Neandertals and Modern Humans in Western Asia

Neandertals and Modern Humans in Western Asia

Edited by

Takeru Akazawa

International Research Center for Japanese Studies
Kyoto, Japan

Kenichi Aoki

University of Tokyo
Tokyo, Japan

and

Ofer Bar-Yosef

Harvard University
Cambridge, Massachusetts

Plenum Press • New York and London

Library of Congress Cataloging-in-Publication Data

Neandertals and modern humans in Western Asia / edited by Takeru Akazawa, Kenichi Aoki, and Ofer Bar-Yosef.
p. cm.
Includes bibliographical references and index.
ISBN 0-306-45924-8
1. Neanderthals--Middle East. 2. Neanderthals--Asia. 3. Hominids--Middle East. 4. Hominids--Asia. 5. Middle East--Antiquities. 6. Asia--Antiquities. I. Aoki, Kenichi, 1948- II. Akazawa, Takeru, 1938- . III. Bar-Yosef, Ofer.
GN285.N45 1998
569.9--dc21 98-35581
CIP

ISBN 0-306-45924-8

A Division of Plenum Publishing Corporation
233 Spring Street, New York, N.Y. 10013

http://www.plenum.com

10 9 8 7 6 5 4 3 2 1

Printed in the United States of America

PREFACE

Research in recent years placed Western Asia in a central position concerning the debate on the origin of modern humans. It is not surprising that this region occupies a special place in human evolution. The particular geographic position at the crossroads of Africa and Eurasia and the mixed landscape of hilly areas and intermontane valleys next to the Mediterranean Sea created an ecosystem suitable for human habitation since the Early Pleistocene. Numerous prehistoric sites are already known and more are discovered every year. Intensive and extensive field and laboratory research, especially in the Levant, was and is targeted to clarify issues of interest to many researchers. The increasing number of investigations contribute to the ongoing search into the origins of modern humans and the demise of the Neandertals. On the basis of mitochondrial and nuclear genetics, modern humans are considered to have originated in sub-Saharan Africa. Human remains from Qafzeh and Skhul are often cited as supportive evidence of this idea. At the same time, queries concerning the date of Levantine Neandertals are not yet fully resolved. Although no scholars propose that Neandertals are native to western Asia, a few suspect that they never even reached the region. Hence, it appears that the human fossils and their archaeological contexts, as revealed by Levantine sites, play a major role in our understanding of the intricate relationships between Neandertals and their contemporaries.

Since the last century, it has been the more recent historical past that has attracted European scholars from different countries to the Levant. Archaeological projects first began in the region during the early part of the nineteenth century, but systematic study of the prehistoric remains was initiated only after the First World War. During the "Golden Age" of Near Eastern archaeology, between the two world wars, numerous large-scale excavations of prehistoric sites were carried out. This work has increased significantly since the Second World War. During the last four decades, foreign archaeologists have been joined by local ones, creating a mosaic of academic approaches still unmatched in other regions of world archaeology. In addition, despite the changing geopolitical situations in this region, field research continues without interruption.

Sites attributed to the Lower, Middle, and Upper Paleolithic have been excavated and reported, and many more are known from surveys. Already in the 1930s, Middle Paleolithic sites produced a large number of human remains. Mousterian assemblages were studied, and the excavations during the 1960s and 1970s produced additional fossils, especially in Qafzeh and Kebara. The renewed excavations at Tabun raised numerous important questions concerning the chronology of Middle Paleolithic industries. At the same time excavations at Sefunim and Geula caves, both in Mt. Carmel, as well as Rosh Ein

Mor and Ain Aqev in the Negev, provided additional information. In Syria and Lebanon excavations in Ras el Kelb and the Adlun sites were followed by Jerf Ajla, Douara, Keoue, Nahr Ibrahim, and Masloukh. Renewed excavations in Yabrud rockshelter produced anew a long sequence of Acheulo-Yabrudian and Mousterian layers. Technological and typological studies developed the sequence of the Mousterian industries in Tabun Cave as a basis for relative cultural chronology. Other investigations stressed the potential of recognizing settlement patterns that include hunting camps, quarry sites, and the relationship between the time-trajectory of metrical attributes of artifacts and human evolution.

However, it was the debate concerning the origin of modern humans and the arguments about the dating of the local Neandertals and the various Mousterian industries that have turned a number of sites into targets for intensive research.

The new era of dynamic field and laboratory investigations, initiated in the early 1980s, led to a series of discoveries of new human fossils. Other important advances include the accumulation of new observations concerning site formation processes and past subsistence strategies, and the dramatic increase in number and quality of radiometric dates. Among the new field projects are the excavations in Dederiyeh Cave in northern Syria, where the skeleton of a Neandertal child was uncovered; the excavations in the El-Kowm basin and especially in Umm et Tlel, where a series of Mousterian living floors was unearthed; the new series of excavations at Kebara Cave in Mt. Carmel and the renewal of the excavations at Hayonim Cave; the completion of the excavations at Amud Cave and the discovery of additional human remains there; and the excavations of the Mousterian rockshelters and caves in southern Jordan such as Tor Faraj, Tor Sabiha and Ain Difla. During the same period ESR, TL, and Amino Acid Racemization readings were obtained in sites such as Qafzeh, Kebara, Tabun, Amud, Tor Faraj, Tor Sabiha, and Ain Difla. Through these projects it became obvious that the Mousterian chronology was longer than had ever been suspected.

Other key achievements in recent years deserve to be mentioned as well. These include in-depth lithic studies carried out in several sites, the search for biological indications for seasonal occupation of the sites, and the study of site formation processes that, in addition to micromorphology, included the investigation of ashes and hearths as well as bone preservation. Faunal analyses produced improved data sets concerning past human diets and hunting techniques, in addition to the traditional reconstruction of past environments.

Both the new fossil relics and the ones discovered in the past raised numerous issues on two fronts: the approximate taxonomic status of these humans; and the possible relationships between those traditionally classified as Neandertals and the modern humans that were also named as "proto-Cro-Magnons." Disagreements concerning taxonomic and paleobehavioral aspects made it obvious that investigators from different countries and various schools needed to meet to discuss the most updated results of their scientific inquiries. The result was the organization of the first meeting at the University of Tokyo in 1990 (Akazawa et al. 1992).

The success of this conference and the rapidly accumulating data sets after 1990 made it clear that another meeting was desired. We therefore invited a large number of colleagues to take part in a meeting that convened at the University Museum of the University of Tokyo in November 1995. In the meantime, it had become evident that the implications of the current debate on the evolution of modern humans and the disappearance of the Neandertals reach far beyond the geographic boundaries of western Asia and indeed the disciplinary boundaries of traditional anthropology and archaeology. Therefore, since we also wished to enlarge the scope of the conference, we invited a few scholars to contribute surveys of adjacent regions in order to enable readers to place the information from western Asia in a global, Eurasian context. Also invited were scholars with a theoretical interest in the relationship between taxonomically and culturally determined aspects of paleobehavior.

Unfortunately, a few colleagues could not attend the conference and several were unable to submit their papers. We thank all those who attended the conference for their lively participation as well as the contributors to the volume for their dedication and patience.

While editing the volume we faced the need to homogenize certain terminologies and spellings that recur throughout the various contributions. We hope that both the authors and the readers will understand the constraints and difficulties involved in this task. One of the more common issues is the spelling of Neandertals, which was originally written with "th" but has been shortened in recent years to only "t." We also favored the term "Near East" rather than "Middle East," and have chosen to spell "Paleo" as such rather than "Palaeo." We have standardized the references to dates of thousands and millions of years by using ka for thousands and Ma for millions.

The organizers would like to thank the University of Tokyo, the International Research Center for Japanese Studies, and the American School of Prehistoric Research (Peabody Museum, Harvard University) for providing the funds for the editing of this volume. We would also like to thank David Pilbeam for his comments on many of the papers and Naomi Ornstein for her work completing the task.

Takeru Akazawa
Kenichi Aoki
Ofer Bar-Yosef

CONTENTS

PART I

Issues of Evolution and Chronology

The study of human evolution involves intricately combining two lines of research. One is the taxonomical definition of the human fossils and their phylogenetic relationships. The second is the chronological framework, without which we cannot reconstruct the history of the human lineages. Given that chronology is crucial in reevaluating the place of known fossils and their cultural material remains, we decided to open this volume with two general evolutionary overviews and a range of papers treating the chronology of the human fossils, fauna, and lithic industries.

Asia is a large continent, and the amount and scope of prehistoric research done across this vast territory is undoubtedly unequaled in other parts of the world. During the first part of the twentieth century, European scholars, after making prehistoric discoveries and acquiring knowledge of prehistory in their home countries, began research in other regions. Major discoveries were made in South Africa, China, and the Near East. The second part of this century witnessed numerous finds of hominids in East Africa and then the expansion of research into many other parts of the Old World. Our current knowledge of prehistoric sequences is far from evenly distributed across Africa, Eurasia, and Australia. However, the overall picture is more complete than it ever has been and it allows for more vivid discussions.

Most readers are probably familiar with the general trends in the interpretation of human evolution. The bulk of the fossils that are being discussed were discovered since the nineteenth century in Africa, Europe, and eastern Asia. This is not to say that the human remains uncovered in the Levant since the first excavations by F. Turville-Petre in Zuttiyeh cave in 1925, and the following ones in Tabun, Skhul, Qafzeh, Kebara, and Amud, or the skeletal remains from Shanidar cave were not taken into account. However, western Asia rarely played a major role in most synthetic overviews, although it is true enough that, geographically speaking, this region comprises only a small portion of an immense continent. Biases also emerge from linguistic barriers, and these affect the facility with which available fossils can be reexamined.

Historically, the Levantine human remains were classified according to the then prevailing conventional wisdom of the profession. Found in Middle Paleolithic contexts, they were first seen as Neandertals. While compiling their report on the Mt. Carmel skeletal remains, Keith and McCown (1939) suggested viewing them as part of a local population with mixed characteristics. It was only later that the entire collection was subdivided into

Neandertals and "proto-Cro-Magnons." Given that Neandertals were considered as a grade in human evolution by authorities such as Hrdlicka, von Koenigswald, and Tobias, it seemed reasonable to place them in a period earlier than those that looked like the forerunners of modern humans from Qafzeh and Skhul. Moreover, in reference to the European Neandertals, those of western Asia were named Asian Neandertals after the region in which they were discovered.

The Levant gained additional attention immediately after the onset of the current debate concerning the origins of modern humans. Molecular and nuclear genetic evidence indicate that western Asia probably served as one of the most important corridors for the dispersals of humans into Eurasia. Although this point had been recognized earlier, when the movements of *Homo erectus* "out of Africa" were considered, it has received further attention since the early 1980s. A series of international conferences (e.g., Otte 1988; Trinkaus 1989; Mellars and Stringer 1989; Bräuer and Smith 1992; Aitken et al. 1993; Nitecki and Nitecki 1994) demonstrated that this region may have a particular place in the evolution of modern humans. These led to the organization of the first conference in Tokyo (Akazawa et al. 1992), which dedicated a major portion of its papers to the Levant.

The questions raised in each of these conferences were essentially similar: (1) How are the available human fossils classified? (2) Do the fossils of the late Middle and Upper Pleistocene demonstrate concrete evidence for the "out of Africa" model? (3) What is the archaeological evidence for modern behavior that is revealed, if any at all, through the excavations of Middle Paleolithic contexts? (4) Where, when, and, if possible, why did the transition from the Middle to the Upper Paleolithic occur? (5) What was the nature of this "revolution"? (6) Was this rapid cultural shift a local Levantine invention or was it the work of the new people coming from Africa?

As time goes by, site reports are published, faunal collections are analyzed, and issues of cognition and language are discussed. Every conference brings new information, ideas, and interpretations. This volume follows the same track, although it is more focused on one region with a few insights from other parts of the Old World (Part IV). This type of continent-wide summary opens the book.

The first paper by F. Clark Howell provides a basis for assessing the evolutionary trajectories and phylogenetic affinities by looking at prehistoric populations. He paints a wide canvas of human evolution in Eurasia, from Dmanisi to the Atapuerca-Gran Dolina fossils, Mauer, Arago, and Petralona, with more particular treatment of the western Asian remains. As will be seen through other studies, this picture is incomplete if the North African fossils are not taken into account. Recognizing the relationship of the Jebel Irhoud specimens to the Levantine Skhul and Qafzeh fossils is essential. All are forerunners of the modern forms and producers of Mousterian industries. The Neandertals, well defined in the European world, are found in an older time zone in the west and in the late Mousterian in the Levant. The survey continues with the presence across the land of Cro-Magnons and their cultural attributes, namely the Aurignacian. In addition to the fossils, Howell deals with various aspects of cultural evolution, types of industries, and possible dispersals, including how the latter could have been triggered by climatic/environmental changes. Questions pertaining to our perceptions and evaluations of the differences between the Middle and the Upper Paleolithic are discussed. The conclusions state the importance of evolutionary studies as both historical and processual and the discoveries of fossils and other archaeological remains would continue to serve for building and refuting hypotheses.

Chris Stringer also opens his chapter with the wide scope of human evolution and he looks at Neandertals and modern humans as clades. Both Eurasian and African fossils form the basis for a discussion of morphological and metrical traits and for recognizing the patterns in the Levant. He stresses how the introduction of the new dating techniques such as ESR and TL have changed the geochronological framework, emphasizing the role that the new dates play in reconstructing the tempo of human evolution. In this domain, the stratigraphic ambiguities concerning the position of the woman from Tabun Cave are justifiably mentioned as a crucial chronological problem. Resolving conflicting dates will have major implications, as until now it seems that most of the Levantine Neandertals are associated with the late "Tabun B-type" industry (see also Chapter 3). The genetic evidence leads to the final conclusions that the rise of the modern *Homo sapiens* occurred in sub-Saharan Africa, and this idea seems to be supported by the Levantine evidence.

In Chapter 3, Ofer Bar-Yosef summarizes issues including definitions of lithic industries, chronology, and the relationship between the available dates and the interpretation of site formation processes. The definitions of the various Mousterian industries make use of the newly dominant concept of *chaîne opératoire*, which prevails in lithic studies. The main premise in such an analysis is that the artisan has a technical knowledge that is based on commonly held notions in his or her society. Reconstructed past patterns of foragers of Levantine Middle Paleolithic, as well as the high degree of availability and good quality of raw materials, are additional elements that are introduced in reviewing the industries. Hence, changes in the reduction sequences in the Levantine Mousterian are essentially seen as dictated to a greater degree by cultural concepts and to a lesser degree by mobility, tool function, and raw material.

The chronology of the known Middle Paleolithic sites and assemblages is discussed and a list of available dates are given. Discrepancies between dating techniques are pointed out. However, available readings are employed in order to estimate the rate of accumulation in Kebara and Hayonim caves, taking into account the effects of diagenesis. The reader is introduced to some of the achievements of the current mineralogical studies of ashes and bones and their implications for the paleoanthropological interpretations.

Henry P. Schwarcz and W. J. Rink bring the results of the ESR and U-series contributions to the Levantine Paleolithic. They discuss the independence of each of the techniques with special reference to TL, and they explain the limitations of what archaeologists often refer to as "absolute" dates. Following a brief description of the ESR and U-series techniques, they discuss the dates from Hayonim, Tabun, and Amud caves, as well as Farah II. The latter is an open air site in the northern Negev that contained a Mousterian industry. The ESR dates at Hayonim are in general accordance with the TL dates (see the paper by Valladas et al. which follows) and indicate that "Tabun D-type" industries in Mt. Carmel and the Galilee are of the same age, falling within Isotope Stages 7 and 6. The authors discuss the discrepancy between the TL and ESR dates from Tabun and make some useful suggestions for future research that will aim to resolve this problem. The ESR dates for Amud onform to the expectations for an overall "Tabun B-type" industry.

Thermoluminescence and ESR revolutionized the chronology of the Middle Paleolithic, and not only in the Levant. In the next chapter, Valladas and associates briefly describe the principles of the technique and then discuss the dates from Kebara, Qafzeh, Skhul, Tabun, Hayonim, Amud, and Zuttiyeh caves.

Finally, this section ends with Eitan Tchernov's overview of the faunal sequence of the Middle Paleolithic and the way it can be related to human dispersals. The author demonstrates the nature of the intricate ecological mosaic of the region and how climatic changes can be related to animal and human dispersals. He points out how hominids were no less susceptible to ecological stresses and ameliorations than their contemporary mammalian communities. The dating of the sites, assemblages, and industries is, like in the other papers, based on the available TL and ESR readings. Among the faunal assemblages, the most sensitive to ecological change are often the micromammals. Tchernov builds upon his earlier publications on the sequence of the biozones and indicates the absence and presence of various species. The Late Acheulian faunal assemblage from the cave site of Oum Qatafa, excavated many years ago by R. Neuville (1951), serves as the baseline. It is followed by the entities of the Acheulo-Yabrudian and then the three Mousterian phases called here "early," "middle" and "late." Faunal replacements, such as the disappearance of the *Ochotona*, an archaic lagomorph close to *Lepus*, are considered major events. Other older species of rodents are also replaced at that time by newcomers. In the second phase, the archaic rodents that persist in Hayonim Cave, including a few Palearctic elements, were replaced by Afro-Arabian species at the time of the Qafzeh Mousterian. The uniqueness of the Qafzeh assemblage receives attention, and is considered to indicate the dominance of savannah conditions (for a detailed report see Rabinovitch and Tchernov 1995). The larger mammals exhibit a mixed environment with high frequencies of Red and Fallow deer, followed by aurochs and rhino, and low presence of gazelle and Roe deer. Ostrich is represented by eggshells like at Hayonim, and the wild camel and the equids are worth mentioning. An additional faunal shift is recorded after Qafzeh by the mammals and micromammals from Kebara and other late Mousterian sites. In conclusion, the arrival of the modern humans, often referred to as the "proto-Cro-Magnons," precedes the late appearance of the Neandertals in the Levant.

REFERENCES

Aitken, M. J., C. B. Stringer, and P. A. Mellars, eds. 1993. *The Origin of Modern Humans and the Impact of Chronometric Dating*. Princeton: Princeton University Press.

Akazawa, T., K. Aoki, and T. Kimura, eds. 1992. *The Evolution and Dispersal of Modern Humans in Asia*. Tokyo: Hokusen-sha.

Bräuer, G., and F. H. Smith, eds. 1992. *Continuity or Replacement: Controversies in* Homo sapiens *Evolution*. Rotterdam: A. A. Balkema.

McCown, T. D., and A. Keith. 1939. *The Stone Age of Mount Carmel*. Vol. 2. Oxford: Clarendon Press.

Mellars, P. A., and C. B. Stringer, eds. 1989. *The Human Revolution*. Edinburgh: Edinburgh University Press.

Nitecki, M. H., and D. V. Nitecki, eds. 1994. *Origins of Anatomically Modern Humans*. New York: Plenum Press.

Otte, M., ed. 1988. *L'Homme de Néandertal*. Liège: ERAUL.

Rabinovich, R., and E. Tchernov. 1995. Chronological, paleoecological and taphonomical aspects of the Middle Paleolithic site of Qafzeh, Israel, 5–44. In *Archaeozoology of the Near East II*, ed. H. Buitenhuis and H. P. Uerpmann. Leiden: Backhuys.

Trinkaus, E., ed. 1989. *The Emergence of Modern Humans: Biocultural Adaptations in the Later Pleistocene*. Cambridge: Cambridge University Press.

1

EVOLUTIONARY IMPLICATIONS OF ALTERED PERSPECTIVES ON HOMININE DEMES AND POPULATIONS IN THE LATER PLEISTOCENE OF WESTERN EURASIA

F. Clark Howell

Laboratory for Human Evolutionary Studies/Museum of Vertebrate Zoology
University of California
Berkeley, California 94720

1. INTRODUCTION

This contribution outlines my own current perspective on some important aspects of the hominine fossil record of western Eurasia with particular reference to demes and populations. Such aggregates comprise essential components for the recognition of lower taxonomic categories (species). They afford the necessary basis for assessments of phylogenetic affinities, for recognition of evolutionary trends, for measurement of evolutionary tempo, and for study of questions of dispersal, isolation, provincialism, and extinctions.

The term deme was introduced by Gilmour and Gregor (1939) for a local interbreeding population or community ("any assemblage of taxonomically closely related individuals" [Gilmour and Gregor 1939:333]), and distinguishable by reproductive (genetic), geographic and ecological (habitat) parameters. Carter, who urged employment of the concept, considered the deme "as the unit of natural history and of evolutionary divergence," and defined it as "a communal interbreeding population within a species" (1951:142). A modern definition is, for example, that of Mayr and Ashlock: "a local population of a species; the community of potentially interbreeding individuals at a given locality" (1991:413). Together, of course, demes (even as isolates, sometimes) constitute subspecies, that is the "aggregate of local populations of a species inhabiting a geographic subdivision of the range of the species" (Mayr and Ashlock 1991:413). Trinkaus has also considered that "the best approach . . . is probably one that regards the available fossil samples (or specimens) as representative of prehistoric populations or lineages acting as portions of dynamic evolutionary units" (1990:9). Past hominine demes are denominated as paleo-demes (p-demes) (Howell 1996). Familiar examples of significant p-deme sam-

Neandertals and Modern Humans in Western Asia, edited by Akazawa *et al.*
Plenum Press, New York, 1998.

ples are well known among several *Australopithecus* species, *Paranthropus* species, Asian *Homo erectus*, *Homo neanderthalensis* including antecedent populations, and *Homo sapiens palestinus*. These constitute of course individual, spatio-temporally disparate p-deme samples. Since there is disjunctive distribution of late Cenozoic continental sediments, within a variety of sediment traps and in association with the vagaries of accumulation, fossilization, and preservation processes, it is rare indeed to have much spatial (geographic) documentation of subspecies distribution and variation from a suitable series of p-deme samples. It is of course essential to bridge the gap between individual (often incomplete) specimens and such p-demic samples to achieve an appreciation of the subspecies and its variation. Such can only be achieved, or sought for, through requisite comparative studies and appropriate morphometric analyses, ultimately updated as further relevant specimens come to light.

The known hominine p-demes of western Eurasia are briefly considered, by decreasing age, in the following section.

For years hominine evolutionary studies have been confounded and obfuscated by epistemological and conceptual difficulties. This is epitomized in historical treatments of western Eurasia and its paleoanthropological documentation. Clearly major difficulties have attended the prevalent state of science and empirical knowledge, including: inadequacy of the fossil record and of its scanty samples; brevity of the fossil record; common reliance on insecure (e.g., Alpine) Pleistocene stratigraphic frameworks; difficulties and mistakes in application of biostratigraphic principles to elucidate relative ages of fossil localities; lack of reliable geochronological frameworks consequent upon absence of isotopic dating methods. However, other hindrances reflect conceptual and related weaknesses: undue reliance or emphasis on fragmentary, often ambiguous fossil specimens; inadequate assessment and comparative analysis of individual and larger samples as reflective of former (paleo-) demes of populations; employment of 'cultural' associations as reflective of biological affinity among fossil samples; inappropriate usage of ascriptions such as 'archaic,' 'intermediate,' 'transitional,' 'anatomically modern,' and the like; confusion of grade and clade, and adherence to ill-founded or outmoded precepts that reflect gradistic, progressivist, and adaptationist perspectives; and surprisingly little concern with fundamental evolutionary processes—adaptive radiation, diversification (including cladogenesis), species recognition, divergence, stasis, endemism, range extensions and dispersals, and extinctions.

Fortunately the last several decades have witnessed major developments in Quaternary studies and in geochronology that have revolutionized conceptions of the Eurasian late Cenozoic. Central has been elucidation of past paleoclimatic events, and thus elaboration of a formal global stratigraphic framework, from isotopic and associated analyses of deep-sea cores of the world ocean basins. The development and refinement of isotopic geochronological methods—including the potassium/argon system and more recently U-series, ESR, TL and OSL, and AMS procedures—afford a battery of approaches relevant to direct age assessments of paleoanthropological localities and their residues. Approaches to the hominine fossil record on the one hand, and toward understanding of the meaning of associated paleoanthropological residues on the other, have similarly been extensively transformed. The former includes: (a) extensive evaluation of extant human population skeletal samples, their variabilities and phenetic resemblances, through multivariate morphometric analyses, thus affording essential perspectives for realistic appraisal of a past hominine fossil record; (b) a rejection of typological and idealistic perspectives, and a reoriented concern with characters/character states within an ontogenetic, functional and phylogenetic framework; (c) application of cladistic analytic approaches toward character state evaluation, ascertainment of polarities, and the employment of derived (novelties) versus shared primitive conditions as informative of phylogenetic affinities and suggestive of evolutionary divergences and trends; (d) reduction of emphasis on incomplete/ambiguous specimens as phylogenetically pivotal or informative; (e) recognition of individual or associated samples as reflective of p-demes of larger population aggregates; (f) substantially decreased emphasis on gradistic, stage of evolution, and related progressivist concepts. In respect to the latter are wide-ranging refinements in exposure and recovery procedures in controlled excavations; elaborated approaches toward the analytical treatment of lithic assemblages and the recognition and evaluation of cultural/industrial stratigraphic units/complexes; intensive employment of much refined approaches in zooarcheology and in taphonomy toward the identification and evaluation of faunal residues from paleoanthropological contexts; and, innovative analytical approaches appropriate toward elucidation of behavioral and adaptive contents of artifacts, cultural features and faunal residues in past occupational contexts. All such developments and shifted emphases are pertinent with respect to the status of hominine evolutionary studies in western Eurasia.

2. WESTERN EURASIAN HOMININE P-DEMES

2.1. Dmanisi

For the moment this constitutes the oldest known p-deme in western Eurasia. The single specimen, an adult mandible with full dentition, derives from fluvio-lacustrine sediments of the eponymous locality in the Georgian Caucasus, and apparently dates from the post Olduvai (N) sc interval of the Matuyama (R) chron. It is associated with an upper Villafranchian type fauna (Palaearctic aspect) and a lithic industry within the Mode 1 technological stage and specifically of Oldowan Industrial Complex affinity. A comparable occurrence of broadly similar (or maybe just younger) age is that of ‘Ubeidiya (Jordan Rift) at which over 60 artifact-bearing horizons, of a further derived Oldowan/cum Acheulean industry, and rich faunal associations of mixed Palaearctic/Ethiopian aspect, are extensively documented. The Dmanisi hominin has a mosaic of characters, in part shared with the (African) Nariokotome p-deme (1.85–1.5 Ma), in part with more derived hominines, and others of which are seemingly autopomorphically unique.

2.2. Atapuerca-Gran Dolina

From very recent excavations of the Gran Dolina (GD) karstic cavity at Atapuerca (Spain) hominin occupation is documented in at least two lower levels, TD-4 (rarely) and overlying TD-6 (abundantly). These lower reaches of ~20 m infilling fall within the upper Matuyama (R) chron and antedate the B/M geomagnetic inversion recorded slightly higher in the sequence. The TD-6 level includes a substantial lithic assemblage (lacking Acheulean hallmarks), abundant faunal residues, and hominine cranial, gnathic, dental, and postcranial elements (of several individuals). The overall morphological pattern, as thus far revealed, is distinctive and is definitely less derived and unlike that of less ancient European counterparts. Comparison with the antecedent Dmanisi p-deme is restricted due to the scant comparability of skeletal parts. However, the GD sample clearly merits recognition as a new p-deme. A partial adult calvaria, recently recovered serendipitously from paludal sediments in the Ceprano basin (southern Latium, Italy), antedates a mid-Pleistocene volcanic succession and is underlain by fluvial sands yielding Galerian mammal-fauna and non-Acheulean industry (Castro dei Volsci facies). The calvaria exhibits morphological development analagous to that of the Atapuerca-Gran Dolina p-deme and thus may be parsimoniously subsumed within it.

2.3. Mauer/Arago

Hominin remains from these localities once constituted the oldest documented hominine samples in Europe proper. The isolated Mauer mandible (with dentition) from the Grafenrain quarry (Rhineland) is the type of the nomen, *heidelbergensis* Schoetensack. The Arago (e. Pyrénées) sample, probably over a dozen individuals, and comprising cranio-facial elements, partial mandibles (2), dentitions, and various postcranial elements, is best taken as paratype and basis of comparison. Each occurrence has rich faunal associations, and in the case of Arago a (largely) non or biface poor flake-dominated lithic industry of distinctive (Taubachian-like) facies. Overall this p-deme is distinguished by an idiosyncratic (regional) mosaic of some (sym)plesiomorphic cranial, gnathic, and postcranial features coupled with derived (apomorphic) features of other aspects of cranium (fronto-parietal elements, facial skeleton) and mandible (symphysis and ramus, dentition) which may approximate and foreshadow structure characteristic of subsequent Neandertals. There is no direct evidence as yet to attribute the Boxgrove (Sussex) tibial diaphysis to this p-deme, although that attribution is not inappropriate. It has both faunal and industrial (Acheulean) associations. The extent to which fragmentary hominin remains from Italy—including Fontana-Ranuccio (Latium) (dentition), Visogliano-2 (Trieste) (mandible fragment, upper premolar) and Venosa-Notarchirico (Basilicata) (femoral diaphysis)—might reflect this or another related p-deme remains to be evaluated. The M/A p-deme encompasses an ill-defined mid-Pleistocene time span, broadly within the mid-Brunhes (N) chron, perhaps 0.45–0.60 Ma (about ^{18}O stages 16–12). There is a notable temporal gap between the M/A and antecedent A/GD p-demes.

2.4. Petralona/Atapuerca (Sima)

This p-deme is among the best known and characterized in the European mid-Pleistocene, initially as a consequence of the recognition of the distinctive Petralona (Khalkidhiki) cranium, the type of the nomen, *petralonensis* Murrill. More recently, the extensive hominin assemblage of all major skeletal elements from dozens of individuals, subadult and adult, recovered from Sima de los Huesos (SH) Atapuerca has afforded the largest single hominin sample from the mid-Pleistocene of Europe, and it is enlarged yearly as fieldwork progresses. This remarkable sample affords also a basis for fuller evaluation and attribution of isolated and less complete/damaged specimens (Montmaurin—La Niche mandible, Vértesszölös occipital, and dental elements, and Steinheim, Swanscombe, and Bilzingsleben partial crania) about which considerable discussion and indeed some controversy prevailed over too many years. The Steinheim specimen is the type of the nomen, *steinheimensis* Berckhemer and this nomen has priority over *petralonensis*. These examples, along with the still ill-known Apidima Diros (southern Greece) crania and, perhaps, the recently found and remarkably complete Altamura (Italy) skeleton, may ultimately prove to be attributable to the same demic group. Similarly, a diversity of more or less fragmentary hominin specimens from Italy—in Liguria, Prince Cave (innominate), in the Latium, Castel di Guido (cranial, femoral fragments), Cava Pompi (postcrania, cranial fragments), Casal dé Pazzi (cranial fragment), Sedia del Diavolo (post cranial/cranial fragments), and Ponte Mammolo (postcranial fragment)—require comparative examination from this perspective. This p-deme reveals persistence of some plesiomorphic features, but substantially still stronger and widespread expression of Neandertal synapomorphies throughout the cranial, gnathic, and postcranial skeleton. The timespan of the p-deme is broadly estimated as about 400–300 ky (^{18}O stages 11–9). All referred representatives have faunal associations, and known industrial associations vary from advanced or Middle Acheulean (as at Swanscombe, Castel di Guido) to several manifestations of flake-based (Taubachian-like) industries (Vértesszölös, Bilzingsleben) in largely open air situations.

2.5. P-Deme Indeterminate—Western Asia

The apparently oldest mid-Pleistocene hominin occurrence in the Levant is that of Gesher Benot Ya'aqov (GBY, Jordan Valley), associated with an African-style Middle Acheulean assemblage. It most likely falls within the (lower) Brunhes (N) chron and is at least <800 Ky (based on the age of the underlying Yarda Basalt). The p-deme (hominin femoral diaphyses) represented is indeterminate on available evidence. This is equally the case for the gnathic fragment from Azykh Cave (southern Azerbaidzhan, Transcaucasus), in an Acheulean association context and perhaps of broadly similar if not rather younger age. Kudaro-I Cave (Osetia, southern Caucasus) is another (reportedly Middle) Acheulean occupation with a single, indeterminate hominine (I) specimen. In view of the demonstrable African character of the GBY (industrial) occurrence it would be premature to consider its attendant hominin to be of strictly European as opposed to African demic affinity (and particularly as the latter merit distinction at the p-demic level).

2.6. Irhoud

This p-deme is well-typified by adult (no. 1) cranium and subadult (no. 2) calvaria and mandible (no. 3); there is also an imperfect infant humerus (no. 4), all from the eponymous cave filling in southern coastal Morocco. ESR age estimates (90–125 ky, EU: 105–190 ky, LU) suggest an age within ^{18}O stage 6. Partial cranial and gnathic-dental remains from Kébibat (Mifsud-Giudice quarry, Morocco) may well represent the same or closely-related p-deme.

To this distinctive p-deme should be attributed the fronto-facial fragment from Zuttiyeh Cave (Galilee). It occurred in the basal infilling in an Acheuleo-Yabrudian (so-called Mugharan Tradition or [MT]) industrial context. That, and distinctive micro-faunal associations indicative of substantial antiquity, have led to its long overdue reevaluation. (The Irhoud occurrence is perhaps less old and the industrial association comparable to Levantine Mousterian, perhaps equivalent to LM-D). The estimated (isotopic) age range at Zuttiyeh is >150–250 ky, based on several U-series determinations on stalactites or travertines. However, such age estimates are not well grounded, and further isotopic dating is sorely needed. Elsewhere, TL ages on this industry have been minimally (without corrections) 160 ± 22 ka (Hummal, Syria) and 195 ± 15 ka (Yabrud 1b, Syria). The MT industry is now well documented only in the central and northern Levant. The only other such hominine fossil association, a femoral fragment, of indeterminate demic affinity, is that of Tabun (Eb) (Mt. Carmel), considered of ^{18}O stage 6 age. Similarly, the demic affinities of the fragmentary hominine remains from Sel'Ungur (Kirgizia, central Asia) cave infilling, in an industrial association with some resemblances to the above, are also unresolved; a final mid-Pleistocene age is estimated.

Table 1. Isotopic age determinations on S/Q p-deme localities

Locality/Bed(s)	U-series (ka)	ESR (ka) EU	ESR (ka) LU	T/L (ka)
Qafzeh				
(level 19)	89.0 ± 3	96.0 ± 13	115.0 ± 15	av. 92 ± 5 (isochron 88 ± 9)
(levels 17–23)	106.0 ± 2	100.0 ± 10	120.0 ± 8	
Skhul B	80.3 ± 6	81.0 ± 15	101.0 ± 12	av. 119 ± 18 (99–167)
Tabun C	101.5 ± 2.9	102.0 ± 17	119.0 ± 11	134–184 (171 ± 17)
		118.0 ± 5	127.0 ± 10	195–263

2.7. Skhul/Qafzeh

This p-deme (S/Q) is well and only surely known from the eponymous cave localities of es-Skhul (level B, Mt. Carmel) and Djebel Qafzeh (levels 17–23) in the southern Levant. These samples comprise partial or semi-complete skeletons, with associated skull parts, of 10 and 13 individuals, respectively, and including both adults and infantile (3, 6) individuals. (An infant individual, no. 1, from the es-Skhul site is designated lectotype for the species/subspecies nomen *palestinus* McCown and Keith). In cranial vault and facial morphology, gnathic features, some dental characteristics, both ontogenetically and in adults, and in varied aspects of postcranial (both axial and appendicular) morphology and proportions, these samples differ consistently and substantively from Neandertals and more closely approximate *Homo sapiens* condition, without however replicating it. A very likely if not firmly established attribution to the S/Q deme is the Tabun C2 adult mandible. At the great cave of Haua Fteah, in Cyrenaican Libya, two partial human mandibles from an industrial association (in beds XXXII–XXXIII) comparable to that of S/Q in the Levant very likely represent the same p-deme. Some phenetic and thus presumably genetic affiliations with the Maghrebian p-deme of Irhoud, of which a representative (Zuttiyeh) is known in the southern Levant, have been established.

This p-deme is associated with the Levantine Mousterian (of Levallois-facies) (or LM) at these three localities and in particular with the LM-C industrial manifestation (epitomized by the Tabun Cave sequence of D, C, and B subdivisions according to upwards successive beds). All these occurrences have afforded age estimations (Table 1) by three isotopic methods: U-series, ESR, and T/L. In all instances the ages are substantial and fall broadly within the mid- to lower reaches of ^{18}O stage 5, and perhaps in the (decreasing) age order of Tabun C, Qafzeh/17–23, and Skhul B. T/L ages for Tabun are uniformly greater than ages afforded by other methods, and such is so throughout that succession.

2.8. Neandertal

This p-deme was the first premodern human form known in the fossil record and is now, overall, the most fully and extensively represented hominin from western Eurasia. The first such recognized example, a partial skeleton from Feldhofer Cave, Neandertal, is the type of the nomen *neanderthalensis* King. It is most extensively documented in the European Last Glacial (^{18}O stages 4–3), from central (Teshik-Tash) and western Asia (Shanidar) into the Levant (Kebara, Amud, Tabun C1, Dederiyeh), the Crimea, and throughout the reaches of central and western Europe into the Apennine and Iberian peninsulas.

Its earlier history is documented in ^{18}O stage 5 (Krapina, Saccopastore, La Chaise-Bourgeois-Delaunay, Gánovce, Taubach, Salzgitter-Lebenstedt), stage 6 (in France, Lazaret, La Chaise-Suard, Fontéchevade, Biache; and in Wales, Pontnewydd), and even into stage 7 (Ehringsdorf). Partial or largely complete skeletons, some in definite interment circumstances, enable a very full if not absolutely exhaustive elucidation of skeletal paleobiology. The distinctiveness of the p-deme, with extensive autapomorphic character states throughout its spatio-temporal range, and including variability, aspects of dimorphism, and ontogenetic development, is well-established and extensively documented. There are overall phenetic resemblances, and sharing of particular discrete traits with antecedent demes, particularly with Petralona/Atapuerca p-deme and, but substantially less so, with that of Mauer/Arago.

The geographic distribution of the Mousterian Industrial Complex (MIC) is extensive in western Asia, the adjoining Caucasus, Crimea and environs, and in portions of (southern) central Asia, whereas the hominine fossil record is largely concentrated in the southern Levant and, to an extent, in the Transcaucasus and Crimea. There are as well important individual site outliers in the Iraqi Zagros range (Shanidar) and in southern central Asia (Teshik-Tash).

There are three major temporal variants of the Levantine Mousterian (of Levallois facies), epitomized by the Tabun (Mt. Carmel) succession (beds with, successively, LM-D, -C and -B). These variants exhibit quantitative differences in some technological practices, particularly core preparation and blank production, and in the types and forms of finished tool products, and their frequencies. From some localities there are evidences to suggest certain differences in landscape usage and settlement pattern between occupations of LM-C and those of LM-B. Respectively, such apparently include circulating versus radiating mobility strategies (or 'foragers' versus 'collectors') and seasonally-occupied residential locales in contrast to multi-seasonal bases and seasonally-employed logistical situations. No hominine remains are as yet known in association with the oldest, LM-D facies. As aforementioned, the distinctive Skhul/Qafzeh p-deme is associated, at the eponymous localities, and at the Tabun type site (individual C2), with the LM-C facies. Representatives of the Neandertal p-deme are associated with the LM-B facies at sites in the southern Levant (Kebara, Tabun B = C1 individual, Hayonim, Amud) and in the northern Levant (Ksar 'Akil, Lebanon lev. xxv, individual KA-2; Dederiyeh Cave, Syria). Another sample of the west Asian Neandertal population is represented in the Shanidar Cave (D, upper and middle levels, 9 individuals in all) of Iraqi Kurdistan, in association with a distinctive Zagros Mousterian industry.

All Levantine and perhaps other west/central Asian Neandertals as well fall within the earlier pleniglacial of the Last Glacial. Thus, isotopic ages from Kebara (ESR/TL) approximate 60–65 ky and from Amud (T/L) fall between 50–60 ky. There is a range of ages for Tabun B, from which the individual Tabun C1 very likely derived, depending on the method employed and the appropriate correction factor applied: U-series between ~43 and 79–80 ky; T/L 119 ± 18 ky (elevated values as in the case of all such measures at this locality); and ESR, between 56 ± 9 (EU) and 76 ± 19 (LU), an average of five analyses, and 81 ± 15 (EU) and 101 ± 12 (LU), according to two different laboratories. These values either overlap or exceed those from the two aforementioned caves, and it is difficult to resolve the discrepancies at this juncture. The earlier obtained ^{14}C dates for the Shanidar (level D) occurrence are 46,900 ± 1500 and 50,000 ± 3000/4000, and thus fall in the aforementioned younger age range, but may very well only constitute minimum values.

The Mousterian, in several expressions, is becoming increasingly well known in parts of central Asia and in the adjacent Siberian Altai. The first-known skeletal association, a juvenile, at Teshik-Tash Cave (southern Uzbekistan) occurs with Levallois-Mousterian Variant (sometimes termed 'Mountain Mousterian'). It remains, unfortunately, undated. At two, among a number of Altai Mousterian sites (and which in fact extend well into adjacent Mongolia), fragmentary (dental) remains of Neandertal morphology occur. These include: Peschera Denisova (2 teeth), in a Levallois-Mousterian association, having an age (^{14}C) of ~39,390; and Peschera Okladnikov (6 teeth, of several individuals) in a typical Mousterian (non-Levallois) association, having ages (AMS) of 33,500 ± 700 (level 3) and 37,750 ± 750 (level 2).

There is no fully convincing evidence for the presence of representatives of a Neandertal p-deme in the Levant, or the greater west Asian region overall, until the earlier Upper Pleistocene (^{18}O stage 4 and perhaps upper 5). This is in marked contrast to the situation in Europe proper and even into the Pontic area and, apparently, the Caucasus. Thus Neandertals are seemingly allochthonous and immigrants into west Asia, from their autochthonous source and broad distributions over greater Europe. Such shifts are still ill-appreciated and were perhaps a consequence of range displacement attendant on natural, paleoclimatic factors (onset of last glacial), population factors, or even aspects of intergroup competition and aggression. European p-demes of the Neandertal lineage (*Homo neanderthalensis*) constituted ecological vicars just as did Asian p-demes of the Zhoukoudian lineage (*Homo erectus*), and each in their own different ways.

In Europe Neandertals are always associated with the Mousterian technocomplex. There are known associations in western Europe with most conventionally recognized facies, including Charentian (both Ferrassie and Quina variants), Typical (TM), Denticulate (DM), and, in the Latium area (Italy), the so-called 'Pontinian.' There are none, however, demonstrably associated with the Cantabrian Vasconian or (Olha-type) Mousterian or with the MTA (Mousterian of Acheulean tradition, in either A or B subgroup). In central Europe associations are known with the c/e Micoquian at the open site of Salzgitter, and the cave of Wildscheuer, Klausennische, and Kulna 7a. There are associations with the Taubachian at Taubach, Ehringsdorf in Thüringia. Subalyuk (level 11), a Bükk mountain cave is a rare association with the Balkan/Pannonian area Charentian. The c/e Typical Mousterian (both Levallois and non-Levallois) are at the Sipka 5, Svédul stul; Krapina; Vindija G3, G1 localities. In respect to circumstances at the Krapina (Croatia) shelter it should be noted that the best age assessment (by ESR) suggests, as long claimed on other grounds, an (early) Eemian interglacial (^{18}O stage 5e) age, about 130 ± 10 ka. Neandertal fossils, with extensive evidence of deliberate breakage, cut marks, and other such damage, representing multiple individuals, derive there from levels (descendingly) 7–3, and mostly 3–4. However, Neandertals also persist into level 8 (skull parts of two juvenile individuals), as demonstrated by their distinctive morphology; claims made as to their anatomical 'modernity' are completely ill-founded. The Kulna 7a occurrence has an age of ~45,600 (^{14}C) and 50 ± 5 (ESR). Similarly, at Vindija, also extensively fragmented Neandertal skeletal parts derive from the lower G3 and the upper G1 levels. Neandertal teeth occur in Remete Felsö Cave (Hungary) in association with the Jankovichian industry (ex-Transdanubian Szeletian), a distinctive Mousterian with bifacial leaf points found at a cluster of sites, below the great southward bend of the Danube. Another similar leaf point occurrence of this industry is that of Dzeravá skála (or Pálffy, western Slovakia) from which a juvenile's second molar, attributed to a Neandertal, was recovered long ago.

A diversity of expressions of the Mousterian technocomplex elsewhere in central and in eastern Europe are well documented archeologically, but there are also only scant and rarely associated human remains. In the Crimea, such a diversity is well documented. At two of the Zaskalnaya (White Rock) sites (Va, VI, adolescent and child remains, respectively) fragmentary remains occur with the c/e Micoquian-like 'Ak-Kaya' industry, in the latter instance having a quite young age (AMS) of 30–31 ka. Other adult (I) and child/infant (II, III, IV) remains occur at other caves in the Ak-Kaya massif. At Kiik-Koba the (lower) association is with a Taubachian-like industry, and a fragmentary infant may, possibly, have occured with the subsequent (upper level) Kiik-Kobian, a 'para'-Micoquian-like industry. In the northwest Caucasus, south of the Kuban river, both open air and rather numeous cave localities with Mousterian occupations are becoming better known. Four caves have afforded Neandertal remains: Monasheskaya; Barakaevskaya (cranial/gnathic parts of an infant, in level 3); Matuzka (in levels with Typical Mousterian, non-Levallois and in overlying Levallois/Mousterian). In Mezmaiskaya Cave there are four Mousterian occupation horizons, the lower (3, 2B) of perhaps c/e Micoquian affinity and the upper (2, 2A) of perhaps TM affinity. Neandertals occur in both lower (3) and upper (2) levels, represented by cranial parts (2) and a partial child's skeleton (B). The latter occurrence is >45 ky (^{14}C) and succeeding levels have ages of 40,660 ± 1600 (2B, ^{14}C), and 35,760 ± 400 and 36,280 ± 540 (2A, AMS). Again, these are ages within the span of the earlier (Aurignacian) Upper Paleolithic.

Two occurrences in France demonstrate Neandertal presence in a Chatelperronian industrial context. At Grotte du Renne, Arcy-sur-Cure (Yonne), the lower (X) of five levels of this industry afforded nine teeth of indisputable Neandertal morphology. It has an age (AMS) of 33,820 ± 720; a higher level (VIII) has an age (^{14}C) of 33,000 ± 1400. The succeeding Aurignacian (II) occupation has an age (^{14}C) of 31,800 ± 1240. The upper of some 8–10 antecedent (Denticulate and other) Mousterian levels have (^{14}C) ages of (XI) 33,700 ± 1400 and of (XII) 37,500 ± 1600 (or 39,400 ± 1450 by AMS). Neandertal remains (3 teeth) are known from such Mousterian occupations (in Schoeplin gallery) as well as from other levels at adjacent Grotte du Loup and Grotte de l'Hyène. The Chatelperronian of Grotte du Renne (GR) is among the best known and most culturally elaborated of any such occurrence in western Europe, of which some three dozen are documented. That at GR demonstrates most remarkably, even uniquely, the Upper Paleolithic aspect and composition of the industry. It is well distinguished by its differential exploitation of raw (lithic) materials, veritable blade (prismatic core) technology, new and diverse lithic tool kit, use of and tool production from bone, personal ornaments produced in shell, bone, tooth, and ivory, elaborated utilization and modification of cave front habitation space, distinctive patterns of exploitation of both prey and predator species, and their accumulation, dismemberment, and utilization.

At Saint Cèsaire (Charente Maritime) cranial and postcranial parts of an adult Neandertal individual occurred in the upper of two Chatelperronian occupations. The age (T/L) of the occurrence is 36,300 ± 2700. Those industrial occurrences succeeded two MTA and three DM horizons (the uppermost = 40,900 ± 2500 BP). The Chatelperronian levels were overlain by four Aurignacian occupations, the upper two developed, and the lower two archaic (age of Ejo 6 = 32,100 ± 3000), and early (I) Aurignacian, respectively.

Younger still than these Neandertal occurrences are several persistences of Mousterian into mid-Last Glacial times in southern Spain. Thus, at Zafarraya Cave (Málaga), in which there are multiple Mousterian occupations, Neandertal remains (mandible, dentition, etc.) are found in lower (E, D) and middle (B) levels multiply dated 31,800 BP (AMS) and 31,700 BP (Th/U), and 33,400 BP (Th/U), respectively.

The total body of current evidence quite clearly indicates the unexpectedly late persistence in certain situations of (some) Neandertals, temporal overlap of some Neandertal and (earliest) Cro-Magnon populations, and most importantly, the ultimate displacement/replacement of the Neandertals by Cro-Magnon peoples.

2.9. Cro-Magnon

This p-deme subsumes all human populations of western Eurasia of late Upper Pleistocene age and, apparently, always associated with Upper Paleolithic (sensu strictu) industries. (It thus effectively excludes the Chatelperronian of southwestern Europe, which has both technological and some typological links with both Middle and Upper Paleolithic). In other instances, for example, the Uluzzian of the southern Apennine peninsula, the Bohunician of Moravia, and the Szeletian/Jerzmanowician of central/east European uplands, scant or no directly associated human skeletal remains are as yet known.

The oldest representatives of this p-deme in Europe are all associated with the Aurignacian technocomplex. The Aurignacian is now known to date substantially earlier than was formerly envisioned. Ages (conventional ^{14}C, AMS, and some T/L determinations) in excess of 40 ky BP have been obtained from such occurrences in Bulgaria (Temnata 4, Bacho Kiro 11, Samulica) and in Hungary (Istallosko 9). Some such occurrences (in Bulgaria) have been attributed to a distinctive industry, Bachokirian. The lower (of two) Aurignacian levels at the Trou Magrite Cave (Wallonia) is now known also to have an age in excess of 40 ky BP. A number of occurrences have slightly younger ages, between 35–36 and 40 ky BP, in the northern Iberian peninsula (Romani 2, Arbreda B1, Castillo 18, Reclau Viver III/27). A later group has isotopic dates from around 34–35 ky in Hungary (Peskö), in Austria (Krems-Hundsteig, Willendorf 3), in southern Germany (Geissenklösterle 15/16), in Belgium (Trou Magrite 2), and in (southern) France (Pataud 14, Ferrassie K6, Flageolet 11, St. Césaire-Ejo, and Esquicho SLC 1b). The former group of occurrences might be (informally) termed Proto-Aurignacian and the latter group Basal Aurignacian. The oldest known occurrences yield (very fragmentary) human remains only at three localities (Bacho Kiro 11, Istallosko 9, Castillo 18), and the next younger localities all lack human remains.

Most human remains with Aurignacian industrial associations are from situations (caves, shelters) commonly referred to as the (so-called) 'Middle' Aurignacian, having isotopic ages between about 31–35 ky BP. Cranial remains, largely from central Europe, with Aurignacian associations are those from Vogelherd-V (Lontal), Velika Pecina (Croatia), Koneprusy (Bohemia) and Mladec or Lautsch (Moravia); all derive from early ('middle') Aurignacian contexts. Associated ^{14}C ages are 30,162–31,900 BP (Vogelherd-V) and >33,850 ± 520 BP (Velika Pecina). Other individual specimens, without industrial associations, but appropriately considered of broadly comparable age are those of (perhaps) Svitavka (Moravia) and the frontal from Hanöfersand (Elbe R.) (age 35–36.3 ky BP). Of many such sites 20 (France), 1 (Italy), 3 (Germany), 1 (Croatia), and 2 (Moravia) have human associations, usually fragmentary, unfortunately. These all represent *Homo sapiens sapiens*. And, although sometimes claimed otherwise, there is nothing 'transitional' or 'intermediate' with respect to their morphology in any instance.

Our conception of the Aurignacian technocomplex has been altered in recent years from previous linear and stadial frameworks proposed in the first half of this century. This has come about as a consequence of greatly enhanced knowledge of its (geographic) distributional aspects, of its composition and variability (from substantial, diverse assemblages from well-controlled stratigraphic contexts), and from new or re-excavation of long occupation sequences (e.g., the Périgord shelters of Abri Pataud and La Ferrassie), their isotopic dating and elucidation of often rich paleoenvironmental records. In those situations in which Mousterian occupations are also evidenced, Aurignacian *always* succeeds Mousterian of whatever facies, and occurrences of the latter are almost always of (at least somewhat) greater antiquity. There are now well over two dozen localities in western Europe in which the Chatelperronian industry is adequately defined and stratigraphically delineated. Many (19) have antecedent Mousterian occupations, and many (some dozen and a half) also afford Aurignacian (AUR) occupation(s) as well. In nearly all (15) Aurignacian overlies the Chatelperronian (CH) occurrence. There are, however, three situations—two in Aquitanian France, one in Cantabrian Spain—in which industrial interstratification occurs. Thus at Roc de Combe (Lot) there are successive CH (10) to AUR "O" (9) to CH (8) to AUR I (7) occupations. The Chatelperronian of level 8 has an age (^{14}C) of 38,000 ± 2,000 BP. At Le Piage (Lot) there are AUR (K, J, G–I levels) to CH (F1) to AUR (F) occupations. At El Pendo (Cantabria), analogously, the sequence is AUR "O" (VIII b, a) to CH (VIII) to AUR I (VI). Unfortunately, none of these caves yield associated human skeletal remains in these provocative occupational circumstances. In the neighboring Morin Cave there is a long sequence (7 levels of two facies) of Mousterian occupations. These are succeeded by a well-defined Chatelperronian level (^{14}C = 36,950 ± 6,777). There are a total of five successive Aurignacian occupations thereafter, the upper three representing AUR I (with an age of 32,415 ky) and AUR II. The underlying levels are at least Basal Aurignacian (typologically), with misleadingly young ^{14}C ages (~28 ky). The upper of these contains traces ('ghosts') of a disintegrated human corpse, of modern human size and proportions.

In the Apennine peninsula some 21 Uluzzian localities (8 caves, 13 surface locales) are recorded. This industry apparently always underlies Aurignacian in situations in which multiple occupations are represented. No instances of interstratification are reported. Although dated occurrences are rare the duration of the Uluzzian has been suggested as between 33–34.5 ky and 33–31.5 ky. Only two (upper) deciduous molars are known in association from the Grotte del Cavallo locality.

A complex of Upper Paleolithic (UP) industries, first numerically distinguished, then denominated as spatio-temporal industries or industrial phases, succeed the West Asian Mousterian subsequent to about 42–47 thousand years ago. In fact there is some evidence, particularly in the Negev and in the northern Levant, for progressive technological shifts in core-reduction processes and in the production of blade blanks, as well as in the introduction and elaboration of newer tool types and variants, all from a dominantly Mousterian/Levallois facies base. There is, unfortunately, only a scant, poorly preserved human fossil record associated with this important industrial succession. These roots of the UP complex are best illuminated in the northern Levant (Ksar 'Akil, levels xxv–xxi) and in the Negev of the southernmost Levant (Boker Tachtit, level 4). It is appropriate to denominate these occurrences as an Initial Upper Paleolithic, fundamentally not so dissimilar from that once termed Emiran (recognized in occupations from the Emireh, el-Wad F/G, Qafzeh E, and Tabban B caves). In each instance, and subsequently (levels xx–xviii) at Ksar 'Akil, significant additions to the (lithic) technological repertoire are manifest as are also certain important innovations in tool kit variety and composition. Human skeletal associations are unknown with these oldest (=earliest known) Levantine UP occurrences.

The p-demic identity of subsequent Levantine population samples is specifically *Homo sapiens*, with particular affinities, so far as evidence affords a basis for judgment, with broadly contemporaneous European populations, the Cro-Magnons. The oldest occurrences of human cranial remains are apparently those at the caves of Erq el Ahmar E and at Qafzeh D in (unnamed) 'UP-II' contexts (to employ a former terminology). An occurrence of a juvenile cranium at Ksar 'Akil (level xvii, specimen KA-1) is in an Ahmarian-like UP context, and purportedly >35–40 ky in age. Most other occurrences, as at el-Wad D, Kebara E and D, and Hayonim D are in Levantine Aurignacian (formerly Antelian) contexts, all <40 ky. in age. It is commonly held that this distinct industry, which both persists for some 15–16 thousand years and is seemingly penecontemporaneous with the (typologically) distinctive Ahmarian, was intrusive into west Asia from a southeast European source in which it evidently had deep roots. The best preserved, most complete Levantine example, a largely complete (burial) skeleton from Nahal 'En-Gev, is apparently a late manifestation of this same UP industry. Final and immediately post-Pleistocene populations would appear to have their ancestral roots largely in the late Pleistocene of this same region. The Upper Paleolithic is much less well known elsewhere in southwest and also in central Asia. The occurrence at Shanidar C of the Baradostian in Iraqi Kurdistan, is minimally dated ~35 ky BP, and the only slightly, if at all younger Kara-Kamar (Afghanistan) occurrence are notable exceptions. Human remains are unknown in either such instance.

3. WHERE ARE WE?

Empirical documentation of hominid fossil and archeological records of western Eurasia has increased markedly and substantively in recent decades. Such documentation has been achieved within the context of new and significant developments in elucidation and comprehension of late Cenozoic stratigraphy, paleoclimates, and isotopic geochronology of events, formations and situations on global, regional, and local scales. The interface between more continuous marine records and largely discontinuous continental instances of paleontological and/or archeological preservation has often been bridged. The contexts and co-associations represented in such instances, whether of basin-wide or particular locality extent, have been elucidated at more elaborated and ever finer scales through the practice of more systematic and refined observational and recovery procedures and the application of a vastly enhanced battery of analytical techniques incorporated within the armamentarium of modern, scientific archeology, paleontology, and paleoanthropology. Thus the framework within which such investigations are conducted and the perspectives from which results obtained are viewed have changed profoundly and broadened enormously. Such transformations are surely in the realm of the paradigmatic, but probably not so much at the level of the metaphysical as at certainly the level of the construct (methodological) paradigm and, perhaps, verging on the sociological, particularly within prehistoric (behavioral) archeology proper. Distinctions between "processual" and "historicist" perspectives have, at times, been emphasized, their roots purportedly lying at national and disciplinary-affiliated levels. However, such essential differentiations are both overly simplistic and unduly and misleadingly exclusional. Past events and circumstances are by definition historical and necessarily demand conjoint historical documentation and processual analysis requisite to insight into issues of causation.

The primary and essential focus in west Eurasian paleoanthropology must of necessity be an elucidation and sufficient comprehension of hominin populations, as sampled by p-demes. This is a matter of at least (a) demonstration of initial population emplacement through immigration consequent to dispersion or dispersal; (b) explication of timing, spatial extent and magnitude of this emplacement process; (c) evaluation of the broader natural context(s) within which such occurred; (d) establishment of phylogenetic affinities of dispersed population(s); (e) undertakings to gain insight into subsequent population history in respect to demographic parameters, nature/extent of p-demic (or other, lesser units) of interaction, re-dispersal(s), extinctions; and (f) evaluation of adaptive strategies, and attendant technological practices, capabilities, and innovations (through establishment of appropriate analytical units and employment of suitable methodological procedures). This last is particularly troublesome in light of some professed disaffection, disdain, and even outright rejection of some valued culture-stratigraphic categories (including tradition, industry, technocomplex, etc.), both traditionally and currently the focus of comparative and analytical concern among prehistorians. Such is perhaps an unjust attribution of normative (and typological) thinking, as well as undue linkages of culture and ethnicity of Pleistocene space-time systematics.

For the oldest (yet) known west Eurasian p-demes, those of Dmanisi and Atapuerca/Gran Dolina, hardly any such parameters are either capable of being addressed or immediately promising of rewarding resolution. Not the least among the perhaps intractable problems is that of sample size of documentation, in each instance limited (now) to two relevant occurrences. And the affinities, if any, between these successive demic units, separated temporally some $5–7 \times 10^5$ years, is still to be elucidated. The former has some marked African affinities (to *Homo ergaster* hypodigm) whereas the latter is apparently uniquely distinct and most probably merits recognition as a new species' taxon. Both p-demes reveal African linkages via their associated Mode 1 industrial technology.

How is the appearance in Europe of the biface technological tradition to be envisaged? Is this (so-called Mode 2) an autochthnous emergence within Europe from an antecedent Mode 1 technocomplex, itself having demonstrable links through western Asia with the Oldowan complex of eastern equatorial Africa? Or is the Acheulean technocomplex, seemingly appearing belatedly and then disparately in parts of eastern, southern, and western Europe within the earlier Brunhes (N) chron, an allochthonous imposition into the European realm, and thus a consequence of cultural diffusion, hominin dispersal, or both? It is questionable if any such (European) occurrences much (if at all) antedate oxygen isotope stage (OIS) 15; most convincing documentation is rather younger still. The hominin p-deme of this approximate interval is that of Mauer/Arago (MA/ARA) separated substantially in time from the antecedent Atapuerca/Gran Dolina (ATA/GD) p-deme. Their interdemic affinities are still to be adequately elucidated, although such is now feasible. Similarly, there is now the possibility of examining affinities between ancient hominin populations of the African Maghreb (Tighenif p-deme) and their penecontemporaries in peri-Mediterranean Europe.

Were hominin demographics in Europe such that demic transformations, displacements and even replacements might be envisioned between ATA/GD and MA/ARA demic units? The former antedated, and the latter were coincident with the major cyclic, orbitally-induced paleoclimatic deteriorations/rebounds initiated ~800 ka. The former was also perhaps broadly coincident with the Galerian Faunal event of the late Lower Pleistocene, effecting extensive faunal turnover and replacement throughout Europe (from Asia). Further, but quite different faunal change, amidst extensive, repeated habitat transformations, is evidenced subsequently in the span of the Matuyama/Brunhes (=L/M Pleistocene) boundary interval, and in part broadly coincident with MA/ARA p-deme presence.

Another relevant issue is that of the introduction into west Eurasia of Levallois preform-flake blank technology. This has been variously conceived as a consequence of biological (populational) movement, diffusion of a technological concept, and/or a reflection of raw material availability, size specificity, or intensity of such utilization. Much new research devoted to operational (reduction) sequences, to the elucidation of knapping signatures, and appropriate analyses of lithic assemblages, in intact stratified contexts, has substantially altered perspectives on this problem. There is, for example, some demonstrable interrelationship between biface preparation, production and treatment procedures, and purported occurrences of Levallois technical modes, particularly in northwestern Europe and in the Levant. Insight into this linkage is evident as early as OSI-12, at Cagny-la-Garenne (Somme), although the fullest documentation of an elaborated Levallois technology is substantially later, around OSI 8/7. In the Levant such technology is questionably expressed in earlier (i.e., so-called Middle) and subsequent Acheulean industrial contexts, but seemingly appears emergent in Final Acheulean assemblages of late mid-Pleistocene antiquity. Nonetheless it remains a fact that a 'proto'-Levallois preform-flake production technology is apparently widespread in east equatorial, and northern (including Saharan) Africa, has roots and expression quite deep into the mid-Pleistocene, and may exhibit—perhaps in part as a reflection of raw material utilization—discernible variability, development, and elaboration in certain localities in which stratigraphic situations and contexts are favorable for such analysis. This is exemplified in the Baringo basin (East African Rift) wherein the lower reaches of the Kapthurin Formation yield flake-based, and non-biface assemblages based on both radial and opposed platform cores in a lower Brunhes (N) chron situation. Although independent (polyphyletic) development of such predetermined flake-blank production, and its attendant operational (reduction) sequences, has been proposed by several workers, this suggestion minimizes both an African priority and thus potential reality of conceptual diffusion regardless of any consideration of biological population factors. Nonetheless some oft-remarked resemblances between widely-separated African (Kabwe) and European (Petralona/Atapuerca-Sima, or P/A-S) p-demes suggest that perhaps biological factors should not be too readily dismissed out of hand.

The P/A-S is at, or close to the node of differentiation of the Neandertal clade, the distinctiveness of which surely warrants recognition at the species level, as *Homo neanderthalensis*. The roots of the Mousterian technocomplex (so-called Mode 3), in which the Levallois system flourished in several and diverse manifestations, extend well into the upper mid-Pleistocene throughout much of the extent of hominin-occupied west Eurasia. The Mousterian also occurs, perhaps discontinuously, along the Mediterranean littoral of Africa and westward to the Atlantic littoral of the Maghreb. In Europe proper only populations of this (Neandertal) clade are known in association with components of this technocomplex. Some older representatives of the clade are associated with the Taubachian industry (Central and Crimean Europe) and, at times, with (upper) Acheulean industry (western and southern Europe).

The Mousterian technocomplex is well known to manifest very substantial variability in respect to both technological processes and typological composition of lithic assemblages. Efforts at delineating and partitioning this variability led to concepts of Mousterian 'facies' or variants, variously recognized on technological/functional (economic) and/or presumed stylistic grounds. The extent to which analytical procedures and typological systematics have been appropriate to any such delineation, and hence the extent to which any such categories reflect natural, or cognitively/behaviorally real categories has come under intense scrutiny and, not surprisingly, severe criticism. And, in particular, attribution of such presumptive categories to identity-conscious social units (e.g., ethnic groups) has been repeatedly questioned. Sampling factors alone can and do affect assemblage make-up (representativeness), as also do such primary factors as: diverse aspects of raw material resources, their availability, abundance, size, form, and quality; operational modes and sequences employed in blank production, and intensity and extent of reduction/utilization treatment of particular lithic categories; the functional linkage of particular artifact classes/types with specific patterns of human activity; nature, size, duration, and overall intensity of occupational context; and, nature, extent, and setting of the occupational locality itself. Hence it is possible, to an extent, to dissect assemblage composition, variability, and pattern with respect to such processual factors and, in conjunction with other analytical procedures appropriate to study of the composition of associated residues, both organic and inorganic, and attendant natural factors to elucidate such occurrences in far fuller, more holistic fashion.

A seemingly ever-increasing number of distinguishable spatio-temporal variants (manifestations), listed below, are now commonly recognized to be represented within the Middle Paleolithic (Mousterian) technocomplex. The star (*) indicates hominin remains.

- * Charentian Mousterian (CM) complex, with Ferrassie (CM-F) and Quina (CM-Q) groups: pan-Europe
- * Typical Mousterian (TM) complex, including (European) Levalloiso-Mousterian: pan-Europe; strongly laminar components at some (mostly) open-air sites, in northwest, central, eastern Europe, and Crimea
  Levantine Mousterian (LM): west Asian Levant
- * Typical Mousterian/Crvena Stijena type (TM/CS): Balkans
- • Mousterian/Vasconian type (Mva): northern Iberia
- * Denticulate Mousterian (DM): pan-Europe
- • Mousterian/Acheulean tradition (MTA): southwestern Europe
- * Mousterian/Chatelperronian type (M-Ch): southwestern Europe
- • Mousterian/Cambresian type (Mca): northwestern Europe
- * Mousterian/Pontinian type (MP): greater Latium
- * Mousterian/Kartstein type (M-Ka): central Europe
- • Mousterian/Tata type (M-Ta): Bükk mountains
- • Mousterian/Staroselje type (M-St): Crimea
- • Mousterian/Tsutskhvatskaya type (M-Tv): Crimea, western Caucasus
- • Mousterian/Khostinskaya type (M-Kh): western Caucasus
- • Mousterian/Kudaro type (M-Ku): western Caucasus, Georgia
- * Mousterian/Zagros type (M-Z): greater Zagros mountains
- • European Micoquian (MQ), including subgroups *Bocksteinian (MQ-B) (central Europe), Altmühlian (MQ-A) (upper Danube),*Jankovician (MQ-J) (trans/lower Danube), Babonyian (MQ-Ba) (Bükk mountains), and *Kiik-Koban (MQ-KK) and *Ak-Kayan (MQ-AK) (Crimea)
- * Acheuleo-Yabrudian (A-Y), or Levantine Mugharan Tradition (L-MT): Levant

These industrial groups and variants are distinguished on diverse and now much refined technological grounds (raw material exploitation, utilization, reduction sequences, and procedures of blank production) and typological composition (nature, diversity, relative frequencies of retouched pieces, and shaped ‘tools’). They may vary in duration; in local, provincial, or broader regional distribution; and seemingly (sometimes) in stratigraphic superposition or, more broadly, temporal occurrence. Although a matter of some controversy, they do not, apparently readily or even *directly* reflect subsistence patterns and attendant activities. The extent, contrariwise, to which they reflect or somehow mirror resource exploitation, land use practices, and extent and patterns of mobility has been a subject of enhanced interest and recent investigation. This promising development may thus afford linkages between otherwise disparate matters of technology, settlement, subsistence, resource use, provisioning practices, mobility, and group, local, and regional demographics. Such matters are of course of direct concern in the understanding of past populations, their adaptations, and dynamics. And they are central and critical for unraveling the disappearance of Mousterian cultural practices in conjunction with Neandertal extinction.

The mid-portion of OIS-3, between ~45–32 ka, appears now as a critical interval in recent hominin evolutionary history in western Eurasia. The lower and upper limits of the interval are approximately bounded, respectively, by the H5 and H4 (Heinrich) events, episodes of massive enhanced iceberg discharges, attendant surface water cooling and strong oceanic influxes of fresh meltwater, and attendant perturbation of the North Atlantic conveyor circulation. These H-events constitute the repetitive crescendos of antecedent Dansgaard/Orschger (D-O) oscillations, pulses of global cooling reflected in high ^{18}O depletions in North Atlantic marine cores and in Greenland (GISP2, GRIP) ice core records. In the latter instance, a series of milder interstadials (i. 12 through i. 5) of decreasing age and varying duration occur over this interval between the variably spikey or rectangular-shaped D-O oscillations. In this respect i. 12 and i. 8 are particularly noteworthy in their greater duration and lesser ^{18}O depletions (shorter, however, but otherwise comparable to the antecedent i. 14–13 complex, between ~52–46 ka). The ~45–32 ka interval contrasts, rather substantially, with the subsequent pleniglacial conditions of the Late Glacial and those antecedent pre-50 ka conditions of the early Pleniglacial.

Terrestrial records only imperfectly mirror these ameliorations, largely due to preservational circumstances, e.g., erosional truncation, non-sequences, and relative rapidity of sedimentation. However, the Moershoofd (Netherlands) interstadial corresponds, in part, to the earlier reaches of the interval, and the Hengelo (Netherlands) to its later part. Pollen sequences, particularly from cores, at a number of localities in western and southern Europe now document the nature and extent of floral change, including other comparable milder ameliorations, within this interval of the Late Pleistocene. Such evidence indicates substantial reduction of mean annual temperatures and of precipitation, steep north-south vegetational gradients, extensive and poorly and sparsely wooded areas, limited distribution of deciduous taxa, and distinctive mixed evergreen and deciduous, often open parkland circumstances in steppic landscapes in diverse north Mediterranean situations.

The Mousterian technocomplex, in its varied techno-typological manifestations, is well-documented within the earlier half of this interval throughout western Eurasia. This is established on both stratigraphic and on associated palynological grounds in a large number of instances. Increasingly geochronometric methods (some conventional ^{14}C, and more recently AMS, T/L and, to an extent, U-series and ESR) afford isotopic age assessments. (However, their reliability must be appreciated within the perspective of probabilities, error limits, and, particularly, requisite correction factors consequent upon demonstrable changes in geomagnetic field intensity). Thus within the subsequent (40–35 ka [BP]) span isotopic dating already provides ages of a number of Mousterian occurrences in Iberia, France, Italy, Hungary, Rumania, Greece, the Crimea, Caucasus, and the Siberian Altai, unknown until only recently. And, equally interestingly, the persistence of Mousterian (and Neandertals) into the 35–30 ka (BP) range is similarly assessed in the case of cave/shelter occupations in Iberia (Arbreda, Ermitons, Beneito, Zafarraya, probably Cariguela), in the Crimea (Zaskalnaya 6, Kabazi 2) and in the Caucasus (Láska, Akhshtyr). Although perhaps premature, and only a partially substantiated conclusion, there is some suggestion of fragmentation and attendant peripheralization and isolation of once prevailing ranges and associated habitat utilization by Neandertal populations. Various hypotheses or scenarios might, and probably should now, be proposed to account for and to explicate the demographic and adaptational consequences of these altered circumstances within known areal paleoecological frameworks.

Within this overall time span are also manifest new, distinctive, and provincially characteristic industries or industrial variants, not manifest (as such) previously. Some have, obscurely and misleadingly in my view, been termed at times as "transitional" industries, a premature preceptual commitment that tends in fact to preempt understanding and explanation. Two such manifestations appear, overall, rather alike, in part overlapping in time; they have been considered by some as manifestations of a similar (if not the same) industrial complex. These are the Szeletian (of Moravia, southern Poland, and the Bükk mountains), and the Jerzmanovician (of eastern Germany, southern Poland). The latter, once considered distinct because of its unifacial foliates, is only known at some half dozen localities. Both have some (but few) demonstrable Upper Paleolithic-like components in addition to a fundamentally Middle Paleolithic configuration. The Bohunician, restricted to Moravia (and the environs of Brno) and largely known from demonstrably workshop occurrences, occurs also over a span of some 15 ka, at least until about 35 ka. It is essentially a Mousterian/Levallois technique industry, even approximating Upper Paleolithic crested, prismatic-like core forms, with some bifacial foliates and some (not large) number of Upper Paleolithic tool types, particularly in steep-scraper and burin categories (and some dorsally retouched elements). Both local and some more distant raw material sources were exploited. It has been proposed by some (not all!) workers that these particular industrial manifestations correspond, most probably, to an "acculturation" phenomenon between practitioners of an autochthonous Mousterian technocomplex and those of an intrusive early Upper Paleolithic (specifically Aurignacian) technocomplex.

Falling within the last third of the aforesaid interval are occurrences in some dozen open-air localities and in four stratified cave occupations in Italy attributed to a distinctive industry, the Uluzzian. These occur in Tuscany, Campania, and Apulia. The open-air occurrences are problematic, perhaps variously representing mixed (composite) assemblages or Mousterian of some sort. The cave occupations, eight levels in all, differ to an extent among themselves although the distinctiveness seems real enough overall. Stratigraphic evidence demonstrates a post-Mousterian and pre-Aurignacian stratigraphic position at several sites. The Uluzzian combines, variously and minimally, Middle Paleolithic forms (and technology)—(many) denticulates and notches, side-scrapers, scaled pieces—and Upper Paleolithic (specifically Aurignacian) elements—various burins, end-scrapers (including steep and nosed forms), retouched blades (including Aurignacian kinds), some backed pieces, and the distinctive (and eponymous), angular to curved backed knives. Coloring minerals and worked bone (points) also occur. This industrial manifestation may be strongly Upper Paleolithic in aspect, but if and when particular Aurignacian components are represented it is not always inadmissable to reject contamination or other comingling factors in assemblage makeup.

The Chatelperronian, considered here as Mousterian/Chatelperronian (M-Ch) type, has a limited distribution in southern France (south of the Seine, and largely south of the Loire) and, infrequently, in Cantabrian (and probably Catalonian) Spain. Some three dozen valid occurrences are known to date. A few isotopically age-determined occupations (only 4) fall between ~37–32 ka (BP). Thus, by inference, this industrial type falls within the last half of the aforesaid interval, and (most) often in an interstadial amelioration (Denekamp or Cottés interstade). However, some localities in the north (Renne Cave in Yonne, Grand-Roche in Vienne) as well as several others in the south and near-Pyrenean, occur in circumstances indicative of the subsequent, early Pleniglacial. This industrial type always succeeds other Mousterian types/facies—including TM, DM and MTA—in many localities having such antecedent occupations. It has close overall fundamental linkages, both technological and even typological, with MTA-B, a perspective long recognized by some workers. (The geographic distribution of MTA and M-Ch has long been recognized as remarkably coincident.) It is laminar, exhibits soft hammer technique of blank production, and contains (variously) tool types of both Middle and Upper Paleolithic typology, although the latter are not specifically Aurignacian. Chatelperron points, with some geometric variability, and often trimmed for hafting, are distinctive. Occurrences which reflect workshop activities, hunting stations, and habitation occupations at least are all reflected among the known occurrences, which include open sites as well as cave and shelter localities. There is some evidence to suggest more substantial Mousterian techno-typological features in those M-Ch situations which directly succeed other Mousterian occupations. This might, conceivably, reflect industrial commingling and admixture in some such instances. Four localities (at least) are known in which M-Ch occurs interstratified in relation to a specifically Upper Paleolithic industry, the Aurignacian, either in an 'archaic' or so-called 'typical' phase—in Lot (Roc-de-Combe, Le Piage), in Allier (Grotte des Fees), and in Cantabria (El Pendo). Such occurrences demonstrate the independence of these industries and, as well, the close penecontemporaneity of such social (and biological) entities responsible for their production. A single locality, Renne Cave (Arcy-sur-Cure, Yonne) with multiple (3) inhabitations, documents an occupational situation with structures, hearths, and an uncommon elaboration in exploitation of animal (mammal, avian) skeletal resources—bone, teeth, and ivory—for utilitarian as well as ornamental and, perhaps, other decorative purposes. The extent, if any, to which this occurrence (and some half dozen other instances of decorative items) reflects particularly an 'acculturation' phenomenon is both controversial, on the one hand not wholly convincing of such, and on the other, not readily resolvable in any case.

An enhanced consensus has emerged in respect to the distinctive nature and fundamental significance of the Aurignacian in the establishment of west Eurasian Upper Paleolithic (UP) industrial complex. The Aurignacian technocomplex, almost a century after its incontrovertible recognition and stratigraphic fixation, appropriately stands at the base of the UP complex. Such acceptance has been accelerated and forced as a consequence of intensified and well-founded techno-typological analytical procedures, elaboration in excavation and recovery procedures and in associated microstratigraphic analyses, the application of refined methods of isotopic age assessment, and, overall, in a greatly enhanced geographic perspective that oftentimes previously was more narrowly focused at the site-specific or solely local and provincial levels. As a consequence the distinctive character and the role of diverse innovations that typify the Aurignacian are now fully manifest.

a. The Aurignacian (sensu lato) exhibits a remarkably extensive longitudinal distribution—from northern central Asia, southwestern Asia, and throughout continental Europe, including into the southern British Isles. It is notably absent in northern Africa, except potentially for outliers in Cyrenaican Libya (the Dabban industry of the eponymous Hagfet ed Dabba and, especially, in the great Haua Fteah Cave). This distribution apparently altered through time, leading to instances of intensive provincial settlement and resource exploitation in several instances.
b. On the basis of (mean) isotopic age assessments Aurignacian occurrences in Europe span minimally some 15 Ky. There is a demonstrable, if imperfect, cline in ages, older to younger, of occurrences from southeast Europe to south-central Europe to northern Iberia and thence to northern Italy and to southwest France. There is demonstrable age overlap (and interstratification) with M-Ch type occurrences, as well as other evidence indicative of Mousterian industrial groups (including surely TM and DM) contemporaneous with some Aurignacian occur-

rences, particularly in western Europe. Occurrences in northern central Asia and possibly in Cyrenaica apparently have an antiquity seemingly as great as those of eastern Europe.

c. The Aurignacian lacks suitable industrial antecedents within most (or all) of Europe. Hence its roots are often, even generally, considered to have flourished elsewhere, and perhaps, most plausibly, within greater western Asia. One likely source is an 'early' manifestation of Levantine Mousterian (Levallois facies), as epitomized in the ancient occurrence at Mt. Carmel (Tabun D-type or LM-D), and by similar manifestations of much lesser antiquity known at localities in the northern Levant (Ksar 'Akil, Abou Halka in Lebanon), in the southern Levant (Boker Tachtit), and in Jordan as well. The successive occupations revealed at the Ksar 'Akil shelter (levels xxv–xx, in particular) and at Boker Tachtit (levels 1–4) document the nature and process of autochthonous technotypological transformation from a type of M. Paleolithic (LM-D) to a form of Upper Paleolithic (Ahmarian, or early manifestation of Aurignacian). The timing of this emergence, as evidenced by these instances, can only be imperfectly estimated, perhaps between ~50–40 ka BP. Hence there is only slight precedence of UP in the Levant in respect to its otherwise oldest (known) appearance in Balkan Europe. Unfortunately there is a paucity of relevant information from the whole of Anatolia. On the other hand early occupation (>49 ka BP) is evidenced at several sites in west Siberia and the Altai.

d. The Aurignacian complex evinces substantial 'cultural' homogeneity over its very extensive distribution. There is remarkably scant distinguishable regional differentiation. However, some temporal differentiation is manifest. Although former attributed stages (I–IV, V), based on both bone and some lithic types/frequencies, have tended to break down, investigators have proffered schema reflective of space/time delineations for Aquitanian France, central Europe and even much of west Eurasia. Such is evidenced in the increasing employment of different terminologies for earlier-known occurrences (dubbed as 'basal,' 'archaic,' 'proto,' and including also Balkan Bachokirian and 'Olschewian,' the last of which often constitute transient hunting and prey processing stations) in comparison with younger age 'typical' Aurignacian manifestations. The latter were first recognized in southern France and in Cantabrian Spain. Similarly, the Ahmarian and Aurignacian are (largely) successively distinctive industries in the Levant, and the Baradostian is a not dissimilar manifestation in the Zagros foothills and environs. All exhibit some number of distinctive UP technotypological innovations, coupled with their increased presence and diversity, coincident with successive reductions and/or loss of MP elements.

e. The Aurignacian complex is consistently and uniquely found in association with human populations representative of *Homo sapiens*. In Europe over 50 such occurrences are substantiated. The majority are associations with the geologically younger, and so-called 'typical' Aurignacian, the well-known Cro-Magnon people. Only four occurences (in Bulgaria, Croatia, Ligurian Italy, and Cantabrian Spain), and then often fragmentary cranial parts, are demonstrably associated with the antecedent, geologically older Aurignacian; in all instances their known morphology is also that of *Homo sapiens*. Of broadly comparable antiquity are Ahmarian occurrences, three in number and each with fragmentary cranial parts, in the Levant similarly referable to *Homo sapiens*. Consequently the Aurignacian complex, allochthonous to much of Europe and quite abrupt in its appearance there, is a manifestation of population growth (inferentially) and of expansion and colonization, in association with social and technological change (empirically). Such a perspective is not wholly new, but is now supported by an overwhelming body and diversity of evidence.

f. Aurignacian lithic technology exhibits persistence of some traditional (MP) production procedures and few tool types associated with significant (UP) production innovations and attendant diversification of artifact groups and of particular new tool types. The extensive employment of prismatic core platforms, common use of the (soft) punch technique, and attendant production of blades exemplifies this transformation (in addition to flake production employing hard hammer methods). The margins of several types/shapes of blanks are submitted to diverse retouches—varyingly steep, scalariform, high lamellar, lateral blunting, terminal and marginal burin blow(s)—according to preferred end-product and requisite use. The expansion of and standardization within lithic categories is epitomized in the scraper and burin groups, in the varied development and treatment of (marginally) retouched blades, and in the proliferation in retouched bladelet categories and in associated point production (Dufour, Font-Yves, Krems, el-Wad, etc., types). There are attendant hafting and mounting procedures, by inference as such organic materials are rarely preserved.

An extensive employment of organic hard parts, including bone, ivory, teeth and particularly antler is well evidenced. This technology produces a diversity of (5–6 types) distinctive and replicative point forms, mostly in antler, according to treatment of base and tip and form of cross-section. These constitute a significant, innovative projectile armamentarium, also employing hafting and binding methods. Other products include spatulate pieces, polishers, chisels, wedges, and (bone) awls. Diverse activities and actions are employed in such (and other) productions—cutting, incising, splitting, wedging, controlled percussion, grinding, polishing as well as perforation, gouging, drilling in other instances. Such are utterly unknown in MP industries.

g. Artistic and decorative expressions are demonstratively first commonly practiced within Aurignacian industrial contexts. Such instances are found largely, though not exclusively clustered in Aquitanian France and the Alpine forelands of central Europe; still earlier occurrences are known in the Balkans (Temnata Cave). Mobiliary items are most often expressed in media afforded by animal hard parts, notably teeth, ivory and bone, and more rarely antler. Perforated teeth, both ungulate and carnivore species, as well as exotic shell may be abundant. Ivory (and shell) beads, the former produced through an elaborate, multistage process employing well-defined steps of manufacture, are locally common. Other productions comprise statuettes and figurines, both zoo- and anthropomorphic, zoomorphic pendant-like objects, incised, notched and punctated bone shafts and segments (plaques), with both abstract motifs and repetitive, time-transgressive patterns only explicable within the framework of notational schema(s), some purportedly and likely calendrical in function. Shelter walls or derivative fallen blocks may exhibit geometric or stylized quasi-zoomorphic outlines and images created through pecking, engraving, and/or imposed coloration. At least one instance (Chauvet Cave) of expression of realistic, figurative animal art specifically dates from main or 'typical' Aurignacian times. Very probably some half dozen others of similar themata and convention are also of such affinity.

Any and all such expressions have no incontrovertible MP antecedents. There are expressions of individual and group social identity, perhaps also denotations of hierarchically-structured social systems hitherto unevidenced. And the manifestation of already complex symbolic ideational systems are often interpreted in the context of ritual and ceremonial behavior.

Fundamental to evaluation of the Aurignacian, as earliest expression of the European Upper Paleolithic, is an elucidation of settlement system, land and resource use, subsistence practices within a foraging spectrum, and demographics. In all respects assertions and evaluations have been put forward as to the transformation of such adaptations and lifeways in respect to a purported, antecedent Middle Paleolithic state.

Thus early Upper Paleolithic organizational patterns have been considered 'qualitatively different,' and constitute the initial manifestation of a 'creative explosion.' Sites, whether cave, shelter or open-air, show an overall increase in number, and are of different types and sizes (in contrast to 'fewer,' 'small,' 'amorphous,' and 'homogeneous' in the Middle Paleolithic). Moreover, settlement may occur in different, including additional regions (in contrast to more limited, and 'strict regionalization'). There is expansion in exploitation, extraction, and utilization of exotic in addition to local resources (stone, shell, etc.), in contrast to almost exclusive employment of local, particularly near by resources. In respect to provender there is enhanced transport and, presumptively, development of storage methods/facilities (in contrast to minimal transport, and little or no demonstrable storage). Prey procurement involves 'organizationally complex' and 'variable' hunting, with more 'specialized' and 'strategic' practices, and employment of innovative projectile technologies (in contrast to [often] non-confrontational scavenging, and broadly based, eclectic hunting strategies, presumably opportunistic as well as [some] ambush hunting). Demographic parameters are reflected in overall increased population size, greater population densities, and larger, more permanent social groupings (in contrast to smaller, more local even partially isolated and habitat 'bound' aggregates, extensive group [and subgroup] mobility, and frequent practice of seasonal rounds and associated shifts in annual range).

Are such frameworks mere phantasma or inference to the best explanation and, hence, some reflection of verisimilitude? Or, is the null hypothesis of 'no difference,' and largely overlapping social organization capabilities and subsistence practices more correct? The roots of such concern extend well past a century, and are epitomized in, for example, volumes by Wolfgang Soergel (1922) and Kurt Lindner (1950). The extent to which the prehistoric record affords documentation appropriate toward elucidation of such inquiry has often been considered moot. However, the application of taphonomic approaches to archeological residues, and the ever-expanding practice of more intensive zooarcheological investigations, would seemingly promise resolution of traditionally obtuse and even intractable issues. Nevertheless a certain prescriptive treatment of faunal residues from archeological contexts is mandatory to maximize any hope of progress toward such resolution. These include, minimally, (a) full-scale taphonomic evaluations of the roles (if any) of natural and predator agencies in faunal residues accumulation; (b) full taxonomic assignment of assemblage components; (c) presentation of full NISP and MNI values, and estimation of NBP, MNE, and MAU values; (d) full presentation of skeletal representation and frequency (by taxon), evaluation of age classes, and derivation of mortality profiles; (e) estimation of seasonality of death. Unfortunately basic data relevant to (a) are often incomplete or cursory, and that of (d) are still too rarely sufficiently collected or presented in extenso. Such fundamental information bearing on animal resource exploitation is ultimately only revealing of adaptation and behavior when integrated within the context of local habitat; locality setting; site types, particularities and sizes; magnitude, intensity, and frequency of occupation and utilization; composition, diversity, and density of associated artifactual residues; and the nature, situation, and association of cultural features with respect to organic residues; and full explication of the condition, utilization, and potential human employment of animal body parts.

The presence of mammal species reflects of course biogeography, appropriate biotic zones, and suitable provincial and even quite local habitats. Absence, as prey, may mirror regional environmental transformation under glacial regimes. This is further compounded by the demonstrable existence of provincial and local refugia, singularly attractive of predatory species (human and otherwise) as well as some diversity of potential prey.

Examination of a very substantial data base reflecting quantitative inventories of faunal residues of diverse Middle Pleistocene and Upper Pleistocene occupation localities is only somewhat informative. As long known there is substantial and broad overlap in total prey spectrum and often similar predominance of a limited number of ungulate taxa between various culture-stratigraphic or technocomplex categories. Provincial or more local differences are manifest in the exploitation of certain taxa, and these distinctions have not only a biogeographic basis, but may also reflect local factors, conditions, and seasonality. Open-air occupation localities may often afford a limited number of taxa, but in a number of instances notable to greatly enhanced numbers of an individual species. In some such instances evidence is sufficient to demonstrate utilization of such locales for the (sometimes) repetitive, (sometimes) directed, or contrived demise of social groups or herds in favorable natural features and settings, presumably through collective efforts of directed driving, coralling, and employment of natural traps, and (sometimes) in seasonally appropriate circumstances. Uncommonly have appropriate and explicit analytical procedures been employed to elucidate predation practices, faunal exploitation methods, and processing procedures. This is a research focus of singular importance critical to fuller, more comprehensive elucidation of past human lifeways.

4. CONCLUDING REMARKS

Western Eurasia affords evidence of a succession of recognizably distinctive hominin p-demes since colonization of its eastern reaches, by populations of African origin well back in the Lower Pleistocene. Inexplicably, hominin penetration into Europe proper was seemingly retarded by some half million years in comparison with that of western Asia. The nature of this earliest occupation is still poorly documented in Europe, and no hominin is surely associated at any locality. A greatly abbreviated chronology for hominin presence in Europe is not supported by currently available evidence.

The west Eurasian archeological record becomes substantial and fairly widespread geographically only with the Middle Pleistocene (Brunhes [N] chron). Three p-demes are distinguishable prior to that of the Neandertals, established in the upper Middle Pleistocene. Neandertals are linked with the preceding M/A and P/A p-demes as a demonstrable clade (phylogenetic lineage). An earlier, A/GD p-deme seemingly has African (not Asian) affinities, but any immediate relationship with subsequent (M/A and P/A) p-demes remains as yet to be firmly established. The successful, wide-ranging, and long-lasting Neandertal p-deme persisted through the initial Pleniglacial of the Upper Pleistocene and was demonstrably contemporaneous with and even (sometimes and in some places) broadly sympatric with initial (Cro-Magnon, or CM) *Homo sapiens*, intrusive (with the Aurignacian technocomplex) into Europe between 30–40 ka ago. No incontrovertible instance of hybridization between the Neandertal clade (*Homo neanderthalensis*) and *Homo sapiens* has been documented. Phylogenetic affinities are established between the CM and the more ancient S/Q p-deme of the east (and possibly south?) Mediterranean lands. S/Q has antecedents in both the Levant and in the Magreb, as well as affinities with penecontemporaneous populations in (eastern) sub-Saharan Africa. However, human populations linking S/Q and CM p-demes are unknown over an interval encompassing some 50 ka (in spite of substantial archeological documents in successions and sites within the greater Levant, and in parts of northeastern and northwestern Africa). This is an important problem worthy of further intensive research, particularly in largely unknown parts of western Asia (Turkey, Iran, Afghanistan) and adjacent southern central Asia.

Human evolutionary studies are equally historical and processual. The historical documentation afforded by fossil and archeological records affords the basic stuff from which hypotheses are generated, tested through analysis and controlled comparisons, and ultimately eventuate in development of sustainable and productive theoretical frameworks. Some over-arching theoretical perspectives of hominin evolutionary history and process are congruent with neither the fossil record as known nor the increasingly extensive archeological record as testimony of past human distribution, behavior, and adaptation. This latter record, ever enhanced and increasingly elaborated through sounder analytical perspectives and techniques, offers a wealth of insight into the human past. Students of human evolution ignore that record and the insights it affords at their own peril.

REFERENCES

Carter, G. S. 1951. *Animal Evolution: A Study of Recent Views of its Causes.* London: Sidgewick and Jackson.

Gilmore, J. S. L., and J. W. Gregor. 1939. Demes: A suggested new terminology. *Nature* 144:333.

Howell, F. C. 1996. Thoughts on the study and interpretation of the human fossil record, 1–45. In *Contemporary Issues in Human Evolution*, ed. W. E. Meikle, F. C. Howell, and N. G. Jablonski. San Francisco: California Academy of Sciences.

Lindner, K. 1950. *La Chasse Préhistorique.* Paris: Payot.

Mayr, E. 1951. Taxonomic categories in fossil hominids. *Cold Spring Harbor Symposia on Quantitative Biology* 15:109–118.

Mayr, E., and P. D. Ashlock. 1991. *Principles of Systematic Zoology.* New York: McGraw Hill.

Soergel, W. 1922. *Die Jagd der Vorzeit.* Jena: Fischer.

Trinkaus, E. 1990. Cladistics and the hominid fossil record. *American Journal of Physical Anthropology* 83:1–11.

2

CHRONOLOGICAL AND BIOGEOGRAPHIC PERSPECTIVES ON LATER HUMAN EVOLUTION

Chris Stringer

Department of Palaeontology
The Natural History Museum
London SW7 5BD, United Kingdom

1. INTRODUCTION

The last decade has seen remarkable advances in our knowledge of the distribution of early humans in time and space. In Africa, the earliest appearance of possible members of the genus *Homo* has been pushed back to at least 2.3 Ma in Ethiopia, and perhaps 2.5 Ma based on evidence from Kenya and Malawi (Kimbel et al. 1996). However, it is now unclear where the earliest examples of more advanced humans first appeared. Specimens attributed to *Homo erectus* or *Homo ergaster* are recorded from Koobi Fora at 1.8–1.9 Ma, while sites yielding specimens attributed to *Homo erectus* in Indonesia have been dated, more controversially, to 1.6–1.8 Ma (Culotta 1995). There is also a claim for a very early human presence in China, but the hominine nature of the main fossil, a mandibular fragment, is in dispute (Culotta 1995; Huang et al. 1996; Schwartz and Tattersall 1996a). From Europe there have been very controversial claims for a similarly early human presence, both from archaeology and fossils, at Orce in southern Spain (Dennell and Roebroeks 1996), while an undoubted hominid of *Homo erectus* affinities is known from Dmanisi in Georgia, close to the eastern borders of Europe (Gabunia and Vekua 1995). Here a possible age of 1.6–1.8 Ma is also claimed. Of more recent antiquity, perhaps close to the Lower-Middle Pleistocene boundary, a larger fossil sample has been excavated from the Gran Dolina locality at Atapuerca (Carbonell et al. 1995), while a calvaria reconstructed from cranial fragments has been recovered from Ceprano, in central Italy (Ascenzi et al. 1996). The taxonomic attribution of these later European fossils is still unclear, although the Ceprano specimen shows the more immediate affinities to *Homo erectus*. No comparably ancient hominid material has yet emerged from southwest Asia, although it must exist, while archaeological evidence does take a human presence there to about 1.4 Ma at 'Ubeidiya (see various papers this volume).

Neandertals and Modern Humans in Western Asia, edited by Akazawa *et al.*
Plenum Press, New York, 1998.

While some of the data remain controversial, these recent developments, taken together, show that humans had probably dispersed across the lower latitudes of the Old World by, or during, the earliest Pleistocene (about 1.6 Ma), and were probably capable of at least a sporadic human presence in southern Europe by that time. This provides a time scale for consideration of the evolution of our own species. Under the multiregional model, this dispersion would represent the beginning of the evolution of *Homo sapiens* and its regional variants. Under single origin models, these early dispersals were radiations which probably led to cladogenesis, and the evolution of separate human lineages in different regions, only one of which ultimately gave rise to *Homo sapiens*.

2. THE EVOLUTION OF THE NEANDERTAL AND MODERN HUMAN CLADES

Claims for the existence of separate species of humans in the later Pleistocene have been strengthened considerably by recent developments in dating and morphological studies. From this perspective, the evolution of *Homo neanderthalensis* and its probable sister species, *Homo sapiens*, are intimately linked, and dating the time of origin of one clade will date the origin of the other. Recognition of the Neandertals as a separate species seems increasingly justified as their derived morphology becomes more apparent. Computerized tomography (CT) has allowed detailed study and reconstruction of Neandertal crania and comparisons with other hominines. These have confirmed the distinctive morphology and robusticity of Neandertal children, such as the Devil's Tower individual (Zollikofer et al. 1995), while examination of Neandertal temporal bones has revealed completely unsuspected idiosyncrasies in the shape of the semi-circular canals of the inner ear, which serve to distinguish all known Neandertals from earlier and later humans (Hublin et al. 1996). The functional basis of these differences is not yet understood, but the characteristic Neandertal morphology is present even in young children and must be a feature under genetic control, since the form of the labyrinth is determined before birth. Recent research on the nasal form of Neandertals has also produced evidence of a derived morphology of the nasal rim and conchae, where these are well preserved (e.g., in the Forbes' Quarry Neandertal from Gibraltar), and it may well be that these distinctions are related to the respiratory demands of a high metabolic rate combined with the relatively cold and dry environments of glacial times (Schwartz and Tattersall 1996b).

In the postcranial skeleton, there is also evidence of Neandertal specializations. While their general pattern of robusticity was undoubtedly an inheritance from Middle Pleistocene ancestors, Neandertal body shape was distinctive, paralleling that of recent cold-adapted populations. It is unclear to what extent their shortened distal extremities reflect adaptation to reduce skin surface area and hence heat loss (Trinkaus 1981; Ruff 1993), or the effects of cold on growth through factors such as vasoconstriction (Trinkaus 1981; L. Aiello and P. Wheeler, personal communication), but the impact of temperate-boreal environments and the absence of strong cultural buffering remains the same. In fact, the apparent presence of the distinctive Neandertal body shape in a child as young as the Dederiyeh infant (Akazawa et al. 1995) certainly seems to emphasize the importance of a genetic component in its development. Their distinctive pelvic morphology is also notable, but here the polarity of the traits concerned and the factors behind the anatomy are less clear. While gestation-length and obstetric considerations seem to be excluded, the functional anatomy of a hip joint which operated differently from our own remains a possible underlying consideration (Rak 1993). The known differences in orientation and implied functional anatomy of the ilium in fossils attributed to *Homo ergaster*, *Homo erectus*, and *Homo heidelbergensis* from those of *Homo sapiens* would suggest that the distinctiveness of the pelvis in *Homo neanderthalensis* might be plesiomorphous. However, this growing body of new data about the apomorphous nature of the Neandertals, which amplifies those characters already known from the cranium and mandible, leads us on to the difficult question of recognizing the first hominines to show significant numbers of such traits, and the seemingly even more difficult question of recognizing primitive members of an equivalently early *Homo sapiens* clade. This latter clade, of course, would need to be diagnosed not only through a lack of Neandertal apomorphies (which would also be expected in more plesiomorphous forms), but additionally through the presence of a comparable body of derived *sapiens* traits.

A trend is developing for an ever earlier recognition of the Neandertal clade. The Swanscombe "skull" is generally regarded as a primitive Neandertal, based on its occipital features (Stringer et al. 1984), and recent research on the geomorphology of the Thames terraces suggest that it may be from deposits of oxygen isotope stage 11 age (about 400 ka) (Bridgland 1994). The Arago material is similarly seen as displaying proto-Neandertal features, and may be of approximately the same antiquity (Stringer et al. 1984). The more ancient Mauer mandible is seen by some workers as showing characters which anticipate those of the Neandertals, while some workers have hinted that even the early Gran Dolina sample from Atapuerca shows such characters (Carbonell et al. 1995). However, as indicated above, recognition of a Neandertal clade stretching back to the early Middle Pleistocene would imply that the modern human clade must be of similarly great antiquity. But where are the correspondingly early members of the *Homo sapiens* clade? The Bodo cranium, now dated at about 600 ka, hardly seems appropriate with its host of archaic features (Rightmire 1996), and even the presumably younger Elandsfontein and Broken Hill crania show plesiomorphous or arguably (for example in the face of Broken Hill) apomorphous characteristics which differ strongly from those of *Homo sapiens*. Perhaps a better case can be made for more gracile specimens like Ndutu, but the complicating factor of probable marked sexual dimorphism in these Middle Pleistocene populations means that its facial gracility might merely reflect its derivation from a small female.

At about this time in Europe, we can at last get a better hold on intrapopulation variability from the remarkable Atapuerca Sima de los Huesos sample. This site has so far yielded a sample of over 30 individuals of both presumed sexes, and of various developmental ages. If we can assume that the sample is penecontemporaneous, then its degree of variation is very instructive (Arsuaga et al. 1993). For example, there is a large variation in cranial size, with endocranial capacity virtually spanning that already known from the entire European Middle Pleistocene (about 1150–1400 ml). Facial shape ranges from specimens reminiscent of Petralona, with inflated cheekbones and clear midfacial prognathism, in this case surprisingly associated with one of the smallest individuals, to much larger facial fragments which appear to show a concave and non-Neandertal zygomaxillary form. There is similar variation in many other morphological features such as brow ridge form, mandibular size and shape, and temporal morphology, but overall the sample does display an incipient Neandertal character, especially in features such as a suprainiac region which already shows stages in the development of the idiosyncratic Neandertal suprainiac fossa. However, the team describing the Atapuerca specimens points out that individually they display almost haphazard combinations of plesiomorphies and Neandertal apomorphies. This supports the view that many Neandertal characters probably began as uncommon polymorphic variants, but gradually increased in frequency as selection operated on populations, increasing their frequency until some reached fixation in the later "classic" Neandertals. The Atapuerca specimens illustrate perfectly a morphological transition taking place between populations showing characters found in (predominantly or assumed) earlier European hominids assigned to *Homo heidelbergensis* and those showing the full suite of Neandertal apomorphies. Whether they actually represent ancestors of succeeding Neandertals or not, they certainly match theoretical expectations of the character of such ancestors. However, they support an accretionary model of "neandertalisation," not one where each Neandertal character evolves in turn and reaches fixation. If this was the course of Neandertal evolution to the north of the Mediterranean, then it may provide clues to the possible course of *Homo sapiens* evolution to the south.

My view up to now has been that a specimen should display a certain minimum number of morphological or metrical characters found across modern human populations before it could be assigned to *Homo sapiens* (Stringer et al. 1984). Thus the more complete early late Pleistocene specimens from Skhul and Qafzeh (Israel) could be so assigned, as well as the Omo Kibish 1 partial skeleton from Ethiopia. But more fragmentary specimens, such as those from Klasies River Mouth and Florisbad (South Africa) or Guomde (Kenya), have insufficient diagnostic characters preserved to properly test the relevant criteria. However, it is evident that an accretionary pattern of evolution within the *Homo sapiens* clade would mean that earlier specimens might lack the majority of the diagnostic characters of *Homo sapiens* today, but would nevertheless be more closely related to *Homo sapiens* cladistically than to any other taxon. Applying a similar test of Neandertal apomorphies to individuals within the Atapuerca sample, most would lack a majority of Neandertal traits, yet across the sample, most Neandertal derived characteristics are represented in it. A related point has been raised by Lahr (1996). She suggested that using only the derived characteristics of recent *Homo sapiens* biases the recognition of the species in the fossil record, because there have been major reductions in skeletal robusticity since the end of the Pleistocene, and a consequent decrease in overall patterns of cranial variation. Specimens such as Ngaloba, Omo Kibish 2, and Jebel Irhoud should be included in a more widely defined *Homo sapiens* clade. If her views are accepted, then we can certainly extend the age of this clade back beyond 130 ka (the approximate age of the Omo Kibish 1 skeleton, based on uranium series dating). The Irhoud material may well date from oxygen isotope stage 6 (about 190–130 ka), based on ESR dating of fauna (Grün and Stringer 1991), while the Singa calvaria is of comparable age based on ESR/U-S dating (McDermott et al. 1996). The Ngaloba cranium has an estimated minimum age of about 200 ka based on ostrich egg shell amino acid correlations with sites dated by argon-argon (Manega 1995), while direct ESR dating of the Florisbad upper molar has recently produced an unexpectedly great age estimate of about 260 ka (Grün et al. 1996). Direct gamma ray dating on the Guomde KNM-ER 999 femur and KNM-ER 3884 partial cranium, both of which show combinations of archaic and more modern features, points to a similar antiquity (Bräuer et al. 1997).

Thus taking a rather conservative view of the recognition of members of the Neandertal clade, which excludes the most ancient claimed members such as Atapuerca Gran Dolina and Mauer, but a more radical view of the recognition of the modern human clade, which includes specimens such as Ngaloba and Florisbad, we can arrive at a compatible estimate for their ages of origin. Using current estimates of the ages of the Atapuerca Sima de los Huesos and Ehringsdorf early Neandertal samples, together with the age estimates already discussed for the proposed early African members of the *Homo sapiens* clade, a likely minimum date for the beginning of their evolutionary divergence is about 250–300 ka. To what extent gene flow between these two clades continued after this date is uncertain, and thus evidence from any region of geographical overlap or contact between the respective areas of evolution of the two clades is of critical importance for our understanding of the course of later human evolution. Thus a proper understanding of the human fossil record of the Levant must remain a primary objective of current palaeoanthropological research. If the Neandertal and *Homo sapiens* clades maintained their separate identity through a long period of overlap in their geographical ranges, as certainly seems to be the case for the Levant in the late Pleistocene, then this provides powerful evidence in support of a specific difference between them.

3. THE SIGNIFICANCE OF THE LEVANTINE RECORD

Although the discovery of the enigmatic Zuttiyeh fossil heralded the beginning of the fossil human record in the Levant, it was the Mount Carmel excavations at Skhul and Tabun which really established the lasting importance of the region to palaeoanthropology. Remarkable advances in our knowledge of the Pleistocene chronological framework of the region have been made, and yet there are still many uncertainties which hinder further progress. The chronological and evolutionary status of the Zuttiyeh fossil itself is still unclear, and it is difficult to align it convincingly with either the Neandertal or *Homo sapiens* clades—indeed perhaps it belongs to neither, although it would require the application of a technique like direct gamma ray dating to attempt to establish its antiquity. However, we now know that primitive modern humans were buried at Skhul and Qafzeh, while Neandertals were buried in the Tabun, Kebara, Amud, and Dederiyeh Caves. The Tabun archaeological sequence remains a key one for the whole Levant, and we have seen a remarkable change in views concerning the probable time range of the Tabun deposits. From Jelinek's chronology, which placed the main sequence entirely in the late Pleistocene, and the Tabun Neandertal skeleton at probably about 50 ka (Jelinek 1982), we have moved through an approximate doubling of the estimated ages for Beds B–E with an ESR chronology based on dating Garrod's fauna (Grün et al. 1991), and an approximate further 50% increase in the possible ages of Beds C–E from TL dating of burnt flints from Jelinek's excavations (Mercier et al. 1995). It is certainly premature to judge which of the various chronologies will prove to be the most accurate, but it seems certain that it will be a long, rather than a short, time-scale, in keeping with the estimated ages of comparable archaeological sequences in neighboring regions.

But even given a reliable chronology for the Tabun archaeological sequence, there remain major uncertainties about the stratigraphic or taxonomic position of the most important hominids from Tabun (see various papers in this volume). The Tabun skeleton derived from a lateral position close to the cave wall, where Garrod's units thinned and were difficult to recognize, so it is unclear whether it derives from her Beds B, C (the usual attribution), or even D. These various assignments could provide an enormous range of possible ages, and it will probably be necessary to rely on direct dating to establish a realistic age for this important specimen. Current attempts center on gamma ray dating of the mandible (Simpson, Schwarcz, and Stringer, in progress) and direct ESR dating of an enamel flake from a mandibular molar (Grün and Stringer, in progress). While the skeleton is quite reasonably attributed to a Neandertal, its evolutionary position remains uncertain, but CT reconstruction on the whole skeleton (Zollikofer, Ponce de Leon, and Stringer, in progress) should help place the specimen more precisely within the known range of early and late Neandertal variation. The Tabun isolated mandible, which can be more definitely attributed to Bed C, has recently been the subject of several restudies, with conflicting results (this volume). Multivariate studies by Smith and colleagues suggest modern affinities, as does Rak's morphological analysis, while Trinkaus and colleagues find that its affinities are with Neandertals and other archaic hominids. Of course, if the most ancient estimates for Bed C are accurate (about 170 ka; Mercier et al. 1995), then the mandible might represent a primitive Neandertal, a primitive *Homo sapiens*, or even some other, currently poorly known, population.

The site of Skhul, and its Middle Paleolithic artifacts and human burials, has similarly undergone major reevaluations in the last few years. From age estimates based on inferred archaeological correlations with the Middle Paleolithic sequence of Tabun of about 45 ka (Jelinek 1982), we have moved on to ESR age estimates from fauna of about 80–100 ka (Stringer et al. 1989) and TL dates on burnt flints of about 120 ka (Mercier et al. 1993). Recent work raises the possibility that the site has a more complex depositional history, covering quite a wide time range (McDermott et al. 1993). Corruccini (1992, 1993) has reemphasized the morphological variation in the Skhul sample, arguing that specimens such as Skhul 4 and 9 are markedly more Neandertal-like and less "modern" than the better-known Skhul 5. There may be some validity in this view, at least for Skhul 9, provided we substitute "archaic" for Neandertal-like, for it is possible that this specimen is the earliest of the Skhul burials, and perhaps (allowing for the uncertainties about the age of the Tabun skeleton and the range of the Qafzeh age estimates) the oldest known of all intentional burials. We (Stringer, Grün, and McDermott) are currently attempting to date the Skhul 5 and 9 specimens by ESR/U-S dating of putatively associated fauna. Skhul 5 was buried with the mandible of a pig in his arms, and a date for a tooth of that mandible should provide a reasonable age estimate for the burial itself. In contrast, McCown (1937) argued that the bovid skull found with Skhul 9 was inserted into the burial at a later date, destroying the legs, and must therefore post-date it. Hence, dating a tooth of the bovid skull should provide a *minimum* age for the Skhul 9 burial. However, as Skhul 9 was buried virtually on the rock floor of the cave, there remains the possibility that the bovid skull was actually excavated during the digging of the grave, and was then added to the burial as part of the fill, in which case it would instead provide a *maximum* age for the burial. Thus it will be necessary to compare uranium levels in the various Skhul specimens to check possible associations, and in the longer term, carry out direct gamma ray dating of human bones, or ESR dating of enamel flakes where human teeth are preserved (but unfortunately, Skhul 9 has no associated teeth).

The early modern burial site of Jebel Qafzeh has less evident dating problems than Tabun and Skhul, but most dating work has been conducted on fauna or flints from the terrace, and it would certainly be valuable to have age estimates for the internal cave deposits which yielded the Middle Paleolithic human material found in the earlier excavations as well. In particular, the two most complete adult crania, Qafzeh 6 and Qafzeh 9, contrast markedly in morphological features and shape, especially in dental size and face shape. These differences cannot be attributed solely to sexual dimorphism, particularly as the supposed female specimen, Qafzeh 9, has the larger teeth. My suspicion is that both the Skhul and Qafzeh Middle Paleolithic human skeletal samples will turn out to have longer time ranges than currently believed.

For the Neandertal sites apart from Tabun, it is still possible to argue that they are associated with a particular Middle Paleolithic variant (Tabun B; see Bar Yosef, this volume), and available dating evidence for Kebara and Amud does not contradict such a scheme, although we await age estimates for the Dederiyeh site. Perhaps Neandertal occupation of the Levant was indeed episodic, and it would be valuable to identify the environmental factors that might have controlled this either directly, or indirectly, by affecting the geographical range of early modern humans, who would perhaps have competed for the available resources in the region.

4. GENETIC ANALYSES

While we await the exciting possibility of recovering meaningful ancient DNA from fossil hominids, a great deal of inferential information continues to be reconstructed from the DNA of living humans (Nei 1995). For example, analysis of short tandem repeat polymorphisms and Alu alleles shows patterns of variation which suggest that non-African populations are sampled from a small sub-set of North or East African populations, with an estimated age for an "Out of Africa" dispersal of about 100 ka (Tishkoff et al. 1996). Takahata (1995), using coalescence analyses from nuclear and mitochondrial DNA, argued that the effective (breeding) human population size in the Pliocene-early Pleistocene was about 100,000 individuals using nDNA, but from the low variation inherent in modern human mtDNA, this must have dropped to about 10,000 individuals in the later Pleistocene. In my view, this contraction or bottleneck may in fact reflect a partitioning of the widespread common ancestral population of Neandertals and modern humans by climatic change in the Middle Pleistocene into smaller isolates. Only one of these isolates, about 10,000 in number and restricted to Africa, then gave rise to living *Homo sapiens*. It seems most likely that the critical biogeographic barrier was the Sahara, increasing in scale through each cold stage of the later Pleistocene, but this still remains to be established.

5. CONCLUDING REMARKS

In this review, I have tried to show why accurate chronological data can be as important as morphological study of the fossils in developing meaningful evolutionary scenarios in human evolution. Recent intense debate over ESR dating of fauna apparently associated with *Homo erectus* fossils from Ngandong and Sambungmachan (Gibbons 1996; Swisher et al. 1996) shows that while dating is not essential for morphological or systematic studies, it is of vital importance in the testing of evolutionary scenarios. We still have a long way to go in both morphological and geochronological studies, but it seems certain that the astonishing progress we have made in the last decade will continue over the next one, as we enter the new millenium.

ACKNOWLEDGMENTS

I would like to thank Takeru Akazawa, Ofer Bar-Yosef, and all their colleagues who made the symposium "Neanderthals and modern humans in West Asia" such a success in Tokyo in November, 1995.

REFERENCES

Akazawa, T., S. Muhesen, Y. Dodo, O. Kondo, and Y. Mizoguchi. 1995. Neanderthal infant burial. *Nature* 377:585–6.

Arsuaga, J. L., I. Martinez, A. Gracia, J. L. Carretero, and E. Carbonell. 1993. Three new human skulls from the Sima de los Huesos Middle Pleistocene site in Sierra de Atapuerca. *Nature* 362:534–7.

Ascenzi, A., I. Bidditu, P. Cassoli, A. Segre, and E. Segre-Naldini. 1996. A calvarium of late *Homo erectus* from Ceprano, Italy. *Journal of Human Evolution* 31: 409–23.

Bräuer, G., Y. Yokoyama, C. Falguires, and E. Mbua. 1997. Modern Human oringins backdated. *Nature* 386: 337–8.

Bridgland, D., ed. 1994. *Quaternary of the Thames*. London: Chapman and Hall.

Carbonell, E., J. M. Bermudez de Castro, J. L. Arsuaga, J. C. Diez, A. Rosas, G. Cuenca-Bescos, R. Sala, M. Mosquera, and X. P. Rodriguez. 1995. Lower Pleistocene hominids and artefacts from Atapuerca - TD6 (Spain). *Science* 269:826–30.

Corruccini, R. S. 1992. Metrical reconsideration of the Skhul IV and IX and Border Cave 1 crania in the context of modern human origins. *American Journal of Physical Anthropology* 87:433–45.

Corruccini, R. S. 1993. Reagonomics and the fate of the progressive Neandertals, 697–708. In *Integrative Paths to the Past*, ed. R. Corruccini and R. Ciochon. New Jersey: Prentice Hall.

Culotta, E. 1995. Asian hominids grow older. *Science* 270:1116–7.

Dennell, R., and W. Roebroeks. 1996. The earliest colonisation of Europe: The short chronology revisited. *Antiquity* 70:535–42.

Gabunia, L., and A. Vekua. 1995. Plio-Pleistocene hominid from Dmanisi, East Georgia, Caucasus. *Nature* 373:509–12.

Gibbons, A. 1996. *Homo erectus* in Java: A 250,000 year anachronism. *Science* 274:1841–2.

Grün, R., and C. Stringer. 1991. Electron spin resonance dating and the evolution of modern humans. *Archaeometry* 33:153–99.

Grün, R., C. Stringer, and H. Schwarcz. 1991. ESR dating of teeth from Garrod's Tabun cave collection. *Journal of Human Evolution* 20:231–48.

Grün, R., J. Brink, N. Spooner, L. Taylor, C. Stringer, R. Franciscus, and A. Murray. 1996. Direct dating of Florisbad hominid. *Nature* 382:500–1.

Huang, W., Y. Gu, R. Ciochon, R. Larick, Q. Fang, J. de Vos, H. Schwarcz, W. Rink, and C. Yonge. 1996. Whose teeth? *Nature* 381:201–2.

Hublin, J. J., C. Barroso Ruiz, P. Medina Lara, M. Fontugne, and J. L. Reyss. 1995. The Mousterian site of Zafarraya (Andalucia, Spain): Dating and implications on the Palaeolithic peopling processes of Western Europe. *Comptes Rendus de l'Académie de Sciences* 321(IIa):931–7.

Hublin, J. J., F. Spoor, M. Braun, F. Zonneveld, and S. Condemi. 1996. A late Neanderthal associated with Upper Palaeolithic artefacts. *Nature* 381:224–6.

Jelinek, A. 1982. The Tabun Cave and Palaeolithic man in the Levant. *Science* 216:1369–75.

Kimbel, W. H., R. Walter, D. C. Johanson, and 14 others. 1996. Late Pliocene *Homo* and Oldowan tools from the Hadar formation (Kada Hadar Member), Ethiopia. *Journal of Human Evolution* 31:549–62.

Lahr, M. 1996. *The Evolution of Modern Human Diversity: A Study of Cranial Variation*. Cambridge: Cambridge University Press.

Manega, P. 1995. *New Geochronological Results from the Ndutu, Naisiusiu and Ngaloba Beds at Olduvai and Laetoli in Northern Tanzania: Their Significance for Evolution of Modern Humans*. Paper presented at the conference "Preservation and use of Olduvai Gorge, Laetoli, Rock art and other palaeoanthropological resources in Tanzania." Bellagio, Italy.

McCown, T. 1937. Mugharet es-Skhul. Description and excavations, 91–107. In *The Stone Age of Mount Carmel*, Vol. 1, *Excavations at the Wadi-el-Mughara*, ed. D. A. E. Garrod and D. M. A. Bate. Oxford: Clarendon.

McDermott, F., C. Stringer, R. Grün, and C. Hawkesworth. 1993. Mass-spectrometric U-series dates for Israeli Neanderthal/early modern hominid sites. *Nature* 363:252–5.

McDermott, F., C. Stringer, R. Grün, C. T. Williams, V. Din, and C. Hawkesworth. 1996. New Late-Pleistocene uranium-thorium and ESR dates for the Singa hominid (Sudan). *Journal of Human Evolution* 31:507–16.

Mercier, N., H. Valladas, O. Bar-Yosef, B. Vandermeersch, C. Stringer, and J. L. Joron. 1993. Thermoluminescence date for the Mousterian burial site of Es-Skhul, Mt. Carmel. *Journal of Archaeological Science* 20:169–74.

Mercier, N., H. Valladas, G. Valladas, J. L. Reyss, A. Jelinek, L. Meignen, and J. L. Joron. 1995. TL dates of burnt flints from Jelinek's excavations at Tabun and their implications. *Journal of Archaeological Science* 22:495–509.

Nei, M. 1995. Genetic support for the out-of-Africa theory of human evolution. *Proceedings of the National Academy of Science USA* 92:6720–2.

Rak, Y. 1993. Morphological variation in *Homo neanderthalensis* and *Homo sapiens* in the Levant: A biogeographic model, 523–36. In *Species, Species Concepts and Primate Evolution*, ed. W. Kimbel and L. Martin. New York: Plenum Press.

Rightmire, G. P. 1996. The human cranium from Bodo, Ethiopia: Evidence for speciation in the Middle Pleistocene? *Journal of Human Evolution* 31:21–40.

Ruff, C. 1993. Climatic adaptation and hominid evolution: The thermoregulatory imperative. *Evolutionary Anthropology* 2:53–9.

Schwartz, J., and I. Tattersall. 1996a. Whose teeth? *Nature* 381:201–2.

Schwartz, J., and I. Tattersall. 1996b. Significance of some previously unrecognized apomorphies in the nasal region of *Homo neanderthalensis*. *Proceedings of the National Academy of Science USA* 93:10852–4.

Stringer, C., J. J. Hublin, and B. Vandermeersch. 1984. Western Europe, 51–135. In *The Origins of Modern Humans*, ed. F. Smith and F. Spencer. New York: Alan R. Liss.

Stringer, C., R. Grün, H. Schwarcz, and P. Goldberg. 1989. ESR dates for the hominid burial site of Es Skhul in Israel. *Nature* 338:756–8.

Swisher, C., W. Rink, S. Anton, H. Schwarcz, G. Curtis, A. Suprijo, and Widiasmoro. 1996. Latest *Homo erectus* of Java: Potential contemporaneity with *Homo sapiens* in Southeast Asia. *Science* 274:1870–4.

Takahata, N. 1995. A genetic perspective on the origin and history of humans. *Annual Review of Ecology and Systematics* 26:343–72.

Tishkoff, S., E. Dietzsch, W. Speed, A. Pakstis, J. Kidd, K. Cheung, B. Bonne-Tamir, A. Santachiara-Benerecetti, P. Moral, M. Krings, S. Paabo, E. Watson, N. Risch, T. Jenkins, and K. Kidd. 1996. Global patterns of linkage disequilibrium at the CD4 locus and modern human origins. *Science* 271:1380–7.

Trinkaus, E. 1981. Neanderthal limb proportions and cold adaptation, 187–224. In *Aspects of Human Evolution*, ed. C. Stringer. London: Taylor and Francis.

Yokoyama, Y., C. Falgueres, and R. Bibron. 1988. Direct dating of Neanderthalian remains and animal bones by the non-destructive gamma-ray spectrometry: comparison with other methods, 135–141. In *L'Homme de Neandertal* Vol. 1, ed. H. P. Schwarcz. Liège: ERAUL.

Zollikofer, C., M. Ponce de Leon, R. Martin, and P. Stucki. 1995. Neanderthal computer skulls. *Nature* 375:283–5.

3

THE CHRONOLOGY OF THE MIDDLE PALEOLITHIC OF THE LEVANT

Ofer Bar-Yosef

Department of Anthropology
Peabody Museum
Harvard University
Cambridge, Massachusetts 02138

1. INTRODUCTION

In 1980, following the first conference on the "Prehistory of the Levant" (Cauvin and Sanlaville 1981), we (the author and B. Vandermeersch) realized that the accepted Middle Paleolithic chronology was too short and the ordering of sites and industries was not well-founded. We therefore initiated a joint project to re-excavate and date key cave sites, including Kebara, Qafzeh, and Hayonim, with the hope of establishing a more sound chronology for the Levantine Mousterian. This volume contains many of the contributions that have been made to this on-going project.

By now, the difficulties in establishing the chronology of the Middle Paleolithic assemblages in the Levant are fairly well-known. The earlier chronological model was based on the European concept that prevailed through the 1960s, namely, that the Mousterian industries date solely to the Last Glacial Age. Different opinions first emerged in Western Europe when assemblages bearing Mousterian characteristics were securely dated to the Penultimate Glacial cycle (or Riss III in the French Alpine-based time scale). In addition, dates from the MSA sequence in East and South Africa indicated a similar tendency towards increasing antiquity. However, the major shift began when new radiometric dating techniques became available (e.g., Aitken 1990). These included thermoluminescence (TL), electron spin resonance (ESR), Uranium series, and amino acid racemization (AAR).

Middle Paleolithic sites of the Levant (Figure 1) have produced numerous assemblages since the inception of systematic excavations in the 1920s. In the 1930s, both Garrod and Neuville used French terminology to define the industries. Garrod recognized the inadequacy of this terminology which, at her time, distinguished between Levalloisian and Mousterian industries based on the proliferation of classical Levallois products in one entity and retouched pieces in the other. Her subdivision (Garrod and Bate 1937) into two stratified entities named Lower and Upper Levalloiso-Mousterian was employed by others for many years.

Neandertals and Modern Humans in Western Asia, edited by Akazawa *et al.*
Plenum Press, New York, 1998.

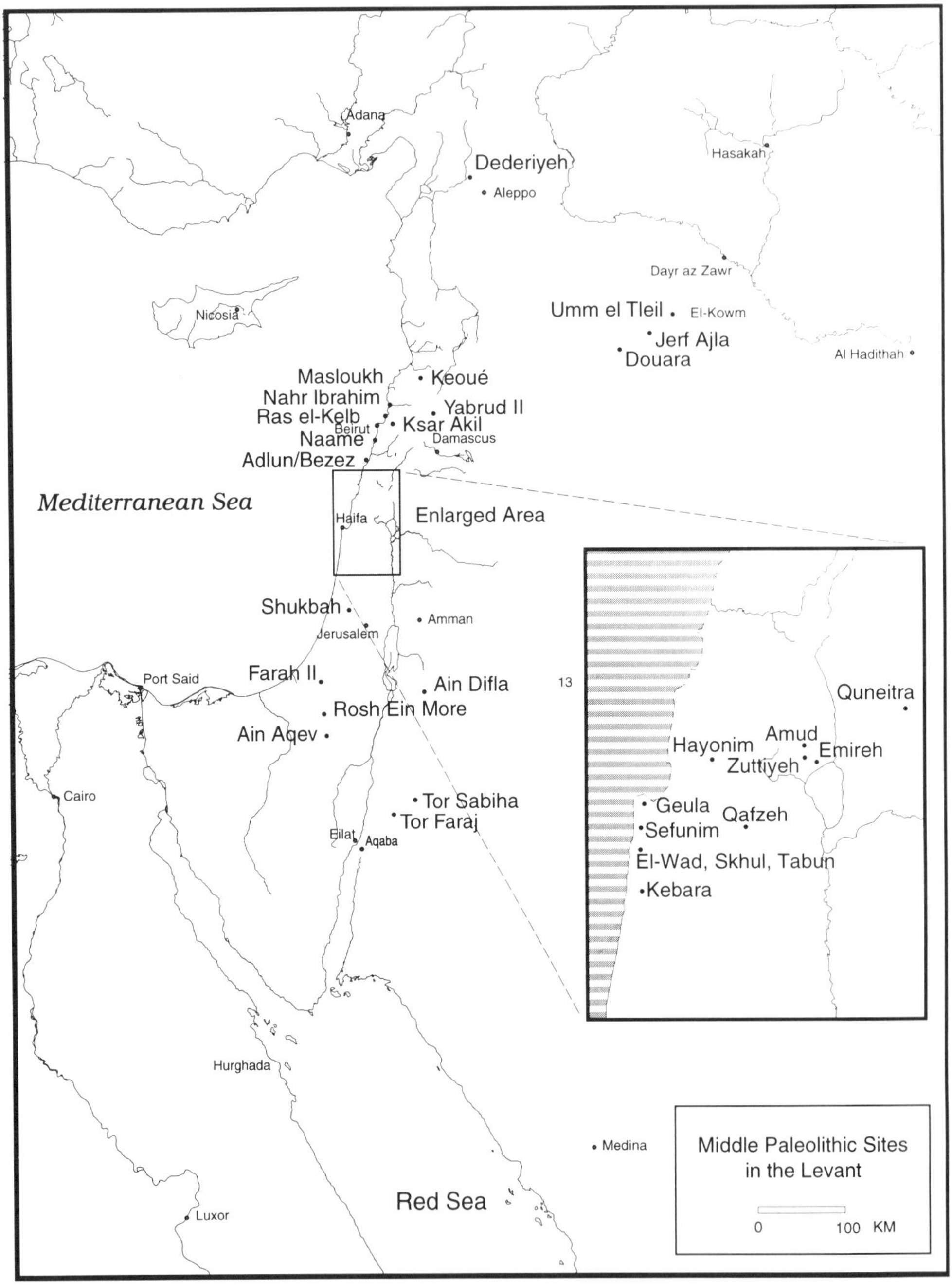

Figure 1. The Middle Paleolithic sites in the Levant.

New classification followed Bordes' studies of the Yabrud I assemblages (Bordes 1956). Employing the Bordesian type-list (Bordes 1961), Perrot (1968) named Mousterian assemblages as follows: Mousterian of Acheulian Tradition (now called Acheulo-Yabrudian); Mousterian of Levallois facies (in which he included most known assemblages such as Tabun D, Tabun B, Kebara, Shukbah D); Mousterian of elongated points (the Abu Sif and Sahba assemblages); Typical Mousterian (including Tabun C, Um Naqus C, Qafzeh F, Shovakh Cave, and Yabrud I, 7). Denticulate Mousterian was identified according to this subdivision only in Yabrud I and Turkish sites south of the Taurus. Such a cultural scheme includes implicitly the possible contemporaneity of the various entities.

A return to an explicitly linear model began with Copeland (1975). She suggested a tripartite subdivision of "Mousterian 3, 2, 1" on the basis of the Tabun sequence that includes three Mousterian layers, defined by Garrod (B, C, and D).

A slightly different view was expressed by Jelinek (1981, 1982a,b,c). While summarizing the early discoveries he insisted on three main points. First, deeply stratified caves (Yabrud I, Adlun, Tabun) contain the same sequence beginning with the Late Acheulian and followed by the Acheulo-Yabrudian, which is overlain by the Mousterian of "Tabun D-type." In some cases the younger layer contains a "Tabun C-type" or "Tabun B-type" industry. Secondly, it was then and still remains unknown how many sites have a sequence where "Tabun C-type" is overlain by "Tabun B-type," and Jelinek (1982c) wondered whether the differences between the two are functional rather than stylistic. Finally, Jelinek expressed that the need for radiometric dating was both obvious and urgent.

The ensuing decade and a half demonstrated that, despite a number of uncertainties, radiometric techniques are helpful in building a more sound chronology. However, we are still far from having a fully established chronology that will enable us to concentrate solely on the cultural issues. The current state of indecision over definitions and interpretations is quite evident in regional or site summaries (e.g., Bar-Yosef 1989, 1994, 1996; Marks 1990, 1992; Henry 1995; Marks and Monigal 1995; Meignen 1995; Hovers 1997, this volume). This situation is exemplified by the proposal that the "Tabun D-type" industry persisted in the semi-arid belt of the Negev and southern Jordan for a very long time and was contemporary with "Tabun C-type" and "Tabun B-type" in the Carmel-Galilee area (Henry 1995; Marks and Monigal 1995; and see Klein 1993 for a different view).

In this paper I briefly discuss the definitions still employed in the Levant, the available dates, and the implications of these dates and artifact densities on the interpretation of site formation processes.

2. THE DEFINITION OF THE LEVANTINE MOUSTERIAN LITHIC INDUSTRIES

Using the basic lithic sequence of Tabun Cave as a rough scale for ordering the Middle Paleolithic industries, at least for the central and southern Levant, was advocated by this author (Bar-Yosef 1989, 1992, 1994) following previous scholars (e.g., Copeland 1975; Ronen 1979; Jelinek 1981, 1982a). It includes the basic notion that a degree of technological and typological variability is expected within each entity in a similar fashion to accepted definitions of Upper Paleolithic entities. This point will be explained further through discussion of the novel approach of using *chaîne opératoire* for studying lithic industries. This approach, translated as 'operational sequence,' was first proposed by A. Leroi-Gourhan and developed by French social anthropologists who studied *chaîne opératoire* as recorded in ethnographic studies, historical documents and currently used technical methods (e.g., Creswell 1983; Lemonnier 1992).

Chaîne opératoire is not studied purely in its descriptive sense. It is an integral part of what is known as the anthropology of technology. The basic notion is that "technology embraces all aspects of the process of action upon material" (Lemonnier 1992:1–2). The idea that it "is a social production" (Lemonnier 1992:1–2) is also accepted by others (e.g., Basalla 1988; Pfaffenberger 1992). In contrast to today's ethnographers, but like historians of technology, archaeologists are entangled in deciphering the ways in which a given technology is a social production. Such an endeavor is more difficult to accomplish perhaps because one cannot observe the past operational sequences in action. The reasons for adopting one manufacturing technique of stone artifacts rather than another are often related to constraints imposed by the available raw material, its mechanical and physical properties, and procurement energy expenditures. In addition, the artisan is limited by the knowledge (*savoir faire*) of knapping methods. The possiblity that these constraints emerged not only from the nature of the given environment, the functional needs, and the knowledge of the producer, but also from the social system within which this *chaîne opératoire* was practiced, are rarely given a second thought (but see Goren-Inbar and Belfer Cohen, this volume; Hovers, this volume). Research among living populations indicates that technological traits as observed in tool making and usage may carry symbolic meanings. Unfortunately, when one can only study the knapping methods and describe tool types, it is impossible to decipher from these mute stones the symbols conveyed by the social system. By reducing the interpretation of the artifacts solely to the availability and nature of raw material, the practiced reduction sequence, resharpening, and microscopic microwear, we lose the rare hints of past social systems but gain insights into the *savoir faire* of the prehistoric artisans.

The basic definition and the descriptive language for the study of *chaîne opératoire*, aiming to adaquately describe the phases in lithic operational sequence, was developed in recent years (e.g., Boëda et al. 1990; Geneste 1990; Perlès 1992; Pigeot 1991; Meignen 1995; Schlanger 1996). It encompasses a detailed description of the various stages of tool production and use: from the procurement of raw material, to the shaping of the core and systematic production of blanks, and finally to secondary modification of selected blanks, their use, and discard. The descriptive terminology is detailed for each of the main core reduction steps, with additional terms that define the various methods. The basic determinations include the 'initialization' of the nodule, the shaping of the surfaces (process of decortication and the preparation of the striking platform), and the formation of one surface as the striking surface, from which flakes/blades are removed (e.g., Boëda 1995; Meigenen 1995; Schlanger 1996).

Most Middle Paleolithic sites in the Levant contain Levallois-dominated industries, but there are sites where a different *chaîne opératoire* is present (e.g., Boëda and Muhesen 1993; Gilead 1995; Meignen, this volume). Tracing the different phases in core and blank exploitation is done best when blanks that have been refitted to cores are available. However, even without the more expensive, and not always realizable, refitting process, researchers can identify core reduction phases through detailed examination of blanks alone (e.g., Meignen 1995, this volume; Marks and Monigal 1995; Hovers, this volume).

The reconstructed operational sequence provides insights into the cognitive abilities of the prehistoric flint knapper (e.g., Schlanger 1996). Most informative are the initial phases, when the selected and often-tested nodule is being exploited. The first blank removals following decortication reflect the desired products. For example, in the method known as Levallois *récurrent*, a series of blanks are removed successively. Except for special ones that reshape the core convexities, most other blanks, either flakes or blades, conform to the definition of the desired products (for details see Meignen 1995; Schlanger 1996).

In identifying the operational sequence, it may be difficult to decipher the process that resulted in the types of discarded cores. Well-used cores, in their final stage, may inform us about changes in a knapper's conceptualization. It could occur, for example, that unidirectional recurrent detachment of blanks is replaced by centripetal (also called 'radial') exploitation. This change in the organization of flake removals was implemented by either the original knapper or someone else. It is a problem that has yet to be studied. One way of examining the various options is when the evidence for the use of the core comes from refitted blanks and small debitage collected from various concentrations that are spatially separated. In such a situation, one may conclude that the same person had different knapping sessions across the space occupied by him/her and the group, or that different knappers used the same core during its life history.

In sum, *chaîne opératoire* is basically a system of technical skills that is learned by the user and does not require the use of language (although talking while knapping does help). On the basis of ethnographic observations, we assume that a certain *chaîne opératoire* employed in the manufacture of a product from one or more raw materials represents the technical tradition of a specific group of humans. The teaching of this skill, either through instructional sessions or modeling the technique, ensures the passage of this knowledge from one generation to the next. Therefore the time depth during which a certain *chaîne opératoire* operated would indicate how long a particular technical tradition lasted. This information may raise the question of biological and/or linguistic continuity of a given population in a particular region. Dates obtained by radiometric techniques indicate this time span. Therefore, in combining dated deposits and their archaeological assemblages we may investigate the history of a social group or closely-tightened interaction sphere. Similar notions concerning the steps that characterize each *chaîne opératoire* during the Upper Paleolithic and in later times provide a sound basis for equating industries with human populations spread over vast teritories.

If this conclusion is not upheld, then the study of *chaîne opératoire* is only of limited value in the analysis of a particular assemblage, as it does not convey the information required for inter-assemblage comparisons. Indeed, some researchers regard *chaîne opératoire* as a purely descriptive procedure and not as a valid measure that provides common cultural denominators for clustering assemblages into industries. Such an approach diminishes the value of this analytical tool. For those who justified the rejection of "guide fossils" as a tool for identifying industries or "cultures," it is perhaps neither pleasing nor appropriate to see *chaîne opératoire* as a replacement for the typologically-oriented *fossile directeur*.

In the historical rivalry that developed between the functional approach for the interpretation of the Mousterian variability and the equation of lithic industries with human populations (represented by the now-famous Bordes-Binford debate), the latter viewpoint prevailed. Thirty years of microwear and edge damage studies have indicated that while tool forms may sometimes relate to function, the same activity was certainly carried out by various tool types (e.g., Beyries 1988). Hence the degree of correlation between polish identified as resulting from butchery or wood working and type of artifacts from the Middle Paleolithic through the Neolithic demonstrated that more than one tool form was employed in these activities.

Finally, it should be emphasized that I make a rather schematic use of the available studies of *chaîne opératoire* in the Levantine Middle Paleolithic (Meigen and Bar-Yosef 1991; Bar-Yosef and Meignen 1992; Meignen 1995). Assemblages for which *chaîne opératoire* has been recorded and which seem to be rather similar are clustered together. This is not to say that a certain amount of variability does not exist among these assemblages. However, current research does not yet indicate the sources for the observable variability such as that seen when comparing units XII–IX with VIII–VII in Kebara or examining the various layers at Qafzeh and Amud (for details see Meignen and Bar-Yosef

1991; Hovers 1997, this volume). Possible interpretations include the following: the individual variability among flint knappers who were members of the same group; situations when expediency needs ruled over systematic core reduction; a short period (a season?) of raw material shortage; training children as future artisans by using cores or thick flakes that adult knappers would consider to be unusable.

In addition, along with the enthusiasm that accompanies a new approach in lithic studies, we should not ignore the significance of the retouched pieces, traditionally called tools. Identifying phases in the *chaîne opératoire* reveals the kind of blanks that were made, selected, and used, and in particular those that ended up being retouched. Hence the identification and classification of retouched pieces is a critical aspect of the specific pattern belonging to a given chaîne opératoire.

The current level of research in the Levant allows us to recognize several *chaînes opératoires* that serve as the basis for identifying the different Mousterian industries. As mentioned above, the description of the Levantine sequence is still based on the three stratified assemblages in Tabun Cave. Given the stratigraphic gaps in this site (Jelinek 1982a) and the rapidly increasing number of other assemblages that are dated and published, one may consider the use of this scheme to be only temporary.

The three main entities are described as follows:

1. "Tabun D-type:" Typical blanks were obtained from essentially unipolar convergent cores with evidence for bidirectional flaking that is often predominantly Levallois but could be non-Levallois in certain assemblages. The bidirectional flaking often addressed the need for shaping the opposite end of the core in order to secure the removal of elongated, pointed blades. Minimal preparation is evident on the striking platforms. The blanks are classified as blades and elongated points. In some cases the presence of crested blades indicates a change in the volumetric concept of the reduction sequence to one that corresponds to the prismatic volume that characterizes the Upper Paleolithic blade industries (Meignen 1994).

   In general, the "Tabun D-type" assemblages contain higher frequencies of retouched pieces than the later industries. The retouched tools include elongated points, racloirs, and burins. It has been suggested that the central Levantine sites be grouped under the term Abu Sifian in order to differentiate them from the Hummalian sites in El-Kowm. The "Tabun D-type" or the Abu Sifian industry is found in a number of sites: Tor Abu Sif and Sahba in the Judean Desert (Neuville 1951; Gordon 1993; Meignen 1994, 1995); Rosh Ein Mor and Nahal Aqev 3 in the Negev highlands (Crew 1976; Marks and Monigal 1995); Hayonim Lower Layer E in the Galilee (Meignen, this volume); Yabrud I (Solecki and Solecki 1995); the Palmyra basin in Jerf 'Ajla (Schroeder 1969); and Douara Cave layer IV (Akazawa 1979; Nishiaki 1989). The Hummalian from the El-Kowm basin is a similar industry and is considered to be non-Levalloisian (Copeland and Hours 1983; Boëda and Muhesen 1993; Meignen 1995), but it resembles the Abu Sifian in terms of its core reduction strategy and some of the shapes of the desired blanks.

   Dating the "Tabun D-type" industry proved to be a complicated issue (see Table 1, Figure 2). TL dates in Tabun indicate a time span from 270 ka through 170 ka (Mercier et al. 1995). Preliminary TL and ESR readings from Hayonim Cave support the evidence from Tabun (Mercier and Valladas personal communication; Rink personal communication; Figure 2).

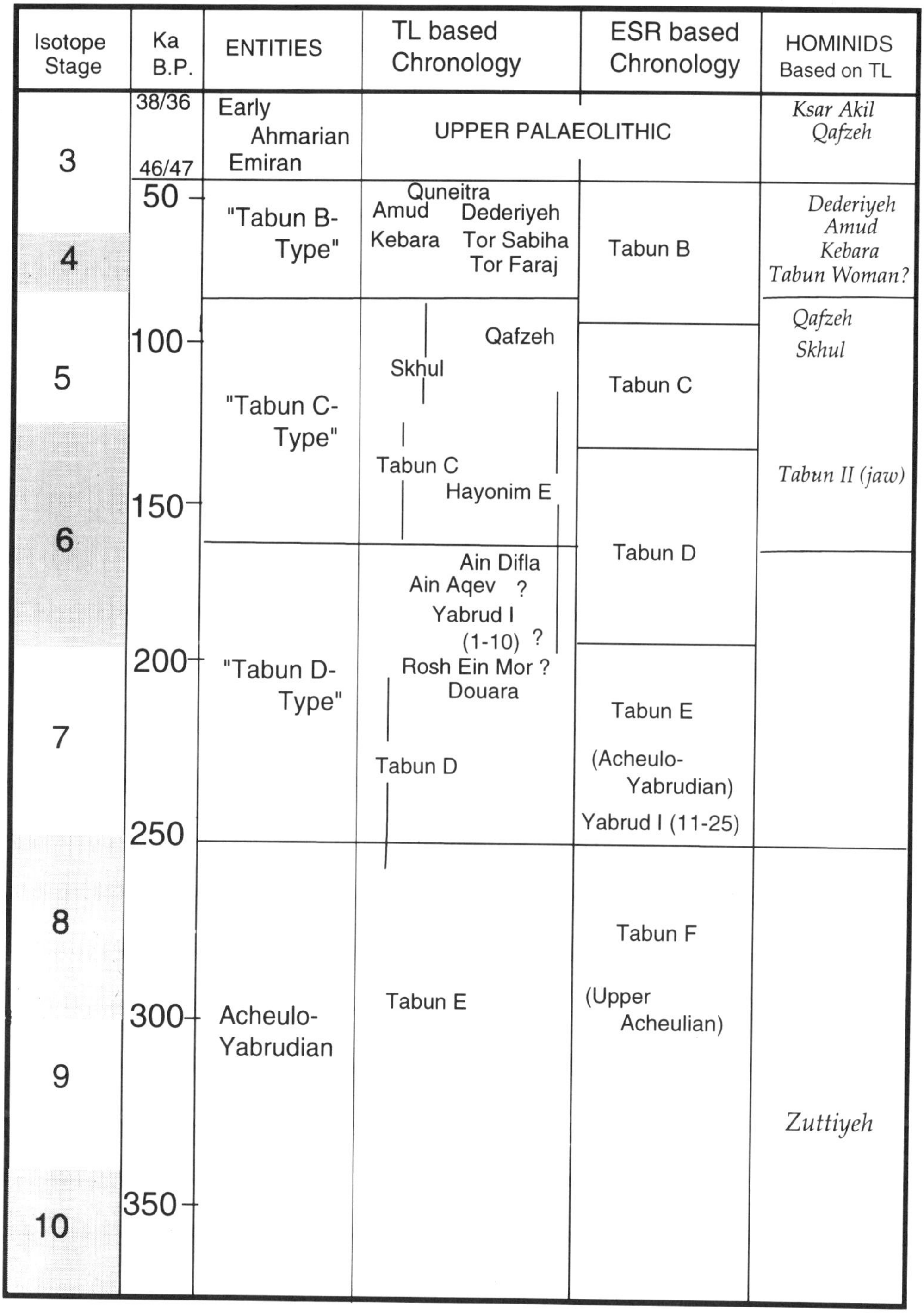

Figure 2. The current suggested chronology of Mousterian industries in the Levant. The ESR based chronology relies solely on the dates from Tabun Cave. As mentioned in the text, new ESR dates from Hayonim Cave corroborate the TL chronology.

Table 1. Th/U, TL, ESR dates by site documenting the available readings for the Acheulo-Yabrudian, Mousterian, and early Upper Paleolithic sites

Site/Layer	Industry	Th/U	TL	ESR/EU	ESR/LU	Ref
El Kowm						
Hummal 2	Yabrudian	156±16				1
Oumm et Tleil 3	Yabrudian	139±16				1
Hummal Well	Layer Ib		180±25			1
			145±20			
			160±23			
	Yabrudian		Avg. 160±22			1
	Layer 6b		97±8			1
			105±9			
			112±10			
	Mousterian (abraded)		Avg. 104±9			1
Oumm et Tleil 5	pre-Yabrudian	245±16				1
Oumm et Tleil 4	inter Yabrudian-Mousterian	76±16				1
Tell 6	Yabrudian	99±16				1
Zuttiyeh						
76ZU 1a	Acheulo-Yabrudian	164±21				2
76ZU 4	Acheulo-Yabrudian	148±6				2
76ZU 1	inter Yabrudian-Mousterian	95±10				2
76ZU 6	inter Yabrudian-Mousterian	97±13				2
	Mousterian	106±7				3
	Mousterian	157±13				3
Naamé						
Strombus level	Enfean II *	90±20				4
Strombus level	Enfean II *	93±5				5
vermet level	Naamean	90±10				4
Nahal Aqev						
Layer B	pre-Mousterian	Avg. 211±19				6
76NZ6d-4	Layer D below Mousterian	85.2±10				6
76NZ1	Layer D below Mousterian	74±5				6
Tabun						
Unit I (=C)	Mousterian		171±17			7
Unit II (=D)	Mousterian		212±22			7
Unit V (=D)	Mousterian		244±28			7
Unit IX (=D)	Mousterian		265±27			7
Unit X ("Transitional" E-D)			270±22			7
Unit XI (=E)	Acheulo-Yabrudian		306±33			7
Unit XII (=E)	Acheulo-Yabrudian		350±33			7
Unit XIII (=E)	Acheulo-Yabrudian		331±30			7
B	Mousterian			86±11	103±18	8
C	Mousterian			102±17	119±11	8
D	Mousterian			122±20	166±20	8
Ea	Acheulo-Yabrudian			154±34	188±31	8
Eb	Acheulo-Yabrudian			151±21	168±15	8
Ec	Acheulo-Yabrudian			176±10	199±76	8
Ed	Acheulo-Yabrudian			182±31	213±46	8
E	Acheulo-Yabrudian			215±22	290±36	16

Table 1. (*Continued*)

Site/Layer	Industry	Th/U	TL	ESR/EU	ESR/LU	Ref
Hayonim						
Upper E (base?)	Mousterian	163±3				2,16
Upper E	Mousterian			164±15	171±17	16
Lower E	Mousterian			241±11	257±6	16
Ain Difla						
Test A, level 5	Mousterian		105±15			17
Upper Tufa	Mousterian	141±20				17
Test A, Avg.	Mousterian			102.9±12.9	162±18.2	17
Qafzeh						
XVII–XXII range of dates	Mousterian		Avg. 92±5			9
XV–XXI	Mousterian			Avg. 96±13	115±15	10
Kebara						
Unit VI	Mousterian		48.3±3.5			11
Unit VII	Mousterian		51.9±3.5			11
Unit VIII	Mousterian		57.3±4.0			11
Unit IX	Mousterian		58.4±4.0			11
Unit X	Mousterian		61.6±3.6	60±6	64±4	11,12
Unit XI	Mousterian		60.0±3.5			11
Unit XII	Mousterian		59.9±3.5			11
Quneitra						
	Mousterian			39.2±4.2	53.9±5.9	13
Skhul						
	Mousterian		Avg. 119±18			14
	Mousterian			81±15	101±12.6	15
Amud						
B1/6	Mousterian			43±5	48±6	16
B2/8	Mousterian			59±8	70±8	16
B4	Mousterian			68	113±8	16
Farah II						
	Mousterian	72.5±1.5		49.1±4.1	62.2±7.0	16

References: 1) Henning and Hours 1982; 2) Schwarcz 1980; 3) Valladas et al., this volume; 4) Stearns and Thurber 1965; 5) Leroi-Gourhan 1980; 6) Schwarcz et al. 1979; 7) Mercier et al. 1995; 8) Grün et al. 1991; 9) Valladas et al. 1988; 10) Schwarcz et al. 1988; 11) Valladas et al. 1987; 12) Schwarcz et al. 1989; 13) Ziaei et al. 1993; 14) Mercier et al. 1993; 15) Stringer et al. 1989; 16) Schwarcz and Rink, this volume; 17) Clark et al., in press

Legend: EU=Early uptake; LU=Linear uptake; All are averages as published.

*marine transgression

2. "Tabun C-type:" The blanks are often oval-rectangular, sometimes large flakes, struck from Levallois cores through centripetal and/or bidirectional preparation. Triangular points appear in small numbers and in definite horizons, such as the top of Layer C in Tabun and layer XV in Qafzeh. This industry is common in Qafzeh (Boutié 1989; Hovers 1997), Tabun Layer C, Skhul, and Naamé (Fleisch 1970) as well as at Hayonim Cave Upper E. The proposed TL dates range from 170 ka to 90/85 ka while ESR readings suggest a similar time range (Figure 2).
3. "Tabun B-type:" The blanks were removed mainly from unipolar convergent Levallois cores (Levallois *récurrent*). Typical products are broad-based Levallois points, commonly with a *chapeau de gendarme* striking platform, and with a special tilted profile (called *en concorde* in the French terminology; e.g., Meignen 1995). Flakes and blades were also produced. This industry is characterized by the assemblages of Tabun B (Copeland 1975), Bezez B (Copeland 1983), Sefunim (Ronen 1984; which resembles unit V in Kebara), layer H at Erq

el Ahmar (Neuville 1951), possibly Dederiyeh (Akazawa et al. 1995) as well as Units XII-IX in Kebara, Amud, Tor Faraj, Tor Sabiha, and Ksar Akil XVIII (Hovers et al. 1991; Meignen and Bar-Yosef 1991; Henry 1995; Meignen 1995).

A later phase within this industry has a tendency for increasing centripetal preparation exemplified in Kebara Units VIII–VII, Quneitra, and Ksar Akil XXVI (Goren-Inbar 1990; Meignen and Bar-Yosef 1991; Meignen 1995). If a technological transition to the early Upper Paleolithic took place locally, it will be difficult to argue that it emerged from the centripetal core preparation produced by the latest Mousterian in the Levant (see below for further discussion of this point).

In sum, there are common elements among all the Levantine Mousterian industries. The first similarity is the use of the Levallois technique in the volumetric sense (Boëda 1995). Second is the appearance of the *récurrent* aspect, which, when applied to the production of points, produces certain resemblances between the "Tabun D-type" assemblages and those of "Tabun B-type." Thus sites such as Tor Faraj and Tor Sabiha, originally thought to resemble the "Tabun D-type," are probably of the "Tabun B-type." Perhaps the best indicator that an assemblage should be formally included in the "Tabun D-type" industry is the presence of cores and debitage products, aside from the flatish Levallois one, that express the prismatic volumetric concept (e.g., Tabun D, Hayonim Lower E; Meignen, this volume).

3. THE CHRONOLOGY OF THE LEVANTINE MOUSTERIAN

In the last ten years, various new radiometric techniques have enabled us to construct a tentative chronology for the Middle Paleolithic assemblages and the human fossils (Figure 2). The newly obtained dates indicated that the time span is longer than was previously perceived, and they improved the biogeographical interpretations, which were originally based on biozones within a relative chronological framework (Tchernov 1981, 1988, 1992, 1994, this volume). The current version still does not provide the level of knowledge one would desire, such as that displayed in the recently available radiocarbon chronology of the Upper Paleolithic and Neolithic. The latter dating technique benefited from five decades of research and numerous radiocarbon laboratories. We therefore have to struggle with a mixture of relative chronology as based on sites' stratigraphies, correlations with Pleistocene shorelines, and geomorphic sequences, while taking into account the readings of various radiometric techniques. Major discrepancies still exist between certain dating techniques when applied to the same site (e.g., Tabun; see Bar-Yosef 1995, 1996), but the situation has improved considerably since the 1990 conference in the University of Tokyo (e.g., Jelinek 1992).

Table 1 lists most of the available published dates including those obtained by older laboratory techniques and those which are clearly aberrant due to problems in calculating the paleodose and identifying the proveniences of the samples. The detailed TL and ESR dates for each layer in the included sites are not given due to limited space. The averages, as reported, are taken as the best chronological indicators.

Even a cursory examination of the dates in these tables demonstrates certain inherent irregularities and it will not be surprising if doubts linger concerning the validity of the results. In recent years we realize that a better understanding of site formation processes, especially those in which variable diagenetic effects are recorded, may help in sorting the available dates. Hence, the present limitations to chronometric readings require that we do not consider the entire set of published dates to be of equal precision.

Figure 2 was created in order to compensate for the inconsistencies. Dates chosen for this figure all agree with observable stratigraphic relationships and inter-site correlations. Three main considerations guided the selection. First, there is a good correlation between TL dates from sites with "Tabun B-type" and "Tabun C-type" industries in the central Levant (Bar-Yosef 1995, 1996). Secondly, there is a discrepancy between the TL and ESR dates from Tabun Cave (Grün et al. 1991; Mercier et al. 1995). The TL dates for Tabun are preferred for at least two reasons. First, there is, unfortunately, poor stratigraphic control over the dated teeth from Garrod's excavations. The second reason is a good correspondance between the TL dates of the industries of Tabun D and C with similar assemblages in Upper and Lower Layer E at Hayonim Cave (Mercier and Valladas unpublished data) and a general agreement with ESR dates (Rink personal communication). Finally, "Tabun C-type" assemblages in other sites fall within the projected time range that includes both Tabun C and Qafzeh, namely from about 170/150 ka through 90/85 ka.

On the other hand, we should note that the overlap between the ESR dates from Tabun B and C supports the contention that there is no erosional gap between the two depositional entities (e.g., Farrand 1979; Jelinek 1982a,b).

While discussing the correlation of sites and/or assemblages and suggesting an overall contemporaneity, it is worth mentioning the possibilities of identifying geographic boundaries between entities. A good case is the Acheulo-Yabrudian, which extended from the Taurus foothills to the central Levant (Bar-Yosef 1989). Its southern limit cuts across various ecozones and cannot be corroborated with a natural geographic or climatic boundary. This situation probably indicates the co-existence between the Acheulo-Yabrudian and the Late Acheulian in the southern Levant, which lasted for about 100 ka or more (about 400–300/270 based on TL chronology). This kind of contemporaneity among Mousterian entities was part of the Bordes-Binford debate (see Mellars 1996 for updated discussion). The evidence for such a situation during the time of the Levantine Mousterian is not available to date.

Two cautionary comments should be made when speculating on shifting Mousterian boundaries in the Levant. First, sites located close to one another could be expected to have been used by groups who practiced the same *chaîne opératoire* over and over again, as demonstrated in the deep deposits dated to time spans in the order of 50 to 100 ka. In evaluating the archaeological situation, we need to take into account the size of a biologically viable population. In the world of historically known foragers such a population carries the same technological tradition. If the estimates made for modern humans are applicable to the past (Wobst 1974; Zubrow 1989), then the minimum number was 200–400 people. Hence, we assume that several small mobile bands that occupied each of these cave sites and probably a number of open air camps, they were members of the same mating syatem, and therefore they shared the same technical tradition. Thus, sites situated within short distances from one another in the central Levant could have been seasonally, ephemerally, or permanantly inhabited by humans who made their artifacts through the same *chaîne opératoire*. It is important to keep in mind that the distance between Hayonim, Qafzeh, and Kebara Caves is about 30–35 km, and Amud Cave is located 25 km east of Qafzeh (Figure 1). Seasonal mobility patterns for Levantine geographic situations were proposed by Marks and Friedel (1977), Marks (1981), Shea (1991), Henry (1994, 1995), and Lieberman (1993).

Secondly, there could be a difference in the dates obtained from open-air and cave sites, particularly given the ambiguities and uncertainities that accompany dating cave assemblages (Bar-Yosef 1995, 1996). Such ambiguities were caused in part by post-depositional diagenetic processes, especially when the layers are attributed to older glacial cycles (e.g., Isotope Stages 6 and 8). Hence, systematic dating of open-air sites is needed before contemporaneity between two different industries can be claimed. For example, it is not impossible that the "Tabun D-type" industry lasted into the Last Glacial in the Negev (Marks and Monigal 1995). This would make it contemporary with the "Tabun C-type" in the caves 150–200 km further north or with the "Tabun B-type" industry in Tor Faraj and Tor Sabiha, which are the same distance southward (Henry 1995). Acceptance of such a proposal requires dating by various techniques and, in particular, dating stratified sites such as Ain Aqev (D35; Goldberg 1976).

In sum, the parsimonious approach adopted in this paper, as in previous ones (Bar-Yosef 1989, 1992, 1995, 1996), is that the main Levantine Mousterian industries are seriated through time (Figure 2). This viewpoint involves the understanding that each entity incorporates a certain degree of variability in core reduction strategies, as mentioned above.

4. THE IMPLICATIONS OF DATING AND CURRENT KNOWLEDGE OF SITE FORMATION PROCESSES ON BEHAVIORAL INTERPRETATIONS

Recent investigations of cave sediments demonstrate that a major portion of the observable volume in Levantine sites accumulated due to anthropogenic and biogenic activities (Laville and Goldberg 1989; Goldberg and Laville 1991; Schiegl et al. 1995; Goldberg and Bar-Yosef, this volume). Given that the deposits bear no clear paleoclimatological signals, it is difficult to make correlations with the Isotope Stages or even with major Pleistocene geomorphic phases such as marine regressions and transgressions. On the other hand, the study of these biogenic accumulations opens new avenues for paleoanthropological research. The following comments are based on a few preliminary observations.

In a previous study, the particular distribution of animal bones in Kebara Cave was suggested to be linked to human activities, and the disappearance of bones was related to diagenetic processes (Bar-Yosef et al. 1992; Weiner et al. 1993). Mineralogical investigations revealed the effects of the latter processes: within a well-preserved area, the spatial distribution of hearths, ashes, and lithics indicates that the cave inhabitants some 60–55 ka maintained a certain separation between their activities. Their actions included the removal and dumping of ashes near the cave wall and discarding a considerable percentage of the large debitage products (unused blanks and exhausted cores) in the same general area. In the central space, animal bones and small debitage products accumulated adjacent to the hearths but were restricted in their distribution. In the main ash lenses (perhaps intentionally spread for sleeping), only dispersed artifacts were recovered.

The role of human activities in Kebara and Hayonim Caves is clearly expressed through the study of ashes. These and other organics brought in by humans and animals are the major depositional component (see Goldberg and Bar-Yosef, this volume). The effects of ashes resulting from in-cave camp-fires were explored in both sites (Schiegl et al. 1996; Weiner et al. 1995). The study began by demonstrating, through use of experimental hearths, that fresh ash from a variety of trees in the Galilee and Mt. Carmel area contains between 0.3 and 3.7 weight percent of insoluble fraction. This material is composed mainly of fine-grained crystalline minerals embedded in an amorphous mineral matrix rich in silica, aluminum, iron, and potassium (Schiegl et al. 1996; Weiner et al. 1995). Earlier studies described these as siliceous aggregates of which large accumulations have been identified in the deposits of Hayonim and Kebara Caves (Schiegl et al. 1996; Weiner et al. 1995).

Thus it seems that hearth building began by digging or scraping an often rounded or oval pit, some 40–60 cm in diameter (Meignen et al. 1989). The main combustible was wood in most cases and the reasons for lighting the fire were to parch, roast, and produce warmth. The remains of such campfires are described by Courty et al. (1989). The burned wood produced calcitic ashes. This deposit was altered to a phosphate mineral, and when the original hearth structure is well-preserved, it overlies a reddened sediment. Charred remains of plant tissues and burnt bone fragments are observed in thin sections. Actual carbonized plant remains were collected through floatation from the layers at Kebara (Lev and Kislev 1993).

The hearths in Kebara, well-preserved in comparison to other sites, as well as the hearths in Hayonim show many signs of having been altered post-depositionally, both chemically and mechanically. Through analysis of the sediments, it was recognized that altered, reworked, and redeposited ashes are quantitatively the most important constituent of the deposits in both caves.

The next element of the interpretation takes into account the frequencies of microvertebrates, the number of artifacts per cubic meter, and the suggested rate of accumulation based on the available TL and ESR dates. Preliminary calculations in Kebara Cave indicate that each cubic meter contains about 1000 pieces (larger than 2 cm), accumulated over about 3,000 TL years. There are meager quantities of microfauna. In Hayonim Cave, each cubic meter contains about 300 pieces (larger than 2 cm), accumulated during about 10/15,000 TL years. Microfauna are relatively abundant.

Therefore, as most of the sediments are biogenic, we may compare them either by "squeezing" or "stretching" one of the stratigraphies. Using the latter approach, the number of artifacts at Hayonim would be reduced to about 100 per cubic meter. The microfauna indicates repeated habitations by Barn Owls and other small predators. In terms of settlement pattern, this means that human occupation at Hayonim Cave some 200–100 Ka ago was ephemeral when compared to Kebara, which dates to the later part of the Levantine Mousterian sequence. Hence different types of mobility were practiced by the humans who used these caves, as suggested by Lieberman and Shea (1994). However, the "central place foraging" (Isaac 1984) attributed to either Kebara or Tabun B inhabitants, based on cementum analysis and microwear, was not the only model that was practiced by the bearers of "Tabun B-type" industry. Henry (1995) demonstrated that a higher degree of mobility, probably based on seasonal movements along altitudinal transect, is indicated by the similarity between the Mousterian assemblages in Tor Faraj and Tor Sabiha in southern Jordan. A similar pattern is proposed for the Zagrosian sites such as Shanidar, a lowland cave, and Kunji Cave in the highlands (Solecki and Solecki 1993; Baumler and Speth 1993). It is essentially expressed in the large amount of curated artifacts and high degree of resharpening in the Kunji assemblage.

An additional conclusion relates to the length of time during which an identifiable *chaîne opératoire* was practiced. The stratified assemblages at Hayonim Cave demonstrate that the core reduction strategy for the production of elongated blades/points (Meignen 1995, this volume) was employed over a very long time. However, a taxonomic determination of the manufacturers of this industry remains unknown. Aside from that point, this observation raises the possibility of a long-held mental template by a population that maintained the way in which they made their stone tools, despite climatic changes and shifts in resource availability and accessibility. This is, however, a coarse-grain conclusion that intentionally ignores a certain amount of variability in lithic production, for reasons explained above.

5. CLOSING REMARKS

The long-term and tedious process of dating Mousterian assemblages in the Levant and neighboring regions in Western Asia is an integral aspect of prehistoric research. Time, as is well known, is an important element in cultural evolution. Changes and shifts in subsistence systems incorporating hunting, gathering, group mobility, the use of sites, the production of artifacts, and the emergence of lasting elements of symbolic behavior, expressed within a given period of known calendrical length, are amenable to different interpretations. The mosaic of Middle Paleolithic cultural components within a regional or on a global scale, as known from the archaeological records, is often interpreted as directly related to biological aspects of human evolution. In the world of *Homo sapiens sapiens*, one does not face taxonomical disputes concerning the relationships between human fossils. However, there is still a lack of consensus concerning the taxonomic status of Middle Paleolithic human fossils and not only those found in the Levant. The disgareement is part and parcel of scientific debates that have accompanied discoveries of such fossils in the Old World since the inception of human paleontology as a field of investigation. Somewhat similar arguments arise concerning lithic industries. However, the basic morpho-typology of artifacts is often agreed upon, and most scholars use F. Bordes' full or shortened type-list. Metrical attributes are also commonly employed. It is the interpretation of typological counts, the composition of assemblages, the formation of cultural entities, and their meaning in terms of behavioral aspects that are constantly argued. The Levantine Mousterian is just one additional example in world prehistory of competing interpretations. However, in our efforts to understand the changes and shifts in cultural evolution, we need to re-examine the meaning of our observations when modifications in dating techniques make available improved time tables.

ACKNOWLEDGMENTS

This paper has benefited from numerous discussions since 1982 with my colleagues with whom the research at Kebara, Qafzeh, and Hayonim caves was jointly carried out, including B. Arensburg, A. Belfer-Cohen, P. Goldberg, S. Kuhn, the late Henri Laville, L. Meignen, N. Mercier, Y. Rak, J. Rink, E. Tchernov, A. M. Tillier, H. Schwarcz, J. Speth, M. Stiner, S. Weiner, H. Valladas, and B. Vandermeersch. Last but not least, Mr. M. Chech was instrumental is our successful field work which was generously sponsored by the L. S. B. Leakey Foundation, the French Ministry of Foreign Affairs, and the NSF (grants to the author). Needless to say, I must stress that I alone am responsible for this version of the Middle Paleolithic chronology.

REFERENCES

Aitken, M. J. 1990. *Science-based Dating in Archaeology*. New York: Longman.

Akazawa, T. 1979. Middle Paleolithic assemblages from Douara Cave. *Bulletin of the Museum of the University of Tokyo* 16:1–30.

Akazawa, T. 1987. The ecology of the Middle Paleolithic occupation at Douara Cave, Syria. *Bulletin of the Museum of the University of Tokyo* 29:155–66.

Akazawa, T., S. Muhesen, Y. Dodo, O. Kondo, Y. Mizoguchi, Y. Abe, Y. Nishiaki, S. Ohta, T. Oguchi, and J. Haydal. 1995. Neanderthal infant burial from the Dederiyeh Cave in Syria. *Paléorient* 21:77–86.

Bar-Yosef, O. 1989. Geochronology of the Levantine Middle Palaeolithic, 589–610. In *The Human Revolution*, ed. P. A. Mellars and C. B. Stringer. Edinburgh: Edinburgh University Press.

Bar-Yosef, O. 1992. Middle Paleolithic human adaptations in the Mediterranean Levant, 189–216. In *The Evolution and Dispersal of Modern Humans in Asia,* ed. T. Akazawa, K. Aoki, and T. Kimura. Tokyo: Hokusen-Sha.

Bar-Yosef, O. 1994. The contributions of southwest Asia to the study of the origin of modern humans, 23–66. In *Origins of Anatomically Modern Humans*, ed. M. H. Nitecki and D. V. Nitecki. New York: Plenum Press.

Bar-Yosef, O. 1995. The Lower and Middle Palaeolithic in the Mediterranean Levant: Chronology, and cultural entities, 247–63. In *Man and Environment in the Palaeolithic*, ed. H. Ullrich. Liège: ERAUL.

Bar-Yosef, O., and L. Meignen. 1992. Insights into Levantine Middle Paleolithic cultural variability, 163–82. In *The Middle Paleolithic: Adaptation, Behavior, and Variability*, ed. H. Dibble and P. Mellars. Philadelphia: The University Museum of the University of Pennsylvania.

Bar-Yosef, O., B. Vandermeersch, B. Arensburg, A. Belfer-Cohen, P. Goldberg, H. Laville, L. Meignen, Y. Rak, J. D. Speth, E. Tchernov, A. M. Tillier, and S. Weiner. 1992. The excavations in Kebara Cave, Mount Carmel. *Current Anthropology* 33:497–550.

Bar-Yosef, O., M. Arnold, A. Belfer-Cohen, P. Goldberg, R. Housley, H. Laville, L. Meignen, N. Mercier, J. C. Vogel, and B. Vandermeersch. 1996. The dating of the Upper Paleolithic layers in Kebara Cave, Mount Carmel. *Journal of Archaeological Science* 23:297–306.

Basalla, G. 1988. *The Evolution of Technology*. Cambridge: Cambridge University Press.

Baumler, M. F., and J. D. Speth. 1993. A Middle Paleolithic assemblage from Kunji Cave, Iran, 1–74. In *The Paleolithic Prehistory of the Zagros-Taurus*, ed. D. I. Olszewski and H. L. Dibble. Philadelphia: The University Museum of the University of Pennsylvania.

Beyries, S. 1988. Functional variability of lithic sets in the Middle Paleolithic, 213–24. In *Upper Pleistocene Prehistory of Western Eurasia,* ed. H. L. Dibble and A. Montet-White. Philadelphia: The University Museum of the University of Pennsylvania.

Binford, S. R. 1968. Early Upper Pleistocene adaptations in the Levant. *American Anthropologist* 70:707–17.

Boëda, E. 1995. Levallois: A volumetric construction, methods, a technique, 41–68. In *The Definition and Interpretation of Levallois Technology*, ed. H. L. Dibble and O. Bar-Yosef. Madison: Prehistory Press.

Boëda, E., J. M. Geneste, and L. Meignen. 1990. Identification de chaînes opératoires lithiques du Paléolithique ancien et moyen. *Paléo* 2:43–80.

Boëda, E., and S. Muhesen. 1993. Umm El Tlel (El Kown, Syrie): Etude préliminaire des industries lithiques du Paléolithique moyen et supérieur 1991–1992. *Cahiers de l'Euphrate* 7:47–91.

Bordes, F. 1955. Le Paléolithique inférieur et moyen de Jabrud (Syrie) et la question du Pré-Aurignacien. *L'Anthropologie* 59:486–587.

Bordes, F. 1961. *Typologie du Paléolithique Ancien et Moyen*. Bordeax: Delmas.

Boutié, P. 1989. Etude technologique de l'industrie Moustérienne de la grotte de Qafzeh (près de Nazareth, Israël), 213–29. In *Investigations in South Levantine Prehistory: Préhistoire du Sud-Levant*, ed. O. Bar-Yosef and B. Vandermeersch. Oxford: British Archaeological Reports International Series 497.

Cauvin, J., and P. Sanlaville, eds. 1981. *Préhistoire du Levant.* Paris: CNRS Editions.

Clark, G. A., et al. In press. Chronostratigraphic contexts of Middle Paleolithic horizons at the Ain Difla rockshelter (WHS 634), west-central Jordan. In *The Prehistory of Jordan II*, ed. H. G. Gebel, Z. Kafafi, and G. O. Rollefson. Berlin: Ex Oriente.

Copeland, L. 1975. The Middle and Upper Palaeolithic of Lebanon and Syria in the light of recent research, 317–50. In *Problems in Prehistory: North Africa and the Levant*, ed. F. Wendorf and A. E. Marks. Dallas: SMU Press.

Copeland, L. 1983. The Palaeolithic industries at Adlun, 89–366. In *Adlun in the Stone Age: The Excavations of D. A. E. Garrod in the Lebanon, 1958–1963,* ed. D. Roe. Oxford: British Archaeological Reports International Series 159.

Copeland, L., and F. Hours. 1983. Le Yabroudien d'El Kowm (Syrie) et sa place dans le Paléolithique du Levant. *Paléorient* 9:21–37.

Courty, M. A., P. Goldberg, and R. Macphail. 1989. *Soils and Micromorphology in Archaeology.* Cambridge: Cambridge University Press.

Cresswell, R. 1983. Transferts de techniques et chaînes opératoires. *Techniques et Cultures* 2:143–63.

Crew, H. L. 1976. The Mousterian site of Rosh Ein Mor, 75–112. In *Prehistory and Paleoenvironments in the Central Negev (Israel)*, Vol. 1, *The Avdat/Aqev Area*, Part 1, ed. A. E. Marks. Dallas: SMU Press.

Farrand, W. R. 1979. Chronology and paleoenvironment of Levantine prehistoric sites as seen from sediment studies. *Journal of Archaeological Science* 6:369–92.

Fleisch, S. J. 1970. Les habitats du paléolithique moyen à Naamé (Liban). *Bulletin du Museé Beyrouth* 23:25–98.

Garrod, D. A. E. 1954. Excavations at the Mugharet Kebara, Mount Carmel 1931: The Aurignacian Industries. *Proccedings of the Prehistoric Society* 20:155–92.

Garrod, D. A. E., and D. M. Bate. 1937. *The Stone Age of Mount Carmel.* Oxford: Clarendon Press.

Garrod, D., and G. Henri-Martin. 1961. Rapport préliminaire sur la fouille d'une grotte au Ras el-Kelb, Liban. *Bulletin du Museé Beyrouth* 16:7–46.

Geneste, J. M. 1990. Dévelopment des systèmes de production lithique au cours du paléolithique moyen en Aquitaine septentrionale, 203–13. In *Paleolithique Moyen Recent et Paleolithique Superieur Ancien en Europe: Ruptures et Transitions*, ed. C. Farizy. Nemours: APRAIF.

Gilead, I. 1995. Problems and prospects in the study of Levallois technology in the Levant: The case of Fara II, Israel, 79–92. In *The Definition and Interpretation of Levallois Technology*, ed. H. Dibble and O. Bar-Yosef. Madison: Prehistory Press.

Gisis, I., and O. Bar-Yosef. 1974. New excavations in Zuttiyeh Cave. *Paléorient* 2:175–80.

Goldberg, P. 1976. Upper Pleistocene geology of the Avdat/Aqev area, 25–55. In *Prehistory and Paleoenvironments in the Central Negev (Israel)*, Vol. 1,*The Avdat/Aqev Area*, Part 1, ed. A. E. Marks. Dallas: SMU Press.

Goldberg, P., and H. Laville. 1991. Etude géologique des dépôts de la grotte de Kébara (Mont Carmel): Campagne 1982–1984, 29–42. In *Le Squelette Mousterien de Kébara 2*, ed. O. Bar-Yosef and B. Vandermeersch. Paris: CNRS Editions.

Gordon, D. 1993. Mousterian tool selection, reduction, and discard at Ghar. *Israel Journal of Field Archaeology* 20:205–18.

Goren-Inbar, N. 1990. *Quneitra: A Mousterian Site on the Golan Heights*. QEDEM 31. Jerusalem: Institute of Archaeology, Hebrew University.

Grün, R., C. B. Stringer, and H. P. Schwarcz. 1991. ESR dating of teeth from Garrod's Tabun Cave collection. *Journal of Human Evolution* 20:231–48.

Henning, G. J., and F. Hours. 1982. Dates pour le passage entre l'Acheuléen et le Paléolithique moyen à El Kowm (Syrie). *Paléorient* 8:81–3.

Henry, D., and F. Servello. 1974. Compendium of C-14 determinations derived from Near Eastern prehistoric sites. *Paléorient* 2:19–44.

Henry, D. O. 1994. Prehistoric cultural ecology in southern Jordan. *Science* 265:336–41.

Henry, D. O. 1995. *Prehistoric Cultural Ecology and Evolution*. New York: Plenum Press.

Hovers, E. 1997. Variability of Levantine Mousterian Assemblages and Settlement Patterns: Implications for the Development of Human Behavior. Ph. D. diss. Hebrew University. Jerusalem.

Hovers, E., Y. Rak, and W. Kimbel. 1991. Amud Cave—1991 season. *Journal of the Israel Prehistoric Society* 24:152–7.

Isaac, G. L. 1984. The archaeology of human origins: Studies of the Lower Pleistocene in East Africa 1971–1981, 1–87. In *Advances in World Archaeology*, ed. F. Wendorf and A. Close. New York: Academic Press.

Jelinek, A. J. 1981. The Middle Palaeolithic in the Southern Levant from the Perspective of the Tabun Cave, 265–80. In *Préhistoire du Levant*, ed. J. Cauvin and P. Sanlaville. Paris: CNRS Editions.

Jelinek, A. J. 1982a. The Middle Palaeolithic in the southern Levant with comments on the appearance of modern *Homo sapiens*, 57–104. In *The Transition from Lower to Middle Palaeolithic and the Origin of Modern Man*, ed. A. Ronen. Oxford: British Archaeological Reports International Series 151.

Jelinek, A. J. 1982b. The Tabun Cave and Paleolithic Man in the Levant. *Science* 216:1369–75.

Jelinek, A. J. 1982c. The Middle Paleolithic of the Levant: Synthesis, 299–302. In *Préhistoire du Levant*, ed. J. Cauvin and P. Sanlaville. Paris: CNRS Editions.

Jelinek, A. J. 1992. Problems in the chronology of the Middle Paleolithic and the first appearance of early modern *Homo sapiens* in southwest Asia, 253–75. In *The Evolution and Dispersal of Modern Humans in Asia*, ed. T. Akazawa, K. Aoki, and T. Kimura. Tokyo: Hokusen-Sha.

Klein, R. G. 1993. The Middle Paleolithic: Adaptation, behavior, and variability. *Journal of Field Archaeology* 20:237–40.

Laville, H., and P. Goldberg. 1989. The collapse of the Mousterian sedimentary regime and the beginning of the Upper Palaeolithic at Kebara, 75–95. In *Investigations in South Levantine Prehistory*, ed. O. Bar-Yosef and B. Vandermeersch. Oxford: British Archaeological Reports International Series 497.

Lemonnier, P. 1992. *Elements for an Anthropology of Technology*. Ann Arbor: University of Michigan Museum of Anthropology.

Leroi-Gourhan, A. 1980. Les analyses polliniques au Moyen-Orient. *Paléorient* 6:79–91.

Lev, E., and M. Kislev. 1993. *The Subsistence and the Diet of the "Neanderthal" Man in Kebara Cave, Mt. Carmel.* Ramat Ha'nadiv Project Research Report Series no. 9.

Lieberman, D. E. 1993. The rise and fall of seasonal mobility among hunter-gatherers. *Current Anthropology* 34:599–631.

Lieberman, D. E., and J. J. Shea. 1994. Behavioral differences between archaic and modern humans in the Levantine Mousterian. *American Anthropologist* 96:300–32.

Marks, A. E. 1981. The Middle Paleolithic of the Negev, 287–98. In *Préhistoire du Levant*, ed. J. Cauvin and P. Sanlaville. Paris: CNRS Editions.

Marks, A. E. 1990. The Middle and Upper Palaeolithic of the Near East and the Nile Valley: The problem of cultural transformations, 56–80. In *The Emergence of Modern Humans*, ed. P. Mellars. Ithica: Cornell University Press.

Marks, A. E. 1992. Upper Pleistocene archaeology and the origins of modern man: A view from the Levant and adjacent areas, 229–52. In *The Evolution and Dispersal of Modern Humans in Asia*, ed. T. Akazawa, K. Aoki, and T. Kimura. Tokyo: Hokusen-Sha.

Marks, A. E., and D. A. Freidel. 1977. Prehistoric settlement patterns in the Avdat/Aqev area, 131–58. In *Prehistory and Paleoenvironments in the Central Negev, Israel*, Vol. 2, *The Avdat Aqev Area, Part 2 and Har Harif*, ed. A. E. Marks. Dallas: SMU Press

Marks, A., and K. Monigal. 1995. Modeling the production of elongated blanks from the Early Levantine Mousterian at Rosh Ein Mor, 267–78. In *The Definition and Interpretation of Levallois Technology*, ed. H. Dibble and O. Bar-Yosef. Madison: Prehistory Press.

McKown, T., and A. Keith. 1939. *The Stone Age of Mt. Carmel*, Vol. 2, *The Fossil Remains from the Levalloiso-Mousterian*. Oxford: Clarendon.

Meignen, L. 1994. Le Paléolithique moyen au Proche-Orient: Le phénomène laminaire, 125–59. In *Les Industries Laminaires au Paléolithique Moyen*, ed. S. Révillion and A. Tuffreau. Paris: CNRS Editions.

Meignen, L. 1995. Levallois lithic production systems in the Middle Paleolithic of the Near East: The case of the unidirectional method, 361–80. In *The Definition and Interpretation of Levallois Technology*, ed. H. Dibble and O. Bar-Yosef. Madison: Prehistory Press.

Meignen, L., O. Bar-Yosef, and P. Goldberg. 1989. Les structures de combustion moustériennes de la grotte de Kébara (Mont Carmel, Israël), 141–6. In *Nature et Fonction de Foyers Préhistorique*, ed. M. Olive and Y. Taborin. Nemours: APRAIF.

Meignen, L., and O. Bar-Yosef. 1991. Les outillage lithiques moustériens de Kébara, 49–76. In *Le Squelette Mousterienne de Kebara 2*, ed. O. Bar-Yosef and B. Vandermeersch. Paris: CNRS Editions

Mellars, P. 1996. *The Neanderthal Legacy*. Princeton: Princeton University Press.

Mercier, N., H. Valladas, O. Bar-Yosef, C. Stringer, and J. L. Joron. 1993. Thermoluminescence date for the Mousterian burial site of Es-Skhul, Mt. Carmel. *Journal of Archaeological Science* 20:169–74.

Mercier, N., H. Valladas, G. Valladas, J. L. Reyss, A. Jelinek, L. Meignen, and J. L. Joron. 1995. TL dates of burnt flints from Jelinek's excavations at Tabun and their implications. *Journal of Archaeological Science* 22:495–510.

Neuville, R. 1951. *Le Paléolithique et le Mésolithique de Désert de Judée*. Paris: Masson.

Nishiaki, Y. 1989. Early blade industries in the Levant: The placement of Douara IV industry in the context of the Levantine Early Middle Paleolithic. *Paléorient* 15:215–29.

Perlès, C. 1992. In search of lithic strategies: A cognitive approach to prehistoric chipped stone assemblages, 223–47. In *Representations in Archaeology*, ed. J. C. Gardin and C. S. Peebles. Bloomington: Indiana University Press.

Perrot, J. 1968. La Préhistoire palestinienne, cols. 286–446. In *Supplément au Dictionnaire de la Bible 8*. Paris: Letouzey and Ané.

Pfaffenberger, B. 1992. Social anthropology of technology. *Annual Review of Anthropology* 21:491–516.

Pigeot, N. 1991. Reflexions sur l'histoire technique de l'homme: De l'evolution cognitive à l'evolution culturelle. *Paléo* 3:167–200.

Roe, D., ed. 1983. *Adlun in the Stone Age*. Oxford: British Archaeological Reports International Series 159.

Ronen, A. 1979. Paleolithic industries, 296–307. In *The Quaternary of Israel*, ed. A. Horowitz. New York: Academic Press.

Ronen, A. 1984. *Sefunim Prehistoric Sites, Mount Carmel, Israel*. Oxford: British Archaeological Reports International Series 230.

Rust, A. 1950. *Die Höhlenfunde von Jabrud (Syrien)*. Neumünster: Karl Wacholtz Verlag.

Sanlaville, P. 1977. *Etude géomorphologique de la région littorale du Liban*. Beirut: Lebanese University.

Sanlaville, P. 1981. Stratigraphie et chronologie du Quaternaire continental du Proche Orient, 21–32. In *Préhistoire du Levant*, ed. P. Sanlaville and J. Cauvin. Paris: CNRS Editions.

Schick, T., and M. Stekelis. 1977. Mousterian assemblages in Kebara Cave, Mount Carmel, 97–150. In *Moshe Stekelis Memorial Volume*, ed. B. Arensburg and O. Bar-Yosef. Jersualem: Israel Exploration Society.

Schiegl, S., P. Goldberg, O. Bar-Yosef, and S. Weiner. 1996. Ash deposits in Hayonim and Kebara Caves, Israel: Macroscopic, microscopic and mineralogical observations, and their archaeological implications. *Journal of Archaeological Science* 23:763–81.

Schlanger, N. 1996. Understanding Levallois: Lithic technology and cognitive archaeology. *Cambridge Archaeological Journal* 6:231–54.

Schroeder, B. 1969. The Lithic Industries from Jerf Ajla and their Bearing on the Problem of the Middle to Upper Paleolithic Transition. Ph. D. diss. Columbia University. New York.

Schwarcz, H. P., B. Blackwell, P. Goldberg, and A. E. Marks. 1979. Uranium series dating of travertine from archaeological sites, Nahal Zin, Israel. *Nature* 277:558–60.

Schwarcz, H. P., P. D. Goldberg, and B. Blackwell. 1980. Uranium series dating of archaeological sites in Israel. *Israel Journal of Earth-Sciences* 29:157–65.

Schwarcz, H. P., R. Grün, B. Vandermeersch, O. Bar-Yosef, H. Valladas, E. Tchernov. 1988. ESR dates for the hominid burial site of Qafzeh in Israel. *Journal of Human Evolution* 17:733–7.

Schwarcz, H. P., W. M. Buhay, R. Grün, H. Valladas, E. Tchernov, O. Bar-Yosef, and B. Vandermeersch. 1989. ESR dating of the Neanderthal site, Kebara Cave, Israel. *Journal of Archaeological Science* 16:653–9.

Shea, J. J. 1991. The Behavioural Significance of Levantine Mousterian Industrial Variability. Ph D. diss. Harvard University. Cambridge.

Solecki, R. L., and R. S. Solecki. 1995. The Mousterian industries of Yabrud Shelter I: A reconsideration, 381–98. In *The Definition and Interpretation of Levallois Technology*, ed. H. Dibble and O. Bar-Yosef. Madison: Prehistory Press.

Solecki, R. S., and R. L. Solecki. 1993. The Pointed Tools from the Mousterian Occupations of Shanidar Cave, Northern Iraq, 119–46. In *The Paleolithic Prehistory of the Zagros-Taurus*, ed. D. I. Olszewski and H. L. Dibble. Philadephia: The University Museum of the University of Pennsylvania.

Stearns, C., and D. Thurber. 1965. Th230-U234 dates of Late Pleistocene marine fossils from the Mediterranean and Moroccan littorals. *Quaternaria* 7:29.

Stringer, C. B., R. Grün, H. P. Schwarcz, and P. Goldberg. 1989. ESR dates for the hominid burial site of Es-Skhul in Israel. *Nature* 338:756–8.

Suzuki, H., and F. Takai, ed. 1970. *The Amud Man and His Cave Site*. Tokyo: Academic Press of Japan.

Tchernov, E. 1981. The biostratigraphy of the Levant, 67–97. In *Préhistoire du Levant*, ed. J. Cauvin and P. Sanlaville. Paris: CNRS Editions.

Tchernov, E. 1988. Biochronology of the Middle Paleolithic and dispersal events of hominids in the Levant, 153–68. In *L'Homme de Néandertal*, Vol. 2, *L'Environment*. ed. M. Otte. Liège: ERAUL 34.

Tchernov, E. 1992. Evolution of complexities, exploitation of the biosphere and zooarchaeology. *Archaeozoologia* 5:9–42.

Tchernov, E. 1994. New comments on the biostratigraphy of the Middle and Upper Pleistocene of the southern Levant, 333–50. In *Late Quaternary Chronology and Paleoclimates of the Eastern Mediterranean*, ed. O. Bar-Yosef and R. S. Kra. Tucson and Cambridge: Radiocarbon and the Peabody Museum of Archaeology and Ethnology, Harvard University.

Turville-Petre, F. 1927. *Researches in Prehistoric Galilee (1925–1926)*. London: The Council of the School.

Valladas, H., J. L. Joron, G. Valladas, B. Arensburg, O. Bar-Yosef, A. Belfer-Cohen, P. Goldberg, H. Laville, L. Meignen, Y. Rak, E. Tchernov, A. M. Tillier, and B. Vandermeersch. 1987. Thermoluminscence dates for the Neanderthal burial site at Kebara in Israel. *Nature* 330:159–60.

Valladas, H., J. L. Reyss, J. L. Joron, G. Valladas, O. Bar-Yosef, B. Vandermeersch. 1988. Thermoluminescence dating of Mousterian 'Proto-Cro-Magnon' remains from Israel and the origin of modern man. *Nature* 331:614–15.

Vandermeersch, B. 1981. *Les Hommes Fossiles de Qafzeh (Israel)*. Paris: CNRS Editions.

Watanabe, H. 1970. Chapter VI: A Paleolithic Industry from the Amud Cave, 77–114. In *The Amud Man and His Cave Site*, ed. H. Suzuki and F. Takai. Tokyo: Academic Press of Japan.

Weiner, S., P. Goldberg, and O. Bar-Yosef. 1993. Bone preservation in Kebara cave, Israel using on-site Fourier Transform Infrared Spectrometry. *Journal of Archaeological Science* 20:613–27.

Weiner, S., S. Schiegl, P. Goldberg, and O. Bar-Yosef. 1995. Mineral assemblages in Kebara and Hayonim caves, Israel: Excavation strategies, bone preservation and wood ash remnants. *Israel Journal of Chemistry* 35:143–54.

Weinstein, J. M. 1984. Radiocarbon dating the southern Levant. *Radiocarbon* 26 (3):297–366.

Wobst, M. H. 1974. Boundary conditions for Paleolithic social systems: A simulation approach. *American Antiquity* 39:147–78.

Ziaei, M., H. P. Schwarcz, C. M. Hall, and R. Grün. 1990. Radiometric dating of the Mousterian site at Quneitra, 232–5. In *Quneitra: A Mousterian Site on the Golan Heights*, ed. N. Goren-Inbar. Jerusalem: Hebrew University.

Zubrow, E. 1989. The demographic modelling of Neanderthal extinction, 212–31. In *The Human Revolution*, ed. P. A. Mellars and C. B. Stringer. Edinburgh: Edinburgh University Press.

4

PROGRESS IN ESR AND U-SERIES CHRONOLOGY OF THE LEVANTINE PALEOLITHIC

Henry P. Schwarcz and W. J. Rink

Department of Geology
McMaster University
Hamilton, Ontario L8S 4M1, Canada

1. INTRODUCTION

Any study of the origins of humans must be set in a chronological context. Prehistorians now have available to them a reasonably large set of tools with which to assign ages to events of the past. Radiocarbon is essentially useless as a tool for the time range under discussion here: In samples older than about 40 ka, the ^{14}C atoms originally deposited with the sample have decayed to less than 1% of their initial value, while equal or greater amounts of contaminant ^{14}C atoms may be present. Newer dating methods now exist which are based on the measurement of doses of radioactivity trapped in natural or artificial materials. While these methods can span the ^{14}C time range, they can also extend much further back in time, and are applicable to a wide variety of archaeological deposits. While their normal precision of ±10% is much less than that of ^{14}C dating, they do not require independent calibration.

These methods encompass the following: thermoluminescence (TL) dating of burned flint artifacts (Valladas et al., this volume); optical- and infrared-stimulated luminescence (OSL, IRSL) dating of wind-blown sand and silt (e.g., desert loess) which has been zeroed by solar bleaching; and electron spin resonance (ESR) dating of tooth enamel of larger mammals. Some of the physical parameters used to determine the age are shared by all these methods (external dose rate, water content of sediment) while others are unique to each method and sample type (e.g., internal dose rate) and assure that these various methods provide dates that are largely independent of one another. For example, ESR ages at a site may largely depend on the internal uranium content of the tooth sample and its uptake history, whereas TL dates on flint at the same site may depend equally on internal and external dose rates.

Over the past decade there has also been an order-of-magnitude increase in the precision of uranium-series (US) dating through the advent of thermal ionization mass spec-

Neandertals and Modern Humans in Western Asia, edited by Akazawa *et al.*
Plenum Press, New York, 1998.

trometric (TIMS) analyses of U and Th isotope ratios (Edwards et al. 1986). Using TIMS/US dating of stalagmitic calcite from cave deposits, we can in principle determine ages with a precision of ±1% of the age or better (Li et al. 1989). This method is applicable to some of the cave deposits in Israel and we are now beginning to refine dates previously obtained on these deposits using alpha-spectrometric methods (Schwarcz et al. 1980). However, many of these deposits are either contaminated with detritus or altered post-depositionally, so that the additional precision of TIMS dating may not significantly add to the accuracy of their dates. In spite of this, we have been able to obtain new, high-precision dates at Hayonim Cave, as summarized here.

The broader features of the time scale of biological and cultural evolution in this region have been worked out through the application of TL, ESR, and alpha-spectrometric US dating of the key sites of Qafzeh, Tabun, es Skhul, Kebara, and Zuttiyeh (see Schwarcz 1994 for summary). While the TL, ESR, and U-series records are generally concordant over these sites, a major discordance has arisen between ESR and TL dates for the earliest part of the record at Tabun. In this review we shall shed some new light on this question, although the major conflict remains for the moment. The transition from a Lower Paleolithic Acheulean industry to a distinctive Levantine Middle Paleolithic Mousterian industry appears to have taken place some time near or just before 200 ka, as recorded at Tabun. Levallois-Mousterian lithics found at Tabun, Skhul, and Qafzeh attest to a series of stone-tool industries being present throughout the Levant for about 150,000 years. The appearance of an upper Paleolithic, Aurignacian in the northern Negev is dated at around 48,000 y by U-series analyses of associated spring-deposited travertines (Schwarcz et al. 1979).

In this paper we shall summarize ESR and U-series results that we have obtained over the past few years on four sites in Israel, as a part of the ongoing program of dating of Paleolithic sites in the Old World being carried out at the McMaster Geochronology Center. The dates presented here, for Hayonim, Tabun, Amud, and Far'ah II, are given without supporting radiometric or other analytical data. Where problems with these data arose during the dating procedures, we have noted these in the text. Papers presenting these details and the specific methods used and problems encountered at each site are in preparation.

Before presenting the new data it is important to consider the purposes and limitations of "absolute" or "chronometric" dating as applied to prehistoric sites. Prehistorians are engaged in the definition of the stages in the evolution of hominids and their artifactual record, and the assignment of a chronology to these stages. We commonly assume that at any given time in the past, and within a restricted geographic region, particular stages of cultural or biological evolution are coeval. In other words, particular lithic industries can be taken as relative time markers, in the fashion of extinct fossil species. One of the chief functions of absolute dating is to test that assumption. Indeed, the specific dates that are obtained at a site are less important than the question as to whether they are the same, older, or younger than dates on supposedly equivalent levels in other archaeological sites. Eventually, as we extend the scope and resolution of absolute dating around the world, we will be able to construct a single, multiregional time frame for all hominid evolution, and to test the limits of this assumption of coevality. At least in a limited region such as the Levant, it would seem to be safe to make this assumption. It is, however, contradicted by some existing data, suggesting that at least some lithic industries have regionally distinct time ranges. The usual attitude of prehistorians when confronted with such data is to assume that the dates at some sites are wrong (while the dates at other sites are assumed to be correct). The task of the geochronologist is to sort out these conflicts and to develop and apply more accurate chronometers to these problems. But, clearly, the existence of such seemingly contradictory data may also be an impetus for archaeologists to reevaluate the supposed equivalence of cultural data from different sites and regions, in order to test whether supposed equivalences are valid.

2. ESR AND U-SERIES DATING

The theory and methods of electron spin resonance dating have been summarized in several articles (e.g., Grün and Stringer 1991; Schwarcz and Grün 1992; Rink 1997). Briefly, the method is based on the measurement of the intensity of a characteristic signal, produced by trapped electronic charges in samples of tooth enamel, which is detectable on an electron spin resonance spectrometer. The height of the signal increases with radiation dose. We calibrate the sample's sensitivity to dose by exposing it to additional doses of artificial gamma rays, permitting us to convert the peak height to an equivalent dose (De) in grays. The dose rate (i.e., the annual dose) is determined from the natural radioactivity of the sample and its surrounding sediment, as well as the calculated cosmic dose. The cosmic + sediment gamma and beta ray dose rate is assumed to have remained constant through time or, at most, to have varied with water content as a function of climate. The internal dose rate of the enamel and the dose of beta-rays from adjacent dentine and cementum is attributed to uranium absorbed by these materials. This dose rate is assumed to have increased through time from an initial zero value, as a result of: a) uptake of U; and b) growth of the daughter isotopes of ^{238}U. The increase in U content is assumed to be a regular function of time. Two possible limiting cases are early uptake (EU), or acquisition of present-day U content soon after burial, and linear uptake (LU), where the present U content has been acquired at a constant rate through time. To some extent we can define the true uptake history through U-series analysis of the enamel and dentine (Grün et al. 1988; Grün and McDermott 1995).

Part of the beta dose from both sediment and dental tissues adjacent to the enamel (cementum or dentine) is attenuated by the enamel, due to the short (about 2 mm) range of beta rays in enamel. A model to describe this attenuation is needed in order to calculate the beta dose rate delivered to the enamel by these external sources. This calculation was formerly done assuming a quasi-exponential attenuation (Grün 1987). We have developed a new method of computing this attenuation, called Rosy, which is based on "one-group" theory (Brennan 1997a). This method is used to calculate all the dates presented in this paper.

The methods and theory of U-series dating are described by Schwarcz and Blackwell (1991). Most stalagmitic calcite formed in caves is initially devoid of thorium (including ^{230}Th). The age of such a deposit can thus be determined by observing the extent of growth of ^{230}Th from its parent and grandparent, ^{234}U and ^{238}U respectively. Mass spectrometric measurement of the ^{234}U/^{238}U and ^{230}Th/^{234}U ratios allows us to determine the age with a precision of 1% over the time range up to 400 ka. The reliability of the dates depends on the assumption of closed-system behavior, which appears to be valid for most cave-deposited calcites. Tooth enamel and dentine can also be dated, but the significance of their "ages" (as calculated from the isotope ratios) depends on the U uptake history, which can be inferred by comparison between ESR and U-series data for the same materials (enamel, dentine). According to the model presented by Grün et al. (1988), a tooth exhibiting early uptake, should give a TIMS/US age equal to the EU ESR age, whereas a tooth which had undergone continuous (e.g., linear) U uptake would give an apparent U-series age less than the EU age.

3. HAYONIM CAVE

This site in the Upper Galilee region of Israel is currently under excavation by an Israeli-French-American team. Thin layers of Natufian and Kebaran deposits cap a deep sequence of Middle Paleolithic deposits which form the main filling of the cave. No Lower Paleolithic deposits have yet been encountered. The following is a preliminary report on ESR and U-series dating of the Middle Paleolithic layers. Hayonim presents an interesting challenge to the application of radiation-dose dating at a site. The applicability of ESR dating at this site is limited by the variable availability of skeletal remains. In some parts of the site, both bones and teeth are absent, apparently as a result of post-depositional alteration of the sediments in the site (Schiegl et al. 1996). This diagenetic process has affected the chemical composition of the sediments (Mercier et al. 1994), including the concentrations of the radioactive elements U, Th, K, etc., which contribute to the external dose rate. We suppose that there has been a progressive change in the dose rate experienced by teeth (and burned flint) in some, but not all, parts of the cave. The areas of best tooth preservation appear to have been essentially unaffected by these changes, whereas areas of high concentration of ash in the central part of the cave are mostly strongly affected, resulting in almost total disappearance of bones and teeth (Stiner, this volume).

Hayonim presents an interesting opportunity to compare U-series dates on stalagmites with ESR and TL dates on teeth and burned flint. A series of stalagmitic deposits extends from the lowest exposed levels of the cave to above the Natufian levels, and apparently represents accumulations from a single area of quasi-continuous dripping from the roof of the cave, in the central part of the cave. A large fallen stalagmite was exposed in the Middle Paleolithic levels of the cave prior to present excavations and was dated by alpha-spectrometric U-series (Schwarcz et al. 1980), giving a date of 163^{+63}_{-40} ka. The newer excavations revealed that the base of this stalagmite lay in the upper part of stratigraphic level E, at a level of about 240 cm below datum, in square K 20. A number of attempts were made to date lower parts of this stalagmite, resulting in poor yields or anomalously low ages, apparently as a result of alteration and partial replacement of the stalagmite by phosphate minerals (S. Weiner, personal communication 1996). A sample of the fallen stalagmite, collected from a point estimated to be about 50 cm below the former cap of the deposit, gave a good yield and appeared to be free of any alteration effects. This sample gave a date of 161 ± 3 ka (2σ), in surprisingly good agreement with the earlier date (considering the large error in the latter). We assume that this is a minimum estimate for the age of deposition of the –240 cm horizon since the dated level of the stalagmite is approximately 1 meter above the base of stalagmite. Further analyses of other levels of these stalagmitic deposits are in progress.

ESR dates have been obtained on teeth collected in two areas of the cave: the central part (squares H–K, 20–24) and the front of the cave (squares J21, K21). The teeth from all sites contained relatively low internal U contents and their dates are therefore strongly dependent on the external gamma ray + cosmic dose rates. These dose rates were obtained in two ways: a) by neutron activation analysis (NAA) of U, Th, and K in sediment from near the points in the section where the teeth had been recovered; and b) using thermoluminescence dosimeters (TLD). A third method, using a portable gamma-ray scintillometer, could not be employed here because of the need to drill unacceptably large holes for the emplacement of the detector probe.

In the central sites, further survey by TLDs are still underway in some areas. At the remaining three sites, TLD-based dose rates were consistently higher than NAA-based rates: the average ratio of NAA to TLD dose rates was 0.75 ± .09. The NAA dose rates are based on the assumption that U in the sediment is in equilibrium with its daughter radioisotopes. These data suggest that there has been some diagenetic addition of daughter isotopes leading to an overestimate of the effective dose from uranium (measured in grays per microgram of U). Note that more than half the dose attributable to U is generated by daughter isotopes. We plan to test the bulk activity of the sediment.

Tooth samples were collected in the central area from levels between 328 and 435 cm below datum within stratigraphic level E. Three samples that were located close to TLDs gave dates of 164 ± 15 (EU) and 171 ± 17 (LU) ka (Table 1). These lay at levels between 150 and 185 cm below the base of the dated stalagmite, and they suggest that the LU date is a better estimate of the true age. Five other teeth collected between 328 and 427 cm below datum, at sites for which we have so far only NAA-based dose rates, gave dates ranging from 159 to 252 ka with very little difference between EU and LU dates. The average for this group is 196 ± 34 ka (EU), 200 ± 32 (LU), which is about 20% higher than the average for the teeth for which TLD dose rates were used and is consistent with the difference in dose rates reported by the two methods. U-series analysis of these tooth enamels would have little effect on the ages due to the low internal U contents of enamel and dentine.

There is no apparent tendency for an increase in age with depth in this series of samples spanning more than 1 meter of stratigraphic thickness, and suggesting rather high sedimentation rates at this point in the history of the deposit. This is also consistent with the small difference between the TIMS/US age of the stalagmite and the ESR dates on teeth.

Near the front of the cave, dose rates determined using TLDs and NAA of sediments were in excellent agreement. Four teeth were analyzed, from square F26, between –385 and –395 cm. They are also assigned to level E, although the upper sample (94884) was originally designated as level "C/E" in excavation field notes. These teeth gave ages of 241 ± 11 and 257 ± 6 ka (average of EU and LU), which are significantly older than ages for level E at comparable depths below datum in the central part of the cave. Two other teeth from this area, from stratigraphically deeper levels, gave ages of 70 and 115 ka.

Table 1. ESR dates for Hayonim

Sample no.	Depth (cm)	Square	Dosimetry	EU (ka)	LU (ka)
a. Central area					
95605	–403	I 23c	TLD	175 ± 17	186 ± 19
95603	–417	I 23a	TLD	147 ± 20	152 ± 21
95606	–435	I 24b	TLD	171 ± 22	176 ± 23
Mean				164 ± 15	171 ± 17
94881[a]	–328	J 21	NAA	242 ± 30	245 ± 31
95902[a]	–341	H 23d	NAA	174 ± 22	183 ± 24
95901[a]	–344	H 23b	NAA	159 ± 20	164 ± 22
94878[a]	–388	I 20d	NAA	218 ± 28	220 ± 29
95602[a]	–427	K 21c	NAA	186 ± 24	187 ± 24
b. Entrance to cave					
94884	–385	F 26	TLD	249 ± 31	261 ± 34
94886	–395	F 26	TLD	233 ± 27	252 ± 32
Mean				241 ± 11	257 ± 6
94882	–431	G 27	TLD	69 ± 8	70 ± 8
95601	–538	D 28b	TLD	113 ± 14	117 ± 15

[a]The following dates are based on neutron activation analyses (NAA) of sediments from central area; discrepancies between TLD and NAA-based dose rates in nearby areas suggest that these dates may be approximately 25% too old (see text).

There is some suggestion of rodent burrowing activity in this part of the cave, which could have intruded teeth from higher levels in the sediments. One of these teeth (95601) was described by its excavator as having been cemented in breccia, but this is not inconsistent with intrusion during an earlier episode of burrowing.

Overall, the ESR and U-series dates suggest that the end of deposition of Layer E must be close to 161 ka, and at least some of the deposition of the level took place between 240 and 170 ka. Finer temporal resolution of the deposition of these deposits awaits further dating work.

It is interesting to compare the chronology of Hayonim with other sites in Israel. The upper part of level E is characterized by the production of Levallois flakes on short blanks, in a fashion resembling that seen in level C at Tabun. In the lower part of E we find the use of more elongated blanks leading to the production of blades and retouched points resembling Tabun D (L. Meignen, personal communication 1996). Our dates for level E are significantly older than those obtained by ESR at Tabun for levels C and D (Grün et al. 1991), and they are somewhat younger than ages for this transition as obtained by TL on burned flint (Mercier et al. 1995).

4. TABUN

The site of Tabun, about 20 km south of Haifa at the entrance to Wadi Mughara, has provided some classic problems in the dating of archaeological sites. This cave was filled with at least 25 m detrital sediments that were rich in Lower and Middle Paleolithic artifacts and contained in their upper parts some well-preserved hominid remains. A large part of the sedimentary fill of this cave was removed by the original excavators, Garrod and Bate (1937). As at Hayonim, the sediments of this cave have been subjected to a great deal of diagenetic alteration, with the result that bones and teeth were lost from a large part of the deposits. In addition, there is a possibility that sediment-derived radiation dose rates have changed through time in some parts of the cave. The deepest deposits with good bone-tooth preservation were found in the central part of the cave, an area that was almost totally emptied by Garrod and Bate. Most of the teeth from their excavation are stored in the Museum of Natural History, London. Some of these teeth were used in an ESR dating study of the cave by Grün et al. (1991), who used NAA analyses of attached sediment to estimate the external gamma and beta dose rates to their teeth samples.

In 1994 the well-preserved mandible of a cervid (red deer) was found exposed in one of the witness sections left by Garrod and Bate, near the west wall of the cave, in an area shown by them to have been excavated to a depth of –8.26 m below datum. Preliminary typological analysis of lithics from the sediments surrounding this site by A. Ronen (University of Haifa) suggests that they correspond to the Tabun E industry. These are the first samples of teeth from Tabun found in situ to be subjected to ESR analysis. We emplaced several TLDs near the site of the mandible, and also collected samples of sediment for NAA. These analyses showed a very marked variation in the concentration of uranium in sediment near the teeth, which made it difficult to estimate the average expected dose rate; TLDs also gave variable values for dose rates. The average ages of the teeth (using TLD data only) were 215 ± 22 ka and 290 ± 36 ka (LU). The large difference between LU and EU ages is due to the significant internal dose rate from enamel and dentine. McDermott et al. (1993), using TIMS/US dates for teeth from Tabun, found that U uptake had occurred relatively early at this site. Therefore we might expect that the true age of these teeth is close to the lower figure of about 220 ka. However, our EU age is significantly older than Grün et al.'s EU age estimate for the Tabun E industry of between 151 ± 34 and 182 ± 61, and closer to their LU estimate of 188 ± 31 to 213 ± 46 ka. On the other hand, the present ages are significantly younger than the TL ages for the Tabun E industry of 278 ± 34 to 386 ± 33 ka (Mercier et al. 1995). In order to further resolve this question, we are now analyzing these teeth by TIMS/US. Note that if our LU ages for this level prove to be correct, then this age for level E would be consistent with the ESR ages which we obtained for layer E at Hayonim which contains a "Tabun D-type" industry.

5. AMUD

This site in Wadi Amud is located about 4 km northwest of the Sea of Galilee. Originally a karstic cave, progressive collapse of the roof has left most of the archaeological deposits exposed in front of a small cave remnant. The site was originally excavated in the 1960s by a Japanese team led by Suzuki and Takai (1970) who found a well-preserved, almost complete Neandertal skeleton (Amud I). It was buried in the uppermost layers of the cave and, as a result, it would be difficult to assign any date other than a terminus post quem to this specimen. Fifteen other individuals have been subsequently recovered from the cave, mainly represented by fragmentary remains (Hovers et al. 1995). The stratigraphy of the site is complex, and it is an extremely "lumpy" site, as defined by Rink et al. (1994), see also Schwarcz 1994 and Brennan et al. 1997b), due to the presence of large, angular blocks of bedrock limestone that have fallen into the deposit as it accumulated. The limestone contains much less U, Th, and K than the fine sediment of the site. This leads to a large degree of spatial variation in the external dose rate. Currently the site is being dated by TL of burned flint and ESR on faunal teeth.

Due to the lumpiness of the sediments at the site, we depended on TLD measurements as well as in situ gamma-ray scintillometry to determine the gamma + cosmic dose rate. Unfortunately, three of the critical TLDs for level B2/8 were stolen and we were forced to use the scintillometer readings exclusively. Notably, we found that dose rates calculated from NAA analyses of sediment were systematically higher than TLD values by as much as 2 ×, which we attribute to the much lower radioactivity of the limestone blocks ("lumps").

The new ESR ages generally increase in age with increasing stratigraphic depth (Table 2). The youngest age, 43 ± 5 ka (EU), 48 ± 6 ka (LU), was obtained on a tooth from level B1/6, which is believed to be the same stratigraphic level that contained Amud I (Hovers et al. 1995). Even though the sample was collected 8 m away from the

Table 2. ESR dates for Amud

Sample no.	Depth (cm)	Square	Level	EU (ka)	LU (ka)
95505	−282	L 3b	B 1/6	43 ± 5	48 ± 6
95506	−364 −369	O 2d	B 2/8	52 ± 4	66 ± 5
95507	−364 −369	O 2d	B 2/8	68 ± 5	79 ± 7
95508	−364 −369	O 2a	B 2/8	56 ± 5	65 ± 6
Mean			B 2/8	59 ± 8	70 ± 8
95501	−534	P 9b	B 4	68 ± 7	72 ± 8
95504	−423	O 15d	B 4	110 ± 8	113 ± 8
Range			B 4	68–113	

locus of the Amud I burial, this date on a single tooth may have some relevance to the age of the Amud I Neandertal. A group of three teeth has been dated from underlying level B2/8. Their average ages are 59 ± 8 (EU) and 70 ± 8 (LU). These constitute a post quem date for Amud I, and would also provide a post quem age for the Neandertal infant burial, Amud 7 (Rak et al. 1994). These ages are in excellent agreement with preliminary TL dates of 50–60 ka by Valladas and Mercier for the stratigraphic layer containing Amud 7 (Rak et al. 1994). It is apparent that they are older than the age for level B1/6, and tend to support the previously cited age. Dates for underlying level B/4 range from 68 to 113 ka. Theoretical analysis of the dose distributions at lumpy sites (Brennan et al. 1997) suggests that the true age may be a few percent higher than quoted mean ages for each set of dates.

In conclusion, our present data for this site suggest that Neandertals were living at this site as recently as 43,000 ± 5,000 years ago, but more results would be needed to confirm this estimate. The LU estimate of 48 ± 6 ka for this same level is consistent with preliminary data from Valladas suggesting that the hominid levels date to about 50 ka (cited in Hovers et al. 1995), and with ESR ages for the top of level B of about 42 ka (EU) and 50 ka (LU) (Grün and Stringer 1991). This extends the range of Neandertal activity in Israel closer to the end of the known time range of this subspecies, but these dates are still much older than the lowest ages reported for Neandertals in southwestern Europe at about 35–30 ka (Hublin et al. 1995). We also note that Mousterian tools were being made by Neandertals at Amud at this date, close to the time when an Upper Paleolithic industry is first recognized at about 48 ka in the northern Negev at Boker Tachtit.

6. FAR'AH II

This Mousterian site consists of two thin occupation levels that lie below a loess cap in partly eroded fluvial sediments of Nahal Besor, an intermittent stream flowing out of the northern Negev (the locality is marked on most maps as En Sharuhen). A high density of flakes, bone fragments, and teeth occurs over an area of about 20 square meters, mainly in the rich upper horizon which is only a few tens of centimeters thick. Gilead and Grigson (1984) showed that the site had been used as a hunting and butchery site and may have been occupied over the dry season. Goldberg (1981) argues from geomorphological evidence that the site was occupied within the last glacial cycle, at about 50 ka.

Teeth were obtained from the collection in the Hebrew University of Jerusalem, Department of Zoology. We used in situ gamma scintillometry, TLDs, and NAA to determine the external gamma dose rate. All three methods agreed to within 5% of one another. The moderately high U content of the teeth (16 ppm in dentine; 1.2 ppm in enamel) resulted in significantly different average EU and LU ages of 44 ± 4 and 59 ± 7 ka, respectively (Table 3). In order to resolve differences between these and to determine the correct model of U uptake, we have begun TIMS/US analyses of the enamel. The first data, however, gave an age of 76.4 ± 1.5 ka (2σ), greater than the EU or even the LU age for the tooth. The Grün et al. (1988) U-uptake model cannot be used to compute an age for such a sample. A possible explanation for this behavior is that the tooth has experienced early or continuous U uptake, followed by a brief period of U loss late in its history. At this stage, after a substantial amount of 230Th had grown as a daughter of U in the sample, then leaching of the tooth could easily remove some uranium while leaving the less soluble thorium in the tooth. This would tend to increase the apparent age. In such a case it is not generally possible, however, to determine the true U-uptake behavior before the period of U loss. At the

Table 3. ESR dates for Far'ah II

Sample no.	EU	LU	U-series
95366	45.6 ± 2.7	57.1 ± 4.1	—
95367A	46.2 ± 2.7	57.1 ± 4.1	74.5 ± 1.5 (2s)
95368A	54.4 ± 3.2	72.0 ± 4.9	—
95370A	50.1 ± 3.1	62.7 ± 4.7	—
Mean	49.1 ± 4.1	62.2 ± 7.0	

moment we suppose that the age of Far'ah II lies somewhere between 50 to 60 ka, and is therefore roughly coeval with part of the Mousterian industry at Kebara. Gilead and Grigson (1984) note that the lithics from this site resemble those from Tabun level IX (=Tabun D). The dates obtained here show that this example of a "Tabun D-type" industry is much younger than its occurrence at Tabun Cave.

7. SUMMARY

We have summarized some new data on the ESR and U-series chronology of the Levantine Mousterian. In presenting these data we have shown how each site presents special problems that limit the precision and accuracy of the dates. Nevertheless, we can now define the time range of occupation of many of these sites with a precision of between 10 and 20%. This allows us, in most cases, to address the main issue of archaeological chronology as outlined earlier: to determine the possible coevality or sequence of sites in this region and in relation to sites at greater remove (e.g., Europe). The seemingly early dates for the Levallois Mousterian of Hayonim Cave appear to set back in time the period at which this industry appears in the archaeological record. In an independent study by this laboratory, Kleindienst et al. (1997) have shown that a rather poorly defined but distinctly Levalloisian industry was being produced in the Western Desert of Egypt, near the Kharga oasis, dated by U-series analyses of travertines to between 200 and 240 ka. This would suggest that the use of prepared cores for manufacture of stone flakes was already widespread by the time of the penultimate interglacial (isotope stage 7), and may perhaps have spread through northern Africa and the Middle East during a period of wetter climate that allowed hominids to traverse the normally hyperarid Sahara.

The chronology of the arrival and dispersal of Neandertals in the Levant appears to be further constrained by the data presented here for Amud. Neandertals are known for three other sites in the Levant: Kebara, Tabun, and the newly described (but as yet undated) Syrian site of Dederiyeh. The Neandertal burial at Kebara is securely dated to 60 ka (Valladas et al. 1987; Schwarcz et al. 1989). The Tabun Neandertal (Tabun C1) would appear to be an outlier, as Grün et al. suggest an age of about 120 ka for layer C. However, preliminary gamma-spectrometric analyses of the mandible of this skeleton by Simpson, Schwarcz, and Stringer (submitted to *Journal of Human Evolution*) give an EU U-series age of about 60 ka. If these data are confirmed, then we would have a remarkably uniform chronology for the appearance (and disappearance) of this taxon in the Levant.

ACKNOWLEDGMENTS

Funding for this research was provided by a grant to HPS and WJR from the U.S. National Science Foundation (DBS-9210469) and travel grants from the L. S. B. Leakey

Foundation. We are grateful to H. K. Lee, J. Rees-Jones, Karen Goodger, and Jean Johnson for analytical assistance. We thank Ofer Bar-Yosef, Itzhak Gilead, Paul Goldberg, Caroline Grigson, Erella Hovers, Liliane Meignen, Norbert Mercier, Rivka Rabinovich, Yoel Rak, Avraham Ronen, Mary Stiner, Eitan Tchernov, Hélène Valladas, and Steve Weiner for assistance in the field and helpful discussions. We thank Professors Takeru Akazawa, Kenichi Aoki, and Ofer Bar-Yosef for organizing the Tokyo University Museum symposium "Neanderthals and Modern Humans in West Asia," where a rather different version of this paper was presented.

REFERENCES

Brennan, B. J., W. J. Rink, E. L. McGuirl, H. P. Schwarcz, and W. V. Prestwich. 1997a. Beta doses in tooth enamel by "One Group" theory and the Rosy ESR dating software. *Radiation Measurements* 27:307–14.

Brennan, B. J., H. P. Schwarcz, and W. J. Rink. 1997b. Simulation of the gamma radiation field in lumpy environments. *Radiation Measurements* 27:299–305.

Edwards, L. R., J. H. Chen, and G. J. Wasserburg. 1986. 238U-234U-230Th-232Th systematics and the precise measurement of time over the past 500,000 years. *Earth and Planetary Science Letters* 81:175–92.

Garrod, D. A. E., and D. M. A. Bate. 1937. *The Stone Age of Mount Carmel*, Vol. 1, *Excavations at the Wadi El-Mughara*. Oxford: Clarendon.

Gilead, I., and C. Grigson. 1984. Far'ah II: A Middle Paleolithic open air site in the northern Negev. *Proceedings of the Prehistoric Society* 50:71–98.

Goldberg, P. 1981. Late Quaternary stratigraphy of Israel: An eclectic view, 55–66. In *Préhistoire du Levant*, ed. J. Cauvin and P. Sanlaville. Paris: CNRS Editions.

Grün, R., and C. B. Stringer. 1991. Electron spin resonance dating and the evolution of modern humans. *Archaeometry* 33:153–99.

Grün, R. 1987. Beta attenuation in thin layers. *Ancient TL* 4:1–8.

Grün, R., and F. McDermott. 1995. Open system modelling for U-series and ESR dating of teeth. *Quaternary Science Review (Quaternary Geochronology)* 13:121–6.

Grün, R., and C. B. Stringer. 1991. Electron spin resonance dating and the evolution of Modern Humans. *Archaeometry* 33:153–99.

Grün, R., C. B. Stringer, and H. P. Schwarcz. 1991. ESR dating of teeth from Garrod's Tabun Cave collection. *Journal of Human Evolution* 20:231–48.

Grün, R., H. P. Schwarcz, and J. Chadam. 1988. ESR dating of tooth enamel: Coupled correction for U-uptake and U-series disequilibrium. *Nuclear Tracks and Radiation Measurement* 14:237–41.

Hovers, E., Y. Rak, R. Lavi, and W. H. Kimbel. 1995. Hominid remains from Amud cave in the context of the Levantine Middle Paleolithic. *Paléorient* 21/2:47–61.

Hublin, J. J., C. Barroso Ruiz, P. Medina Lara, M. Fontugne, and J. L. Reyss. 1995. The Mousterian site of Zafarraya (Andalucia, Spain): Dating and implications on the Palaeolithic peopling processes of Western Europe. *Comptes Rendus de l'Académie de Sciences* 321(IIa):931–7.

Kleindienst, M., H. P. Schwarcz, K. Nicoll, C. S. Churcher, J. Frizano, R.W. Giegengack, and M. Wiseman. 1997. *Pleistocene geochronology and paleoclimates at Dakhleh and Kharga oases, Western Egypt, based on uranium-series determinations from water-laid tufas*. Paper presented at 13th Biennial meeting, Society for African Archaeology. Poznan, Poland.

Li, W., J. Lundberg, A. Dickin, D. C. Ford, H. P. Schwarcz, and D. Williams. 1989. High-precision mass-spectrometric U-series dating of cave deposits and implications for paleoclimate studies. *Nature* 339:534–6.

McDermott, F., R. Grun, C. B. Stringer, and C. J. Hawkesworth. 1993. Mass-spectrometric U-series dates for Israeli Neanderthal/early modern hominid sites. *Nature* 363:252–5.

Mercier, N., H. Valladas, J. L. Joron, S. Schiegl, O. Bar-Yosef, and S. Weiner. 1995. Thermoluminescence dating and the problem of geochemical evolution of sediments. A case study: The Mousterian of Levels at Hayonim (1994), Israel. *Journal of Chemistry* 35:137–41.

Mercier, N., H. Valladas, G. Valladas, A. J. Jelinek, L. Meignen, J. L. Joron, and J. L. Reyss. 1995. TL dates of burnt flints from Jelinek's excavations at Tabun and their implications. *Journal of Archaeological Science* 22:495–509.

Rak, Y., E. Hovers, and W. H. Kimbel. 1994. A Neandertal infant from Amud Cave, Israel. *Journal of Human Evolution* 26:313–24.

Rink, W. J. 1997. Electron spin resonance (ESR) dating and ESR applications i Quaternary science and archaeolmetry. *Radiation Measurements* 27:975–1025.

Rink, W. J., H. P. Schwarcz, R. Grün, I. Yalçinkaya, H. Taskiran, M. Otte, H. Valladas, N. Mercier, O. Bar-Yosef, and J. Kozlowski. 1994. ESR dating of the Last Interglacial Mousterian at Karaïn Cave, Southern Turkey. *Journal of Archaeological Science* 21:839–49.

Schiegl, S., P. Goldberg, O. Bar-Yosef, and S. Weiner. 1996. Ash deposits in Hayonim and Kebara caves, Israel: Microscopic, macroscopic, and mineralogical observations and their archaeological implications. *Journal of Archaeological Science* 23:763–83.

Schwarcz, H. P., B. Blackwell, and P. Goldberg. 1979. Uranium series dating of travertine from archaeological sites, Nahal Zin, Israel. *Nature* 277:558–60.

Schwarcz, H. P., P. Goldberg, and B. Blackwell. 1980. Uranium series dating of archaeological sites in Israel. Israel *Journal of Earth Sciences* 29:157–65.

Schwarcz, H. P., and R. Grun. 1992. ESR dating and the origin of modern man. *Philosophical Transactions of the Royal Society of London B* 337:145–8.

Schwarcz, H. P., W. M. Buhay, R. Grün, H. Valladas, E. Tchernov, O. Bar-Yosef, and B. Vandermeersch. 1989. ESR dating of the Neanderthal site, Kebara Cave, Israel. *Journal of Archaeological Science* 16:653–61.

Schwarcz, H. P., and B. Blackwell. 1991. Archaeometry, 513–52. In *Uranium Series Disequilibrium: Application to Environment Problems in the Earth Sciences*, 2nd edition, ed. M. Ivanovich and R. S. Harmon. Oxford: Oxford University Press.

Schwarcz, H. P. 1994. Chronology of modern humans in in the Levant, 21–31, In *Late Quaternary Chronology and Paleoclimates*, ed. O. Bar-Yosef and R. Kra. Tucson and Cambridge: Radiocarbon and the Peabody Museum of Archaeology and Ethnography, Harvard University.

Schwarcz, H. P. 1994. Current challenges to ESR dating. *Quaternary Science Reviews (Quaternary Geochronology)* 13:601–5.

Suzuki, H., and F. Takai. 1994. *The Amud Man and His Cave Site*. Tokyo: Academic Press of Japan.

Valladas, H., J. L.Joron, G. Valladas, B. Arensburg, O. Bar-Yosef, A. Belfer-Cohen, P. Goldberg, H. Laville, L. Meignen, Y. Rak, E. Tchernov, A. M. Tillier, and B. Vandermeersch. 1987. Thermoluminescence dates for the Neanderthal burial site at Kebara cave in Israel. *Nature* 330:159–60.

5

GIF LABORATORY DATES FOR MIDDLE PALEOLITHIC LEVANT

Helene Valladas,[1] Norbert Mercier,[1] Jean-Louis Joron,[2] and Jean-Louis Reyss[1]

[1]Centre de Faibles Radioactivités
Laboratoire Mixte CNRS-CEA
Gif sur Yvette Cedex 91198, France
[2]Groupe des Sciences de la Terre
Laboratoire Pierre Süe, CEN Saclay
Gif sur Yvette Cedex 91191, France

1. INTRODUCTION

From modest beginnings twenty years ago, the TL dating of burnt flint has emerged as a major and reliable dating technique, especially at sites that are too old for radiocarbon dating. During the last two decades, the "Centre des Faibles Radioactivités" (CNRS/CEA) at Gif sur Yvette has played a leading role in perfecting this technique, thereby enabling it to make some major contributions to the elaboration of the chronology of the Middle Paleolithic Levant (Aitken and Valladas 1993; Mercier and Valladas 1994).

2. PRINCIPLES OF TL DATING

Since the theory and practice of TL dating have been discussed at some length in several recent publications, the technical discussion will be confined to a few key points that illustrate the strengths and weaknesses of this dating method (Valladas 1992; Mercier et al. 1995a).

Anything buried in the ground is exposed to ionizing radiation because natural radioactivity is all around us. Many minerals, among them flint, store the energy they receive and release it only when heated to over 450°C. The intensity of light emitted during heating, or *thermoluminescence*, is proportional to the total energy accumulated during burial, or *paleodose* (Aitken 1985).

When a flint falls into a fire, it loses the energy it received during prior burial and this event sets the time clock to zero. After the hearth is abandoned, the flint can again

Neandertals and Modern Humans in Western Asia, edited by Akazawa *et al.*
Plenum Press, New York, 1998.

store the energy received from within and without. To calculate the time elapsed since the flint was last burned, we need to know two things: the paleodose, or the total amount of radiation energy received during burial and released as TL; and the *dose-rate*, or the amount of radiation received per year.

The age of the abandoned hearth is calculated using a simple linear equation:

$$\text{Age} = \text{paleodose}/\text{dose-rate}$$

The dose-rate is divided into two components: *internal* and *external*. The internal radioactivity of dated flint core depends exclusively on the concentrations of radioisotopes present in it. So the internal dose-rate remains virtually unchanged over hundreds of thousands of years. This stability contrasts with the situation in tooth enamel, where internal radioactivity is a function of two external factors: the chemical composition of the surrounding sediment and the rate of radioisotope uptake. This point is sometimes overlooked when dates obtained by TL for burnt flints are compared with ESR dates for teeth (Grün 1989).

The external dose-rate received by the flints comes mainly from γ-rays produced by the radioisotopes present in the environment, and includes a small contribution of cosmic rays. This total external dose is usually measured in situ by dosimeters buried in the sediment in the vicinity of the flint findspots. The relative γ dose-rates contributed by various radioisotopes can be calculated from α and γ spectrometric measurements done on sediment samples in the laboratory. The results are used to determine if the uranium and thorium series are in a state of secular equilibrium or not. Significant disequilibrium indicates that the sediment underwent geochemical changes at some time in the past. Such changes in the concentrations of radioisotopes might make the modern external dose-rates different from those prevailing in the past.

Since internal dose-rates can be determined with precision, the major potential source of error comes from bad estimates of past external dose-rates. One way to test how well the measured dose-rate approximates what the flint experienced during its burial is to use the *isochron analysis* (Aitken and Valladas 1993). This test can be used when flints from adjacent findspots (same age and external dose) have different internal dose-rates due to different radioisotopic compositions. Such variability of internal dose-rates allows one to compute the mean age of an archaeological level without explicit use of the measured external dose rate.

3. THERMOLUMINESCENCE DATING OF SOME LEVANTINE MIDDLE PALEOLITHIC SITES

Using this technique TL age-estimates were made for seven sites in Northern Israel: three in the Mt. Carmel area (Kebara, Skhul, and Tabun); and four in the Galilee (Qafzeh, Amud, Zuttiyeh, and Hayonim).

3.1. Kebara Cave

Age-estimates ranging from 48.3 ± 3.5 ka to 61.6 ± 3.6 ka were obtained for 38 burnt flints from Mousterian occupation units VI–XII (Valladas et al. 1987). The relatively narrow span of ages suggests rapid sedimentation rate. Of particular interest was unit XII because it contained a Neandertal skeleton (Arensburg et al. 1985). The estimated age of this unit was 59.9 ± 3.5 ka. Isochron analysis of flints from units XI and XII gave an age of 57 ± 8 ka, close to the 60 ka mean of individual flints.

The flint TL and similar tooth enamel ESR dates of Schwarcz et al. (1989) indicate that a Neandertal population occupied the cave about 60 ka ago.

3.2. Qafzeh Cave

Age-estimates ranging from 102 to 85 ka were obtained for 20 burnt flints from strata XVII–XIX and XXI–XXIII, which contain the remains of early modern humans (Vandermeersch 1981). The ages exhibit little systematic variation with depth (Valladas et al. 1988). This is consistent with geological evidence of rapid sedimentation rates for the Mousterian strata and with biostratigraphic analysis of microvertebrate remains (Haas 1972; Bar-Yosef and Vandermeersch 1981; Tchernov 1981). Isochron analysis confirmed that the skeletal remains are about 90 ka old (Aitken and Valladas 1993). Comparable ESR dates for teeth were subsequently reported by Schwarcz et al. (1988).

Qafzeh was the first site that demonstrated the presence of modern humans long before the Neandertals disappeared from the area. This situation stands in contrast with that in Europe, where none of the Cro-Magnon skeletons found so far antedate the Neandertal.

3.3. Skhul Rock Shelter

This rock shelter yielded the skeletal remains of at least ten Proto-Cro-Magnon hominids, found in layer B, which was rich in Mousterian industries of the Levallois type (McCown and Keith 1939). When six burnt flints from layer B were dated by TL, an average age of 119 ± 18 ka was obtained (Mercier et al. 1993), which stands in reasonable agreement with several ESR dates (Stringer et al. 1989).

3.4. Tabun Cave

The Tabun Cave is most impressive by its size and depth of accumulated strata (Garrod and Bate 1937). The great variety of industries makes Tabun a reference point for the Late Pleistocene cultures of the Levant (Bar-Yosef 1989, 1992, 1993).

Eight units in the intermediate chamber excavated by Jelinek (1981, 1982) yielded 39 burnt flints suitable for dating. These flints gave age estimates listed in Table 1 (Mercier et al. 1995b).

Table 1. Calculated mean ages of the investigated units at Tabun

Jelinek's major unit	Jelinek's bed	Garrod's layer	Age in ka*
I	17–26	C	171 ± 17
II	27–32	D/C	212 ± 22
V	41–42	D/C	244 ± 28
IX	62–69	D	263 ± 27
X	70–72	Ea-D	270 ± 22
XI	73–77	Eb-Ea	306 ± 33
XII	78–80	Ec(?)-Eb	350 ± 33
XIII	81–85	Ed	331 ± 30

*The error accompanying each age is the arithmetic average of errors in individual flint ages.

Examination of the table reveals the extent to which the TL chronology increases the antiquity of the site (Jelinek 1982; Bar-Yosef 1993). While one can see an overall age increase with depth, no sharp separations can be made between individual units due to the magnitudes of standard deviations. This is particularly true for units XI to XIII, all parts of Garrod's Layer E.

The situation at the lower strata might become clarified after we have finished dating flints from Garrod's Layers F and G, which were kindly provided by the current excavator, Avraham Ronen. At this point one can only say that the few dates obtained so far tend to confirm the great antiquity of Garrod's Layer E.

The α and γ spectrometric analyses detected no evidence of significant past radioisotopic mobility. So, the external dose-rate measured by 49 dosimeters cannot be dramatically different from what it was in the past. This was confirmed by an isochron test on flints from unit XI which yielded an age of 287 ± 20 ka, a value not significantly different than the mean of individual flints from this unit, 306 ± 33 ka. The external dose-rate computed from the isochron plot was close to the dosimetric mean confirming the conclusions of radioisotopic analysis.

When our TL ages are compared with linear uptake ESR dates for teeth (Grün et al. 1991), the result is that the latter and some published uranium series measurements (McDermott et al. 1993) give a much shorter chronology. Possible explanations for this discrepancy are discussed by Mercier et al. (1995b).

The new chronology raises some interesting questions about the hominid remains labeled Tabun I and II. The former shows affinities with some early European Neandertals. If it belongs to the top of layer C, where it was found (Garrod and Bate 1937), it would be 170,000 years old and the population represented by this hominid would have settled in the Levant during oxygen isotope stage 6 or earlier. In this case, the occupants of Kebara would have been preceded by another Neandertal population who arrived in the area more than 100 ka before. However, Garrod wondered if the remains were not intrusive from layer B, in which case the hominid would be more recent.

This Neandertal would have been preceded in the cave by a *Homo sapiens sapiens*, if we accept Vandermeersch's suggestion (1989) that the Tabun II mandible found in Layer C belonged to a hominid of the type found at Skhul and Qafzeh. In that case the presence of modern humans would have to be pushed back in time at least to oxygen isotope stage 6.

3.5. Hayonim Cave

This cave is of great interest because the Mousterian lithic industries of the Upper and Lower layers E show characteristics associated with Tabun C and D lithic assemblages (Meignen, this volume), respectively. So TL dates obtained for these layers can provide more information on the antiquity of these industries.

Preliminary ages for seven flints from Upper E and five from Lower E averaged to about 150 and 200 ka, respectively. The Hayonim results for levels containing Tabun C and D industries make the unusually long TL chronology proposed for Tabun more convincing.

3.6. Amud Cave

Preliminary age-estimates for 19 flints from levels B2 and B4 fall in the 50–70 ka range. When the dating is completed we will be able to assign a more precise age to the

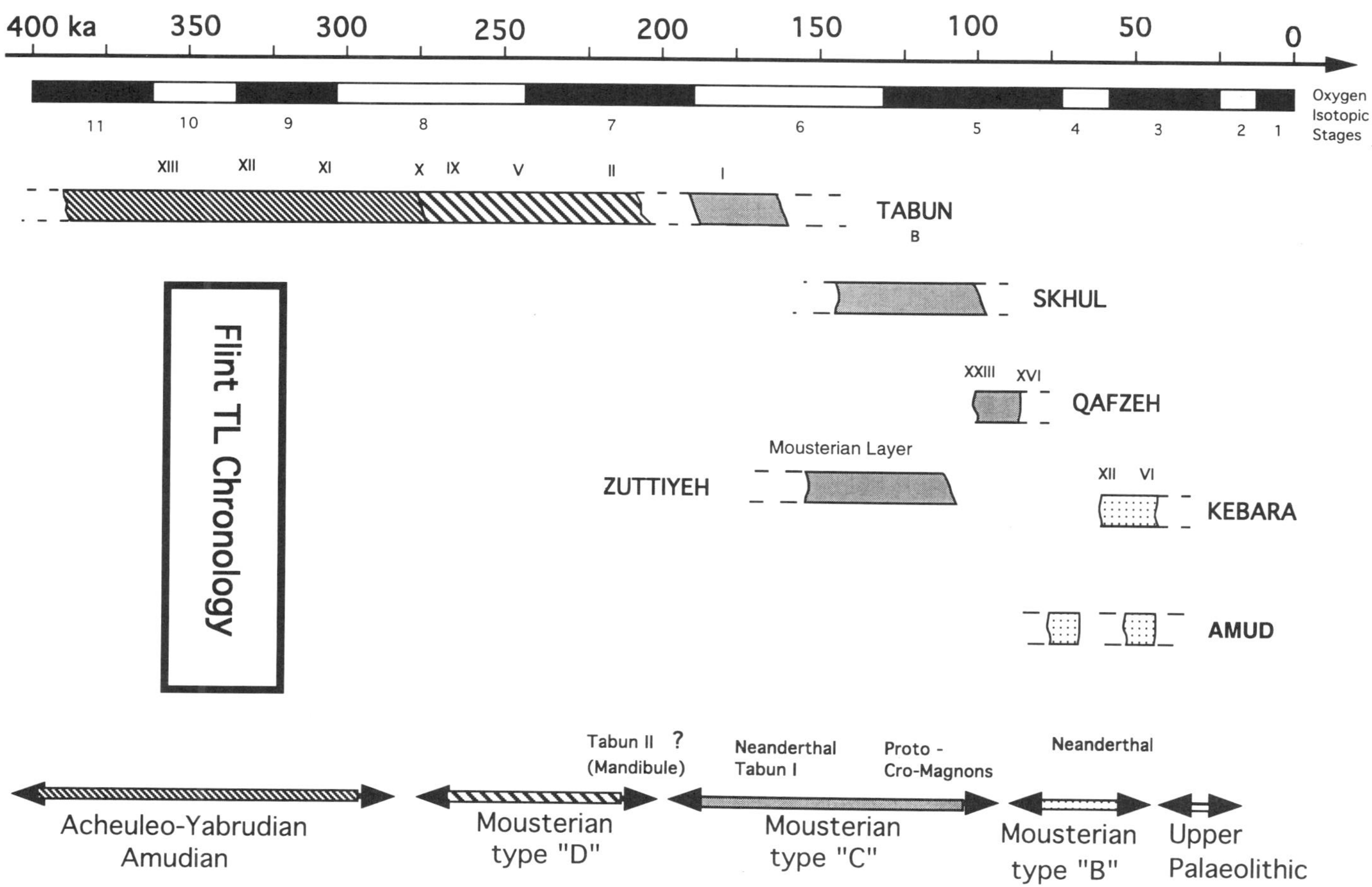

Figure 1. Oxygen isotope scale (Imbrie et al. 1982; Martinson et al. 1987) and the TL flint chronology of selected Near Eastern sites. Darkest grey shading designates Lower Paleolithic, while several lighter shades are used to indicate several types of Mousterian industries represented in the Near East. The numerals or letters to the left of each bar locate the archaeological levels.

skeletal remains of several Neandertals excavated by Japanese archaeologists in the 1930s and to others discovered more recently (Rak et al. 1994; Hovers 1995).

3.7. Zuttiyeh Cave

Part of the Mousterian levels at Zuttiyeh, excavated by Gisis and Bar-Yosef (1974), consisted of breccia of very low radioactivity. They yielded only five dateable flints. Three had a mean age of 106 ± 7 ka and two had a mean age of 157 ± 13 ka. Flints from the underlying Acheuleo-Yabrudian level, from which part of a human skull had been excavated by Turville-Petre (Turville-Petre 1927), had not been heated high enough to be dateable by TL.

4. CONCLUSION

The TL dates (Figure 1) shed new light on the evolution of Middle Paleolithic industries in the Near East. The new chronology assigns a much longer lifespan to specific types of Levantine Mousterian lithic industries than previously believed. For example, the Tabun types C and D remained in use for about 80 thousand years each. These dates suggest that the Mousterian industries were introduced at about the same time in the Near East as they were in Europe. The difference is that in the Near East this industry was used by early modern humans as well as by Neandertals.

The Qafzeh and Skhul dates indicate that a population of modern humans inhabited the Near East 100 ka ago and maybe even much earlier if the Tabun Layer C mandible belonged to a modern human (Vandermeersch 1989). That a Neandertal population was present in the Near East about 60 ka ago is proven by the Kebara and Amud dates, but tis arrival in the area will remain uncertain until an independent date has been obtained for the Tabun skeleton. Whether the two populations were present continuously or intermittently and precisely when they first appeared are unknown at present.

ACKNOWLEDGMENTS

Numerous individuals have provided burnt flint specimens and shared knowledge about the sites discussed in this article. We wish to thank, in particular, O. Bar-Yosef, E. Hovers, A. Jelinek, L. Meignen, Y. Rak, A. Ronen, A. M. Tillier, and B. Vandermeersch. We are also grateful to M. Chech for assistance at most of the studied sites. Contribution no. 1929 from the Centre des Faibles Radioactivités (CFR), Gif-sur-Yvette.

REFERENCES

Aitken, M. J. 1985. *Thermoluminescence Dating*. London: Academic Press.

Aitken, M. J., and H. Valladas. 1993. Luminescence dating relevant to human origins, 27–39. In *The Origins of Modern Humans and the Impact of Chronometric Dating*, ed. M. J. Aitken, C. B. Stringer, and P. A. Mellars. Princeton: Princeton University Press.

Arensburg, B., O. Bar-Yosef, P. Goldberg, H. Laville, L. Meignen, Y. Rak, E. Tchernov, A. M. Tillier, and B. Vandermeersch. 1985. Une sépulture néandertalienne dans la grotte de Kébara (Israël). *Comptes Rendus de l'Académie des Sciences* 300:227–30.

Bar-Yosef, O. 1989. Geochronology of the Middle Levantine Paleolithic, 589–610. *The Human Revolution*, ed. P. A. Mellars and C. B. Stringer. Edinburgh: Edinburgh University Press.

Bar-Yosef, O. 1992. Middle Paleolithic human adaptations in the Mediterranean Levant, 189–215. In *The Evolution and Dispersal of Modern Humans in Asia*, ed. T. Akazawa, K. Aoki, and T. Timura. Tokyo: Hokusen-Sha.

Bar-Yosef, O. 1993. The role of western Asia in modern human origins, 132–47. In *The Origins of Modern Humans and the Impact of Chronometric Dating*, ed. M. J. Aitken, C. B. Stringer, and P. A. Mellars. Princeton: Princeton University Press.

Bar-Yosef, O., and B. Vandermeersch. 1981. Notes concerning the possible age of the Mousterian layers in Qafzeh cave, 282–6. In *Préhistoire du Levant*, ed. J. Cauvin and P. Sanlaville. Paris: CNRS Editions.

Garrod, D. A. E., and D. M. A. Bate. 1937. *The Stone Age of Mount Carmel*, Vol. 1, *Excavations at the Wadi-El-Mughara*. Oxford: Clarendon.

Gisis, I., and O. Bar-Yosef. 1974. New excavations in Zuttiyeh cave, Wadi Amud, Israel. *Paléorient* 2:175–80.

Grün, R. 1989. Electron Spin Resonance (ESR) dating. *Quaternary International* 1:65–109.

Grün, R., C. B. Stringer, and H. P. Schwarcz. 1991. ESR Dating of teeth from Garrod's Tabun cave collection. *Journal of Human Evolution* 20:231–48.

Haas, G. 1972. The microfauna at Djebel Qafzeh cave. *Paleovertebrata* 5:261–70.

Hovers, E. 1995. Renewed excavations at Amud cave: Implications for the end of the Levantine Middle Paleolithic. Paper presented at the conference on "Neanderthals and Modern Humans in West Asia," Tokyo.

Imbrie, J., J. D. Hays, D. G. Martinson, A. McIntire, A. C. Mix, J. J. Morley, N. G. Pisias, W. G. Prell, and N. J. Shackleton. 1982. The orbital theory of Pleistocene climate: support from a revised chronology of the $\delta^{18}O$ record, 269–305. In *Milankovitch and Climate*, ed. A. Berger. Dordrecht: Reidel.

Jelinek, A. 1981. The Middle Palaeolithic in the Southern Levant from the perspective of the Tabun Cave, 265–80. In *Préhistoire du Levant*, ed. J. Cauvin and P. Sanlaville. Paris: CNRS Editions.

Jelinek, A. J. 1982. The Tabun cave and Paleolithic man in the Levant. *Science* 216:1369–75.

Martinson, P. M., N. G. Pisias, J. D. Hays, J. Imbrie, T. C. Moore, and N. J. Shackelton. 1987. Age dating and the orbital theory of the ice ages, development of a high-resolution 0 to 300,000-year old chronostratigraphy. *Quaternary Research* 27:1–29.

McCown, T. D., and A. Keith. 1939. *The Stone Age of Mount Carmel*. Oxford: Clarendon.

McDermott, F., R. Grün, C. B. Stringer, and C. J. Hawkesworth. 1993. Mass-spectrometric U-series dates for Israeli Neanderthal/early modern hominid sites. *Nature* 363:252–5.

Mercier, N., and H. Valladas. 1994. Thermoluminescence dates for the Paleolithic Levant, 13–20. In *Late Quaternary Chronology and Paleoclimates of the Eastern Mediterranean*, ed. O. Bar-Yosef and R. S. Kra. Tucson and Cambridge: Radiocarbon and the Peabody Museum of Archaeology and Ethnology, Harvard University.

Mercier, N., H. Valladas, O. Bar-Yosef, B. Vandermeersch, C. Stringer, and J. L. Joron. 1993. Thermoluminescence date for the Mousterian burial site of es-Skhul, Mt. Carmel. *Journal of Archaeological Science* 20(2):169–74.

Mercier, N., H. Valladas, and G. Valladas. 1995a. Flint thermoluminescence dates from the CFR laboratory at Gif: Contributions to the study of the chronology of the Middle Palaeolithic. *Quaternary Science Reviews (Quternary Geochronology)* 14:351–64.

Mercier, N., H. Valladas, G. Valladas, J. L. Reyss, A. Jelinek, L. Meignen, and J. L. Joron. 1995b. TL dates of burnt flints from Jelinek's excavations at Tabun and their implications. *Journal of Archaeological Science* 22:495–509.

Rak, Y., W. H. Kimbel, and E. Hovers. 1994. A Neandertal infant from Amud Cave, Israel. *Journal of Human Evolution* 26:313–24.

Schwarcz, H. P., W. M. Buhay, R. Grün, H. Valladas, E. Tchernov, O. Bar-Yosef, and B. Vandermeersch. 1989. ESR dating of the Neanderthal site, Kebara Cave, Israel. *Journal of Archaeological Science* 16:653–9.

Schwarcz, H. P., R. Grün, B. Vandermeersch, O. Bar-Yosef, H. Valladas, and E. Tchernov. 1988. ESR dates from the hominid burial site of Qafzeh in Israel. *Journal of Human Evolution* 17:733–7.

Stringer, C. B., R. Grün, H. P. Schwarcz, and P. Goldberg. 1989. ESR dates for the hominid burial site of es-Skhul in Israel. *Nature* 338:756–8.

Tchernov, E. 1981. The biostratigraphy in the Middle East, 67–97. In *Préhistoire du Levant*, ed. J. Cauvin and P. Sanlaville. Paris: CNRS Editions.

Turville-Petre, F. 1927. *Researches in Prehistoric Galilee (1926–1927)*. Oxford: British School of Archaeology in Jerusalem.

Valladas, H. 1992. Thermoluminescence dating of flint. *Quaternary Science Reviews* 11:1–5.

Valladas, H., J. L. Joron, G. Valladas, B. Arensburg, O. Bar-Yosef, A. Belfer-Cohen, P. Goldberg, H. Laville, L. Meignen, Y. Rak, E. Tchernov, A. M. Tillier, and B. Vandermmeersch. 1987. Thermoluminescence dates for the Neanderthal burial site at Kebara in Israel. *Nature* 330:159–60.

Valladas, H., J. L. Reyss, G. Valladas, O. Bar-Yosef, and B. Vandermeersch. 1988. Thermoluminescence dating of Mousterian "Proto-Cro-Magnon" remains from Israel and the origin of modern man. *Nature* 331:614–6.

Vandermeersch, B. 1981. *Les Hommes Fossiles de Qafzeh (Israël)*. Paris: CNRS Editions.

Vandermeersch, B. 1989. The evolution of Modern Humans: Recent evidence from Southwest Asia, 155–64. In *The Human Revolution*, ed. P. A. Mellars and C. B. Stringer. Edinburgh: Edinburgh University Press.

6

THE FAUNAL SEQUENCE OF THE SOUTHWEST ASIAN MIDDLE PALEOLITHIC IN RELATION TO HOMINID DISPERSAL EVENTS

Eitan Tchernov

Department of Evolution, Systematics, and Ecology
Hebrew University
Jerusalem 91904, Israel

1. INTRODUCTION

A realistic appraisal of Southwest Asia during the late Middle and Upper Pleistocene is hampered by the difficulties involved in understanding two key issues. The first is the taxonomical position and phylogenetic relationship of Southwest Asian hominids and the correct sequence of their evolutionary events. The second issue is the biogeographical position of this region as a crossroads between two continents. Misinterpretation of the morphological, taxonomical, and evolutionary status of the southern Levantine hominids is essentially due to the inadequacy of the fossil record. Furthermore, the reliance on insecure stratigraphic frameworks, samples taken from undated layers, and non-matching radiometric methods have done little to provide clear insight into the real sequential biotic and anthropological events during this period in Southwest Asia. In particular, the difficulties of correlating the biotic changes with the geochronological and archaeological succession prevent the establishment of a more solid framework between the different chronologies, faunal turnovers, and anthropological events.

A central question in archeobiology is whether a causal relationship exists between cultural events and environmental changes. This question is magnified with respect to Southwest Asia, a region where the climatic fluctuations were relatively mild but where significant socioeconomic transformations and hominid interchanges took place during the Quaternary. One of the most conspicuous yet highly debatable events in Southwest Asia concerns the emergence of anatomically modern humans and Neandertals. During recent years, a sufficient amount of anthropological evidence has been accumulated to indicate that each of the two types of humans probably occupied the region, either alternatively or sympatrically: Proto-Cro-Magnons and Western Asian Neandertals. At present, the main disagreements concern the dating of the fossils, the definition of the various morphotypes

Neandertals and Modern Humans in Western Asia, edited by Akazawa *et al.*
Plenum Press, New York, 1998.

in terms of phylogenetic relationships, and the place of the fossils within the evolutionary and cultural sequence. The correct sequence of their appearance and disappearance, or whether they were sympatric or exclusive of each other, is still not resolved. Thus efforts to establish a reliable bio-cultural chronology of the Levantine Middle Paleolithic have become inseparable from efforts to understand the evolution of modern humans. The study of mammal faunal spectra and the new established radiometric dating techniques, mainly ESR and TL, have provided (in spite of the discrepancy between these results) some preliminary clues for a basic sequence in which late Acheulian, Acheulo-Yabrudian, and Mousterian industrial complexes are ordered according to faunal turnover and are correlated with the cultural chronostratigraphy.

2. PROVINCIALIZATION OF SOUTHWEST ASIA

Southwest Asia is neither simply a bridge between two continents nor a trivial corridor for biotic and hominid intercontinental exchanges. The biogeographical position of this region and its relation with other biogeographic provinces is extremely complex and can be explained only by examining global environmental fluctuations. *De facto* during the later Neogene and Quaternary periods, Southwest Asia acts more as a barrier than a corridor; only intermittently did biotic and hominid exchanges occur between Africa and Eurasia. During the Miocene, the Afro-Arabian plate continued to move northward and its northern edge drew up against the margin of the Eurasian plate by the subduction zone along the present Anatolian-Iranian tectonic suture line. The result was the formation of the Taurus-Zagros mountain chain, which steadily became a biogeographic barrier. This barrier increased with time, not only because of the recurrent tectonic movements, but also due to the intensification of the climatic oscillations during the Pleistocene. Ultimately, the eastern Mediterranean biota grew ever more isolated from the rest of Palearctica.

Although the climate fluctuated significantly, the entire region of the Levant is marked principally by a progressive desiccation (Rognon 1987, 1996). This increasing aridity was a major cause of the extinction of many Afro-tropical and Palearctic elements as well as the increased separation between tropical Africa and Eurasia. The impact of the glacial period and the close proximity of a major desert belt have played a very complicated role in the distribution of the Levantine plants and animals and have constantly reshaped the biotic gradients between Palearctica and the eremian belt (Tchernov 1988). Thus southward-northward shifts of the Palearctic regime fluctuated throughout the Pleistocene. Northern elements occasionally migrated southward deep into the Saharo-Arabian belt, apparently in correlation with the colder phases, with the partial exclusion of the eremian and xerotropical species, which temporarily dominated the region. Rognon argues for a "near disappearance of the Saharan desert" (1996:553) between 40 to 25 ka. Forty-five ka ago the earliest Upper Paleolithic cultures in the southern region of the Levant emerged, where many Upper Paleolithic sites were recorded (Marks 1983; Goring-Morris 1987). During warmer, drier, or interglacial periods, the reverse process occurred and desertic areas were abandoned.

Another cause of the partial isolation of Southwest Asia, which was no less important, was a deep penetration of the Mediterranean sea into the lower Nile (Messinian) canyon, as far south as Aswan, some 1200 km inland (Said 1975; Williams and Williams 1980). It is thus obvious that the southern Levantine region and part of eastern Egypt created a peninsula-like regime which remained biogeographically quite isolated from the rest of the world during the Pliocene. The fragmentation of the circum Mediterranean region, in particular its eastern part, into practically isolated biogeographic provinces is a direct consequence of raised sea-level during the Pliocene.

This biogeographical provincialism also characterizes the eastern Mediterranean communities at the dawn of the Pleistocene. During cold periods, the shallow shelf of the eastern Mediterranean allowed the arms of the Nile Delta to nearly reach and almost interfinger with the southern Levantine coastal rivers (Tchernov 1988). On these occasions, Nilotic elements were spread into the eastern Mediterranean coastal riverine system. It seems that during the Quaternary there is a general trend of further fragmentation of the region into smaller biotic units and an accentuation of the climatic-ecological belts, both of which created abrupt clinal transitions and caused major changes in the plant and animal lives over relatively short distances. Biotic interchanges between Africa and Eurasia took place mainly through the Afro-Arabian land masses. The opening of the Red Sea trough, which occurred only later, eliminated this geographical corridor. During the later Pliocene, when the Red Sea became firmly established, the only other possible passageway was the coastal plain of northern Sinai. At the end of the Pliocene and early Pleistocene, insemination of biotas into the Levant was possible only for a selected number of species and it was mainly restricted to this region. Thus the southern Levant, surrounded by the Saharo-Arabian desert belt in the south and the high Alpine mountain chain in the north, with the ever increasing aridity of the Syrian-Transjordanian plateau in the east, became quite isolated. This isolation was sufficient to hamper gene flow with other regions and consequently encourage endemism. Only major geological and climatic fluctuations could initiate the intrusion of new flora and fauna into this province, either over the Taurus-Zagros mountain range, across the Saharo-Arabian arid belts, or through the Red Sea.

3. THE ECOLOGY OF HOMINID DISPERSALS

The amplitude of climatic fluctuations as reflected in the deep-sea oxygen-isotope stages increased during the last 5 my, particularly the last one Ma, and caused extreme geographic shifts in the ecological belts. Understanding the response of biotic communities to environmental fluctuations is fundamental to creating models of cultural change during human history. Despite the indirect and complicated response of mammalian communities to climatic fluctuations, a rather rigid correlation between biotic changes, abiotic parameters, and socio-cultural events is possible. This can be done by studying the relative rapidity and high frequencies of such changes during the Pleistocene, as evidenced in the rich fossil record and the available radiometric data.

Ecological fluctuations, though extremely complicated, are essentially an indirect reflection of the environment. The more vigorous the action of natural agencies, the more simple and direct the ecological response. According to the same Darwinian logic, under low level environmental changes, the interplay between and within the biotic components in a relatively stable environment is minimal. Hence in the southern Levant, in comparison with higher latitudes, the magnitudes of Quaternary environmental changes were not severe enough to cause large scale biotic turnovers. Therefore the large mammals of this province cannot be used for biostratigraphic correlation. It is mostly the micromammals that have shown dynamic changes and therefore may be more efficiently used for a comprehensive paleoecological consideration and biochronology.

During the course of human history, we witness unidirectional changes in the human biosocial structure and a widening of niche space. These changes eventually became detached from both environmental factors and biological causes, and recently humans have become the main forces of change on both the biotic and abiotic environments. For most of their history, however, humans' impact on the ecosystem was negligible. Hominids were indeed an integral component of their community system, playing a particularly important role within the large-mammal communities (Isaac 1980, 1983; Brain 1981; Foley 1984; Turner 1984). As such, hominids were certainly susceptible to more or less the same ecological selective pressures as all other members of their community and they took part in any of the dispersal events that affected mammalian communities. Turner has shown (1984) that migration of carnivores occurs in the same direction as the major dispersal of the hominids and within the same general time span. It is probably true that any hominid dispersal was a natural part of any emigration until the late Upper Pleistocene period. Given that hominids play an integral part in the ecological events, hominid dispersals and faunal changes are actually part of the same phenomenon.

4. DATING THE SOUTHWEST ASIAN MIDDLE PALEOLITHIC

It is now clearly understood that the Middle Paleolithic of the eastern Mediterranean actually lasted much longer than previously thought, especially if we rely on TL-based chronology. Yet we have to be aware of two factors: not only is this series of dates far from complete, as we need more reliable dates for some critical archaeological layers, but there is also a substantial discrepancy between ESR, TL, and Th/U-based chronologies. Moreover, confusion was caused concerning several important strata due to unreliable sampling of teeth, mainly for ESR dating. The stretch of time suggested at present for the southern Levantine Middle Paleolithic is from 270 to 45 ka, based on TL chronology (Mercier et al. 1995). The ESR-based chronology provides a span of time from 170 to 45 ka. It seems that the younger the dated period, the fewer discrepancies in the datings.

The sequence of industries that emerges for the Levantine Middle Paleolithic seems to be more clearly divided into three complexes. The later one, known as "Tabun B-type," is well-represented in level B of Tabun and in Amud and Kebara Caves (as well as Sefunim and Geula Caves, Mt. Carmel). It is dated by both ESR and TL methods to around 65 to 45 ka, well within oxygen isotope stages 4 and early stage 3.

The excavations in Kebara (Bar-Yosef et al. 1992) exposed successive and intensive occupations by humans, with an increased rate of sedimentation and an industry that resembles "Tabun B-type" (Meignen and Bar-Yosef 1988, 1989). A series of TL dates indicates that Units XII through VI span the range of 60 to 48 ka (Valladas et al. 1988). ESR dates (Schwarcz et al. 1988; Grün et al. 1991) suggested a range of 60 to 64 ka for Units X–XI. If the estimated age of the first Neandertal occupation is around 65 ka, then the occupation of Kebara began at the end of stage 4, when Neandertals were present in the region.

We intend to show in this paper that the "Tabun B-type" industry corresponds well with the renewal of the fauna during the early part of the last glaciation in the Levant.

A "Tabun C-type" industrial assemblage is represented in Skhul and Qafzeh. Microfaunal remains from the "Tabun C-type" industry are known mainly from Qafzeh and not from Skhul. Furthermore, because the list of mammals from layer C at Tabun is only known from Bate's publications (1937a,b, 1942, 1943), in which some of the described material came from mixed D+C layers and some from C+B layers, we can only rely on the Qafzeh fossils for this period. We will also show that the faunal assemblage from Hayonim Upper E cannot be associated with the unique fauna of Qafzeh. The faunal assemblages exposed until now from Hayonim Upper E predate Qafzeh. The ESR and TL datings for Qafzeh agree satisfactorily and yield a time span of about 92 to 110 ka (Table 1).

Table 1. Comparison of micromammalian assemblages from late Acheulian, Acheulo-Yabrudian, and Mousterian deposits from Israel. When available, dates and references are added to each site. The faunal assemblages of the late Middle Paleolithic sites, Amud (Galilee), Sefunim, and Geula (Mt. Carmel), are identical with the Mousterian complex of Kebara. The faunal description for Tabun D is only based on Bate (1937a, 1937b, 1944, 1943), and includes remains from Tabun E and C in a few layers

OUMM QATAFA (Judean Desert)	TABUN E (a) (=unit XI)② (Mt. Carmel)	TABUN D (unit IX) ②	HAYONIM E (Western Galilee)	QAFZEH (XV-XXV) (Lower Galilee)	TABUN B (Mt. Carmel)	KEBARA (MT. CARMEL)
Late Acheulian	Acheulo - Yabrudian	Lower Mousterian ("Tabun D - type")		Upper Mousterian ("Tabun C - type")	("Tabun B - type")	
	TL = 306 (±33) ka ③	ESR = 140-170 ka ④ TL = 263 (±27) ka ②	U-TH = 164±60/40 ka ⑤ ESR = 120-17 ka TL = 115-225 ka ⑥	ESR = 115±15 ka⑦ TL = 92(±)5 ka ⑧	ESR (EU) = 85-105 ka ④ TL = not directly dated	ESR = 45-60 ka ⑨ TL = 47 - 61 ka ✸
RODENTIA	RODENTIA	RODENTIA	RODENTIA	RODENTIA	RODENTIA	RODENTIA
Hystrix indica	*Hystrix indica*	*Hystrix indica*	*Hystrix indica*	*Hystrix indica*	*Hystrix indica*	*Hystrix indica*
Sciurus anomalus	*Sciurus anomalus*	*Sciurus anomalus*	*Sciurus anomalus*	*Sciurus anomalus*	*Sciurus anomalus*	*Sciurus anomalus*
	Mesocricetus auratus	*Mesocricetus auratus*	*Mesocricetus auratus*		*Mesocricetus auratus*	*Mesocricetus auratus*
Allocricetus jesreelicus	*Allocricetus jesreelicus*	*Allocricetus jesreelicus*				
Allocricetus magnus	*Allocricetus magnus*	*Allocricetus magnus*	*Allocricetus magnus*			
					Cricetulus migratorius	*Cricetulu smigratorius*
Rattus haasi						
Mastomys batei	*Mastomys batei*	*Mastomys batei*		*Mastomys batei*		
Arvicanthis ectos	*Arvicanthis ectos*	*Arvicanthis ectos*		*Arvicanthis ectos*		
Apodemus mystacinus	*Apodemus mystacinus*	*Apodemus mystacinus*	*Apodem. mystacinus*	*Apodemus mystacinus*	*Apodemus mystacinus*	*Apodemusmystacinus*
Apodemus sylvaticus	*Apodemus sylvaticus*	*Apodemus sylvaticus*	*Apodemus sylvaticus*	*Apodemus sylvaticus*	*Apodemus sylvaticus*	*Apodemus sylvaticus*
	Apodemus flavicollis	*Apodemus flavicollis*			?	
Mus macedonicus	*Mus macedonicus*	*Mus macedonicus*	*Mus macedonicus*	*Mus macedonicus*	*Mus macedonicus*	*Mus macedonicus*
Gerbillus dasyurus				*Gerbillus dasyurus*		
Meriones tristrami	*Meriones tristrami*	*Meriones tristrami*	*Meriones tristrami*	*Meriones tristrami*	*Meriones tristrami*	*Meriones tristrami*
Ellobius fuscocapillus	*Ellobius fuscocapillus*	*Ellobius fuscocapillus*	*Ellobius fuscocapillus*			
Microtus guentheri	*Microtus guentheri*	*Microtus guentheri*	*Microtus guentheri*	*Microtus guentheri*	*Microtus guentheri*	*Microtus guentheri*
Spalax ehrenbergi	*Spalax ehrenbergi*	*Spalax ehrenbergi*	*Spalax ehrenbergi*	*Spalax ehrenbergi*	*Spalax ehrenbergi*	*Spalax ehrenbergi*
Myomimus judaicus	?					
		Myomimus qafzensis	*Myomimus qafzensis*	*Myomimus qafzensis*		
					Myomimus roachi	*Myomimus roachi*
LAGOMORPHA	LAGOMORPHA	LAGOMORPHA	LAGOMORPHA	LAGOMORPHA	LAGOMORPHA	LAGOMORPHA
	Lepus capensis	*Lepus capensis*	*Lepus capensis*	*Lepus capensis*	*Lepus capensis*	*Lepus capensis*
Ochotona sp.						
INSECTIVORA	INSECTIVORA	INSECTIVORA	INSECTIVORA	INSECTIVORA	INSECTIVORA	INSECTIVORA
Crocidura leucodon	*Crocidura leucodon*	*Crocidura leucodon*	*Crocidura leucodon*	*Crocidura leucodon*	*Crocidura leucodon*	*Crocidura leucodon*
Crocidura russula	*Crocidura russula*	*Crocidura russula*	*Crocidura russula*	*Crocidura russula*	*Crocidura russula*	*Crocidura russula*
			Suncus etruscus	*Suncus etruscus*	*Suncus etruscus*	*Suncus etruscus*
				Suncus cf. murinus		
	?	*Talpa chthonia*	*Talpa chthonia*			
STAGE 8	STAGE 7	STAGE 6		STAGE 5	STAGES 4 + 3	

① The bone bearing bed of Oumm Qatafa is estimated here as roughly contemporaneous to Tabun G-Ed (unit XIII of Jelinek, 1982, 1992) which were dated in Tabun to 331±30 Ka (Mercier et al., 1995).

② Jelinek, 1982, 1992; Mercier and Valladas, 1994; Mercier et al., 1995.

③ Mercier et al., 1995; Mercier and Valladas, 1994.

④ Grün et al., 1991. Teeth Used for dating Tabun B, were probably belong to Tabun C.

⑤ Schwarcz; this symposium.

⑥ Valladas and Mercier; this symposium.

⑦ Schwarcz et al., 1988.

⑧ Valladas et al., 1988.

⑨ Schwarcz et al, , 1989.

✸ Valladas et al., 1987; Rak et al., for Amud.

The small rockshelter of Skhul (Garrod and Bate 1937) actually contains two complexes: a lower unit C, which is overlaid unevenly by unit B (the latter contained the hominid burials). The erosional phase between B and C was correlated by Bar-Yosef (1989) to the erosional phase that truncated the top of Tabun D. ESR datings for this site (Stringer et al. 1989; Grün et al. 1991) gave an average age of 81 ± 15 ka (linear uptake) and 101 ± 12 ka (early uptake), and hence it may be partially contemporaneous with Qafzeh. No microfauna was described from Skhul.

The "Tabun D-type" industry is known from Tabun D, Rosh Ein Mor, Nahal Aqev 3, Abu Sif, Douara 3, and a few other sites. It is worth noting that in Hayonim, Lower E was associated with "Tabun D-type" and the industry of Upper E was linked with "Tabun C-type." Nonetheless, the faunal assemblages of the whole complex of Hayonim E, which stretches over a relatively long span of time, antedate all the faunal assemblages of Qafzeh.

The Acheulo-Yabrudian macrofauna of Zuttiyeh Cave (Bate 1932) might be correlated with at least part of Tabun F and E, but the paucity of the faunal remains does not permit a finer correlation. The industry shows some similarity with Tabun E (Gisis and Bar-Yosef 1974). The layers above the Acheulo-Yabrudian in Zuttiyeh provided U-series dates of 95 ± 10 ka and 97 ± 13 ka, while dates of 148 ± 6 ka and 164 ± 21 ka were recorded below these layers by Schwarcz et al. (1980). This layer was associated with the fragmentary hominid remains.

5. THE FAUNAL SEQUENCE OF THE SOUTHWEST ASIAN MIDDLE PALEOLITHIC

The osseous beds of Oumm-Qatafa seem to predate all Acheulo-Yabrudian lithic accumulations in the Levantine region. This can be asserted on the basis of the relative antiquity of its micromammalian assemblage including *Rattus haasi*, *Arvicanthis ectos*, *Mastomys batei* (Muridae), *Myomimus judaicus* (Gliridae). In particular, the unique appearance of *Ochotona* sp. (Ochotonidae, Lagomorpha), the remains of which were never recovered from later deposits (Tchernov 1981, 1984) (Table 1), is of great importance. Moreover, several genera, all of a Palearctic origin, are not yet represented in the Southern Levant during this period, such as *Lepus* (Leporidae, Lagomorpha), *Talpa* (Talpidae, Insectivora), *Sciurus* (Sciuridae, Rodentia) (Figure 1). This assemblage may be best correlated with Tabun G (Farrand 1979; Jelinek 1981, 1982a,b), preceding Yabrud rockshelter I (Copeland and Hours 1983; Bar Yosef 1989), and perhaps of the same age as Birket Ram in the Golan Heights (Goren-Inbar 1985). Hence Oumm-Qatafa may delineate a biostratigraphic baseline for all the later Acheulo-Yabrudian and Mousterian faunal sequences in this region (Table 1; Figure 1). In addition, the closely related species of *Rattus* and *Ellobius* found at Yarimburgaz Cave (Turkey) may indicate that these sites are roughly contemporaneous and should be dated to the late Middle Pleistocene (Koenigsvald, personal communication).

It should be stressed that the spatio-chronological distribution of the lagomorphs in the Southern Levant indicates a short event during late Acheulian times when the genus *Ochotona* is recorded at Oumm-Qatafa. The post-Acheulian replacement of *Ochotona* by the genus *Lepus* occurred during the time represented by Tabun F+E (Tchernov 1981, 1994). The disappearance of *Ochotona* together with a few other ancient forms (e.g., *Myomimus judaicus* and *Rattus haasi*), and the first appearance of new elements (e.g., *Lepus capensis*, *Sciurus anomalus* and *Myomimus qafzensis*) (Haas 1972, 1973; Daams

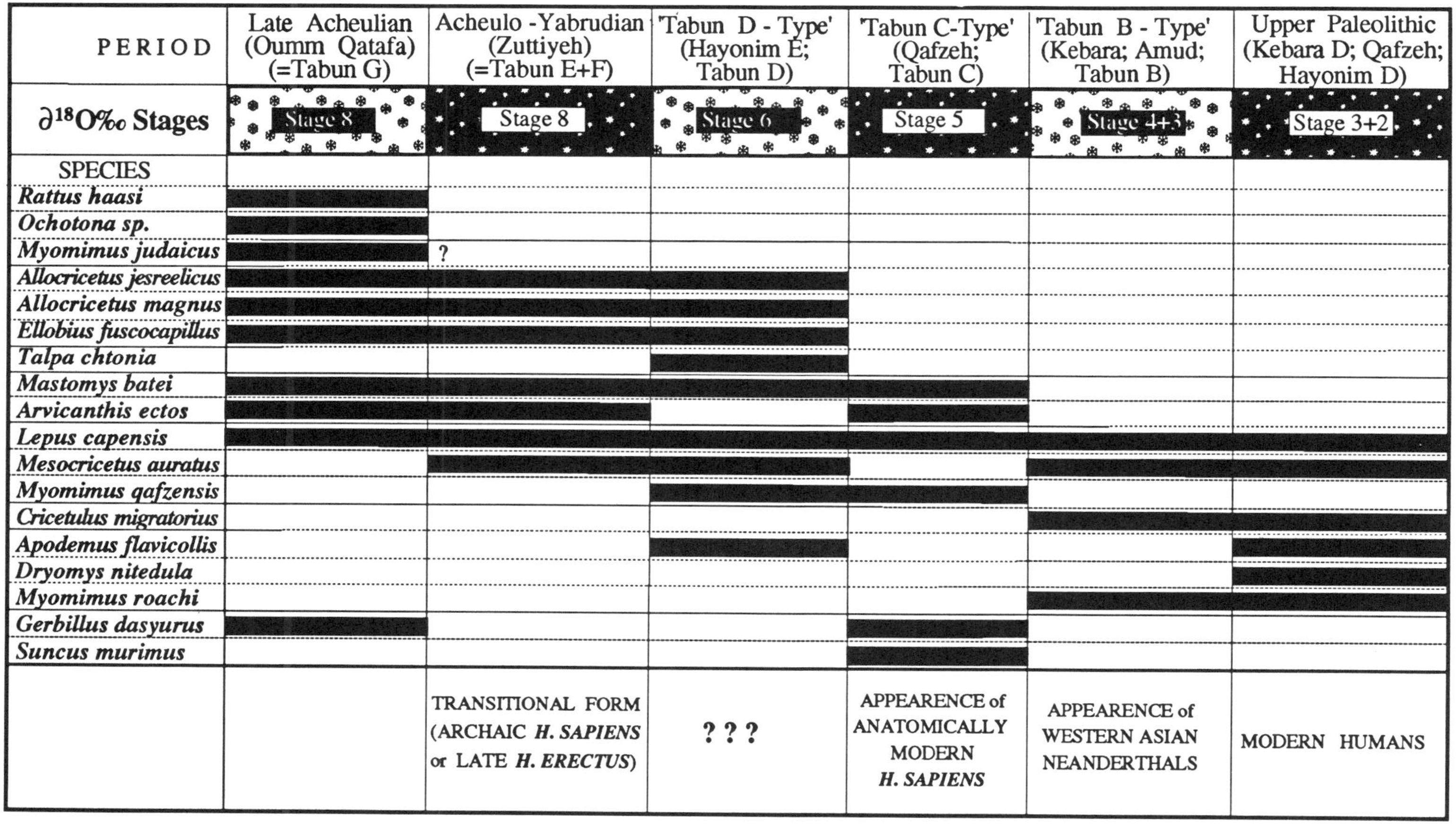

Figure 1. The essential microfaunal turnover events in late Acheulian, Acheulo-Yabrudian, and early, middle, and late Mousterian communities from the main bone-bearing sites in the southern Levant, correlated with the late Middle and Upper Pleistocene oxygen isoptope stages and human dispersal events in the Southwest Asia.

1981), suggests that the complex of Tabun F+E as well as Zuttiyeh, in particular layer E, should be placed chronologically well after the fauna of Oumm-Qatafa (Table l; Figure 1).

A later significant faunal change is evidenced when the assemblage of Hayonim E is compared to that of Qafzeh. The fauna with Mousterian of "Tabun D-type" in Hayonim still includes earlier Pleistocene elements, such as representatives of the fossil genus *Allocricetus* (*A. jesreelicus* and *A. magnus*) and the predominance of Palearctic mammals (e.g., *Talpa chthonia*). All of these were abruptly replaced in the assemblage of Qafzeh, mainly by Afro-Arabian elements. Biostratigraphically and ecologically, the fauna of Hayonim E can be fit best to isotope stage 6, which in this case corresponds more or less to the ESR-based chronology (Table 1; Figure 1) but less so to the TL-based chronology. We therefore suggest that Tabun D (unit IX) and Hayonim Upper and Lower E antedate the complex of Qafzeh.

The unique situation of the Qafzeh assemblage is clear (Tchernov 1981, 1994), with the total absence of cricetines being the most obvious characteristic. This is the only known cave site in the southern Levant where no cricetines were retrieved. This phenomenon cannot be due to ecological circumstances given that cricetines, such as *Mesocricetus auratus* and *Cricetulus migratorius*, do occur in the Upper Paleolithic deposits of this cave. *Ellobius fuscocapillus*, a Euro-Siberian Microtinae, and *Talpa chthonia*, a typical European insectivore, are also missing from this micromammalian community. All of them, however, exist in the Acheulo-Yabrudian of Tabun E and in the Mousterian of Tabun D. None of these groups, in particular the Microtinae and the Cricetinae, have ever developed any adaptations to withstand arid conditions in the Old World, and they are completely absent from the Saharo-Arabian arid belt.

The complete absence of these Palearctic elements from the assemblage of Qafzeh can only be explained by the unique ecological circumstances that prevailed during this period in this area (Rabinovich and Tchernov 1995). This impression is reinforced by the salient increase of East African savanna and Arabian elements: *Arvicanthis ectos*, *Mastomys batei*, *Gerbillus dasyurus* and *Suncus murimus*, the giant Indo-Arabian shrew. The latter two species are unknown from the Acheulo-Yabrudian material and mark the drastic change in the presence of rodent species in Hayonim E, the Middle Paleolithic deposits of

Table 2. Distribution of rodent species in three Middle Paleolithic cave sites in Israel

	HAYONIM (E) (Galilee)		QAFZEH (XV-XXV) (Galilee)		KEBARA (Units VI-XII) (Mt. Carmel)	
Species	MNI	%	MNI	%	MNI	%
Mesocricetus auratus	9	0.29	0	0	3	0.13
Allocricetus magnus	5	0.16	0	0	0	0
Cricetulus migratorius	0	0	0	0	17	0.72
Mastomys batei	5	0.16	1,172	4.18	0	0
Arvicanthis ectos	0	0	200	0.71	0	0
Apodemus spp.	128	4.18	92	0.33	105	4.42
Mus macedonicus	326	10.64	496	1.77	74	3.12
Gerbillus dasyurus	5	0.16	19	0.07	0	0
Meriones tristrami	54	1.76	332	1.18	124	5.22
Ellobius fuscocapillus	4	0.13	0	0	0	0
Microtus guentheri	2,476	80.86	25,157	89.66	1,955	82.35
Spalax ehrenbergi	20	0.65	125	0.45	49	2.06
Myomimus spp.	22	0.72	923	3.29	25	1.05
Sciurus anomalus	17	0.56	39	0.14	22	0.93
	3,062		28,059		2,374	

Qafzeh, and the Upper Mousterian layers of Kebara as shown in Table 2. The microfaunal assemblage of Qafzeh is dominated by open-country, steppe, or savanna species and is associated with the African antelope *Alcelaphus buselaphus*, and possibly an African ass *Equus africanus* (Eisenmann 1992; Rabinovich and Tchernov 1995), a North African Equid, *Equus* cf. *tabeti* (Eisenmann 1992), as well as *Camelus* cf. *dromedarius* (according to Bouchud (1974), and *Struthio camelus*. This is further manifested by the complete absence of all the cricetines as well as *Ellobius fuscocapillus* and *Apodemus flavicollis*. There is thus a strong Afro-Arabian stamp upon the fauna of Qafzeh, indicating that savanna conditions were represented in the immediate environment of the site.

As demonstrated in Table 2, a significant increase in the commensal rat, *Mastomys batei*, is shown in the deposits of Qafzeh compared with Hayonim E. A complete disappearance of the cricetines characterizes the savannization of the southern Levant during stage 5, as represented by the fauna of Qafzeh, and their recovery at the dawn of stage 4 (Kebara), followed by a drastic decrease in the distribution of forest and wood dwellers (*Apodemus* spp. and *Sciurus anomalus*) in Qafzeh as compared with Hayonim E and Kebara.

Species distribution of ungulates in Qafzeh shows that *Cervus elaphus* and *Dama mesopotamica* are the most common species, followed by large mammals such as bovids (*Bos primigenius* and *Dicerorhinus hemitoechus*). Mountain gazelles (*Gazella gazella*; N = 91, 6.2%) and roe deer (*Capreolus capreolus*; N = 6, 0.4%) are not very frequent in the assemblage. Wild goat remains (*Capra aegagrus*) outnumber the gazelle remains, probably due to the craggy and rocky area around the site. The frequencies approach the *Capra/Gazella* ratio of the rocky region near Ksar 'Akil (Lebanon) (Hooijer 1961; Marks and Volkman 1986). The hartebeest (*Alcelaphus buselaphus*) is rare, and wild boar (*Sus scrofa*) are represented by a few bones, mainly from layer XV (Rabinovich and Tchernov 1995).

Worth noting is the relatively large number of ostrich (*Struthio camelus*) egg shells retrieved from the Mousterian deposits of Qafzeh, indicating the abundance of this typical savanna species in the area. *Camelus dromedarius* is an additional steppic element that was described from this site. Equids constituted 5.26 percent (N = 78) of the Qafzeh remains (following the study by Eisenmann [1992] on the Middle Paleolithic equids of Israel). The relative abundance of equids in the Middle Paleolithic complex in the site mentioned above is given in Table 3. Notice that the North African *Equus* cf. *tabeti* is dominant in Qafzeh and is almost totally replaced by the Palearctic *Equus hydruntinus* in Kebara. Among the carnivores, which were studied by Dayan (1989), the spotted hyena (*Crocuta crocuta*) is the most common and the fox (*Vulpes vulpes*) is remarkably absent. The carnivores comprise only 1.8 percent (N = 26) of the Mousterian sequence. In conclusion, the fauna of Qafzeh provide clear evidence for a significant northward expansion of Afro-Arabian biotic elements among which African populations of anatomically modern *Homo sapiens* were also introduced.

Table 3. Distribution of equids in Middle Paleolithic sites in Israel. Based on data given by Eisenmann (1992)

	Tabun E		Hayonim E		Qafzeh (iv-xxiv)		Kebara E		Tabun B	
Species	NISP	%	NISP	%	NISP	%	NISP	%	NISP	%
Equus hydruntinus	0	0	(0 or 1)	(?100)	(1 or 0)	(3.125)	24	96	9	100
Equus caballus	13	81.25	0	0	4	12.5	(1 or 0)	(4)	0	0
Equus tabeti	3	18.75	0	0	27	84.375	(0 or 1)	(4)	0	0
Equus africanus	0	0	0	0	(0 or 1)	(3.125)	0	0	0	0

Table 4. Distribution of commensal species of rodents in three Middle Paleolithic cave sites in Israel

	HAYONIM (Galilee)		**QAFZEH** (XV-XXV) (Galilee)		**KEBARA** (Units VI-XII) (Mt. Carmel)	
Species	MNI	%	MNI	%	MNI	%
Mastomys batei	5	1.09	1172	60.10	0	0
Arvicanthis ectos	0	0	200	10.26	0	0
Apodemus spp.	128	27.88	92	4.72	105	58.66
Mus macedonicus	326	71.02	496	25.43	74	41.34
	459		1950		179	

Hovers et al. argue that "there are several subtle nuances in the archaeological observable mortuary practices" (1995:58). They also indicate a higher number of burials and longer occupation of the sites such as Skhul and Qafzeh by the anatomically modern humans who inhabited them. I have already pointed out (Tchernov 1984) that the predominance of commensal mice and African rats (*Mus musculus*, *Mastomys batei*) indicates a long-term occupation of Qafzeh (Table 4). The fauna and dates of Qafzeh support the notion that a significant northward shift of Afro-Arabian species, associated with the first introduction of modern humans, occurred on Asian terrain during this period. This is clearly demonstrated by the unique faunal community and the earliest modern human burials in Qafzeh during late stage 5, some 100 ka ago. The southern Levant was in a state of quasi-isolation, sealed from the north by the Alpine mountain chain of the Taurus-Zagros as well as from the east by the eastern deserts. Therefore this Afro-Arabian dispersal episode, which also included the introduction of an early form of modern humans, was confined to this region.

As shown in Table 4, *Mus* is very common in Hayonim Cave ("Tabun D-type"), but the type of human that occupied the Mousterian layers of this cave is not yet known. In Kebara ("Tabun B-type;" occupied by Neandertals), the relative abundance of *Mus* is much lower, while *Mastomys* became extinct. The commensal species in Qafzeh predominate the rodent community of this cave, suggesting long-term occupation of the site (Tchernov 1984).

In considering the microfauna and its paleoecological implications, a distinct faunal change between the assemblages of Qafzeh and Kebara and other "Tabun B-type" sites such as Amud becomes evident. A total elimination of almost all the Afro-Arabian micromammals and a significant decrease in the predominance of large open-land mammals is very clear in the fossil record. The sharp faunal turnover and the ecological shift between the faunal assemblage of Qafzeh and those of Kebara and Amud remains in accordance with the ESR and TL radiometric datings, and it fits the transition from isotope stage 5 to stage 4.

In sum, the cold and dry period, stage 4, caused a conspicuous faunal turnover in the region when a swarm of Palearctic-European elements including the Neandertals invaded the southern Levant. Several archaic species of mammals became extinct or were eliminated from the Levantine scenario at this time: *Myomimus qafzensis*, *Mastomys batei*, *Arvicanthis ectos*, *Suncus murimus*, and temporarily also *Gerbillus dasyurus* and *Camelus* cf. *dromedarius*. On the other hand, the cricetines reappear in the southern Levant, together with a new Euro-Siberian species similar to *Myomimus roachi*. This significant faunal break of a large-scale southward dispersal of Palearctic elements drove many of the Afro-Arabian species southward.

If this reconstruction is verified, then the arrival of Neandertals into the Southwest Asia occurred around 70 ka ago. Early stage 4 and its faunal turnover were associated in the Levant with both a lowering of sea level, particularly from Mt. Carmel southward onto

Sinai, and with increased dune activity, which resulted in the development of *kurkar* ridges along the coast.

6. FAUNAL CHANGES AND HUMAN EMIGRATIONS: CONCLUDING REMARKS

The regional sequence, although poorer in fossils in its early part, demonstrates that most Neandertal remains appear relatively late in the Mousterian sequence (e.g., Kebara, Amud, Tabun I remains from Tabun C, which possibly belong to Tabun B, and the earlier group of Neandertals from Shanidar). A cautious age estimate would be around 70–50 ka BP. Thus the arrival of the Neandertals in Southwest Asia could be explained as the result of a swift dispersal into the Mediterranean coastal ranges or along inland steppic routes leading eastward from European populations. This population expansion could have happened with the onset of the cold and dry conditions of stage 4 in western Europe, which evidently caused drastic changes in the availability, reliability, and accessibility of basic food resources (Bar-Yosef 1988).

The southward and eastward expansion of continental glaciers as well as the advent of mountain glaciers enlarged the surface of periglacial lands. Furthermore, they shifted the boundaries of vegetational belts, as regions such as Belgium turned into polar deserts (Cordy 1984). In southwestern France, environmental changes led to the replacement of red deer, wild horse, and roe deer mainly by reindeer herds in the area of Combe Grenal (Delpech and Prat 1980; Chase 1986), challenging the abilities of Mousterian people to adapt to a more precarious meat resource (Burch 1972). A somewhat similar faunal change is also observed in central Europe (Gabori 1976). In sum, the new environmental configuration across Europe challenged the foraging technologies, the social structure, and the spatial organization of Mousterian populations, forcing them to move into the Mediterranean lands. However, discussing in detail all the possible environmental effects on the Eurasian Mousterian adaptations is beyond the scope of this paper. If this reconstruction is verified, then the arrival of Neandertals or neandertaloid morphological features in Southwest Asia occurred around 70 ka (Figure 2).

During this relatively long time span, some very important events in human evolution took place within this region. During the early Middle Paleolithic in the southern Levant, a late form of *Homo erectus* or an early stage of *Homo sapiens*, possibly similar to what is known from the site of Irhoud in Morocco, could have been around. However, no human fossils from the time span of the "Tabun D-type" industry are available (Figure 2).

Some 100 ka ago, however, a more evolved form of *Homo sapiens*, or Proto-Cro-Magnons, were already present in Southwest Asia and occupied the region before any Neandertaloid type first appeared there. There is a tendency to regard these two forms as genetically isolated, but other interpretations persist. Our working hypothesis is that modern humans and Neandertals are the products of vicariant evolution of *Homo erectus*; the former evolved in Afro-Arabia (the remains from Qafzeh may represent one of the oldest paleodeme of a seemingly indigenous *Homo sapiens*), the latter in western Europe. No solid and indisputable evidence is at hand as yet to show that Neandertals predated stage 4 in Southwest Asia; the date of Tabun D is as yet not final. That Neandertals could have been late immigrants to the Levant, as a direct consequence of the abrupt and severe climatic change during early stage 4, is still illappreciated by many scholars. Our understanding of whether these two forms of hominids coexisted or overlapped with each other in a certain period in this region is entirely dependent on the fossil record.

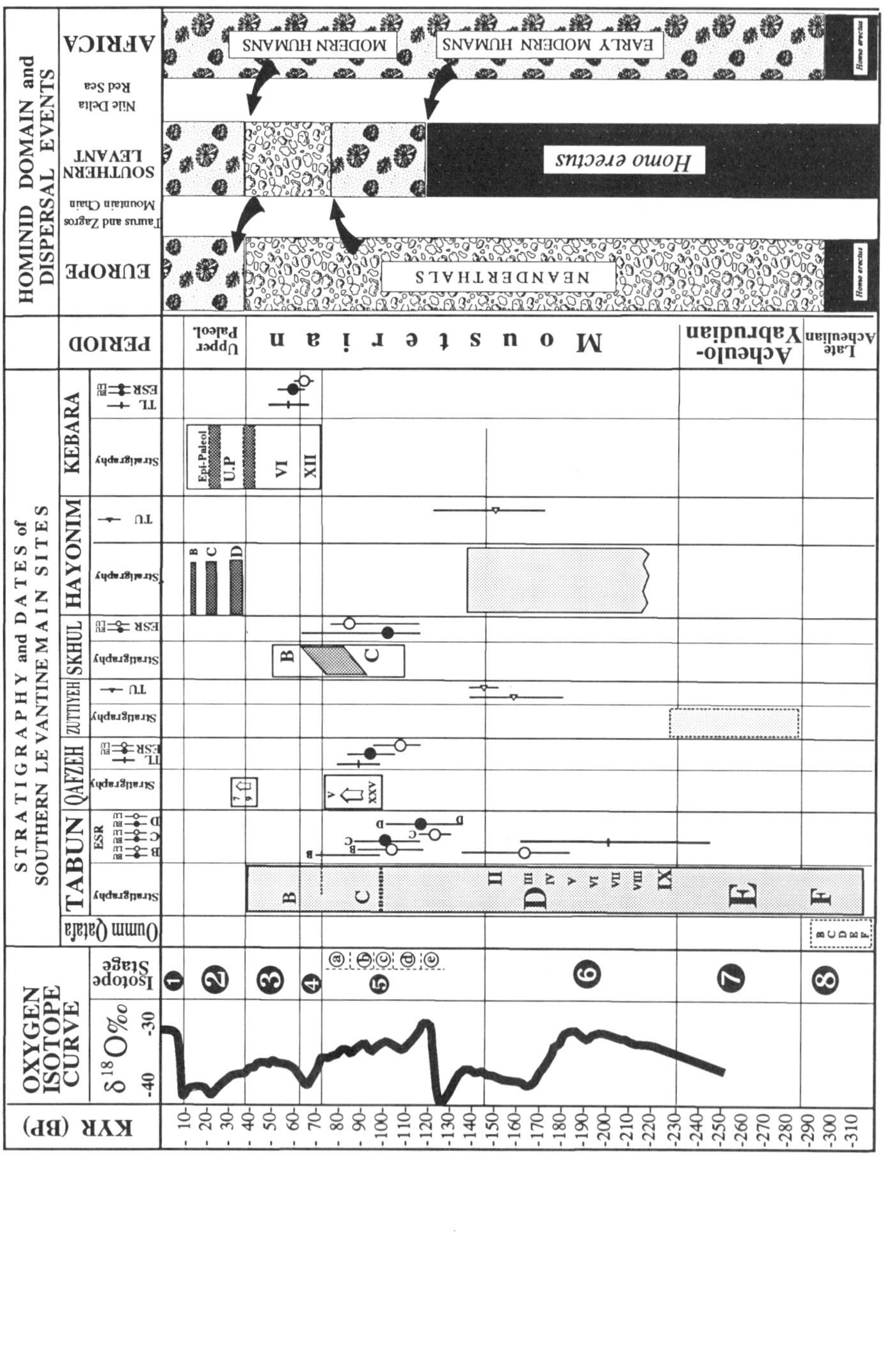

Figure 2. The chronostratigraphy and absolute dating of the main Middle Paleolithic sites, correlated with the hominid dispersal events in Southwest Asia.

REFERENCES

Bar-Yosef, O. 1988. The date of Southwest Asian Neanderthals, 31–38. In *L'Homme de Néandertal*, Vol. 3, ed. M. Otte. Liège: ERAUL.

Bar-Yosef, O. 1989. Geochronology of the Levantine Middle Palaeolithic, 589–610. In *The Human Revolution*, ed. P. A. Mellars and C. B. Stringer. Edinburgh: Edinburgh University Press.

Bar-Yosef, O., B. Vandermeersch, B. Arensburg, A. Belfer-Cohen, P. Goldberg, H. Laville, L. Meignen, Y. Rak, J. D. Speth, E. Tchernov, and A. M. Tillier. 1986. The Excavations in Kebara Cave, Mt. Carmel. *Current Anthropology* 33:497–550.

Bate, D. M. A. 1932. On the animal remains obtained from the Mugharet-el-Zuttiyeh in 1925, 27–52. In *Prehistoric Galilee*, ed. F. Turville-Petre. London: British School of Archaeology.

Bate, D. M. A. 1937a. New Pleistocene mammals from Palestine. *Annals and Magazine of Natural History* 20:397–400.

Bate, D. M. A. 1937b. Palaeontology: The fossil fauna of the Wadi el-Mughara Caves, 139–237. In *The Stone Age of Mount Carmel*, Vol. 1, *Excavations at Wadi el-Mughara*, ed. D. A. E. Garrod and D. M. A. Bate. Oxford: Clarendon.

Bate, D. M. A. 1942. Pleistocene Murinae from Palestine. *Annals and Magazine of Natural History* 9:465–86.

Bate, D. M. A. 1943. Pleistocene Cricetinae from Palestine. *Annals and Magazine of Natural History* 11:823–38.

Bouchud, J. 1974. Etude preliminaire de la faune provenant de la grotte du Djebel Qafzeh, près de Nazareth, Israel. *Paléorient* 2:87–102.

Brain, C. K. 1981. *The Hunter or the Hunted? An Introduction to African Taphonomy*. Chicago: University of Chicago Press.

Burch, E. S. Jr. 1972. The caribou, wild reindeer as a human resource. *American Antiquines* 37:339–68.

Chase, P. G. 1986. *The Hunters of Combe Grenal: Approaches to Middle Paleolithic Subsistence in Europe*. Oxford: British Archaeological Reports International Series 286.

Copeland, L., and F. Hours. 1983. Le Yabrudien d'El Kowm (Syrie) et sa place dans le Paleolithique du Levant. *Paléorient* 9:21–38.

Cordy, J. M. 1984. Apport de la paleontologie à la paleoecologie et à la chronostratigraphie en Europe du nord-occidentale, 55–64. In *L'Homme de Néanderthal*, Vol. 2, ed. M. Otte. Liege: ERAUL.

Daams, R. 1981. The dental pattern of Dormice Dryomys, Myomimus, Microdyromys and Peridyromys. *Utrecht Micropaleontological Bulletin* (Special Publ.) 3:1–115.

Dayan, T. 1989. The Succession and Community Structure of the Carnivores of the Middle East in Space and Time. Ph. D. diss. (in Hebrew). Tel Aviv University. Tel Aviv.

Delpech, F., and F. Prat. 1980. Les grands mammifères Pleistocene de sud-ouest de la France, 268–297. In *Problèmes de Stratigraphies Quaternaire en France et dans les Pays Limitrophe*, ed. J. Chaline. Supplement au Bulletin de l'AFEQ, no. 1

Eisenmann, V. 1992. Systematic and biostratigraphical interpretation of the equids from Qafzeh, Tabun, Shkull, and Kebara (Acheulo-Yabrudian to Upper Paleolithic of Israel). *Archaeozoologia* 5:43–62.

Farrand, W. R. 1974. Observations on the Kurkar-Hamra succession in the Carmel coastal plain. *Tel Aviv* 1:45–54.

Foley, R. 1984. Early man and the red queen: Tropical African community evolution and hominid adaptation, 85–110. In *Hominid Evolution and Community Ecology*, ed. R. Foley. London: Academic Press.

Gabori, M. 1976. *Les Civilisations du Paléolithique Moyen Entre les Alpes et l'Oural*. Budapest: Akadémiai Kiadó.

Garrod, D. A. E., and D. Bate, 1937. *The Stone Age of Mount Carmel*, Vol. 1, *Excavations at the Wadi el-Mughara*. Oxford: Clarendon.

Gisis, I., and O. Bar-Yosef. 1974. New excavations in Zuttiyeh cave, Wadi Amud, Israel. *Paléorient* 2:175–80.

Goring-Morris, A. N. 1987. *At the Edge: Terminal Pleistocene Hunter-Gatherers in the Negev and Sinai*. Oxford: British Archaeological Reports International Series 361.

Grün, R., C. B. Stringer, and H. Schwarcz. 1991. ESR dating from Garrod's Tabun cave collection. *Journal of Human Evolution* 20:231–48.

Haas, G. 1972. The microfauna of Djebel Qafzeh cave. *Paleovertebrata* 5:261–70.

Haas, G. 1973. The Pleistocene glirids of Israel. *Verhandlungen der Naturforschenden Geselshaft in Basel* 83:76–110.

Hooijer, D. A. 1961. The fossil vertebrates of Ksar 'Akil, Paleolithic rock shelter in the Lebanon. *Zoologische Verhandlanger* 48:1–67.

Hovers, E., Y. Rak, R. Lavi, and H. Kimbel. 1996. Hominid remains from Amud Cave in the context of the Levantine Middle Paleolithic. *Paléorient* 21(2):47–61.

Isaac, G. L. 1980. Casting the net wide: A review of archaeological evidence for early hominid land-use and ecological relations, 226–51. In *Current Argument on Early Man*, ed. L. Konigson. Oxford: Pergamon Press.

Isaac, G. L. 1983. Bones in contention: Competing explanations for the juxtaposition of early Pleistocene artifacts and faunal remains, 1–20. In *Animals and Archaeology*, Vol. 1, *Hunters and Their Prey*, ed. J. Clutton-Brock and C. Grigson. Oxford: British Archaeological Reports International Series 163.

Jelinek, A. J. 1981. The Middle Paleolithic in the Southern Levant from the perspective of the Tabun cave, 265–80. In *Préhistoire du Levant*, ed. J. Cauvin and P. Sanlaville. Paris: CNRS Editions.

Jelinek, A. 1982a. The Tabun cave and Palaeolithic man in the Levant. *Science* 216(4553):1369–75.

Jelinek, A. J. 1982b. The Middle Paleolithic in the southern Levant with comments on the appearance of modern *Homo sapiens*, 57–104. In *The Transition from the Lower to the Middle Paleolithic and the Origin of Modern Man*, ed. A. Ronen. Oxford: British Archaeological Reports International Series 151.

Jelinek, A. J. 1992. Problems in the chronology of the Middle Paleolithic and the first appearance of early modern *Homo sapiens* in Southwest Asia, 253–76. In *The Evolution and Dispersal of Modern Humans in Asia*, ed. T. Akazawa, K. Aoki, and T. Kimura. Tokyo: Hokusen-Sha.

Marks, A. 1983. The middle to upper Paleolithic transition in the Levant, 51–98. In *Advances in World Archaeology*, ed. F. Wendorf and A. Close. New York: Academic Press.

Marks, A., and P. Volkman. 1986. The Mousterian of Ksar Akil: Levels XXVIA through XXVIIIB. *Paléorient* 12(1)5–20.

Meignen, L., and O. Bar Yosef. 1989. Nouvelles recherches sur le Paleolithique moyen d'Israël: La grotte de Kebara, unites VII à XII, 169–184. In *Investigations in South Levantine Prehistory*, ed. O. Bar Yosef and B. Vandermeersch. Oxford: British Archaeological Reports International Series 497.

Mercier, N., and H. Valladas. 1994. Thermoluminescence dates for Paleolithic Levant, 13–20. In *Late Quaternary Chronology and Paleoclimates of the Eastern Mediterranean*, ed. O. Bar-Yosef and R. Kra. Tucson and Cambridge: Radiocarbon and the Peabody Museum of Archaeology and Ethnology, Harvard University.

Mercier, N., H. Valladas, G. Valladas, and J. L. Reyss. 1995. TL dates of burnt flints from Jelinek's excavations at Tabun and their implications. *Journal of Archaeological Science* 22:495–509.

Rabinovich, R., and E. Tchernov. 1996. Chronological, paleoecological and taphonomical aspects of the Middle Paleolithic site of Qafzeh, Israel, 5–44. In *Archaeozoology of the Near East*, ed. H. Buitenhuis and H. P. Uerpmann. Leiden: Backhuys.

Rognon, P. 1987. Late Quaternary climatic reconstruction for the Maghreb (North Africa). *Palaeogeography, Palaeoclimatology and Palaeoecology* 58:11–34.

Rognon, P. 1996. Climatic change in the African deserts between 130,000 and 10,000 y BP. *Comptes Rendus de l'Académie des Sciences* 323:549–61.

Said, R. 1975. The geological evolution of the river Nile, 7–44. In *Problems in Prehistory: North Africa and the Levant*, ed. F. Wendorf and A. E. Marks. Dallas: SMU Press.

Schwarcz, H. P., P. Goldberg, and B. Blackwell. 1980. Uranium series dating of archaeological sites in Israel. *Israel Journal of Earth Sciences* 29:157–65.

Schwarcz, H. P., R. Grun, B. Vandermeersch, O. Bar-Yosef, H. Valladas, and E. Tchernov. 1988. ESR dates for the hominid burial site of Qafzeh in Israel. *Journal of Human Evolution* 17:733–7.

Schwarcz, H. P., W. M. Buhay, R. Grün, H. Valladas, E. Tchernov, O. Bar-Yosef, and B. Vandermeersch. 1989. ESR dates for the Neanderthal site of Kebara, Israel. *Journal of Archaeological Science* 16:653–61.

Tchernov, E. 1981. The biostratigraphy of the Middle East, 67–97. In *Préhistoire du Levant*, ed. J. Cauvin and P. Sanlaville. Paris: CNRS Editions.

Tchernov, E. 1984. Commensal animals and humans in the Middle East, 91–115. In *Animals in Archaeology: Early Herders and their Flocks*, Vol. 3, ed. J. Clutton-Brock and C. Grigson. Oxford: British Archaeological Reports International Series 202.

Tchernov, E. 1988. The biogeographical history of the southern Levant, 159–250. In *The Zoogeography of Israel*, ed. Y. Yom-Tov and E. Tchernov. Dordrecht: Dr. Junk Publishers.

Tchernov, E. 1994. New comments on the biostratigraphy of the Middle and Upper Pleistocene of the southern Levant, 333–50. In *Late Quaternary Chronology and Paleoclimates of the Eastern Mediterranean*. ed. O. Bar-Yosef and R. S. Kra. Tucson and Cambridge: Radiocarbon and the Peabody Museum of Archaeology and Ethnology, Harvard University.

Turner, A. 1984. Hominids and fellow travelers: Human migration into high latitudes as part of a large mammal community, 193–217. In *Hominid Evolution and Community Ecology*, ed. R. Fole. London: Academic Press.

Valladas, H., J. L. Joron, G. Valladas, B. Arensburg, O. Bar-Yosef, A. Belfer-Cohen, P. Goldberg, H. Laville, L. Meignen, Y. Rak, E. Tchernov, A. M. Tillier, and B. Vandermeersch. 1987. Thermoluminescence dates for the Neanderthal burial site at Kebara in Israel. *Nature* 330:159–60.

Valladas, H., J. L. Reyss, J. L. Joron, G. Valladas, O. Bar-Yosef, and B. Vandermeersch. 1988. Thermoluminescene dating of Mousterian "Proto-Cro-Magnon" remains from Israel and the origin of modern man. *Nature* 331:614–16.

Williams, M. A. J., and F. M. Williams. 1976. Evolution of the Nile basin, 207–24. In *The Sahara and the Nile*, ed. M. A. J. Williams and H. Fauve. Rotterdam: A. A. Balkema.

PART II

Archaeological Assemblages: Cultural Interpretations and Subsistence Strategies

The archaeology of western Asia has been researched since the nineteenth century for a variety of reasons, not the least of which was the desire of European and American scholars to study the origins of western civilizations. The Levant, in particular, attracted travelers who sought antiquities from the time of the Bible and the New Testament. Thus the presence of prehistoric sites was noticed, but it was not until the first excavations of F. Turville-Petre in Wadi Amud in 1925 that more systematic research began. He was followed by D. Garrod who excavated in Mt. Carmel (1928–1934) and also did fieldwork in Iraq (Zarzi and Hazar Merd caves). R. Neuville, one of their contemporaries, dug in rockshelters in the Judean Desert (1928–1935) as well as in Qafzeh Cave with M. Stekelis. A. Rust made his contribution to these early efforts by digging a series of rockshelters near the village of Yabrud in Syria. During the same period, excavations began at Ksar 'Akil in Lebanon. Surface surveys in these countries pointed to the presence of numerous Middle Paleolithic sites, both in caves and open-air localities.

Much more fieldwork has been done since the 1950s. The series of Paleolithic excavations are well-known and include, among others, Shanidar (Iraq), Jerf 'Ajla (Syria), Ras el Kelb and Adlun caves (Lebanon), Kebara, Geula, Sefunim, Tabun, Qafzeh, Amud (Israel), and a host of open-air sites in the Negev. Paleolithic research in Jordan began a little later but is rapidly catching up.

From the papers in this part and those in Part III, it is clear that some sites are considered more often than others in our efforts to build a sound chronology and to reconstruct Middle Paleolithic lifeways. Other sites make additional contributions, however, and it would be unreasonable to state that one can attain the full picture by studying only a limited number of sites. This notion is especially true for Lower and Middle Paleolithic occurrences. When performing research aimed at achieving resolutions to currently asked questions, several factors need to be taken into account.

The first factor that should be considered is the topographic and climatic variability that characterizes western Asia, and the Levant in particular. Within short distances (e.g., 50–100 km), one may encounter different types of vegetation (from the Mediterranean type through the Saharo-Arabian belt) and uneven distribution of sources of water, such as ponds and lakes. Even with vague estimations of the carrying capacities of various habitats, variability across adjacent territories is quite prominent. Settlement patterns can therefore be tentatively reconstructed on the basis of the spatial distribution of resources, and these can be tested with the archaeological records. However, one must keep in mind the changing climatic conditions of the late Middle and Upper Pleistocene before launching hypotheses concerning the kinds and degrees of hunter-gatherer mobility. We cannot assume that the availability, predictability, and reliability of food sources that mark the current landscape was the same in the past. Uncovering this past should be one of our research goals. Hence, several lines of evidence are required for reconstructing mobility patterns and comparing the local Neandertals with modern humans.

The second factor that must be considered in this research is preservation, which unfortunately plays an important role: It often dictates whether bones and other organic matters are present in the site. It is not surprising, therefore, that many studies that address human behavior derive their conclusions essentially from lithics.

The papers in this section begin with the early part of the prehistoric sequence. Sultan Muhesen describes the geographic background of Syria and the Pleistocene sequence. Coastal and inland Acheulian sites are reported and the presence of core and flake industries (also known as the core-chopper industries) as well as the Acheulian are noted. The overall chronocultural framework of the Lower Paleolithic of the Syrian and Lebanese sequences provides the basis for the ensuing Middle Paleolithic entities in the Levant.

The importance of understanding the role of site formation processes seems to have gained more support in recent years. Paul Goldberg and Ofer Bar-Yosef provide a basic description of the regions and the caves, with particular attention to the contributions of geogenic and biogenic agencies to the depositional history of Kebara and Hayonim. The authors provide the readers with a more comprehensive discussion on the basis of the most recent analysis of sites, where various aspects of diagenesis were recognized through the study of ashes and bone mineralogy (mostly conducted by Steve Weiner). Through their examination of different caves, the authors detail the role and nature of the different geogenic processes. Biogenic processes are treated similarly; the anthropogenic aspects are summarized separately. It has become obvious in recent years that only through such systematic analyses of mineralogy and micromorphology can one disclose site formation processes. And without this information we have very few foundations for paleoanthropological interpretations. This is undoubtedly only the first step in this direction. More chemical studies are needed as well as further research into the causes and effects of bioturbation. However, in order to perform these studies one must clearly pose the paleoanthropological questions.

One of the aspects that reflects human behavior is the spatial patterning of the archaeological remains. The paper by Donald O. Henry addresses this goal. In the search for signs of modern behavior which, on the basis of the European notions, is missing in Middle Paleolithic contexts, the Levantine sites with their modern looking fossils (the Qafzeh-Skhul group) are often compared to the local Neandertals (Kebara, Amud, Dederiyeh). Henry discusses the expected correlations between behavior and archaeological remains. Patterns of exploitation, such as residential moves, logistical task groups, or opportunistic strategy, are explained. Models of raw material and tool provisioning, whether individuals or places, as well as the resulting intrasite patterns are taken into account when comparing Middle to Upper Paleolithic records. Henry then proceeds to examine the information from his excavations at Tor Faraj, where layers were uncovered containing a "Tabun B-type" assemblage dated to about 70 ka. Looking at hearths, the distribution of artifacts, and phosphorous concentrations, he concludes that it hardly differs from patterns created by modern foragers.

The lithic industries are the main subject of the next four papers. Erella Hovers examines in detail the lithic assemblages from Amud Cave, which generally fall into the category of the "Tabun B-type" industries and are dated to the same time span. Core modifications, blank production, core reduction, and tool manufacture are examined. Hovers stresses the variability within the assemblages at Amud and provides an overall review of other sites and assemblages that would fall within the late Mousterian. She concludes by pointing out that a known technological system recovered from sites across the Levant indicates the flow of technical information. She views the apparent continuity between the late Middle Paleolithic and the early Upper Paleolithic as evidence against replacement.

An earlier Mousterian industry from Hayonim Cave is the subject of Liliane Meignen's paper. The lithic assemblages in the Mousterian deposit, which was temporarily subdivided into Lower and Upper Layer E, are compared to other sites. While Upper Layer E resembles the "Tabun C-type" from various sites, Lower Layer E falls within the "Tabun D-type." The author stresses the bladey character of the lower assemblage and compares it to other known Middle Paleolithic sites in Africa and Eurasia.

The paper by Eric Boëda and associates describes and discusses the unique finds that demonstrate that bitumen was employed by Mousterian people. The preservation in the site of Umm el Tlel in the El Kowm basin in northern Syria seems exceptional when one remembers that this is an open air site. The detailed report on this unusual find provides the readers with a significant amount of information. Each of the artifacts is analyzed and placed within the core reduction sequence (the elaborate *chaîne opératoire* concept) that is currently employed in studies of the Levant. The authors conclude that the bitumen was used as an adhesive and the various hafting options are deliberated.

Evaluating the technological aspects of the Levantine Mousterian industries as indicators of cultural and mental capacities, a highly debated subject, is the topic of the paper by Naama Goren-Inbar and Anna Belfer-Cohen. The authors open the paper by enumerating the hallmarks of the Middle Paleolithic. These serve as a basis for discussing the departure of their producers from the previous Acheulian mode. The raw materials concerned, typology, and technology (the various technical traits of operational sequences) form the basis for a general discussion that centers on cores and blanks that exhibit inter-assemblage variability. The authors believe that regarding these as expressions of group traditions is better than other alternative models. They stress that Fodor's theory of mental modules and social learning would explain the modularity as well as the flexibility recorded in the cultural remains.

Subsistence systems and their relationship to the environment and human behavior are the subjects of the last four papers in this part. The role of hunting and scavenging as reflected in the faunal collections from Kebara Cave is the topic of John Speth's and Eitan Tchernov's work. The Mousterian hunters who inhabited this cave site, even temporarily, were effective hunters who went after prime adults including some dangerous species such as the aurochs and the wild boar. The authors raise issue with the perception of the Neandertals by some scholars as in some sense prehumans, with no art objects, no style in stone tools, and no grave goods, who were possibly opportunistic scavengers. They employ the faunal remains collected during the excavations by M. Stekelis and the more recent project to explore the presence or absence of hunting by both Middle and Upper Paleolithic humans. The models used are those developed by Stiner for Middle Paleolithic assemblages in Italy. Examining the taphonomic characteristics of the Kebara faunal assemblages, they conclude that the animal bone accumulations were essentially the results of human activities and transport. This does not imply that hyenas did not use the cave. Hyenas were in fact the occupiers sporadically, during periods when the site was abandoned and, in particular, during the Upper Paleolithic times. Ungulate mortality patterns for gazelle, Fallow deer, and Red deer demonstrate that Mousterian hunters did not differ from the Upper Paleolithic ones.

In the next paper, the faunal assemblages from Hayonim Cave are used by Mary Stiner and Eitan Tchernov to test the effects of the economic approaches of the Mousterian, Kebaran, and Natufian foragers. The shifts among ungulate species between large and small game are best seen as the results of environmental shifts caused by climatic changes. The entire array of species uncovered at Hayonim is reported. Following the description of taxonomic representation in the three archaeological entities, the authors focus on the Mousterian. The intrasite variability is seen especially in terms of body size, and the proportional abundance of ungulates is interpreted, as other scholars such as Garrard, Uerpmann, and the earlier work by Tchernov have done, as reflecting the availability across the landscape during each time span. The presence of small game species seems to indicate the role played by females and children in a foraging society.

The aspect of seasonal site occupation, whether base camps or transitory stations, has triggered a lot of research in recent years. Revealing the mode of mobility among prehistoric foragers at a given time in a certain territory is necessary for testing hypotheses concerning the evolution of societal organization. Daniel Lieberman examines this aspect by explicitly examining the question of differences between Neandertals and modern humans. He uses cementum increment analysis of gazelle teeth retrieved in Mousterian sites to uncover the patterns of seasonal hunting. This technique, calibrated on the basis of controlled modern samples, was also employed to check the seasonality of Natufian sites. In addition to examining the biological information, the author studies the differences between the human fossils and states that the available data does not allow the conclusion that Neandertals and modern humans had an identical pattern of lifeways.

The prospects for stable isotope analysis of human remains as indicators for prehistoric diets is the subject of the final chapter in this section by Stanley Ambrose. The author mentions the sources that may tell us about past diets, such as faunal and floral remains, bone pathologies, as well as the chemical and isotopic analysis of bones and teeth. He describes how the diet can be reconstructed by the analysis of stable isotopes of bones and discusses the effects of preservation of their signature. Samples of bones from Europe, Kenya, and the Levant were tested. The author concludes by stating that poor preservation hampers further studies in the Levant.

7

THE EARLIEST PALEOLITHIC OCCUPATION IN SYRIA

Sultan Muhesen

Director-General of Antiquities and Museums
Damascus, Syrian Arab Republic

1. INTRODUCTION

Information on the Paleolithic of Syria dates to the beginning of this century, though the first systematic research began with the excavation of Yabrud Rockshelter (Rust 1950; Solecki 1966, 1987–88). From that time on, many excavations and surveys took place in different parts of the country. Among the most important were those along the coastal region (Sanlaville 1979), the Orontes valley (Clark 1966a,b; Besançon et al. 1978; Muhesen 1985; Sanlaville et al. 1993), the Euphrates valley (Besançon et al. 1980a,b) and Douara Cave (Akazawa and Sakaguchi 1987). Important excavations are currently conducted by a Syro-Swiss expedition at Nadaouiyeh I (Le Tensorer et al. 1993), a Syro-French expedition in Umm el-Tlel (Böeda and Muhesen 1993), both in the el-Kowm basin, and a Syro-Japanese mission in Dederiyeh Cave in the region of Afrin (Muhesen et al. 1988).

To date, results of field work demonstrate the rich, varied, and extensive Paleolithic occupation all over the country. Some regions proved to be a center of early and long Paleolithic occupations and produced the best examples for a continuous Paleolithic cultural evolution in the Levant. We will focus here only on the earliest Paleolithic occupation as uncovered by recent research.

2. GEOGRAPHIC BACKGROUND

Syria, with its rich resources and varied environments, has played a crucial role since the beginning of humankind due to its strategic geographical intermediate position between the three main continents of Africa, Asia, and Europe. During the Pleistocene, the country experienced important climatic fluctuations as warm and dry periods alternated with colder and wetter ones, having major effects on the floral cover of the country and

Neandertals and Modern Humans in Western Asia, edited by Akazawa *et al.*
Plenum Press, New York, 1998.

years time scale	Oxygen Isotope core stage	Continental Formation	Marine Formation	N. el-Kebir	Orontes	Euphrates	General	Stone industries	.European Equivalance
150-40	2 3 4	QFI		el-Chir	Sarout	Abu Chahri	el-Chir	Middle Paleolithic	Würm
250-150	5a 5d 5e		QM I	Banyas				Transitional Industries (Yabrudian.....)	Riss -Würm
450-250	6 7 8 9 10 11 12	QFII		Jinndiryeh	Jraibiyat	Abu Jumaa	Abu Jumaa	Evolved upper Acheulean	Riss
600-450	13 14 15		QMII	Khellaleh				Evolved middle Acheulean	Mindel - Riss
800-600	16 17 18 19 20	QFIII		Berzin	Latamne	Chnineh	Latamne	Middle Acheulean	Mindel
1000-800	21 22 23 24 25		QMIII	Baksa				Evolved Lower Acheulean	Günz - Mindel
1400-1000	26 27 28 29 30 31	QFIV		Sitt-Markho	Khattab		Khattab	Lower Acheulean Flakes Industries	Günz
			QMIV	Micherfet el Somouk					
		QFV				Qara Yakoub			

Figure 1. Pleistocene of Syria.

the faunal assemblages. The best field evidence of these changes was observed in the valleys of the main rivers and the coastal region. Cold periods are represented by series of river terraces, while warm ones are better documented by the sequence of shore lines.

Both fluvial and marine formations were studied and given local names in the different geographic regions. The obtained chrono-archaeological-stratigraphical scheme is presented in Figure 1.

The oldest Pleistocene formation (QFV) was distinguished at Qara Yakoub in the Sajour valley, an upper tributary of the Euphrates River. It was followed by a marine transgression (QMIV) documented at Micherfeh in the valley of Nahr el-Kebir. Both QFV and QMIV formations were poorly represented and they did not contain any stone artifacts, which indicates that they were formed prior to the arrival of humans in Syria. It was the QFIV formation which contained the oldest artifacts in both Nahr el-Kebir valley (the Sitt-Markho Formation) and in the valley of the Orontes (the Khattab Formation).

Early Paleolithic industries were found also in the QMIII (Baksa-Idriss formation) along the coast, while the QFIII and later marine and fluvial formations (QFII, QMI, QFI) produced Middle and Upper Acheulean and Middle Paleolithic industries.

In the following pages we will describe the sites embedded in the QFIV and QMIII formations, in which stone industries attributed to the early Lower Paleolithic phase were uncovered (old Acheulean) in both the coastal and the interior regions (Muhesen 1988) (Figures 2, 3).

3. THE COAST

The most complete sequence of Quaternary formations is exposed in the coastal region, including the valley of Nahr el-Kebir. The valley of Nahr el-Kebir lies between two geologically different areas: the mountainous area of Bayer and Bassit, of Maestrichtian age to the north; and the Jurassic and Lower Cretaceous coastal range to the south. The Quaternary formations, in the forms of butts and hills, occur mainly in the lower part of the river valley at its junction with the sea.

The stepped fluvial terraces consist of coarse and silty material. Their altimetric position, relative stratigraphy, depositional facies, surface alteration, and stone artifacts allowed us to establish their relative chronology. Sitt-Markho formation (QFIV) occurred at an outcrop of 110 m above the river level. It is several meters thick consisting of a cemented conglomerate of limestone pebbles, green rocks, and flint cobbles. The important presence of decalcified flint, the geochemical alteration of the limestone, the altitude, and the morphology of the stone artifacts suggested that this terrace can be correlated with the Günz glaciation according to Alpine chronology.

4. OLD ACHEULEAN ASSEMBLAGES

4.1. Sitt-Markho

The most important stone assemblage was found in QFIV at Sitt-Markho village. The artifacts were collected from a relatively small surface. There were 89 pieces whose physical condition permits us to divide them into two series. The first contained 56 artifacts which were heavily rolled and show archaic features, both typologically and technologically. The assemblage consists of four cores, 41 flakes, generally large and thick, and

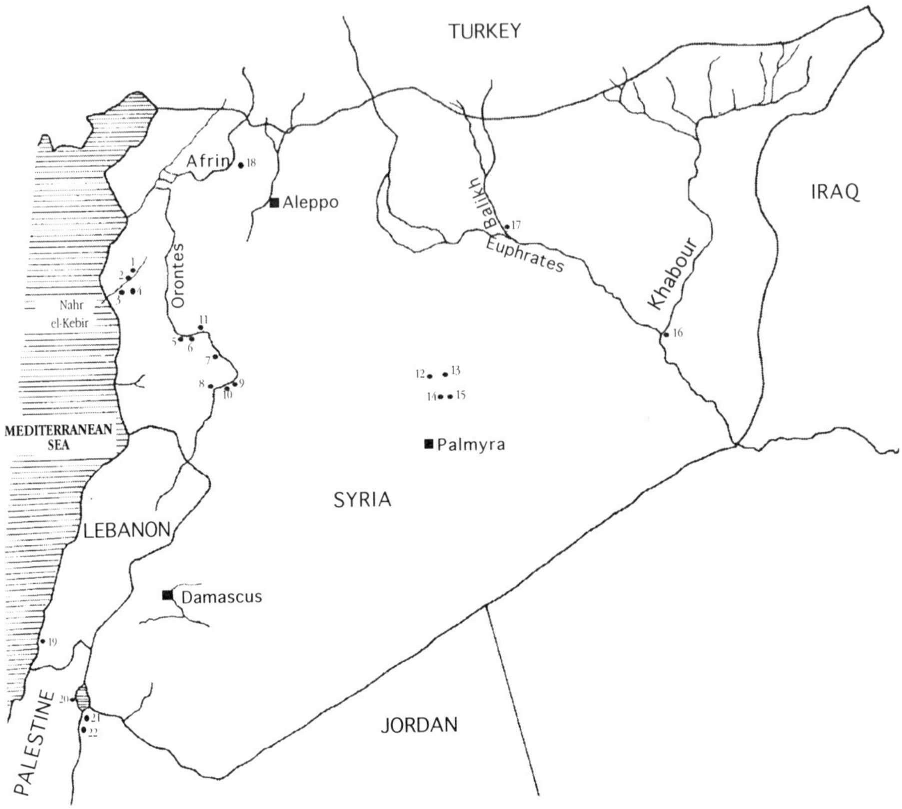

1	Sitt-Markho	12	Nadaouiyeh I
2	J. Idriss I, II	13	Um el-Tlel
3	Fidio II	14	el-Kowm
4	Cheikh Mohammed	15	el-Meirah
5	Abu Obeida	16	Deir ez-Zor
6	Mahardeh 2	17	Chniné
7	Khattab 2	18	Afrin (Deideriyeh)
8	Khor el-Aassi	19	Borj Qinnarit
9	Ard Habibe	20	Ubeidiyeh
10	el-Farché I	21	Abu el-khass
11	Latamne	22	Abu Habil

Figure 2. Early Paleolithic sites in Syria.

11 tools. The tools are massive and show significant use. They consisted of five chopping tools, one chopper, and two primitive bifaces which can be classified as of Abbevilian type, two cleavers on flakes, and one pick; there were no light duty tools in this series (Figures 4, 5). The second series contained 33 artifacts which were slightly rolled and seemed to bear less archaic forms than those of the first series. They consist of nine cores with one or two striking platforms, and 18 flakes which were smaller than those of the first series. The tools included one chopper, one biface, one heavy denticulated piece, and three light duty tools (a side-scraper, end-scraper, and a backed knife). Although both se-

Site	region	Geomorphlogical Formation	Debitage		Tools	Total Artifacts (214)
			Flakes	Cores		
Sitt-Markho	N. el-Kebir	QF IV	59	13	17	89
Cheik Mohammad	N. el-Kebir	QM III	15	2	18	35
J. Idriss I	N. el-Kebir	QM III	3		3	6
J. Idriss II	N. el-Kebir	QM III	23	4		27
Fidio II	N. el-Kebir	QM III	15	2	3	20
Khattab 2	Orontes	QF IV	7	1	4	12
Mahardeh 2	Orontes	QF IV	1	1	6	8
Khor el-Aassi	Orontes	QF IV	2		1	3
Ard Habibe	Orontes	QF IV	1	1		2
Abu obeida	Orontes	QF IV			7	1
El-Farche I	Orontes	QF IV	1			1

Figure 3. Inventory of early Paleolithic sites in Syria.

ries were contemporary with QFIV formation, it seems that the first series is older than the second one. However, both series form a homogenous assemblage with an interesting composition of tools: bifaces, picks, cleavers, choppers, and light duty tools. It was therefore placed in the very beginning of the Acheulean, thus being the oldest stone industry as yet known in Syria, dated to around one million years ago (Copeland and Hours 1979).

4.2. Cheikh Mohammed, Fidio II, and Jebel Idriss I, II

The QMIII formation represents the transgression of the Mediterranean sea, which followed the QFIV regression of the Sitt-Markho formation. It was relatively well-preserved at some places, such as the lower valley of Nahr el-Kebir mainly at Baksa, where it was purely marine, and at Jabal Idriss, where it is of fluvio-marine facies.

The QMIII formation is well-presented on both sides of Nahr el-Kebir, with altitudes ranging between 90 m and 134 m. It consists of marine pebbles and sand with some limestone and green rocks. This formation corresponds largely with the post-Günz interpluvial (correlated with the Cromerian complex in Europe). The QMIII contained stone industries, which were more evolved than those of QFIV at four spots.

4.3. Cheikh Mohammed

This site is located on the left bank of Nahr el-Kebir. An assemblage of 35 artifacts was found, including two cores, 15 flakes and 18 tools. Two series could be distinguished: an older one, rolled with brownish thick patina; a younger one, slightly rolled with reddish patina. The 12 bifaces were the most characteristic tools. Typologically they are more

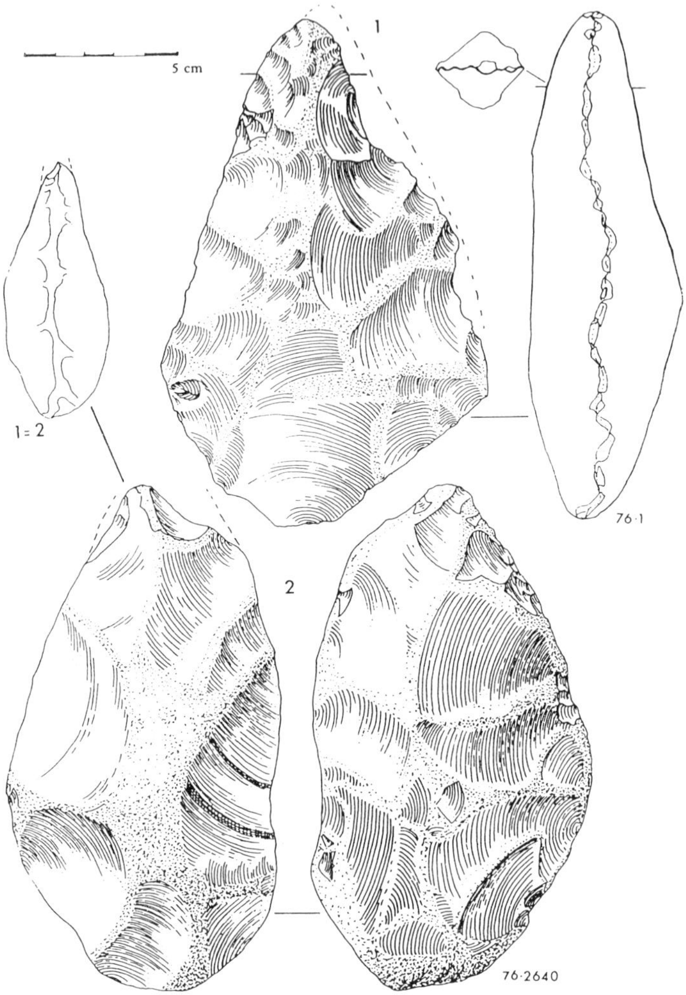

Figure 4. Bifaces from Sitt-Markho (after Sanlaville 1979).

evolved than those of Sitt-Markho, and bear the forms of amygdaloids, lanceolates, and ovates. There were also four choppers, one pick, and one cleaver (Figure 6).

4.4. Fidio II

In this site, a little to the north of Cheikh Mohammed, stone industries of different periods were found embedded within a marine pebble conglomerate. An assemblage of 20 artifacts rolled with red-brown patina seems to be of almost the same age as that of the younger series of Cheikh Mohammed. They contained one biface, one pick, one side-scraper, two cores, and 15 flakes.

4.5. Jebel Idriss I, II

On the right bank of Nahr el-Kebir, in a fluvio-marine deposit contemporary with QMIII formation, different stone industries were found, some of which resemble those of

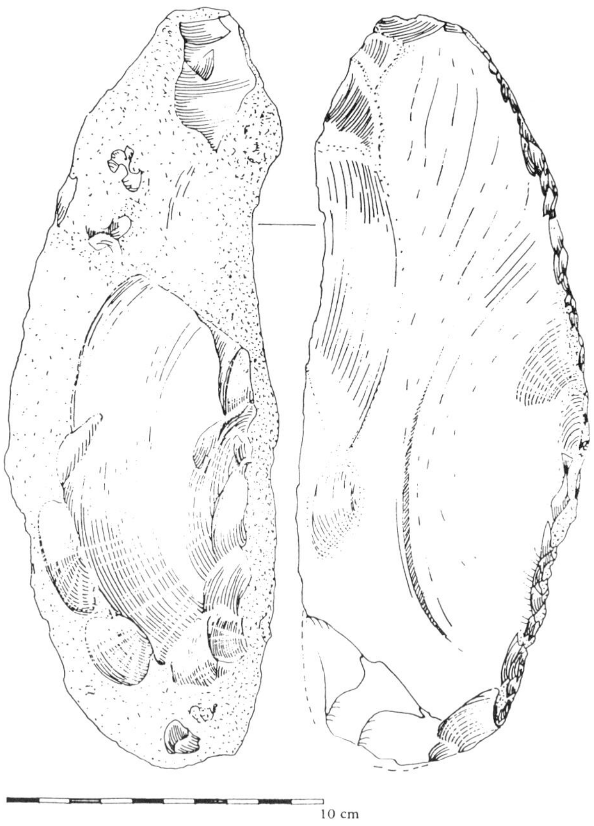

Figure 5. Flake from Sitt-Markho (after Sanlaville 1979).

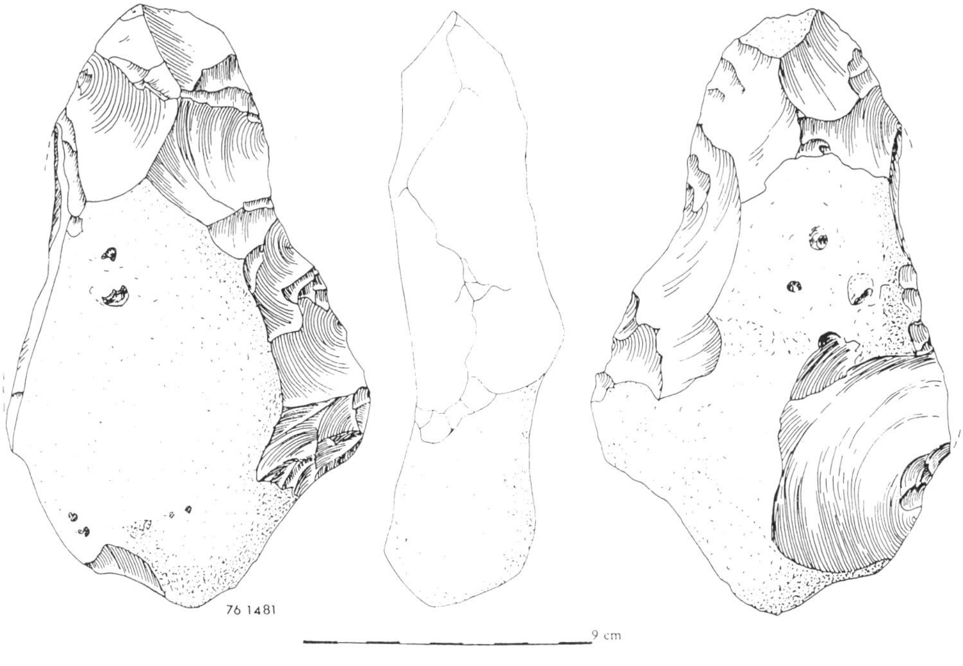

Figure 6. Biface from Cheikh Mohammed (after Sanlaville 1979).

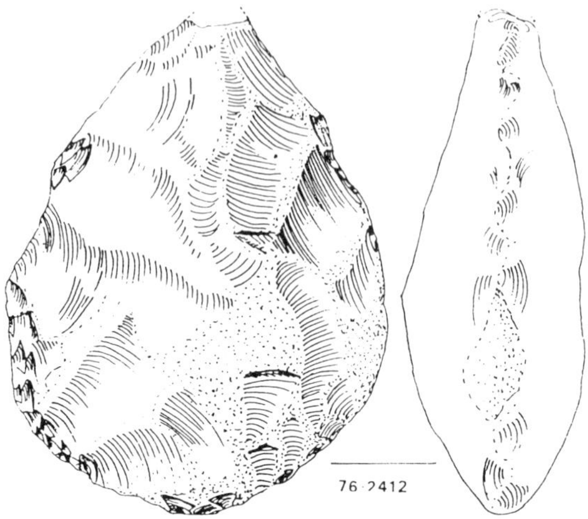

Figure 7. Biface from Jebel Idress I (after Sanlaville 1979).

Cheikh Mohammed. Artifacts from Jebel Idriss I included six patinated artifacts, consisting of one biface, two choppers, and three flakes. Jebel Idriss II contained 27 artifacts bearing brown patina, morphologically more recent than those of Jebel Idriss I. They contained no characteristic tools but included four cores and 23 flakes. All artifacts are contemporary with QMIII formation (Figure 7).

5. THE INLAND

Stone industries of early Paleolithic age in Syria's interior are poor and less characteristic than those of the coastal region. The most important studied area is the valley of the Orontes river, in which at least four terraces of Pleistocene age are exposed. The oldest is QFIV (Khattab formation), which was the best-preserved one upstream of the Khattab village in the middle Orontes valley. Here, deposits about 10 m thick were found, consisting of a conglomerate of sand, limestone, and flint. In the Khattab formation and in other contemporary sites, a total of 27 artifacts were collected from six places: Khattab 2, Mahardeh 2, Khor el-Aassi, el-Faraché I, Ard Habibe, and Abou Obeida (Sanlaville et al. 1993).

Because the samples were very small, they were clustered and classified into the following categories: three cores, 12 flakes, 12 tools, seven choppers, one polyhedron, one spheroid, one rabot, one bifacial item, and one denticulate. All artifacts except the limestone spheroid were made of flint. They had a very light patina with grayish and beige color. Also, they were well-preserved and had no traces of chemical alternation. They were slightly rolled. In brief, their physical condition indicated that they were in situ and not transported by the river over a large distance. Regardless of the small number, this collection, with its typical tools, attests for the human occupation of the Orontes valley at an early Paleolithic age (Figures 8, 9).

On the other hand, scarce and dubious stone artifacts were east of the Orontes indicating a possible early Paleolithic human existence in these regions. These were the few flakes collected from the surface of QFV formation in the valley of the Sajour (Qara Yak-

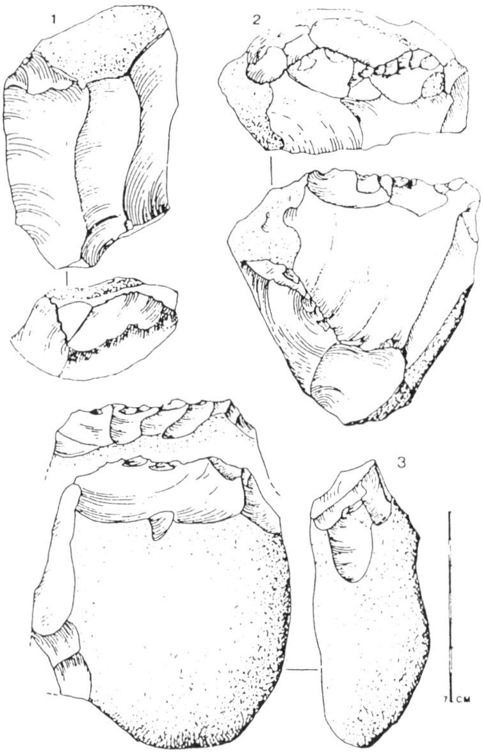

Figure 8. Khattab 2. 1: prismatic core; 2: chopper on a slab; 3: chopper on a river pebble (after Sanlaville et al. 1993).

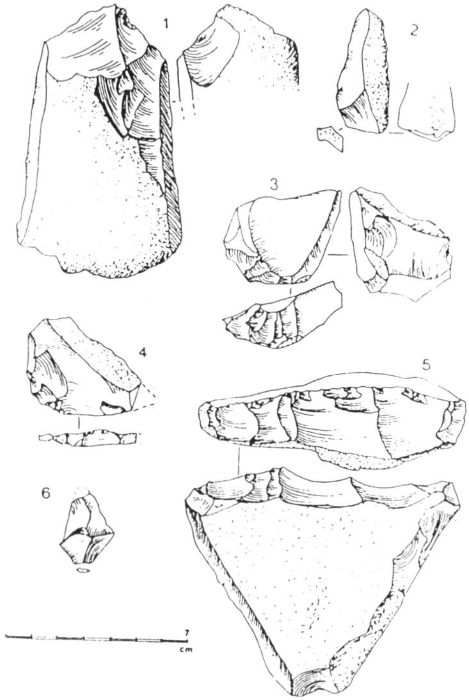

Figure 9. Artifacts from the Khattab formation: 1–4 from Khattab 2; 5–6 from Ard Habibe. 1: pointed chopper or rough pick; 2: offset cortex-flake; 3: core fragment; 4: core-ridge flake; 5: heavy-duty end-scraper on a slab or broken flake; 6: trimming flake (éclat de taille) (after Sanlaville et al. 1993).

oub) and flakes collected in the QFIII formation in el-Kowm (Besançon et al. 1981). It seems that human occupations in eastern parts of Syria began with a Middle Acheulean phase as attested by finds from the el-Kown by el-Meira site (Boëda et Muhesen 1993) and in the region of Deir ez-Zor in the middle Euphrates valley (Besançon, personal communication).

6. CONCLUSION

The data collected to date permit us to state that *Homo erectus* arrived in Syria during the Lower Pleistocene. The well-preserved evidence in the valley of Nahr el-Kebir is in the form of early Acheulean industry, and that in the Orontes valley is as a flake industry, a kind of core-chopper industry. The assemblages of these two regions were embedded within the QFIV formation. However, we are not yet sure if these Acheulean and core and flake industries were contemporary or successive, and we still do not know the relationships that could have existed between both regions at that early phase of the Paleolithic sequence.

If we take into consideration the length of time represented by the accumulation of the QFIV deposits, and the lack of radiometric dates, it is possible that the non-bifacial flake industry of the Orontes could correspond to the first half of the QFIV formation, while the early Acheulean of Nahr el-Kebir could be related to the second half of the same QFIV formation. If so, Khattab 2 will be the early phase of the Sitt-Markho Formation.

In view of the small quantity of artifacts, any statistical comparison is meaningless. The general features of the discussed assemblages indicate that the Orontes valley produced a core-chopper industry with chopping tools made of limestone, spheroids, and polyhedrons and is currently named Khattabian by F. Hours and L. Copeland (Sanlaville et al. 1993). However, more evidence is needed in order to maintain this conclusion and to justify the validity of the Khattabian as a typical industry of this region.

The industries of Nahr el-Kebir are quantitatively and qualitatively more important. They contained typical tools such as choppers, primitive bifaces, cleavers, picks, chopping tools, and light duty tools, but no polyhedrons. Their debitage was coarse and the Levallois technique is completely absent. This early Acheulean and the core chopper industry are not known in eastern Syria, in the desert, or the Euphrates Valley. Their absence could be explained either by erosion, which caused the disappearance of old sites, or the view that this region was only later occupied, which seems to be more likely.

There are very few early Paleolithic sites in the Levant to compare with those of Syria. Borj Qinnarit, on the southern Lebanese coast, gave a small assemblage of flake industries, which was found in QMIII formation (the Zaqrounian of Lebanon). This assemblage resembles that of the Orontes valley, which is older (Hours 1992). The Orontes sites can also be compared to Kefar Menahem on the southern coastal plain of Palestine, where a core-chopper industry was succeeded by an early Acheulean one (Gilead and Israel 1975).

'Ubeidiya is another early Paleolithic site in the Jordan valley. Several successive archaeological layers were exposed in the context of the immediate environment of a Paleo-lake. The assemblage layers II-24 and K-20 were dominated by choppers, chopping tools, rare picks and trihedrals, and limestone spheroids (Bar-Yosef and Goren-Inbar 1993). This assemblage renders 'Ubeidiya both technically and typologically comparable to materials from the coastal and Orontes regions though without indicating a precise chronological relation between the discussed sites.

The fluvial member Fi at 'Ubedeiya, which contained the early Acheulean assemblages, could correspond with the QFIV of Syria, estimated at around one million years (Sanlaville et al. 1993). This date could agree with the faunal interpretation that places 'Ubeidiya between 1 and 1.4 Ma (Tchernov 1988).

Early Paleolithic industries are also known from other sites, which were connected with the QFIV in the Jordan valley. These are Abu Habil, which is assigned to the early Acheulean with its primitive bifaces (Muhesen 1988), and Abu el-Khass, which produced a less important flake industry (Villier 1983).

In conclusion, the early Paleolithic sites of Syria and the Levant correspond with the out-of-Africa expansion of *Homo erectus*, who, while moving up north, used the natural and favorable passage along the coast and through the Rift Valley. Without a well-founded chronological control, it will be difficult to follow the evolution and cultural scheme of the early Paleolithic of the Levant and compare it with that of Africa. As the first Acheulean manifestations in Africa are dated at around 1.5 million years ago, we may conclude that it took about a half-million years before the arrival of the first Acheuleans in the Levant.

REFERENCES

Akazawa, T., and Y. Sakaguchi, eds. 1987. *Palaeolithic site of Douara cave and paleography of Palmyra Basin in Syria, Part IV, 1984 excavation*. Tokyo: University of Tokyo Press.

Bar-Yosef, O., and N. Goren-Inbar. 1993. *The Lithic Assemblages of Ubeidiya: A Lower Palaeolithic Site in the Jordan Valley*. QEDEM 34. Jeruselem: Institute of Archaeology, Hebrew University of Jeruselem.

Besançon, J., L. Copeland, F. Hours, and P. Sanlaville. 1978. Morphologie et Préhistoire de la vallée de l'Oronte entre Rastan et le Ghab (Syrie). *Comptes Rendus de l'Académie des Sciences* 287:857–60.

Besançon, J., H. de Contenson, L. Copeland, F. Hours, S. Muhesen, and P. Sanlaville. 1980a. Étude géographique et prospection archéologique des plateaux de la région de Menbij (Syrie du Nord). *Travaux Recherche Co-operatif Programme* 438 (11 pages) and *Les Annals Archeologiques Arabes Syriennes* 32:211–32.

Besançon, J., L. Copeland, S. Muhesen, and P. Sanlaville. 1980b. Géomorphologie et préhistoire de la vallée moyenne de l'Euphrate. Essai de chronologie du Pléistocène et du Paléolithique de Syrie. *Comptes Rendus de l'Académie des Sciences* 290D:167–70.

Boëda, E., and S. Muhesen. 1993. Umm el-Tlel (el-Kowm, Syrie): Étude préliminaire des industries lithiques du Paléolithique Moyen et supérieur 1991–1992, 47–92. In *Cahiers de l'Euphrate*, ed. J. Cauvin. Paris: CNRS Editions.

Clark, D. 1966a. The middle Acheulean occupation site at Latamné, northern Syria. *Les Annals Archeologiques Arabes Syriennes* 16(2):31–74 and *Quaternaria* 9:1–68.

Clark, D. 1966b. The middle Acheulean occupation site at Latamné, northern Syria, further excavation (1965). *Les Annals Archeologiques Arabes Syriennes* 16(2):75–120 and *Quaternaria* 10:1–76.

Gilead, D., and M. Israel. 1975. An early Paleolithic site at Kafer Menahem: Preliminary report. *Tel Aviv* 2:1–12.

Le Tensorer, J. M., S. Muhesen, and R. Jagher. 1993. Nadaouiyeh I Aïn Askar: Une grande séquence Paléolithique du bassin d'el-Kowm (Syrie), premiers résultats, fouilles 1989–1992, 11–36. In *Cahiers de l'Euphrate*, ed. J. Cauvin. Paris: CNRS Editions.

Muhesen, M. 1988. Le Paléolithique et l'Epipaléolithique en Jordanie. Ph. D. diss. University of Bordeaux I. Bordeaux.

Muhesen, S. 1985. *L'Acheuléen Récent Evolué de Syrie*. Oxford: British Archaeological Reports International Series 248.

Muhesen, S. 1988. Le Paléolithique inférieur de la Syrie. *L'anthropologie* 93(3):863–82.

Muhesen, S., T. Akazawa, and A. Abdul Salam. 1988. Prospections préhistoriques dans la région d'Afrin, Syrie. *Paléorient* 14(2):145–53.

Rust, A. 1950. *Die Höhlenfunde von Yabrud (Syrien)*. Neumünster: Karl Wachbaltz Verlag.

Sanlaville, P., ed. 1979. *Quaternaire et Préhistoire du Nahr el-Kebir Septentrional, les Débuts de l'Occupation Humaine dans la Syrie du Nord et du Levant*. Paris: CNRS Editions.

Sanlaville, P., J. Besançon, L. Copeland, and S. Muhesen. 1993. *Le Paléolithique de la vallée moyenne de l'Oronte (Syrie): Peuplement et Environnment*. Oxford: British Archaeological Reports International Series 587.

Solecki, R. S., and R. L. Solecki. 1966. New data from Yabroud, Syria. *Les Annals Archeologiques Arabes Syriennes* 16(2):121–54.

Solecki, R. S., and R. L. Solecki. 1987–88. Archaeological researches at Yabroud and vicinity, Syria. Report of the summer 1987, Yabroud researches. *Les Annals Archeologiques Arabes Syriennes* 37/38:9–49.

Tchernov, E. 1988. The age of the Ubediyeh formation (Jordan Valley, Israel) and the earliest hominids in the Levant. *Paléorient* 14:63–5.

Villiers, L. E. 1983. Final report on Paleolithic sampling at Abu al-Khas, North Jordan. *Annual of the Department of Antiquities of Jordan* 27:27–43.

8

SITE FORMATION PROCESSES IN KEBARA AND HAYONIM CAVES AND THEIR SIGNIFICANCE IN LEVANTINE PREHISTORIC CAVES

Paul Goldberg[1] and Ofer Bar-Yosef[2]

[1]Department of Archaeology
Boston University
Boston, Massachusetts 02215
[2]Department of Anthropology
Peabody Museum
Harvard University
Cambridge, Massachusetts 02138

1. INTRODUCTION

Paleolithic caves and rockshelters have long been the subject of archaeological and geological research in the Levant. Caves are particularly noteworthy in comparison to open-air sites, in which organic remains are rarely preserved or were subject to numerous post-depositional processes that obliterated their association with the lithics. In contrast, caves tend to preserve the context of material culture in clear stratigraphic entities that also provide local and regional paleoenvironmental data derived from sedimentological, faunal, and floral analyses. Levantine caves, in particular, are known for their numerous human relics and their bearing on the debates concerning the origin of modern humans and the demise of the Neanderthals.

As in Western Europe, early archaeologists in the Levant paid attention first to caves because of their long stratified sequences. Pioneering archaeologists such as Zumoffen, Turville-Petre, Garrod, Neuville, Stekelis, Rust, Ewing, and others were keen observers of the sedimentary context of the deposits. They were rarely conscious of the agents responsible for these thick accumulations, and studied these deposits in a very eclectic manner (e.g., Brammall and Leech in Garrod and Bate 1937). Only since the 1960s and early 1970s have researchers begun to identify the detailed aspects of the geological context and overall site formation processes in caves (e.g., Wright 1962; Farrand 1971, 1979; Goldberg 1972, 1973, 1979; Jelinek et al. 1973; Hanihara and Sakaguchi 1975; Hanihara

Neandertals and Modern Humans in Western Asia, edited by Akazawa *et al.*
Plenum Press, New York, 1998.

1976). Yet, even with this increasing attention, the details of depositional and post-depositional processes—be they geogenic, biogenic, or anthropogenic—were never the focus of in-depth investigations. Neither the role of these latent processes in significantly influencing the archaeological record, nor their implications for the interpretation of the archaeological contexts, was fully appreciated.

The aim of this paper is to provide an overview of the site formation processes occurring in Kebara and Hayonim Caves. We have chosen these two caves because the former has been recently excavated and the latter is currently under investigation. In both cases, a principle goal of the excavation was to reconstruct a comprehensive history of the cave, including recognition of all depositional and post-depositional processes. As outlined below, cognizance of these processes is crucial if we are to appreciate the true context of the artifacts and to understand fully the archaeological record in all its aspects, including inferences concerning human behavior and evolution. By using data and observations collected using modern excavation methods at these two cave sites, we will attempt to show how such information can reflect on similar types of site formation processes that might have been overlooked at other Middle Paleolithic cave sites in the region. Thus, we will commonly refer to what we believe are similar or comparable situations or processes that appear to have escaped the attention of the original excavators.

We begin with a brief characterization of the region and continue with a description of the sediments and stratigraphy of Kebara and Hayonim. This section is followed by a discussion of the types of processes operating in each cave.

2. THE REGION AND THE CHARACTERISTICS OF CAVES

Most of the well-documented cave sequences in the Levant (Figure 1) are situated either within the Mediterranean climate belt (including Kebara, Hayonim, as well as Tabun, el-Wad, Skhul, Qafzeh) or in the drier, steppic and desertic regions (e.g., Jerf ‘Ajla, Douara in Syria). The Mediterranean region is characterized by warm, dry summers and cool, wet winters, with marked climatic gradients such as increased continentality and aridity from west to east. Thus the coastal ranges receive more precipitation than the eastern flanks at the edge of the Jordan-Beqa’a-Orontes Rift valley. The same pattern is repeated along the Anti-Lebanon mountains and the Trans-Jordanian plateau.

Kebara and Hayonim, like most caves in the region, are limestone cavities that formed under phreatic conditions, commonly along vertical and horizontal joints. Subsequent vadose enlargement has in many cases (e.g., Kebara, Hayonim, Tabun, Qafzeh, Bezez, Nahr Ibrahim, Dederiyeh) resulted in vaulted or domed ceilings, which have been prolonged into chimneys that extend to the hillside surface above the caves. In the same way, many of the chimneys are mirrored by depressions or sinkholes that are developed in the underlying bedrock. These depressions have resulted in undulating and irregularly shaped floors, which indicates that the earliest deposits were not horizontal. Moreover, long-term continuity of karstic activity, even on a minor scale, is revealed by cm scale offsets, tilted layers, and collapse of sediments into the sinkholes. Such phenomena in turn indicate the evacuation of the sinkhole, followed by collapse of sediments into the depression (Garrod and Kirkbride 1961; Bar-Yosef and Vandermeersch 1972; Goldberg 1973; Farrand 1979; Kirkbride 1983; Goldberg and Laville 1989). Modern-day karstic action in most of these caves is minimal, however, and the remnants of formerly more extensive and well-developed dripstone signifies karstic conditions that are markedly different from those of today, where dripping water is hardly active.

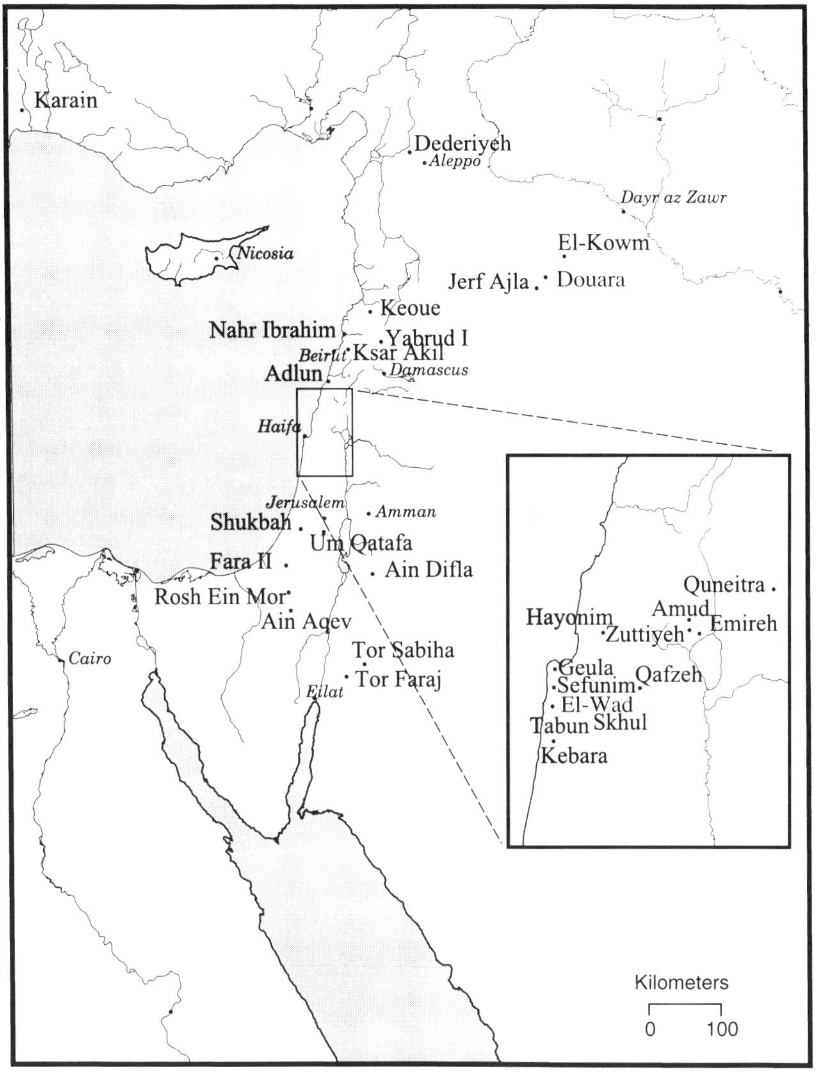

Figure 1. Map showing location of caves and rockshelters discussed in the text.

3. KEBARA

Kebara (Figure 1) is located along the western slope of Mt. Carmel, about 30 km south of Haifa, two km from the coastal shore-line, at an elevation of about 60 m asl (Valladas 1987; Laville and Goldberg 1989; Goldberg and Laville 1991; Bar-Yosef et al. 1992, 1996; Weiner et al. 1993). Initial excavations by Turville-Petre (1932) revealed deposits with Natufian, Kebaran, and Upper Paleolithic industries (Layers B, C, and D, respectively). Later excavations by Stekelis (Schick and Stekelis 1977) exposed Upper and Middle Paleolithic deposits (Layers D and E; Figure 2). Recent excavations concentrated on the Mousterian layers (units XIV through V), reaching bedrock (Bar-Yosef et al. 1992) and verifying the stratigraphy of the remaining Upper Paleolithic deposits (Bar-Yosef et al. 1996). Burnt Mousterian flints have yielded thermoluminescence (TL) dates between

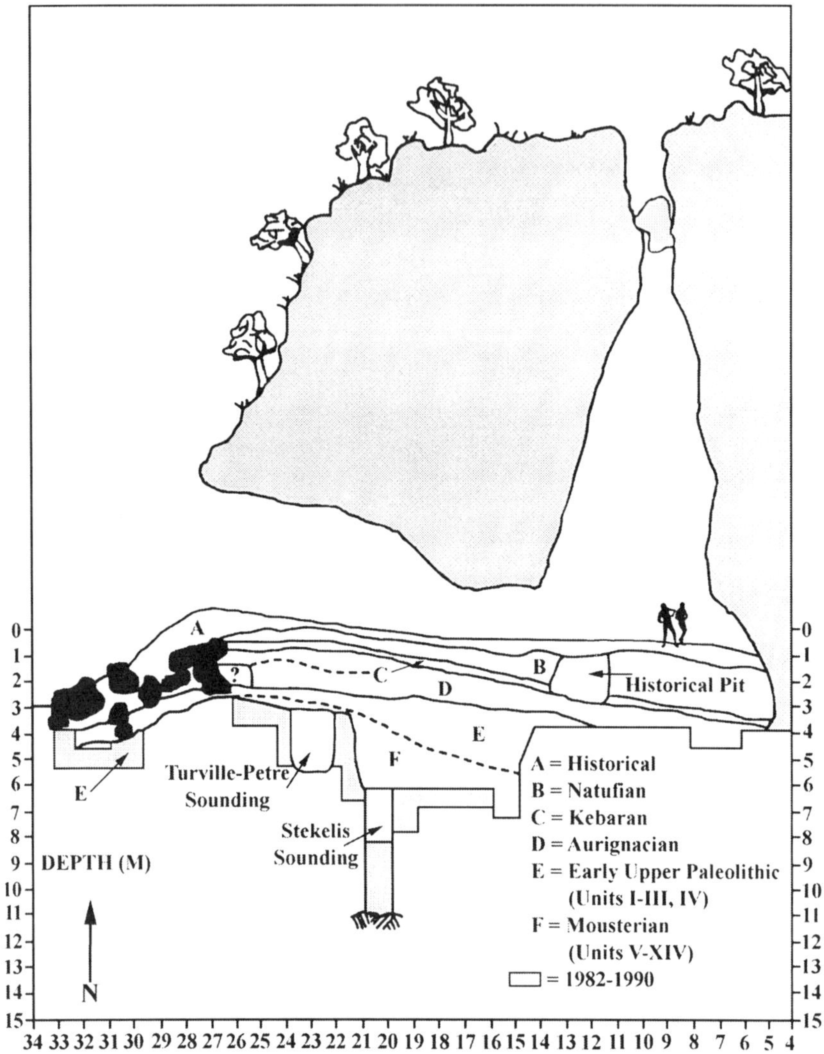

Figure 2. Cross section of Kebara showing overall stratigraphy and cultural layers.

60,000 to 48,000 B.P. (Valladas et al. 1987). These were supported by ESR dates from the lower Mousterian occupations (Schwarcz et al. 1989), whereas charcoal from overlying Upper Paleolithic sediments was dated by radiocarbon to about 43,000 to 28,000 B.P. (Bar-Yosef et al. 1996).

The sediments at Kebara vary from the interior to the exterior and from bottom to top. Inside the cave, the lowermost, sterile sediments closely follow the plunge of the bedrock, and in general consist of poorly sorted, somewhat silty, angular shiny sand, and well-bedded, grayish yellow, and light brown silts (Figure 3); all appear as though they were drawn into an underlying swallow hole. Overlying this sequence are massive, horizontally bedded silts, with numerous siliceous seed coats and centimeter thick darker bands that are finer grained and presumably richer in organic matter. These appear to have been water laid.

The Mousterian sequence is a succession of superimposed organic rich silty layers that preserved laminar, lenticular, and basin-shaped hearths and bands (Figure 4) composed

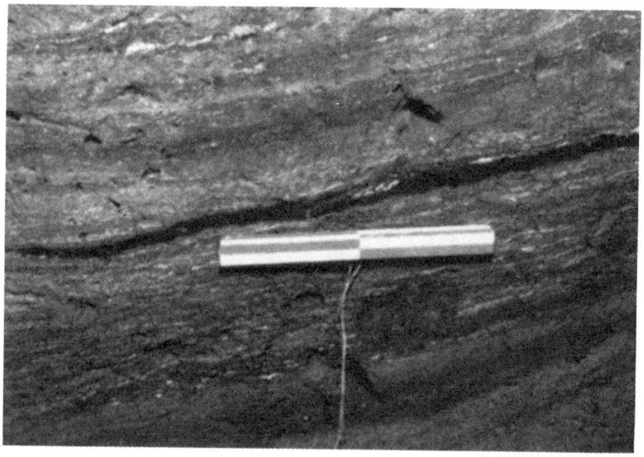

Figure 3. Kebara Cave. Finely laminated, basal sediments resting on bedrock (just below the bottom of the photograph). These sterile sands and silts are intercalated with darker, organic-rich bands and lighter colored siliceous seed coats. Scale is 20 cm across.

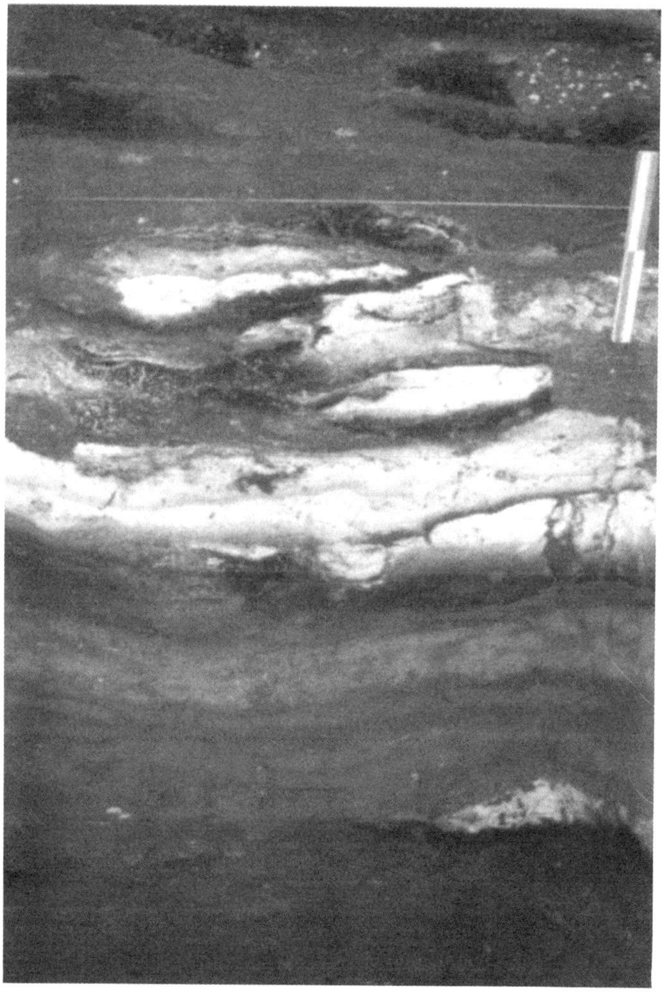

Figure 4. Kebara Cave. Massive, lenticular, and laminar ashy hearths overlying sterile silts and sands that are continuations of sediments shown in Figure 3. Scale is 20 cm long.

of fresh and diagenetically altered ashes and generally organic-rich silts (Laville and Goldberg 1989; Meignen et al. 1989; Bar-Yosef et al. 1992; Weiner et al. 1995; Schiegl et al. 1996). The lower stratigraphic units tend to dip at different angles into the basal swallow hole, whereas the upper ones tend to be more or less horizontal.

The importance of the diagenetic processes is expressed by the decalcification of many pieces of rock fall, and by the formation of numerous secondary phosphate minerals. The crucial role that these diagenetic transformations played is in the disappearance of bones in the southern portion of the cave (Weiner et al. 1993). Similar effects are known from Tabun Cave (e.g., Jelinek et al. 1973; Goldberg and Nathan 1975). Their main impact is in reducing the totality of animal and human bones, thus causing a biased spatial as well as diachronic picture.

In addition, the sediments display conspicuous evidence of biological reworking, in the form of simple and complex rodent burrows (*térrier* in French), ranging from a few cm to several dozens of cm in diameter (Figure 5). Field and micromorphological data show that burrowing was essentially penecontemporaneous with individual occupations and thus artifacts collected from these *térrier* are roughly coeval with those from surrounding, undisturbed sediments.

The uppermost Mousterian layers are exposed not only in the interior sections but also in excavated areas at the cave entrance. As with the earlier sediments, these also consist of silty layers with evidence of hearths, often much reduced in their thickness. These layers are capped by homogeneous, calcareous, reddish silty clays that contain numerous blocks of bedrock roof fall, particularly prominent near the walls and across the entrance to the cave. Overall, this silty clay is lithologically similar to that from the Upper Paleolithic sediments in the entrance, suggesting that deposition beneath the brow of the cave represents a continuous accumulation of material derived from the soil-covered slopes of Mt. Carmel throughout much of the history of the cave. In fact, the sediments beneath the brow can be considered a talus-like accumulation of derived terra rosa soils interdigitated with archaeological remains.

Visible in these entrance Mousterian deposits, and also the Upper Paleolithic ones, are numerous signs of earthworm activity (casts), which is responsible for blurring the boundary between the Upper and Middle Paleolithic sediments, especially since they are lithologically similar in this portion of the cave. However, the Upper Paleolithic layers do exhibit some localized, mm-size impregnations of $CaCO_3$ that is presumably derived from water dripping beneath the brow of the cave. These secondary accumulations are not apparent in the Middle Paleolithic sediments.

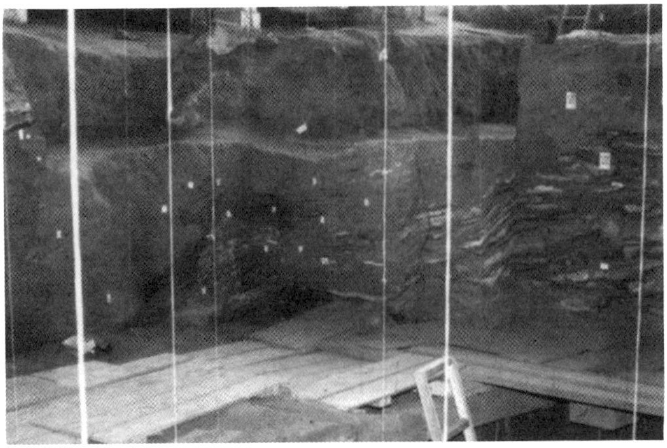

Figure 5. Kebara Cave. Southwestern corner of excavations showing well-bedded, Middle Paleolithic ashy hearth sequence at right. These wedge out to the left (South Profile) where they are overlain by extensively bioturbated and slumped Middle and Upper Paleolithic deposits. The remains of an isolated Upper Paleolithic hearth is shown along the left-hand edge of the photograph.

Post-depositional chemical modifications of the Mousterian deposits at Kebara are widespread and take the form of extensive phosphate diagenesis (Laville and Goldberg 1989; Bar-Yosef et al. 1992; Weiner et al. 1993; Schiegl et al. 1994, 1996; Weiner et al. 1995), although this is rare in areas adjacent to the bedrock walls where precipitation of calcite is common. Physical modifications include localized slumping and faulting. The most significant slumping event affected the entire Mousterian sequence and took place at the end of the Middle Paleolithic (Figure 5). It was followed by inward erosion, truncating especially the upper Mousterian units (VIII-V). This subsidence, coupled with intensive biological activity, continued into the early Upper Paleolithic. Biogenic effects take the form of mammal and insect burrowing, similar to those at Hayonim Cave (see below).

In the interior of the cave, the Upper Paleolithic sediments, like the Mousterian ones, exhibit a noticeably different character as one moves from the entrance to the interior of the cave. In the interior, they are typically composed of compact to cemented, finely laminated and bedded sandy silts (Figures 5 and 6). The number of hearths is much reduced in comparison to the Middle Paleolithic units (Bar-Yosef et al. 1996). Many of these seem to have been obliterated by low energy water flowing from the entrance inward.

The major rock-fall across the entrance (in part removed by Stekelis) is embedded within the red silty clay unit. The collapse of the blocks from the cliff above the brow of the cave occurred during the early Upper Paleolithic. Its correlation with the gap between the Middle and Upper Paleolithic sequences is still unknown. The depositional hiatus, well recorded in the southern section inside the cave (Figure 6), was probably not a long one. The TL dates indicate an age of 48,000 B.P. for Unit VI (while V, the latest Mousterian, is yet undated) and the early radiocarbon readings for the Upper Paleolithic indicate an age of 43,000 B.P., which is possibly a minimum age (Bar-Yosef et al. 1996).

The Upper Paleolithic sediments in the eastern profile, in the back of the cave and under the chimney, are diffused, irregularly bedded yellowish brown layers. In one unit microscopic observations have revealed the presence of diatoms indicating the presence of standing water at one time.

The Upper Paleolithic sediments also have been modified by post-depositional processes that include: (a) local cementation by phosphate (possibly silica) and concomitant destruction of bones in certain parts of the cave. Bones were well-preserved near the northern wall, which is in accordance with the post-Mousterian and perhaps post-Upper Paleolithic diagenesis. In addition, bones are well preserved near the entrance in the clay

Figure 6. Kebara Cave. South profile and lateral continuation of the photograph shown in Figure 5. These finely bedded sediments are clearly water lain and likely represent a markedly wet episode during the Upper Paleolithic occupation of the cave. A 20 cm long scale is in the lower left.

rich deposits and near the northern wall in an area that was not excavated but just tested over a very limited surface; (b) numerous burrows are particularly evident in the direction of the rear of the cave (the eastern section); and (c) cutting and filling by a semi-circular channel measuring about 1 m across.

4. HAYONIM

Hayonim Cave is situated in the Western Galilee at an elevation of about 250 m, about 13 km from the Mediterranean Sea (Figures 1 and 7; Bar-Yosef and Goren 1973; Goldberg 1979; Bar-Yosef 1991; Schiegl et al. 1996).

The established stratigraphic sequence is as follows (Bar-Yosef 1991):

- Layer A: Ashy deposits (200 A.D. to recent)
- Layer B: Built-up rooms and graves of Early through Late Natufian age (radiocarbon dated to 12,500 to ?11,000 B.P.)
- Layer C: Mostly re-worked Mousterian silty deposits, containing Kebaran industry (?17,000–?14,500 B.P.), accumulated at the entrance to the cave.
- Layer D: Levantine Aurignacian kitchen midden (with four sub layers) radiocarbon dated to 29–27,000 B.P. (Belfer-Cohen and Bar-Yosef 1981).
- Layer E: Mousterian. Layer E is a thick deposit (although bedrock has not yet been reached) and is the earliest Mousterian as well. The current excavations (1992–1998) target mainly the Mousterian deposits which occupy the deeper portions of the sequence (Figure 8). The available exposures are situated near the cave entrance (the Deep Sounding) and in a larger area in the central part inside the cave.

As at Kebara, the sediments in the entrance section—similarly situated beneath the brow of the cave—differ from coeval units excavated in the interior. The basal 3.7 m of

Figure 7. Hayonim Cave from afar looking to the West. Much of the reddish silty clay sediment found in the entrance section is reworked terra rosa material that presumably covered much of the terrain in this area. The soil material was removed from the mountain slope and deposited in the cave environment either through the chimney or in the front part of the cave. In any case, the volume of this reworked soil material is significantly less than the volume of excavated sediments within the cave, the balance being made up primarily by anthropogenically derived ashes, organic matter, mineral matter, and aeolian additions.

deposits exposed in the entrance deep sounding (Figure 9) contain Middle Paleolithic artifacts within stratified, generally light colored (brown to reddish brown), soft, powdery silts containing dispersed and powdery flecks of charcoal. These dip slightly toward the exterior and are partially disturbed by extensive burrowing visible in exposed sections. Toward the top or laterally near the walls of the cave these silts are locally cemented by calcium carbonate (Figure 8), and appear to be associated with water dripping off the brow of the cave (see below), or with the formation of dripstone/travertine along the cave walls.

Although the effects of bioturbation in the Deep Sounding are conspicuous in the form of circular burrows about 9 cm in diameter (Figure 9), numerous remains of hearths can be observed, usually taking the form of brown circular features about 30–40 cm across. These hearths are recovered typically without white ashy components (originally calcareous), as these apparently have been eroded, dissolved, or transformed into phosphates. Similarly, much of the lower sediments have been extensively modified by siliceous and phosphatic diagenesis that is currently being investigated by S. Weiner (Weizmann Institute, Rehovot). Interestingly, however, the clearly defined burrows do not show signs of having been phosphatized after their formation, suggesting that the burrows post-date the process of phosphatization.

Figure 8. Excavations at Hayonim Cave looking outward toward entrance. Note the excavations at the base of the photograph known as the Central Area, the right part of which exhibits stalagmitic accumulations of calcite.

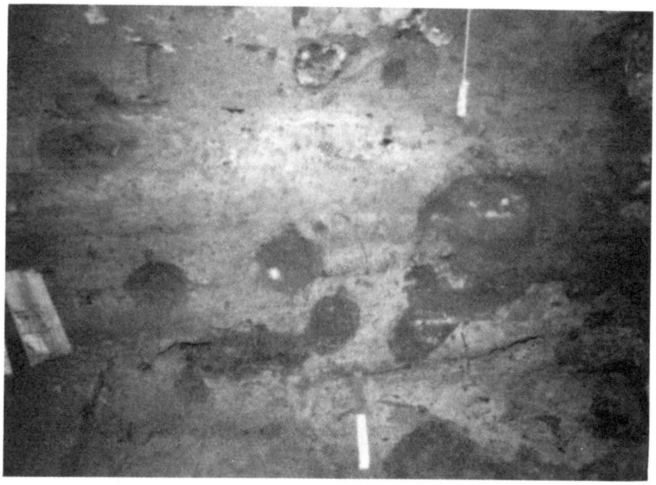

Figure 9. Deep entrance section at Hayonim Cave showing diagenetically altered, finely bedded, water-laid silts and clay perforated by a number of circular rodent holes. The scale is 20 cm long.

Kebaran and Natufian deposits (2.5 m thick in total) directly overly those of the Mousterian, taking the form of an elongated, channel-like depression whose lower surface is inclined toward the exterior. The presence of this depression is likely related to its position directly under the drip line, where runoff from strong winter rains could readily produce chutes of water cascading along the flat brow of the cave. This erosion, similarly to the evidence of inward erosion, truncated the top of the Mousterian deposits. Under the brow of the cave, the excessive amount of winter precipitation and summer dryness turned the top of the Mousterian layers into calcite-cemented, hard breccia.

In the interior, Middle Paleolithic sediments are presently exposed over a total thickness of about 4.5 m and are noticeably more anthropogenic and diagenetically transformed than those at the entrance. The basal sediments in the central section are characterized by anthropogenic accumulations of bedded, lenticular or tabular ashy burnt layers and hearths that are set into a mass of homogeneous brown, yellow or reddish silty clay, commonly mixed in a chaotic arrangement with fragments of charcoal and bones (Figures 10 and 11). The massive accumulation of these sediments appears to result from penecontemporaneous reworking of Middle Paleolithic hearths by the actions of humans (e.g., trampling) and possibly other animals (e.g., insect and rodent burrowing). Moreover, some of the lower

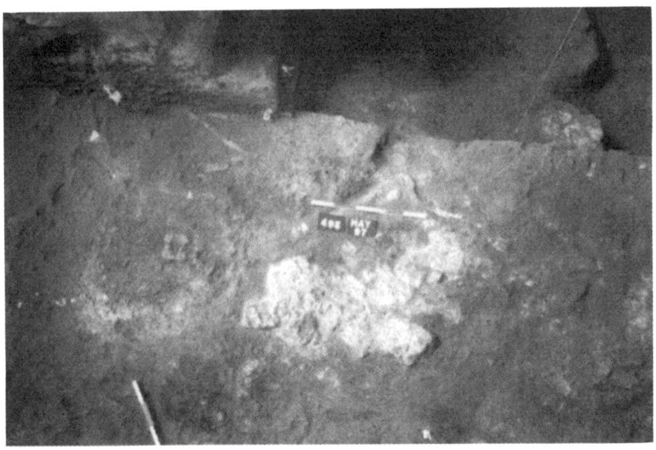

Figure 10. Hayonim Cave. Remains of irregular white ashes (about 10–15 cm thick) from hearth at base of Central Area. These ashes, over a meter in diameter, are surrounded by redder silts and clay rich in charcoal, ash, and organic matter, essentially representing trampled and reworked hearth materials (including cleanouts). Scale in center is 50 cm across.

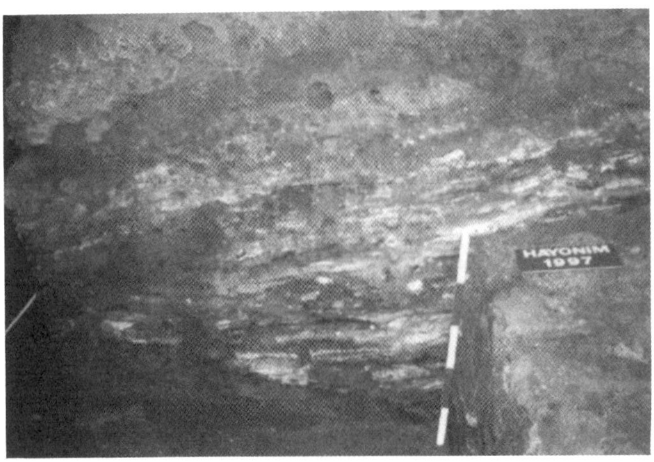

Figure 11. Hayonim Cave. North wall of Central Area showing finely bedded ashes and hearths, with some disturbance by rodent holes. The preservation of hearths becomes worse in an upward direction, mirroring the increase in phosphate diagenesis and reduction in bone preservation. Phosphates at the base of the profile are typically apatitic; these give way in an upward direction to more complex phosphates such as taranakite, leucophosphite, crandallite, and montgomeryite. Scale is 50 cm.

layer E hearths tend to be larger and more massive than overlying ones, attaining over 110 cm in diameter and 10–12 cm in thickness. Many of the hearths, particularly those exposed in the 1997 season, are intact, and consist of slightly cemented, but otherwise unaltered, calcareous ashes (Figure 10). Most of the lower layer E deposits are cut diagonally (relatively to the grid system) by diagenetic front, which destroyed the bones only in a band across the central area.

This situation contrasts with hearths from overlying Mousterian units, in upper layer E, where the original calcite has been diagenetically altered to apatite and other phosphatic minerals. Such increased diagenesis in the younger Mousterian sediments seems to represent a major change in the geological regime of the cave, which is also expressed by concomitant poorer bone preservation, and the relatively sudden occurrence of travertines, fallen stalagmites, and calcareous cave breccia concentrated along the walls. This increase in phosphatization, calcite cementation, and flowing water seems to mark a period of increased wetness which commenced at the end of the Middle Pleistocene, as evidenced by the U/Th date of 163±3 ka for the fallen stalagmite (Schwarcz and Rink, this volume). An additional wet period probably occurred during the Late Pleistocene and was responsible for the erosion of the top of the Mousterian layers.

In sum, depositional processes at Hayonim range from predominantly geogenic ones at the entrance, where contributions from runoff, colluviation, and possible aeolian inputs gave way to largely anthropogenic ones in the interior. These anthropogenic effects represent both depositional operations (hearths) but also penecontemporaneous ones (e.g., trampling, hearth clean-out). Post-depositional physical modifications are clearly present as burrowing in both the interior and entrance sections, and as cementation by calcite and phosphate. Both cementation, as well as bioturbation, are responsible for, or at least associated with, both destruction and effacing the original character of the deposits and its stratification, either geogenic or anthropogenic.

5. DISCUSSION

It should be apparent from the descriptions above that Kebara and Hayonim Caves share a number of characteristics that relate to depositional and post-depositional processes. In this section, we discuss these processes and attempt to show how similar or different processes may be acting in other caves and rockshelters in the area. We subdivided these processes into geogenic (sensu lato, i.e., including pedogenic processes) or anthropogenic in nature.

5.1. Geogenic Processes

5.1.1. Clastic Sediments. At Kebara the deposition of the earliest well-sorted sands resting upon bedrock appear to be associated with phreatic water flow that took place before the cave was occupied. As such, it represents a rare instance in which the effects of the phreatic environment—in contrast to vadose effects in the form of travertine and calcareous cementation—can be documented. Such conditions have not been revealed at Hayonim, although the finely bedded and laminated silts in the profile of the deep sounding at the entrance clearly point to standing water. In the case of Hayonim, this could be pooled water accumulating below the brow. At Qafzeh Cave, on the other hand, phreatic water is likely responsible for the extensive cementation by manganese and calcite of most of the Mousterian deposits in front of the cave's brow, known as the 'terrace' (Vandermeersch 1981). It should be noted that, like in Hayonim, the cementation solely affected the upper part of the sequence. Similar phreatic processes probably occurred at el-Wad, where the earliest deposit, layer G, accumulated in a pit at the base of the first chamber, and was comprised of light grey to greenish-yellow gritty, siliceous material that Garrod and Bate (1937:24) ascribed to the effects of subterranean drainage. The overlying layer F, containing both early Upper Paleolithic as well as Middle Paleolithic artifacts "was made up of a vast number of flint flakes and implements without earth between" (Garrod and Bate 1937:22). In addition, inside the cave, in Chamber II, most of the flints were abraded and locally rolled.

The bulk of the geogenic sediments at both Kebara and Hayonim is represented by those occurring at the entrance, where excavations reveal a mass of red silty clay. These deposits are essentially terra rosa soils that have been colluvially reworked from the surface above the caves. Although some of this material was also washed into the caves through the chimneys at the rear, most accumulated at the entrance. Similarly derived "entrance facies" occur at or near the entrances of Sefunim, Rakefet, Amud, and possibly Umm Qatafa, although the last case is difficult to evaluate because of extreme diagenesis of the deposits.

These accumulations, derived from outside the cave, are mostly associated clays and silts of slope-derived soils and sediments (colluvium), or aeolian sand and silt. Soil materials vary according to locality (particularly climate and parent material). In the Mediterranean zone, typical soils (e.g., terra rosa, hamra, rendzina, grumosols; Dan et al. 1972) are composed of clay and some silt (usually imbedded aeolian dust), with little organic matter. These can be calcareous or not, depending on local conditions. These soil materials are typically deposited in the cave or rockshelter either by slopewash or runoff, either directly through a chimney or through the entrance (e.g., Tabun layers B and C, Kebara, Hayonim, Qafzeh [mainly the Upper Paleolithic layers], Ksar 'Akil), where they can temporarily accumulate as a small talus cone beneath the brow and are later washed into the cave (e.g., Kebara, Hayonim).

In addition to colluvial accumulations at the entrance, we note for Kebara the inwashing of finely laminated silty clays by sheet wash, particularly during the Upper Paleolithic. The slope of the sediments toward the rear of these caves clearly demonstrates that the source of these water laid sediments is from the entrance. Accumulations of organic-rich layers at the back of the cave and associated diatoms (Barbara Winsborough, personal communication 1994) point to wetter conditions at this time. Flowing surface water is also indicated by a channel developed within the Upper Paleolithic deposits along the south section (Goldberg and Laville 1989; Bar-Yosef et al. 1996). The fresh, red silty clay aspect of the entrance facies, however, is lost as we move to the interior of the caves, where much of the sediments have undergone extensive diagenesis in the form of decalcification and phosphatization. As a result, much of the interior sediments tend to be darker, more gritty, and cemented.

Internally derived, autochthonous sediments generally take the form of clasts of bedrock (limestone or dolomite) that have been detached from the roofs and walls of the caves. Large, meter size blocks are evident at the entrances to both Kebara and Hayonim where they tend to occur at the top of the Mousterian layers or in the Upper Paleolithic ones. Their origin could be caused by increased dissolution of the bedrock, possibly associated with different climatic conditions, or they could be tied to other mechanisms, such as detachment by earthquakes (Ronen 1971). The fact that many caves in the area (e.g., Sefunim, Umm Qatafa, Zuttiyeh, Jerf 'Ajla, Yabrud; Figure 1) exhibit such large blocks at the entrance suggests that the brows and overhangs are likely to be unstable loci, and subject to collapse simply because of the effects of gravity.

Less striking, although more invariable, is the occurrence in most of the sediments of cm to dm size pieces of roof fall that are scattered throughout the sediments. Such clasts are present in most sites, including Jerf 'Ajla and Yabrud Rockshelter I (Goldberg 1972), Tabun layers B and C (Jelinek et al. 1973), Qafzeh (both the interior and the terrace deposits (Vandermeersch 1981), Amud (Suzuki and Takai 1970), Skhul (Garrod and Bate 1937), Sefunim (Ronen 1984), Kebara, and Hayonim (Goldberg and Laville 1989).

The accumulation of these smaller blocks of roof fall is likely tied to detachment of bedrock blocks by dissolution along joints. The fact that in most cases these clasts are generally sprinkled throughout the sediments suggests that bedrock dissolution has produced a regular supply of these clasts from the ceiling and walls throughout the depositional history of these sites. Massive accumulations of *éboulis* are best known from the sequence of Ksar Akil (e.g., Tixier 1974; Copeland 1975; Bergman 1987), where they probably depict the effects of frost shattering. In extreme cases, such as in Yabrud and Jerf 'Ajla, the bulk of the fill is composed of this type of *éboulis*. In the case of the latter sites, which are situated in a drier, more steppic environment than those of the Galilee-Mt Carmel area, increased freeze-thaw activity in a climate colder than today's has been suggested as the principal agent for detaching these pieces of rock fall (Goldberg 1972; Farrand 1979). Interestingly, however, the rockfall in caves from these drier environments (Yabrud, Jerf 'Ajla, and Amud) is quite fresh, indicating the necessity of water to effect or enhance diagenetic alterations.

It is worth noting here that in the caves of the Mediterranean coastal ranges, many of the smaller clasts of limestone have been diagenetically modified and commonly dissolved, often leaving hollow ghosts. Consequently, the amount of roof fall observed in the sediments today is clearly less than what was originally present. The dissolution of these blocks is particularly noteworthy in the interior portions of the caves, where more moisture is present and thus diagenesis is more prevalent. Large blocks of roof fall occurring at cave entrances tend to be unaltered.

Aeolian-derived sediment is generally rare or volumetrically limited in the caves discussed here, although much of the sediment at Tabun Cave (layer G-E) is composed of aeolian sand and silty sand that seems to have accumulated over a long period of time probably before 200,000 ka (Mercier et al. 1995). These sands were likely blown into the cave from beaches and coastal dunes situated along the Mediterranean littoral. Although originally thought to be associated with one glacio-eustatic retreat of sea level, current age assessments indicate that sands must have accumulated during more than one fluctuation of sea level. On the other hand, stratigraphic breaks in the sedimentary sequence are not evident and have likely been overlooked.

Tabun is rather unique in its coastal-derived sandy accumulations. Interestingly, Kebara Cave, situated only 13 km south of Tabun, on the whole contains only traces of aeolian sand. Preliminary observations of the lowermost units indicate thin beds of sand situated just above bedrock. Their polished nature suggests that these are possibly traces of phreatically-derived sand (Laville and Goldberg 1989). It seems that when most of the Middle and Upper Paleolithic sequence accumulated in Kebara, the narrow coastal plain already accommodated one of the kurkar (sandstone) ridges, where a Mousterian assemblage was deposited inside the hamra (red loam layer) (Ronen 1977).

Most aeolian derived sediment, however, takes the form of finer, silt size accumulations of dust that has been blown in the caves and settled out during dust storms or during dusty days, even in the northern parts of the country (Ganor 1975). Such dusty accumulations are manifest in the caves by the quartz silt fraction that is readily obvious in thin sections of these sediments. Some of this quartz silt, however, is likely recycled with the input of soil-derived material (see below) that already contains imbedded aeolian quartz silt (Yaalon and Ganor 1973). In soil-derived sediments (e.g., terra rosa) it is not possible to differentiate the silts originating with the soils from those that have been deposited directly into the cave. Micromorphological analysis reveals the presence of quartz silt in anthropogenic ashes, which could suggest direct aeolian accumulation of silt.

5.1.2. Chemical Sedimentation. Primary chemical sedimentation is not volumetrically abundant in these Levantine caves, and is usually represented by calcitic deposits of travertine (both stalactites and stalagmites, as well as more horizontal, tabular deposits of speleothems, known in French as *plancher*). These calcium carbonate accumulations, moreover, grade laterally into detrital sediments which they cement, yielding what is known in the archaeological literature as "brecciated deposits" or just "breccia." As such, they represent a form of water activity that can considerably obliterate the details of past events. Accumulations of precipitated calcite are absent in Kebara but more prominent in Hayonim, as well as in Umm Qatafa, Tabun, Zuttiyeh, and Skhul. They are very important because of their ability to be dated by uranium-series techniques (Schwarcz et al. 1980). Moreover, their accumulation may reflect wider scale phenomena, such as past environments and climates. At Hayonim, for example, travertine formation seems to be localized stratigraphically near the upper part of the Middle Paleolithic deposits that have been dated at 163±3 Ka. This occurrence of travertine seems to indicate notably wetter conditions possibly during Stage 6 with a renewal during stage 5 d-a, as can be inferred from the latest TL dates for Mousterian pieces (Mercier and Valladas, personal communication).

Brecciated detrital deposits, on the other hand, are quite abundant in these caves, particularly in locations adjacent to the walls of the caves where carbonate-rich solutions can easily migrate. It is worth nothing that speleothems, travertines, and calcitic "breccias" are virtually absent from deposits associated with artifacts younger than the Upper Paleolithic, implying that the later part of the Late Pleistocene was overall drier than the earlier half associated with Middle Paleolithic artifacts.

5.2. Biogenic Processes

Biogenic processes occur in Kebara and Hayonim, as well as in other caves in the area, whether recognized as such during the excavation or later. The most striking biogenic effect is expressed in the form of burrows (*térriers*), which occur typically as circular to elliptical features about 6–10 cm across. They are generally filled with softer material than the surrounding matrix and micromorphologically are clearly characterized by a loose, aggregated fabric. Where repeated burrowing has taken place, larger, decimeter size areas occur which lack bedding or other indications of primary sedimentation.

To date, the animal(s) responsible for the formation of these biogenic features has not been determined since they are characteristically devoid of bones of an animal of suitable size to make such a burrow. Modern burrowing was observed outside of Kebara Cave, where an experimental hearth was burrowed the following year by a gerbillid (E. Tchernov, personal communication 1988). Thus, we have yet to find a smoking gun, and we can only surmise that the originator of such features will be uncovered during future excavations. Nevertheless, the issue of disturbance is clear and is likely more widespread than commonly appreciated or stressed. In Kunji cave, in the Zagros, extensive bioturbation was attributed to the burrowing effects of porcupines (Baumler and Speth 1993). Similarly, burrowing was clearly recognized at Douara (Hirai 1987:52), although its implications on the interpretation of the faunal analysis, which showed a generally uniform assemblage, was never taken into account. We suspect that such burrowing processes are more widespread than commonly realized.

Other bioturbation processes are more subtle, such as the activity of earthworms and insects. Both groups of animals were observed at Kebara, particularly at the entrance to the cave. Although because of their small size it is unlikely that their activities were responsible for movement of artifacts (a constant dread), they are clearly responsible for microscale obliteration of original stratigraphic features, such as bedding and contacts between lithological units (Wood and Johnson 1979).

A more subtle biogenic process involves the contribution of organic matter (guano) by bats and pigeons. The present-day effects of the latter in particular can be demonstrated by the presence of pigeons in many caves in Israel. Recognition of past contributions, however, survive in the sediments as mm size, well-rounded and polished grains of limestone and chert that can be ascribed to gizzard stones. At Tabun, such grains made up about 1% of the sediment by weight (Goldberg 1973), and identical grains were observed in the wet screened material from both Kebara and Hayonim. The presence of such biogenic inputs provides additional sources of phosphate and organic matter for evaluating processes of phosphate diagenesis. Phosphate from certain minerals, such as taranakite, have consistently been interpreted to be derived from bat guano and not bones (Murray and Dietrich 1956). The past activity of bats in these caves is less demonstrable because caves in Israel are currently thought to have been occupied by insectivorous bats whose remains have yet to be identified during micromorphological analysis of the sediments.

Carnivore contributions from hyaenas and their phosphatic coprolites is more demonstrable in many caves where complete or fragmented parts of coprolites have been recovered. Their presence is also reflected in bone gnawing and the presence of hyaena bones. More subtly, however, the occurrence of hyaenas and their denning activities is likely another source of bioturbation of the sedimentary substrate. It is perhaps not coincidental that in Kebara the concentration of hyaena bones near the north side of the entrance mirrors the bioturbated nature of the sediment, as seen both in the field and in thin section.

5.3. Anthropogenic Processes

One of the main anthropogenic features found in well-preserved contexts are hearths and ashy layers. Middle and Upper Paleolithic hearths and ash deposits appear in various forms. Hearths are often oval- to circular-shaped lens of material distinct from the associated sediment. Ash mantles or distributions sometimes are spread over 2–3 meters across a large and not always clearly delineated lens. In cases of good preservation, it has become clear that ashes were spread outward from the hearth area to cover a larger surface, which would perhaps be adequate for sleeping. Hearths, ashes, and burned bones can provide important information on human ways of controlling and using fire and ashes. Little has been published concerning the compositions and microscopic structures of prehistoric ashes. Courty et al. (1989) provide the most detailed descriptions of calcite-containing fossil wood ash, and recent studies in Kebara and Hayonim Caves have produced a wealth of information (e.g., Courty et al. 1989; Weiner et al. 1995; Schiegl et al. 1996; see also Meignen et al. 1989). Among the main discoveries was that unrecognized portions of ash, namely the relatively insoluble siliceous fraction, is a quantitatively important constituent. Similar types of studies should be undertaken in other Levantine caves, where in situ hearths, reworked hearths, and ashy deposits are evident or suspected (e.g., Amud, Qafzeh, Douara, Jerf 'Ajla). A comparison between the Kebara and Hayonim burned layers and those from layers B and C from Tabun would be particularly relevant.

Anthropogenic tracking of debris into the cave is an elusive process and one that is suspected but virtually non-testable. It is difficult to imagine that such a process did not occur, particularly for those who have been in Northern Israel during the winter where muddy terrains are typical. Nevertheless, this process must remain under suspicion until it can be adequately demonstrated.

Synsedimentary anthropogenic effects are represented by hearth cleaning and trampling. Both these activities effectively rework hearths and burnt areas and result in chaotic mixtures of dismantled hearth components (ashes, charcoal, phytoliths) that lack bedding or integrity. Such homogenized sediments can be observed both in the field and in thin section. At Kebara, micromorphological study of hearths, still in progress, clearly demonstrates areas of bedded ashes and organic matter that likely represent hearth rake-out. Elsewhere, toward the rear of the cave, thin section analysis of crudely bedded, decimeter thick accumulations of ash are manifestly the result of Middle Paleolithic dumping of ash.

Bones that are black in color may certainly have been burned, at least at relatively low temperatures (less than 750°C). Bones may also be darkly colored due to staining by various metal ions. Above that temperature, the burned bones are white as all the organic matter is oxidized and the mineral also recrystallizes (Shipman et al. 1984). These calcined bones lose much of their mechanical integrity, however, and are rarely preserved (Oakley 1955; Stiner et al. 1995). The presence in burned bones of carbonized remains of organic material (Brain and Sillen 1988) is certainly consistent with burning, although it is probably not definitive proof of burning. Low temperature diagenesis also causes severe alteration of the organic material. Stiner et al. (1995) observed that infrared spectra of burned bone organic remnants in Hayonim Cave are distinct from those obtained from diagenetically altered organic matrices. Shahack-Gross et al. (1997) have developed the means for separating burned from stained bones. These methods should be useful as a means of identifying burned bones in other sites.

6. CONCLUSIONS

A number of trends appear to immerge from the detailed observations and analysis of the Kebara and Hayonim sediments and their comparison with situations from other caves and rockshelters in the Levant.

Anthropogenic sedimentation is a significant depositional agent and is responsible for supplying inputs of charcoal, ash, and organic matter. These anthropogenic additions make up the sedimentary imbalance produced by geogenic sediments, which are not sufficient to fill up the caves to the levels presently observed. Individual events stemming from human activities such as burning are generally recognizable in most Mediterranean caves such as Tabun, Kebara, Hayonim, Amud, Tor Faraj, and Douara. The composition and types of anthropogenic units is a subject that needs further attention and detailed study should provide information that is compatible and no less important than data collected from other material sources such as lithics, bone, and botanical remains.

Human activities are not limited simply to net additions of materials to the sedimentary record but also are responsible for penecontemporaneous modification of existing sediments in the form of trampling, hearth rake-out, and dumping of ash.

Bioturbation can be a significant process in displacing artifacts to varying extents. Such activities are already well documented in Kebara, Hayonim, and Douara, and by analogy could have taken place in other sites. Systematic refitting, although both expensive and a time consuming effort, will indicate the amount of vertical movement that has taken place. It still remains to be determined which animal(s) produce the observed bioturbation.

Secondary phosphatization of the sediments tends to be localized in caves with Mediterranean climates such as Tabun, Kebara, Hayonim, Qafzeh, Umm Qatafa, and Zuttiyeh. Other caves in drier areas (Yabrud, Jerf 'Ajla, Douara) are generally calcareous and drier and appear to have remained so. They also appear to have overall less biological activity associated with them which, coupled with reduced water supply, would manifest itself in less phosphate accumulation and diagenesis.

We hope that our overview of the types of site formation processes acting in these important sites has provided insights into the nature of the real context of these important sites, many of which provide the keys to understanding cultural and biological evolution in this part of the globe.

ACKNOWLEDGMENTS

We would like to thank all of our colleagues in the project on the Origins of Modern Humans who have, over the years, provided us with insights, discussions and criticisms of some of the points made in this paper. Not all of them might agree with what we have written here.

REFERENCES

Bar-Yosef, O. 1992. The role of western Asia in modern human origins. *Philosophical Transactions of the Royal Society of London* B 337:193–200.

Bar-Yosef, O. 1993. Site formation processes from a Levantine point of view, 11–32. In *Formation Processes in Archaeological Context*, ed. P. Goldberg, D. T. Nash and M. D. Petraglia. Madison: Prehistory Press.

Bar-Yosef, O. 1994. The contributions of southwest Asia to the study of the origin of modern humans, 23–66. In *Origins of Anatomically Modern Humans*, ed. M. H. Nitecki and D. V. Nitecki. New York: Plenum Press.

Bar-Yosef, O., and N. Goren. 1973. Natufian remains in Hayonim Cave. *Paléorient* 1:49–68.

Bar-Yosef, O., B. Vandermeersch, B. Arensburg, A. Belfer-Cohen, P. Goldberg, H. Laville, L. Meignen, Y. Rak, J. D. Speth, E. Tchernov, A. M. Tillier, and S. Weiner. 1992. The excavation in Kebara Cave, Mt. Carmel. *Current Anthropology* 33:497–550.

Bar-Yosef, O., M. Arnold, A. Belfer-Cohen, P. Goldberg, R. Housley, H. Laville, L. Meignen, N. Mercier, J. C. Vogel, and B. Vandermeersch. 1996. The dating of the Upper Paleolithic layers in Kebara Cave, Mount Carmel. *Journal of Archaeological Science* 23:297–306.

Baumler, M. F., and J. D. Speth. 1993. A Middle Paleolithic assemblage from Kunji Cave, Iran, 1–74. In *The Paleolithic Prehistory of the Zagros-Taurus*, ed. D. I. Olszewski and H. L. Dibble. Philadelphia: The University Museum of the University of Pennsylvania.

Bergman, C. A. 1987. *Ksar Akil, Lebanon: A Technological and Typolological Analysis of the Later Palaeolithic Levels of Ksar Akil.* Oxford: British Archaeological Reports International Series 329.

Brain, C. K., and A. Sillen. 1988. Evidence from the Swartkrans Cave for the earliest use of fire. *Nature* 336:464–6.

Brammal, A., and J. G. C Leech. 1937. Report on samples of matrix from the Mugharet es-Skhul and et-Tabun, 126–7. In *The Stone Age of Mount Carmel*, ed. D. A. E. Garrod and D. M. Bate. Oxford: Clarendon Press.

Copeland, L. 1975. The Middle and Upper Palaeolithic of Lebanon and Syria in the light of recent research, 317–50. In *Problems in Prehistory: North Africa and the Levant*, ed. F. Wendorf and A. E. Marks. Dallas: SMU Press.

Courty, M. A., P. Goldberg, and R. I. Macphail. 1989. *Soils and Micromorphology in Archaeology*. Cambridge: Cambridge University Press.

Dan, J., D. H. Yaalon, H. Koyumdjisky, and Z. Raz. 1972. The soil association map of Israel. *Israel Journal of Earth-Sciences* 21:29–49.

Farrand, W. R. 1979. Chronology and paleoenvironment of Levantine prehistoric sites as seen from sediment studies. *Journal of Archaeological Science* 6:369–92.

Farrand, W. R. 1970. Geology, climate and chronology of Yabrud Rockshelter I. *Fundamenta* (Institut für UR- und Frühgeschichte der Universität zu Köln, Reihe A) 2(2):212–23.

Ganor, E. 1975. Atmospheric Dust in Israel—Sedimentological and meteorological analysis of dust deposition. Ph. D. diss. (unpublished). Hebrew University. Jerusalem.

Garrod, D., and D. Kirkbride. 1961. Excavation of the Abri Zumoffen—A Palaeolithic Rock-Shelter Near Adlun, South Lebanon, 1958. *Bulletin de Musée de Beyrouth* 16:7–45.

Goldberg, P. 1972. Analyses of sediments from Jerf 'Ajla and Yabrud Rockshelters, Syria, 747–54. In *Compte Rendu Officiel du VIIIe Congres INQUA* (Paris 1969). Paris.

Goldberg, P. 1973. *Sedimentology, Stratigraphy and Paleoclimatology of et-Tabun Cave, Mount Carmel, Israel.* Ann Arbor: University Microfilms.

Goldberg, P., and Y. Nathan. 1975. The phosphate mineralogy of et-Tabun Cave, Mt. Carmel, Israel. *Mineralogical Magazine*, 40:253–8.

Goldberg, P. 1979. Micromorphology of sediments from Hayonim Cave, Israel. *Catena* 6(1):60–7.

Goldberg, P. 1979. Micromorphology of sediments from Hayonim Cave, Israel. *Catena* 6(2):167–81.

Goldberg, P., and H. Laville. 1991. Etude géologiques des dépôts de la grotte de Kébara (Mont-Carmel): Campagnes 1982–1984, 29–41. In *Le Squelette Moustérien de Kébara 2*, ed. O. Bar-Yosef and B. Vandermeersch. Paris: CNRS Editions.

Goldberg, P., and H. Laville. 1989. Le contexte stratigraphique des occupations paléolithiques de la grotte de Kebara (Israël). *Paléorient* 14:117–23.

Hanihara, K. 1976. Excavations at the Douara Cave, Palmyra, Syria, 1974: A preliminary report. *Annales Archéologique Arabes Syriennes* 27:120–35.

Hanihara, K., and Y. Sakaguchi. 1978. *Palaeolithic site of Douara Cave and palaeogeography of Palmyra Basin in Syria*, Part I. Tokyo: Bulletin of the University Museum.

Hirai, Y. 1987. Stratigraphy and sedimentological analysis of the Douara Cave deposits, 1984 excavations, 49–60. In *Paleolithic site of Douara Cave and Paleogeography of Palmyra Basin in Syria. Part IV: 1984 Excavations*, ed. T. Akazawa and Y. Sakaguchi. Tokyo: The University Museum, The University of Tokyo.

Jelinek, A. J., W. R. Farrand, G. Haas, A. Horowitz, and P. Goldberg. 1973. New excavations at the Tabun Cave, Mount Carmel, Israel, 1967–1972: A preliminary Report. *Paléorient* 1:151–83.

Kirkbride, E., S. De Saint-Mathurin, and L. Copeland. 1983. *Results, Tentative Interpretation and Suggested Chronology*. Oxford: British Archaeological Reports International Series 159.

Laville, H., and P. Goldberg. 1989. The collapse of the Mousterian regime and the beginnings of the Upper Palaeolithic in Kebara Cave, Mount Carmel, 75–95. In *Investigations in South Levantine Prehistory*, ed. O. Bar-Yosef and B. Vandermeersch. Oxford: British Archaeological Reports International Series 497.

Meignen, L., O. Bar-Yosef, and P. Goldberg. 1989. Les structures de combustion moustériennes de la grotte de Kébara (Mont Carmel, Israel), 141–6. In *Nature et Fonction des Foyers Prehistoriques*, ed. M. Olive and Y. Taborin. Nemours: APRAIF.

Mercier, N., H. Valladas, G. Valladas, J. L. Reyss, A. Jelinek, L. Meignen, and J. L. Joron. 1995. TL dates of burnt flints from Jelinek's excavations at Tabun and their implications. *Journal of Archaeological Science* 22:495–510.

Murray, J. W., and R. V. Dietrich. 1956. Brushite and taranakite from Pig Hole Cave, Giles County, Virginia. *American Mineralogist*. 41:616–26.

Oakley, K. P. 1955. Fire as a Palaeolithic tool and weapon. *Proceedings of the Prehistoric Society* 21:36–48.

Ronen, A. 1971. Post-Pleistocene Stony Layers in East Mediterranean Sites. *Quartär* 22:73–93.

Ronen, A. 1977. Mousterian sites in red loam in the coastal plain of Mount Carmel, 183–90. In *Eretz Israel 13* (Moshe Stekelis Memorial Volume), ed. B. Arensburg and O. Bar-Yosef. Jerusalem: Israel Exploration Society.

Ronen, A. 1984. *Sefunim Prehistoric Sites, Mount Carmel, Israel*. Oxford: British Archaeological Reports International Series 230.

Schick, T., and M. Stekelis. 1977. Mousterian Assemblages in Kebara Cave, Mount Carmel, 97–150. In *Eretz Israel 13* (Moshe Stekelis Memorial Volume), ed. B. Arensburg and O. Bar-Yosef. Jerusalem: Israel Exploration Society.

Schiegl, S., S. Lev-Yadun, O. Bar-Yosef, A. El Goresy, and S. Weiner. 1994. Siliceous aggregates from prehistoric wood ash: A major component of sediments in Kebara and Hayonim caves (Israel). *Israel Journal of Earth Science* 43:267–78.

Schiegl, S., P. Goldberg, O. Bar-Yosef, and S. Weiner. 1996. Ash deposits in Hayonim and Kebara Caves, Israel: Macroscopic, microscopic and mineralogical observations, and their archaeological implications. *Journal of Archaeological Science* 23:763–81.

Schwarcz, H. P., P. Goldberg, and B. B. Blackwell. 1980. Uranium-series dating of archaeological sites in Israel. *Israel Journal of Earth-Sciences* 29:157–65.

Schwarcz, H. P., W. M. Buhay, R. Grün, H. Valladas, E. Tchernov, O. Bar-Yosef, and B. Vandermeersch. 1989. ESR dating of the Neanderthal site, Kebara Cave, Israel. *Journal of Archaeological Science* 16:653–9.

Shahack-Gross, R., O. Bar-Yosef, and S. Weiner. 1997. Black-coloured bones in Hayonim Cave, Israel: Differentiating between burning and oxide staining. *Journal of Archaeological Science* 24:439–46.

Shipman, P., G. Foster, and M. Schoeninger. 1984. Burnt bones and teeth: An experimental study of color, morphology, crystal structure and shrinkage. *Journal of Archaeological Science* 22:223–37.

Stiner, M., S. Kuhn, S. Weiner, and O. Bar-Yosef. 1995. Differential burning, recrystallization, and fragmentation of archaeological bone. *Journal of Archaeological Science* 22:223–37.

Suzuki, H., and F. Takai, ed. 1970. *The Amud Man and His Cave Site*. Tokyo: Academic Press of Japan.

Tixier, J. 1974. Poinçon décoré du paléolithique supérieur à Ksar Aqil (Liban). *Paléorient* 2:187–92.

Turville-Petre, F. 1932. Excavations in the Mugharet el-Kebarah. *Journal of the Royal Anthropological Institute* 62:271–6.

Valladas, H., J. L. Joron, G. Valladas, B. Arensburg, O. Bar-Yosef, A. Belfer-Cohen, P. Goldberg, H. Laville, L. Meignen, Y. Rak, E. Tchernov, A. M. Tillier, and B. Vandermeersch. 1987. Thermoluminescence dates for the Neanderthal burial site at Kebara in Israel. *Nature* 330:159–60.

Vandermeersch, B. 1981. *Les Hommes Fossiles de Qafzeh (Israel)*. Paris: CNRS Editions.

Weiner, S., P. Goldberg, and O. Bar-Yosef. 1993. Bone preservation in Kebara cave, Israel using on-site Fourier Transform Infrared Spectrometry. *Journal of Archaeological Science* 20:613–27.

Weiner, S., S. Shiegl, P. Goldberg, and O. Bar-Yosef. 1995. Mineral assemblages in Kebara and Hayonim Caves, Israel: Excavation strategies, bone preservation, and wood ash remnants. *Israel Journal of Chemistry* 35:143–54.

Wood, W. R., and D. L. Johnson. 1978. A survey of disturbance processes in archeological site formation. *Advances in Archaeological Method and Theory* 1:315–81.

Wright, H. E., Jr. 1962. Late Pleistocene geology of coastal Lebanon. *Quaternaria* 6:525–39.

Yaalon, D., and E. Ganor. 1973. The influence of dust on soil during the Quaternary. *Soil Science* 116(3):146–55.

9

INTRASITE SPATIAL PATTERNS AND BEHAVIORAL MODERNITY

Indications from the Late Levantine Mousterian Rockshelter of Tor Faraj, Southern Jordan

Donald O. Henry

Department of Anthropology
University of Tulsa
Tulsa, Oklahoma 74104

1. INTRODUCTION

Not only do anatomically modern fossils appear much earlier in Africa and the Near East, but so do important elements of the lithic technologies that were formerly thought to be diagnostic of the Upper Paleolithic traditions in Europe. Furthermore, the abrupt transition between Middle and Upper Paleolithic traditions that characterizes the European sequence is not evident in the Levant. The Levantine Mousterian/Upper Paleolithic boundary is not only chronologically hard to pinpoint, perhaps encompassing some 15,000 years, but it is equally blurred from a techno-typologic perspective. Had the pioneering efforts in defining the archaeological signatures and fossil associations of the Middle and Upper Paleolithic taken place in the Levant and southern Africa rather than Europe, our view of the situation likely would be quite different. Given such a perspective, I think we would be puzzled by the apparent tardiness of the transition in Europe as well as its abruptness.

Even though the fossil and artifactual evidence recovered from outside Europe has caused us to abandon the traditional equation that linked anatomically modern populations with a time after about 40,000 years ago and with Upper Paleolithic artifacts, there persists a notion that these traditional markers are important in distinguishing modern from archaic behaviors. Beyond the questions tied to symbolic cognition as reflected in works of art and ornamentation, archaic behaviors are thought to differ from modern ones in the organizational areas of planning depth and flexibility. In general, archaic hominids are thought to have followed settlement and procurement strategies that relied largely upon routinized, expedient responses to unanticipated opportunities which re-

Neandertals and Modern Humans in Western Asia, edited by Akazawa *et al.*
Plenum Press, New York, 1998.

quired limited planning ahead. In contrast, the hallmarks of modern behaviors are flexibility and anticipation.

Efforts to trace these basic differences in behavior within the archaeological record have focused mainly on comparisons of artifact inventories between sites (Munday 1976; Roebroeks et al. 1988; Geneste 1989; Shea 1989; Kuhn 1992; Henry 1992; Lieberman 1993; and others). Studies have also approached the problem from an examination of intrasite spatial patterns (Hietala and Stevens 1977; Hietala and Marks 1981; Gilead and Grigson 1984; Simek 1987; Saragusti and Goren-Inbar 1990). And a few research efforts have attempted to develop comprehensive reconstructions of settlement and procurement strategies by meshing pattern searches both between and within sites (Marks and Friedel 1977; Henry 1995a). The results of these studies differ principally along geographic lines. Studies undertaken in Europe largely point to a shift in behavioral organization coincident with the Middle to Upper Paleolithic transition about 40,000 years ago, but the Levantine research fails to identify such changes, at least as traditionally held. This raises the interesting possibility that fully modern forms of behavioral organization emerged earlier in the Levant than Europe, paralleling the precocious appearance of anatomically modern humans and Upper Paleolithic technotypological elements in lithic assemblages. Although this is a reasonable interpretation of the available evidence, it should be underscored that our understanding of the problem comes from only a handful of research efforts in the Levant. We clearly need more studies, especially those that integrate intersite and intrasite data, before credible conclusions can be reached.

2. BACKGROUND

Research of Late Levantine Mousterian sites situated in the mountains of southern Jordan has produced evidence that points to the sites having been inhabited during different segments of an annual cycle of transhumance. The rockshelters of Tor Faraj and Tor Sabiha have been the focal sites of the study, but the recent discovery of Levantine Mousterian deposits at six other sites in the area furnishes additional support for the transhumance model (Figure 1). Patterned variability in the natural settings of the sites, the sizes of their occupational areas, the thicknesses of their cultural deposits, and their artifact inventories suggests that their occupants adjusted their behavioral organization in concert with seasonal shifts in their catchments, residential mobility levels, and group sizes (Henry 1992, 1994, 1995a, 1995b; Henry et al. 1996). These intersite data indicate that logistical and opportunistic provisioning strategies, following Binford's (1980) terminology, were employed alternatively on a seasonal basis by the Late Levantine Mousterian groups inhabiting the region. Long-term winter camps appear to have been maintained logistically, in part through the acquisition of critical resources from outside their immediate catchments. In contrast, ephemeral, high elevation, summer camps were provisioned opportunistically from resources found within their catchments.

The importance of this reconstruction extends beyond that of better defining Middle Paleolithic cultural ecology. If accurate, it implies that the occupants of the sites had the capacity to both anticipate their resource needs and to make significant seasonal adjustments in their provisioning strategies. In that both planning depth and behavioral flexibility are viewed as hallmarks of the behavioral organizations of modern hunter-gatherers (Binford 1979, 1989), the evidence from southern Jordan points to the presence of a fully modern strategy some 20,000–40,000 years earlier than is generally suspected (Binford 1989; Klein 1992:5, 1995; Stringer and Gamble 1993:174–6).

3. EXPECTED CORRELATIONS: INTERSITE AND INTRASITE PATTERNS

The traditional view of archaic behavioral organization (i.e., short planning depth and inflexible, routinized behaviors) has been interwoven with dichotomous models of hunter-gatherer settlement and procurement strategies. These were not initially related to contrasting archaic with modern behavioral organization, but were developed to better understand the variability in archaeological site types. In an effort to better grasp why sites of the same period displayed patterned variation in their sizes, settings, structures, and artifact inventories, polar models were created that linked these variates largely to differ-

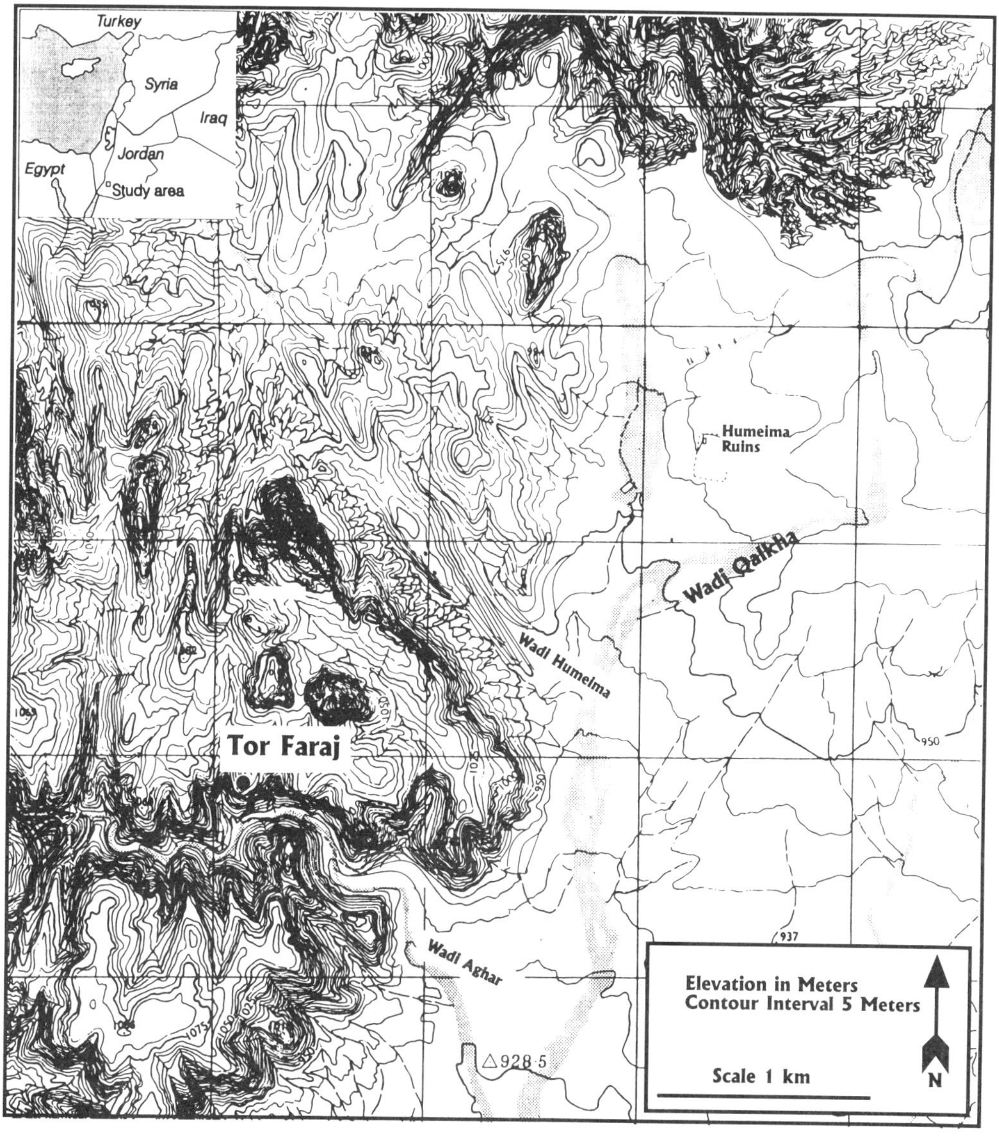

Figure 1. Map showing the location of Tor Faraj and nearby Levantine Mousterian sites in southern Jordan.

ences in residential permanence, activities, and acquisition of resources. The specific configurations of the models and their linkages were inspired mainly by ethnographic information. Although the models differ in their terminology and emphasis, they have in common a fundamental notion that in the range of hunter-gatherer adaptive strategies there are two radically different ways in which groups distribute themselves across the landscape and secure their resources. Within the context of contemporary hunter-gatherers, it should be noted that these two contrastive settlement-procurement patterns appear as polar points along a continuum. They also appear to be followed situationally, dependent upon certain ecological conditions.

In contrasting opportunistic and logistically based encampments, scholars have emphasized their differences relative to residential permanence, activity diversity, and acquisition of resources. These factors form the linkage between our abstract notions of behavioral organization and the tangible evidence within the archaeological record (e.g., site size, site structure, variability in artifact inventories). And when employed as measures of behavioral modernity, these factors have been used to infer fundamental cognitive or adaptive differences.

3.1. Opportunistic and Logistical Strategies

An opportunistic strategy relies upon high levels of individual and group mobility. Resources are acquired primarily from within the catchments of hunter-gatherer residential encampments largely through daily, opportunistic encounters. The settlement and procurement pattern generated by the strategy consists of two basic kinds of sites: residential base camps and non-residential task specific sites, such as kill sites. Base camps are generally occupied briefly by a small number of residents. They may be occupied for several days to a few weeks by a basic social unit numbering some 11–31 (average 22) persons (Hassan 1981:92). In contrast, kill, quarry, and other task specific sites are utilized over a few hours to a few days by small, often gender specific, adult task groups numbering fewer than 10 persons. The range of resources processed at base camps may differ depending upon the catchment, but the overall range of subsistence and related extractive activities remains common to most residential encampments. Although task specific sites introduce the greatest variability to the opportunistic settlement-procurement strategy, the residue of these highly ephemeral, mostly non-residential sites is largely invisible to the archaeological record. Most resources processed within residential base camps come from the immediate site catchment, rarely exceeding a 10 km radius (Vita-Finzi and Higgs 1972; Jochim 1976:138; Butzer 1982:237; Griffin 1984). Less common, highly portable materials such as certain lithic tools may be transported between catchments by curation or trade, however.

Unlike an opportunistic strategy, a logistical strategy is based upon bringing resources to consumers rather than the other way around. Such a strategy is required when a group is tethered to a specific locus and unable to reside near certain critical resources (Binford 1980). These critical resources are acquired by specialized task groups who establish camps and exploit satellite catchments beyond the immediate catchment of the base camp. The resources utilized within a logistically supported base camp come from both within and beyond a radius of about 10 km. Moreover, they are likely to show considerable diversity because of their disparate origins. By funneling resources back to a base camp from distant settings, greater residential permanency is typically maintained in logistically supported settlements than in those supported by an opportunistic strategy. And from a different perspective, it is, in fact, the residential permanency that makes the logistical strategy practical. Such sites tend be occupied over periods ranging from several weeks to several months by groups numbering from some 23–150 persons (Hassan 1981; Butzer 1982).

3.2. Technological Provisioning

Recently Steven Kuhn has introduced a model of "technological provisioning" in an effort to examine the "nature and extent of anticipatory organization within a technological system" (1990:70). As a synthesis of Binford's (1973, 1977, 1979) concepts of curation and expediency, Kuhn's provisioning model is based upon the assumption that a principle objective of a technological system is to "make tools available when and where they are needed" (1990:70). He recognizes three modes of provisioning that have to do with organizing technologies for activities, individuals, and/or places. The *provisioning of activities* requires the least amount of planning depth in that in this context needs are anticipated only as they occur. Making tools "expediently" as an immediate need arises and abandoning them when that need ceases is only efficient, however, in those contexts where raw material is immediately available and where manufacturing time does not interfere with meeting the goals of the task. In those situations where raw material and time are not available, anticipation or planning depth plays a stronger role in organizing the technology. The *provisioning of individuals* with "personal gear" frees individuals and groups from the constraints imposed by raw material availability and enables them to engage their technology more or less immediately when a need arises. This requires a considerable amount of anticipation or planning ahead in organizing the technology with the fundamental constraint being that of portability. Depending upon the specificity by which needs can be anticipated, individuals may be provisioned with specialized tools, generalized tools, or even raw materials for making tools. The *provisioning of places* is an alternative strategy for organizing a technology which also depends on a substantial anticipation of needs. In contrast to the provisioning of activities or individuals, the provisioning of places is not constrained by the distribution of raw materials, manufacturing time, or portability. The major constraint in organizing a technology dependent upon the provisioning of a place is the duration of residence. As the duration of an occupation increases, there is an attendant reduction in the importance of raw material availability, manufacturing time, and portability of implements in organizing a technology.

There is a strong correlation between the different modes of technological provisioning and the contrastive (opportunistic-logistical) settlement and procurement strategies. The provisioning of activities would most likely accompany an opportunistic settlement-procurement pattern, while the provisioning of places would be necessarily tied to logistically supported settlements. The provisioning of individuals would be expected to occur in both contexts.

4. INTRASITE CORRELATIONS

Ethnoarchaeological studies have shown that campsites which were provisioned logistically display more complex site structures than those in which resources were acquired opportunistically from within their catchments (Binford 1983:191; O'Connell 1987:104). This complexity appears to be linked primarily to the longer durations of residence at logistically supported camps and the tendency for the occupants of such sites to engage in a greater amount of site modification (e.g., features) and to more formally segregate their activities (e.g., discrete activity areas). It is interesting to note that anticipated length of residence has been found to provide a stronger predictor of site size and complexity than actual length of residence (Kent 1991:49).

To understand the ways in which the prehistoric occupants of sites organized their behaviors, archaeologists have focused on defining the contextual relationship of features (e.g., hearths), artifacts, and other occupational residue (e.g., bone) (Leroi-Gourhan and Brezillon 1972; Binford 1978, 1983:144–92; Kind 1984; Simek 1987). When concerned with sheltered sites, these relationships have been extended to include the wall and dripline in the assumption that these natural features also would have influenced the distributions of behaviors within sites (Thomas 1983; Flannery 1986).

The structure of a site is thus reflected in the pattern produced by the contextual relationship of natural and cultural features, artifacts, and other behavioral residue. Binford (1983:144) argues that the relationship of cultural features creates a basic framework from which to examine both the range of activities undertaken at a site and how these activities were organized.

Among foragers, in both ethnographic and archaeological contexts, research suggests that the most important cultural feature in shaping the structure of a site is the hearth plan (Binford 1983:149–60; Gamble 1991:11–12). Given that hearths provide a focal area for all of those activities needing heat and light, it is not surprising that the organizational plan for many activities within a site are referenced to the arrangement of hearths. Binford (1978, 1983) has observed, for example, that behaviors tethered to a hearth form an area of intensive activity which is found within a radius of 2–3 m from the hearth center. Within this area, concentrically arranged "drop" and "toss" zones are defined by the size-sorting of artifacts and other residue. A fan-shaped swath, found immediately peripheral to the hearth, is used for seating and this is where ordinary behaviors (e.g., cooking, eating, talking, maintenance tasks) most often occur. This zone is typically linked to small residue (e.g., bone and flint chips) that has been dropped in place. Another swath forming a concentric band beyond the drop zone contains larger residue that has been tossed beyond the limits of the immediate hearth area. Outside of this area of intensive activity, an area of extensive activity is defined where specialized tasks such as butchery and other primary extractive activities are most likely to occur.

4.1. Comparison of Middle with Upper Paleolithic Structures

In an attempt to test propositions linking archaic behaviors with opportunistic settlement-procurement strategies, Simek (1987) compared the intrasite structures of the Upper Paleolithic horizons of Le Flageolet rockshelter with a Mousterian level of Abri Vaufrey (Rigaud 1978). He emphasized the relative proportions of the living floors devoted to intensive versus extensive activities and the degree of variability between the activity areas. He argued that in an opportunistic settlement-procurement system, one would expect to find the site structures of base camps to be relatively simple, dominated by an intensive activity area tied to a central hearth. In contrast, the intrasite structures of sites in a logistically organized system would be expected to display greater proportions of extensive activities and greater between area diversity. Simek (1987) found the Mousterian intrasite structure, although without a hearth, to display an "occupation 'core' area" largely lacking spatial variability. The Upper Paleolithic horizons of Le Flageolet, on the other hand, revealed several distinct artifact concentrations tethered to hearths and these revealed significant diversity between concentrations.

5. SITE STRUCTURE OF TOR FARAJ

Within the upper part of the >3.5 m deposit of Tor Faraj two near-level bedded floors (Floors I and II) were identified on the basis of concentrations of artifacts and hearth horizons. The bedding of the deposit, dipping off at an angle of 5–10° from the backwall to the dripline, was created by sediments being entrapped behind a natural wall formed by the collapsed brow of the shelter (Figure 2). The identification of the site structure of the Tor Faraj occupations rests in the recognition of the contextual and spatial relationships of the natural features of the shelter with the spatial arrangements of cultural features (hearths), artifacts, and occupational residues (phosphates). Also, the presence of stratified living floors offers the opportunity for determining if a common site structure persisted throughout the occupations of the shelter. Once defined, the site structure of Tor Faraj furnishes a basis for (a) assessing the degree to which the occupants of the shelter followed modern forms of behavioral organization and (b) reconstructing specific activity areas within the shelter.

5.1. Hearths

Although Floor I contains less than half the number of hearths recorded for Floor II, the arrangement of hearths for the two floors is remarkably similar (Figures 3 and 4). Both floors show a primary distribution of hearths that forms a gentle arc following the wall of the shelter. Also, in both floors these are accompanied by another hearth-line found nearly equidistant from the backwall and dripline (in units C3–C5).

The quantitative attributes related to the sizes, shapes, and workspaces of hearths show no significant statistical differences between the two floors (Table 1). Moreover, a comparison of these data with similar data obtained from a cross-cultural survey of hearths from ethnographic and late Paleolithic contexts also shows there are no significant

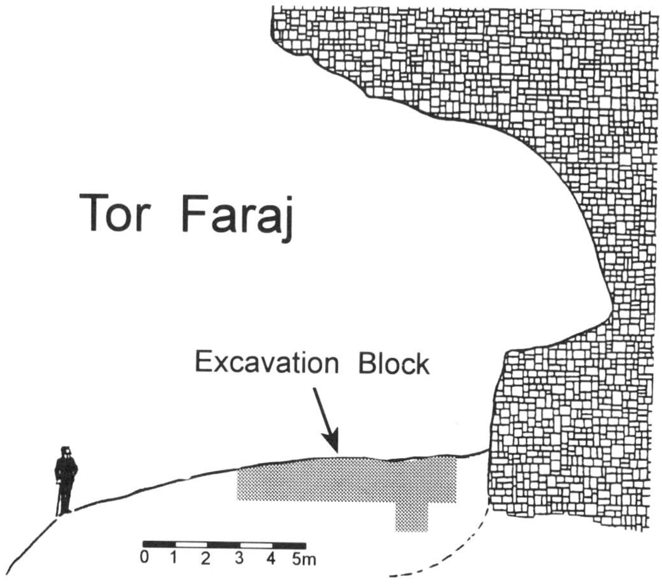

Figure 2. Schematic cross-section showing the location of the excavation block within the Tor Faraj rockshelter.

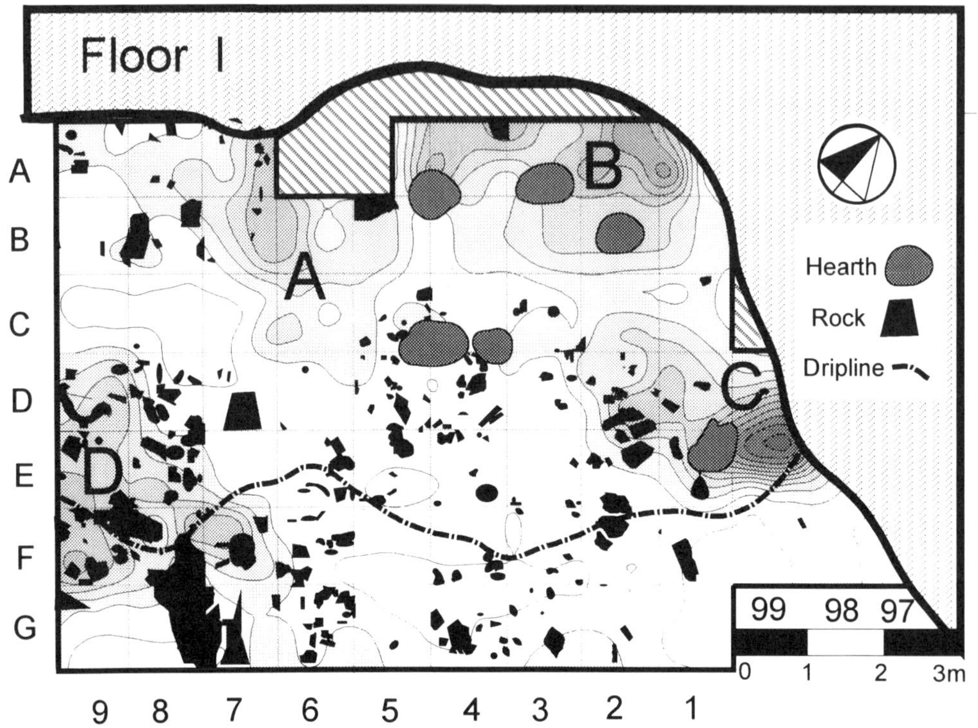

Figure 3. Plan of Floor I showing the distributions of chips (N=4,085) by density contours of 20 specimens, hearths and rocks. Note that the activity areas (A, B, C, and D) are defined by the presence of high chip densities and hearths.

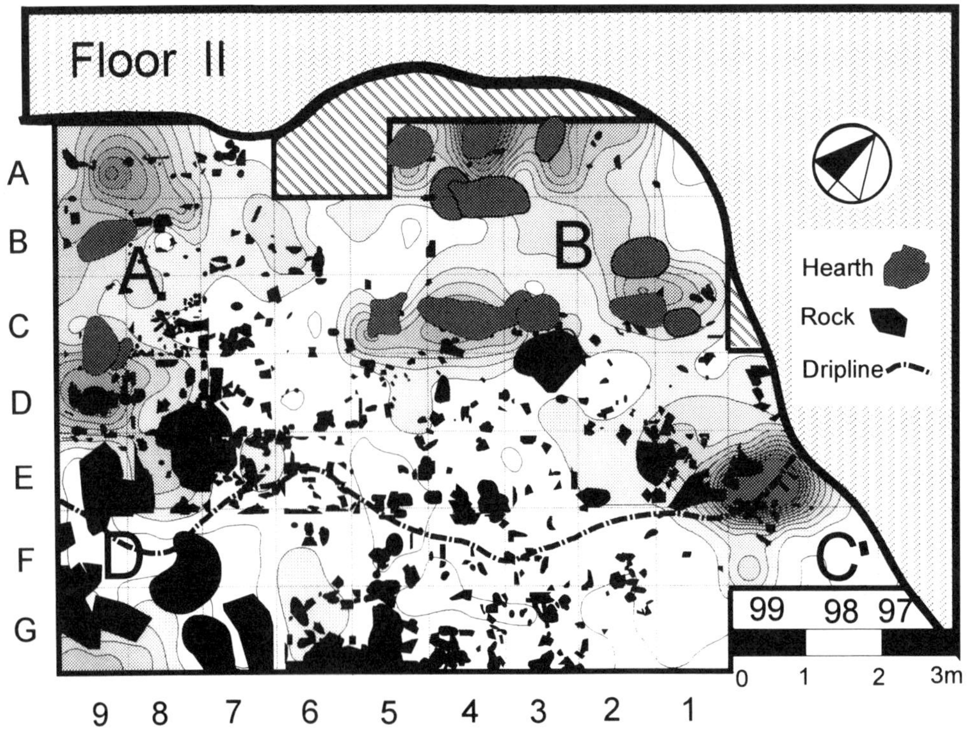

Figure 4. Plan of Floor II showing the distributions of chips (N=4,953) by density contours of 20 specimens, hearths and rocks. Note that the activity areas (A, B, C, and D) are defined by the presence of high chip densities and hearths.

Table 1. Comparison of positional data for hearths from Tor Faraj with hearths from late Paleolithic and ethnographic contexts[a]

Site/floor	No. of hearths	Shape of hearth	Diameter of hearth (cm)	Distance (m)					
				1st hearth	2nd hearth	3rd hearth	Near wall	Back wall	Drip line
Tor Faraj									
Floor I	6								
1a		round	50	1	2	2.5	1.5	1.8	3.5
2		round	60	1	1.5	2.1		1.3	4.5
3		round	50	1.5	2.1	2		1.6	4.4
6		oval		1	2	2.3		3.6	2
7		round	50	1	2.1	2.2		3.5	
8		round	80	3	3.1	4	1	4.5	
Floor II	14								
1		oval		0.5	0.8	1.3	1.5	2.5	3.3
4		round	50	1.5	1.8	2		3.1	2
5		round	75	0.8	2	2.1		1.5	4.5
9		round	50	0.8	1.5	1.6		0.8	5.5
10		oval		1.5	1.7	3.8		1.5	3.5
11		oval		0.8	1	1.1		1.5	5
12		round	60	1	1.2	0.8		0.8	5
14		round	50	1	1.4	1.5	2.8	3	2.8
15		oval		1	1.5	2.3		3	2.9
17		round	50	0.3	1.5	3.8		3	2
18		round	50	0.8	1	2	1	2.5	2.4
19		round	50	0.9	1	1.5		1	4.5
20		round	75	0.8	1.1	1.6	1	2	3.1
21		oval		0.5	1.5	3.5		3	2
Comparison of mean data from Tor Faraj with mean cross-cultural data									
Tor Faraj	20	Mean	57.1	1	1.6	2.2	1.3	2.3	3.5
		Std. dev.	10.8	0.5	0.5	0.9	0.8	1	1.1
Cross-cultural	75	Mean	51.7	2.2	3.1	3.2	2	2.5	2.4
		Std. dev.	19.8	1.2	1.8	1.3	1.2	1.5	2.1

[a]The cross-cultural data come from 15 sites containing 25 living floors found within exogene rockshelters located in North America (Thomas 1983; Flannery 1986), Australia (Nicholson and Cane 1991), Africa (Clark and Walton 1962), and Europe (Hahn 1984; Kind 1984).

statistical differences between the data sets. As in the cross-cultural sample, the Tor Faraj hearths are mostly (70%) round in shape with diameters averaging about one-half meter. All of the Tor Faraj hearths were kindled in shallow depressions dug in the floor of the shelter and none were rock-lined. Although the absence or paucity of rock-lined hearths in Mousterian contexts is often interpreted as a characteristic of archaic behaviors, these cross-cultural data indicate that rock lining is also rarely employed (3% of the sample) by modern foragers.

While not statistically significant, the mean spacing between hearths and between hearths and the near wall in Tor Faraj is less than in the cross-cultural sample. This difference, however, would appear to be an expression of the inverse relationship that exists between the numbers of hearths and the distances by which they are separated. When the cross-cultural sample is limited to only those floors with six or more hearths (thus resembling the Tor Faraj floors), the mean distance to the first hearth is 1.5 m (std .7 m), compared to a mean distance of 1 m (std .5 m) for Tor Faraj. The other spacing attributes (i.e., distances to back wall and dripline) recorded at Tor Faraj are also remarkably similar to the cross-cultural sample.

The arrangement of hearths across the Tor Faraj living floors indicates that (a) the basic site structure remained largely the same for the two occupations and (b) the structure is statistically indistinguishable from that reported for modern foragers. This structure, accompanied by the remarkable regularity in spacing, is more than merely an expression of the working space required by the human body. The mechanical dimensions of the human body undeniably dictate the amount of free space necessary for conducting work, but it was only through forethought that the hearths were placed in specific positions such that they were roughly equidistant from obstacles or conditions that would intrude upon this workspace.

The regularity in the spacing of hearths at Tor Faraj clearly reflects the anticipation of the shelter's occupants that 1–3 m of working space would be needed around each hearth. While the directions of wind shifts or the numbers of those who would join a hearth might not be known, such changes were anticipated and accommodated by a very specific hearth placement.

5.2. Artifact Distributions

The distributions of artifacts across the two floors show a common pattern in the general arrangements of their concentrations relative to the natural constraints (walls and dripline) of the shelter. Although most of the artifact categories show a similar pattern, chips (i.e., flakes <25 mm maximum dimension) offer the finest resolution in defining the concentrations because they form such a large sample, numbering over 9,000 specimens. Trend-line analysis of chips shows the consistent presence of three to four major clusters (A, B, C, and D; Figures 3 and 4). They are positioned within an approximately 3 m wide swath that parallels the wall of the shelter and rests behind the dripline. This swath is nearly identical to that defined by the hearth line.

The clusters of chips also show redundancy between levels and floors in their relative sizes and positions. The largest cluster (B) is always found centrally positioned in the shelter's arc, followed by a smaller cluster (A) to its west and an even smaller, often very compact, cluster (C) to the east. Throughout the occupations of Floor II, Cluster A shows a clear separation from Cluster B, but during Floor I times Cluster A had come to adjoin Cluster B. Cluster C, although always the smallest of the clusters, is typically defined by very high chip densities. Cluster C also shows the least variability in positioning, for its peak concentration is normally centered within a 2 m area at the juncture of the dripline and backwall. Cluster D, lacking the artifact density and definition of the other clusters, appears in the area of roof-fall in the southwest corner of the excavation.

The distribution of artifact clusters parallels the "drop and toss" model as advanced by Thomas (1983) for exogene rockshelter occupations. Chips, representing the kinds of small residue expected in the drop zone, are virtually restricted to that portion of the shelter found inside the dripline. And the chip concentrations also show strong associations with individual hearths and with the swath forming the hearth line.

At a finer level of analysis, the clusters show clear and consistent differences in the proportions of Levallois cores and points contained within them (Figures 5 and 6). In Floor I, points dominate Clusters A, B, and C, whereas only cores are found within Cluster D. In Floor II, points again dominate Clusters B and C, but cores are dominant along the western edge of the excavation block within Cluster A.

These differences in the proportionate representations of cores and points indicate fundamental differences in the activities undertaken within the clusters. While caution needs to be exercised when attempting to define the spatial loci of certain behaviors from the find-spots of artifacts, the artifacts found within the clusters at Tor Faraj do appear to be

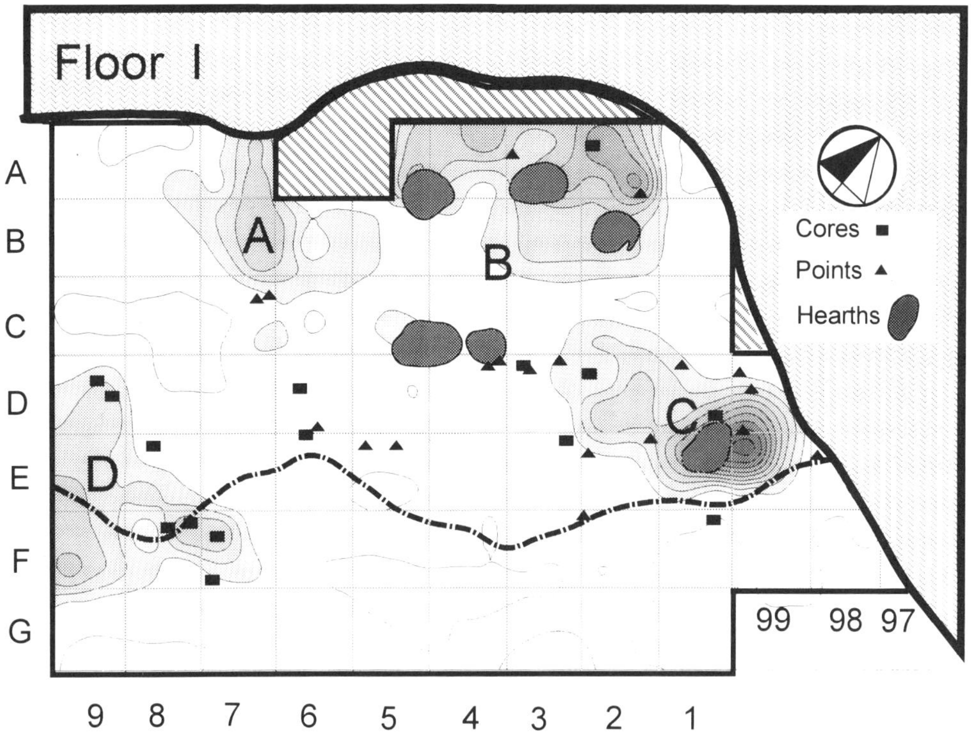

Figure 5. Plan of Floor I showing the relative distributions of Levallois point cores and Levallois points. Note the dominance of points in activity areas A, B, and C and the dominance of cores in area D.

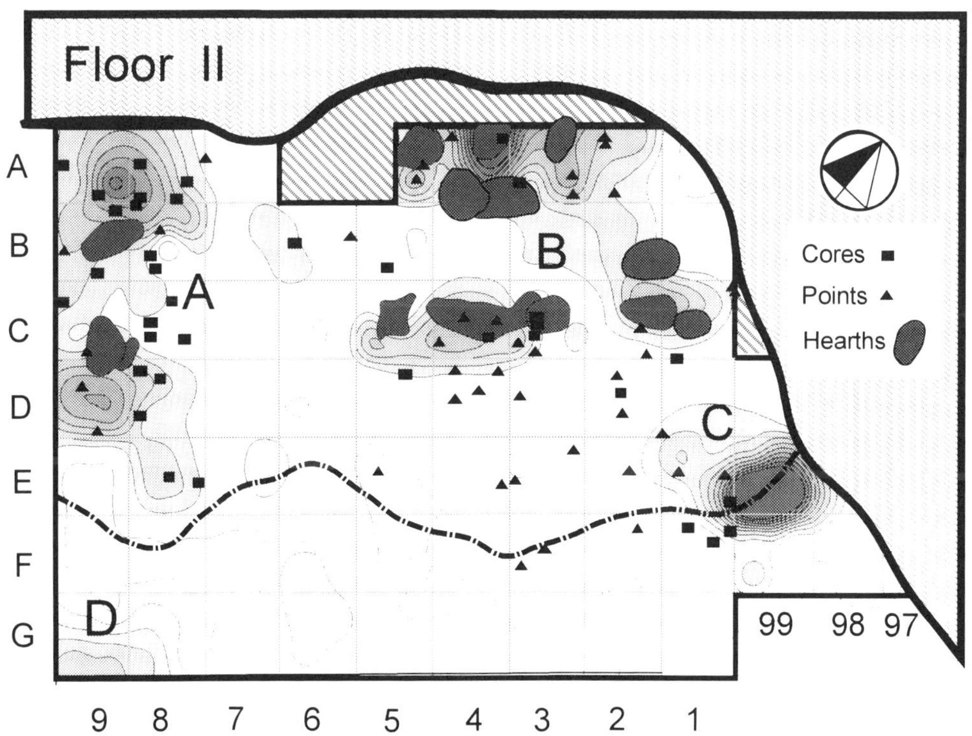

Figure 6. Plan of Floor II showing the relative distributions of cores and points. Note the dominance of points in activity areas B and C and the dominance of cores in area A.

behaviorally linked for several reasons. First, the cores and points are spatially associated with clusters that are defined on the basis of concentrations of chips that are often accompanied by hearths. This implies that the clusters are indeed in primary context. Secondly, the cores and points rest in what is widely recognized as a drop zone. This implies that they were unlikely to have been discarded there as refuse or residue from another activity area. Thirdly, the spatial pattern associated with the cores and points is remarkably redundant, not only between floors, but between the 5 cm levels composing the floors. If the points and cores are in primary context as argued, Cluster D (Floor I) and Cluster A (Floor II), both positioned peripheral to the central hearth areas of the floors, would have been associated largely with the fabrication of cores and the primary production of points. The points, however, would have been used, modified, and refurbished mainly in the central part of the shelter in the areas of Clusters B and C and of course off-site.

5.3. Phosphorus Concentrations

Only a few scraps of bone and tooth enamel have been preserved in the Tor Faraj deposit and these are confined to a thin zone rich in carbonates that veneers bedrock. Ostrich eggshell fragments and opal phytoliths furnish the only other physical evidence of organic remains. In an effort to gain some additional information on the spatial distribution of organic residue within the shelter, sediment samples from Floor II were analyzed for phosphorus following the methods developed by Eidt (1984). Eidt has found phosphorus values to reflect the degree to which organic residue was introduced to sediments. The sediment samples from the Tor Faraj deposit were collected from the center of each excavation unit for two levels (180 and 185) within Floor II. The phosphorus values (total P) for the samples ranged from 500–4,880 ppm. A trend-line analysis of these data identifies areas of phosphorus concentrations that strongly match the clusters as defined by chip concentrations (Figure 7). The highest values form a swath along the eastern and central portions of the shelter's wall, largely encompassing artifact Clusters B and C. Peak phosphorus values also occur in the area of artifact Cluster A, but these are much lower than those recorded in the eastern and central portions of the shelter.

During collection of the sediment samples, care was taken to avoid contamination, especially from hearths. Although many of the areas of phosphorus concentrations do overlap with hearths, only one peak in phosphorus values corresponds to a hearth (unit C1). Furthermore, several hearths occur in areas of low phosphorus values. The areas of high phosphorus values indicate that organic materials were used or discarded to a much greater extent in Clusters B and C than elsewhere within the shelter.

6. INFERENCES BEYOND HUMAN MODERNITY

The patterns observed for hearths, artifact distributions, and phosphorus concentrations indicate a site structure at Tor Faraj that differs little from the structures observed for modern foragers inhabiting rockshelters. The spacing of the hearths relative to other hearths and to the natural features of the shelter is statistically indistinguishable from that recorded for modern foragers. Artifact distributions reveal concentrations of fine residue (chips) indicative of intensive activity areas around hearths, resembling the drop zones recognized for modern foragers. Beyond the dripline, artifacts as a whole decline in frequency, but fine residue is virtually absent. This again resembles that aspect of the site structure of modern foragers inhabiting shelters as reflected in a drop zone inside the dripline and a toss zone outside the shelter.

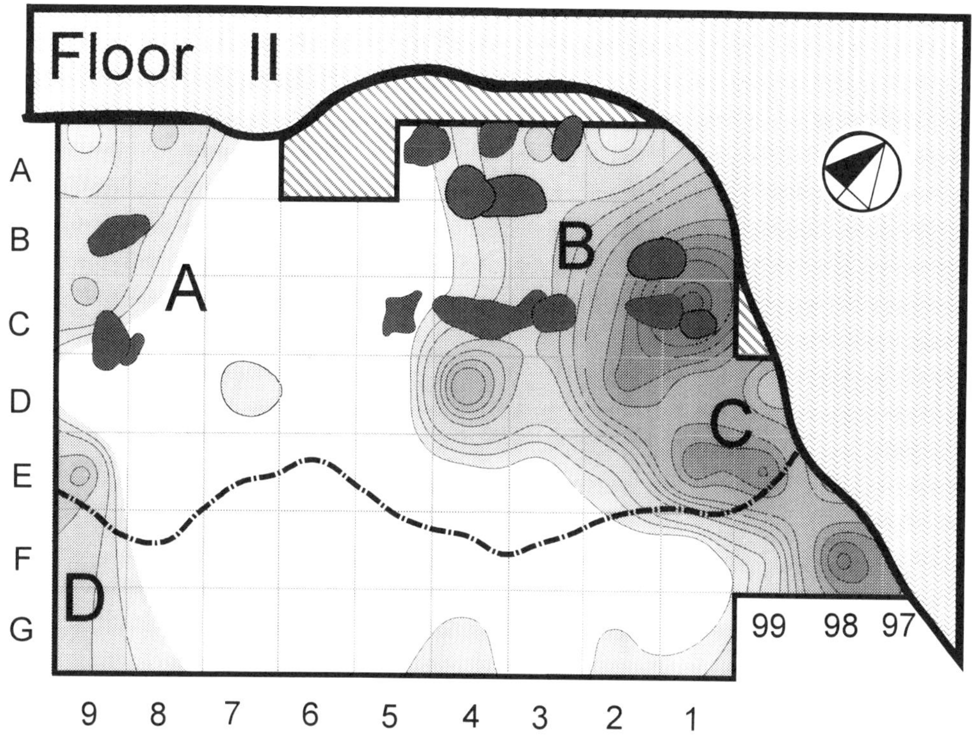

Figure 7. Plan of Floor II showing the distribution of phosphorus values relative to hearths and activity areas. The phosphorus values are reflected in contour intervals of 250 ppm and a base contour of 1,250 ppm. Note that the highest concentrations of phosphorus, found in areas A and C, show strong spatial correlations with the highest chip densities of the floor (see Figure 4).

In light of cautions regarding the uncritical use of ethnographic or modern analogues as aids in the interpretation of the prehistoric record, especially when potentially linked to archaic behaviors, the question emerges as to whether the site structure of Tor Faraj can be interpreted accurately at an even finer resolution. I think it can. Given that the basic site structure of Tor Faraj is not distinguishable from that of modern foragers, it seems reasonable that more specific relationships within this structure might be compared to similar relationships observed among modern foragers. These specific relationships are largely associated with both the behavioral organization of the occupation as a whole and the nature of the specific activity areas encompassed within this organization.

The complexity of the site structures of Floors I and II points to behaviors that supported a logistically organized base camp. In contrast to a site structure composed of a hearth surrounded by an area of intensive but largely redundant activities, the occupational floors of Tor Faraj point to at least three and perhaps four major areas of intensive activities that differed from one another (Table 2). The archaeological expressions of these differences are seen in the variability between clusters as related to their sizes, number of hearths, numbers of micro-clusters, core:point ratios, and phosphorus concentrations. Here, it is important to recall that the intrasite structures of occupations that had been organized logistically would be expected to display a greater number of activity areas and greater diversity between these areas than occupations that were organized opportunistically (Binford 1983:191; O'Connell 1987:104; Simek 1987:3).

Table 2. A comparison of the artifact clusters at Tor Faraj for Floors I and II

	Cluster			
	A	B	C	D
Floor I				
Size	Small	Large	Small	Medium
No. of hearths	0	5	1	0
Micro-clusters	1	2	2	2
Phosphorus	no data			
Core:point ratio	0:2	0.4:1	0.4:1	7:0
Wear functions	no data			
Floor II				
Size	Medium	Large	Small	Smallest
No. of hearths	0–2	5–11	0–1	0
Micro-clusters	1–2	3–5	1–2	1
Phosphorus	Moderate	Low-high	High	Low
Core:point ratio	3.5:1	0.4:1	0.6:1	0
Wear functions	1	5	0	0

A closer look at the characteristics of the artifact clusters or activity areas suggests that the largest, most diverse of these (Cluster B) is likely to have served as the focus for domestic tasks related to food preparation, cooking, maintenance activities, and sleeping. The greater number of hearths and high concentrations of organic residue is consistent with the hearth-side activities of cooking, eating, and sleeping. Preliminary results of a microscopic wear analysis study show striking differences between the clusters and furnish additional evidence for the activity diversity of Cluster B. Overall there is a low intensity of utilization, but all of the Levallois points show wear and traces of hafting. Artifacts showing wear in Cluster A are associated exclusively with butchery, whereas diverse tasks including green-wood working, hide-working, bone-working, and butchery were undertaken in Cluster B.

7. SUMMARY

The distributions of hearths, artifacts, and occupational residue across two stratified living floors at Tor Faraj reveal similar, complex site structures that are positioned nearly identically within the shelter. Each of the structures contains a central large artifact concentration (Cluster B) associated with numerous hearths and 2–3 outlying concentrations (Clusters A, C, D) with few or no hearths. The central concentration also encompasses several smaller concentrations or micro-clusters, while the other concentrations show only one or two peaks in artifact density. Phosphate concentrations largely parallel the artifact distributions. The highest phosphate values are found in the area of Cluster B and imply that this was an area most used in processing and/or discarding organics and for bedding. Artifact analysis identifies fundamental differences within the site's structure. Cluster A stands out as an area for core preparation, blank production, and butchery. In contrast, Cluster B served primarily as an area of tool use and maintenance accompanied by a wide range of activities including butchery and the working of wood, bone, hide, and tough green plant such as reed or palm.

The complexity of the site structure of Tor Faraj, as expressed in numerous and diverse activity loci, suggests a logistically organized encampment rather than one organized opportunistically. Comparison of the site structures of Tor Faraj and its smaller, upland counterpart of Tor Sabiha underlines these differences. It would appear then that within the Late Levantine Mousterian of southern Jordan, both logistical and opportunistic strategies were adopted situationally as adaptive responses to different environmental settings. Clearly, the hominids who occupied these sites some 70K years ago had the capacity to both anticipate needs beyond their immediate future and to significantly alter their adaptive strategies in response to differences in their environment. The rather detailed reconstructions of their inter- and intrasite patterns fail to show any significant differences with those of fully modern humans. While this does not necessarily mean that they thought as we do, it does mean that they used an extended planning depth, a flexibility in behavior, and a logistical settlement-procurement strategy to cope. This suggests that these characteristics are either not useful indicators of modern humanness, or that modern humans were extant in southern Jordan some 70K years ago.

REFERENCES

Binford, L. 1973. Interassemblages variability: The Mousterian and the "functional" argument, 227–54. In *The Explanation of Cultural Change: Models in Prehistory*, ed. C. Renfrew. London: Duckworth.

Binford, L. 1977. Forty-seven trips: A case study in the character of archaeological formation processes, 24–36. In *Stone Tools as Cultural Markers*, ed. E. Hall. Canberra: Australian Institute of Aboriginal Studies.

Binford, L. 1978. *Nunamiut Ethnoarchaeology*. New York: Academic Press.

Binford, L. 1979. Organization and formation processes: Looking at curated technologies. *Journal of Anthropological Research* 35:255–73.

Binford, L. 1980. Willow smoke and dogs' tails: Hunter-gatherer settlement systems and archaeological site formation processes. *American Antiquity* 45:4–20.

Binford, L. 1983. *In Pursuit of the Past*. New York: Thames and Hudson.

Binford, L. 1989. Isolating the transition to cultural adaptations: An organizational approach, 18–41. In *The Emergence of Modern Humans*, ed. E. Trinkaus. Cambridge: Cambridge University Press.

Butzer, K. 1982. *Archaeology as Human Ecology*. Cambridge: Cambridge University Press.

Clark, J. D., and J. Walton. 1962. A late stone age site in the Erongo mountains, South West Africa. *Proceedings of the Prehsitoric Society* 28:1–16.

Eidt, R. 1984. Theoretical and practical considerations in the analysis of anthrosols, 155–90. In *Archaeological Geology*, ed. G. Rapp, Jr., G. Gifford, and J. Gifford. New Haven: Yale University Press.

Flannery, K. 1986. Spatial analysis of Guila Naquitz living floors: An introduction to Part VI, 319–30. In *Guilá Naquitz*, ed. K. Flannery. New York: Academic Press.

Gamble, C. 1991. An introduction to the living spaces of mobile peoples, 1–24. In *Ethnoarchaeological Approaches to Mobile Campsites*, ed. C. Gamble and W. Bosmier. Ann Arbor: International Monographs in Prehistory.

Geneste, J. M. 1989. Economie des ressources dans le Moustérien du Sud-Ouest de la France, 75–97. In *L'homme de Néandertal* Vol. 6, ed. M. Otte. Liège: ERAUL.

Gilead, I., and C. Grigson. 1984. Far'ah II, a Middle Palaeolithic open-air site in the Northern Negev, Israel. *Proceedings of the Prehistoric Society* 50:71–97.

Griffin, P. 1984. Forager resource and land use in the humid tropics: The Agta of northeastern Luzon, the Phillippines, 95–121. In *Past and Present in Hunter-Gatherer Studies*, ed. C. Schrire. Orlando: Academic Press.

Hahn, J. 1984. Spatial organization and occupation of the Helga-Abri, near Schelklingen, Swabian Jura, 79–88. In *Upper Palaeolithic Settlement Patterns in Europe,* ed. H. Berke, J. Hahn, and C. J. Kind. Tübingen: Verlag Archaeologica Venatoria, Institut für Urgeschichte, Der Universität Tübingen.

Hassan, F. 1981. *Demographic Archaeology*. New York: Academic Press.

Henry, D. 1992. Transhumance during the late Levantine Mousterian, 143–62. In *The Middle Paleolithic: Adaptation, Behavior, and Variability*, ed. H. Dibble and P. Mellars. Philadelphia: The University Museum of the University of Pennsylvania.

Henry, D. 1994. Prehistoric cultural ecology in southern Jordan. *Science* 265:336–41.

Henry, D. 1995a. *Prehistoric Cultural Ecology and Evolution: Insights From Southern Jordan*. New York: Plenum Press.

Henry, D. 1995b. The influence of mobility levels on Levallois point production, Late Levantine Mousterian, Southern Jordan, 185–200. In *The Definition and Interpretation of Levallois Technology*, ed. H. Dibble and O. Bar-Yosef. Madison: Prehistory Press.

Henry, D., S. Hall, H. Hietala, Y. Demidenko, V. Usik, A. Rosen, and P. Thomas. 1996. Middle Paleolithic behavioral organization: 1993 excavation of Tor Faraj, southern Jordan. *Journal of Field Archaeology* 23:1–23.

Hietala, H., and A. Marks. 1981 Changes in spatial organization at the Middle to Upper Paleolithic transitional site of Boker Tachtit, Central Negev, Israel, 305–18. In *Préhistoire du Levant*, ed. J. Cauvin and P. Sanlaville. Paris: CNRS Editions.

Hietala, H., and D. Stevens, 1977. Spatial analysis: Multiple procedures in pattern recognition studies. *American Antiquity* 42:539–59.

Jochim, M. 1976. *Hunter-Gatherer Subsistence and Settlement: A Predictive Model*. New York: Academic Press.

Kent, S. 1991. The relationship between mobility strategies and site structure, 33–60. In *The Interpretation of Archaeological Spatial Patterning*, ed. E. Kroll and T. Price. New York: Plenum Press.

Kind, C. J. 1984. The Felstalle, a rockshelter near Ehingen-Mühlen, Alb-Donau-Kreis (Southwest Germany) with particular emphasis on the occupational structures from the Magdelenian horizon IIIb, 89–97. In *Upper Palaeolithic Setttlement Patterns in Europe*, ed. H. Berke, J. Hahn, and C. J. Kind. Tübingen: Verlag Archaeologica Venatoria, Institut für Urgeschichte, Der Universität Tübingen

Klein, R. 1992. The archaeology of modern human origins. *Evolutionary Anthropology* 1:5–14.

Klein, R. 1995. Anatomy, behavior, and modern human origins. *Journal of World Prehistory* 9:167–98.

Kuhn, S. 1990. Diversity within Uniformity: Tool Manufacture and Use in the 'Pontinian' Mousterian of Latium (Italy). Ph. D. diss. (unpublished). University of New Mexico. Albuquerque.

Kuhn, S. 1992. On planning and curated technologies in the Middle Paleolithic. *Journal of Anthropological Research* 48:185–214.

Leroi-Gourhan, A., and M. Brezillon. 1972. *Fouilles de Pincevent d'Analyse Ethnographique d'un Habitat Magdalénien*. Paris: CNRS Editions.

Lieberman, D. 1993. The rise and fall of hunter-gatherer seasonal mobility: The case of the southern Levant. *Current Anthropology* 34:599–631.

Marks, A., and D. Friedel. 1977. Prehistoric settlement patterns in the Avdat/Aqev area, 131–58. In *Prehistory and Paleoenvironments in the Central Negev, Israel*, Vol. 2, ed. A. E. Marks. Dallas: SMU Press.

Munday, F. 1976. Intersite variability in the Mousterian occupation of the Avdat/Aqev area, 113–40. In *Prehistory and Paleoenvironments in the Central Negev, Israel*, Vol. 1. ed. A. E. Marks. Dallas: SMU Press.

Nicholson, A., and S. Cane. 1991. Desert camps: Analysis of Australian Aboriginal proto-historic campsites, 263–354. In *Ethnoarchaeological Approaches to Mobile Campsites*, ed. C. Gamble and W. Bosmier. Ann Arbor: International Monographs in Prehistory.

O'Connell, J. 1987. Alyawara site structure and its archaeological implications. *American Antiquity* 52:74–108.

Rigaud, J. P. 1978. The significance of variability among lithic artifacts: A specific case from southwestern France. *Journal of Anthropological Research* 34:299–310.

Roebroeks, W., J. Kolen, and E. Rensink. 1988. Planning depth, anticipation and the organization of Middle Paleolithic technology: The "archaic natives" meet Eve's descendants. *Helinium* 28:17–34.

Saragusti, I., and N. Goren-Inbar. 1990. Conjoinable pieces from the lithic assemblage of Quneitra, Area B, 173–88. In *Quneitra: A Mousterian Site on the Golan Heights*, ed. N. Goren-Inbar. QEDEM 31. Jerusalem: Institute of Archaeology, Hebrew University of Jerusalem.

Shea, J. 1989. A functional study of the lithic industries associated with hominid fossils in the Kebara and Qafzeh Caves, Israel, 611–25. In *The Human Revolution*, ed. P. A. Mellars and C. B. Stringer. Edinburg: Edinburgh University Press.

Simek, J. 1987. Spatial order and behavioural change in the French Paleolithic. *Antiquity* 61:25–40.

Stringer, C., and C. Gamble. 1993. *In Search of the Neanderthals*. London: Thames and Hudson.

Thomas, D. 1983. *The Archaeology of Monitor Valley: Gatecliff Shelter*. Anthropological Papers, Vol. 59, Part 1. New York: American Museum of Natural History.

Vita-Finzi, C., and E. Higgs. 1972. Prehistoric economy in the Mount Carmel area of Palestine: Site catchment analysis. *Proceedings of the Prehistoric Society* 36:1–37.

10

THE LITHIC ASSEMBLAGES OF AMUD CAVE

Implications for Understanding the End of the Mousterian in the Levant

Erella Hovers

Institute of Archaeology
Hebrew University
Mt. Scopus, Jerusalem 91905, Israel

1. INTRODUCTION

The Middle-Upper Paleolithic transition in the Levant is a cultural phase of short duration. Indeed, its actual occurrence as reflected in the region's archaeological record is debated. The blade industries that mark the onset of the Levantine Upper Paleolithic were viewed for a long time as a local development, stemming either directly out of the later stages ("Tabun B-type") of the Levantine Middle Paleolithic (Garrod 1951, 1955, 1957; Copeland 1986:11), or from an early lithic tradition ("Tabun D-type") that had been retained over the millennia (Marks 1993, and references therein). More recently it was suggested that transitional assemblages are virtually absent from the Levant, as even the lowest level of Boker Tachtit, the type locality for the alleged transitional assemblages, is argued to contain a full-fledged Upper Paleolithic occupation (Bar-Yosef et al. 1996). The latter scenario suggests that the fundamental change in lithic technology occurred during an extremely brief time span, between about 48 ky ago, the latest known date for a Mousterian assemblage in the Levant (Kebara VI), and about 45 ky, the date of Boker Tachtit. According to this view, the abrupt change in lithic technology between the Late Mousterian and the Early Upper Paleolithic may have been the result of "a rapid acculturation" rather than an autochtonous process (Bar-Yosef 1994:41; and cf. Klein 1995, this volume).

Because "[I]nnovations involve . . . little more than new ways of doing old things" (Renfrew 1978), 'pre-adaptations' for accepting such new ways are thought to have occurred among Late Mousterian populations, permitting the swift shift in lithic technology between the Middle and Upper Paleolithic periods. These pre-adaptations would have enabled the assimilation of less familiar technological traits through either cultural contact with extraneous groups or autochthonous groups. In the former case, such pre-adaptations would allow easy incorporation of external innovations into the normative technological

Neandertals and Modern Humans in Western Asia, edited by Akazawa *et al.*
Plenum Press, New York, 1998.

repertoire of a population. In the latter, they might have facilitated selection for technological inventions which had already risen within the local population itself (ibid.).

The present article examines aspects of the Middle-Upper Paleolithic transition in the Levant from the perspective of the earlier period. The assemblages of interest are therefore those that are the immediate predecessors of the early Upper Paleolithic, namely, those derived from Late Mousterian contexts as determined by radiometric dates. The newly-recovered lithic assemblages from Amud Cave, dated to about 65–55 ky BP (Rink et al., in prep.; Valladas et al., in press), are a recent addition to this group.

The original stratigraphic framework of the Amud Cave sequence (Chinzei 1970), while refined, was nonetheless confirmed in the course of recent excavations at the site (Hovers et al. 1991; Hovers et al. 1995). The Paleolithic deposit, designated as unit B, is divided into four layers. The matrix of each layer consists mainly of fine-grained silty ashes of anthropic origins, previously misinterpreted as alternating black soils and gray concretions (Chinzei 1970). In several instances, mainly in the older layer designated as B4, the outline of hearths could be discerned. This unit is separated from the overlying artifact-bearing sediments of layers B2 and B1 by an archaeologically sterile horizon (B3) composed mainly of angular *éboulis*, which originated from initial stages of wall and ceiling collapse. Layer B3 thus represents an occupational gap between layer B4 and layers B2–B1. This hiatus is also documented by the TL dates (Valladas et al., in press), which suggest a cessation in the occupation of the site for possibly several thousand years.

Hominid remains from Amud Cave were proclaimed to be a transitional form between Neandertals and modern humans due to the occurrence of a few traits that are uncharacteristic of the former group (Suzuki 1970). Such identification was, at the time, compatible with the assignment of a transitional Middle-Upper Paleolithic age to the occupation, based on characteristics of the lithic assemblage. These include relatively high frequencies of UP tools, the alleged occurrence of chamfered pieces, and lack of "classical" (i.e., centripetal) Levallois cores and products (Watanabe 1964, 1965, 1970; Copeland 1975). Later research has suggested that Amud Cave was a Neandertal occupation (e.g., Rak 1993; Lavi 1994) and that the industry was Mousterian (Akazawa 1979; Ohnuma 1992).

2. THE LITHIC ASSEMBLAGES OF AMUD CAVE

Based on typological grounds, Watanabe (1970:94) concluded that the lithic samples from the various archaeological horizons at Amud represented "a single industrial cycle." This statement is not surprising given that typological homogeneity characterizes Levantine Mousterian assemblages, all of which fall within the parameters of Bordes' (1972) 'typical Mousterian.' However, because retouched tools represent only the last stage of raw material manipulation, typological classification does not normally yield information about the dynamic manufacture process preceding this last stage (but see Dibble 1987). Moreover, it has long been established that the main dimension of lithic variability in the Levantine Mousterian is expressed through variations in knapping methods (but see Marks 1992a on typological variability), viewed as corresponding with the three technological facies represented in the Tabun sequence (Copeland 1975, 1986; Ronen 1979; Marks 1992a; Meignen and Bar-Yosef 1992; and see below).

The Amud lithics were not easily accommodated within the tripartite techno-chronological model of the Levantine Mousterian. Jelinek (1982:99) argued that the assemblage was of a "Tabun D-type," due to its elongated aspect. Other researchers have characterized the assemblage as being of "Tabun B-type" due to the intensive use of unipolar convergent

Levallois flaking methods, which are argued to be typical of this variant (Ohnuma 1992; Meignen 1995). Combined with the evidence for an occupational hiatus at the cave, such opposing interpretations beg the question of technological homogeneity of the Amud lithics. This issue was addressed by analysis of the operational sequences (*chaînes opératoires*), which reflect both the mental templates of their producers and the specific knowledge (*savoir-faire*) of the series of mechanical actions necessary to obtain the desired results (Leroi-Gourhan 1943, 1945; Pelegrin 1990; Karlin et al. 1991; Perlès 1992). The analysis made use of Boëda's (1993, 1995) scheme for characterizing the diversity encountered within Levallois flaking systems.

2.1. The Samples

A restricted sample (n=423) from a single square of layer B4 was described by Ohnuma (1992). His sample from layer B2 (n=1,253), obtained from the same square, was also re-analyzed, along with a small sample (n=474) recently excavated from an adjacent square (Shimron 1996). A larger sample (n=1,248), derived from two adjacent squares, was obtained from layer B1 (Goder 1997; Goder and Hovers in prep.). Initially retrieved as separate samples from two sedimentological horizons, the samples from the two sets proved highly similar and are combined here into one sample.

Not all the quantitative data discussed below appear in Ohnuma's report, hence some of those mentioned here are pertinent only to layer B1 and to the recently obtained, partial sample from layer B2. Whenever possible, however, data from the two separate samples of layer B2 were agglomerated. Such instances are specified in the text.

2.2. Core Reduction Sequences

Lithic raw material was obtained mostly from wadi pebbles and/or from sources further away from the site at all times of occupation. Flint beds are common within the limestone cliffs surrounding the cave, but they seem not to have been extensively exploited by the cave's occupants due to both the low quality of the local flint and the difficulties of quarrying it from the cliff face. The compositions of all three assemblages indicate that raw material exploitation was similar throughout time. All the samples share the high ratio of débitage and tools per core, which is primarily the result of low absolute frequencies of the latter (Table 1). Fully cortical (100% cortical cover) elements are rare in all three assemblages (Table 1), suggesting that initial decortication of raw material nodules took place outside the cave.

2.2.1. Core Modification. Semi-cortical elements, among which naturally backed elements are prominent, occur in all three samples. The latter tend to be elongated (the average length/width ratio of the group is 2.1 in both B1 and B2 samples, with about 60% being of blade proportions). Over half of them exhibit cortex on their lateral and distal regions and have unipolar scar patterns. Overall, these characteristics suggest the use of unidirectional flaking in the first stages of core reduction. The elongated pieces are sometimes plunging (*outrepassants*) as a means to obtain the distal convexities that are necessary for Levallois removals already at the early stages of core reduction. *Eclats débordants* were rarely, if at all, detached during these stages, as indicated by the practical absence of cortical cover on their dorsal faces, characteristics of scar patterns, and their size distributions (see below).

2.2.2. Blank Production. The identical values of IL in the three samples reflect similarities in the extent of application of Levallois flaking. Levallois débitage is never quantitatively dominant in any of the samples (Table 2). Still, the production of Levallois débitage appears to have been the focus of lithic reduction in all the Amud assemblages, given that part of the non-Levallois elements present in Mousterian assemblages are by-products of Levallois flaking rather than products of independent reduction sequences (cf. Geneste 1985; Hovers 1997).

Ohnuma's (1992) 'discoidal cores,' as well as some of the alleged 'prismatic cores' (Ohnuma 1992), exhibit the geometric characteristics of Levallois technology (e.g., ibid.: figs. 7.3:2–3, 7.5:1; Boëda 1993), so that Levallois cores on nodules are relatively more abundant in layer B2 than in layer B4 (Ohnuma 1992: table 7.2) or in B1 (Goder 1997). However, the frequent use of cores-on-flake (Solecki and Solecki 1970; Newcomer and Hivernel-Guerre 1974; Goren-Inbar 1988) in the two latter assemblages (36.4% and 60% of the cores, respectively) is well within the conceptual framework of the Levallois system, as flakes exhibit a priori the two sub-parallel surfaces prerequisite for Levallois production (Boëda 1991, 1993; Delagne 1995; Hovers 1997). Thus, the majority of cores-on-flake were used to produce Levallois débitage. In layer B1, this particular *chaîne opératoire* was extended to the degree that the convex ventral faces of large, cortical hinged flakes, among the

Table 1. Composition of the lithic samples from the various layers at Amud Cave

	Layer B1[a]	Layer B2[b]		Layer B4[c]
Cores	26 (1.95)	[25[1]+6[2]]	31 (2.35)	9 (3.46)
Cortical pieces[d]	36 (2.70)	[28[1]+9[2]]	37 (2.80)	12 (4.61)
Levallois débitage[e]	282 (21.22)	[233[1]+105[2]]	338 (25.66)	86 (33.07)
Non-Levallois débitage	748 (56.28)	[420[1]+287[2]]	707 (53.68)	111 (42.69)
'Core Management Pieces'[f]	112 (8.42)	[70[1]+49[2]]	119 (9.03)	30 (11.53)
Retouched pieces[g]	125 (9.40)	[67[1]+18[2]]	85 (6.50)	12 (4.61)
Total	1329 (99.97)	[843[1]+474[2]]	1317 (100.02)	260 (99.98)
Flakes+tools:cores[h]	50.1:1	42.5:1		40.5:1

[a]Data after Goder 1997. Both complete and broken pieces are included.
[b]Compiled from Ohnuma 1992: tabs. 7.1, 7.9, and Shimron 1996 (see text). Ohnuma's counts (marked [1]) include only unbroken pieces; Shimron's numbers (marked [2]) include all pieces.
[c]After Ohnuma 1992: tabs. 7.1, 7.9. Only unbroken artifacts.
[d]Pieces with over 75% of surface covered with cortex (Goder 1997; Shimron 1996); in B4-over 80% (Ohnuma 1992).
[e]Excluding cortical elements and 'Core Management Pieces'
[f]The category of 'Core Management Pieces' consists of pseudo-Levallois flakes, core trimming elements (including ridge blades), naturally-backed débitage, *éclats débordants*, and *éclats outrepassés* (Hovers 1997). In Ohnuma's (1992) terminology these are referred to as 'naturally-backed débitage,' pseudo-Levallois flakes, and 'crested débitage,' the latter corresponding to '*éclat débordant*' (Ohnuma 1992:94).
[g]Excluding pieces with signs of use.
[h]Calculation includes the broken artifacts.

Table 2. Relative frequencies of unretouched Levallois products and values of IL in each sample

	Flakes		Blades		Points		Total		
	N	%	N	%	N	%	N	%	IL
B1	159	56.32	101	35.81	23	8.16	282	100.28	30.9
B2[1]	81	34.76	56	24.03	96	41.20	233	99.99	31.0
B2[2]	76	58.91	26	20.16	27	20.93	129	100.00	33.3
B2 (combined)	157	43.37	82	22.65	123	33.97	362	99.99	31.8
B4	27	31.39	26	30.23	33	38.37	86	99.98	31.4

[1]after Ohnuma 1992
[2]after Shimron 1996; counts of Levallois products do not include 'Core Management Pieces.'
IL calculated from Ohnuma's data is based only on complete pieces.

first to have been removed off a nodule, were used as the flaking surface for Levallois removals (Figure 1). Cores are also characterized by limited preparation, with the length of striking platforms usually less than half of the core circumference. The tendency for minimal investment in platform preparation was brought to extremes in cores-on-flake, where dorsal ridges of the blank were used as nervures guides for removals, without any further investment in shaping the core's geometry.

The distributions of various scar patterns on cores from layers B4 and B1 are more similar to each other than to B2. The final stages of core reduction were carried out by multiple (recurrent) and often convergent removals from each flaking surface (Figure 2), with detachment of Levallois pieces offset to the core's center. The similarity in core utilization is echoed in the distribution of scar patterns on Levallois products in the two layers, where convergent patterns are dominant. However, whereas flakes were the dominant

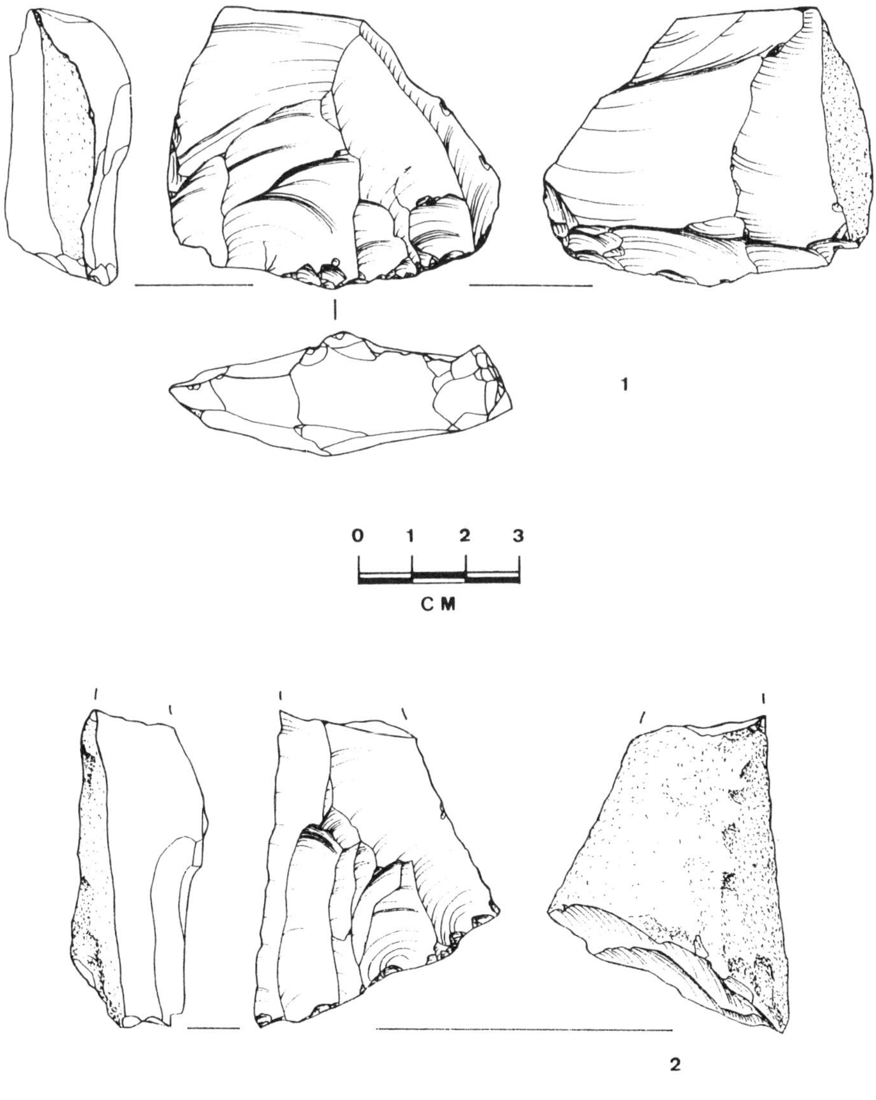

Figure 1. Cores on hinged flakes—layer B1.

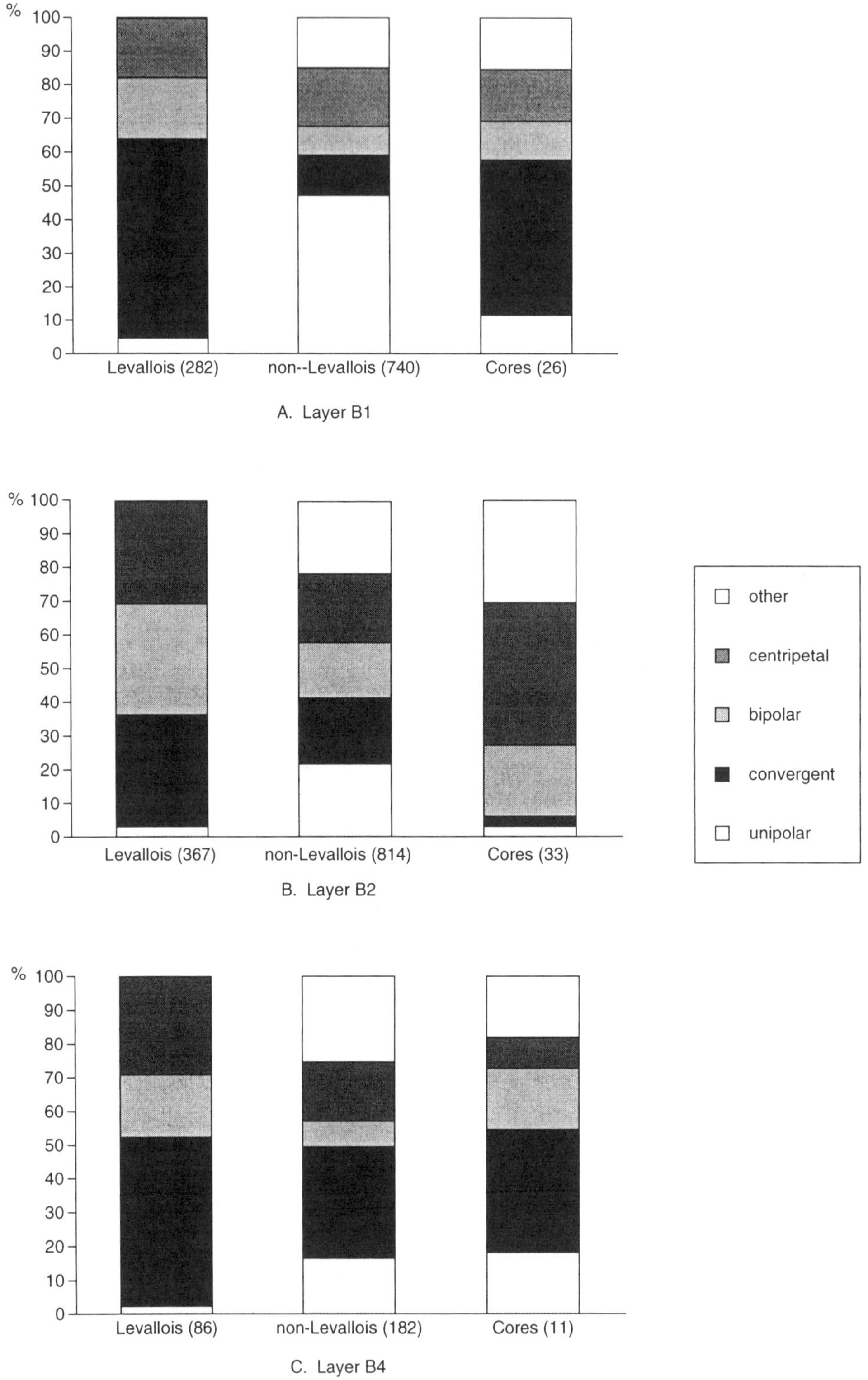

Figure 2. Distribution of scar patterns of various artifact classes of the Amud lithic samples (excluding 'core management pieces').

débitage type (of all flaking systems) produced throughout the sequence, the relative emphasis of Levallois production shifted from points and triangular flakes in layer B4 to a more pronounced preference for rectangular blades (and flakes) in layer B1 (Table 2). In the latter sample, too, Levallois points differ from the typical products of convergent recurrent methods (Meignen and Bar-Yosef 1992:139): arched 'Concorde' profiles are practically absent, whereas impact points appear to have been placed closer together in the course of core reduction. Hence dorsal face scar patterns, while convergent, tend towards more parallel patterns. Points are thus usually elongated (the mean length/width ratio being 1.85; n=18), with over 30% exhibiting blade proportions. Broad-based points, with well-prepared *chapeau de gendarme* platforms, are extremely rare in layer B1 (Table 3), where many of the pointed artifacts are 'leaf shaped' (Watanabe 1965). The former are more common in layer B4 (cf. Ohnuma 1992: fig. 7.4:8–9), where they tend to be squat.

In contrast to layers B4 and B1, relatively high frequencies of centripetal removals on the cores are encountered in the (combined) sample of layer B2 (Figure 2). Nevertheless, the frequencies of scar patterns in the débitage are more evenly distributed within both the Levallois and non-Levallois subassemblages of this layer (Figure 2), and about 30% of all débitage are pointed or triangular in shape. As indicated by the arrangements of scar patterns on the dorsal faces of débitage, centripetal removals were achieved mainly by recurrent flaking methods. In this sample, too, Levallois points were not centered on the core and they tend to be shorter than in layer B1 (mean length/width ratio is 1.63; n=16).

The differences in flaking modes, aimed at obtaining different end products, are clearly depicted by scar patterns on the *non*-Levallois débitage. In layer B1 these tend to be unipolar, while in layer B4 the frequencies of convergent flaking are significantly higher than those of any other (Figure 2). Contrary to expectations, however, frequencies of scar patterns on Levallois cores and débitage do not co-vary directly with the shifting emphasis on the morphology of Levallois products (Figure 2).

2.2.3. Advanced Stages of Core Reduction. The progressive use of convergent recurrent methods (Boëda et al. 1990) throughout the process of core reduction is reflected in the quantitative dominance of asymmetrical Levallois products (in layers B1 and B2; data for B4 are not available). It is also seen in the abundance of asymmetrical striking platforms of Levallois products in all the layers (Figure 3; Ohnuma 1992: figs. 7.4:3–6, 9, 7.6:2–3, 5–6), which is the result of blank removal from the lateral part of the core rather than its center, in all of the layers. However, pieces with lateralized scar patterns (i.e., '*enlèvement* II'), a typical product of the second and later phases of blank production in a

Table 3. Distribution of types of striking platforms in the Amud Cave assemblages[a]

	B1[a]			B2[b]			B4[c]		
Platform type	[d]Lev. (n=333)	[d]n. Lev. (n=592)	All (n=925)	[d]Lev. (n=333)	[d]n. Lev. (n=663)	All (n=990)	[d]Lev. (n=333)	[d]n. Lev. (n=167)	All (n=247)
Plain	30.3	44.9	39.7	22.3	45.8	38.0	26.3	52.7	44.1
Dihedral	10.5	7.8	8.8	31.8	26.9	28.0	32.5	28.7	32.0
Faceted	51.5	33.4	39.8	38.5	20.2	26.3	38.8	10.8	17.8
c.d.g.	0.6	—	0.2	5.2	0.2	1.8	1.3	—	0.4
Other	7.5	13.8	11.5	9.2	7.7	8.2	1.3	7.8	5.6

[a]After Goder 1997
[b]Combined sample of Ohnuma 1992 and Shimron 1996.
[c]After Ohnuma 1992 (calculation on unbroken pieces only).
[d]Lev.=Levallois; n. Lev.=non-Levallois

recurrent method, are not very frequent in the two recently obtained samples. During the late stages of core reduction and re-arrangement, lateral convexities were reshaped by removals of *éclats débordants,* which are short and have squat proportions (length to width ratio is 1.6). In the two recently obtained samples, about 60% of these artifacts exhibit centripetal dorsal face scar patterns and do not bear any cortex. When cortex occurs, it is usually along the lateral edges and on less than a quarter of their dorsal surface area.

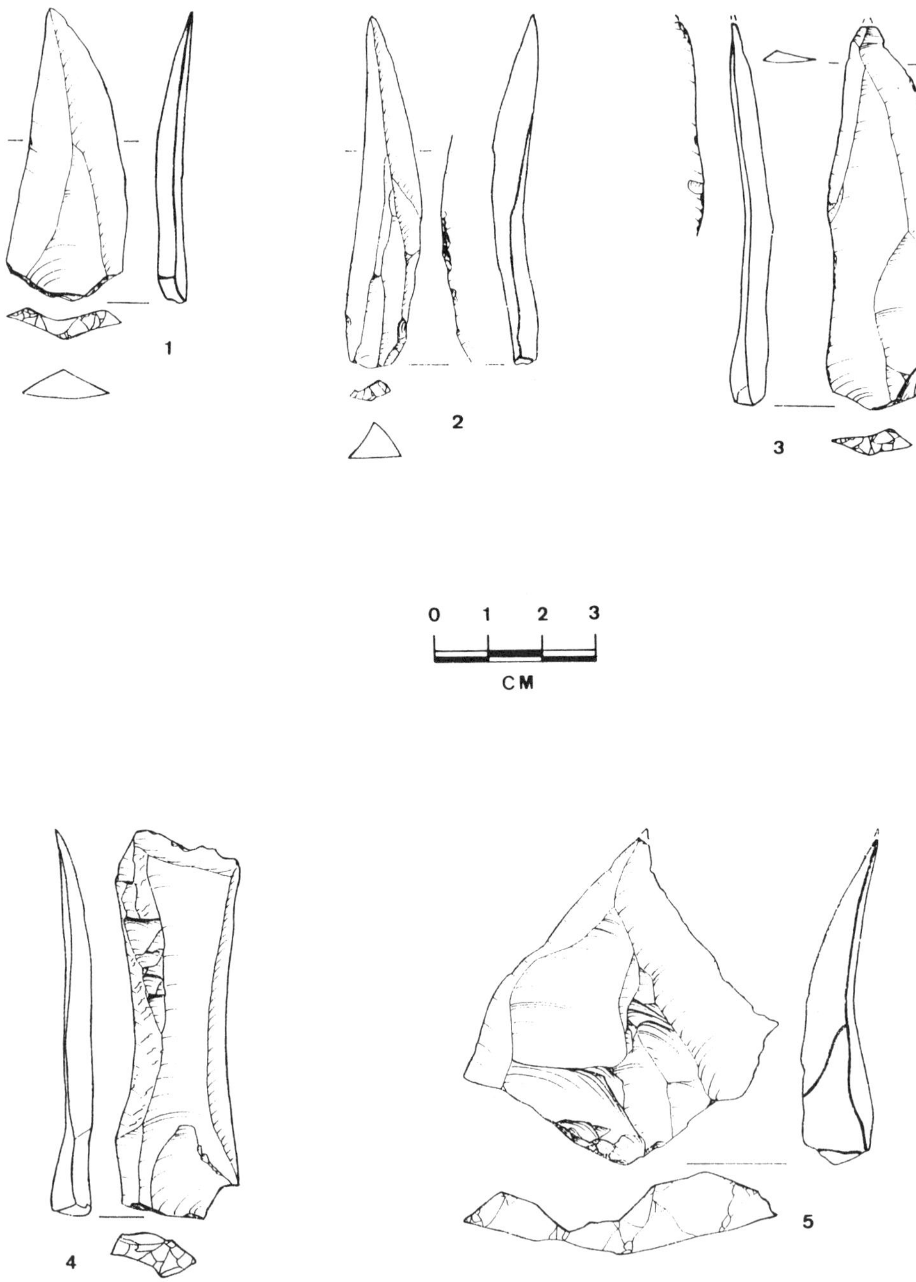

Figure 3. Levallois products. 1–2, 5: layer B2; 3–4: layer B1. Note keeled cross-section of number 2.

These traits suggest that the recurrent convergent mode of flaking was preferred throughout the process of core reduction in all of the layers. However, a comparison of metric attributes reveals several differences among the various assemblages during the advanced stages of the core's use. The average size of Levallois products in layer B4 and in the combined sample from layer B2 (56 mm and 51 mm, respectively) exceeds the size of discarded/abandoned cores (42 mm and 45 mm, respectively). The pattern is different in layer B1, where cores and Levallois débitage are almost the same average size (51 mm and 49 mm, respectively).* Naturally backed pieces in layer B2 (new sample) are often longer than the cores and their mean length far exceeds that of all the other pieces in the sample (58 mm versus 43 mm). In layer B1, the mean dimensions of naturally backed flakes fall within the size range of the cores, and metric differences between those and all other artifacts are less marked, average lengths being 46 mm versus 41 mm. These patterns suggest fewer use cycles for cores in layer B1, in compliance with the extensive use of flakes as blanks for cores.

2.2.4. Tool Manufacture. There is a clear selection in the samples from layers B1 and B2 for larger blanks, hence for Levallois blanks, to be retouched, except in the case of denticulates. While many pieces bear signs of use on their lateral edges, formally retouched tools are scarce in all the layers (Ohnuma 1992; Shimron 1996; Goder 1997). Notches are the dominant group in layer B1, followed by the Mousterian group, which consists mainly of simple scrapers. In layers B2 and B4, Upper Paleolithic tools are more abundant. Chamfered pieces are absent from all the assemblages.

2.3. Variability of Core Reduction Methods in the Amud Samples

Convergent recurrent Levallois flaking was the dominant method in the Amud assemblages, but it was not the only one. Smaller proportions of asymmetrical Levallois pieces and higher frequencies of Levallois elements with well-prepared, faceted *chapeau de gendarme* platforms in layer B2 (Table 3) might reflect more frequent use of fully convergent and centripetal modes of flaking. The relatively high frequencies of centripetal scar patterns on both Levallois and non-Levallois débitage in this assemblage, accompanying the frequent application of convergent flaking, imply that two different reduction sequences, relying on both centripetal and convergent flaking modes, were used synchronously. However, frequencies of small (up to 30 mm in length) 'centripetal' débitage are high compared to those of débitage bearing other scar patterns, suggesting that in some instances convergent flaking might have shifted toward core exploitation through centripetal flaking in the final stages of a core's use (cf. Meignen and Bar-Yosef 1992; Bar-Yosef 1994).

Centripetal flaking, reflected in distinct scar patterns on both cores and débitage (Figure 2), was also used in layers B1 and B4, sometimes to the extent that preferential flakes (Boëda 1986) were produced, mainly at the end of a core's use life. The similarities between size distributions of centripetal versus convergent and bipolar débitage in layer B1 and in layer B2 suggest that in the B1 assemblage, too, a change occurred toward more intensive use of centripetal flaking at the end of a core's use life. However, in contrast to layer B2, a consistent core reduction by centripetal modes of flaking is less likely to have appeared in the initial phases of the process.

Finally, the presence of miscellaneous and prismatic cores (Ohnuma 1992:table 7.2), amorphous core trimming elements, and some sporadic ridge blades, as well as the occur-

* The mean length of non-Levallois débitage is smaller than that of cores in all of the samples.

rence of blades with "keeled cross-sections" (Figure 3:2; cf. Marks and Monigal 1995; Meignen 1995:375), suggests the use of non-Levallois technologies in general and blade technologies in particular, together with the dominant Levallois system. This is evident in layer B1 by the high frequency (23%) of "Jerf 'Ajla" type, i.e., "change of orientation" cores (Figure 1; cf. Schroëder 1969; Crew 1976; Meignen and Bar-Yosef 1992: fig. 9.4:3), flaked on both surfaces. Still, these cores never achieve the bi-conical shapes typical of discoidal cores. The latter morphology occurs on some eight percent (n=2) of the cores of the B1 sample. In both cases there is cortex on one of the surfaces, suggesting that the specimens are not exhausted Levallois cores and indeed were exploited by true discoidal flaking (Boëda 1993).

Variability within the lithic assemblages of Amud Cave is thus discerned on two distinct levels: intra- and inter-assemblage variability. Intra-assemblage variability within each stratigraphic unit is expressed by two patterns. One is the use of several core reduction sequences in any assemblage. In addition to a preferred reduction strategy—recurrent, unidirectional convergent Levallois flaking—other Levallois and non-Levallois production processes were applied to lithic raw material. At the same level of variability, a second pattern consists of changes of reduction methods in the course of exploiting any single core. These changes mainly involve shifts from recurrent to preferential methods, regardless of the specific mode (centripetal or convergent) employed, towards the end of a core's use life.† These shifts demonstrate the cognitive flexibility of Middle Paleolithic hominids (cf. Goren-Inbar and Belfer-Cohen, this volume), in this case of Neandertals. At the same time, they appear to be the means for extracting serviceable blanks from a core which had reached the threshold of minimal useful size for Levallois débitage (i.e., they reflect rational materialistic and economic considerations). Given that the morphology of Levallois products is not directly correlated with the specific reduction sequence employed (see below), intra-assemblage variations in reduction methods cannot be viewed as responding to concrete requirements for specific functional morphologies. Indeed, to a certain degree the patterns seen at Amud refute the notion that "convergent pattern often gives the products a triangular . . . shape" (Bar-Yosef et al. 1992:516).

The second level of variability revealed in the Amud Cave samples is inter-assemblage variability throughout the occupation sequence at the site. The variability of lithic production among the layers of Amud Cave is demonstrated by the fact that the relative dominance of the single preferred reduction sequence varies across stratigraphic units. Moreover, within this preferred reduction method there are subtle technological changes from one period of occupation to another. Varied choices as to the preliminary form of raw material (e.g., flake versus nodule), made at the initial stages of the manufacture process, result in different longevity of cores and affect modifications in the technological choices in the course of core exploitation.

3. VARIABILITY OF LITHIC PRODUCTION IN THE LEVANTINE LATE MIDDLE PALEOLITHIC

The recent dating of the Amud Cave assemblages to about 65–55 ky BP renders them contemporaneous with the Late Mousterian sequence of Kebara Cave (Valladas et al. 1987; Schwarcz et al. 1989), as well as with the sites of Quneitra (Ziaei et al. 1990), Tor Sabiha and Tor Faraj (Henry and Miller 1992; Mercier and Valladas, in press), and Far'ah

† Whether such changes also occurred in earlier stages of reductions is a moot point in the absence of refittings.

II (Schwarcz and Rink, this volume), all of which date to the time span between 70 and 48 ky. The lithic assemblages of Rosh Ein Mor and Nahal Aqev were originally characterized as Early Mousterian ("Tabun D-type") and later as Late Early Mousterian (Marks 1993 and references therein). Still, Rosh Ein Mor is now tentatively dated to the late Middle Paleolithic, while at Nahal Aqev a *pre*-Mousterian travertine was dated to about 80 ky BP (Schwarcz et al. 1979). Thus the lithic assemblages from these two sites are discussed here in a Late Mousterian context. The lithic assemblages from these sites have been published in sufficient detail to allow the identification of reduction sequences,[‡] notwithstanding the fact that in some cases the technological analyses concentrated only on the Levalloisian component. Comparison of the assemblages reveals that during the short time span directly preceding the transition to the Upper Paleolithic, the types of technological variability which characterize the Amud Cave assemblages, if not the specific patterns, are by no means unique to this site.

Far'ah II was recently interpreted as a short-term fortuitous occupation with truly expedient lithic production (Hovers 1997), hence its lithic assemblage does not appear to reflect rigid, structured reduction sequences. In the other assemblages under consideration, intra-assemblage variability is evident in the number of flaking systems applied to lithic raw material in each assemblage. Core morphologies and scar patterns indicate that at least in the final stages of reduction, arrays of flaking technologies were used. These comprised non-Levallois reduction sequences (i.e., blade technologies resulting in prismatic cores, and authentic discoidal techniques), in addition to a wide range of Levallois methods (Crew 1976; Munday 1977; Goren-Inbar 1990; Meignen and Bar-Yosef 1991; Marks 1993; Marks and Monigal 1995), incorporating considerable frequencies of cores-on-flakes. 'Derived' core forms (e.g., 'change of orientation cores'), achieved by slight modification of the geometric constraints of Levallois concepts, also occur in all the assemblages.

Levallois cores exhibit unidirectional, centripetal, and bi-directional scar patterns. In the Negev, "centripetal preparation cannot be regarded even as the norm" (Crew 1976:89). This is the case also at Tor Faraj and Tor Sabiha, where the emphasis of lithic reduction was upon the production of Levallois points from both unidirectionally and bi-directionally prepared cores, and centripetal preparation never occurred (Henry 1995a,b). Still, some cores that occur in the Negev sites "come closest to the classic western European variety" (Munday 1976a:129, fig. 6–6, 6–7b; Crew 1976:fig. 5–8b), namely, they bear lineal, centripetal removals.

At Quneitra, on the other hand, the use of lineal methods appears to have been much more common (Goren-Inbar 1990:128, tab. 28, fig. 84), complementing equally well-represented centripetal recurrent methods of flaking. Centripetal flaking is not exclusive, however, and other modes of Levallois flaking as well as non-Levallois reduction (e.g., Goren-Inbar 1990:128: fig. 88) are also encountered, albeit in much lower frequencies.

In units XI–IX of Kebara Cave, over half of the cores exhibit unipolar scar patterns, among which convergent scar patterns are the majority. Still, in the very same assemblages there are high frequencies of cores with centripetal patterns (Meignen, personal communication). While specimens produced by the two flaking modes indicate the use of recurrent methods, broad scars of preferential flakes sometimes occur, which probably indicate a shift from recurrent convergent methods of flaking to lineal ones at the end of the core's use (Bar-Yosef et al. 1992:513). The latter tendency does exist in the assemblage from Rosh Ein Mor (see above), but seems to be less frequent, as small Levallois cores for

‡ Other important sites that are not securely dated (e.g., Ksar Akil) are not discussed here in detail, as the emphasis of the discussion is on lithic variability within a specific time span.

flakes, at the end of their use lives, still maintain the characteristic scar patterns of unidirectional recurrent flaking (Crew 1976:91).

Unidirectional patterns are common among the non-Levallois blanks (hence dominant within the complete assemblage) in all instances. However, in most of the assemblages under discussion, characteristics of scar patterns of Levallois débitage complement those seen on the cores, as in layers B1 and B4 (but unlike layer B2) at Amud Cave. Levallois artifacts at Quneitra often bear indications of centripetal preparations (Goren-Inbar 1990:fig. 61), with the lineal method used at the end of a core's use life. In contrast, the most frequent scar pattern seen on Levallois débitage in the Negev assemblages exhibit unidirectional (often recurrent and convergent, to judge by published drawings) scar patterns. Whereas this trait of Late Mousterian assemblages led to the notion that the convergent recurrent method of flaking is typical of Levantine Mousterian assemblages (Meignen and Bar-Yosef 1992; Meignen 1995), it is never the only method of Levallois production in any assemblage. As in the case of layer B2 at Amud Cave, the data seem to indicate that both parallel use of several Levallois reduction sequences and constant reorganization of core reduction methods took place in the course of the knapping of any single core. The high frequency of points among the unretouched Levallois products (23%) in area A of Quneitra (cf. Goren-Inbar 1990:table 17), combined with the occurrence of several Levallois cores for points (Goren-Inbar 1990:table 24), is a clear example of such parallel use of two distinct reduction sequences.

The high frequency of points in area A of Quneitra is similar to that seen in Kebara units X–IX, in spite of the marked differences in the predominant reduction sequence. Moreover, in both these assemblages the production of points was emphasized more than at layer B1 of Amud Cave, although a variant of the recurrent convergent method is prevalent at that site. These patterns suggest that even where the recurrent convergent flaking method was selected over other flaking methods, it was not a monolithic, rigid choice. Indeed, patterns of inter-assemblage variability are apparent even within the favorable reduction sequence. For instance, morphometric characteristics of naturally backed pieces at Kebara (Bar-Yosef et al. 1992; Meignen and Bar-Yosef 1992) and Rosh Ein Mor (Crew 1976:86) strongly imply that they were removed in the early stages of core reduction, as was shown to be the case in all the Amud assemblages. However, whereas in the two cave sites these artifacts were knapped in order to shape the distal convexities of Levallois cores, this may not necessarily have been the case at Rosh Ein Mor (Marks and Monigal 1995).

By the same token, Amud, Rosh Ein Mor, and Kebara are similar in that a particular scar pattern is clearly exhibited on the dorsal faces of the vast majority of Levallois points. This pattern reflects sequential, spatially-adjacent preparatory removals, which create the *nervures guides* prior to the detachment of the point itself (Crew 1976:73–86, fig. 5–5). After the removal of the first point from the specific surface, the knapper progressed to remove additional points from the same flaking surface by rotating the core slightly after each point removal until the flaking surface lost its convexities (Meignen 1995: fig. 25.5). This characteristic, combined with high investment in the preparation of the striking platforms, is argued to be the technological hallmark of recurrent convergent Levallois methods. Still, the investment in platform preparation seen at Amud B1 and B2 hardly matches the careful shaping of cores' platforms reported from Kebara and from the two Negev sites, where *chapeau de gendarme* platforms are argued to be an important technological aid for producing elongated pieces and Levallois points (Munday 1979). Another case in point is the site of Tor Faraj where, as in the Amud assemblages in general and layer B1 in particular, point production was achieved by two variants of the recurrent convergent method (Henry 1995b). The result was broad-based points with *chapeau de gendarme* on the one hand, and 'leaf shaped' pointed artifacts on the other. Morphometric properties, too, are only weakly correlated with specific technological traits. A comparison between the layers with the highest frequencies of laminar products at Kebara (units XI–XII) and at Amud (layer B1) shows essentially opposing trends: in Kebara the tendency towards laminarity appears to be correlated with longer débitage (Meignen and Bar-Yosef 1992:134; and see also Crew 1976:84–6), but the reverse is true in Amud, where débitage in layer B1 is on average shorter than in the other layers of the site.

Nor does technological variability in the last stage of the Levantine Mousterian display a clear chronological patterning. The pattern of temporal changes in the values of the laminar index is an illustrative example. If a gradual, developmental process favoring blade production was the operative agent affecting the morphometric properties of blanks (e.g., Marks 1993:7–8), then a tendency toward increased blade production over time is to be expected. Figure 4 refutes this model (and see Bar-Yosef 1994:39). A comparison of the trends seen in the stratified assemblages of Kebara and Amud Caves also shows that the directions of variation differ among sites: at Kebara the laminar index is highest in the oldest units, XII–XI (Meignen and Bar-Yosef 1991), whereas the trend observed in the assemblages of Amud is practically the opposite, with the youngest assemblages being the most laminar.

The kinds of variability within and among Late Mousterian assemblages clearly echo the patterns of variation observed in the sequence of Amud Cave. Whereas one

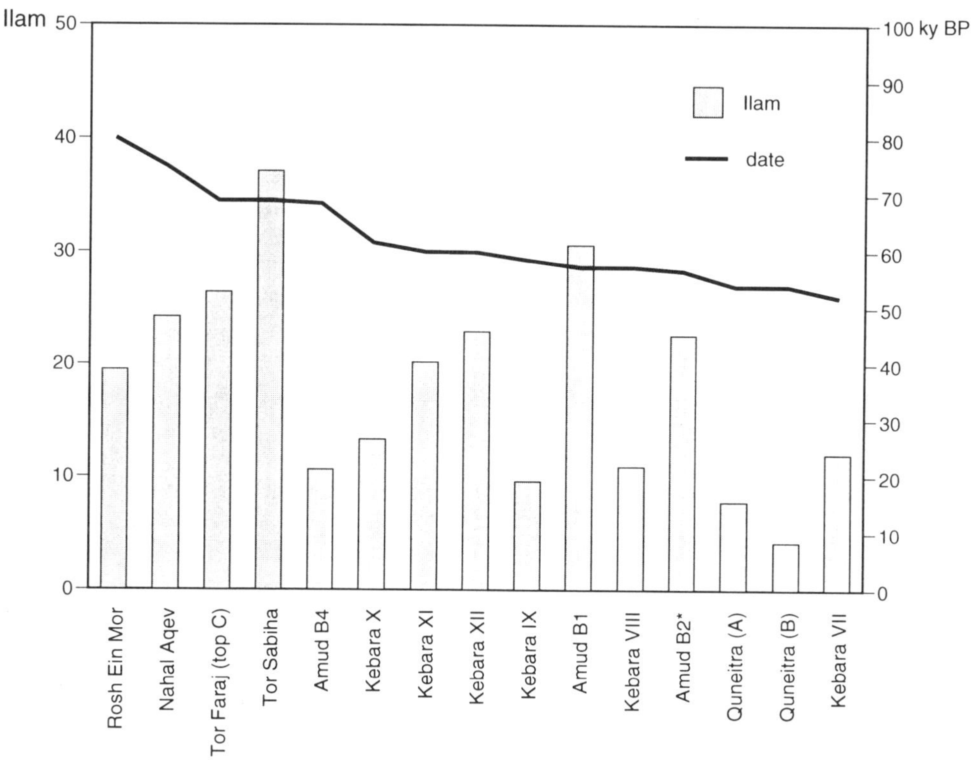

Figure 4. Values of laminar indices in various Late Mousterian assemblages as a function of their ages. Note that dates for Far'ah II, Nahal Aqev, and Rosh Ein Mor are tentative. Amud B2* relates to the sample from the new excavation; the index for Amud B4 was compiled from Ohnuma 1992: tab. 7.9 and is an estimate.

reduction method was usually preferred, parallel use of other methods is clearly documented in each assemblage. Most notably, a high degree of variability is manifested in nuances of *savoir-faire* (i.e., in the knowledge of gestures and their implications for the physical properties of the final products; Pelegrin 1990:118). Archaeologically, this translates into the degree of platform preparation, positioning of the hammer in the process of blank removal, and symmetry of the products. These vary as a means for attaining diversified morphological products while conforming with the dominant reduction system. This facet of variability is complementary to the commonly noted phenomenon of identical products that were achieved through diverse reduction systems (Marks and Volkman 1983; Boëda 1995).

4. DISCUSSION

Following Copeland (1975) and Ronen (1979), the range of variability observed among Middle Paleolithic lithic assemblages has usually been described as adhering to the tripartite scheme of lithic technological variants observed in the Tabun Cave sequence. This sequence, albeit only cursorily studied, formed the model for change during the Levantine Middle Paleolithic which led to a view of time-trajectoried trends of variability in the Levantine Mousterian. This in turn implied that the specific *chaînes opératoires* typical of each technological variant became a 'guide fossil', i.e., represented a specific time span whose limits were to be defined by radiometric dating (cf. Bar-Yosef 1991). Several workers have defied the chronological significance of the technological variations throughout the Tabun sequence (cf. Munday 1979; Marks 1992b, 1993; Marks and Monigal 1995; Meignen 1995; Hovers 1997; Goren-Inbar and Belfer-Cohen, this volume), although some still submit that they represented the full range of Middle Paleolithic lithic variants in the region.

There is no clear-cut fit between patterns seen in the Amud Cave assemblages and either of these two models. Jelinek (1982:99), followed by Ohnuma and Akazawa (1988), suggested that the assemblage of Amud be designated as "Tabun D-type." The notion may not have been too far-fetched, since the earliest Tabun phase is characterized by the use of unipolar Levallois flaking (Bar-Yosef et al. 1992:531). Specific Levallois methods used at Amud for obtaining blades and elongated points, especially those seen in Amud B1, are similar to those documented in Jelinek's unit IX (the equivalent of Garrod's layer D) at Tabun (cf. Meignen 1995:374–5). Another feature of Tabun IX which is argued to be a typical property but is missing from the Amud assemblages is the quantitative dominance of Levallois blades over flakes (Bar-Yosef et al. 1992:531). This feature might well be the result of differential transport (Jelinek 1975, 1982; Hovers 1997) rather than a direct outcome of specific reduction modes.

A believer in the linear techno-chronological scheme, Jelinek argued that such characteristics placed the Amud lithics (and the Neandertal remains associated with them) at "...an early point in the technological chronology of the Mousterian" (1982:99). By the same token, if the Tabun sequence is accepted as a pan-Levantine techno-chronological yardstick, then the recent dating of the various stratigraphic units at Amud to 65–55 ky BP requires that the assemblages be assigned to "Tabun B-type." While several assemblages from various sites were placed within this technological phase (Bar-Yosef 1992, 1994, 1995), the term "Tabun B-type" has become in recent years synonymous with the Kebara industries (Meignen and Bar-Yosef 1988, 1989, 1991, 1992; Bar-Yosef et al. 1992; Meignen 1995).

None of the Amud assemblages falls indisputably into the parameters of this variant. Amud B1 indeed differs from "Tabun D-type" assemblages in the absence of the typical Abu Sif points and in that elongated points are less common. But similarities also abound, including the fact that over a third of the points are of blade proportions. Furthermore, whereas layers B2 and B4 exhibit technological traits that are closer to those characteristics identified as the telltale properties of the "Tabun B-type" industries, they are not wholly similar. Most notably, reduction sequences indeed involve the use of both "one-axis or radially prepared cores" and the occurrence of flakes among which laminar and narrow forms prevail over oval ones (Copeland 1975:332). The production of Levallois points is not as monotonous and standardized as at Tabun B (Copeland 1975) or at Kebara (Meignen and Bar-Yosef 1991, 1992; Bar-Yosef et al. 1992).

Even without radiometric dates, the order of appearance of the particular technological traits in the Amud assemblages would considerably weaken the validity of a techno-chronological model based on the Tabun sequence. The technological traits at Amud are reminiscent of more than one Tabun type, but they occur in a sequence that is opposite the one predicted from the Tabun stratigraphic section. That the various traits are confined to a short time span at the final stages of the Middle Paleolithic, rather than to a long-term sequence like that of Tabun (Grün et al. 1991; Mercier et al. 1995), further undermines the use of Tabun as a techno-chronological scale for the Levantine Middle Paleolithic. Even more striking differences are observed between the various units of Kebara and Amud sequences and the contemporaneous assemblages of Quneitra and, possibly, Far'ah II (see above). It is possible that the Tabun model holds more firmly within a constrained geographical area, whereas technological homogeneity within any given time span decreases towards the periphery of that region (Hovers 1997). The term "Tabun phase," insinuating chronological significance, is obviously unwarranted (see also Copeland 1983).

The differences observed among the Amud assemblages and "Tabun B-type" assemblages also undermine the alternative model of Tabun as a "catalogue" encompassing all the Levantine Mousterian technological variants. Under the tenet that each Levantine Mousterian assemblage must fit into only one of three acceptable slots of technological variants, this "lumpers" model becomes a vicious circle. The Amud assemblages, as well as those of Tor Faraj, Tor Sabiha, and Rosh Ein Mor, serve as an example of this process. They are assigned to a specific Tabun type based on discrete, sometimes tenuous, technological criteria, the interpretation of which is disputed in some cases (Ohnuma 1992; Marks 1992b; Bar-Yosef 1994:41; Henry 1995a,b; Marks and Monigal 1995; Meignen 1995). At the same time, many points of dissimilarity with each other and with the type-assemblage as described by Copeland (1975) are overlooked. To be sure, many (but not all) Late Mousterian assemblages conform with a unifying conceptual framework associated with one prevalent technological system (namely, the use of unipolar Levallois flaking). This system, however, must have been flexible, as diversity in the finer details of core reduction tactics exceeds the constraints set by the more rigid tripartite Tabun model and results in a higher degree of cultural complexity.

Several studies related Mousterian lithic variability (as expressed by reference to Tabun types) to ecological conditions such as climatic changes and raw material availability (Munday 1976a,b, 1979; Marks and Freidel 1977; Marks 1988), or to functional requirements following specific behavioral adaptations (Shea 1991; Lieberman and Shea 1994). Recent analyses indicate that such factors are less significant in governing the degree of technological variability than had been assumed previously (Hovers 1997). The technological variation of the Levantine Mousterian may be better understood as specific production techniques which, like other cultural behaviors, were transmitted vertically by social learning (Rogers 1988; Laland, this volume). Technological traits reconstructed through the *chaîne opératoire* approach are the visible sum of individual choices, hence they depict societal choices based on several equally-adequate available options. As such, these traits are material manifestations of the social infrastructure and world views of the societies within which the objects were produced (Levi-Strauss 1976; Pfaffenberger 1992; Dobres and Hoffman 1994). Be they mundane (e.g., stone tool production) or spectacular (e.g., the creation of parietal art), technical choices are socially meaningful (Lemonnier 1992:17–24) and are related to social identities and differences.

Application of this ethnological stance to the study of Levantine Late Mousterian lithic assemblages reveals the operation of social processes on two different scales. A fixed technological system occurred over the larger geographic scale, so that the use of unipolar (convergent) Levallois flaking became the technological paradigm in the Levant at this time. This is attributed to information flow among the human groups in the region. On the local scale, changes might be due to a group's wish to incorporate marks of its independent identity into the process of lithic production, i.e., stylistic manifestations (cf. Dobres and Hoffman 1994:217–18). Another possible cause for local changes might have been a decrease in communication quality over distance, causing random slight changes in the technical "recipes" received at nodes along the communication chain. Those local, particular reduction tactics, still within the paradigmatic concept, were later perpetuated into the local norm following socially and politically dominant group members with whom other members tend to conform (cf. Boyd and Richerson 1995). This complemented the process of innovation within particular groups, in this way keeping at bay (but rarely preventing altogether) the occurrence of "rival" systems (Boyd and Richerson 1992; Dobres and Hoffman 1994; Neiman 1995). This process, which causes repetition of fixed practices over time, culminates in local traditions (Hobsbawm 1983), clearly observed in the archaeological record of the Late Mousterian, especially in the stratified sites (see Hovers et al. 1995).

That punch blade techniques came to dominate the technological repertoire of the Levantine habitants reflects a process of innovation, namely, the widespread adoption of this specific technology (Renfrew 1978). Essentially a process of diffusion, adoption of a technological trait is a product of communication, since "a person becomes more and more inclined to accept an innovation the more often he comes into contact with other persons who have accepted it" (Hagerstrand 1953, cited in Moore 1983:179; and see Spratt 1982:81–3; Davis 1983). The communication networks already established among Late Mousterian groups, serving as the venue by which blade technologies spread, were a prerequisite in the process of the latter becoming the normative lithic production system. The diversity of lithic traditions among Levantine Late Mousterian groups was another prerequisite for accepting a new technological mode. Interpreted here as the result of social processes within each group, the relaxed technological norms would have complemented the spread of innovations, as a trait would be adopted locally if a dominant member of the group accepts it for any reason.

In the specific case of blade technologies, the use of unipolar flaking during the Late Mousterian might have helped their acceptance. While the technological concept is completely different, the products of blade production sequences were morphologically similar, at least in some cases where the laminar index is high, to series of products used by the Late Mousterian hominids (e.g., at Amud B1, Rosh Ein Mor). This similarity would have facilitated the switch to reduction sequences emphasizing blades.

The discovery of punch blade technologies and of the volumetric concept of blade cores (Boëda 1989, 1990), and their consequent invention (i.e., the perception of their practical use; Renfrew 1978; Spratt 1982) need not necessarily be sought outside the Levant. Ample evidence demonstrates that within the geographic context of the Levant it occurred at the latest during the Late Lower Paleolithic/Early Middle Paleolithic (e.g., Meignen 1994, this volume; Goren-Inbar and Belfer-Cohen, this volume), and became a successful innovation for some time (e.g., some "Tabun D-type" assemblages). The appearance of blade technologies at the end of the Mousterian represents a known phenomenon of a long "delay period" between the point of invention and its spread, during which "world outlook was not conductive to this innovation" (Spratt 1982:87, 92; cf. Renfrew 1978:391). Hence it is not necessary to invoke population replacement and external diffusion as the mechanisms responsible for the occurrence of Upper Paleolithic technologies. Indeed, given the degree of continuity between the Late Mousterian and Upper Paleolithic in aspects other than lithic technologies, it is difficult to accept such a scenario. On the other hand, the Levantine archaeological record makes it hard to accept that blade technologies flourished at the end of the Middle Paleolithic because they were retained as such over a period of some 200 ky. Although the invention was kept alive, namely, the technical knowledge survived through time, it did not exist as a full-fledged technological practice.

Although no direct line of technological 'progress' is discerned in the archaeological record of the Middle Paleolithic in general, and of the Late Mousterian in particular, this time span was not a static period, encompassing facets of variability and variation that may be best understood as indications of social processes (Hovers 1997). Rather than discrete technological traits, it is the social/demographic environment that formed the necessary preadaptations for the acceptance, and eventually the dominance, of Upper Paleolithic lithic technologies.

ACKNOWLEDGMENTS

Field work at Amud Cave was supported by the L. S. B. Leakey Foundation, the Irene Levi-Sala CARE foundation, the Institute of Human Origins, currently at the Arizona State University at Tempe, and the Institute of Archaeology, Hebrew University. Grants from the Irene Levi-Sala CARE foundation and the Jerusalem Center for Anthropological Research enabled analyses of the lithic assemblages. Artifact drawings were done by Julia Moskovich, and graphs were done by Guy Hivroni. I thank Anna Belfer-Cohen and Ofer Bar-Yosef for their comments on an early draft of this paper.

REFERENCES

Akazawa, T. 1979. Middle Paleolithic assemblages from Douara Cave, 1–30. In *Paleolithic Site of Douara Cave and Paleography of Palmyra Basin in Syria*, ed. K. Hanihara and T. Akazawa. Tokyo: The University Museum, University of Tokyo.

Bar-Yosef, O. 1991. The search for lithic variability among Levantine Epi-Palaeolithic industries, 319–36. In *25 Ans d'Études Technologiques en Préhistoire: XI Recontres Internationales d'Archéologie et d'Histoire d'Antibe*. Juan-les-Pins: APDCA.

Bar-Yosef, O. 1992. Middle Paleolithic human adaptations in the Mediterranean Levant, 189–216. In *The Evolution and Dispersal of Modern Humans in Asia*, ed. T. Akazawa, K. Aoki, and T. Kimura. Tokyo: Hokusen-sha.

Bar-Yosef, O. 1994. The contribution of southwest Asia to the study of the origin of modern humans, 23–66. In *Origins of Anatomically Modern Humans*, ed. M. H. Nitecki and D. V. Nitecki. New York: Plenum Press.

Bar-Yosef, O. 1995. The role of climate in the interpretation of human movements and cultural transformations in Western Asia, 507–23. In *Paleoclimate and Evolution*, ed. E. Vrba, G. Denton, T. Partridge, and L. Burckle. New Haven: Yale University Press.

Bar-Yosef, O., B. Vandermeersch, B. Arensburg, A. Belfer-Cohen, P. Goldberg, H. Laville, L. Meignen,, Y. Rak, J. D. Speth, E. Tchernov, A. M. Tillier, and S. Weiner. 1992. The excavations in Kebara Cave, Mt. Carmel. *Current Anthropology* 33:497–550.

Bar-Yosef, O., M. Arnold, N. Mercier, A. Belfer-Cohen, P. Goldberg, R. Housely, H. Laville, L. Meignen, J. C. Vogel, and B. Vandermeersch. 1996. The dating of the Upper Paleolithic layers in Kebara Cave, Mt. Carmel. *Journal of Archaeological Science* 23:297–306.

Boëda, E. 1986. Le débitage Levallois de Biache-Saint-Vaast (Pas-de-Calais): Première étude technologique, 209–18. In *Chronostratigraphie et Faciès Culturels du Paléolithique Inférieur et Moyen dans l'Europe du Nord-ouest*, ed. A. Tuffreau and J. Somme. Paris: Société Préhistorique Française.

Boëda, E. 1989. Le concept laminaire: Rupture et filiation avec le concept Levallois, 41–59. In *L'Homme de Neandertal*, Vol. 8, *La Mutation*, ed. M. Otte. Liège: ERAUL.

Boëda, E. 1990. De la surface au volume: Analyse des conceptions des débitages Levallois et laminaire, 63–8. In *Paléolithique Moyen Récent et Paléolithique Supérieur Ancien en Europe,* ed. K. Farizy. Nemours: APRAIF.

Boëda, E. 1991. Approche de la variabilité des systèmes de production lithique des industries du Paléolithique inférieur et moyen: Chronique d'une variabilité attendue. *Technique et Culture* 17–18:37–79.

Boëda, E. 1993. Le débitage discoïde et le débitage Levallois récurrent centripète. *Bulletin de la Société Préhistorique Française* 90:392–404.

Boëda, E. 1995. Levallois: A volumetric construction, methods, a technique, 41–68. In *The Definition and Interpretation of Levallois Technology*, ed. H. L. Dibble and O. Bar-Yosef. Madison: Prehistory Press.

Boëda, E., J. M. Geneste, and L. Meignen. 1990. Identification de *chaînes opératoires* lithiques du Paleolithique ancien et moyen. *Paléo* 2:43–79.

Bordes, F. 1972. *A Tale of Two Caves*. New York: Harper and Row.

Boyd, R., and P. J. Richerson. 1992. How microevolutionary processes give rise to history, 179–210. In *History and Evolution*, ed. M. H. Nitecki and D. V. Nitecki. Albany: State University of New York Press.

Chinzei, K. 1970. The Amud Cave site and its deposits, 21–52. In *The Amud Man and His Cave Site*, ed. H. Suzuki and F. Takai. Tokyo: Academic Press of Japan.

Copeland, L. 1975. The Middle and Upper Palaeolithic in Lebanon and Syria in the light of recent research, 317–50. In *Problems in Prehistory: North Africa and the Levant,* ed. F. Wendorf and A. E. Marks. Dallas: SMU Press.

Copeland, L. 1983. The Palaeolithic industries at Adlun. Section III: The Levalloiso-Mousterian of Bezez Cave level B, 261–366. In *Adlun in the Stone Age: The Excavations of D. A. E. Garrod in the Lebanon, 1958–1963*, ed. D. Roe. Oxford: British Archaeological Reports International Series 159(ii).

Copeland, L. 1986. Introduction to volume I, 1–24. In *Ksar Akil, Lebanon: A Technological and Typological Analysis of the Transitional and Early Upper Palaeolithic Levels of Ksar Akil and Abu Halka*, Vol. 1, *Levels XXV–XII*. Oxford: British Archaeological Reports International Series 289(i).

Crew, H. L. 1976. The Mousterian site of Rosh Ein Mor, 75–112. In *Prehistory and Paleoenvironments in the Central Negev, Israel*, Vol. 1, *The Avdat/Aqev Area*, ed. A. E. Marks. Dallas: SMU Press.

Davis, D. D. 1983. Investigating the diffusion of stylistic innovations, 53–89. In *Advances in Archaeological Method and Theory 6*, ed. M. B. Schiffer. New York: Academic Press.

Delagne, A. 1995. Variability within uniformity: Three levels of variability within the Levallois system, 201–12. In *The Definition and Interpretation of Levallois Technology*, ed. H. L. Dibble and O. Bar-Yosef. Madison: Prehistory Press.

Dibble, H. L. 1987. The interpretation of Middle Paleolithic scraper morphology. *American Antiquity* 52:109–17.

Dobres, A. M., and C. R. Hoffman. 1994. Social agency and the dynamics of prehistoric technology. *Journal of Archaeological Method and Theory* 1:211–58.

Garrod, D. A. E. 1951. A transitional industry from the base of the Upper Palaeolithic in Palestine and Syria. *Journal of the Royal Anthropological Institute* 81:121–34.

Garrod, D. A. E. 1955. The Mugharat el-Emireh in Lower Galilee: Type station of the Emiran industry. *Journal of the Royal Anthropological Institute* 85:141–62.

Garrod, D. 1957. Notes sur le paléolithique supérieur du Moyen Orient. *Bulletin de la Société Préhistorique Française* 54:439–46.

Geneste, J. M. 1985. Analyse lithique d'industries moustériennes du Périgord: Une approche technologique des comportements des groupes humains au Paléolithique moyen. Ph. D. diss. University of Bordeaux. Bordeaux.

Goder, M. 1997. Technological and typological aspects of Layer B1 in the Late Mousterian site at Amud Cave. M. A. thesis. Hebrew University. Jerusalem.

Goren-Inbar, N. 1988. Too small to be true? Reevaluation of cores on flakes in Levantine Mousterian assemblages. *Lithic Technology* 17:37–44.

Goren-Inbar, N. 1990. The lithic assemblages, 61–167. In *Quneitra—A Mousterian Site on the Golan Heights*, ed. N. Goren-Inbar. QEDEM 31. Jerusalem: Institute of Archaeology, Hebrew University of Jerusalem.

Grün, R., C. B. Stringer, and H. P. Schwarcz. 1991. ESR dating of teeth from Garrod's Tabun collection. *Journal of Human Evolution* 20:231–48.

Henry, D. O. 1995a. The influence of mobility levels on Levallois point production, Late Levantine Mousterian, southern Jordan, 185–200. In *The Definition and Interpretation of Levallois Technology*, ed. H. L. Dibble and O. Bar-Yosef. Madison: Prehistory Press.

Henry, D. O. 1995b. The Middle Paleolithic sites, 49–84. In *Prehistoric Cultural Ecology and Diversity: Insights from Southern Jordan*, ed. D. O. Henry. New York: Plenum Press.

Henry, D. O., and G. H. Miller. 1992. The implications of amino acid racemization dates of Levantine Mousterian deposits in southern Jordan. *Paléorient* 18:45–52.

Hobsbawm, E. 1983. Introduction: Inventing tradition, 1–14. In *The Invention of Tradition*, ed. E. Hobsbawm and T. Ranger. Cambridge: Cambridge University Press.

Hovers, E. 1997. Variability of Levantine Mousterian Assemblages and Settlement Patterns: Implications for the Development of Human Behavior. Ph. D. diss. Hebrew University. Jerusalem.

Hovers, E., Y. Rak, and W. H. Kimbel. 1991. Amud Cave—The 1991 season. *Journal of the Israel Prehistoric Society* 24:152–7.

Hovers, E., Y. Rak, R. Lavi, and W. H. Kimbel. 1995. Hominid remains from Amud Cave in the context of the Levantine Middle Paleolithic. *Paléorient* 21:47–61.

Jelinek, A. J. 1975. A preliminary report on some Lower and Middle Paleolithic industries from the Tabun Cave, Mount Carmel (Israel), 279–316. In *Problems in Prehistory: North Africa and the Levant*, ed. F. Wendorf and A. E. Marks. Dallas: SMU Press.

Jelinek, A. J. 1982. The Middle Paleolithic in the Southern Levant, with comments on the appearance of modern Homo sapiens, 57–104. In *The Transition from Lower to Middle Palaeolithic and the Origin of Modern Man,* ed. A. Ronen. Oxford: British Archaeological Reports International Series 151.

Karlin, C., P. Bodu, and J. Pelegrin. 1991. Processus techniques et *chaînes opératoires*: Comment les préhistoriens s'approprient un concept élaboré par les ethnologues, 101–17. In *Observer l'Action Technique: Des Chaînes Opératoires, Pour Quoi Faire?*, ed. H. Balfet. Paris: CNRS Editions.

Klein, R. G. 1995. Anatomy, behavior, and modern human origins. *Journal of World Prehistory* 9:167–98.

Lavi, R. 1994. The Amud I Skull—Taxonomic Assessment and Phylogenetic Status. M. A. thesis. University of Tel Aviv. Tel Aviv.

Lemonnier, P. 1992. *Elements for an Anthropology of Technology*. Ann Arbor: University of Michigan.

Leroi-Gourhan, A. 1943. *Evolution et Techniques. L'Homme et la Matière* (reprinted 1971). Paris: Albin Michel.

Leroi-Gourhan, A. 1945. *Evolution et Techniques. Millieu et Techniques* (reprinted 1973). Paris: Albin Michel.

Levi-Strauss, C. 1976. *Structural Anthropology*, Vol. 2. New York: Basic Books.

Lieberman, D. E., and J. J. Shea. 1994. Behavioral differences between archaic and modern humans in the Levantine Mousterian. *American Anthropologist* 96:300–32.

Marks, A. E. 1988. The Middle to Upper Paleolithic transition in the southern Levant: Technological change as an adaptation to increased mobility, 109–123. In *L'Homme de Neandertal*, Vol. 8, *La Mutation*, ed. M. Otte. Liège: ERAUL.

Marks, A. E. 1992a. Typological variability in the Levantine Middle Paleolithic, 127–42. In *The Middle Paleolithic: Adaptation, Behavior, and Variability*, ed. H. L. Dibble and P. Mellars. Pennsylvania: The University Museum of the University of Pennsylvania.

Marks, A. E. 1992b. Upper Pleistocene archaeology and the origins of modern man: A view from the Levant and adjacent areas, 229–51. In *The Evolution and Dispersal of Modern Humans in Asia*, ed. T. Akazawa, K. Aoki, and T. Kimura. Tokyo: Hokusen-Sha.

Marks, A. E. 1993. The early Upper Paleolithic: The view from the Levant, 5–21. In *Before Lascaux: The Complex Record of the Early Upper Paleolithic*, ed. H. Knecht, A. Pike-Tay, and R. White. Boca Raton: CRC Press.

Marks, A. E., and D. A. Freidel. 1977. Prehistoric settlement patterns in the Avdat/Aqev area, 131–58. In *Prehistory and Paleoenvironments in the Central Negev, Israel*, Vol. 2, *The Avdat/Aqev Area*, ed. A. E. Marks. Dallas: SMU Press.

Marks, A. E., and K. Monigal. 1995. Modeling the production of elongated blanks from the Early Levantine Mousterian at Rosh Ein Mor, 267–78. In *The Definition and Interpretation of Levallois Technology,* ed. H. L. Dibble and O. Bar-Yosef. Madison: Prehistory Press.

Marks, A. E., and P. W. Volkman. 1983. Changing core reduction strategies: A technological shift from Middle to Upper Paleolithic in the southern Levant, 13–33. In *The Mousterian Legacy: Human Biological Change in the Upper Pleistocene*, ed. E. Trinkaus. Oxford: British Archaeological Reports International Series 164.

Meignen, L. 1994. Paléolithique moyen au Proche-Orient: Le phénomène laminaire, 125–159. In *Les Industries Laminaires au Paléolithique Moyen*. Paris: CNRS Editions.

Meignen, L. 1995. Levallois lithic production systems in the Middle Palaeolithic of the Near East: The case of the unidirectional method, 361–80. In *The Definition and Interpretation of Levallois Technology*, ed. H. L. Dibble and O. Bar-Yosef. Madison: Prehistory Press.

Meignen, L., and O. Bar-Yosef. 1988. Kebara et le Paléolithique moyen du Mont Carmel (Israël). *Paléorient* 14:123–30.

Meignen, L., and O. Bar-Yosef. 1989. Nouvelles recherches sur le Paléolithique moyen d'Israël: La grotte de Kebara, unités VII à XII, 169–84. In *Investigations in South Levantine Prehistory*, ed. O. Bar-Yosef and B. Vandermeersch. Oxford: British Archaeological Reports International Series 497.

Meignen, L., and O. Bar-Yosef. 1991. Les outils lithiques moustériens de Kebara (fouilles 1982–1985), 49–85. In *Le Squelette Mousterien de Kebara 2*, ed. O. Bar-Yosef and B. Vandermeersch. Paris: CNRS Editions.

Meignen, L., and O. Bar-Yosef. 1992. Middle Paleolithic variability in Kebara Cave, Israel, 129–48. In *The Evolution and Dispersal of Modern Humans in Asia*, ed. T. Akazawa, K. Aoki, and T. Kimura. Tokyo: Hokusen-Sha.

Mercier, N., and H. Valladas. In press. Chronologie par la thermoluminescence de gisements du Paléolithique moyen du Proche-Orient. In *Actes du Colloque du CTHS, Nice 1996.*

Mercier, N., H. Valladas, G. Valladas, J. L. Reyss, A. Jelinek, L. Meignen, and J. L. Joron. 1995. TL dates of burnt flints from Jelinek's excavations at Tabun and their implications. *Journal of Archaeological Science* 22:495–509.

Moore, J. A. 1983. The trouble with know-it-alls: Information as a social and ecological resource, 173–91. In *Archaeological Hammers and Theories*, ed. J. A. Moore and A. S. Keene. New York: Academic Press.

Munday, F. C. 1976a. Intersite variability in the Mousterian of the Central Negev, 113–40. In *Prehistory and Paleoenvironments in the Central Negev, Israel*, Vol. 1, *The Avdat/Aqev Area*, ed. A. E. Marks. Dallas: SMU Press.

Munday, F. C. 1976b. The Mousterian in the Negev: A Description and Explication of Inter-Site Variability. Ph. D. diss. Southern Methodist University. Dallas.

Munday, F. C. 1977. Nahal Aqev (D35): A stratified, open-air Mousterian occupation in the Avdat/Aqev area, 35–60. In *Prehistory and Paleoenvironments in the Central Negev, Israel*, Vol. 2, *The Avdat/Aqev Area*, ed. A. E. Marks. Dallas: SMU Press.

Munday, F. C. 1979. Levantine Mousterian technological variability: A perspective from the Negev. *Paléorient* 5:87–104.

Neiman, F. D. 1995. Stylistic variation in evolutionary perspective: Inferences from decorative diversity and interassemblage distance in Illiois Woodland ceramic assemblages. *American Antiquity* 60:7–36.

Newcomer, M. H., and F. Hivernel-Guerre. 1974. Nucleus sur éclat: Technologie et utilisation par differentes cultures préhistoriques. *Bulletin de la Société Préhistorique Française* 71:119–28.

Ohnuma, K. 1992. The significance of layer B (square 8–19) of the Amud Cave (Israel) in the Levantine Levalloiso-Mousterian: A technological study, 83–106. In *The Evolution and Dispersal of Modern Humans in Asia*, eds. T. Akazawa, K. Aoki, and T. Kimura. Tokyo: Hokusen-Sha.

Ohnuma, K., and T. Akazawa. 1988. Reexamination of the lithic artifacts from Layer B (square 8–19) of the Amud Cave, Israel. *Paléorient* 14:137–44.

Pelegrin, J. 1990. Prehistoric lithic technology: Some aspects of research. *Archaeological Review from Cambridge* 9:116–25.

Perlès, C. 1992. In search of lithic strategies: A cognitive approach to prehistoric chipped stone assemblages, 223–47. In *Representations in Archaeology*, ed. J. C. Gardin and C. S. Peebles. Bloomington: Indiana University Press.

Pfaffenberger, B. 1992. Social anthropology of technology. *Annual Review of Anthropology* 21:491–516.

Rak, Y. 1993. Morphological variation in *Homo neanderthalensis* and *Homo sapiens* in the Levant: A biogeographical model, 523–36. In *Species, Species Concept and Primate Evolution*, ed. W. H. Kimbel and L. B. Martin. New York: Plenum Press.

Renfrew, C. 1978. The anatomy of innovation, 390–418. In *Approaches to Social Archaeology*, ed. C. Renfrew. Cambridge: Harvard University Press.

Rogers, A. R. 1988. Does biology constrain culture? *American Anthropologist* 90:819–31.

Ronen, A. 1979. Paleolithic industries, 296–307. In *The Quaternary of Israel*, ed. A. Horowitz. New York: Academic Press.

Schroëder, B. 1969. The Lithic Industries from Jerf Ajla and Their Bearing on the Problem of a Middle to Upper Paleolithic Transition. Ph. D. diss. Columbia University. New York.

Schwarcz, H. P., B. Blackwell, P. Goldberg, and A. E. Marks. 1979. Uranium series dating of travertine from archaeological sites. *Nature* 277:558–60.

Schwarcz, H. P., W. M. Buhay, R. Grün, H. Valladas, E. Tchernov, O. Bar-Yosef, and B. Vandermeersch. 1989. ESR dates of the Neanderthal site, Kebara Cave, Israel. *Journal of Archaeological Science* 16:653–9.

Shea, J. J. 1991. The Behavioral Significance of Levantine Mousterian Variability. Ph. D. diss. Harvard University. Cambridge.

Shimron, R. 1996. Layer B2 at Amud Cave: A technotypological analysis of a lithic sample from square S10. Seminar paper. Hebrew University.

Solecki, R. L., and R. S. Solecki. 1970. A new secondary flaking technique at the Nahr Ibrahim cave site. *Bulletin du Musée de Beyrouth* 23:137–42.

Spratt, D. A. 1982. The analysis of innovation processes. *Journal of Archaeological Science* 9:79–94.

Suzuki, H. 1970. Skeletal remains of the Amud Man: General conclusions, 421–2. In *The Amud Man and His Cave Site*, ed. H. Suzuki and F. Takai. Tokyo: Academic Press of Japan.

Valladas, H., J. L. Joron, G. Valladas, B. Arensburg, O. Bar-Yosef, A. Belfer-Cohen, P. Goldberg, H. Laville, L. Meignen, Y. Rak, E. Tchernov, A. M. Tillier, and B. Vandermeersch. 1987. Thermoluminescence dates for the Neanderthal burial site at Kebara in Israel. *Nature* 330:159–60.

Valladas, H., N. Mercier, L. Frojet, J. L. Joron, W. H. Kimbel, and Y. Rak. In press. TL dates for the Neandertal site of Amud Cave. *Journal of Archaeological Science*.

Watanabe, H. 1964. Les éclats et lames à chanfrein et la technique de fracturation transversale dans un horizon paléolithique en Palestine. *Bulletin de la Société Préhistorique Française* 4:84–8.

Watanabe, H. 1965. A Palaeolithic industry from the Amud Cave, Israel: A preliminary report. VIIth International Congress of Anthropological and Ethnological Sciences, Moscow. 15–21.

Watanabe, H. 1970. A palaeolithic industry from the Amud Cave, 77–114. In *The Amud Man and His Cave Site*, ed. H. Suzuki and F. Takai. Tokyo: Academic Press of Japan.

Ziaei, M., H. P. Schwarcz, C. M. Hall, and R. Grün. 1990. Radiometric dating at the Mousterian site at Quneitra, 232–5. In *Quneitra—A Mousterian Site on the Golan Heights*, ed. N. Goren-Inbar. QEDEM 31. Jerusalem: Institute of Archaeology, Hebrew University.

11

HAYONIM CAVE LITHIC ASSEMBLAGES IN THE CONTEXT OF THE NEAR EASTERN MIDDLE PALEOLITHIC

A Preliminary Report

Liliane Meignen

ERA 28 du Centre de Recherches Archéologiques, CNRS
Sophia Antipolis, Valbonne 06560, France

1. INTRODUCTION

The Near East, as a geographical corridor between Africa and Europe, and as a region where both archaic and anatomically modern human remains associated with Middle Paleolithic industries were uncovered, is an area of crucial importance in the discussion of the origins of modern humans. In 1982, an interdisciplinary research project was initiated in order to establish the chronology of this period, acquire greater knowledge of the subsistence strategies, and gain a better understanding of the technical behavior of Mousterian populations living in this region (Bar-Yosef et al. 1992).

This international research program began with the study of two cave sites (Qafzeh and Kebara) which are known for their human remains (*Homo sapiens sapiens* and *Homo sapiens neanderthalensis* respectively) and the associated Mousterian lithic and faunal assemblages (Boutié 1989; Bar-Yosef et al. 1992; Bar-Yosef and Vandermeersch 1992; Meignen and Bar-Yosef 1992; Speth and Tchernov 1995). At the same time, dating by members of the project and others placed the Kebara and Qafzeh sequences within the period of 48 to 100 ka and even confirmed that the long duration of the Middle Paleolithic, aş already known in Western Europe, is valid for the Near East (Valladas et al. 1987; Schwarcz et al. 1988; Valladas et al. 1988; Schwarcz et al. 1989; Mercier et al. 1993; Mercier and Valladas 1994; Mercier et al. 1995).

In the context of this research, Hayonim Cave (Figure 1) was chosen to provide new data concerning human behavior and past technical knowledge over a period that corresponds to the early Middle Paleolithic. Previous excavations at this site directed by Bar-Yosef, Arensburg, and Tchernov (Arensburg et al. 1990; Bar-Yosef 1991; Tchernov 1992) exposed an important Mousterian sequence that, on the basis of both preliminary biostrati-

Neandertals and Modern Humans in Western Asia, edited by Akazawa *et al.*
Plenum Press, New York, 1998.

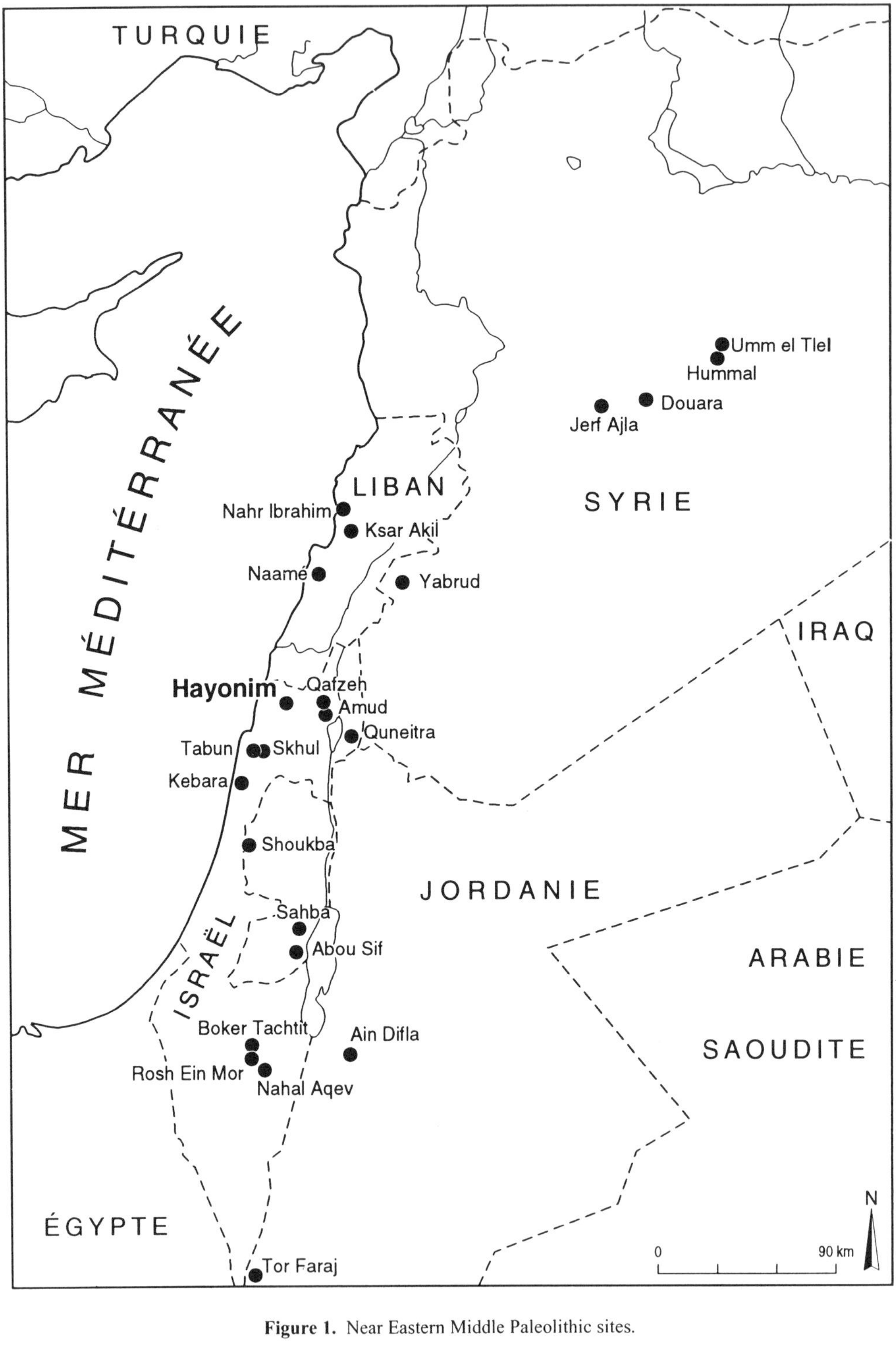

Figure 1. Near Eastern Middle Paleolithic sites.

graphic and lithic studies and a Th/U date, was considered to have preceded the Qafzeh human-bearing layers.

Aside from achieving an improved chronology, the new excavations in this cave, which began in 1992, provided the opportunity to better understand the technical behavior of early Middle Paleolithic populations. The last decade of prehistoric studies has already demonstrated an unsuspected variability in lithic reduction strategies (*chaînes opératoires*) during this period. In particular, blade production, long considered an Upper Paleolithic phenomenon, has recently been discovered in several different Middle Paleolithic contexts (e.g., Meignen 1994 and references therein; Revillion and Tuffreau 1994a,b). The presence of this special blade manufacture in Hayonim prior to 160 ka (Schwarcz et al. 1980) has aroused our interest in the assemblages of Hayonim Cave.

Two mains areas of Hayonim Cave are currently under excavation (Bar-Yosef, this volume). In the central area, the upper part of layer E mainly consists of massive, soft, silty deposits, locally highly phosphatized. Ash accumulations are present as superimposed, thin, whitish lenses or thicker, circularly-shaped hearths, more or less well-preserved. In the northeast corner, a stalagmitic layer covers these Upper Mousterian layers. At the entrance of the cave, the Mousterian deposits are brecciated along the wall and beneath the drip line. A deep sounding penetrates into the lower part of layer E in this area.

Detailed stratigraphical relationships between these two areas are not yet fully established. However, the field observations clearly indicate that the layers in the central area overlie the deposits exposed in the deep sounding at the cave's entrance.

For the purpose of this preliminary study, and in order to obtain an initial understanding of the lithic industries, we selected two large samples of artifacts. One sample is derived from Upper E, collected in the central area. The other comes from Lower E, collected in the deep sounding at the entrance.

The technological approach developed here attempts, in its first phase, to identify the technical lithic systems adopted by the different inhabitants of the cave. Typological studies focus mainly on the final end-products and in general do not take into account the flaking methods employed to manufacture the blanks. The same blank morphology can be obtained by entirely different technical systems (e.g., Marks and Volkman 1983; Boëda 1995; Meignen 1995). It has already been stressed that elongated Levallois points, for example, could have been produced not only through the Levallois flaking methods but also by an Upper Paleolithic-like core reduction. This contention leads us to base the identification of the lithic technical system on the whole dynamic process, from the procurement of raw material to tool production and discard. Our purpose is to decipher the volumetric conceptions adopted by the knappers. This knowledge is reached by understanding the geometric construction of the core and the flintknappers' perception of the volume to be worked, as well as by identifying the steps in the process of knapping that are necessary for obtaining end-products of a desired morphology.

2. LITHIC ASSEMBLAGES

The sourcing of the procured raw materials is currently being investigated by C. Delage. Preliminary results (Delage, personal communication) indicate the presence of four different abundant flint sources of different sizes and qualities within a 10 km radius around the site. Most of the studied artifacts were made of these raw materials collected in the vicinity and transported as blocks and nodules to the cave. However, a few blanks struck from Middle Eocene flints clearly indicate that they came from a source more than 40 km east or south of Hayonim Cave (Delage, personal communication).

2.1. Unit Upper E

This sample was collected from the successive *décapages* excavated on more than 20 m^2 in the central area. In all these levels, lithic densities are relatively low (from 200 to 650 pieces >2.5 cm per cubic meter) in comparison to Kebara Cave. These low densities of artifacts are interpreted as the result of ephemeral occupations, even though numerous cortical elements suggest knapping activities occurred in the cave.

In this unit, the dominant core reduction strategy is clearly linked with the Levallois method, with its now well-known volumetric construction (Boëda 1994, 1995), resulting in more or less flat cores after the last removals.

The debitage was mainly aimed at the production of Levallois flakes (more than 84%), often with faceted butts. Very few true Levallois points and blades are present. The Levallois recurrent method (as defined by Boëda 1986, 1990) was practiced, through which several flakes were struck from each prepared flaking surface. Core management is evidenced by the presence of centripetal and unidirectional removals, rather evenly represented (Figure 2), resulting in quadrangular and triangular blanks.

Retouched tools (Figure 3) include mainly numerous side-scrapers on large, highly-selected Levallois blanks. A series of well-struck typical burins has also been recovered.

In the context of the Levantine Mousterian, if we take into account the high proportion of short Levallois blanks (mainly flakes) and the scarcity of broad-based Levallois points, this lithic assemblage conforms closely with those from part of unit I in Tabun Cave (the so-called "Tabun C-type" industries). This group includes lithic assemblages from Tabun unit I (Jelinek 1981), most layers from Qafzeh (Boutié 1989), Ksar Akil layer XXVI (Marks and Volkman 1986), and Douara layer III (Akazawa 1979). Quneitra could probably be included, although some particularities must be emphasized concerning the retouched tools in this site (Goren-Inbar 1990). All of these sites are located in the northern and central part of the Levant, and Hayonim Cave fits well with this already recognized pattern.

More precise comparisons with Qafzeh industries (from Vandermeersch's excavations), recently studied by E. Hovers, are extremely significant. While the rather elongated flake production is largely predominant in both sites, the Hayonim Upper E assemblage seems to be more diversified in terms of core management methods. Centripetal removals are more common in all layers from Qafzeh, resulting in ovoid flakes, while unidirectional core exploitation is slightly higher in Hayonim Upper E, resulting in more quadrangular and subtriangular blanks. In the same way, the lithic assemblage from Ksar Akil layer XXVI demonstrates some diversity in the Levallois method adopted, directed at the production of subovular and quadrangular flakes.

The Hayonim Upper E industry differs markedly from the lower units (IX-XII) in Kebara Cave (Meignen and Bar-Yosef 1991, 1992) and from Tor Faraj (Henry 1992), where the prevailing unidirectional convergent scheme results in high frequencies of triangular blanks, including the classical broad-based Levallois points. The latter are virtually absent in the upper layers at Hayonim. On the other hand, in the upper units of Kebara (VII-VIII), a slight increase in the centripetal flaking pattern gives a picture closer to Hayonim Upper E, though less unidirectional.

This preliminary examination of Levantine Levallois flake core reduction strategies demonstrates the variability of the Levallois recurrent methods and the difficulty of establishing clear-cut categories. The more we recognize the details of the Levallois components, the more it becomes clear that it is impossible to organize them into completely separate facies. In fact, the Levallois technology is mainly characterized by this flexibility in core management, resulting not in a unique standardized blank production as in the Upper Paleolithic, for example, but in an array of products of controlled morphology.

2.2. Unit Lower E

The sample presented here was selected from the lower levels of the deep sounding. The most striking feature observed in this assemblage—which in fact constitutes a surprising phenomenon in Middle Paleolithic industries—is the high proportion of elongated blanks which are frequently retouched into elongated points.

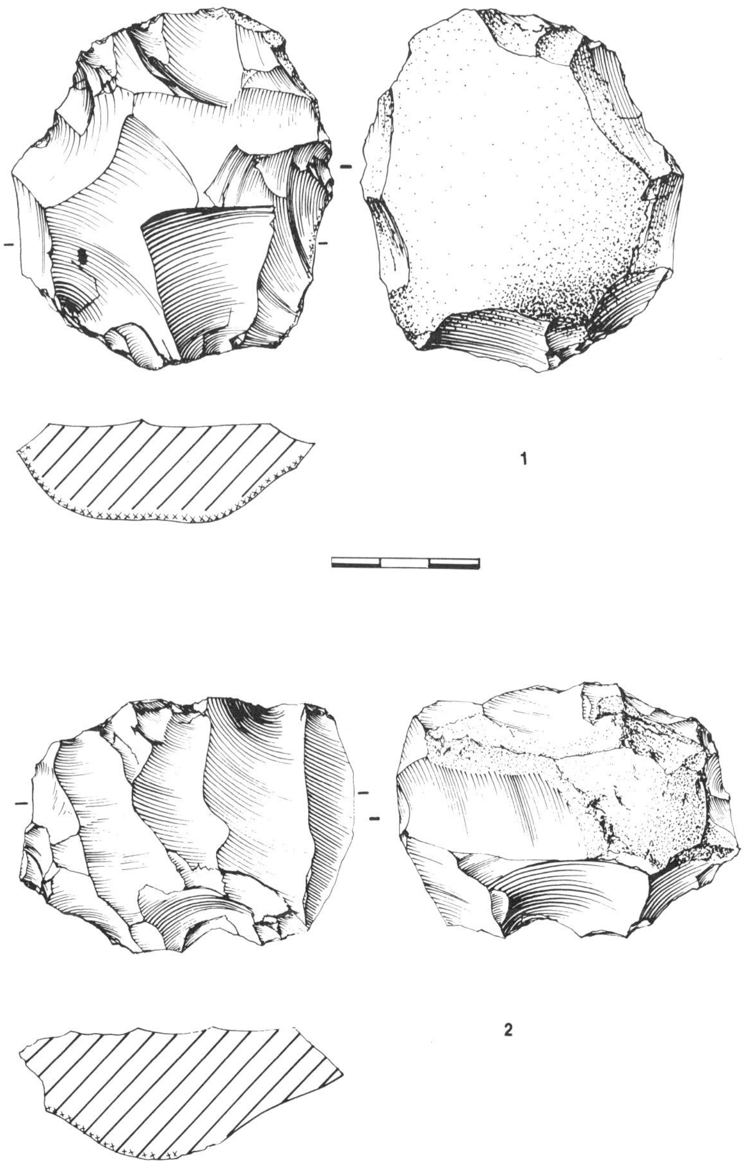

Figure 2. Hayonim Upper E: Levallois recurrent cores.

Our main purpose was to understand which kinds of core reduction strategies were used in the production of this blady industry. Taking into account all the products (blanks, cores, core trimming elements, and tools), we used some specific observations to establish a preliminary reconstruction of the adopted production system:

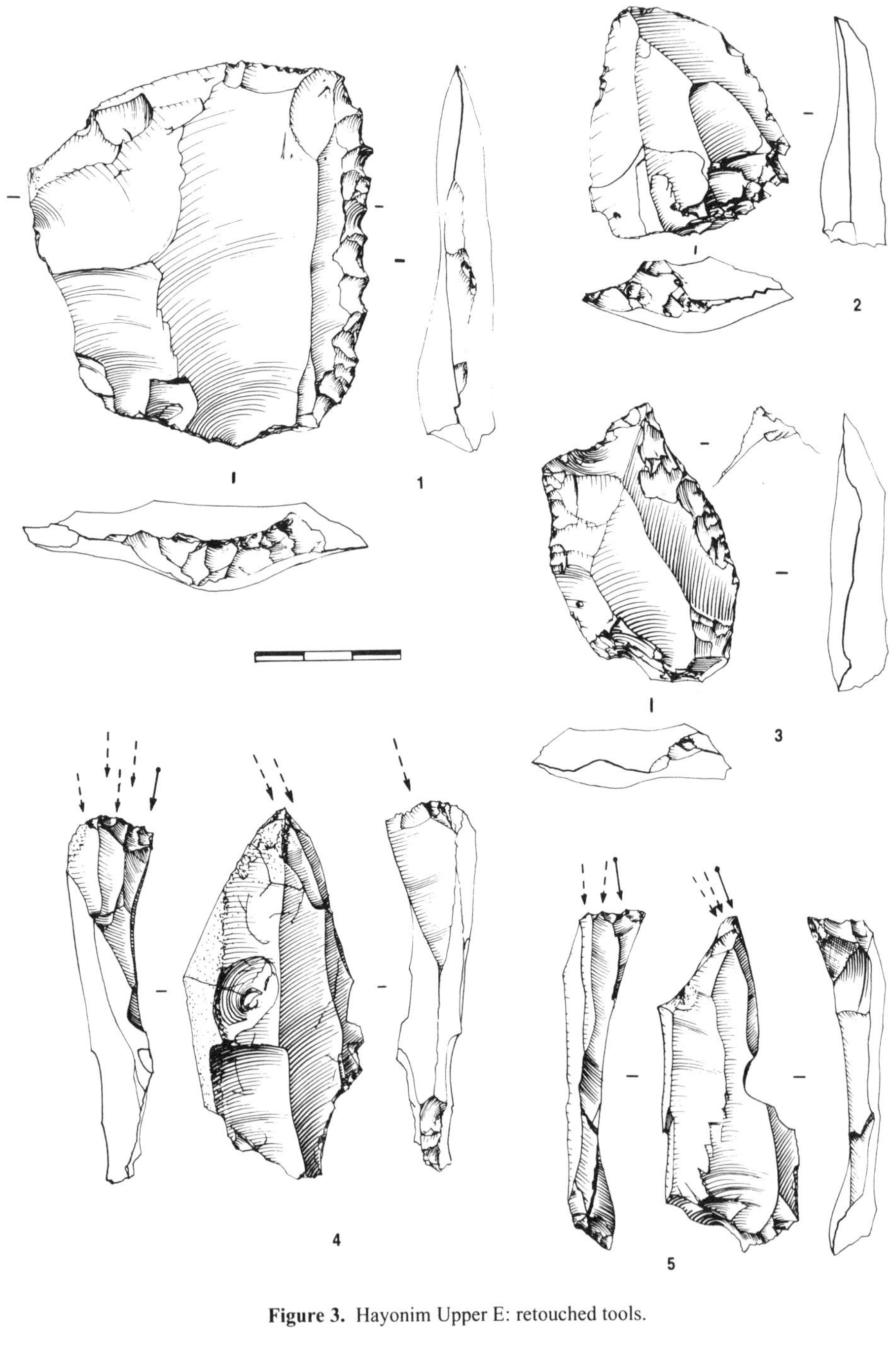

Figure 3. Hayonim Upper E: retouched tools.

1. The desired end-products consist of elongated, narrow, and thick blades, often retouched with triangular or trapezoidal sections (Figures 4, 5). This implies relatively narrow cores with highly oblique lateral sides on the cross section (Figure 6 no. 1), a characteristic which is totally different from a Levallois core section.
2. The elongated products were struck from different kinds of cores. One type includes cores with a single striking platform, where the flaking surface expanded to the lateral edges of the core, resulting in a semi-pyramidal core with a highly convex transverse section (Figure 6 no. 1). Change in the orientation of part of

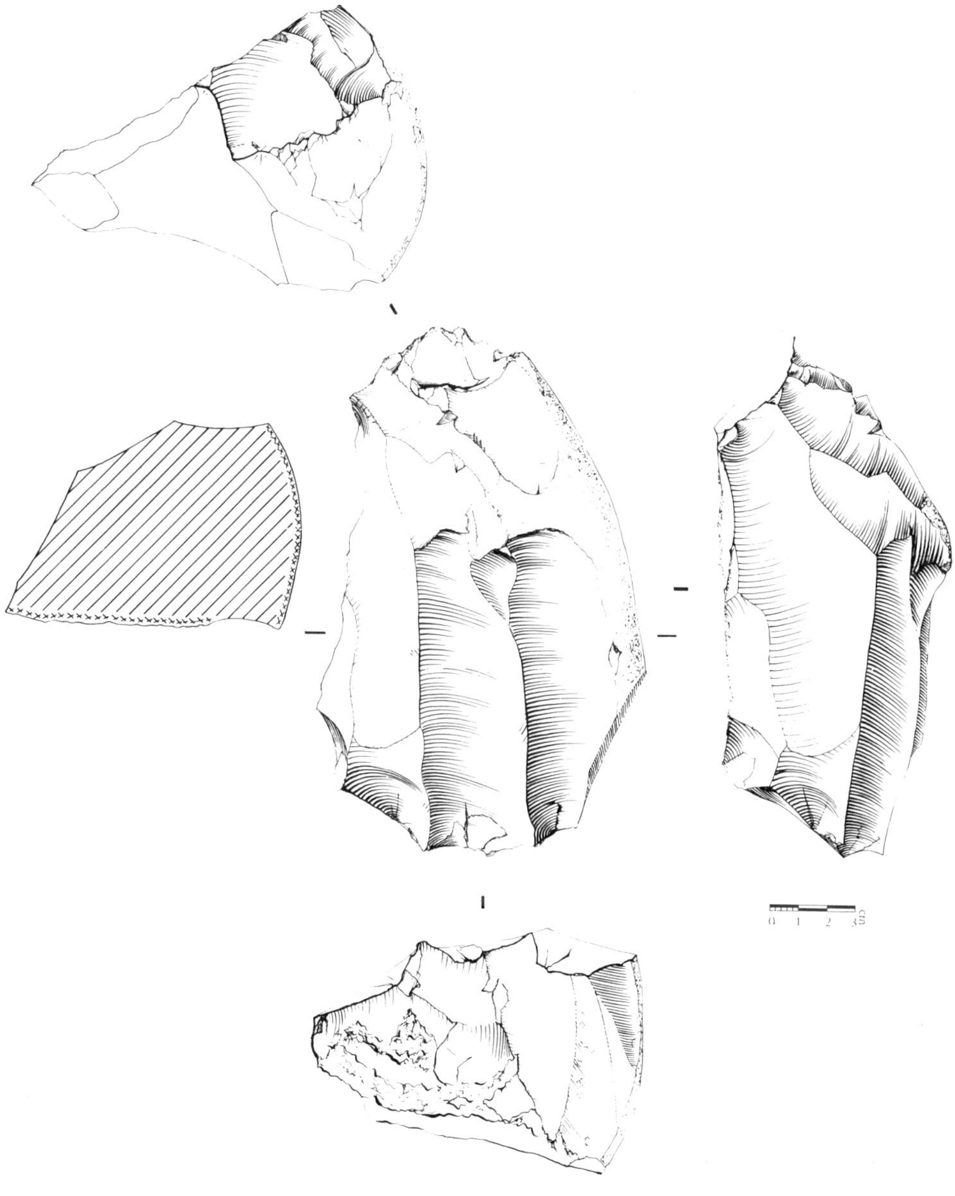

Figure 4. Hayonim Lower E: retouched blades and points.

the striking platform allows the removals from the lateral edges. Another core type consisted of items with two opposed platforms. In some cases, a second striking platform was established with a particular position, not exactly along the same axis, slightly twisted (Boëda 1995). Two reduction surfaces whose intersection creates the necessary convexities are then exploited (Figure 7).

3. Accidental plunging blades (Figure 8) demonstrate the same morphology and confirm the reconstructed volumetric structure as previously identified. The two opposed platforms, slightly twisted, can be observed. These indicate two secant flaking surfaces, not only one convex surface which characterizes the Levallois concept.
4. A few crested blades have been uncovered (Figure 6 no. 2, 3). Classical ones have two prepared sides and their removal allowed the beginning of the core reduction. Lateral and partial ones ensure control of the core shape during the process of knapping and the enlargment of the knapping surface.
5. The characteristics of the striking platform, bulb of percussion, and ventral surface of the products suggest direct percussion with stone hammer, probably "soft" stone (P. Bodu and J. Pelegrin, personal communication).

Thus the laminar debitage system recognized in Hayonim Lower E, which brings to mind the Upper Paleolithic debitage, differs significantly from Levallois core exploitation, and even from the unidirectional or bidirectional recurrent methods for Levallois blade production (Boëda 1988). The way in which the knappers organized the volume to be worked is different. In the case of the Levallois method, the debitage is organized along

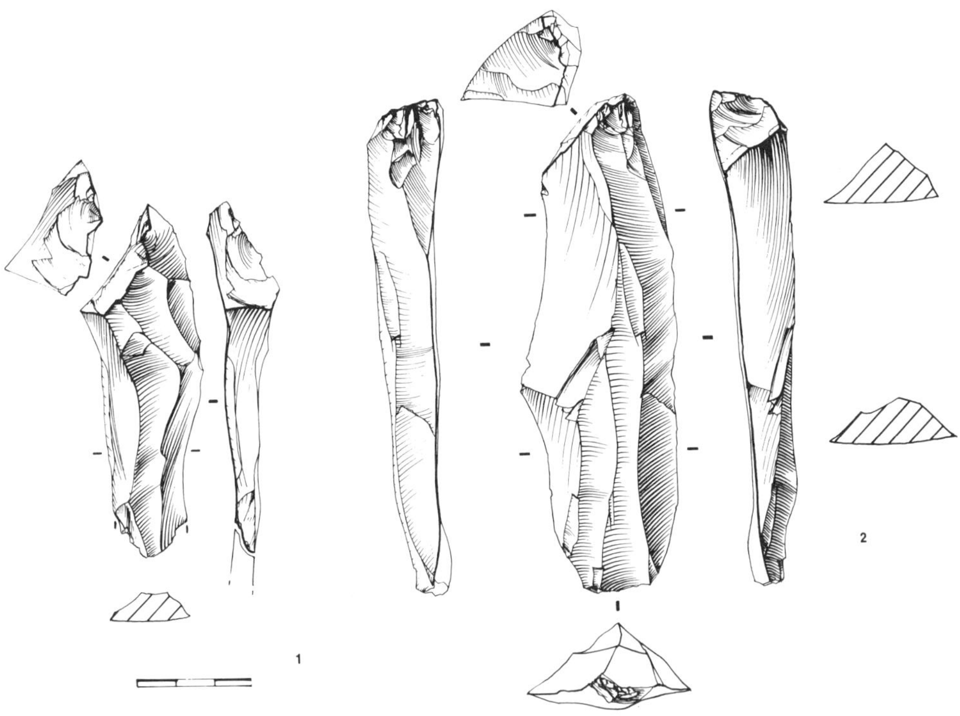

Figure 5. Hayonim Lower E: retouched blades.

the flattened surface of the block, with a series of removals more or less parallel to this plane. In contrast, reduction using the laminar method follows the maximum length, organized along the thickness of the block. All the volume is then exploited in a continuous process, if it is well controlled, without reshaping. The latter conception, which is recognized at Hayonim Lower E, was developed primarily during the Upper Paleolithic.

In Upper E, apart from this specific blade technology, a Levallois core reduction strategy is also carried out and is directed towards the production of wider blanks (flakes

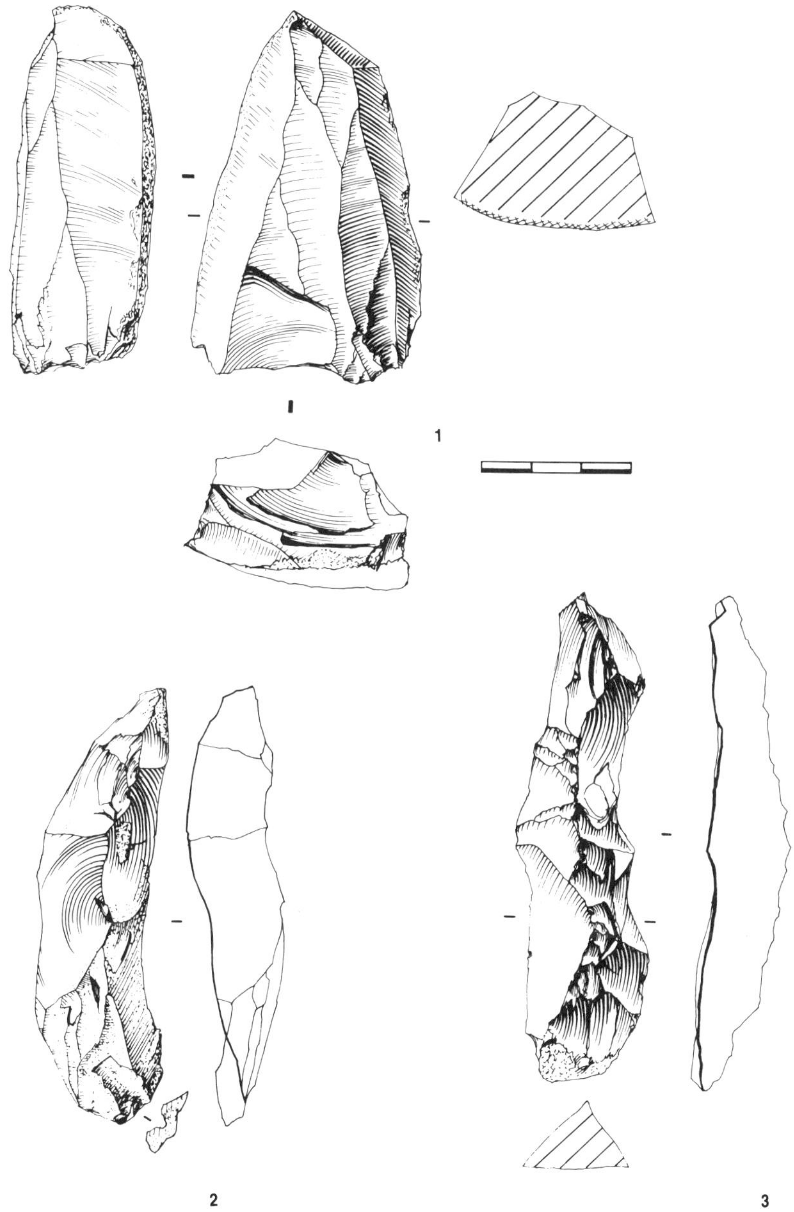

Figure 6. Hayonim Lower E: 1: blade core with a single striking platform; 2, 3: crested blades.

and points), often by unidirectional removals. Whether some elongated products (blades and elongated points) result from the Levallois technology is still debated and must be checked in the near future.

Retouched tools in Lower E are numerous (more than 30%), including many elongated points, retouched blades, sidescrapers (in some levels, often convergent), and typical burins. The largest blanks have been selected for retouch.

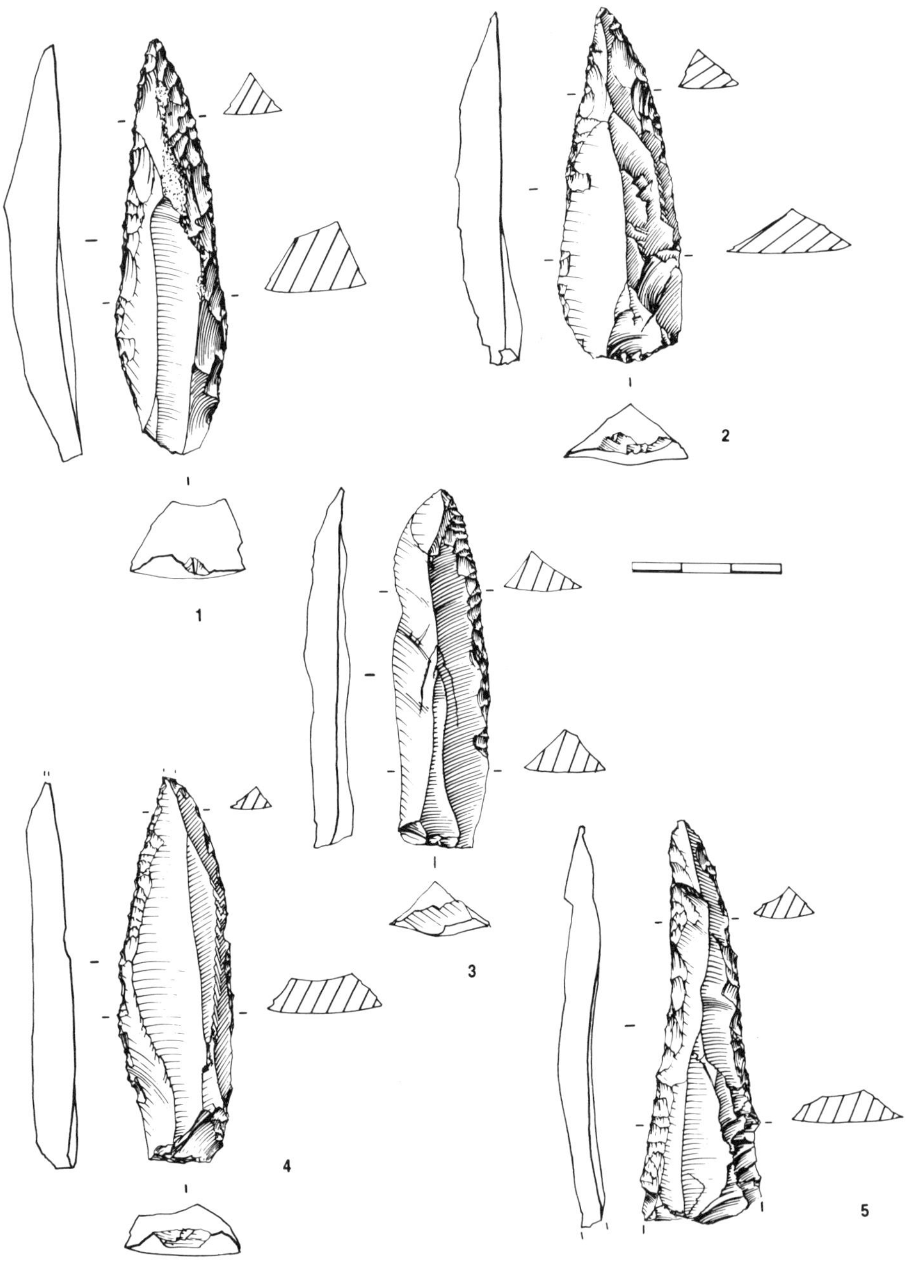

Figure 7. Hayonim Lower E: blade core with two opposed striking platforms slightly twisted.

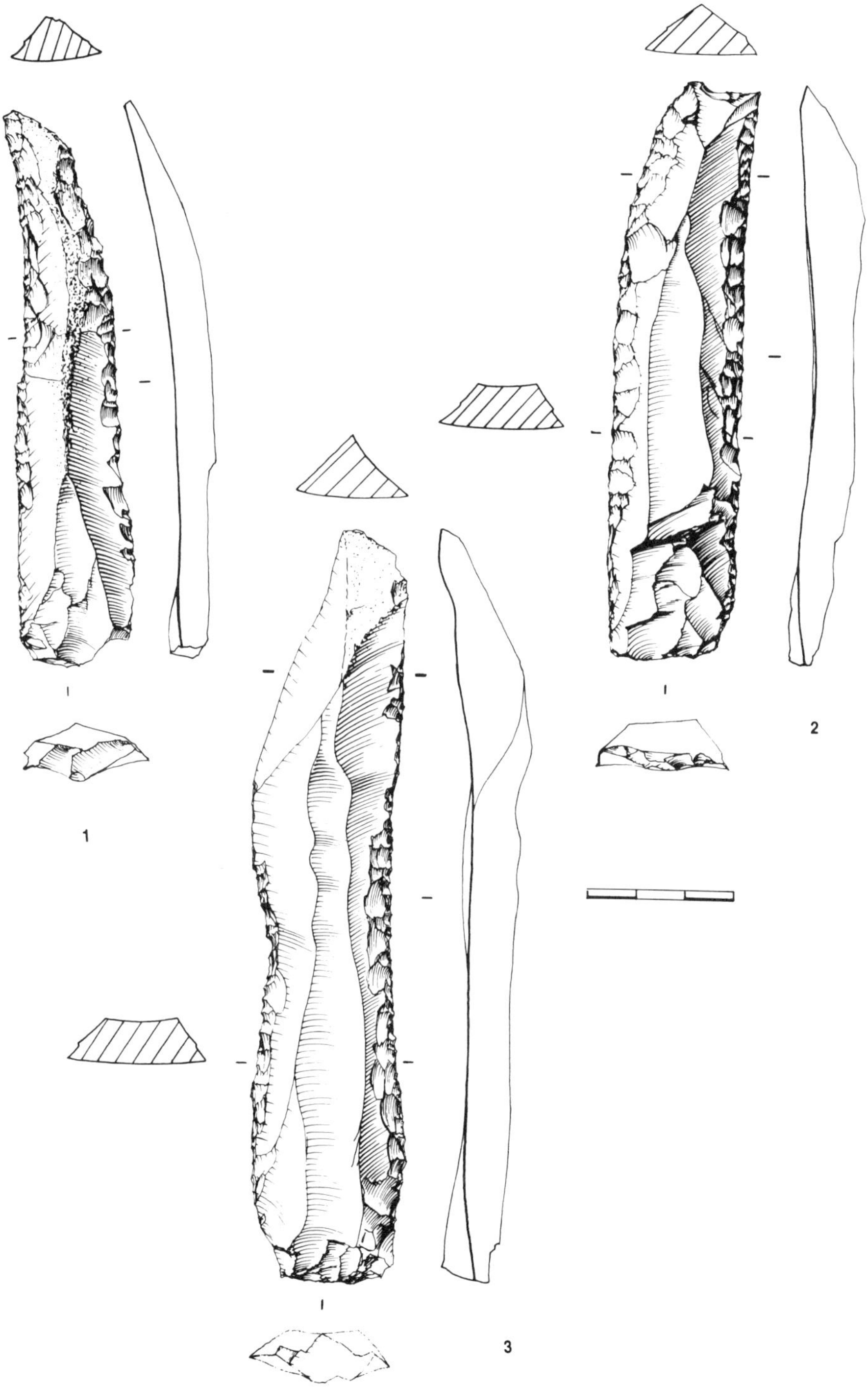

Figure 8. Hayonim Lower E: plunging blades demonstrating the two "twisted" opposed striking platforms.

3. HAYONIM LAYER E AMONG OTHER LEVANTINE SITES

Generally speaking, the assemblage of Hayonim Lower E shares striking common features with the so-called "Tabun D-type" industries, or Early Levantine Mousterian, which are found in several sites throughout the Near East. All these lithic assemblages are grouped under this term, without considering their debitage systems. But careful reading of the reports and preliminary technological studies already indicate diversified core reduction strategies (Levallois and non-Levallois), leading to morphologically identical products (blades and points) in this large techno-complex (Meignen 1994; Marks and Monigal 1995).

Blade manufacture during the Lower and Middle Paleolithic is now a well-known phenomenon that, even though rather infrequent, spread to many areas of the world (Europe, Africa, Western Asia). In the Near East, blade-dominated assemblages (Amudian, Hummalian, pre-Aurignacian, Levallois blade productions) have been discovered in various chronological contexts (from isotopic stages 8 to 5). The primary development of this laminar phenomenon in this part of the world, compared to other areas such as Western Europe, is worth noting.

Most Near Eastern assemblages with elongated blanks, apart from the older blady industries such as the Amudian, Hummalian, and pre-Aurignacian, have been identified as linked with the Levallois technology (see Meignen 1994 for a general review). The recently developed technological approach in lithic studies provides a more complex picture of these assemblages.

A systematic blade production in a volumetric concept close to Hayonim Lower E (let us call it "laminar system" as opposed to "Levallois elongated blank production system"; Meignen 1994) is already recognized in a few sites:

1. In Hummal Ia (Syria), E. Boëda has recently pointed out that thick blades are removed in a series from two opposed striking platforms "not along the same axis" with a hard hammer technique (Boëda 1995).
2. In Rosh Ein Mor (Negev, Israel), some of the elongated blanks were obtained from unidirectional prismatic cores, as recently demonstrated by A. Marks and K. Monigal (Marks and Monigal 1995). In this site, another core reduction strategy, the unidirectional Levallois recurrent method, is also involved in flake and point production.
3. In Douara IV (Syria), thick elongated blanks clearly result from cores whose flaking surfaces expanded onto the lateral sides of the prismatic cores, as described by Akazawa (1979). On the basis of technological attributes, Nishiaki (1987, 1989) has determined closer affinities for the Douara IV assemblage with the early Levantine Mousterian from Tabun IX. In our opinion, most of the available data concerning cores and elongated blanks would indicate the presence of a laminar system as defined above.
4. In Ain Difla (Jordan), Lindly and Clark (1987) identified very narrow elongated blanks that seem to have been knapped from opposed platform narrow cores. Demidenko and Usik (1993) have recently reported the presence of crested blades in this assemblage.

It must be stressed that, just as in Hayonim Upper E, Levallois core reduction strategies coexisted with this laminar system in both Rosh Ein Mor and Douara IV. In contrast, blade production in Tabun unit IX (Tabun D) seems to be linked exclusively to the Levallois concept of core reduction, in a uni/bidirectional modality. However, this assumption must be checked in more detail, taking into account recent discoveries.

In most of these blade-geared industries, the retouched components consist mainly of elongated, retouched points and blades. According to this criterion, the Hayonim Lower E tool-kit (Figures 4, 5) closely resembles the elongated pointed tools from Hummal Ia (Copeland 1985), Abu Sif (Neuville 1951), and even Tabun IX (despite the different geometric construction of the core in the latter). A first glance shows several striking similarities in blanks and in the general composition of tool-kits between Hayonim Lower E and Abu Sif: the presence of numerous thick blades, possibly not retouched; the abundance of elongated pointed tools, many of them with a noticeable asymmetry described as "pointes incurvées" by Neuville (1951); and the presence of short Levallois points, retouched or non-retouched. Only this last characteristic is missing in the Hummal Ia assemblage. The Douara unit IV industry differs significantly from all these blady facies in terms of retouched tools. Here, Upper Paleolithic types dominate, mainly burins and truncated pieces, but the sample collected is very small (Nishiaki 1987).

4. CONCLUSION

This preliminary study provides new data concerning human technical knowledge during the Middle Paleolithic period. The following points summarize the main issues:

1. Elongated blank production precedes short blank production in Hayonim Cave, as already observed in Tabun, Douara, and the basin of El Kowm (Syria). The new TL dates recently published (Mercier et al. 1995; Mercier and Valladas, this volume) indicate a range of 170–190 ka years for this important technical change in the Near East in both Hayonim and Tabun caves.
2. These first results already emphasize the presence of technological variability in the blady core reduction strategies of the so-called Early Levantine Mousterian, considered solely as Levallois until recently. Hence, the "Tabun D-type" industries encompass several technical systems of lithic production that need to be elucidated in detail. Hayonim Lower E is just the first step.
3. It is worth stressing that blade production based on a volumetric concept similar to that developed later by Upper Paleolithic people has already been found in Hayonim Cave. This industry dates to around 190 ka, long before the appearance of morphologically modern humans. The same laminar phenomenon is also reported in Western Europe, especially in northern France, where it is dated to isotope stages 5–4 in several sites. This laminar system and the Levallois technology coexist in the same assemblage in Hayonim, as in other sites from the Near East such as Rosh Ein Mor and Douara, as well as sites in France in Riencourt-lès-Bapaume (Ameloot-Van Der Heijden 1994) and Seclin (Revillion 1994).

Consequently, recent dating suggests a relatively early period for the sporadic appearance of true laminar technical systems in the Levant. The oldest is the Amudian, dated in Tabun XI to around 300 ka (Mercier et al. 1995). Thus, contrary to what was claimed for a long time, the sole presence of true blade-geared industries is not in itself indicative of modern human behavior.

Not represented in Late Mousterian assemblages, it is only around 45–40 ka that the true blade technology greatly increases with the emergence of the Ahmarian industries (e.g., Bergman 1987; Ferring 1988; Ohnuma 1988; Bar-Yosef and Belfer-Cohen 1988; Gilead 1991; Marks 1992). The Upper Paleolithic laminar products obtained within a lithic reduction system comparable to the one described above are thinner, more regular in shape, and more numerous from each core. These are achieved by a more cautious core-shaping and change in the technique of percussion such as the use of soft hammer as opposed to hard hammer (Meignen 1996).

Hence, the concept of blade production developed during the Upper Paleolithic was already part of the Mousterian technical knowledge, even if it was not largely adopted. The most important changes we have observed seem to concern the techniques adopted and the care with which flintknappers shaped the core in the first step of the *chaîne opératoire*. In sum, from the Middle Paleolithic, when the laminar debitage was already known, through the Upper Paleolithic, during which time this lithic production is overwhelmingly practiced, we can observe a change in the general trend rather than a real technical innovation.

ACKNOWLEDGMENTS

This project was sponsored jointly by the National Science Foundation (grant awarded to O. Bar-Yosef) and the French Ministry of Foreign Affairs (grant awarded to B. Vandermeersch and L. Meignen). I am grateful to Erella Hovers for information on the lithics from Qafzeh.

REFERENCES

Akazawa, T. 1979. Middle Paleolithic assemblages from Douara caves. *Bulletin of the Museum of the University of Tokyo* 16:1–30.

Ameloot-Van Der Heijden, N. 1994. L'ensemble lithique du niveau CA du gisement de Riencourt-lès-Bapaume (Pas-de-Calais), 63–75. In *Les Industries Laminaires au Paléolithique Moyen*, ed. S. Révillion and A. Tuffreau. Paris: CNRS Editions.

Arensburg, B., O. Bar-Yosef, A. Belfer-Cohen, and Y. Rak. 1990. Mousterian and Aurignacian human remains from Hayonim cave, Israel. *Paléorient* 16:107–9.

Bar-Yosef, O. 1991. The archaeology of the Natufian layer at Hayonim cave, 81–93. In *The Natufian Culture in the Levant*, ed. O. Bar-Yosef and F. Valla. Ann Arbor: International Monographs in Prehistory.

Bar-Yosef, O., and A. Belfer-Cohen. 1988. The early Upper Paleolithic in Levantine caves, 23–41. In *The Early Upper Paleolithic-Evidence from Europe and the Near East*, ed. J. F. Hoffecker and C. A. Wolf. Oxford: British Archaeological Reports International Series 437.

Bar-Yosef, O., and B. Vandermeersch, eds. 1991. *Le Squelette Moustérien de Kebara 2.* Paris: CNRS Editions.

Bar-Yosef, O., B. Vandermeersch, B. Arensburg, A. Belfer-Cohen, P. Goldberg, H. Laville, L. Meignen, Y. Rak, J. D. Speth, E. Tchernov, A. M. Tillier, and S. Weiner. 1992. The excavations in Kebara cave, Mt. Carmel. *Current Anthropology* 33(5):497–550.

Bergman, C. O. K. 1987. The Upper Paleolithic sequence of Ksar Akil, Lebanon. *Berytus* 35:13–40.

Boëda, E. 1986. Approche Technologique du Concept Levallois et Evaluation de Son Champ d'Application. Ph. D. diss. Université de Paris X. Nanterre.

Boëda, E. 1988. Le concept laminaire: Rupture et filiation avec le concept Levallois, 41–59. In *L'homme de Néandertal*, No. 8, ed. M. Otte. Liège: ERAUL.

Boëda, E. 1990. De la surface au volume: Analyse des conceptions des débitages Levallois et laminaire, 63–8. In *Paléolithique Moyen Récent et Paléolithique Supérieur Ancien en Europe,* ed. C. Farizy. Nemours: APRAIF.

Boëda, E. 1994. *Le Concept Levallois: Variabilité des Méthodes*. Paris: CNRS Editions.

Boëda, E. 1995. Levallois: A volumetric construction, methods, a technique, 41–68. In *The Definition and Interpretation of Levallois Technology*, ed. H. Dibble and O. Bar-Yosef. Madison: Prehistory Press.

Copeland, L. 1985. The pointed tools of Hummal Ia (El Kowm, Syria). *Cahiers de l'Euphrate* 4:177–89.

Demidenko, Y. E., and V. I. Usik. 1993. The problem of changes in Levallois technique during the technological transition from the Middle to Upper Paleolithic. *Paléorient* 19(2):5–15.

Ferring, C. R. 1988. Technological change in the Upper Paleolithic of the Negev, 333–48. In *Upper Pleistocene Prehistory of Western Eurasia,* ed. H. Dibble and A. Montet-White. Philadelphia: The University Museum of the University of Pennsylvania.

Gilead, I. 1991. The Upper Paleolithic period in the Levant. *Journal of World Prehistory* 5(2):105–54.

Goren-Inbar, N. 1990. *Quneitra: A Mousterian Site on the Golan Heights*. QEDEM 31, Jerusalem: Institute of Archeology, Hebrew University of Jerusalem.

Henry, D. O. 1992. Transhumance during the Late Levantine Mousterian, 143–62. In *The Middle Paleolithic Adaptation Behavior and Variability*, ed. H. Dibble and P. Mellars. Philadelphia: The University Museum of the University of Pennsylvania.

Jelinek, A. 1981. The Middle Paleolithic in the southern Levant from the perspective of the Tabun Cave, 265–80. In *Préhistoire du Levant*, ed. J. Cauvin and P. Sanlaville. Paris: CNRS Editions.

Lindly, J., and G. Clark. 1987. A preliminary lithic analysis of the Mousterian site of 'Ain Difla (WHS Site 634) in the Wadi Ali, West-Central Jordan. *Proceedings of the Prehistoric Society* 53:279–92.

Marks, A. E., and P. W. Volkman. 1983. Changing core reduction strategies: A technological shift from the Middle to Upper Paleolithic in the southern Levant, 13–33. In *The Mousterian Legacy: Human Biocultural Change in the Upper Pleistocene*, ed. E. Trinkaus. Oxford: British Archaeological Reports International Series 164.

Marks, A. E., and P. Volkman. 1986. The Mousterian of Ksar Akil: Levels XXVIA through XXVIIIB. *Paléorient* 12(1):5–20.

Marks, A. E. 1993. The Early Upper Paleolithic: The view from the Levant, 5–21. In *Before Lascaux: The Complex Record of the Early Upper Paleolithic*, ed. H. Knecht, A. Pike-Tay, and R. White. Boca Raton: CRC Press.

Marks, A. E., and K. Monigal. 1995. Modeling the production of elongated blanks from the early Levantine Mousterian at Rosh Ein Mor, 267–78. In *The Definition and Interpretation of Levallois Technology*, ed. H. Dibble and O. Bar-Yosef. Madison: Prehistory Press.

Meignen, L., and O. Bar-Yosef. 1991. Les outillages lithiques moustériens de Kebara (fouilles 1982–1985): Premiers résultats, 49–75. In *Le Squelette Moustérien de Kebara 2*, ed. O. Bar-Yosef and B. Vandermeersch. Paris: CNRS Editions.

Meignen, L., and O. Bar-Yosef. 1992. Middle Paleolithic variability in Kebara Cave (Mount Carmel, Israel), 129–48. In *The Evolution and Dispersal of Modern Humans in Asia*, ed. T. Akazawa, K. Aoki, and T. Kimura. Tokyo: Hokusen-Sha.

Meignen, L. 1994. Paléolithique moyen au Proche-Orient: Le phénomène laminaire, 125–59. In *Les Industries Laminaires au Paléolithique Moyen*, ed. S. Révillion and A. Tuffreau. Paris: CNRS Editions.

Meignen, L. 1995. Levallois lithic production systems in the Middle Paleolithic of the Near East: The case of the unidirectional method, 361–80. In *The Definition and Interpretation of Levallois Technology*, ed. H. Dibble and O. Bar-Yosef. Madison: Prehistory Press.

Meignen, L. 1996. Les prémices du Paléolithique supérieur au Proche-Orient, 107–28. In *The Last Neandertals The First Anatomically Modern Humans*, ed. E. Carbonell and M. Vaguero. Tarragona: Universitat Rovira i Virgili, Fondacio Catalana per a la Recerca.

Mercier, N., H. Valladas, O. Bar-Yosef, B. Vandermeersch, C. Stringer, and J. L. Joron. 1993. Thermoluminescence date for the Mousterian burial site of Es-Skhul, Mt. Carmel. *Journal of Archaeological Science* 20:169–74.

Mercier, N., and H. Valladas. 1994. Thermoluminescence dates for the Paleolithic Levant, 13–20. In *Late Quaternary Chronology and Paleoclimates of the Eastern Mediterranean*, ed. O. Bar-Yosef and R. S. Kra. Tucson and Cambridge: Radiocarbon and the Peabody Museum of Archaeology and Ethnology, Harvard University

Mercier, N., H. Valladas, G. Valladas, J. L. Reyss, A. Jelinek, L. Meignen, and J. L. Joron. 1995. TL dates of burnt flints from Jelinek's excavations at Tabun and their implications. *Journal of Archaeological Science* 22:495–509.

Neuville, R. 1951. *Le Paléolithique et le Mésolithique du désert de Judée*. Paris: Masson.

Nishiaki, Y. 1987. Middle Paleolithic Assemblages from the Douara Cave, 1984 Excavations. In *Paleolithic Site of Douara Cave and Paleogeography of Palmyra Basin in Syria, Part IV: 1984 Excavations*. Bulletin of the Museum of the University of Tokyo 29:61–96.

Nishiaki, Y. 1989. Early blade industries in the Levant: The placement of Douara IV industry in the context of the Levantine Early Middle Paleolithic. *Paléorient* 15(1):215–29.

Ohnuma, K. 1988. *Ksar Akil, Lebanon: A Technological Study of the Earlier Upper Paleolithic Levels of Ksar Akil, Vol. III: levels XXV–XIV*. Oxford: British Archaeological Reports International Series 426.

Revillion, S., and A. Tuffreau. 1994a. *Les Industries Laminaires au Paléolithique Moyen*. Paris: CNRS Editions.

Revillion, S., and A. Tuffreau. 1994b. Valeur et signification du débitage laminaire du gisement paléolithique moyen de Seclin (Nord), 19–43. In *Les Industries Laminaires au Paléolithique Moyen*, ed. S. Revillion and A. Tuffreau. Paris: CNRS Editions.

Schwarcz, H. P., P. Goldberg, and B. B. Blackwell. 1980. Uranium-series dating of archaeological sites in Israel. *Journal of Earth Science* 29:157–65.

Schwarcz, H. P., R. Grün, B. Vandermeersch, O. Bar-Yosef, H. Valladas, and E. Tchernov. 1988. ESR dates for the hominid burial site of Qafzeh in Israel. *Journal of Human Evolution* 17:733–7.

Schwarcz, H. P., W. M. Buhay, R. Grün, H. Valladas, E. Tchernov, O. Bar-Yosef, and B. Vandermeersch. 1989. ESR dating of the Neanderthal site, Kebara Cave, Israel. *Journal of Archaeological Science* 16:653–9.

Tchernov, E. 1992. Biochronology, paleoecology, and dispersal events of hominids in the southern Levant, 149–88. In *The Evolution and Dispersal of Modern Humans in Asia*, ed. T. Akazawa, K. Aoki, and T. Kimura. Tokyo: Hokusen-sha.

Valladas, H., J. L. Joron, G. Valladas, B. Arensburg, O. Bar-Yosef, A. Belfer-Cohen, P. Goldberg, H. Laville, L. Meignen, Y. Rak, E. Tchernov, A. M. Tillier, and B. Vandermeersch. 1987. Thermoluminescence dates for the Neanderthal burial site at Kebara in Israel. *Nature* 330:159–60.

Valladas, H., J. L. Reyss, J. L. Joron, G. Valladas, O. Bar-Yosef, and B. Vandermeersch. 1988. Thermoluminescence dates for the Mousterian Proto-Cro-Magnons from Qafzeh Cave (Israel). *Nature* 331:614–16.

12

BITUMEN AS HAFTING MATERIAL ON MIDDLE PALEOLITHIC ARTIFACTS FROM THE EL KOWM BASIN, SYRIA

Eric Boëda,[1] Jacques Connan,[2] and Sultan Muhesen[3]

[1]Department of Ethnology and Prehistory
Université de Paris X
Nanterre 92001, France
[2]Elf Exploration Production
Direction Exploration, CSTJF
Avenue Larribau, PAU 64018, France
[3]Director-General of Antiquities and Museums
Damascus, Syrian Arab Republic

1. INTRODUCTION

The El Kowm basin in Central Syria is located between the Palmyra basin and the valley of the Euphrates River (Figure 1). It is a depression 25 km wide and 80 km long, dominated to the east by Jabal Bicheri (rising to 850 m above sea level) and to the south by Jabal Minshar (879 m) and Jabal Mqaïbara (1110 m). Running down the center of this natural basin is an elongated plateau, the Qdeir plateau, carved out by Quaternary erosion (Figure 2).

During surveys in 1978 directed by T. Fujimoto (1979) and J. Cauvin (Cauvin et al. 1979), numerous lithic assemblages were discovered in the backdirt of many ancient wells situated on the periphery of the plateau or on the surface of tells. Since that time, two Middle Paleolithic sites have been excavated. One is the site of Hummal, situated on the plateau 3 km from the tell of El Kowm (Hours 1982; Le Tensorer and Hours 1989); the other is the site of Umm El Tlel, located on the northern slope of the Qdeir plateau (Molist et al. 1988; Boëda and Muhesen 1993).

Among the material collected at these two sites, we identified 16 artifacts, 15 flint and one limestone, which showed traces of a black substance on one or more surfaces. On

Neandertals and Modern Humans in Western Asia, edited by Akazawa *et al.*
Plenum Press, New York, 1998.

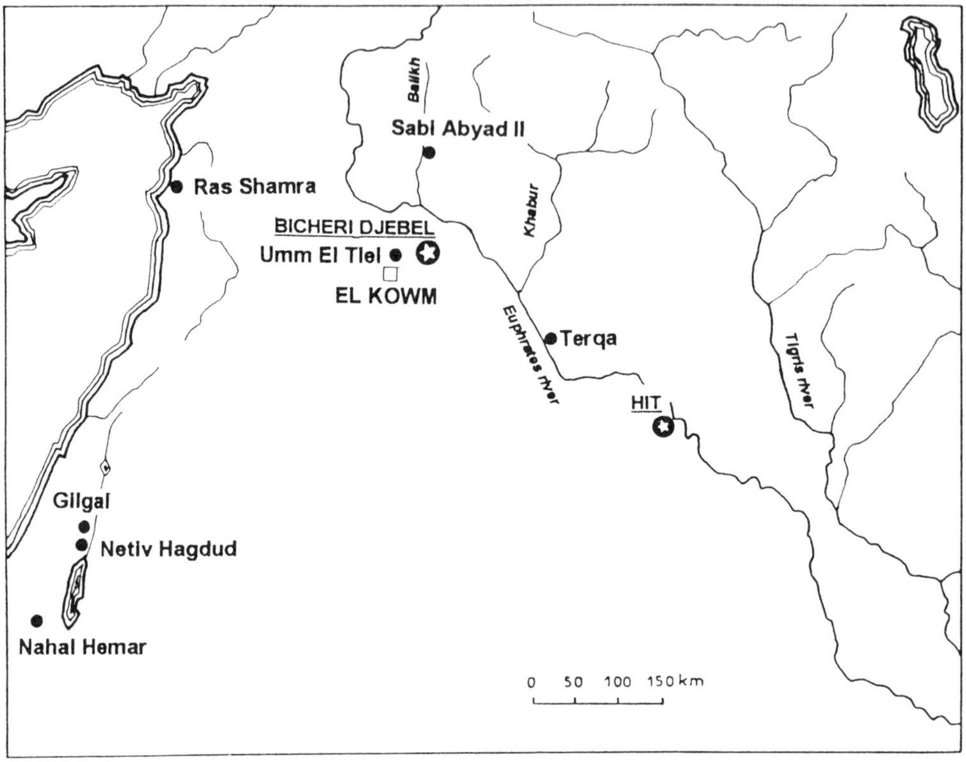

Figure 1. Location of archaeological sites and bitumen sources.

certain pieces this deposit was several millimeters thick, whereas other artifacts contained only a thin film of black substance.

Only one of the pieces mentioned above (number 263) was found at the site of Hummal in the Hummalian level. This piece is the only one among several hundred artifacts found in situ. Unfortunately, most of these pieces had been washed and brushed already, with the exception of those found in a bag containing a few pieces, including the one described below.

Fifteen pieces were found at Umm el Tlel, among more than 30,000 pieces uncovered in the excavations of this site since 1991. No other items with dark marks were found in the 87 archaeological levels exposed to date. Only five of these 15 pieces could be analyzed. These five artifacts show bitumen traces that could be associated with manganese deposits, while the remaining ten either do not have enough material for a chemical analysis or show that the black traces are evidently manganese. These manganese traces were not observed on any other material issued from the 87 levels. From the analysis conducted by one of us (Connan), it seems that manganese accumulated only on artifacts covered with bitumen. This association could possibly explain why black traces are either bitumen or manganese according to the location on the particular piece. In cases where the bitumen has disappeared, only manganese remains. This process has been clearly observed on pieces numbered 840 and 841. Thus bitumen preservation depends on differing surface conditions: bitumen is adhered to the flint on patinated and/or cortical artifacts, while bitumen covers the whole surface on non-patinated artifacts.

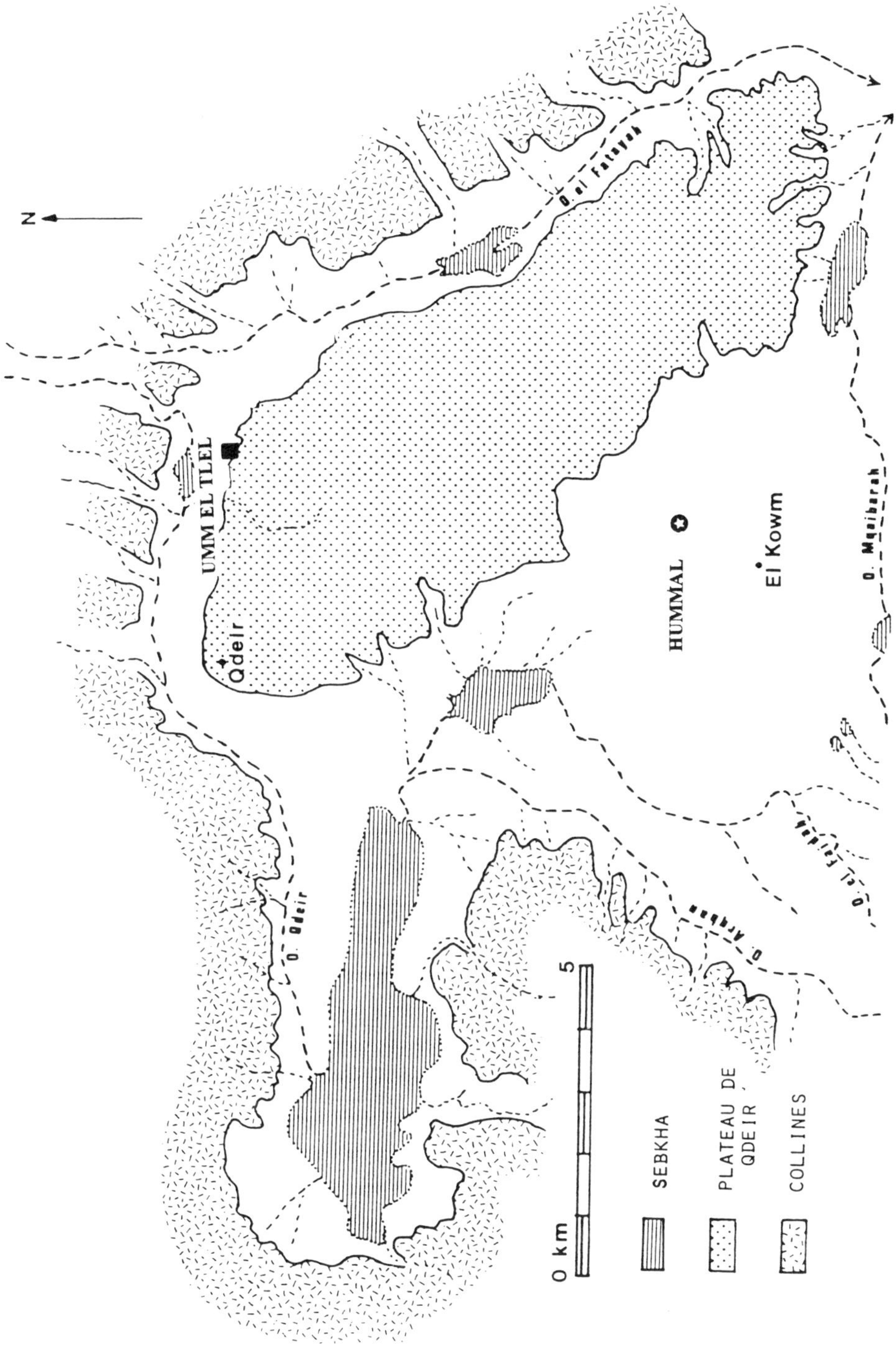

Figure 2. The El Kowm basin.

2. DESCRIPTION OF ARTIFACTS WITH BITUMEN-LIKE TRACES

2.1. Umm El Tlel

Umm el Tlel is an open-air site (Figure 3). Previous research has demonstrated that the site has major potentials for yielding archaeological materials. It was assumed that its stratigraphic sequence preserved a continuum from the Acheulean to the Neolithic. In 1987 and 1989, two soundings were carried out on the north hill. A step-trench revealed the presence of a series of Islamic, Romano-Byzantine, Neolithic (PPNB), Epipaleolithic (Geometric Kebaran), and Upper Paleolithic (Levantine Aurignacian) occupations (Molist et al. 1988). A Yabrudian level was found at more than 12 m deep inside the well (Copeland and Hours 1983). The excavations of the Paleolithic sequence, carried out from 1991 to 1996 (Boëda and Muhesen 1993), resulted in the identification of 87 layers spanning a thickness of more than 4 m. Thirty-one are assigned to the Upper Paleolithic (Levantine Aurignacian and indeterminate Upper Paleolithic industry); three are an Intermediate Paleolithic (Ahmarian (?) and Transitional Industry); and 53 are Middle Paleolithic. Three of these layers, IV2a, IV2b, and IV2cα, have produced 14 special pieces in a typical assemblage of the final Mousterian period.

2.1.1. Archaeological Levels Containing Artifacts with Black Traces. The three successive levels were uncovered from an area less than 5 m^2. Five hundred artifacts were found in level IV2a, including about twenty tools, mainly simple or double scrapers. Core reduction is mainly recurrent Levallois uni- and bi-directional, parallel or convergent, including the production of Levallois flakes of triangular shape. One hundred artifacts were retrieved from level IV2b, among which twenty are various tools. The debitage is preferentially recurrent Levallois, unipolar parallel, and convergent. Finally, Level IV2c contained only about fifty pieces, with a large number of Levallois products including typical Levallois points (Boëda 1994, 1995).

Despite the small sample of artifacts, the technical analyses demonstrate that 47 of the 53 Mousterian levels are different. This observation means that the use of bitumen in the three levels was practiced by more than one human group. This interpretation should be re-examined by subsequent analyses.

Fourteen pieces with black traces were found in the three successive Mousterian levels and are described below.

Level IV2a contained six artifacts. Two of these artifacts had black organic, bitumen-like traces: a convex/concave convergent scraper (number 359); and a quadrangular Levallois flake (number 738) (Boëda et al. 1996). The other four artifacts had traces that were too faint to be subjected to any analysis. The latter group includes a simple scraper, a *débordant* Levallois flake, a flake, and a limestone block (10×6×5 cm) with a black trace on its concave surface.

Level IV2b contained four artifacts. These include three artifacts with black organic, bitumen-like traces: a Levallois flake (number 839); a *débordant* Levallois flake with lateral *coup de tranchet* retouch of the *Prondnickmesser* type (number 841); and a cortical flake (number 840). The fourth artifact in this level was a simple hinged flake with black traces.

Level IV2cα contained five artifacts, but none of these have shown bitumen traces. Some do not display sufficient residue; others show only manganese deposits. New analyses are currently under way in order to identify other possible bitumen traces.

2.1.2. Post-Depositional Contamination? Natural bitumen has not been found within 40 km of the site. The rivers flow in the opposite direction from the site, thus no natural

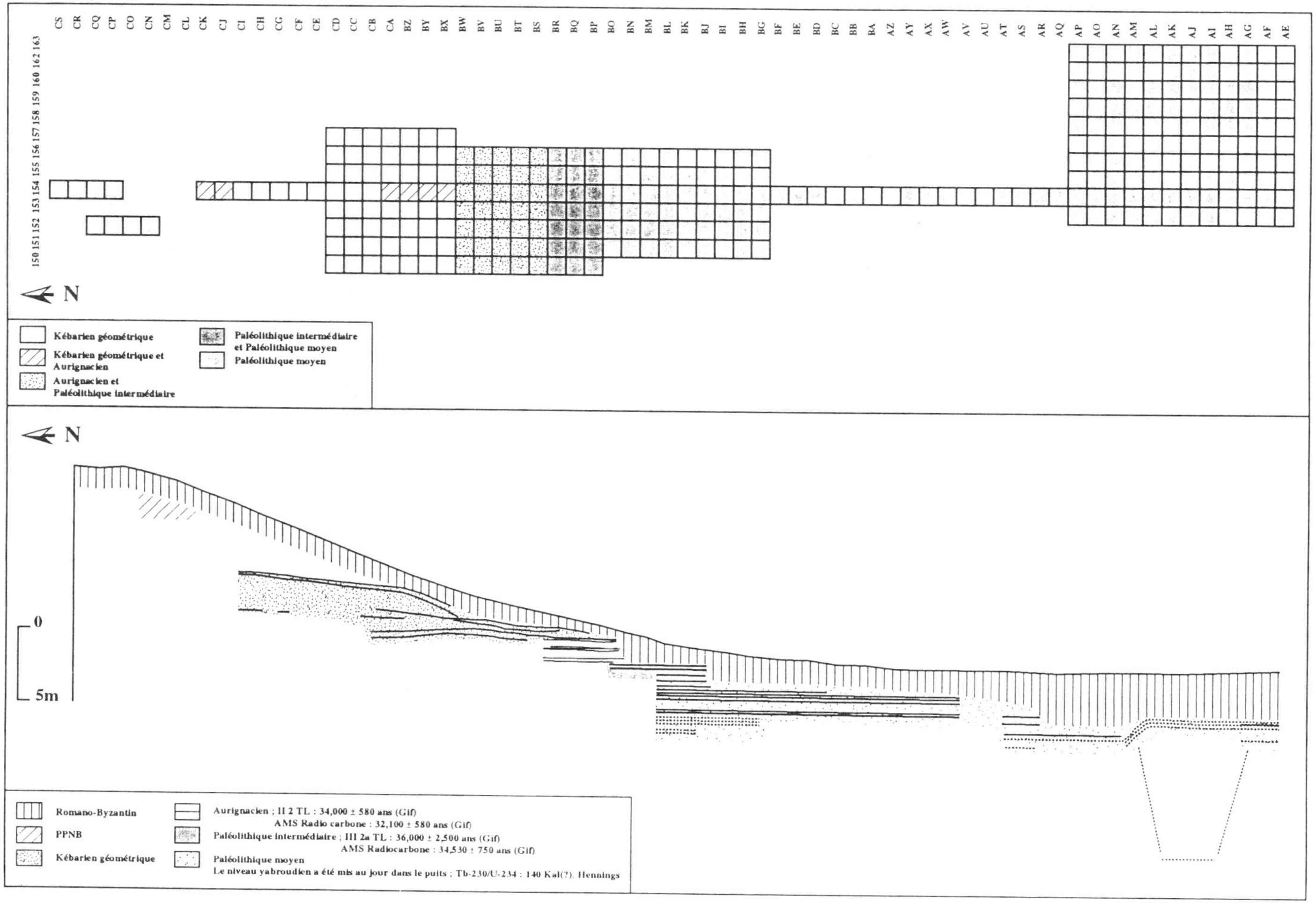

Figure 3. Umm el Tlel site. The excavations carried out from 1991 to 1996 on the Paleolithic sequence have led to the recognition of 87 layers stretching over more than 4 m deep: 31 are allocated to the Upper Paleolithic (Levant Aurignacian and indeterminate); 3 to an Intermediate Paleolithic (Ahmarian (?) and Transitional); and 53 to the Middle Paleolithic.

transport of bitumen is possible. The petrochemical analysis of samples from the various archaeological layers, including sterile deposits, demonstrate that there is no evidence for the presence of fossil fuel traces (numbers 677–680). We may therefore conclude that the black traces did not result from post-depositional contamination.

2.1.3. The Age of the Pieces. Precise dating of layers IV2a, IV2b, and IV2cα has not been possible, but thermoluminescence and ^{14}C readings are available for the overlying layers attributed to the Transitional Industries (layers III) and to the Aurignacian (layers II). The TL of three burnt flints (Mercier et al. 1995) from layer III2a gave a mean age of 36,000±2,500 years. A single AMS radiocarbon date (Arnold et al. 1987) of 34,530±750 years BP (GifA 93216) has been obtained for the same level. The overlying layer II has also been dated. The mean TL age based on four burnt flints is 34,000±2500 years. A single AMS radiocarbon date for the same layer is 32,000±580 years BP (GifA 93212) (Boëda et al. 1996).

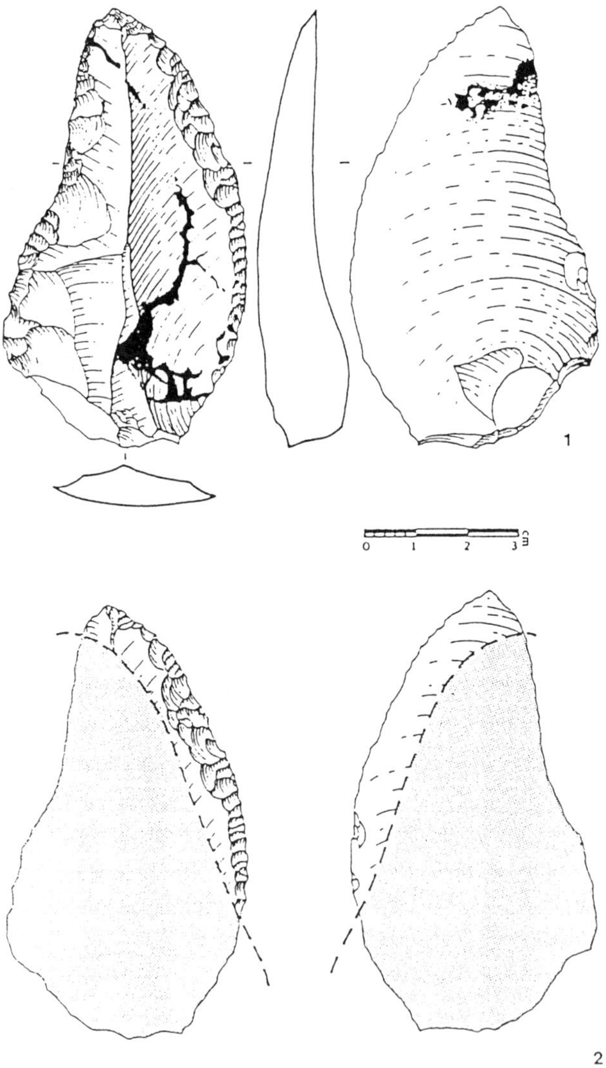

Figure 4. (1) Drawing showing the black bitumen traces on the upper face of the convergent scraper. On this face, the sinuous trace goes across the piece and follows the convex curve of the right edge, at a distance of 1 cm. In the proximal third, the trace swerves to end on the right edge exactly where the retouch stops. (2) Reconstruction of the presumed handle of the convergent scraper.

2.1.4. Techno-Functional Analysis of Specimens with Black Bitumen-Like Traces. 2.1.4.1. Double Side Scraper (Number 359). This is a convex/concave convergent scraper made on an 8 cm long Levallois flake with a triangular transverse cross-section (Figure 4). On the upper face, the sinuous trace traverses the piece, linking the distal end of the left edge to the proximal end of the right edge. This "bitumen" trace steadily follows the convex curves of the right edge at a distance of 1 cm, creating two parallel curves. In the proximal third, the trace swerves to end on the right edge line, exactly where the retouch stops. On the lower face, the black trace is less important and is located at the distal end of the flake, where it starts at the same level as the upper face trace. It is only 1 cm long and is again parallel to the curving right edge.

The black traces on the artifacts suggest the remains of adhesive used to glue the tool into a vegetal or animal haft. This fastening leaves the right edge and the retouched end of the scraper free. The traces facing each other at the distal end of the artifact support this suggested reconstruction. In the same way, we can stress the correlation between the presence of black adhesive and the absence of retouching on the proximal part of the right edge. The latter stops where the fastening began.

2.1.4.2. Levallois Flake (Number 738). This item is a small, subquadrangular Levallois flake (4×4 cm; Figure 5). The left and the distal edges were obtained respectively by only one removal, thus producing cutting edges with a constant angle and a regular shape. The black traces are present only on the upper face. These are small marks creating three lines. The first one starts from the right edge and has an axis parallel to the distal edge. The second one, located along a vein, is parallel to the left edge. The third one is parallel to the proximal edge and starts where the second line ends. A few scattered marks are still

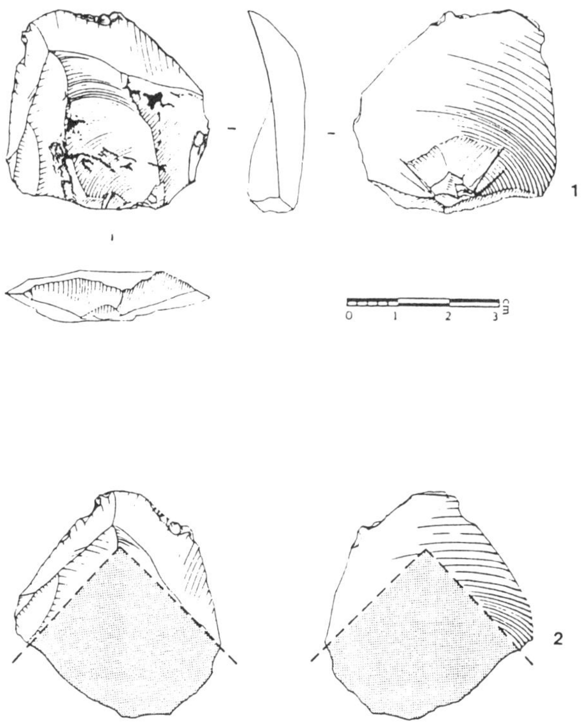

Figure 5. (1) Drawing showing the black bitumen traces on the upper face of the Levallois flake, made on a small, subquadrangular piece 4 cm long and 4 cm wide. The black traces only exist on the face shown (2c). There are small marks, creating three parallel lines to edges at a 1 cm distance. (2) Reconstitution of the presumed handle, only valid for the limits of the handle on the artifact.

detectable within this subquadrilateral formation. These three lines are parallel to one edge and are at a distance of 1 cm from the other edges.

This Levallois flake displays similar traits to the scraper described above. The fastening left the two best edges operational and their convergence produces a projecting edge.

2.1.4.3. *Débordant* Levallois Flake with a Retouch *Coup de Tranchet Lateral* (Number 841). This is a *débordant* Levallois flake which was used as a blank for manufacturing a *Prondnickmesser*, a characteristic tool-type known in central Europe (Figure 6). This type was not reported previously from the Near East.

The technical analysis shows that the flake is issued from a *recurrent* unidirectional Levallois debitage, producing some Levallois removals with edges issued from a unique scar. In the present case, the right edge, made of one removal, faces the non-cutting left edge showing a *débordant* type back. The knapper has arranged a small distal truncation using a reduced plunging area. From this truncation, while employing the lower face of the blank, the knapper obtained a removal called a lateral cutting blow, or *coup de Prondnick*. This removal created a 2 cm-long straight edge, shaped by two surfaces with a 25 degree angle, which is well-suited for cutting.

The positions of the blackened areas are directly related to the preceding technical observations. The hafted or hand-held zone is covered with black traces while the cutting edge is free of these marks. The traces are found mainly on the dorsal face while the ventral face displays less heavy stains, although they start at the same level as those on the upper face. The negative bulb from the *coup de tranchet latéral* cuts through the black traces, as if it was done after the hafting was completed with the intention of sharpening the edge.

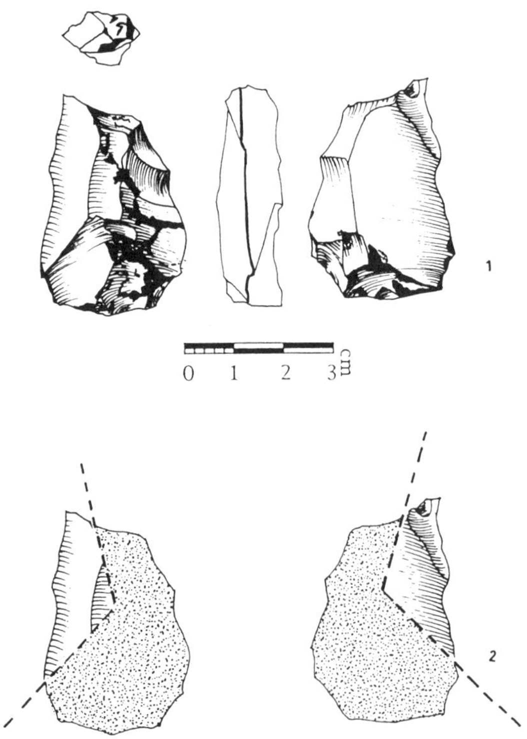

Figure 6. (1) *Débordant* Levallois flake. The right edge, made of one removal, faces the non-cutting left edge showing a *débordant* type back. From this truncation on a distal part, and using the lower face of the blank, the knapper obtained a removal named lateral cutting blow, or *coup de Prondnick*. (2) The analysis of the positions of the black areas is related to preceding technical observations. The entire set of traces agrees with the transformative part of the tool, that is, the right edge of the flake, while the prehensive zone, the back and distal parts, shows black traces. It seems that from the function point of view, the position of the black traces in front of the best cutting planes is not hazardous. If the black traces represent the limits of a hafting, this highlights a cutting zone which has surely been filled up during the work.

It seems that the position of the black traces in front of the best cutting edge is not accidental. If the black traces represent the limits of the hafting, then the sharpening of the cutting edge took place during the activity.

2.1.4.4. Levallois Flake (Number 839). This flake was obtained from unidirectional, *recurrent* Levallois debitage (Figure 7). Distal parts of the right and left edges are converging and a very faint direct and inverse retouch can be observed along the distal left back. This retouch is formed by flat concave removals, thus creating a cutting edge with an angle lower than 25 degrees. Rather than being intentional, it is perhaps due to usage. The right edge is not modified. Black traces are located on the proximal part of the upper

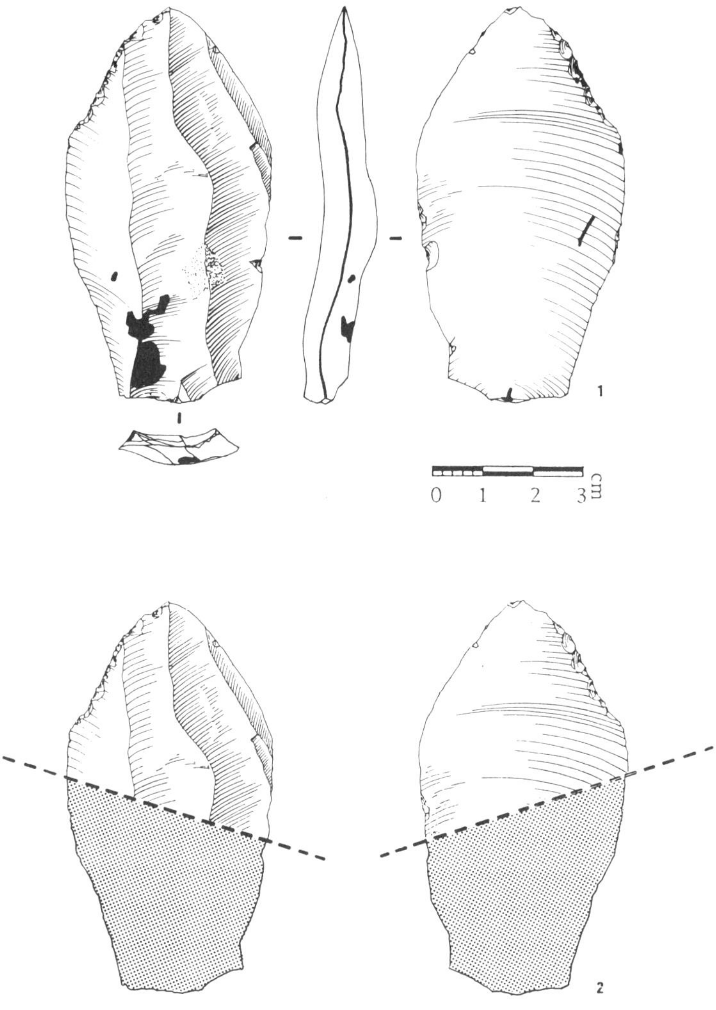

Figure 7. (1) Drawing showing the black bitumen traces on the upper face of the Levallois flake. Distal parts of the right and left edges are converging. A very faint direct and inverse retouch can be observed along the distal left back. The right edge is not modified. Black traces are preferentially located on the proximal part of the upper face and on the butt. (2) Black traces are not sufficient to confirm a hafting hypothesis. Nevertheless, one can again note a black trace distribution outside the *contacts transformatifs* zones of the artifacts. In the present case, the transformative zones are located in the middle part, while black traces are observed in the proximal part.

face and on the butt. Only a limited oblique black trace is observed on the middle part of the flake's lower face.

Black traces are not sufficient to support reconstructed hafting. Nevertheless, the distribution of the black traces is not within the so-called zone of *contacts transformatifs*. In the present case, this zone is located in the middle part of the flake while the black traces are in the proximal portion.

2.1.4.5. Cortical Flake (Number 840). This cortical flake preserves black traces only on the upper faces. On the left part, one notes that the flake removal took place after the black deposit had accumulated (Figure 8). Given that this piece was not used, it seems to confirm that bitumen was in contact with other objects such as non-knapped chert blocks. They were contaminated somehow during activities carried out on site. This suggested interpretation is based on our experimental work. While experimenting with bitumen, we noticed that this is not an easily manageable material and that unintentional contamination may occur, thereby producing stains on surrounding artifacts. We have also discarded the hypothesis that the eventual staining took place at the source locality of this raw material. The condition of the cortical surface of this artifact indicates that it was excavated from a primary source. This type of chert is available in an outcrop more than 50 km away from the source of the bitumen. While using several tons of this particular chert in our experiments, we never noticed traces of black material on any of the cortical areas of the collected cobbles.

2.2. Hummal

The site of Hummal is located on the top of a dome which contains a well more than 20 m deep (Figure 9). No excavation has ever been undertaken at this site, but several archaeological levels have been noted in the stratified exposure. The stratigraphy was recorded by Hours, Copeland, Le Tensorer, and Muhesen and the following assemblages were recognized from bottom to top: Yabroudian, Hummalian, Mousterian, Aurignacian, and Geometric Kebaran (Le Tensorer and Hours 1989; Muhesen 1992).

2.2.1. The Technological Characteristics of the Hummalian Industry. The technological analysis carried out concerns the entire collection, including the in situ assemblage, and allows us to identify the *schema opératoire* of an industry geared almost exclusively towards the production of blade blanks which were made into Middle Paleolithic type tools. However, this cannot be considered as an Upper Paleolithic type of blade reduction sequence. In this respect, the Hummalian industry differs from the Pre-Aurignacian, Yabroudian, or the

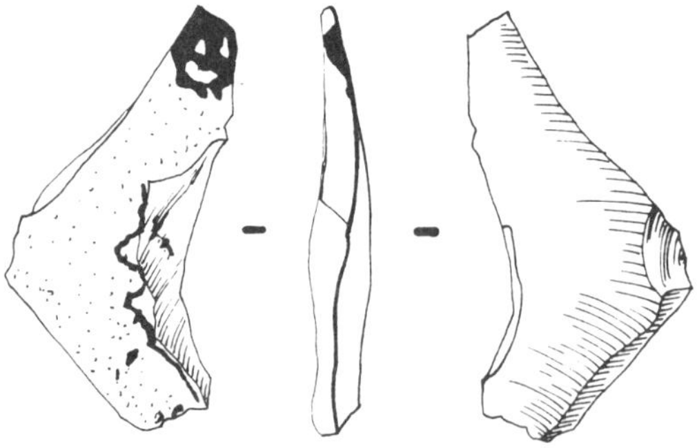

Figure 8. Drawing showing the black bitumen traces on the upper face of the cortical flake. This cortical flake offers evidence of the observation of black traces only on the upper faces. On the left part, one can see that flake removal took place after the black deposit.

Amudian, in which blade blanks (although not the entire production) served as blanks for the fabrication of Upper Paleolithic tool types. Among the Hummalian collection, we found several pieces displaying a deposit of granules of black material on their surfaces, as described below.

2.2.2. Stratigraphic Position. Only one of the stained pieces has been analyzed to date. It was found in situ in the Hummalian level along with many other artifacts. This layer was protected from erosion and contamination by a concrete wall which had been erected to prevent the collapse of the well. Although the deposit is considered to be geologically in primary condition, this is not necessarily the case for the archaeological remains. Hence, the study of the whole collection indicated the presence of many different surface states. One artifact could bear up to three different surface states. Surface conditions ranged from a state of remarkable freshness to advanced alteration, including being eroded, blunted, or polished by water. These variations were not necessarily due to the mixing of artifacts of different ages, as certain artifacts showed two different types of alteration on their various surfaces. In addition, the analysis of the entire collection shows a high level of technological homogenity, making it rather doubtful that mixing with either earlier or later industries occurred. Possibly, this collection represents the mixing of several Hummalian levels.

It is unfortunate that the great majority of the collection had been carefully washed, with the exception of two bags. In one of these, we found the artifact (Hu 82 1a) with black material on its upper surface while its lower side has been partially cleaned in order to mark its provenience.

2.2.3. Post-Depositional Contamination? Natural bitumen is found 40 km from the site. The wadis flow in the opposite direction, thus no natural transport of bitumen is possible. In addition, the analysis of various currently-used fuels for diesel-operated wells nearby shows totally different geochemical compositions which leave different residues on stone artifacts. Therefore, we may conclude that the black traces are not due to post-depositional contamination.

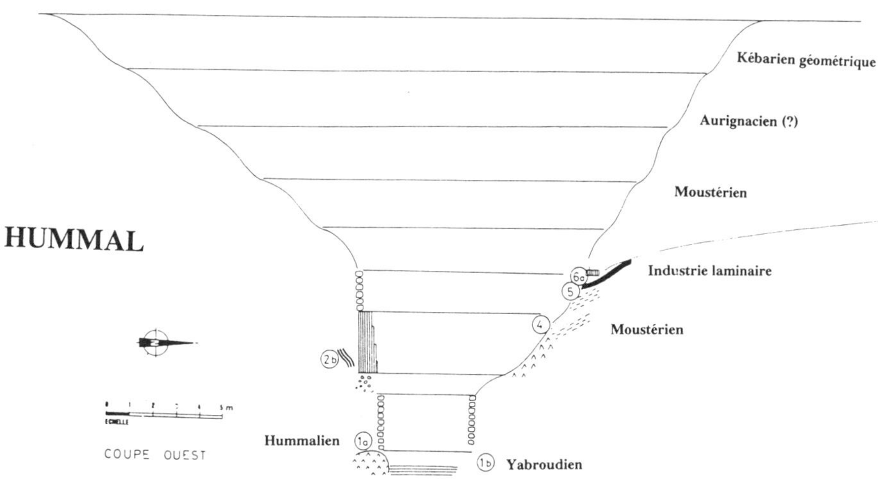

Figure 9. Hummal site.

2.2.4. Description of the Object with Black Traces (Number 263). This is a blade-flake (8.2×2.6×1.0 cm in size) with a triangular cross-section. It fits perfectly in a reduction sequence of blade-flakes (Figure 10). The two edges are subparallel with a central ridge and they converge in the distal quarter to create a point. The proximal end is characterized by

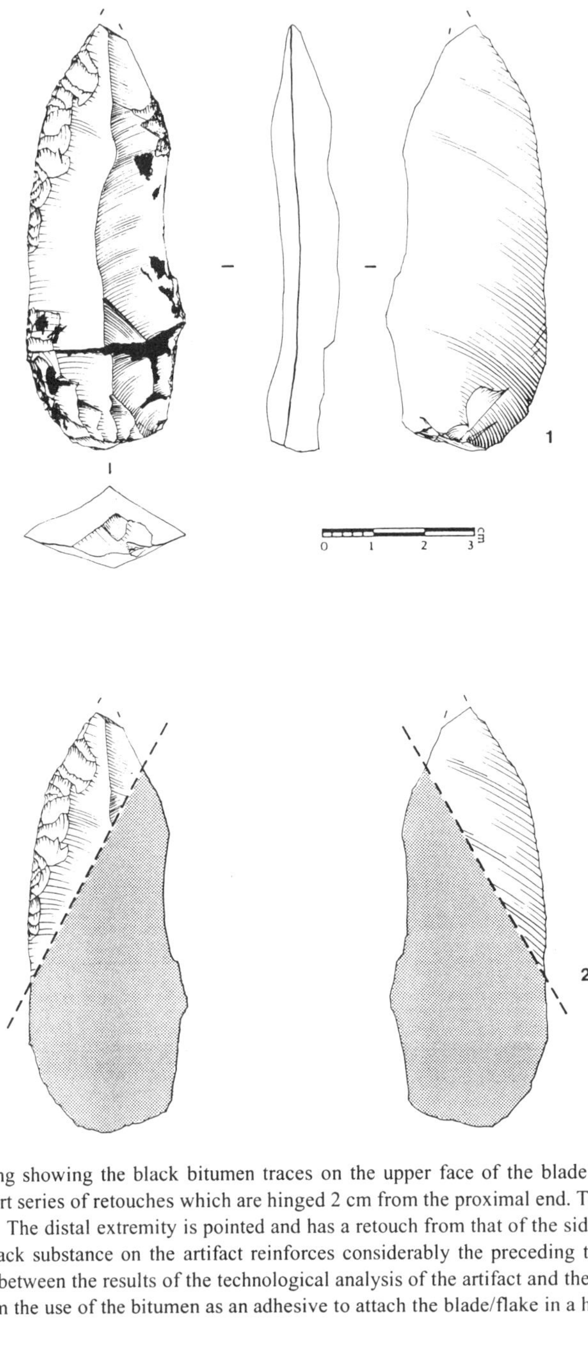

Figure 10. (1) Drawing showing the black bitumen traces on the upper face of the blade. The proximal end is characterized by a short series of retouches which are hinged 2 cm from the proximal end. The left edge presents a discontinuous retouch. The distal extremity is pointed and has a retouch from that of the sides. (2) The analysis of the location of the black substance on the artifact reinforces considerably the preceding technological observations. The correlation between the results of the technological analysis of the artifact and the localization of the bitumen tends to confirm the use of the bitumen as an adhesive to attach the blade/flake in a handle made of vegetal or animal material.

a short series of retouches which stop 2 cm from the proximal end, creating an abrupt uneveness of several millimeters in thickness.

The left edge displays discontinuous retouch. At the proximal end, a shallow irregular retouch along 2.8 cm creates a convex edge and surfaces which define heterogeneous planes and angles. The goal of these modifications seems to be the formation of a more uniform morphology on the left side of the proximal end and not the shaping of an efficient edge. The left edge at the mesial-distal portion has been retouched by the removal of parallel but unevenly invasive mini-flakes. The technical role of this retouch is to smooth the surface and create a uniform cutting plane 4.5 cm long. In addition, the retouch creates a convex cutting edge with an angle of 40 degrees.

The distal extremity is pointed and was retouched from both sides. On the left side, the retouch creates a concave surface, dipping in 1 mm and modifying the original cutting edge. On the right side, the retouch maintains the cutting angle. This retouch may have been created through an intentional strengthening of the point or may have resulted from use. Use-wear study may reveal the nature of this retouch.

The retouch of the mesial-distal portion of the right side was created by a short series of invasive removals from a ridge running perpendicular to the edge. This series of flaking reduces the thickness along the surface and produces the final shape.

2.2.5. Technical Interpretation. We have observed three types of intentions in the modifications described above. First, the modification required for the prehensive and receptive contact is demonstrated by the retouch along the left edge of the proximal end and the right edge of the mesial/distal part. This has the role of creating a convex delineation.

The modification of the transformative contact demonstrates the second and third intentions. On the left side, retouch of the mesial/distal end was intended to create both a techno-functional transformative unit along 4.5 cm of that plane and a particular type of angle, as well as a slightly convex outline. On the extremity, no retouch has been used to form a special modification. Instead, a second functional unit was made operational and shows traces of its activity.

In sum, this object is deceptively simple-looking from a technical perspective, but it is also fully functional as it maintains the proximal end for hafting.

2.2.6. The Distribution of the Black Substance. The location of the black substance on the artifact reinforces the preceding technological observations. Two localities were observed, but as mentioned above, the lower side was washed.

The first trace of black substance constitutes a small mass a few millimeters thick that adheres to the item in the hinged area at the proximal end of the flake. A small quantity of black substance had been preserved, protected in the interstices created by the steep fractures.

The second group of traces is located on the right side of the flake in a segmented line several millimeters thick running along the edge. These spots, 2–5 mm in diameter and 1 mm thick, were found at intervals along the length of the surface from the steep fractures to the retouched area described above.

2.2.7. Interpretation. The correlation between the results of the technological analysis of the artifact and the localization of the bitumen tends to confirm the use of the bitumen as an adhesive to haft the blade/flake in a handle made of vegetal or animal tissue. This hafting leaves free the two technological units which had not been modified, the retouched right edge and the pointed end. The axis of hafting is asymmetrical, encasing the proximal third of the artifact and two-thirds of the right edge.

3. ARCHAEOMETRIC STUDY OF THE BITUMEN-LIKE MATERIAL OCCURRING ON ARTIFACTS

3.1. Introduction

Two Mousterian artifacts from Umm el Tlel from layer IV (numbers 359 and 738) were examined by organic geochemistry methods in order to investigate the nature of the black substance. These imprints were thought to be the remains of the adhesive which was used as a hafting material to affix flint implements to handles. The geochemical techniques included detailed molecular analyses on hydrocarbons, currently applied in oil exploration but also in the study of archaeological bitumens found at excavated sites in the Near East (Connan and Deschesne 1996). These tests already identified the hafting material on the two artifacts as bitumen (Boëda et al. 1996).

Thirteen artifacts from Umm el Tlel, also from the Mousterian layer IV, and one artifact from Hummal, attributed to older layers (Hummalian, see Table 1), were subsequently examined in order to test the previous findings. All of these artifacts showed black stains that were believed to be partly bitumen residue.

Prior to any chemical study, our goals (i.e., to demonstrate the occurrence of bitumen on artifacts) appeared to be difficult to achieve for several reasons. First, the black imprints were generally very thin films, signifying there was little available material. In addition, oxidation of the presumed bitumen film probably generated substantial asphaltene and insoluble residue (Connan et al. 1992), resulting in minute amounts of hydrocarbons for analysis. Moreover, these samples had been collected during previous excavation seasons without the precautions needed to avoid significant organic contamination. Contamination with the paste used for taking pictures of artifacts has been clearly detected on

Table 1. Basic references about the samples examined in the archaeometric study

Elf sample number	Elf laboratory number	Excavation campaign	Area	Layer	Preliminary evaluation	Scanning Electron Microscope	Conclusions
837	B77467	UM 91	BS 156	IV 2a 6	No bitumen		
841	B77471	UM 91	BR 153	IV 2b 13	Possible bitumen	yes	**bitumen**
845	B78476	UM 91	BQ 153	IV 2b 15	Possible bitumen		
876	B78479	UM 94	BR 152	IV 2a 22	Possible bitumen		
877	B78480	UMT 91	BR 154	IV 2c α3	No bitumen		
836	B77466	UMT 91	BR 154	IV 2c α4	Possible bitumen		
838	B77468	UM 91	BR 153	IV 2a 1	Possible bitumen		
840	B77470	UM 91	BR 153	IV 2b 3	Possible bitumen	yes	**bitumen**
839	B77469	UM 94	BQ 153	IV 2b 4	Possible bitumen		bitumen?
844	B78475	UM 94	BX 154	V 2βa 43	Possible bitumen		bitumen?
846	B78477	UM 91	BR 153	IV 2c α11	Possible bitumen		
875	B78478	UMT 91	BR 154	IV 2c α5	Bitumen?		
878	B78481	UMT 91	BR 154	IV 2c α7	Bitumen?		
359	B45164	Um 91	BR 153	IV 2a 2	Possible bitumen		**bitumen**
738	B72934	UM 94	BR 152	IV 2a 26	Bitumen?		**bitumen**
263	B34586	Hu 92	Hu 82	1a	Possible bitumen		**bitumen**
676	B72236	soil		IV	no bitumen/ vegetal contribution		no bitumen
677	B72237	soil		IV 1	no bitumen/ vegetal contribution		no bitumen
678	B72238	soil		IV 1	no bitumen/ vegetal contribution		
679	B72239	soil		IV 2a	no bitumen/ vegetal contribution		
680	B72240	soil		IV 2a	no bitumen/ vegetal contribution		

some samples. Consequently, the geochemical study of the most relevant cases was supplemented by a Scanning Electron Microscope (SEM) analysis in order to establish definitely the organic origin of the black imprints.

3.2. Geochemical Results

The results obtained from a few samples (numbers 839, 840, 841, 359, 738, and 263) confirm the occurrence of bitumen as a potential adhesive on flint artifacts. Evidence of dissolution of the black imprint in chloroform was clearly obtained on three samples: number 839 (Figure 11), number 840 (Figure 12), and number 263. In these cases, the soluble black matter, presumed to be bitumen, was placed in a suitable position to ensure hafting.

Molecular proof was gathered on C15+ alkanes from samples 359, 738 (Boëda et al. 1996), and 263. The distribution patterns of steranes and terpanes, i.e., biomarkers derived from steroids and terpenoids, present in living organisms, were recorded from these samples. Steranes and terpanes are geochemical fossils whose molecular signatures are used as genetic fingerprints to determine the origin of crude oil and bitumen. The artifact distributions, compared to a sample of soil extract (number 677) from layer IV that was used as blank reference, are shown in Figures 15 and 16. Sample number 677 displays a typical pattern of immature organic matter dominated by land-derived debris, characterized herein by a strong predominance of C29, C28, and C27$\alpha\alpha\alpha$R compounds among steranes. The artifact extracts possess different patterns which are typical of fossil fuel fingerprints.

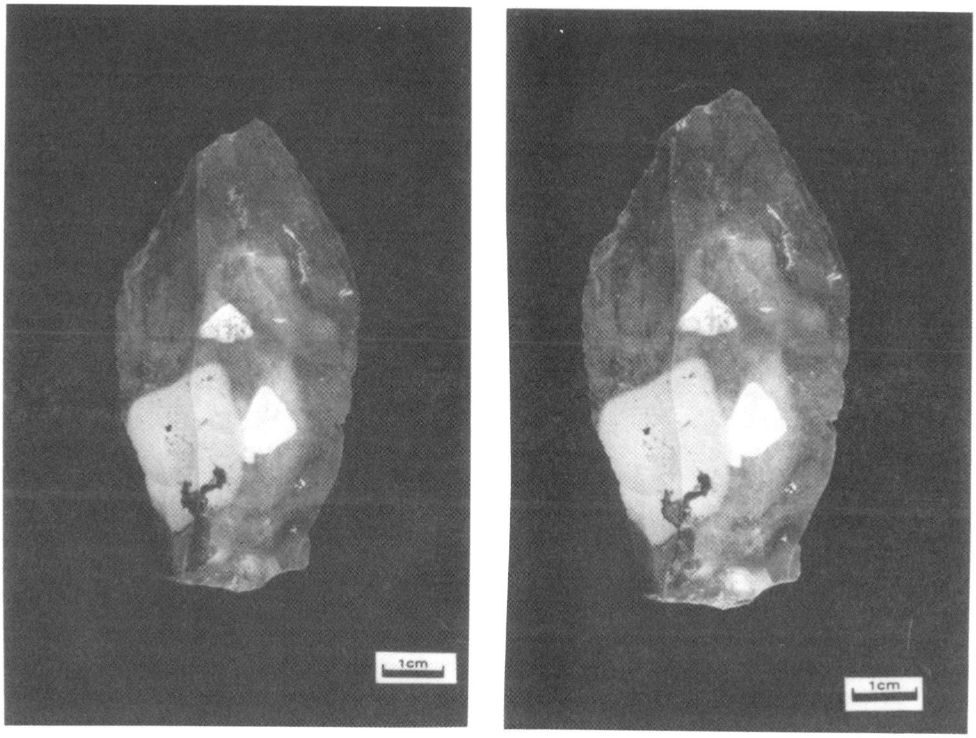

Figure 11. Photograph of sample number 839 (UM 94, Br 153, IV2b 4) before (*left*) and after (*right*) extraction with chloroform. It can be noticed that a significant part of the primary black imprint has been dissolved in chloroform. This soluble part is likely bitumen.

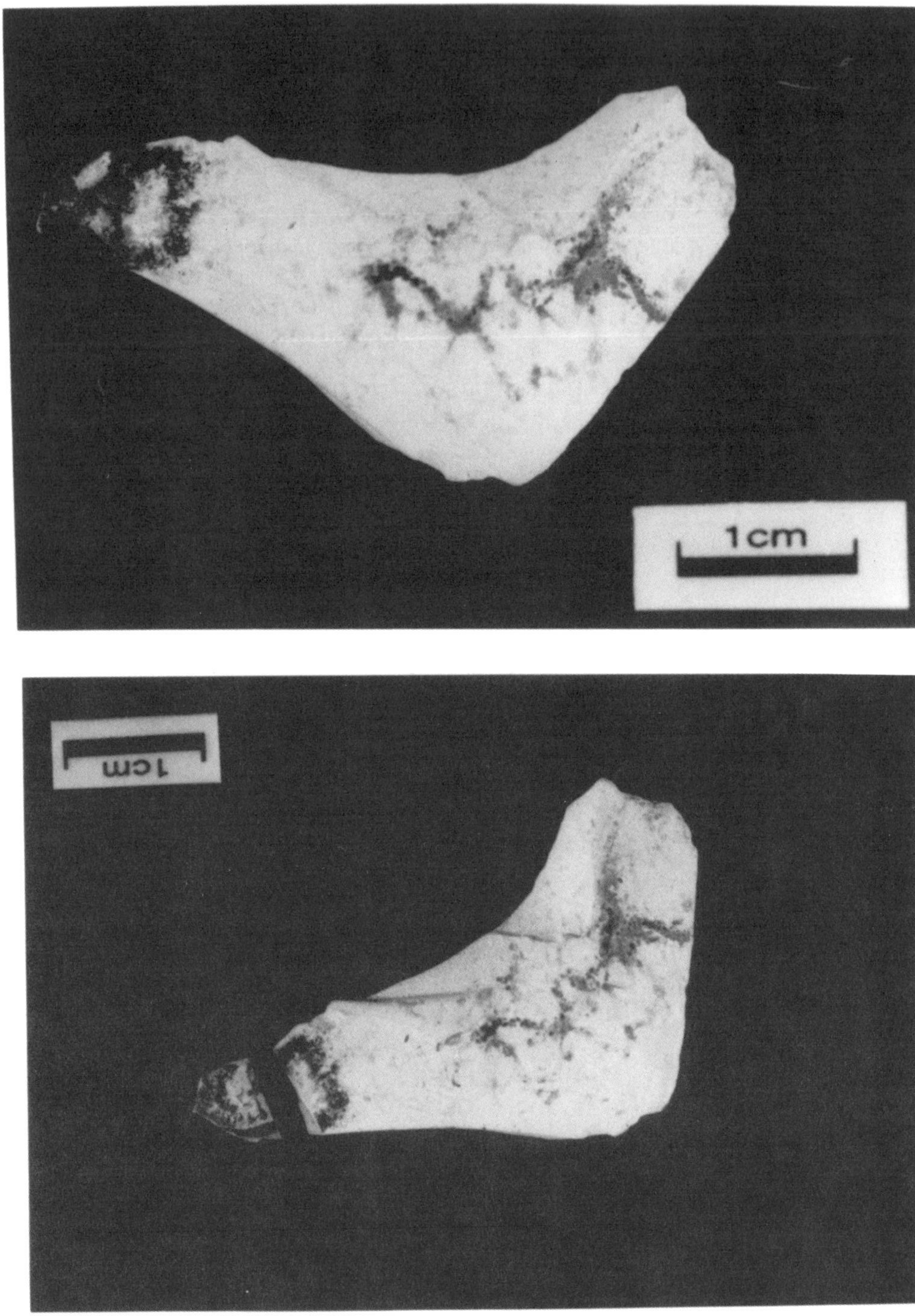

Figure 12. Photographs of sample number 840 (UM 91, BR 153, IV 2b 3), before (above) and after (below) extraction with chloroform. Subsequently, the flint implement has been broken into two pieces to allow their introduction inside the Scanning Electron Microscope (SEM). It can be noticed that a part of the black imprint has been dissolved in chloroform. This soluble part is likely bitumen. The black area, showing solubility features, has been subsequently studied by (SEM) image and X-ray emission analysis (Figures 13 and 14). In addition, one should also note occurrences of rather dark zones that are in fact related to iron and manganese oxide precipitation.

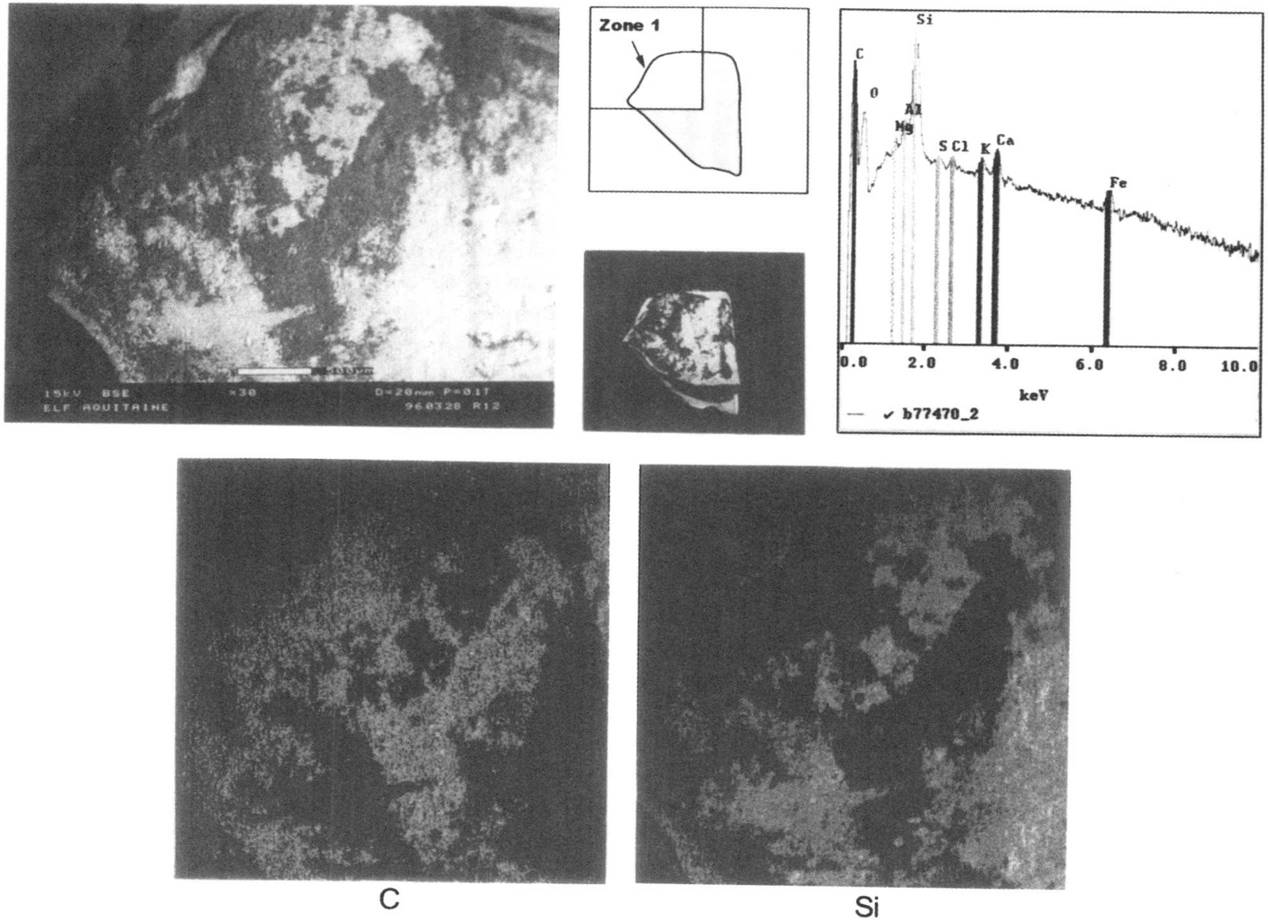

Figure 13. SEM analysis of zone 1 in sample number 840. SEM image in SEI (Secondary Electron Impact) mode with its X-ray spectrum and the related carbon and silicon maps. The black zone on the SEM image is organic-rich, i.e., bitumen-coated.

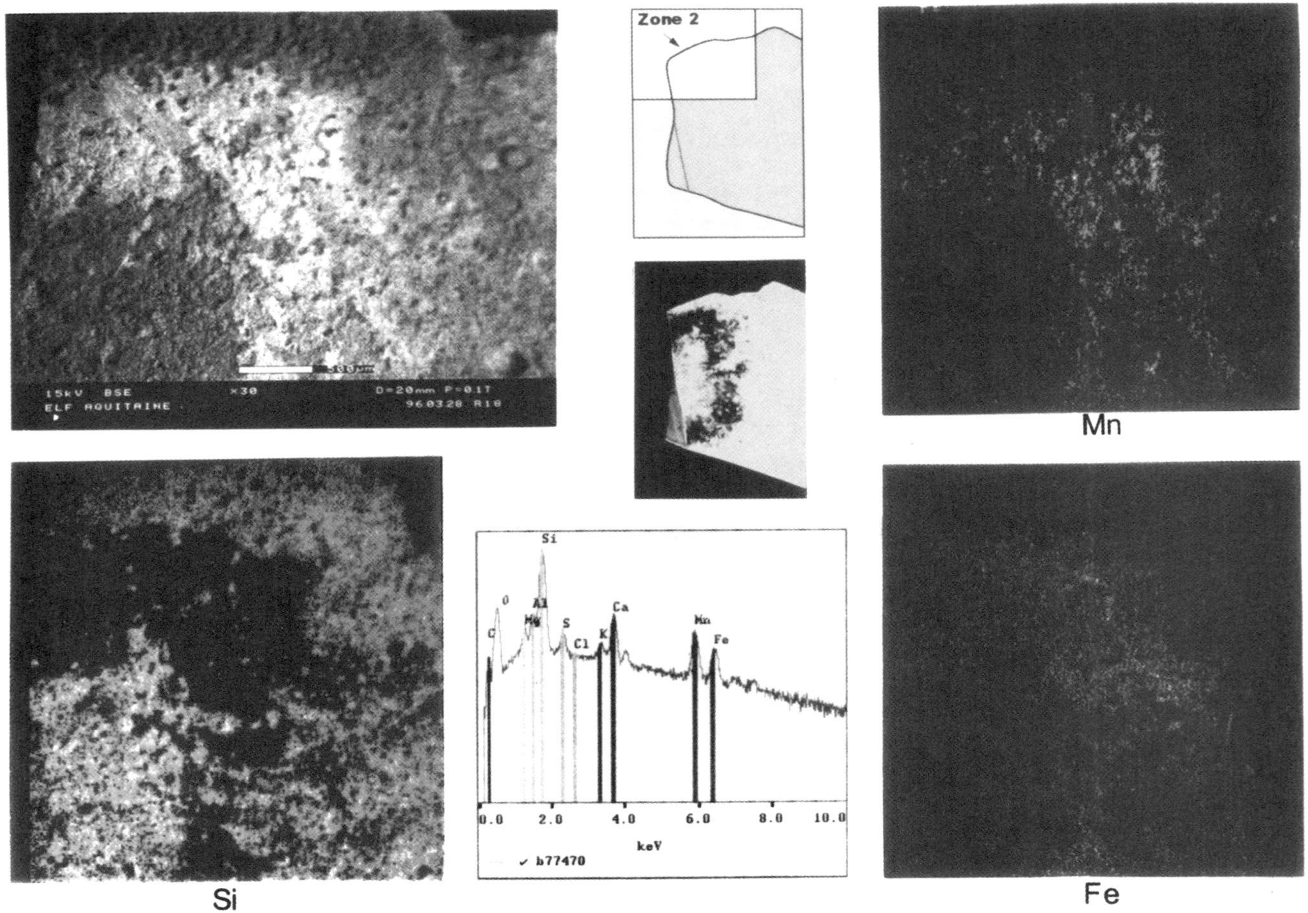

Figure 14. SEM analysis of zone 2 in sample number 840. SEM image in SEI mode with its X-ray spectrum and the related manganese, silicon, and iron maps. The black zone corresponds mainly to manganese and iron oxide.

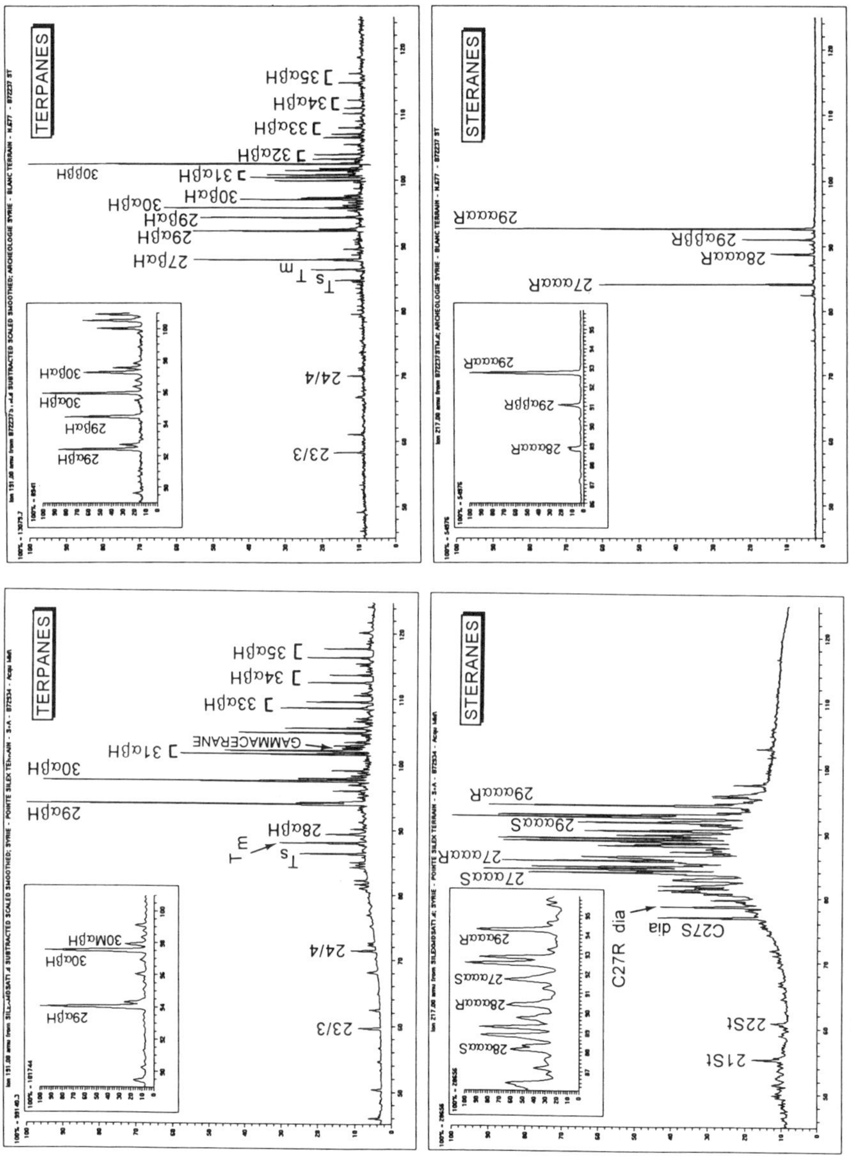

Figure 15. Sterane (m/z 217) and terpane (m/z 191) patterns of samples numbered 738 and 677, obtained by GC-MS analysis of C15+ alkanes. The extracts of the soils surrounding the archaeological pieces show a very different fingerprint which is characteristic of a present-day material and not a fossil fuel.

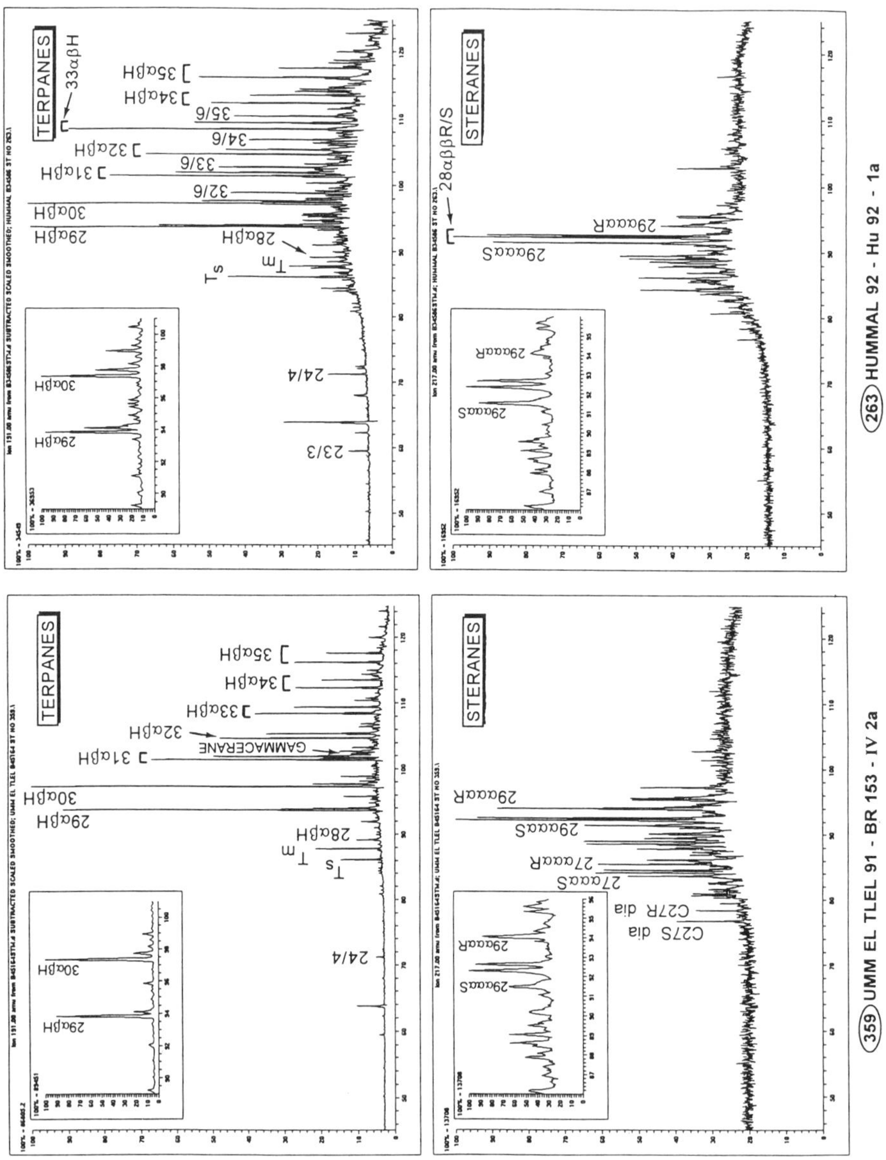

Figure 16. Sterane (m/z 217) and terpane (m/z 191) patterns of samples numbered 359 and 263, obtained by GC-MS analysis of C15+ alkanes. The fingerprint of sample number 263 is typical of a severely biodegraded bitumen.

These patterns are variable as seen in steranes (C27-C29$\alpha\alpha\alpha$ and $\alpha\beta\beta$ steranes) and terpanes (23/3, 24/4, Tm, Ts, C27 to C35$\alpha\beta$ hopanes, etc.).

Of particular interest is the distribution of steranes and terpanes in sample number 263, which features significant degradation of both molecular categories. Short chain steranes (C21-C22) have completely disappeared and C27 steranes have been severely altered. In addition, this alteration preferentially affected C28$\alpha\alpha\alpha$R and C29$\alpha\alpha\alpha$R steranes. Such selective removal of geosteranes with a biological configuration (C28$\alpha\alpha\alpha$R and C29$\alpha\alpha\alpha$R steranes) is characteristic of a biological process which has been reproduced under laboratory conditions using pure strains of Nocardia (Chosson et al. 1991). None of the molecular patterns encountered in the artifact samples fully match the most well-known natural asphalt sources in the area, namely the seepage of the Hit-Abu oil in Iraq and the oil-stained rocks from Jabal Bicheri in Syria. Consequently, the question of the origin of the bitumen is not yet solved.

GC-MS analysis of C15+ aromatics helped identify two classes of molecular structures. The first includes phenylalkanes (Eganhouse 1986; Elis et al. 1995, 1996), fluoranthene, pyrene, phenanthrenes, and chrysenes (White 1986), which are all characteristic pyrolytic products, indicating that the bitumen was subjected to fairly high temperatures. L. Mansuy (1995) has determined that phenylalkanes remain the predominant members of the alkyl-benzene family when a Mahakam delta coal is heated at temperatures in excess of 350°C. The second class includes mono- and tri-aromatic steroids, monoaromatized 8,14-secohopanoids, and benzohopanes, which are common biomarkers in crude oils and natural asphalts from Iraq and Syria.

3.3. Scanning Electron Microscope (SEM) Analysis

As a follow-up to the geochemical study carried out on the chloroform extract, complementary analyses were performed on two chloroform extracted artifacts, using a Scanning Electron Microscope. SEM images, coupled with X-ray spectrum, allowed us to map the major elements and to determine the spatial distribution of the various constituents: silica, organic carbon, iron, and manganese oxide.

An example is presented in Figures 14, 15, and 16. Artifact number 840 was broken so that it could be introduced into the SEM chamber. A comparison of pictures taken prior to and after the chloroform treatment reveals that part of the black imprint was dissolved in chloroform.

The SEM analysis was concentrated on two areas: zone 1 and zone 2. In zone 1 (Figure 15), which corresponds to the extreme end of the artifact, the X-ray analysis was able to detect the following main elements: organic carbon, oxygen, silicon, potassium, calcium, and iron. The most significant elements were mapped to identify organic carbon surrounded by silica associated with subordinate carbonate. In zone 2 (Figure 16), almost unaffected by the chloroform extraction, the X-ray spectrum is dominated by oxygen, silicon, calcium, manganese, and iron. Carbon is present as a very minor element. It is scattered and in low concentration, as confirmed by the mapping of the major elements. The black surface is in fact of mineral origin and corresponds to a coprecipitation of iron and manganese oxide. Yellow-orange surfaces seen on the pictures in Figure 4 are produced by iron oxide only.

Similar results were found by studying artifact number 841. The occurrence of organic material is directly associated with manganese and iron oxide. Carbonates are sometimes detected on the artifacts as result of incomplete scraping of the samples.

3.4. Conclusions

Bitumen has been identified on some artifacts belonging to layer IV of Umm el Tlel (Mousterian) and to the Hummalian layer at Hummal. No definite conclusions about their origin have been drawn due to the low amount of extractable bitumen that was isolated and to the suspected interference of anthropogenic contamination during storage and treatment of the samples.

SEM investigation and X-ray analyses have established that the black residue remaining after the chloroform extraction is not exclusively of organic origin but is sometimes due to coprecipitation of iron and manganese oxide. Black areas on flint implements do not provide accurate maps of the distribution of the organic matter. However, it was observed in the tested cases that organic matter, when present, is generally associated with manganese oxide. This suggests that the bitumen may have played a role in the concentration and subsequently in the precipitation of manganese and iron oxide.

Contamination problems encountered in this study have initiated a more careful sampling procedure to avoid any external contamination. This procedure was applied in the 1996 excavation campaign to prevent mixtures of original molecular signatures with allochthonous contaminants. Forthcoming studies are to be conducted on a new set of samples.

4. DISCUSSION

The black traces on the artifacts suggest the remains of bitumen adhesive used to stick the tool into a vegetal or animal handle. For the scraper, this fastening leaves the right edge and the retouched end of the scraper free. The traces facing each other at the distal end of the artifact confirm this reconstruction. In the same way, on the proximal part of the right edge, we can stress the concordance between the presence of black material and the absence of retouching, the latter stopping where the hafting begins. Similar observations concern both Levallois flakes. The hafting leaves the two best rough edges operational. The *Prondnickmesser* is very interesting since the portion used as the cutting edge was produced as such and is the only edge that does not present any traces of bitumen. There also, we believe, the active edge is deliberately left free. The control flake removal shows bitumen traces only at the periphery of the superior surface. This means that bitumen was spread on the surface of the core and not on the flake after its removal.

If the hafting technique was indeed practiced during the Mousterian period (Anderson-Gerfaud 1981; Beyries 1987; Friedman et al. 1995), the use of bitumen as hafting adhesive during this time period has never been described before. The available literature indicates that the use of bitumen (Lechevallier 1988; Connan and Deschesne 1991, 1992; Barquins 1993) or other adhesive materials (Funke 1969; Binder et al. 1990; Regert 1996) for hafting dates back to the Neolithic. For instance, we can mention a few cases for this period that were recently studied at the PPNA villages of Gilgal (10,000–9,700 BP) and Netiv Hagdud (9,700 BP) in the Jordan Valley and at the PPNB Tell Sabi Abyad II (8,300 BP) in the Balikh Valley in northern Syria (Verhoeven 1994).

Various natural adhesives were used in the Neolithic period and in later times. Birch bark tar is considered the most common one on Neolithic artifacts from Europe (Funke 1969; Binder et al. 1990), but bitumen is the most frequent archaeological tar used in the Near East (Bar-Yosef 1985; Connan and Deschene 1991, 1992; Barquins 1993) and Pakistan (Lechevallier 1988). However, one interesting exception to this wide usage of bitumen should be mentioned. At the Neolithic site of Nahal Hemar Cave (8,800–8,000 BP), close to the Dead Sea (Connan et al. 1995), analysis of a sickle, baskets, and skulls identified the use of collagen instead of the previously reported bitumen (Mills and White 1994). Collagen is the structural fibrous protein of tissues in humans, animals, and fishes. Degraded into gelatin by treatment with hot water, the produced gel acquires efficient adhesive properties already discovered as early as the 9th millenium BP, as documented by the Nahal Hemar finds.

ACKNOWLEDGMENTS

We would like to thank the French Ministry of Foreign Affairs, the Museums and Antiquities directions of Syria, and the Elf Aquitaine Society. I also thank C. Amiot and S. Egna for their assistance.

REFERENCES

Anderson-Gerfaud, P. 1981. Contribution à l'Analyse des Microtraces d'Utilisation. Thèse de 3ème cycle. University of Bordeaux I. Bordeaux.

Arnold, M., E. Bard, P. Maurice, and J. C. Duplessy. 1987. Instrument and methods in physics research. *Nuclear* 29:120–3.

Barquins, M. 1993. Colles naturelles et colles synthétiques. *Pour la Science* 192:8–10

Bar-Yosef, O. 1985. *A Cave in the Desert: Nahal Hemar 9000-year-old Finds*. Jerusalem: The Israel Museum.

Beyries, S. 1987. Quelques exemples de stigmates d'emmanchements observés sur des outils du paléolithique moyen, 55–61. In *La Main et l'Outil: Manches et Emmanchements Préhistoriques*, ed. D. Stordeur. Lyon: Maison de l'Orient.

Binder, D., G. Bourgeois, F. Benoist, and C. Vitry. 1990. Identification de brai de bouleau (Betula) dans le Néolithique de Giribaldi (Nice, France) par la spectrométrie de masse. *Revue d'Archéométrie* 14:37–42.

Boëda, E. 1994. *Le Concept Levallois: Variabilité des Méthodes*. Paris: CNRS Editions.

Boëda, E. 1995. Levallois: A volumetric construction, methods, a technique, 41–68. In *The Definition and Interpretation of Levallois Technology*, ed. H. Dibble and O. Bar-Yosef. Madison: Prehistory Press.

Boëda, E., and S. Muhesen. 1993. Umm El Tlel (El Kowm, Syrie): Etude préliminaire des industries lithiques du Paléolithique moyen et supérieur. *Cahiers de l'Euphrate* 7:47–60.

Boëda, E., J. Connan, D. Dessort, S. Muhesen, N. Mercier, H. Valladas, and N. Tisnerat. 1996. Bitumen as hafting material on Middle Palaeolithic artefacts. *Nature* 380:336–8.

Cauvin, J., M. C. Cauvin, and D. Stordeur. 1979. Recherches préhistoriques à El Kowm (Syrie). Première campagne 1978. *Cahiers de l'Euphrate* 2:80–118.

Chosson, P., C. Lanau, J. Connan, and D. Dessort. 1991. Biodegradation of refractory hydrocarbon blomarkers from petroleum under laboratory conditions. *Nature* 351:640–2.

Connan, J., and O. Deschesne. 1991. Le bitume dans l'antiquité. *La Recherche* 229:152–9.

Connan, J., and O. Deschesne. 1992. Archaeological bitumen: Identification, origins, and uses of an ancient Near Eastern material, 683–720. In *Materials Issues in Art and Archaeology III*, ed. P. B. Vandiver, J. R. Druzik, G. S. Wheeler, and I. C. Freestone. Pittsburgh: Materials Research Society.

Connan J., and O. Deschesne. 1996. *Le Bitume à Suse—Collection du Musée du Louvre*. Paris: Réunion des Musées Nationaux.

Connan, J., A. Nissenbaum, and D. Dessort. 1992. Molecular archaeology: Export of Dead Sea asphalt to Canaan and Egypt in the Chalcolithic-Early Bronze age (4th–3th millenium B.C.). *Geochimica et Cosmochimica Acta* 56:2743–59.

Connan, J., A. Nissenbaum, and D. Dessort. 1995. Geochemical study of Neolithic "bitumen"-coated objects of the Nahal Hemar cave in the Dead Sea area, 699–701. In *Organic Geochemistry: Developments and Applications to Energy, Climate, Environment and Human History*, ed. J. Grimalt and C. Dorronsoro. Donostia-San Sebastian: AIGOA.

Copeland, L., and F. Hours. 1983. Le Yabroudien d'El Kowm (Syrie) et sa place dans le Paléolithique du Levant. *Paléorient* 9(1):21–37.

Eganhouse, R. P. 1986. Long-chain alkylbenzenes: Their analytical chemistry, environmental occurrence and fate. *International Journal Environmental Analytical Chemistry* 26:241–63.

Ellis, L., R. K. Singh, R. Alexander, and R. I. Kagi. 1995. Geosynthesis of organic compounds: III. Formation of alkyltoluenes and alkylxylenes in sediments. *Geochimica et Cosmochimica Acta* 59:5133–40.

Ellis, L., T. A. Langworthy, and R. Winans. 1996. Occurrence of phenylalkanes in some Australian crude and sediments. *Organic Geochemistry* 24(1):57–69.

Friedman, E., N. Goren-Inbar, A. Rosenfeld, O. Marder, and F. Burian. 1995. Hafting during Mousterian times—Further indication. *Journal of the Israel Prehistoric Society* 26:8–31.

Fujimoto, T. 1979. Upper Paleolithic and Epi-Paleolithic assemblages in the Palmyra basin: Site 50 and 74, 77–130. In *Paleolithic Site of Douara Cave and Paleogeography of Palmyra Basin in Syria*, ed. K. Hanihara and T. Akazawa. Tokyo: University of Tokyo Press.

Funke, H. 1969. Chemisch-analytische Untersuchungen verschiedener archäologische Funde. Ph. D. diss. University of Hambourg. Hambourg.

Hours, F. 1982. Une nouvelle industrie en Syrie entre l'Acheuléen supérieur et le Levalloiso-Moustérien, 33–46. In *Archéologie au Levant*, ed. R. Saidah. Lyon: Maison de l'Orient.

Lechevallier, M. 1988. L'industrie lithique de Mehrgarh, 56–7. In *Les Cités Oubliées de l'Indus—Archéologie du Pakistan*, ed. J. F. Jarrige. Paris: Association Française d'Action Artistique.

Le Tensorer, J. M., and F. Hours. 1989. L'occupation d'un territoire à la fin du Paléolithique ancien et au Paléolithique moyen à partir de l'exemple d'El Kowm (Syrie), 107–14. In *L'Homme de Néandertal*, Vol. 6, ed. M. Otte. Liège: ERAUL.

Mansuy, L. 1995. Le rôle du milieu réactionnel au cours de la maturation artificielle de la matière organique. Ph. D. diss. Institut Polytechnique de Lorraine. Nancy.

Mercier, N., H. Valladas, and G. Valladas. 1995. Flint thermoluminescence dates from CFR Laboratory at Gif: Contributions to the study of the chronology of the Middle Palaeolithic. *Quaternary Science Reviews (Quaternary Geochronology)* 14:351–64.

Mills, J. S., and R. White. 1994. *The Organic Chemistry of Museum Objects*. London: Butterworths.

Molist, M., M. C. Cauvin, and A. Taha. 1988. Le gisement d'Umm El Tlel-2 (El Kowm, Syrie): Rapport préliminaire des travaux. *Annales Archéologiques Arabes Syriennes* 37/38:67–77.

Muhesen, S. 1992. The transitional Lower-Middle Palaeolithic industries in Syria, 51–65. In *The Evolution and Dispersal of Modern Humans in Asia*, ed. T. Akazawa, K. Aoki, and T. Kimura. Tokyo: Hokusen-sha.

Regert, M. 1996. Les composés organiques en préhistoire: Nouvelles approches analytiques. Ph. D. diss. Université de Paris X. Paris.

Verhoeven, M. 1994. Excavation at Tell Sabi Abyad II, a later Pre-Pottery Neolithic B village in the Balikh valley, Northern Syria. *Orient Express* 1:9–12

White, R. 1996. Brown and black organic glazes, pigments, and paints. *National Gallery Technical Bulletin* 10:58–71.

13

THE TECHNOLOGICAL ABILITIES OF THE LEVANTINE MOUSTERIANS

Cultural and Mental Capacities

Naama Goren-Inbar and Anna Belfer-Cohen

Institute of Archaeology
Hebrew University
Mt. Scopus, Jerusalem 91905, Israel

1. INTRODUCTION

The present article is an endeavor to understand the technological abilities of the Levantine Mousterian population, from which some insight may be gained into the nature of human cognition at that time. The approach employed here is similar to that adopted in a previous study (Belfer-Cohen and Goren-Inbar 1994) which dealt with issues pertaining to human cognition during the Lower Paleolithic through examinations of Levantine lithic techno-complexes.

Though much of prehistoric research is concerned with the phenomenon of transitions, most efforts have been oriented towards explaining the Middle Paleolithic/Upper Paleolithic transition while the Lower Paleolithic/Middle Paleolithic transition has attracted considerably less attention (but see Clark 1982, 1988; Volman 1984; Klein 1989; McBrearty 1991; Goren-Inbar 1995). The behavior of Middle Paleolithic (i.e., Mousterian) hominids is unquestionably rooted in their past and origins—the Lower Paleolithic patterns. It is nevertheless obvious that a fundamental change must have occurred around the time of the passage from the Lower to the Middle Paleolithic which is reflected in both the typological and the technological aspects of the relevant industries (Figure 1).

When examining traits considered to be the hallmarks of the Middle Paleolithic lithic assemblages, from the Levant as well as from Europe and the Middle Stone Age of Africa, several differences from the Lower Paleolithic industries stand out quite clearly:

1. An emphasis on flake and flake tool production as opposed to an emphasis on production of core tools in the Lower Paleolithic.
2. Dominance of the Levallois technique (Clark 1988; McBrearty 1991).

Neandertals and Modern Humans in Western Asia, edited by Akazawa *et al.*
Plenum Press, New York, 1998.

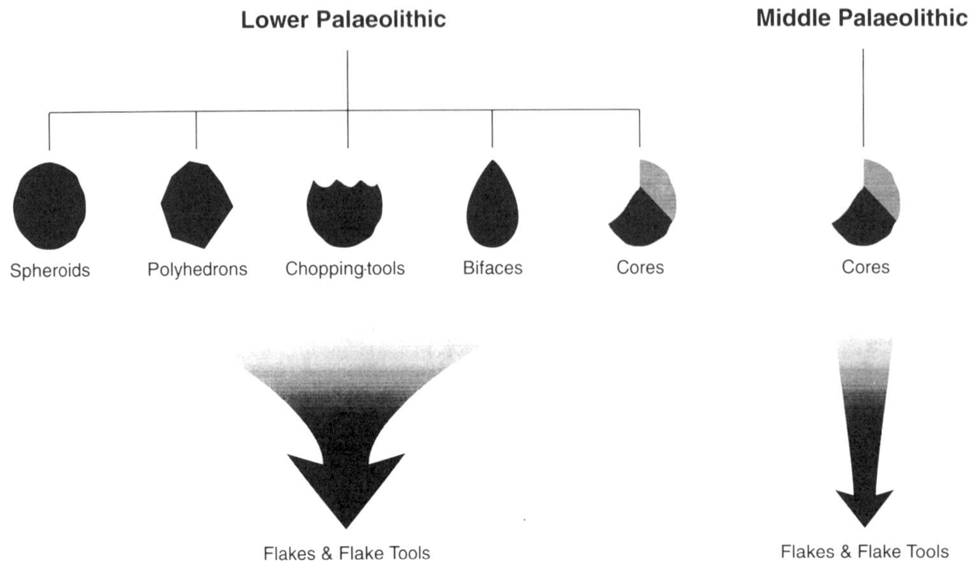

Figure 1. Schematic typological comparison between Lower and Middle Paleolithic morphotypes.

3. The disappearance of tool types that were dominant in Lower Paleolithic industries, i.e., bifaces, spheroids, and chopping-tools (Volman 1984; Clark 1988; McBrearty 1991).
4. Greater standardization of the cores and their by-products, i.e., flakes and flake tools (Clark 1988; Klein 1989).

In summary, the underlying difference between the Acheulean and Mousterian assemblages is the following: in the Lower Paleolithic industries, each morphotype reflects a particular mental template outlining a specific reduction sequence, while in the Mousterian techno-complex all morphotypes derive from a single, though complex, reduction strategy. Accordingly, whereas in the Acheulean techno-complex one observes separate reduction sequences for spheroids, chopping-tools, bifaces, cores, and their products (i.e., flakes), in the Mousterian techno-complex, as a rule one finds reduction sequences involving mainly Levallois cores and their by-products.

There is, nonetheless, a high degree of similarity between the flake/flake tool components of the Late Acheulean and the Mousterian in the Levant. This can be interpreted as indicating continuity, similar to that observed in the Lower to Middle Paleolithic transitional industries in Europe (e.g., Roebroeks et al. 1988; Tuffreau 1995) and in transitional Early to Middle Stone Age lithic assemblages from Africa (Clark 1982, 1988).

Indications of this transitional nature are readily observable in the Levantine archaeological record. Thus the most common Mousterian core technology, the Levallois technique, appears in the Levant as early as the Lower Paleolithic and is used for the production of flakes and flake tools for quite a while before becoming the dominant Mousterian technique. Moreover, spheroids, which are typically absent from Mousterian assemblages just as bifaces are, disappear along with their particular production mode already during the latest stage of the Acheulean (as in Tabun E, Garrod and Bate 1937; and at Gesher Benot Ya'aqov, Goren-Inbar in prep.), though bifaces are still present at that time.

The nature of the mental development processes which may have brought about the observed changes is purely a matter for speculation. Undoubtedly, many complex processes were involved. It seems, however, that the observed phenomena can be adequately elucidated by an analogy derived from the domain of biological evolution (and see Belfer-Cohen and Goren-Inbar 1994). It has been suggested (Gould 1989) that, in the course of biological evolution, the first rush of diversification of life forms brings forth the widest range of anatomical possibilities. This phase is then followed by a lengthy process of restriction, as most of these early 'experimental' forms are 'discarded' while endless variants are generated upon a few surviving models. It can be similarly claimed that in the course of the 'evolution' of lithic production, several production modes appeared initially, each used in a distinct fashion for the manufacture of a specific morphotype. This was followed in later stages (e.g., the Mousterian cultures of the Middle Paleolithic) by a reduction in the number of production sequences alongside an increase in the variability of the products of the 'surviving' strategy.

Stretching the analogy a little further, one may also claim that chance factors like those which fulfilled a crucial role in the survival or extinction of past living phyla (Gould 1989) can be held at least partly responsible for the disappearance or persistence of lithic forms. There seem, however, to be other possible lines of explanation. Indeed, there is firm evidence to suggest that the changes (or at least some of them) observed in the Mousterian techno-complexes can be attributed to the development of greater mental versatility and dexterity in Mousterian hominids, who are known to have possessed cranial capacity well within the modern human range (see below).

If standardization of form and mode of production is one of the criteria used to determine the extent of progress over time with regard to the Lower Paleolithic, then one must consider the criterion of technological flexibility with regard to the Middle Paleolithic. This characteristic appears in opposition to a certain technological *rigidity* exhibited in the preceding Lower Paleolithic techno-complex. Earlier, the very choice of a certain morphotype for production implied application of a particular reduction sequence which allowed for no further alterations once it was underway. The Mousterian production techniques, in contrast, were considerably more versatile, permitting constant modifications to be introduced in the course of the production process.

The present article will focus on the technological characteristics of the Mousterian techno-complex, though other aspects of the lithic industries (raw material, size, typology, etc.) will also be addressed. Attention will be paid also to the inter- and intra-site variability range usually attributed to such factors as ecological constraints and cultural/functional differences. It seems that there is still another line of interpretation to be considered.

2. MOUSTERIAN LITHIC CHARACTERISTICS

2.1. Raw Material

In contrast to the early Lower Paleolithic assemblages, which are characterized by a more sophisticated mode of exploiting diverse raw materials, the Mousterian assemblages are made entirely on a single raw material—flint (Belfer-Cohen and Goren-Inbar 1994; Goren-Inbar 1995). Consequently, various Lower Paleolithic reduction processes associated exclusively with specific raw materials other than flint (e.g., the manufacture of Kombewa basalt flakes for the production of bifaces; Goren-Inbar and Saragusti 1996) disappear altogether. Two open-air Mousterian sites diverge from this pattern: in Quneitra and Far'ah II (Gilead 1980; Gilead and Grigson 1984; Goren-Inbar 1990), basalt and limestone were used to produce small quantities of unique artifacts by a special reduction process, different from that used for the manufacture of flint artifacts (Goren-Inbar in prep.). The overwhelming predominance of flint observed in the long Levantine Mousterian sequences in sites such as the Tabun, Kebara, Qafzeh, and Hayonim Caves (Bar-Yosef et al. 1992; Jelinek 1981; Jelinek et al. 1973) ceases only much later, towards the end of the Epipaleolithic period, with the emergence of mass production of pounding and grinding utensils made of basalt and limestone.

2.2. Size

The Mousterian lithic assemblages show an overall homogeneity of size, probably reflecting standardization. This phenomenon is associated with the dominance of the Levallois technique and the technological shift to the production of cores and their by-products for the most part. The general uniformity in nodule size as discernible in the cores is considered to reflect a well-defined, purposeful process of selection by the manufacturers. This uniformity is manifested in cores as well as in flakes and flake tools (Goren-Inbar 1990). In contrast, Lower Paleolithic tool types such as spheroids, chopping tools, polyhedrons, bifaces, etc., exhibit a size-range far wider than that of Mousterian artifacts (e.g., Bar-Yosef and Goren-Inbar 1993).

2.3. Typology

The observed typological differences between the Lower and Middle Paleolithic assemblages signify a major change the implications of which are dealt with rather cursorily in the published prehistoric research. The disappearance from the Levantine Mousterian assemblages of certain tool types actually reflects the disappearance of complex sets of specific, well-defined typological-technological entities. Chopping tools have been in use for over 2 million years and bifaces for over 1.5 million years, as were also spheroids and sub-spheroids. Each of these tool types was made on a specific raw material and each by a distinct reduction process (Bar-Yosef and Goren-Inbar 1993; Belfer-Cohen and Goren-Inbar 1994). The minimal numbers of chopping tools in Mousterian contexts (Goren-Inbar 1990) may be interpreted as remnants of initial testing and evaluation of flint nodules or of early stages of core modification rather than as products of an intentional process aimed at the manufacture of a specific tool type.

The typological composition of the Levantine Mousterian assemblages is highly consistent throughout the entire duration of the Mousterian (with rare exceptions, such as the limited presence of Abu Zif points in the early part of the Mousterian sequence; Garrod and Bate 1937; Neuville 1951; Bergman and Ohnuma 1983; Copeland 1985). This phenomenon has been observed in most Levantine Mousterian assemblages, such as that from Nahal Aqev in the Negev, in which, "Significantly, . . . there seem to have been no typological shifts through time" (Munday 1977:94). This indicates that technological changes must have occurred against a background of typological stability, in much the same manner as observed in the Mousterian sequences of Tabun (Jelinek 1975) and Qafzeh (Crew 1975).

3. TECHNOLOGY

The technological aspects of the Mousterian lithic assemblages constitute a problematic, multifaceted issue which merits detailed discussion. There is no doubt, however, that the technological abilities of the Mousterian hominids bear direct evidence as to their cultural and mental capacities.

The complexity and range of the lithic variability of the Middle Paleolithic, over time, are best represented in the long sequence of Tabun. Numerous attempts have been made to explain the phenomena observed there, beginning with Garrod's (Garrod and Bate 1937) cultural classification and followed by various others (Jelinek et al. 1973; Copeland 1975; Jelinek 1975, 1981; Meignen and Bar-Yosef 1992; Meignen 1995; Bar-Yosef 1995, in press). It is thus that particular, sequential Levantine Mousterian type-complexes came to be designated "Tabun D," "Tabun C," and "Tabun B" industries (as in Bar-Yosef 1996) after the stratified layers of this type-site with their particular techno-typological configurations.

3.1. Current State of Research

The technological aspects of the Mousterian industries in the Levant have been investigated extensively since the early seventies. Most researchers used the same methodological tool, i.e., the Bordesian system of description and definition (Bordes and Bourgon 1951; Bordes 1953, 1961), which was unfortunately not sufficiently clear-cut. The resulting analyses therefore expressed, at least to a certain extent, subjective interpretations (Copeland 1983; Meignen, this volume).

Most discussions of Mousterian technology revolve around the Levallois technique. The technological variability characterizing both cores and their products is expressed (apart from the size variable) mainly in the scar patterns, the most prominent difference being that between elongated scar patterns and converging ones. Until recently, the main research effort has been focused on the flake and flake-tool components of the Levantine Mousterian assemblages, while the core component was largely neglected.

Indeed, examination of core *typology* has traditionally been considered as the best method for tackling the technological complexity of the Mousterian lithic assemblages. According to the Bordesian typological criteria, the dominant core types in the Levant are those classified as "Levallois" and "Discoidal" as well as "others" (unclassified, "varia," "diversified forms," etc.).

Another analytical approach focused on accurately describing the scar patterns on the dorsal faces of flakes and flake tools, and comparing them with those on the upper faces of cores (e.g., Crew 1975; Munday 1977; Meignen and Bar-Yosef 1988; Goren-Inbar 1990; Boëda and Muhesen 1993). Over the years numerous such studies of the Levantine Mousterian provided a wide array of technological details (i.e., technological attributes and attribute states; see references above). However, it is only with the introduction of a new nomenclature based on a methodological scheme developed mainly by Boëda (1995 and references therein) in the early 1990s that it became possible to analyze all the Mousterian assemblages from the Levant according to detailed and uniform technological criteria.

Admittedly, some of these analyses produced results which merely repeated well-known facts. Others, however, provided new data or rearranged familiar data in different, more coherent configurations. In the past, scarcity of data and inadequacy of the available methodology combined to generate a simplistic view of the Levantine Mousterian techno-complex. Earlier analyses of Mousterian technological data, carried out on material excavated before the 1980s, were consequently somewhat biased. The overall picture was further distorted by a biased choice of material for study and publication. As noted above, technological data were most frequently obtained from analyses of flakes alone while observations made on cores were practically ignored. Failure to integrate the relevant technological data from both sources, cores as well as flakes, led to misconceptions regarding the range of technological variability within the analyzed assemblages. It has indeed been demonstrated (Goren-Inbar 1990) that integrated technological data, obtained from both the flake and the core components, indicate a wider range of technological variability than that suggested by data pertaining to either of the two components alone.

In his methodological scheme, Boëda (1993, 1995; see also Meignen 1995) defines several methods of flaking within the Levallois concept, with the two main variants being the preferential method and the recurrent method (English terminology after: "Definition and Interpretation of the Levallois Technology;" see Dibble and Bar-Yosef 1995). The recurrent method was further divided into the following sub-types: (1) 'unidirectional parallel'; (2) 'unidirectional convergent'; (3) 'bidirectional parallel'; (4) 'bidirectional divergent'; (5) 'centripetal.'

As mentioned earlier, Boëda's scheme of detailed and precise definitions promoted the use of a single, standardized terminology in descriptions of Mousterian Middle Paleolithic technology. This in turn led to recognition of a complex technological pattern within the Levantine Mousterian assemblages. It is now evident that both the preferential method and the recurrent one were used concurrently in the same assemblages (e.g., Bar-Yosef et al. 1992).

The diversity characterizing the Levallois technological pattern became evident to all researchers involved in the study of Levantine Mousterian assemblages (see for example the studies of flake forms and scar-patterning by Crew 1975; Munday 1979; Marks 1988; Meignen and Bar-Yosef 1988; Boutié 1989; Goren-Inbar 1990; Bar-Yosef 1992; Meignen 1994, in press; and others). These assemblages were all found to exhibit the full range of the Levallois reduction sequences. Thus, one finds bipolar, unipolar, and radial cores along with their by-products, as well as evidence of both the preferential and recurrent modes of core exploitation, in every single assemblage. The general awareness of this diversity led to an over-simplified tendency to focus exclusively on the most dominant patterns (and see the example of Kebara Cave below), the most dominant mode of flaking. This tendency soon became the norm in publications discussing Mousterian assemblages and was also expressed in reports of the directionality of scar patterning on the dorsal faces of the Levallois (and non-Levallois) artifacts.

As noted before, technological data are obtained all too frequently through the study of flakes while core data are often not taken into account. It has been shown that, in fact, reliance on both sources of data could lead to conclusions different from those reached on the basis of flake data alone. Thus while observations made on flakes originating from the upper levels of the Mousterian occupation at Kebara would predict a rise in the percentages of the discoidal cores, these percentages are in fact declining. The same is true for the Rosh Ein Mor assemblage (Crew 1975; Munday 1977; Marks 1995), where observations of flakes alone led to the conclusion that all the reduction sequences at the site were of unipolar and convergent cores, whereas the cores themselves present a much more complex and varied picture.

This overall tendency is consistent with the simplistic model which views the Tabun sequence and its range of diversity (defined by the most common pattern in each of its stratigraphic units) as adequately representing the diversity and complexity of all other Levantine Mousterian assemblages.

In order to illustrate the flaws of this approach, examples were selected from recent studies pertaining to such issues as the laminar aspect of the Levantine Mousterian, the "Tabun D" variant (and its Negev manifestation), and the variability in flaking directions observed in the Mousterian lithic assemblages from the cave of Kebara.

3.2. The Laminar Aspect of the Levantine Mousterian

The laminar aspect of the "Tabun D" assemblage stood out very soon after the material was recovered (Garrod and Bate 1937; Neuville 1951) as a unique phenomenon within the Mousterian techno-complex. Evidently, the makers of this Mousterian variety must have been highly skilled craftsmen to have been capable of producing these typically elongated artifacts. As with other Mousterian entities, the technological variability of "Tabun D" assemblages was variously interpreted in terms of climatic, temporal, traditional, environmental, or other differences (Meignen 1994 and references therein).

The Mousterian sites in the Negev were all assigned to the "Tabun D" Mousterian (Marks 1976, 1977). Despite the lack of appropriate radiometric chronology, these sites are regarded as a late phenomenon within the Levantine Mousterian sequence (Marks 1993 and references therein). Referring to certain technological similarities (e.g., the presence of elongated flakes/blades) between the Mousterian Negev assemblages and those from the transitional site of Boker Tachtit, Marks (1988) suggested a model which explained the laminar component at Boker Tachtit as the culmination of a long technological tradition. He claimed that "the second assemblage type belongs technologically within the Early Levantine Mousterian of Tabun D type. . . . Level 1 at Boker Tachtit is a special, terminal expression of this Mousterian" (Marks 1993:7–8) and, "Present evidence, however, strongly indicates that the [Middle to Upper Paleolithic (authors' note)] transition, as defined here, was part of a long-term developmental process within the Tabun D type Mousterian" (ibid). Two Middle Paleolithic sites (Tor Faraj and Tor Sabiha) recently reported from Jordan have also been assigned to the Levantine Mousterian of the "Tabun D" tradition (Henry 1995a and references therein, 1995b).

Laminar components within lithic assemblages are certainly not a Levantine Mousterian novelty: they can be traced back in Levantine lithic industries to as early as Lower Paleolithic times (Meignen 1994; Goren-Inbar 1995). Indeed, laminar components were reported from various Acheulean assemblages where they have been found to comprise cores (either Levallois, prismatique, or pyramidal ones) as well as blanks (Goren 1979; Goren-Inbar 1985). It follows that the Mousterian assemblages from the Negev, or those reported from Jordan, cannot be claimed to unequivocally represent a retention of the "Tabun D" Mousterian tradition. It could be argued just as plausibly that these assemblages represent a facies of their own. The late appearance of laminar-oriented assemblages in the Negev can be regarded as additional proof of the complexity of Mousterian technology.

Marks, in his discussion of the technological variability of the Negev Mousterian, notes that: "During the early Mousterian, a wide morphological range of blanks was produced (Crew 1976) by using a large number of different core-reduction strategies, from Levallois, through discoidal, to true, if rare, blade-core methods (Munday 1976)" (1993:16). Examinations of Mousterian technological variability should thus focus on these observations, rather than on the over-emphasized laminar aspect, which is but a single facet of a much more complex technological pattern. Only when the assemblages from Layer D at Tabun are properly published and the lithic material is adequately presented in illustrations, will it be possible to evaluate the degree of similarity between these and the Negev material.

Despite the strongly emphasized laminar character of the Negev Mousterian assemblages, a very pronounced centripetal pattern of flaking is detectable on both the flakes and the cores from these assemblages. Figures 2 and 3, presenting relevant typological and technological data (obtained from observations on cores), illustrate the fact that the centripetal mode is common. This clearly contradicts previous expectations for predominance of a bipolar or a converging pattern.

3.3. Variability in Flaking Direction

One of the most illustrative cases of the variability in flaking direction observed in Levantine Mousterian assemblages is the lithic assemblages from Kebara Cave. These assemblages show a complex pattern of technological variability characterized by coexistance

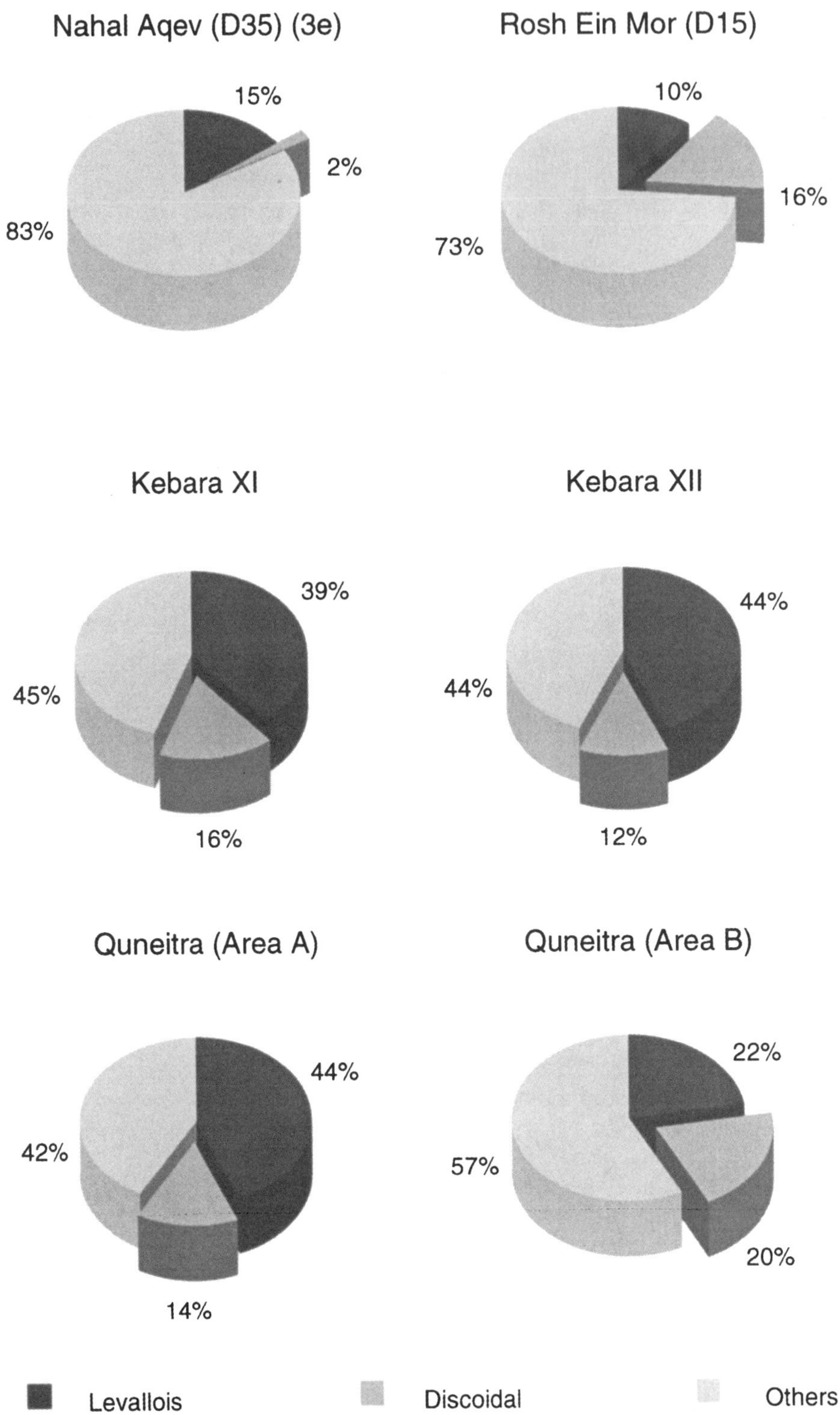

Figure 2. Types of Levallois cores in various Mousterian assemblages.

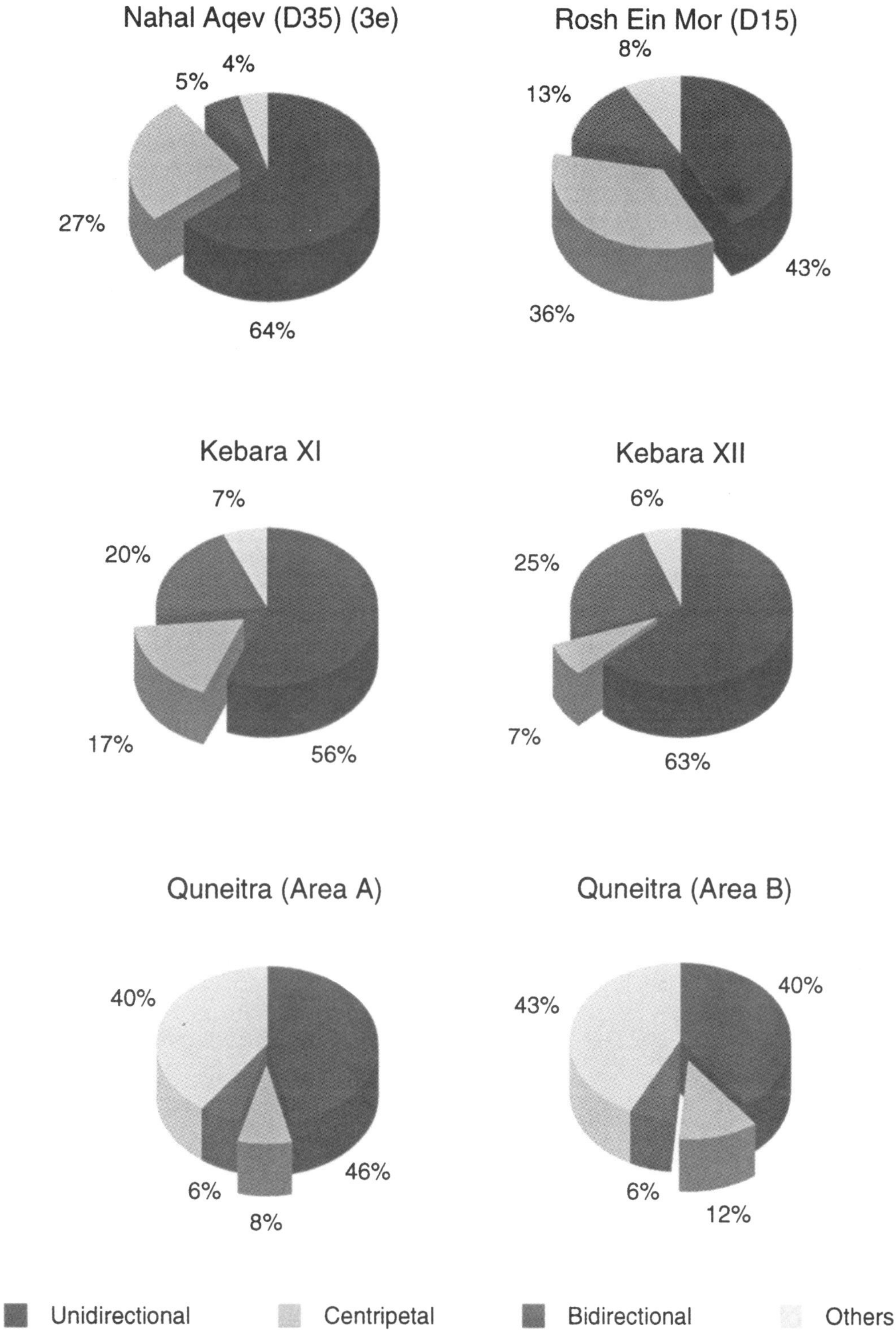

Figure 3. Mousterian cores classified according to scar patterns on the debitage surface.

of both the preferential (lineal) and the recurrent modes of core flaking, with the latter exhibiting a much higher degree of technological variability. The excavators emphasize that "the coexistence throughout the Kebara Mousterian units of the two knapping systems ("recurrent" and "lineal") is noteworthy" (Bar-Yosef et al. 1992:513). However, they proceed to state that the recurrent mode is predominant: "Throughout the levels of Kebara, . . . core reduction was predominantly unidirectional, with converging flake removals. This observation is particularly obvious in units IX and X . . . ; in the upper units (VII and VIII), radial preparation are somewhat more numerous" (Bar-Yosef et al. 1992:513).

The scar pattern most common on cores, the recurrent unidirectional one, is considered by the excavators as the pattern most typical of all Levantine Mousterian industries. Concerning the second mode, the excavators explain: "Throughout the levels, core management is evidenced by the presence of unidirectional removals, a phenomenon that is rather common in the Near East. However, it is important to stress that the convergent pattern often gives the products a triangular or sub-triangular shape. All of these elements constitute a dominant feature of the assemblage. This means that in spite of the availability of other flaking methods, such as the preferred-flake method or radial or bidirectional reduction, the local artisans made a clear choice" (Bar-Yosef et al. 1992:516).

Published core data from Kebara Cave show that even in strata considered to show dominance of elongated blanks and which display a unidirectional (convergent) scar pattern on cores, there is a substantial component of cores with centripetal scar patterns. In the lower strata, this component comprises as much as about 12 percent of the cores, a value similar to that observed in the upper strata (Meignen and Bar-Yosef 1991; for somewhat different values see Meignen and Bar-Yosef 1988). Figures 2 and 3 juxtapose the relevant data from Kebara with those from the Negev sites. The former provide values of "discoidal" cores similar to those encountered in the Negev assemblages as well as in other assemblages considered to have emphasized laminar reduction processes.

4. DISCUSSION

Levantine Mousterian lithic assemblages from different geographical areas of Israel exhibit a high degree of technological variability. This variability is manifested in cores as well as in their by-products, i.e., flakes, blades, and points.

4.1. The Technological Variability

4.1.1. Cores. The fact that the entire range of core variability is represented in nearly every Levantine Mousterian assemblage was noted by many researchers. Thus Crew, for example, reports the presence of non-Levallois prismatic and pyramidal cores in Rosh Ein Mor, an assemblage assigned to the early Mousterian "Tabun D" type (Crew 1976; see also discussion in Meignen 1994). These "Upper Paleolithic" core-types occur in nearly all Levantine Mousterian assemblages in very small frequencies, which accounts for the fact that their presence has been ignored in most reports. These rare types occur in sites from Quneitra in the north (Goren-Inbar 1990: table 24, p. 115) to the Negev sites in the south, and their sporadic appearance is clearly not consistent with any time trajectory trends. As a matter of fact, various types of non-Levallois cores already appear in certain Levantine Acheulean assemblages (i.e., Berekhat Ram; Feraud et al. 1983) and continue throughout the Mousterian sequence up to its most recent occurrences (i.e., Quneitra; Goren-Inbar 1990) dating from some 54,000 years ago.

However, the variability range of the cores is mainly expressed in the co-existence of several, distinct scar-patterns within the Levallois system. This specific variability, expressed in the organization and direction of scar patterns on the 'reduction surface' of cores, was observed in all Mousterian assemblages (Bar-Yosef et al. 1992; Meignen 1995). In the case of the Quneitra Mousterian, for example, this co-existence of different technological approaches was thus summarized: "From a technological perspective, what were assumed to be three distinct Mousterian reduction processes turned to be a single and complex mode. Thus, the three traditionally distinct types of flint cores (Levallois, Discoidal, and Cores-on-flakes) were found to have very similar properties, indicating the presence of a single complex flint reduction process" (Goren-Inbar 1990:237).

Indeed, both the 'preferential' and the 'recurrent' modes of reduction co-exist in many of the lithic assemblages of the Levantine Mousterian (e.g., Meignen and Bar-Yosef 1988, 1991; Goren-Inbar 1990; Meignen 1995).

A possible explanation for this phenomenon, offered by the Kebara excavators, suggests that the reduction sequence of a Levallois core began in the 'recurrent' mode, which was replaced by the 'preferential' mode towards the end of the reduction process, when the core was nearly exhausted, for the terminal removals (Bar-Yosef et al. 1992). On the other hand, one could easily suggest an alternative model whereby both the 'preferential' and the 'recurrent' modes were employed rather randomly and interchangeably.

In order to confirm either suggestion, a Mousterian assemblage would be necessary, with refitting potential similar to that of the Boker Tachtit assemblage, for example (Volkman 1983). Meanwhile, the variability of the available data suggests a very complex situation.

4.1.2. Blanks. The variability range of scar patterns observed on blanks (flakes and blades) is much wider than that observed on cores. Moreover, it has been experimentally demonstrated (e.g., Boëda 1993) that the use of distinctly different Levallois methods could result in morphologically similar products, a fact which causes some confusion as to the nature of the specific reduction sequence employed in any given case. In most cases, however, it is possible to identify the particular Levallois mode employed. Preference of the 'unipolar convergent' mode (as in most of the Kebara stratigraphic units) or the 'centripetal' one (as in most of the Qafzeh levels) is quantitatively evident in various Mousterian assemblages. Nevertheless, ignoring the quantitative aspect, the striking feature of the Mousterian range of variability is the fact that different production modes appear to have co-existed within each of the different Mousterian assemblages (see the case of Quneitra described above).

For example, the production mode for laminar artifacts is evident not only in assemblages where elongated blanks are dominant but also where they are present in minimal numbers. The Mousterian of Quneitra could serve here as an illustrative case for assemblages that are poor in a laminar component. The frequency of knives there is less than 2 percent (Goren-Inbar 1990, table 17) and the Ilam is 7.82 and 4.22 for Areas A and B respectively (Goren-Inbar 1990, table 35). Yet, all the available blades demonstrate excellent craftsmanship (Goren-Inbar 1990, Figure 41).

Analyses of various Levantine Middle Paleolithic assemblages clearly demonstrate that throughout the entire Mousterian sequence hominids were fully capable of producing a wide array of lithic artifacts by different technological methods. This variability of methods is documented in both the core (Goren-Inbar 1990, tables 39–40; Meignen 1994) and the blank components of each of the investigated assemblages. Furthermore, the Mousterian assemblages characteristically show dominance of some of the methods and little evidence for the use of others.

There are indeed technological trends common to all the Levantine Mousterian assemblages. The repetitive pattern of the 'recurrent convergent' mode, for example, is discernible in different sequences, geographical zones, and time trajectory units. Nevertheless, and contrary to Meignen's view (Meignen 1995; Meignen and Bar-Yosef 1988, 1991, 1992), it cannot be considered as the technological mode most typical of the Levantine Mousterian. The 'radial' patterns so common at Qafzeh, and the 'recurrent along axis' pattern typical of Quneitra, are all Mousterian, having in common other technological traits indicative of complex and interactive sets of reduction processes.

The Tabun Mousterian assemblages have been published only cursorily, with the published material consisting mainly of that retrieved from Garrod's old excavations (Garrod and Bate 1937; Jelinek et al. 1973; Jelinek 1975, 1981). As such, attempts to compare their technological characteristics with those of Mousterian assemblages retrieved from various modern excavations (e.g., at Amud, Kebara, Qafzeh, and Quneitra) seem rather presumptuous. Such attempts seem even more precarious in light of new developments in the field of radiometric chronology (Valladas et al. 1987; Schwarcz et al. 1988, 1989; Grün and Stringer 1991; Grün et al. 1991), which enable Mousterian assemblages to be positioned within a more accurate temporal framework. The Tabun sequence was thus shown to be much older than previously envisioned (Mercier et al. 1995; Schwarcz and Rink, in press).

Recent TL dates for Tabun (layers D-C of Garrod's excavations, and units II-IX of Jelinek's excavations; Garrod and Bate 1937; Jelinek 1981) range between 171±17 and 263±27 ky (Mercier et al. 1995 and see discussion therein; Valladas et al., this volume). Thus attempts to correlate various Mousterian assemblages with those of Tabun (e.g., the suggested correlation between the Kebara and the Tabun Mousterian assemblages in Bar-Yosef 1992) are rather dubious in light of the fact that the Tabun sequence is several thousand years older.

Attempts to correlate the undated Negev Mousterian entities (Rosh Ein Mor, Boker Tachtit) with the "Tabun D" assemblages (Marks 1988, 1995), should likewise be suspended until dates are provided for the Negev sites as well.

It should be noted that new dates for different Mousterian assemblages (Amud and Hayonim caves: Hovers, in press; Schwarcz and Rink, in press; Valladas et al., this volume), which place them within a more accurate chronological framework, strongly suggest that the end of the Middle Paleolithic sequence is characterized by inter-assemblage technological variability even more extensive than previously envisioned (for a detailed discussion see Hovers, this volume).

4.2. Explanations of the Observed Technological Variability in the Levantine Mousterian Assemblages

The archaeological record demonstrates the existence of several production modes in each Levantine Mousterian assemblage, one of which is clearly predominant. The reasons for these preferences are a matter for speculation. However, the conservative, traditional view linking particular preferences to "group tradition" rather than to time trajectories or specific geographic locations seems to furnish the most plausible explanation. Still, recent studies indicate that certain technological traits might be correlated with functional demands (Shea 1988; Friedman et al. 1995; Henry 1995a; Boëda et al. 1996). Thus Lieberman and Shea (1994) suggest that the elongated, pointed component of the lithic material they studied is associated with hunting requirements. Similarly, recent studies of Mousterian points, which indicate that some of them have been used as hafted objects, suggest that this function may be correlated with particular dimensions and morphological attributes. The scarcity of these finds prevents the construction of a quantitative or proportional scheme in which the different technological modes identified in a given assemblage may be presented. On the other hand, interpretations focused exclusively on isolated morphotypes, however frequent, inevitably result in generalizations which fail to explain the overall observed variability.

A case at hand is Lieberman's and Shea's (1994) study. They suggest inferring the existence of two distinct human groups (perhaps even species) solely from an inter-assemblage difference, interpreted in functional terms, in the frequency of a single morphotype, namely, elongated pointed artifacts. This approach is clearly invalid since it ignores contemporaneous intra-assemblage technological variability. Even a cursory examination reveals that each Levantine Mousterian assemblage attests to the use of a variety of manufacturing modes. No bimodality exists which can be attributed to or explained by the presence of two different human species. Rather it seems that the occupants of each Mousterian site were capable of manufacturing their artifacts by diverse technological methods.

4.3. Cultural and Mental Capacities

It is generally acknowledged that one of the criteria by which the trajectory of evolution over time is evaluated is the degree of standardization in lithic production. When we compare the degree of standardization observed in Lower Paleolithic industries and that observed in the Mousterian assemblages, the latter is clearly much higher. At the same time, while the Lower Paleolithic is marked by technological inflexibility, i.e., a rigid correlation between the technique employed and the final product, the Mousterian technocomplex is characterized by technological flexibility.

How does all this reflect human capacities? The domain of cognitive archaeology is a problematic one and there are clearly "...reasons for not investing heavily in [it]. The main one...is that it is not archaeology's strong suit—it is not what we can do best" (Hill 1994). All the same, various attempts have been made (for example Karlin 1992; Karlin et al. 1992; Perles 1992; Karlin and Julien 1994, to mention but a few) to draw inferences regarding human intellectual capabilities from evidence of the development of technical skills, trying to evaluate the complexity of the concepts underlying them.

For quite some time, cognitive scientists of different backgrounds have been inclined to regard intelligence as a multi-faceted phenomenon. Gardner (1983) maintains that there are six domains of intelligence: the spatial, musical, logical-mathematical, personal, bodily-kinesthetic, and linguistic ones. Humphrey (1976) speaks of just two: social and non-social intelligence. It is regarded as a uniquely human phenomenon that the "building stones" of the different domains of intelligence are interchangeable. This would imply that each particular aspect of knowledge, acquired in the process of dealing with a particular problem, immediately becomes available to all domains of intelligence with information flowing between the various mental modules (for example, see Fodor 1983, 1987; for further information see Gardner 1985; LeDoux and Hirst 1987; Eccles 1994). This constant flow of information from multiple sources and parallel processors is analyzed simultaneously, which accounts for complex understanding and complex operations and for the speed with which they are achieved (Feldman and Ballard 1982).

On the basis of Fodor's theory of mental modules, it has been suggested (Mithen 1994) that the differences observed between the Middle Paleolithic and Upper Paleolithic cultural entities can be explained in terms of cognitive changes which increased the flexibility of the human cognitive system and improved the flow of information between the different cognitive domains. It seems that this explanation applies equally well to the differences observed between different cultural entities in the Lower Paleolithic.

Indeed, according to many of the researchers cited above, this process of increased accessibility of knowledge may have been a gradual one. At first it may have happened sporadically within a particular society, spontaneously emerging in one individual or another. It may, however, have been acquired by learning, in a process similar to that by which memory abilities are acquired by an individual. It has been suggested that in this process "...repeated stimulation of specific receptors will lead slowly to the formation of an 'assembly' of association-area cells which can act briefly as a closed system after stimulation has ceased; this prolongs the time during which the structural changes of learning can occur and constitutes the simplest instance of a representative process (image or idea)" (Hebb 1949:60). This process of learning may have also been facilitated by mechanisms of social learning (Zentall and Galef 1988).

At present neurologists seem to agree that *Homo erectus* (if not *Homo habilis*) possessed the appropriate gross neuroanatomical configuration to support conceptual structures, which would imply existence of the neural preconditions for language (Gregory 1984; Deacon 1990; Wilkins and Wakefield 1995).

This avenue of thought furnishes a firm basis for explaining the modularity and flexibility observed in human cultural and mental capacities, including those observed in the Mousterian techno-complex. It agrees with the available archaeological evidence, at least that from the Levantine Mousterian, and explains the 'mosaic' nature of various cultural attributes. Endeavors to explain the cultural variability detailed in the present article as reflecting the 'demise' of the Neandertals and their failure to compete with their mentally better-equipped *Homo sapiens sapiens* contemporaries, seem to belong to the realm of fiction rather than to that of science.

The wealth of Levantine Mousterian data and the array of information pertaining to its technological properties await re-examination. Further analyses, both quantitative and qualitative, are bound to shed more light on the evolutionary processes pertaining to the development of humankind during the Middle Paleolithic.

REFERENCES

Bar-Yosef, O. 1992. Middle Palaeolithic human adaptations in the Mediterranean Levant, 189–215. In *The Evolution and Dispersal of Modern Humans in Asia*, ed. T. Akazawa, K. Aoki, and T. Kimura. Tokyo: Hokusen-Sha.

Bar-Yosef, O. 1994. The Lower Paleolithic of the Near East. *Journal of World Prehistory* 8:211–65.

Bar-Yosef, O., and D. Alon. 1988. Excavations in the Nahal Hemar Cave. *'Atiqot* 18:1–30.

Bar-Yosef, O., and N. Goren-Inbar. 1993. *The Lithic Assemblages of Ubeidiya, in a Lower Paleolithic Site in Jordan Valley*. QEDEM 34. Jerusalem: Institute of Archaeology, Hebrew University of Jerusalem.

Bar-Yosef, O., B. Vandermeersch, B. Arensburg, A. Belfer-Cohen, P. Goldberg, H. Laville, L. Meignen, Y. Rak, J. D. Speth, E. Tchernov, A. M. Tillier, and S. Weiner. 1992. The excavations in Kebara Cave, Mt. Carmel. *Current Anthropology* 33:497–550.

Belfer-Cohen, A., and N. Goren-Inbar. 1994. Cognition and communication in the Levantine Lower Palaeolithic. *World Archaeology* 26:144–57.

Bergman, C., and K. Ohnuma. 1983. Technological notes on some blades from Hummal Ia, El Kowm, Syria. *Quartär* 33/34: 171–80.

Besançon, J., L. Copeland, F. Hours, S. Muhesen, and P. Sanlaville. 1981. Le Paléolithique d'el Kowm. Rapport préliminaire. *Paléorient* 7:33–55.

Boëda, E. 1993. Le débitage discoïde et le débitage Levallois récurrent centripète. *Bulletin de la Société Préhistorique Française* 90:392–404.

Boëda, E. 1994. *Le Concept Levallois: Variabilité des Méthodes*. Paris: CNRS Editions.

Boëda, E. 1995. Levallois: A volumetric construction, methods, a technique, 41–68. In *The Definition and Interpretation of Levallois Technology*, ed. H. L. Dibble and O. Bar-Yosef. Madison: Prehistory Press.

Boëda, E., and S. Muhesen. 1993. Umm El Tlel (El Kowm, Syrie): Etude préliminaire des industries lithiques du Paléolithique moyen et supérieur 1991–1992. *Cahiers de l'Euphrate* 7:47–91.

Boëda, E., J. Connan, D. Dessort, S. Muhesen, N. Mercier, H. Valladas, and N. Tisnérat. 1996. Bitumen as a hafting material on Middle Palaeolithic artefacts. *Nature* 380:336–8.

Bordes, F. 1953. Levalloisien et Moustérien. *Bulletin de la Société Préhistorique Française* 4:226–35.

Bordes, F. 1961. *Typologie du Paleolithique Ancien et Moyen*. Bordeaux: Impremeries Delmas.

Bordes, F., and M. Bourgon. 1951. Le complex Moustérien: Moustérien, Levalloisien, et Tayacien. *L'Anthropologie* 55:1–23.

Boutié, P. 1989. Etude technologique de l'industrie de la grotte de Qafzeh (près de Nazareth, Israel): Rapport préliminaire, 147–67. In *Investigations in South Levantine Prehistory*, ed. O. Bar-Yosef and B. Vandermeersch. Oxford: British Archaeological Reports International Series 396.

Clark, J. D. 1982. The transition from Lower to Middle Palaeolithic in the African continent, 235–55. In *The Tansition from Lower to Middle Palaeolithic and the Origin of Modern Man*, ed. A. Ronen. Oxford: British Archaeological Reports International Series 151.

Clark, J. D. 1988. The Middle Stone Age of East Africa and the beginnings of regional identity. *Journal of World Prehistory* 2:235–305.

Copeland, L. 1975. The Middle and Upper Palaeolithic in Lebanon and Syria in the light of recent research, 317–50. In *Problems in Prehistory: North Africa and the Levant*, ed. F. Wendorf and A. E. Marks. Dallas: SMU Press.

Copeland, L. 1983. Levallois/non-Levallois determinations in the early Levant Mousterian problems and questions for 1983. *Paléorient* 9:15–27.

Copeland, L. 1985. The pointed tools of Hummal Ia (El Kowm, Syria). *Cahiers de l'Euphrate* 4:177–89.

Crew, H. 1975. An Examination of the Variability of the Levallois Method: Its Implications for the Internal and External Relationships of the Levantine Mousterian. Ph. D. diss. University of Michigan. Ann Arbor.

Crew, H. 1976. The Mousterian of Rosh Ein Mor, 75–112. In *Prehistory and Paleoenvironments in the Central Negev (Israel), The Avdat/Aqev Area*, Part 1, ed. A. E. Marks. Dallas: SMU Press.

Deacon, T. W. 1990. Brain-language coevolution, 1–35. In *The Evolution of Human Language, SFI Studies in the Science of Complexity*, ed. J. A. Hawkins and M. Gell-Mann. New York: Addison-Wesley.

Dibble, H. L., and O. Bar-Yosef, eds. 1995. *The Definition and Interpretation of the Levallois Technology*. Madison: Prehistory Press.

Eccles, J. C. 1994. *How the Self Controls its Brain*. Berlin: Springer-Verlag.

Feldman, J., and D. Ballard. 1982. Connectionist models and their properties. *Cognition Sciences* 6:205–54.

Feraud, G., D. York, C. M. Hall, N. Goren, and H. P. Schwarcz. 1983. $^{40}Ar/^{39}Ar$ age limit for an Acheulian site in Israel (sample site: Berekhat Ram). *Nature* 304:263–5.

Fodor, A. 1983. *The Modularity of Mind*. Cambridge: MIT Press.

Fodor, A. 1987. *Psychosemantics*. Cambridge: MIT Press.

Friedman, E., N. Goren-Inbar, A. Rosenfeld, O. Marder, and F. Burian. 1995. Hafting during Mousterian times—further indication. *Mitekufat Ha'even* 26:8–31.

Gamble, C. 1986. *The Palaeolithic Settlement of Europe*. Cambridge: Cambridge University Press.

Gardner, H. 1983. *Frames of Mind: The Theory of Multiple Intelligences*. New York: Basic Books.

Gardner, H. 1985. *The Mind's New Science*. New York: Basic Books.

Garrod, D. A. E., and D. M. A. Bate 1937. *The Stone Age of Mount Carmel, Vol. 1, Excavation at Wadi Mughara*. Oxford: Clarendon Press.

Gilead, I. 1980. The Middle Paleolithic open-air site near Tell Fara, western Negev: Preliminary report. *Israel Exploration Journal* 30:52–62.

Gilead, I. 1995. Problems and prospects in the study of the Levallois technology in the Levant: The case of Fara II, Israel, 79–91. In *The Definition and Interpretation of the Levallois Technology*, ed. H. L. Dibble and O. Bar-Yosef. Madison: Prehistory Press.

Gilead, I., and C. Grigson. 1984. Far'ah II a Middle Paleolithic open-air site in Northern Negev, Israel. *Proceedings of the Prehistoric Society* 50:71–97.

Goren, N. 1979. An Upper Acheulian industry from the Golan Heights. *Quartär* 29/30:105–31.

Goren-Inbar, N. 1985. The lithic assemblages of Berekhat Ram Acheulian site, Golan Heights. *Paléorient* 11:7–28.

Goren-Inbar, N. 1990. *Quneitra—A Mousterian site on the Golan Heights*. QEDEM 31. Jerusalem: Institute of Archaeology, Hebrew University of Jerusalem.

Goren-Inbar, N. 1995. The Lower Palaeolithic of Israel, 93–109. In *The Archaeology of Society in the Holy Land*, ed. T. E. Levy. London: Leicester University Press.

Goren-Inbar, N., and I. Saragusti. 1996. An Acheulian biface assemblage from the site of Gesher Benot Ya'aqov, Israel: Indications of African affinities. *Journal of Field Archaeology* 22:15–30.

Gould, S. J. 1989. *Wonderful Life*. New York: W. W. Norton and Company.

Gregory, R. L. 1984. *Mind in Science*. London: Penguin Books.

Grün, R., and C. B. Stringer. 1991. Electron Spin Resonance dating and the evolution of Modern Humans. *Archaeometry* 33:153–99.

Grün, R., C. B. Stringer, and H. P. Schwarcz. 1991. ESR dating of teeth from Garrod's Tabun Cave collection. *Journal of Human Evolution* 20:231–48.

Hebb, D. 1949. *The Organization of Behavior*. New York: Wiley.

Henry, D. O. 1995a. Late Levantine Mousterian patterns of adaptation and cognition, 107–32. In *Prehistoric Cultural Ecology and Evolution*, ed. D. O. Henry. New York: Plenum Press.

Henry, D. O. 1995b. The Middle Paleolithic sites, 49–84. In *Prehistoric Cultural Ecology and Evolution*, ed. D. O. Henry. New York: Plenum Press.

Hill, J. 1994. Interpretation and testability in theories about prehistoric thinking, 15–21. In *The Ancient Mind*, ed. C. Renfrew and E. B. Zubrow. Cambridge: Cambridge University Press.

Humphrey, N. 1976. The social function of intellect, 303–17. In *Growing Points in Ethology*, ed. P. P. G. Bateson and R. A. Hinde. Cambridge: Cambridge University Press.

Jagher, R. 1993. Nadaouiyeh Aïn Askar: Un nouveau site hummalien dans le bassin d'El Kowm (Syrie). *Cahiers de l'Euphrate* 7:37–46.

Jelinek, A. 1975. A preliminary report on some Lower and Middle Paleolithic industries from the Tabun Cave, Mount Carmel (Israel), 297–316. In *Problems in Prehistory: North Africa and the Levant*, ed. F. Wendorf and A. E. Marks. Dallas: SMU Press.

Jelinek, A. 1981. The Middle Palaeolithic in the Southern Levant from the prespective of the Tabun Cave, 265–80. In *Préhistoire du Levant*, ed. J. Cauvin and P. Sanlaville. Paris: CNRS Editions.

Jelinek, A., W. R. Farrand, G. Haas, A. Horowitz, and P. Goldberg. 1973. New excavations at the Tabun Cave, Mt. Carmel, Israel 1967–1972; A preliminary report. *Paléorient* 1:151–83.

Karlin, C. 1992. Connaissance et savoir-faire: Comment analyser un procesus technique en prehistoire. *Technologia y cadenas operativas liticas* Reunion Internacional: 99–124.

Karlin, C., N. Bodu, N. Pigeot, and S. Ploux. 1992. Some socio-economic aspects of the process of lithic reduction among groups of hunter-gatherers of the Paris-Basin area, 318–40. In *The Use of Tools by Humans and Non-human Primates*, ed. A. Berthelet and J. Chavaillon. Oxford: Clarendon Press.

Karlin, C., and M. Julien. 1994. Prehistoric technology: A cognitive science?, 152–64. In *The Ancient Mind*, ed. C. Renfrew and E. B. Zubrow. Cambridge: Cambridge University Press.

Klein, R. G. 1989. *The Human Career: Human Biological and Cultural Origins*. Chicago: University of Chicago Press.

Klein, R. G. 1992. The archeology of modern human origins. *Evolutionary Anthropology* 1:5–14.

LeDoux, J. H., and H. Hirst, eds. 1987. *Mind and Brain*. Cambridge: Cambridge University Press.

Lev, E. 1992. The Vegetal Food of the "Neanderthal" Man in Kebara Cave, Mt. Carmel in the Middle Palaeolithic Period. Ms. C. thesis. Bar-Ilan university, Ramat Gan.

Lieberman, D. E., and J. J. Shea. 1994. Behavioral differences between Archaic and Modern Humans in the Levantine Mousterian. *American Anthropologist* 96:300–32.

Marks, A. E., ed. 1976. *Prehistory and Paleoevnironments in the Central Negev, Israel. the Avdat/Aqev Area*, Vol. 1. Dallas: SMU Press.

Marks, A. E., ed. 1977. *Prehistory and Paleoevnironments in the Central Negev, Israel. the Avdat/Aqev Area*, Vol. 2. Dallas: SMU Press.

Marks, A. E. 1988. The Middle to Upper Paleolithic transition in the Southern Levant: Technological change as an adaptation to increasing mobility, 109–23. In *L'Homme de Neandertal*, No. 35, ed. M. Otte, Liège: Université de Liège.

Marks, A. E. 1993. The Early Upper Paleolithic: the view from the Levant, 5–21. In *Before Lascaux*, ed. H. Knecht, A. Pike-Tay, and R. White. Boca Raton: CRC Press.

Marks, A. E. 1995. Modeling the production of elongated blanks from the Early Levantine Mousterian at Rosh Ein Mor, 267–77. In *The Definition and Interpretation of Levallois Technology*, ed. H. L. Dibble and O. Bar-Yosef. Madison: Prehistory Press.

McBrearty, S. 1991. Recent research in western Kenya and its implications for the status of the Sangoan Industry, 159–76. In *Cultural Beginnings: Approaches to Understanding Early Hominid Life-ways in the African Savanna*, ed. J. D. Clark. Bonn: Dr. Rudolf Habelt GMBH.

Meignen, L. 1994. Paléolithique moyen au Proche-Orient le phénomène laminaire, 125–59. In *Les Industries Laminaires au Paléolithique Moyen*, ed. S. Révillon and A. Tuffreau. Paris: CNRS Editions.

Meignen, L. 1995. Levallois lithic production systems in the Middle Paleolithic of the Near East: The case of the unidirectional method, 361–79. In *The Definition and Interpretation of the Levallois Technology*, ed. H. L. Dibble and O. Bar-Yosef. Madison: Prehistory Press.

Meignen, L. and O. Bar-Yosef. 1988. Variabilite technologique au Proche Orient: L'example de Kebara, 81–95. In *L'Homme de Neandertal*, No. 4, ed. M. Otte. Liège: ERAUL.

Meignen, L., and O. Bar-Yosef. 1991. Les outillages lithiques Moustériennes de Kébara (fouilles 1982–1985). Premiers résultats, 49–75. In *Le Squelette Moustérien de Kébara 2*, ed. O. Bar-Yosef and B. Vandermeersch. Paris: CNRS Editions.

Meignen, L., and O. Bar-Yosef. 1992. Middle Paleolithic lithic variability in Kebara Cave, Mount Carmel, Israel, 129–48. In *The Evolution and Dispersal of Modern Humans in Asia*, ed. T. Akazawa, K. Aoki, and T. Kimura. Tokyo: Hokusen-Sha.

Mercier, N., H. Valladas, G. Valladas, J. L. Reyss, A. Jelinek, L. Meignen, and J. L. Joron. 1995. TL dates of burnt flint from Jelinek's excavations at Tabun and their implications. *Journal of Archaeological Science* 22:495–509.

Mithen, S. 1994. From domain specific to generalized intelligence: A cognitive interpretation of the Middle/Upper Palaeolithic transition, 29–39. In *The Ancient Mind*, ed. C. Renfrew and E. B. Zubrow. Cambridge: Cambridge University Press.

Munday, F. 1977. Nahal Aqev (D35): A stratified open-air Mousterian occupation in the Avdat Aqev area, 35–60. In *Prehistory and Paleoevnironments in the Central Negev, Israel. the Avdat/Aqev Area*, Vol. 2, ed. A. E. Marks. Dallas: SMU Press.

Munday, F. 1979. Levantine Mousterian technological variability: A perspective from the Negev. *Paléorient* 5:87–104.

Neuville, R. 1951. *Le Paléolithique et Mésolithique de Désert de Judée*. Paris: Masson.

Perles, C. 1992. In search of lithic strategies: A cognitive approach to prehistoric chipped stone assemblages, 357–84. In *Representations in Archaeology*, ed. J. C. Gardin and C. S. Peebles. Bloomington: Indiana University Press.

Révillion, S., and A. Tuffreau. 1994. *Les Industries Laminaires au Paléolithique Moyen*. Paris: CNRS Editions.

Roebroeks, W., J. Kolen, and F. Resnik. 1988. Planning depth, anticipation and the organization of Middle Palaeolithic technology: The "archaic natives" meet Eve's descendants. *Hellinium* 38:17–34.

Schwarcz, H. P., W. M. Buhay, R. Grün, H. Valladas, E. Tchernov, O. Bar-Yosef, and B. Vandermeersch. 1989. ESR dating of Neanderthal site, Kebara Cave, Israel. *Journal of Archaeological Science* 16:653–9.

Schwarcz, H. P., W. M. Buhay, R. Grün, B. Vandermeersch, O. Bar-Yosef, and E. Tchernov. 1988. ESR dating for the hominid burial site of Qafzeh in Israel. *Journal of Human Evolution* 17:733–7.

Shea, J. J. 1988. Spear points from the Middle Paleolithic of the Levant. *Journal of Field Archaeology* 15:441–50.

Tuffreau, A. 1995. The variability of Levallois technology in Northern France and neighbouring areas, 413–27. In *The Definition and Interpretation of Levallois Technology*, ed. H. L. Dibble and O. Bar-Yosef. Madison: Prehistory Press.

Valladas, H., J. L. Joron, G. Valladas, B. Arensburg, O. Bar-Yosef, A. Belfer-Cohen, P. Goldberg, H. Laville, L. Meignen, Y. Rak, E. Tchernov, A. M. Tillier, and B. Vandermeersch. 1987. Thermoluminescence dates for the Neanderthal burial site at Kebara in Israel. *Nature* 330:159–60.

Volkman, P. 1983. Boker Tachtit: core reconstructions, 127–90. In *Prehistory and Paleoevnironments in the Central Negev, Israel. The Avdat/Aqev Area*, Vol. 3, ed. A. E. Marks. Dallas: SMU Press.

Volman, T. P. 1984. Early prehistory of southern Africa, 169–220. In *Southern African Prehistory and Paleoenvironments*, ed. R. G. Klein. Rotterdam: A. A. Balkema.

Weiner, S., P. Goldberg, and O. Bar-Yosef. 1993. Bone preservation in Kebara Cave, Israel using on-site Fourier Transform Infrared Spectrometry. *Journal of Archaeological Science* 20:613–27.

Wilkins, W. K., and J. Wakefield. 1995. Brain, evolution and neurolinguistic preconditions. *Behavioral and Brain Sciences* 18:161–226.

14

THE ROLE OF HUNTING AND SCAVENGING IN NEANDERTAL PROCUREMENT STRATEGIES

New Evidence from Kebara Cave (Israel)

John D. Speth[1] and Eitan Tchernov[2]

[1]Museum of Anthropology
University of Michigan
Ann Arbor, Michigan 48109
[2]Department of Evolution, Systematics, and Ecology
Hebrew University
Jerusalem 91904, Israel

1. INTRODUCTION

This paper has two principal goals. The first is to demonstrate through faunal analysis that the Neandertals who inhabited Kebara Cave (Israel) between about 65,000 and 47,000 years ago were effective hunters, repeatedly targeting the prime adults of species as large or dangerous as aurochs and wild boar. The second goal is to show that the Kebara hominids, at least during the seasons of the year when they occupied the cave, did not rely on scavenging as a major component of their animal procurement practices. In so doing, we contribute to the continuing, and often contentious, debate about modern human origins by showing that, at least in these critical behavioral dimensions, the Kebara case reveals nothing demonstrably "archaic" about the procurement practices of Levantine Neandertals.

The origins of anatomically modern humans (henceforth AMH), and the role of Neandertals in human evolution, remain fascinating and hotly debated issues (e.g., Mellars and Stringer 1989; Trinkaus 1989; Wolpoff 1989a,b; Aitken et al. 1992; Bar-Yosef 1992; Frayer et al. 1993; Stringer and Gamble 1993; Templeton 1993; Trinkaus and Shipman 1993; Nitecki and Nitecki 1994; Klein 1995). On one side of this debate are the protagonists of the so-called "Out of Africa" model or "Eve Hypothesis" (see discussions in Stringer 1989; Frayer et al. 1993; Mellars 1996; see also Ayala 1995; Templeton and Ayala 1996). Proponents of this view, citing mitochondrial DNA evidence from modern populations as well as the human fossil record, maintain that AMH arose about 150,000–200,000 years ago in Africa and then spread throughout the Old World, completely replacing Nean-

Neandertals and Modern Humans in Western Asia, edited by Akazawa *et al.*
Plenum Press, New York, 1998.

dertals and other archaic human forms. In this view, Neandertals made no genetic contribution to our own species and represent an extinct side branch in the human family tree. On the other side of the debate are the so-called "multiregionalists," who argue that AMH arose in Europe, Africa, and Asia from preceding archaic forms in each area, maintaining local continuity but considerable inter-regional variability, yet preserving the overall genetic unity of the species through genic exchange and migration (e.g., Frayer et al. 1993). In this view, Neandertals are very much a part of our genetic and cultural heritage.

Both sides of this debate clash head on in the Near East, where fossil remains of both AMH and Neandertals have been found in nearby caves and in broadly similar Middle Paleolithic contexts (AMH at Qafzeh and Skhul; Neandertals at Kebara, Tabun, Amud, and Shanidar). Archaeologists have tried for many years to find evidence of behavioral differences between the two hominid taxa. Neandertals, of course, have been perceived ever since their first discovery as "dim-witted" proto-humans, because they lacked any trace of art, made monotonous stone tools that display little evidence of style, and buried their dead unceremoniously in shallow pits with no grave offerings to indicate a belief in an afterlife (e.g., Chase and Dibble 1987). But the Near Eastern sites that have yielded the remains of AMH have also failed so far to produce clear evidence of grave offerings or art, and the Middle Paleolithic stone tool assemblages from these sites are remarkably similar to those from Neandertal sites (e.g., Chase and Dibble 1987; Bar-Yosef 1992, 1994; Bar-Yosef and Meignen 1992).

Failing to find major behavioral contrasts between the two hominid taxa reflected in their stone tools or burials, some archaeologists have turned to the faunal remains, suggesting that AMH were skilled hunters of large, often dangerous prey, whereas Neandertals, in keeping with their lowly image, were opportunistic scavengers (e.g., Binford 1984, 1988; but for contrasting views see Chase 1989; Jaubert et al. 1990; Farizy et al. 1994; Grayson and Delpech 1994; see also discussion in Rabinovich and Tchernov 1995). Until recently, however, there was very little hard evidence in support of this view, and most discussion remained largely speculative. Fortunately, faunal remains provide a useful archaeological tool for detecting differences in the way ancient hominids organized their economic activities. The potential of faunal research for clarifying the nature of Neandertal subsistence strategies has recently been demonstrated by Stiner (1994) in an analysis of Italian Middle Paleolithic faunal assemblages. Using prey mortality profiles and other data (see below), Stiner (1994:359) showed that there may, in fact, have been a major shift in Neandertal procurement strategies from one that involved a substantial scavenging component to one based far more heavily on the selective hunting of prime adult prey. Surprisingly, however, the shift occurred squarely within the Middle Paleolithic (about 55,000 years ago), well before the Middle to Upper Paleolithic boundary when most proponents of the scavenging view would have anticipated it. Later Neandertals, at least in Italy, apparently were already effective hunters of comparatively large ungulates.

Stiner's approach, in combination with techniques and approaches developed in other areas of faunal and ethnoarchaeological research (e.g., Klein 1978, 1982, 1991; Binford 1981, 1984; Brain 1981; Shipman 1981; Klein and Cruz-Uribe 1983, 1984; Bunn and Kroll 1986, 1988; Blumenschine 1987; O'Connell et al. 1988; Potts 1988; Marean 1991; Marean et al. 1992; Marean and Bertino 1995), offers useful tools for exploring whether there are, in fact, major behavioral differences between Neandertals and their anatomically more modern quasi-contemporaries in the Near East. Ultimately, of course, what is needed are systematic studies of the faunal remains from Near Eastern sites that have produced each of the putative human taxa. This study is a first step toward that goal, examining the ungulate remains from one of the key Neandertal-producing sites in the region—Kebara Cave (Israel).

2. BACKGROUND AND METHODS

The Mugharet el-Kebara (Me'arat Kabara) is a large karstic cave on the western face of Mt. Carmel, about 13 km south of Tabun Cave (e.g., Jelinek et al. 1973; Jelinek 1982) and 2.5 km east of the present-day Mediterranean shoreline (Figure 1). Three major excavations have been conducted in the cave over the past 65 years, the first by Francis Turville-Petre in 1931 (Turville-Petre 1932), the second by Moshe Stekelis between 1951 and 1965 (Stekelis 1956; Schick and Stekelis 1977), and the most recent work by a joint French-Israeli team under the co-direction of Ofer Bar-Yosef and Bernard Vandermeersch between 1982 and 1990 (Bar-Yosef et al. 1986, 1992). These excavations yielded tens of thousands of mostly well-preserved animal bones from a deeply stratified sequence of Middle and Upper Paleolithic deposits.

A number of reports have appeared over the years on various aspects of the Kebara fauna. The bones from Turville-Petre's 1931 excavations were briefly reported by Saxon (1974), Tchernov has published studies of the birds (1962) and microvertebrates (1968), and Dayan (1994) has discussed Kebara's carnivore remains. A major portion of the ungulates recovered by Stekelis were analyzed by Davis (1977, 1980), and preliminary investigations of the ungulates recovered by both Stekelis and the French-Israeli excavations were reported by the present authors (see Bar-Yosef et al. 1992:517–26). This latter study

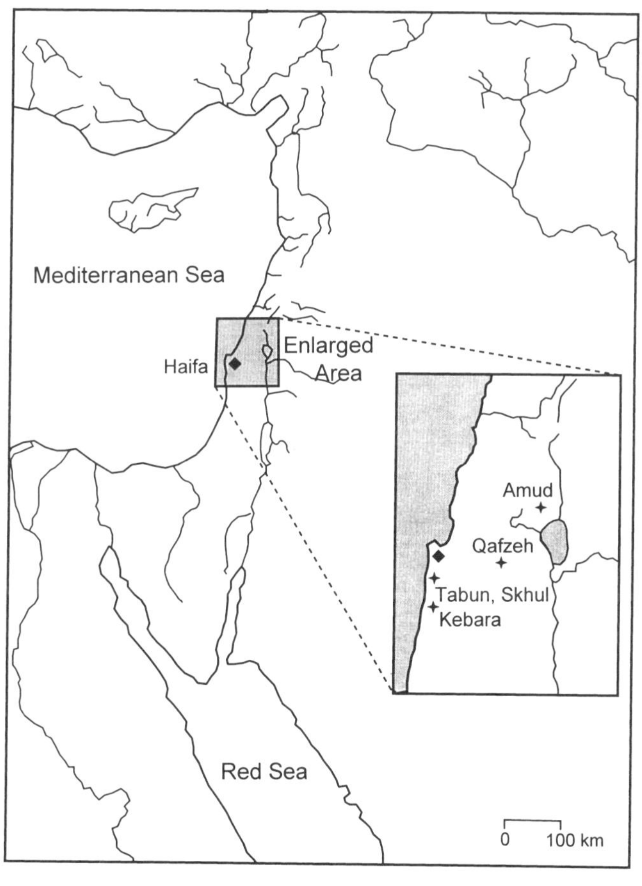

Figure 1. The Near East, showing the location of Kebara Cave.

examined a sample of about 11,000 bones (NISP = 11,375), focusing particularly on the two most abundant species—gazelles (*Gazella gazella*) and fallow deer (*Dama mesopotamica*). The present study continues these earlier analyses, but with a substantially enlarged sample of more than 21,000 specimens (NISP = 21,181). Unlike the previous study, however, which primarily considered taphonomic issues, the principal goal of the present paper is to examine the material for insights into the methods employed by the site's Middle Paleolithic inhabitants to procure medium- to large-sized mammals.

All of the faunal remains from the Stekelis and French-Israeli excavations at Kebara Cave are curated by the Department of Evolution, Systematics, and Ecology at Hebrew University in Jerusalem. The materials that form the basis of the present study are only a small portion of the total collection, however, and constitute anything but a random sample, either in terms of taxa or body parts. Our sampling procedures, therefore, need to be described briefly so that the reader can more readily evaluate our results. We began the analysis in 1986, focusing initially on just the Middle and Upper Paleolithic gazelles and cervids in the Stekelis collections that had been sorted out and identified at least to genus by previous researchers. To date we have coded all of the gazelle remains and all of the Middle Paleolithic *Capreolus*, *Dama*, and *Cervus* remains, but only about half of the Upper Paleolithic cervid bones. Scattered throughout the same drawers there are also many bones of other taxa, including *Capra*, *Sus*, *Bos*, *Alcelaphus*, and various equids. We coded all of these specimens as well, but they represent only a small fraction of the total number of bones of these species that remain to be studied in the Stekelis collections. Thus our present Stekelis sample, totaling nearly 13,000 specimens (NISP=12,852), is clearly biased in favor of gazelles and cervids, and cannot be used reliably to estimate the proportions of other species.

There are also many boxes of "unidentified" specimens in the Stekelis collections that remain to be sorted. Included in these boxes are almost all of the ungulate ribs and vertebrae, as well as thousands of limb shaft fragments that undoubtedly can be identified at least to skeletal element and approximate body size. Aside from scanning these boxes for specific ageable teeth (see below), as well as for all sexable gazelle and cervid pubic elements and horn cores, the remaining materials have not yet been examined. Thus any inferences concerning the relative proportions of skeletal elements or anatomical units must keep these biases in mind.

In 1988 we also began coding the mammal bones from the French-Israeli excavations. So far we have focused primarily on the material recovered from the so-called "décapage," a large horizontal exposure in the central area of the cave that produced several dense concentrations of mammal, bird, and reptile bones (see Bar-Yosef et al. 1992:507, 510). To date we have analyzed slightly over 8,000 bones from the décapage (NISP=8,329), coding every fragment regardless of the taxonomic level to which it could be identified. Nearly 2,000 of these are tiny shaft splinters, however, that contribute little to the present study.

Beginning in 1990, we shifted our research strategy and concentrated our efforts largely on aging the dentitions of five taxa—*Gazella*, *Dama*, *Cervus*, *Sus*, and *Bos*. Our goal here was to examine prey age selection patterns for an array of species of different body sizes using an aging approach that closely paralleled the one used by Stiner (1994) in her study of Italian Middle Paleolithic procurement strategies. For this purpose, we culled all of the deciduous and permanent lower fourth premolars (i.e., dP4, P4) from both the Stekelis and French-Israeli collections (isolated teeth as well as intact tooth rows). While we undoubtedly missed a few ageable specimens, the proportions of these five species should be fairly reliably reflected in the tooth counts.

Like Stiner (1994:289–91), we assigned the lower fourth premolars to three broad age classes—juvenile, prime adult, and old adult. Juveniles are represented by the deciduous premolar. Adults are represented by the permanent tooth, with the boundary between prime and old adult animals placed at the point when half of the tooth crown has been worn away. Stiner (1994:288–92) presents a detailed argument justifying the use of just three age classes of unequal length, and she also outlines the reasons for focusing on the lower fourth premolar rather than on other teeth. These arguments need not be repeated here.

3. TAPHONOMY

In the interests of space, only a brief summary of the taphonomic character of the Kebara fauna is presented here. A more detailed discussion of the taphonomy has been presented elsewhere (Bar-Yosef et al. 1992:517–26). Over much of the Middle Paleolithic sequence, bones of larger mammals, numbering in the tens of thousands, are densely concentrated in a relatively narrow zone along the north wall of the cave, and the location of this concentration appears to have remained relatively stable for millennia. This is strikingly illustrated by the fact that almost 50 percent of the Middle Paleolithic bones coded thus far come from a single four-meter-wide strip adjacent to the north wall (Stekelis's excavation grid unit lines A2 and A3). Smaller, though often extremely dense, concentrations of bones also occur in the central part of the cave, particularly in the area referred to by Bar-Yosef et al. (1992:507, 510) as the "décapage," where several square meters were opened up horizontally and all bones and stone tools were piece-plotted.

Understanding the mechanism(s) by which so many animal bones became concentrated along the north wall, and in the décapage, is an essential first step in the analysis of the Kebara fauna. We obviously cannot simply assume that the human inhabitants of the cave were the primary agents of bone transport and accumulation, and proceed to interpret the remains as though they provided an unbiased and direct record of past human behavior at the site. We must first determine the role that carnivores (and other agents or processes) may have played in the formation and subsequent alteration of these bone concentrations.

With the possible exception of the faunal remains deposited adjacent to the north wall of the cave at the end of unit VII, there is no compelling geological or stratigraphic evidence to suggest that the bulk of the faunal remains in these concentrations represent lag deposits formed as a result of erosion, slumping, or other natural depositional processes. Furthermore, mineralogical studies of the sediments indicate that the highly localized concentrations in the décapage most likely represent the original burial distribution rather than an artifact of differential dissolution of bones following burial (Weiner and Bar-Yosef 1990; Weiner et al. 1993). In contrast, the scarcity of fauna in the southern part of the cave may be due largely to post-depositional losses.

Humans clearly played an important role in the formation of the bone accumulations both along the north wall and in the central area of the cave. Human involvement is indicated, for example, by the presence of many cut-marked and burned bones, as well as ash lenses, hearths, and extremely large numbers of stone tools. However, there are also several lines of evidence that clearly point to the involvement of carnivores in the formation of the Kebara bone assemblages. The most obvious evidence is the presence of many specimens that have been gnawed, punctured, crenulated, or pitted by medium to large carnivores, as well as sharp-edged specimens that have been acid-etched in the gut of a predator (Horwitz 1990). Also noteworthy is the virtual absence in the larger mammal sample of soft, spongy limb epiphyses, such as the proximal humerus and proximal tibia, a possible sign that the assemblage has been modified by attritional processes, almost certainly among them bone-chewing predators (Binford 1981). The assemblage also displays a sharp bias against elements of the upper limb, another possible sign of loss through attrition, again perhaps involving carnivores. In addition, the bones of several different species of carnivore, including those of the spotted hyena (*Crocuta crocuta*), have been found within the north wall concentrations (Dayan 1994), a pattern which taphonomists have often noted as a characteristic feature of carnivore dens (e.g., Klein 1975; Binford 1981; Straus 1982). Finally, there are occasional coprolites in the deposits, attributed by Horwitz and Goldberg (1989; see also Horwitz 1990) to the spotted hyena; these also provide an unambiguous sign of carnivore activity in the cave.

Thus carnivores, not just humans, have contributed to the formation of the Kebara bone concentrations—of this there seems little doubt. The critical question is the nature and extent of their contribution. Did carnivores primarily damage, consume, or remove bones that had already been brought into the site by humans, or did they actually transport significant numbers of bones into the cave themselves? Several lines of evidence lead us to the conclusion that the major bone concentrations in the cave, almost certainly those in the Mousterian and probably also those in the Upper Paleolithic, are in fact largely a product of human, not carnivore, transport. For example, bones and lithics are thoroughly intermingled throughout the north wall concentrations, an unlikely pattern if major portions of the fauna were the products of hyena denning activity. In addition, the north wall bone concentrations actually grade into the dark, ash- and organic-rich cultural horizons that form the core of Kebara's Middle Paleolithic sequence. While this in no way proves absolute contemporaneity between these bone concentrations and human presence in the cave, it does suggest that periods of cultural activity were also periods of bone accumulation. This was not invariably the case, particularly in the Upper Paleolithic, but it certainly was often the case. During periods of intense (and presumably recurrent) human occupation at Kebara, as evidenced by the extremely high lithic densities and superimposed horizons of hearths and ash lenses that characterize much of the four-meter thick Mousterian sequence, hyenas are unlikely to have constructed their dens in the cave. Modern spotted hyena cubs often remain at or close to their den for up to 15 months (e.g., Mills 1990:215–20), a span of time that would not be possible at Kebara if the site's human occupants returned there each year. Under such conditions, hyena activity would have been largely opportunistic, scavenging fresh bones from the floor of the cave when the human occupants were temporarily elsewhere. Many of these scavenged bones would almost certainly have been transported by the hyenas to more protected locations, away from the cave, where they could be consumed in relative security (e.g., Binford et al. 1988). Thus, although bone concentrations that accumulated during periods when Kebara Cave was regularly used or visited by humans were almost certainly ravaged by scavengers, with many bones damaged or destroyed and others removed, most of the contents of these piles nevertheless were probably brought to the cave by humans, not carnivores (other evidence supporting this conclusion is presented later in the paper).

This conclusion in no way implies that hyena feeding and denning did not take place within the cave. In fact, there is convincing evidence that it did, most particularly during the Upper Paleolithic. First, human use of the cave appears to have been more ephemeral during the Upper Paleolithic than during the preceding Mousterian period. Hearths and ash lenses are much less in evidence in the younger deposits, and both lithic and faunal densities are noticeably lower. In addition, carnivore damage to bones is greater in the Upper Paleolithic, an indication that scavengers were more active in the cave during the later period. Both the absolute number of carnivore remains (Dayan 1994) and the ratio (using NISP values) of carnivore to gazelle remains (an approximation of the carnivore to ungulate ratio employed by taphonomists; see for example Klein 1975) are greater in the Upper Paleolithic than in the Mousterian, indirect clues that carnivores frequented the cave more regularly in the later period. In addition, while both periods produced skeletal remains of the hyenas themselves, only the Upper Paleolithic deposits yielded the remains of hyena pups, a tell-tale sign that at least some denning activity took place within the cave (Dayan 1994).

But even during the Upper Paleolithic the faunal remains display many features that suggest they are largely humanly-derived rather than the food remains of hyenas. Several lines of evidence point in this direction. For example, while carnivore damage to bones in the form of gnawing, puncturing, and acid-etching increases in the Upper Paleolithic, the levels are still modest compared to what one might expect if the assemblage were largely or entirely the food remains of carnivores (e.g., Binford 1981; Brain 1981; Stiner 1991a,b). Also striking is the fact that, in both time periods, the proportion of carnivore-damaged bones and of cut-marked and burned bones, as well as the proportion of lower versus upper limb elements, all appear to be comparatively uniform across the site, in contrast to what one would expect if bones accumulating close to the north wall were largely the detritus of feeding or denning hyenas while those out on the cave floor were left there by humans.

Thus the massive north wall concentrations, as well as the smaller concentrations in the "décapage," are accumulations of bones resulting largely from human activities, most probably involving the intentional sweeping, tossing, or dumping of trash into these areas of the site. This conclusion, at least with respect to the Mousterian bone concentrations, fits comfortably with the lithic evidence: the lithic debris close to the north wall is comprised of larger pieces than elsewhere in the cave, and includes an abundance of cores, cortical elements, flakes, and other waste, precisely the kinds of material one might expect to be tossed or dumped along the periphery of the occupation area. Unfortunately, we presently lack comparable data concerning the spatial characteristics of the Upper Paleolithic stone artifacts.

4. UNGULATE MORTALITY PATTERNS

Stiner (1990, 1991a,b, 1994), through a survey of the literature, examined the prey age selection patterns of a wide variety of modern cursorial and ambush predators, including spotted hyenas. In making these comparisons, she divided the ungulate prey into three broad age classes (juvenile, prime adult, and old adult), and displayed these data on simple ternary or triangular diagrams (Figure 2). Her results showed that cursorial predators typically produced prey assemblages biased toward juvenile animals (gray zone on left in Figure 2), whereas non-human ambush predators targeted the three prey age groups in a more random fashion, producing patterns that more closely approximate the age structure of living prey populations (gray zone on right in Figure 2). Interestingly, hyenas and tigers that engaged in a fair amount of scavenging consumed greater numbers of young and old prey, including many individuals that had died of causes other than predation. However, because the bones of young prey were more likely to be consumed entirely, the bone assemblages left behind by scavenging hyenas and tigers were distinctly biased toward old animals.

Stiner then compared the prey age selection patterns of Late Pleistocene and Holocene human hunters with the patterns produced by the modern non-human cursorial and ambush predators. She used the same three broad age classes, determined in the archaeological materials on the basis of the state of eruption and wear of the deciduous and permanent lower fourth premolars. Her results showed very clearly that both Upper Paleolithic and Holocene human hunters were ambush predators, and in fact tended to produce

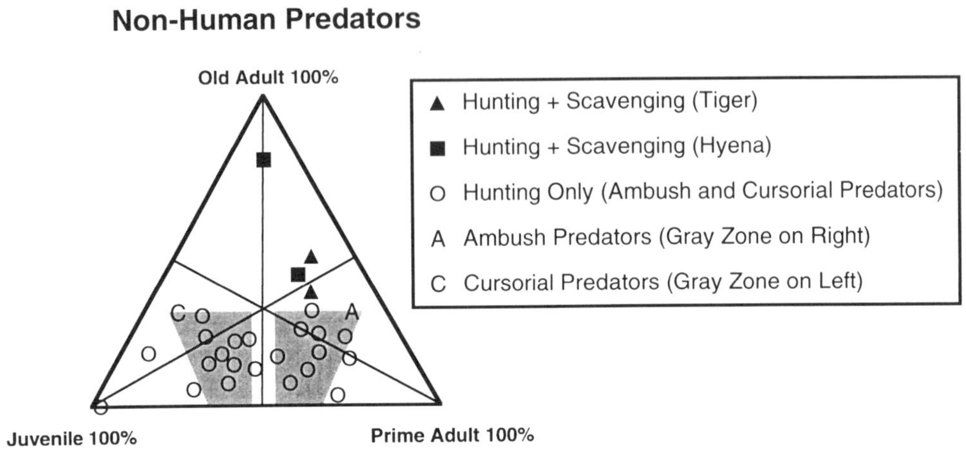

Figure 2. Ternary diagram showing prey age classes taken by modern cursorial and ambush predators (after Stiner 1994).

prey mortality patterns that were distinctly biased toward prime adults. When the ungulate data from several Italian Middle Paleolithic sites were plotted on the same ternary diagram (Figure 3), Stiner found a clear change in prey age selection patterns but, at least in this region, the change fell squarely within the Middle Paleolithic, about 55,000 years ago, not at the boundary between the Middle and Upper Paleolithic, when many archaeologists might have anticipated it (see Stiner 1994 for a detailed discussion of the excavation history, sample sizes, taphonomy, and other aspects of the Italian Middle Paleolithic faunal assemblages). The pattern revealed by the younger Middle Paleolithic assemblages was very similar to the one produced by Upper Paleolithic and Holocene hunters, that is, an ambush pattern biased toward the prime adults of common prey species. This intriguing result suggested that later Neandertals were ambush hunters much like their Upper Paleolithic successors. In contrast, the earlier Middle Paleolithic pattern displayed the same bias toward old animals that was seen in the scavenged hyena and tiger assemblages (see Fig-

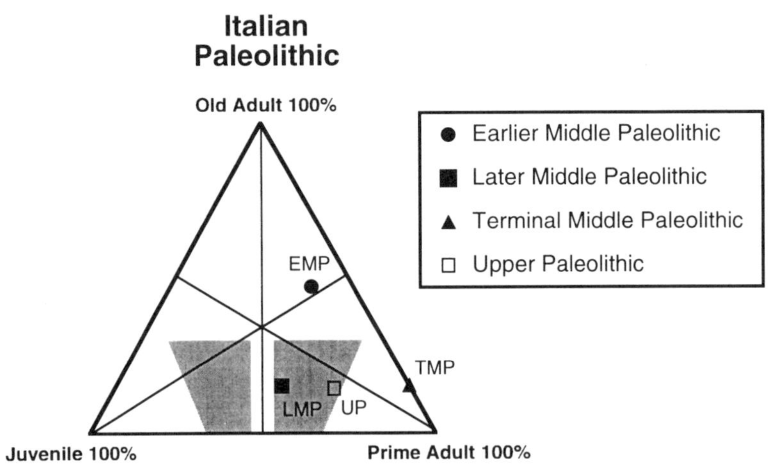

Figure 3. Ternary diagram showing prey age classes taken by Italian Middle and Upper Paleolithic hominids (after Stiner 1994).

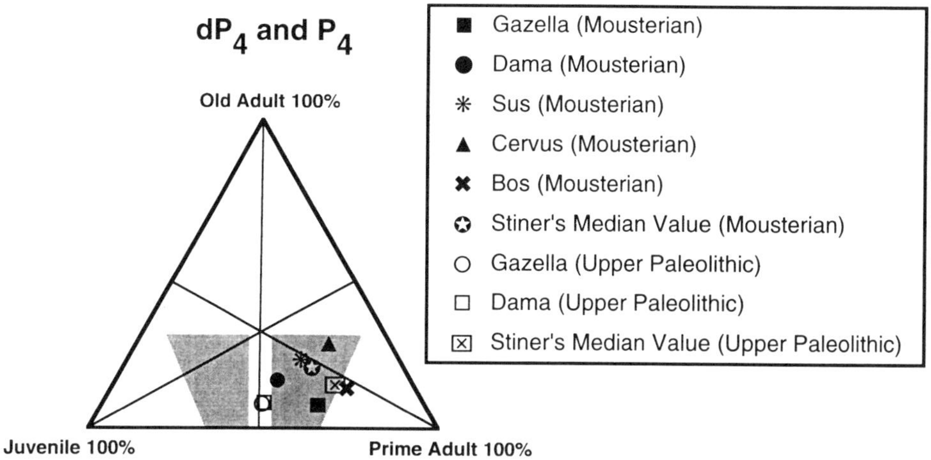

Figure 4. Ternary diagram showing prey age classes taken by Kebara Middle and Upper Paleolithic hominids (based on dP4 and P4).

ure 2), suggesting that scavenging may indeed have been an important component of earlier Neandertal procurement practices.

We have used this same approach in examining the prey age selection patterns of five different mammalian taxa at Kebara—gazelle (*Gazella gazella*), fallow deer (*Dama mesopotamica*), wild boar (*Sus scrofa*), red deer (*Cervus elaphus*), and aurochs (*Bos primigenius*) (Figure 4 and Table 1). As discussed above, we follow Stiner (1994) in distinguishing three broad age classes (juvenile, prime adult, and old adult) using, as she did, the eruption and wear patterns of the deciduous and permanent lower fourth premolars (i.e., dP4 and P4).

Our initial look at prey age selection patterns at Kebara treats the entire four-meter-thick Mousterian sequence as a single block. This is obviously far from ideal, but was necessitated by the extremely small samples of ageable dentitions for the larger-bodied species, and because we lack precise correlations between the bones recovered in the arbitrary horizontal spits employed by Stekelis and those recovered by the recent French-Israeli excavations using natural stratigraphic levels. For the same reasons, we have had to treat the Upper Paleolithic sequence as a single block.

Table 1. Number of juvenile, prime old, and old adult animals in Middle and Upper Paleolithic, based on lower dP4 and lower P4 (age classes after Stiner 1994:288–92)

Taxon	Middle Paleolithic			Upper Paleolithic		
	Juvenile MNE (%)	Prime Adult MNE (%)	Old Adult MNE (%)	Juvenile MNE (%)	Prime Adult MNE (%)	Old Adult MNE (%)
Gazella	96 (30.4)	196 (62.0)	24 (7.6)	39 (45.9)	39 (45.9)	7 (8.2)
Dama	43 (37.7)	53 (46.5)	18 (15.8)	43 (45.3)	44 (46.3)	8 (8.4)
Sus	5 (27. 8)	9 (50.0)	4 (22.2)	--	--	--
Cervus	7 (17.5)	22 (55.0)	11 (27.5)	--	--	--
Bos	9 (19.6)	31 (67.4)	6 (13.0)	--	--	--

As Figure 4 clearly shows, all five taxa fall in the "ambush predator" portion of the triangular diagram, with most taxa clustering around the median values obtained by Stiner in Italy. Since we have lumped the material from all four meters of Mousterian deposits into a single composite sample, these results obviously do not rule out the possibility that procurement strategies changed over time or that they varied seasonally or interannually. But the results do at least suggest that most of the ungulates most of the time were procured at Kebara by hunting, not scavenging, regardless of the prey's body size. Interestingly, the largest species—aurochs (*Bos*)—is the most strongly prime-dominated. Since the *Bos* sample is also the smallest, however, this observation must be viewed with caution.

Figure 5 shows the prey age selection patterns for *Gazella*, *Dama*, *Cervus*, and *Bos* by stratigraphic unit or level (see also Table 2). Because of sample size limitations, we have had to drop *Sus* from this comparison, and we have combined the fauna from the site's many Middle Paleolithic stratigraphic units into just two composite assemblages, an "upper" assemblage that includes material from units VI–VIII, and a "lower" assemblage that includes material from units IX–XII. The decision to split the faunal samples at the boundary between units VIII and IX was made largely because there are significant changes in the techniques of flake production at this point in the sequence (Meignen and Bar-Yosef 1989; Bar-Yosef and Meignen 1992; Bar-Yosef et al. 1992:516). From the faunal perspective, however, the selection of this particular dividing point is arbitrary, since

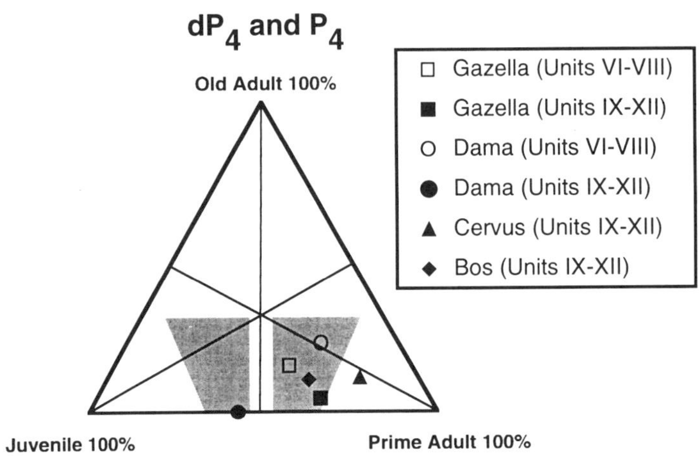

Figure 5. Ternary diagram showing prey age classes, grouped by stratigraphic unit, taken by Kebara Middle and Upper Paleolithic hominids (based on dP4 and P4).

Table 2. Number of juvenile, prime old, and old adult animals, based on lower dP4 and lower P4, in Middle Paleolithic, grouped by stratigraphic unit (age classes after Stiner 1994:288–92)

Stratigraphic Units (Grouped)	Gazella			Dama			Cervus			Bos		
	Juven. MNE (%)	Prime Adult MNE (%)	Old Adult MNE (%)	Juven. MNE (%)	Prime Adult MNE (%)	Old Adult MNE (%)	Juven. MNE (%)	Prime Adult MNE (%)	Old Adult MNE (%)	Juven. MNE (%)	Prime Adult MNE (%)	Old Adult MNE (%)
VI-VIII	7 (35.0)	10 (50.0)	3 (15.0)	2 (22.2)	5 (55.6)	2 (22.2)	-- --	-- --	-- --	-- --	-- --	-- --
IX-XII	36 (31.3)	74 (64.4)	5 (4.4)	8 (57.1)	6 (42.9)	0 (0)	3 (16.7)	13 (72.2)	2 (11.1)	6 (31.6)	11 (57.9)	2 (10.5)

we have no a priori reason to expect changes in the site's fauna to parallel shifts in lithic technology.

Figure 5 shows that the lower and upper assemblages both fall within the "ambush predator" portion of the ternary diagram, the only notable exception being the lower sample for *Dama*. This deviation, however, is almost certainly an artifact of the very small sample for this species. There is certainly no evidence in the upper Mousterian assemblage at Kebara for increased reliance on prime-age animals, nor greater reliance on juvenile and old animals among the largest-bodied species.

5. TRANSPORT PATTERNS

Stiner (1994:242) employed three principal indices to distinguish faunal assemblages transported by scavenging hyenas from those transported by Upper Paleolithic and Holocene human hunters. One of these indices is the ratio of total skeletal elements (tMNE) to the maximum number of individual animals (MNI) represented in the assemblage (tMNE/MNI). This ratio provides a measure of skeletal completeness. The second index is the ratio of heads plus horns and antlers to major limb bones ([H+H]/L), a measure of the degree to which an assemblage is biased toward crania or limbs. The third index is the ratio of prime adults to old adults in the assemblage, a measure of the degree to which the age structure of the assemblage is prime-dominated. In brief, in a scavenged assemblage composed primarily of medium-sized ungulates (in the Italian context these include red deer, fallow deer, and ibex), Stiner (1994:249, 252) expects to find evidence for many animals, each represented by relatively few skeletal elements, a bias toward head parts, and a bias toward old animals.

Again, as in the case with prey age selection patterns, when the Italian Middle and Upper Paleolithic data are considered, all three indices point toward a major transforma-

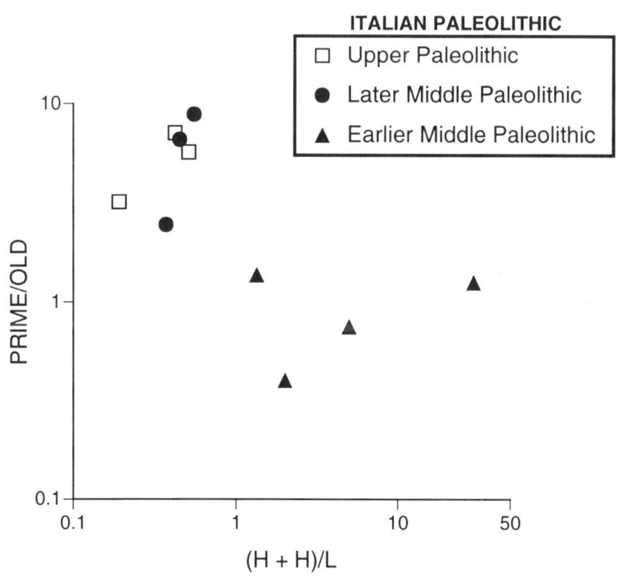

Figure 6. Scattergram of head to limb ratio ([H+H]/L) plotted against ratio of prime to old adults for Italian Middle and Upper Paleolithic assemblages (after Stiner 1994).

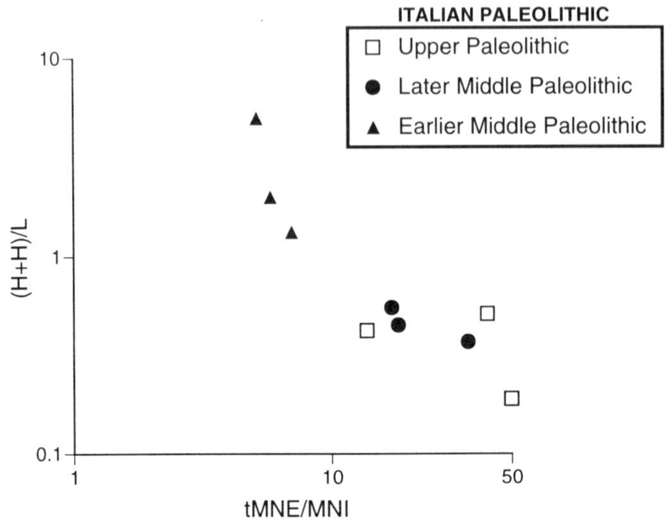

Figure 7. Scattergram of index of skeletal completeness (tMNE/MNI) plotted against head to limb ratio ([H+H]/L) for Italian Middle and Upper Paleolithic assemblages (after Stiner 1994).

tion in procurement practices within the Mousterian, not at the Middle to Upper Paleolithic boundary. Figure 6 shows a scattergram of the head to limb index ([H+H]/L) plotted against the ratio of prime to old animals. The Italian Middle Paleolithic data split very nicely into two groups, the earlier assemblages (to the right in the figure) being distinctly more head-dominated than the younger Mousterian assemblages. The next figure (Figure 7) shows a similar scattergram, this time plotting the index of skeletal completeness (tMNE/MNI) for the Italian assemblages against the head to limb ratio ([H+H]/L). Again, the assemblages split into two groups, the later Mousterian ones falling together with the Upper Paleolithic assemblages toward the right hand side of the figure. These results suggest quite convincingly that scavenging played an important role in the formation of the Italian assemblages dating to the earlier part of the Middle Paleolithic (i.e., prior to about 55,000 years ago).

The Kebara Mousterian faunal data were evaluated using these same indices (Table 3). It should be borne in mind that our analyses include loose teeth, a skeletal category that Stiner excluded in the calculation of these body part indices. Figure 8 shows the relation between the ratio of head to limb parts ([H+H]/L) plotted against the ratio of prime to old individuals. All five species at Kebara, regardless of body size, are limb-dominated, despite our inclusion of loose teeth, and fall together on the graph with Stiner's later Mousterian and Upper Paleolithic assemblages. *Bos* and *Sus* appear to be slightly more head dominated than the other taxa, but this is almost certainly an artifact of our sampling procedures. As noted earlier, in our coding efforts to date we have focused primarily on gazelles and cervids, only coding elements of *Bos* and *Sus* that had been mixed inadvertently into the same storage drawers. Most of the *Bos* and *Sus* remains from Kebara remain to be studied, with the notable exception of the teeth. We systematically culled all of the lower fourth premolars from both the Stekelis and French-Israeli collections so that we could include these two large-bodied taxa in our study of prey mortality patterns. Hence the ratio of prime to old adults for these species should be reasonably reliable (although the sample sizes are small), whereas the other two indices, (H+H)/L and tMNE/MNI, are

Table 3. Total number of skeletal elements (MNE) by major anatomical unit and summary indices discussed in text

Anatomical Unit[1]	Gazella	Dama	Sus	Cervus	Bos
Horn or Antler	170	32	--	0	2
Head[2]	293	111	36	31	49
Neck	27	18	1	7	3
Axial	203	103	2	30	4
Upper Front Limb	272	69	3	9	3
Lower Front Limb	369	321	9	42	25
Upper Hind Limb	149	59	0	5	1
Lower Hind Limb	982	339	15	47	21
Other Limb Elements[3]	207	76	39	9	6
Feet	1099	542	106	163	55
(H+H)/L[4]	0.23	0.17	0.55	0.28	0.91
tMNE/MNI[5]	25.65	29.82	11.72	21.44	6.76
PRIME/OLD[6]	8.17	2.94	2.25	2.00	5.17

[1]Anatomical units follow those used by Stiner (1994:242).
[2]Includes MNE value for most common isolated tooth.
[3]Metapodials not identifiable to front or rear limb.
[4]Total MNE for Horns or Antlers plus Heads divided by total MNE for Limbs, excluding feet (after Stiner 1994:242).
[5]Total MNE for all anatomical units divided by maximum MNI value based on Heads (after Stiner 1994:242).
[6]Number of prime adults (MNE) divided by number of old adults (MNE), based on lower dP_4 and lower P_4 (after Stiner 1994:242).

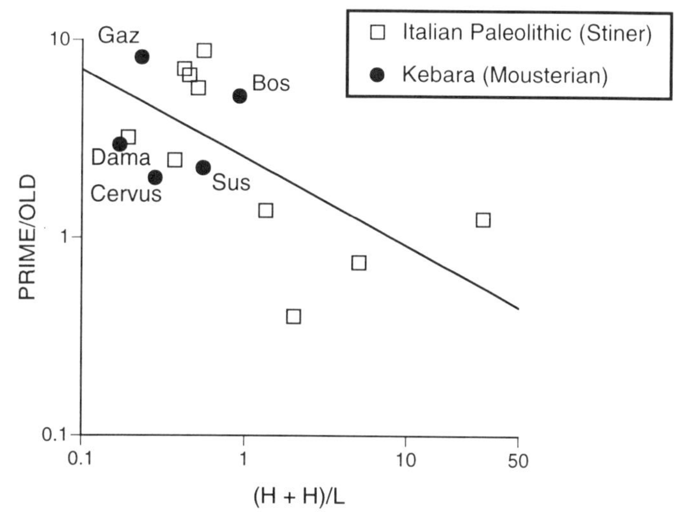

Figure 8. Scattergram of head to limb ratio ([H+H]/L) plotted against ratio of prime to old adults for Kebara Middle Paleolithic assemblages.

undoubtedly biased toward values that one would expect in scavenged assemblages because of our explicit focus on dentitions.

In the next figure (Figure 9), which shows the index of skeletal completeness (tMNE/MNI) plotted against the head to limb ratio ([H+H]/L) for the same five Kebara species, the gazelles and cervids again fall tightly together on the graph with Stiner's later assemblages. As already noted, the indices for carcass completeness and head to limb parts in *Bos* and *Sus* are artificially biased toward many individuals and few postcranial elements per individual by our deliberate focus on ageable dentitions in these taxa. As coding progresses, this bias will disappear, and we suspect that the index values for *Bos* and *Sus* will converge on the values seen in the other species.

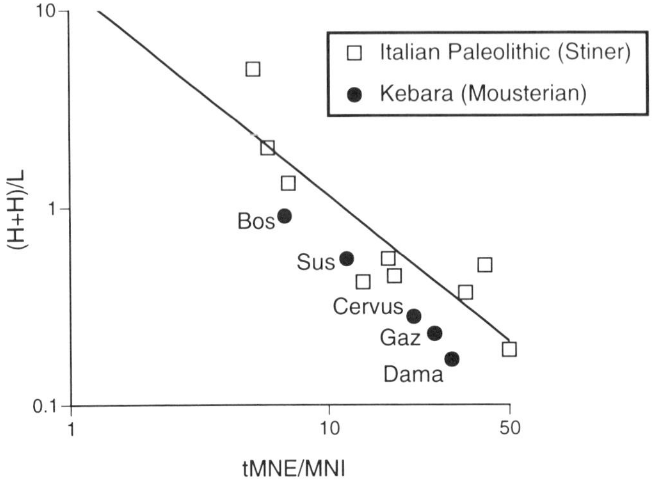

Figure 9. Scattergram of index of skeletal completeness (tMNE/MNI) plotted against head to limb ratio ([H+H]/L) for Kebara Middle Paleolithic assemblages.

6. CONCLUSIONS

Our analyses, though still preliminary in nature, suggest that most ungulates at Kebara were hunted, not scavenged. This conclusion seems applicable not just to small-bodied prey like gazelles, but to animals as large or dangerous as aurochs and wild boar. The body part indices presented here also strengthen our contention that humans, not hyenas, were the major transporters of the Kebara assemblage. In terms of the broader debate about modern human origins, we emphasize one principal conclusion. If the Kebara hominid is accepted as a Neandertal, then Neandertals living in both Europe and the Near East after about 60,000 to 55,000 years ago were already effective hunters, preferentially targeting large, and potentially dangerous, prime adult prey. Thus, at least in this one critical behavioral dimension there is nothing demonstrably "archaic" about the procurement practices of Levantine Neandertals.

ACKNOWLEDGMENTS

We would like to thank Ofer Bar-Yosef, Simon J. M. Davis, Tamar Dayan, Liliane Meignen, Rivka Rabinovich, Mary C. Stiner, and Milford H. Wolpoff for their many helpful comments and suggestions in the preparation of this paper. JDS would also like to express his deep gratitude for the warmth and hospitality of his many Israeli colleagues and friends who have made his many visits to Israel truly enjoyable experiences. Among these he would particularly like to acknowledge Baruch Arensburg, Uri Baruch, Ofer and Danny Bar-Yosef, Anna Belfer-Cohen, Goggy Davidowitz, Tamar Dayan, Avi Gopher, Yuval Goren, Naama Goren-Inbar, Nigel Goring-Morris, Liora Horwitz, Erella Hovers, Gila Kahila, Arlene Miller-Rosen, Dani Nadel, Rivka Rabinovich and Steve Rosen. Partial support for this research was received from the U.S.-Israel Binational Science Foundation, Israel Academy of Sciences and Humanities, L. S. B. Leakey Foundation, Irene Levi-Sala CARE Archaeological Foundation, Hebrew University (Jerusalem), and the University of Michigan's Office of the Vice-President for Research and Horace H. Rackham School of Graduate Studies.

REFERENCES

Aitken, M. J., C. B. Stringer, and P. A. Mellars, eds. 1992. *The Origin of Modern Humans and the Impact of Chronometric Dating*. Princeton: Princeton University Press.

Ayala, F. J. 1995. The myth of Eve: Molecular biology and human origins. *Science* 270:1930–6.

Bar-Yosef, O. 1992. The role of western Asia in modern human origins. *Philosophical Transactions of the Royal Society of London B* 337:193–200.

Bar-Yosef, O. 1994. The contributions of southwest Asia to the study of the origin of modern humans, 23–66. In *Origins of Anatomically Modern Humans*, ed. M. H. Nitecki and D. V. Nitecki. New York: Plenum Press.

Bar-Yosef, O., and L. Meignen. 1992. Insights into Levantine Middle Paleolithic cultural variability, 162–82. In *The Middle Paleolithic: Adaptation, Behavior, and Variability*, ed. H. L. Dibble and P. A. Mellars. Philadelphia: The University Museum of the University of Pennsylvania.

Bar-Yosef, O., B. Vandermeersch, B. Arensburg, A. Belfer-Cohen, P. Goldberg, H. Laville, L. Meignen, Y. Rak, J. D. Speth, E. Tchernov, A. M. Tillier, and S. Weiner. 1992. The excavations in Kebara Cave, Mt. Carmel. *Current Anthropology* 33(5):497–550.

Bar-Yosef, O., B. Vandermeersch, B. Arensburg, P. Goldberg, H. Laville, L. Meignen, Y. Rak, E. Tchernov, and A.-M. Tillier. 1986. New data on the origin of modern man in the Levant. *Current Anthropology* 27(1):63–4.

Binford, L. R. 1981. *Bones: Ancient Men and Modern Myths*. New York: Academic Press.

Binford, L. R. 1984. *Faunal Remains from Klasies River Mouth*. Orlando: Academic Press.

Binford, L. R. 1988. Etude taphonomique des restes fauniques de la Grotte Vaufrey. *Mémoires de la Société Préhistorique Française* 19:535–64.

Binford, L. R., M. G. L. Mills, and N. M. Stone. 1988. Hyena scavenging behavior and its implications for the interpretation of faunal assemblages from FLK22 (the Zinj Floor) at Olduvai Gorge. *Journal of Anthropological Archaeology* 7(2):99–135.

Blumenschine, R. J. 1987. Characteristics of an early hominid scavenging niche. *Current Anthropology* 28(4):383–407.

Brain, C. K. 1981. *The Hunters or the Hunted? An Introduction to African Cave Taphonomy*. Chicago: University of Chicago Press.

Bunn, H. T., and E. M. Kroll. 1986. Systematic butchery by Plio/Pleistocene hominids at Olduvai Gorge, Tanzania. *Current Anthropology* 27(5):431–52.

Chase, P. G. 1989. How different was Middle Palaeolithic subsistence? A zooarchaeological perspective on the Middle to Upper Palaeolithic transition, 321–37. In *The Human Revolution*, ed. P. A. Mellars and C. B. Stringer. Edinburgh: Edinburgh University Press.

Chase, P. G., and H. L. Dibble. 1987. Middle Paleolithic symbolism: A review of current evidence and interpretations. *Journal of Anthropological Archaeology* 6(3):263–96.

Davis, S. J. M. 1977. The ungulate remains from Kebara Cave, 150–63. In *Moshé Stekelis Memorial Volume*, ed. B. Arensburg and O. Bar-Yosef. Jerusalem: The Israel Exploration Society.

Davis, S. J. M. 1980. Late Pleistocene-Holocene gazelles of northern Israel. *Israel Journal of Zoology* 29:135–40.

Dayan, T. 1994. Carnivore diversity in the Late Quaternary of Israel. *Quaternary Research* 41(3):343–9.

Farizy, C., F. David, and J. Jaubert. 1994. *Hommes et Bisons du Paléolithique Moyen à Mauran (Haute-Garonne)*. Paris: CNRS Editions.

Frayer, D. W., M. H. Wolpoff, A. G. Thorne, F. H. Smith, and G. G. Pope. 1993. Theories of modern human origins: The paleontological test. *American Anthropologist* 95(1):14–50.

Grayson, D. K., and F. Delpech. 1994. The evidence for Middle Palaeolithic scavenging from couche VIII, Grotte Vaufrey (Dordogne, France). *Journal of Archaeological Science* 21(3):359–75.

Horwitz, L. K. 1990. The origin of partially digested bones recovered from archaeological contexts in Israel. *Paléorient* 16:97–106.

Horwitz, L. K., and P. Goldberg. 1989. A study of Pleistocene and Holocene hyaena coprolites. *Journal of Archaeological Science* 16:71–94.

Jaubert, J., M. Lorblanchet, H. Laville, R. Slott-Moller, A. Turq, and J. Brugal. 1990. *Les Chasseurs d'Aurochs de la Borde: Un Site du Paléolithique Moyen (Livernon, Lot)*. Paris: Editions de la Maison des Sciences de l'Homme.

Jelinek, A. J. 1982. The Tabun Cave and Paleolithic man in the Levant. *Science* 216:1369–75.

Jelinek, A. J., W. R. Farrand, G. Haas, A. Horowitz, and P. Goldberg. 1973. New excavations at the Tabun Cave, Mount Carmel, Israel, 1967–1972: A preliminary report. *Paléorient* 1:151–83.

Klein, R. G. 1975. Middle Stone Age man-animal relationships in southern Africa: Evidence from Die Kelders and Klasies River Mouth. *Science* 190:265–7.

Klein, R. G. 1978. Stone Age predation on large African bovids. *Journal of Archaeological Science* 5(3):195–217.

Klein, R. G. 1982. Age (mortality) profiles as a means of distinguishing hunted species from scavenged ones in Stone Age archaeological sites. *Paleobiology* 8:151–8.

Klein, R. G. 1991. Why does skeletal part representation differ between smaller and larger bovids at Klasies River Mouth and other archaeological sites? *Journal of Archaeological Science* 16:363–81.

Klein, R. G. 1995. Anatomy, behavior, and modern human origins. *Journal of World Prehistory* 9(2):167–98.

Klein, R. G., and K. Cruz-Uribe. 1983. The computation of ungulate age (mortality) profiles from dental crown heights. *Paleobiology* 9(1):70–8.

Klein, R. G., and K. Cruz-Uribe. 1984. *The Analysis of Animal Bones from Archaeological Sites*. Chicago: University of Chicago Press.

Marean, C. W. 1991. Measuring the post-depositional destruction of bone in archaeological assemblages. *Journal of Archaeological Science* 18:677–94.

Marean, C. W., and L. Bertino. 1995. Intrasite spatial analysis of bone: Subtracting the effect of secondary carnivore consumers. *American Antiquity* 59(4):748–68.

Marean, C. W., L. M. Spencer, R. J. Blumenschine, and S. D. Capaldo. 1992. Captive hyaena bone choice and destruction, the schlepp effect and Olduvai archaeofaunas. *Journal of Archaeological Science* 19:101–21.

Meignen, L., and O. Bar-Yosef. 1989. Nouvelles recherches sur le Paléolithique Moyen d'Israel: La Grotte de Kebara, unites VII à XII, 169–84. In *Investigations in South Levantine Prehistory: Préhistoire du Sud-Levant*, ed. O. Bar-Yosef and B. Vandermeersch. Oxford: British Archaeological Reports International Series 497.

Mellars, P. A. 1996. *The Neanderthal Legacy: An Archaeological Perspective from Western Europe*. Princeton: Princeton University Press.

Mellars, P. A., and C. B. Stringer, eds. 1989. *The Human Revolution*. Edinburgh: Edinburgh University Press.

Mills, M. G. L. 1990. *Kalahari Hyaenas: Comparative Behavioural Ecology of Two Species*. London: Unwin Hyman.

Nitecki, M. H., and D. V. Nitecki, eds. 1994. *Origins of Anatomically Modern Humans*. New York: Plenum Press.

O'Connell, J. F., K. Hawkes, and N. G. Blurton Jones. 1988. Hadza scavenging: Implications for Plio/Pleistocene hominid subsistence. *Current Anthropology* 29(2):356–63.

Potts, R. B. 1988. *Early Hominid Activities at Olduvai*. New York: Aldine de Gruyter.

Rabinovich, R., and E. Tchernov. 1995. Chronological, paleoecological and taphonomical aspects of the Middle Paleolithic site of Qafzeh, Israel, 5–44. In *Archaeozoology of the Near East*, Vol. 2, *Proceedings of the Second International Symposium on the Archaeozoology of Southwestern Asia and Adjacent Areas*, ed. H. Buitenhuis and H. P. Uerpmann. Leiden: Backhuys.

Saxon, E. C. 1974. The mobile herding economy of Mt. Carmel: An economic analysis of the faunal remains. *Journal of Archaeological Science* 1:27–45.

Schick, T., and M. Stekelis. 1977. Mousterian assemblages in Kebara Cave, Mount Carmel, 97–149. In *Moshé Stekelis Memorial Volume*, ed. B. Arensburg and O. Bar-Yosef. Jerusalem: The Israel Exploration Society.

Shipman, P. 1981. *Life History of a Fossil: An Introduction to Taphonomy and Paleoecology*. Cambridge: Harvard University Press.

Stekelis, M. 1956. Nouvelles fouilles dans la Grotte de Kebarah. *Cronica del IV Congreso Internacional de Ciencias Prehistoricas y Protohistoricas, Madrid* (1954) 13:385–89.

Stiner, M. C. 1990. The use of mortality patterns in archaeological studies of hominid predatory adaptations. *Journal of Anthropological Archaeology* 9(4):305–51.

Stiner, M. C. 1991a. Food procurement and transport by human and non-human predators. *Journal of Archaeological Science* 18:455–82.

Stiner, M. C. 1991b. An interspecific perspective on the emergence of the modern human predatory niche, 149–86. In *Human Predators and Prey Mortality*, ed. M. C. Stiner. Boulder: Westview Press.

Stiner, M. C. 1994. *Honor Among Thieves: A Zooarchaeological Study of Neandertal Ecology*. Princeton: Princeton University Press.

Straus, L. G. 1982. Carnivores and cave sites in Cantabrian Spain. *Journal of Anthropological Research* 38(1):75–96.

Stringer, C. B. 1989. Documenting the origin of modern humans, 67–96. In *The Emergence of Modern Humans: Biocultural Adaptations in the Later Pleistocene*, ed. E. Trinkaus. New York: Cambridge University Press.

Stringer, C. B., and C. S. Gamble. 1993. *In Search of the Neanderthals: Solving the Puzzle of Human Origins*. New York: Thames and Hudson.

Tchernov, E. 1962. Palaeolithic avifauna in Palestine. *Bulletin of the Research Council of Israel* 11:95–131.

Tchernov, E. 1968. *Succession of Rodent Faunas during the Upper Pleistocene of Israel*. Hamburg and Berlin: Parey.

Templeton, A. R. 1993. The "Eve" hypothesis: A genetic critique and reanalysis. *American Anthropologist* 95(1):51–72.

Templeton, A. R., and F. J. Ayala. 1996. Gene lineages and human evolution. *Science* 272:1363–4.

Trinkaus, E., ed. 1989. *The Emergence of Modern Humans: Biocultural Adaptations in the Later Pleistocene*. New York: Cambridge University Press.

Trinkaus, E., and P. Shipman. 1993. *The Neandertals: Changing the Image of Mankind*. New York: Alfred A. Knopf.

Turville-Petre, F. 1932. Excavations in the Mugharet el-Kebarah. *Journal of the Royal Anthropological Institute* 62:271–6.

Weiner, S., and O. Bar-Yosef. 1990. States of preservation of bones from prehistoric sites in the Near East: A survey. *Journal of Archaeological Science* 17(2):187–96.

Weiner, S., P. Goldberg, and O. Bar-Yosef. 1993. Bone preservation in Kebara Cave, Israel using on-site Fourier transform infrared spectrometry. *Journal of Archaeological Science* 20(6):613–27.

Wolpoff, M. H. 1989a. The place of the Neandertals in human evolution, 97–141. In *The Emergence of Modern Humans: Biocultural Adaptations in the Later Pleistocene*, ed. E. Trinkaus. New York: Cambridge University Press.

Wolpoff, M. H. 1989b. Multiregional evolution: The fossil alternative to Eden, 62–108. In *The Human Revolution*, ed. P. A. Mellars and C. B. Stringer. Edinburgh: Edinburgh University Press.

15

PLEISTOCENE SPECIES TRENDS AT HAYONIM CAVE

Changes in Climate versus Human Behavior

Mary C. Stiner[1] and Eitan Tchernov[2]

[1]Department of Anthropology
University of Arizona
Tucson, Arizona 85721
[2]Department of Evolution, Systematics, and Ecology
Hebrew University
Jerusalem 91904, Israel

1. INTRODUCTION

The deep faunal series in Hayonim Cave (Upper Galilee, Israel) fills a unique and crucial gap in our knowledge about the Pleistocene sequence of Israel, representing more than 200,000 years of foraging activities by human beings. The bulk of the cultural layers date to the Middle Paleolithic, capped by thinner layers rich in Kebaran and Natufian material (Tchernov 1968, 1981; Goldberg 1979; Belfer-Cohen and Bar-Yosef 1981; Henry et al. 1981; Bar-Yosef and Belfer-Cohen 1988; Belfer-Cohen 1988; Goldberg and Laville 1988; Bar-Yosef 1989, 1991; Valla et al. 1989; Rabinovich n.d.).* What follow are some preliminary results from new investigations of game use by hominids at this cave site. The presentation focuses especially on variation in game species abundances, based on piece-plotted material recovered during the 1992 through 1995 excavation seasons. Though far from finished, the ultimate objectives of the research are to evaluate the taphonomic history of the bone accumulations in Hayonim Cave and to learn about the economic behaviors of the Paleolithic human occupants.

Trends in the prey species consumed by humans here are first examined across the Natufian, Kebaran, and Mousterian (Middle Paleolithic) cultural periods, and then within

* Aurignacian layers were also found in the stratigraphic sequence of Hayonim Cave, but were almost completely removed during the earlier excavation campaigns (pre-1980) at this site. Information on Aurignacian faunas therefore is available only from the old collections and is not represented in most of the preliminary analyses here; the Aurignacian sample ultimately will be examined in its entirety using the same analytical criteria (see also Rabinovich n.d.).

Neandertals and Modern Humans in Western Asia, edited by Akazawa *et al.*
Plenum Press, New York, 1998.

the Mousterian itself. The Natufian occupations span roughly 10–13 ka; the Kebaran occupations 13–19 ka (e.g., Belfer-Cohen 1988; Bar-Yosef 1990, 1991). The time frame represented by the Mousterian is much greater, ending roughly 70 ka and beginning well before 200 ka, based on preliminary uranium-thorium (Schwarcz and Rink, this volume) and thermoluminescence results (Valladas et al., this volume). Because the Mousterian sample certainly cross-cuts significant oscillations in Pleistocene climate, some variation in species representation is also expected within this deep sequence.

What stand out most in the species abundance data for Hayonim Cave are shifts in the nature of large versus small game use by humans over time. Some of the variation, especially in ungulate species harvested, is best explained by climate change. However, variation in the small game species that humans emphasized over the course of the Paleolithic appears to have a different explanation, on grounds that the means required to obtain them and the abilities of the respective species to recover from intensive local harvesting differ greatly. Although the comparisons here must be confined to relative rather than absolute differences, the findings on prey species representation are quite significant and greatly refine the research questions we are asking in our work at Hayonim Cave.

2. CHARACTERISTICS AND TAPHONOMY OF THE FAUNAL SAMPLE

The study sample from Hayonim Cave consists of those faunal remains that clearly represent human refuse, based on processing damage, consistent associations with artifactual material, and, equally important, the consistent absence of damage from other bone collecting agencies (see below). Because the bones are in a good state of macroscopic preservation, details of hominids' processing techniques are readily apparent. The sample therefore includes certain large reptiles, lagomorphs, and birds, along with large game, because it is clear that all of these animals were food to the human occupants. The sample excludes so-called microfauna (e.g., small rodents) that we believe, on the basis of damage patterns, to have been collected by owls.

Carnivore remains and traces of gnawing on all varieties of bones are rare throughout the Natufian, Kebaran, and Mousterian layers excavated thus far. The paucity of carnivore activity in the cave is both encouraging and relatively unusual for Mediterranean sites (Straus 1982; Gamble 1986; Stiner 1991, 1994). Even the damage on carnivore bones is almost always attributable to humans, mainly burning and/or cut marks such as on the leopard phalanx shown in Figure 1a.

The quality of macroscopic skeletal preservation varies among excavation units, ranging from poor to extraordinary. Conditions are especially favorable where the deposits are fully protected by the cave vault. The deeper Mousterian layers contain the greatest quantities of faunal material and the largest fragments on average. Preservation is truly exceptional under the breccia shelves that have formed along the cave walls, as determined by both macroscopic examination and infrared spectrometry analysis of bone mineral. Bones and teeth below these shelves were relatively better protected from percolating water, also shielding them from dissolution effects. Cemented ash, often rich in charred and calcined bone, is also present in these zones. In the zones of good preservation, delicate deciduous teeth in ungulate mandibles are found astride emerging permanent crowns, many maxillary tooth rows are semi- or wholly intact (Figure 1b), and some conjoining bone elements are still in anatomical connection. Large portions of tortoise shells could be reconstructed from fragments found together (Figure 1c), indicating that they were crushed in place rather than scattered once broken (see below). The condition of these

Figure 1. Examples of the condition of faunal specimens from Hayonim Cave: (A) second phalanx of adult leopard (2.1 × 0.8 cm) with cut marks on anterior shaft and proximal epiphysis, found at Middle-Upper Paleolithic interface (F28a, exterior trench); (B) right half of red deer maxilla (12.0 × 4.4 cm) preserving complete cheek tooth row and part of palate, from Middle Paleolithic deposits (I23c, main trench); (C) posterior-right quarter of large tortoise carapace (10.8 × 7.5 cm) restored from fragment concentration in Middle Paleolithic deposits (I23d, main trench); (D) tortoise humeri showing "fresh bone" damage to proximal epiphyses and/or breaks through mid-shafts, from the Mousterian (1.7 to 3.8 cm in maximum dimension); (E) tortoise shell edge fragment with cone fracture, the impact point for which is designated by white arrow, from the Mousterian (specimen is 2.8 cm in maximum length).

specimens suggests that post-depositional disturbance was fairly limited in the main excavation trench where most of the study sample was recovered.

The very dense concentrations of bone found in squares I23-I24 and K21-K22 in 1995 are unprecedented for the Middle Paleolithic layers of this cave. Not far from these concentrations in the main excavation trench are, however, conspicuous voids in the faunal distributions. Figure 2 shows a transparent "view-from-above" of all piece-plotted faunal remains from the Kebaran and Mousterian layers of the main trench. The plot reveals a peculiar empty stripe that runs diagonally through the center, irrespective of culture period. No such gaps are found in the distributions of stone artifacts. This observation raises the possibility of localized in situ bone decomposition, a problem of distinguishing between situations in which bone was never present and those in which bones dissolved in place. Whereas faunal remains near the cave walls are well-preserved, the condition of specimens near the sagittal line of the cave tends to be poorer on the average, as determined by both macroscopic examination and infrared spectroscopy.

The paucity of skeletal material in the midline area may be due to higher rates of water percolation there which, because of the solvent properties of water, is believed to accelerate bone decomposition in sediments also rich in organic material (*sensu* Weiner and Bar-Yosef 1990; Weiner and Goldberg 1990; Weiner et al. 1993; see also Schiegl et al. 1994 on wood ash diagenesis). If this was the case, however, the strikingly heterogeneous pattern of microscopic preservation among the specimens that do occur in the bone-poor central zone also needs to be explained. It is possible that some of the bone fragments were subsequently introduced to the central zone by borrowing rodents, a source of localized (i.e., small-scale) sediment disturbance in this site. If correct, the scale of rodent disturbance was not so extensive as to erase the general spatial distinction between favorable and unfavorable preservation environments within the main trench, nor the stratigraphic integrity of the cultural sequence.

3. METHODS FOR THE SPECIES ABUNDANCE COMPARISONS

The variable, piece-plotted NISP, is the quantitative basis for the species abundance comparisons. NISP stands for the number of identified skeletal specimens, piece-plotted (P-P) on three spatial axes in this case. While the use of NISP for species abundance comparisons is not without its critics, none of the complaints raised in the literature to date approach the computational ills of alternative variables, such as MNE, MAU, and MNI, which in fact are derivatives of this simple counting unit and can introduce serious rounding errors to small samples (Grayson 1984). Virtually all potentially identifiable macrofauna, as well as bone fragments 5 cm or larger, were piece-plotted during the 1992–1995 excavations at Hayonim Cave. The data set does not include bone fragments recovered while sieving; the proportions of small game will certainly increase for all assemblages once bone material recovered in the screens are integrated into the data set. Larger samples naturally are being sought for the future, but neither recovery criteria nor sample size biases can explain the *relative differences* in taxonomic abundances obtained by this preliminary study. The current sample comes from intact cultural strata only—areas of the excavation where significant post-depositional mixing is apparent have been eliminated from consideration. Although the sizes of some of the faunal samples available for comparison are small, Chi-square statistics on raw P-P NISP counts show the contrasts in species representation to be highly significant, with probability values falling far below the .005 level. Because the results diverge substantially from expectations of randomly-dis-

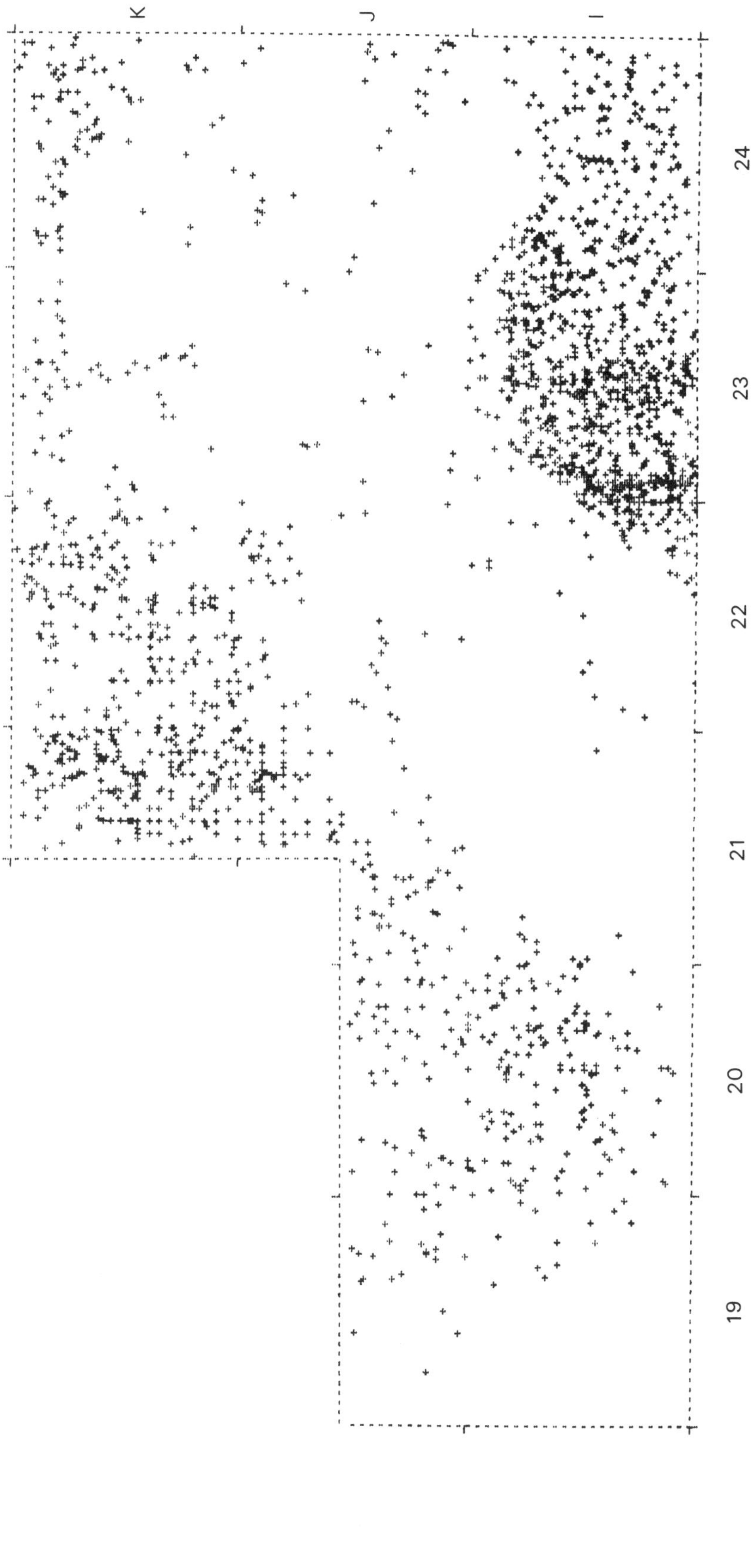

Figure 2. "Transparent" top-view of the distribution of piece-plotted bones and teeth in the Natufian, Kebaran, and Mousterian layers of the main excavation trench as of the close of the 1995 field season (to 455 cm below datum), showing bone-poor diagonal stripe across center.

tributed variation, the trends illuminated in these data must be taken seriously. They are very relevant to future research at this site.

Other characteristics of the study sample further ensure that the comparisons to follow are meaningful: (1) the study sample is from a single site and geological formation, meaning that changes in world climate would be required to alter local species availability; (2) the sample was excavated by a single team of archaeologists using a thorough recovery system, and whatever recovery biases may exist were consistently applied; (3) the skeletal fragment size ranges and means are comparable throughout the sample, owing in large part to the recovery techniques used; (4) the quality of bone preservation is relatively good over much of the excavated area; (5) the comparisons employ exactly the same counting unit throughout (P-P NISP); and (6) the observer was always the same person (the author), using the comparative collections at Hebrew University in Jerusalem, so no substantial variation in criteria of specimen identifiability exists in the sample. Thus while the absolute values used to generate the species abundance graphs are preliminary and subject to revision as sample sizes grow with continued excavation, the *relative* differences in species counts in the 1992–1995 piece-plotted material are informative as they stand.

The question of fragment size comparability across provenience units merits some additional discussion because of its general importance for evaluating species abundance data from discrete assemblages. If fragment sizes are relatively constant and in situ decomposition is limited, one can learn a great deal from NISP counts, including the possibility of ranking species in terms of bulk bone volume. Table 1a shows that the fragment size means for all piece-plotted taxa from the three cultural layers are quite similar, varying between 4.0 and 4.8 cm. T-tests show the means from the Natufian and the Mousterian layers to be indistinguishable statistically (Table 1b). The Kebaran mean is different, although the order of difference is only about half a centimeter.

More important are the data presented in Figure 3, which examines piece-plotted specimen size ranges and means for selected (common) taxon groups representing substantially different body sizes in the Natufian, Kebaran, and Mousterian layers. This more comprehensive approach shows that fragment means and ranges gently increase with original body sizes of the taxon categories compared, but that the overall configuration of means and ranges is essentially the same for the three cultural layers. Only the maximum length for the medium ungulate remains in the Mousterian stands out, due to the vastly larger number of specimens available for comparison, wherein a few whole or nearly whole bones happen to be present; the mean for this category nonetheless is consistent with those for medium ungulate remains in the other cultural layers.

Table 1. Summary comparisons of mean fragment sizes of piece-plotted faunal specimens across cultural layers

a. Fragment size means:

Cultural layer	N specimens	Mean size (cm)	sd
Natufian	105	4.765	2.394
Kebaran	772	4.005	2.216
Mousterian	3213	4.609	2.255

b. T test results for pair-wise comparisons of fragment size means:

Cultural layer pair	T value	p
Natufian-Kebaran	3.265	0.001
Natufian-Mousterian	0.693	0.488
Kebaran-Mousterian	–6.712	0.001

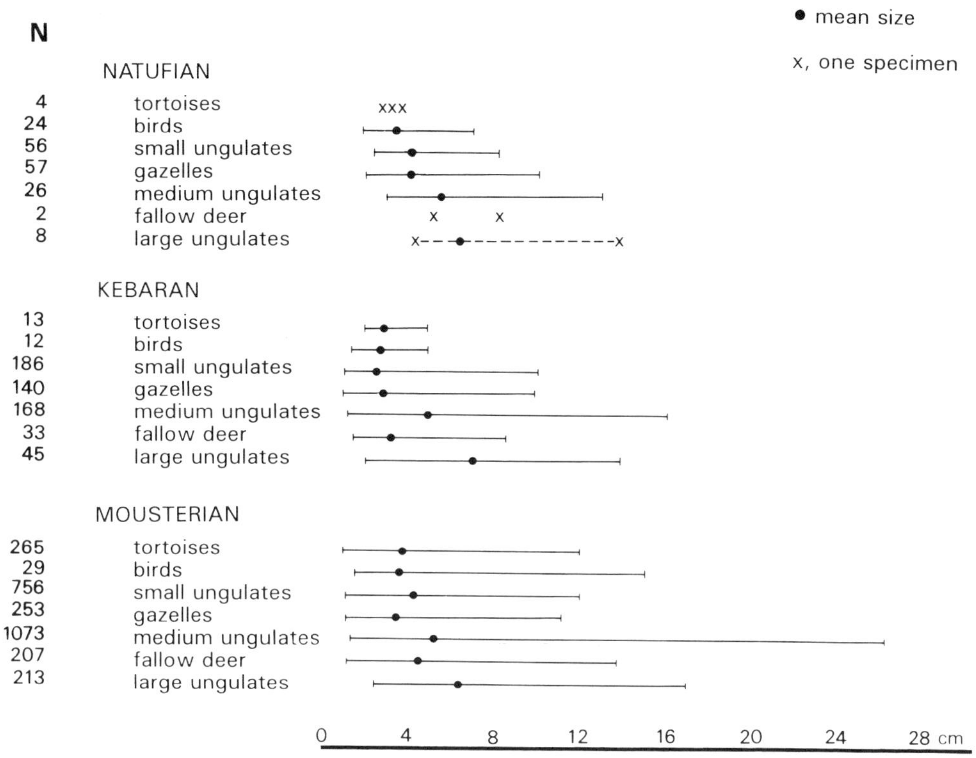

Figure 3. Piece-plotted fragment size means and ranges (cm) for faunal specimens in selected taxon categories, ordered by increasing body size. All taxonomic categories are mutually exclusive. N refers to the number of specimens observed for each taxon.

4. GAME SPECIES ARRAYS IN HAYONIM CAVE

The species utilized by hominids at Hayonim are, in descending order of abundance, gazelle (*Gazella gazella*), fallow deer (*Dama mesopotamica*), red deer (*Cervus elaphus*), aurochs (*Bos primigenius*), wild pig (*Sus scrofa*), roe deer (*Capreolus capreolus*), onager (*Equus hydruntinus* or *E. hemionus*), horse (*E. caballus*), wild goat (*Capra aegagris* and/or *C. ibex*), as well as small game species such as the spur-thighed tortoise (*Testudo graeca*), legless lizard (*Ophisaurus apodus*), hare (*Lepus capensis*), and a variety of passerine, columbine, and galliform birds (see also Tchernov 1992, 1994). Ostrich eggshell fragments (from *Struthio camelus*) are also found in the Mousterian layers.

Small canids (*Vulpes* sp., *Canis aureus*), leopard (*Panthera pardus*), brown bear (*Ursus arctos*), wildcat (probably *Felis sylvestris*), and mustelids (e.g., stone martin, *Martes foina*) occur in much smaller quantities in the Paleolithic cultural layers. Other rare taxa, found mainly in the Aurignacian and Kebaran levels, include an undetermined species of rhinoceros, vulture, and very rarely hyaena (*Hyaena hyaena* and/or *Crocuta crocuta*). These results are consistent with prior observations of Pleistocene archaeofaunas from the Levant (e.g., Davis 1980, 1981, 1985; Garrard 1980, 1982; Henry et al. 1981; Tchernov 1988, 1989, 1992).

4.1. Species Proportions in the Natufian, Kebaran, and Mousterian

Ungulates were the most abundant prey species of hominids at Hayonim Cave, but certain small animals—primarily tortoises—were also important sources of food. Figure 4 compares the percentages of small game in the Natufian, Kebaran, and Mousterian layers (P-P NISP values in Table 2a). Small animals were least abundant in the Kebaran diet, more in the Mousterian, and most abundant in the Natufian diet—at least at Hayonim Cave.

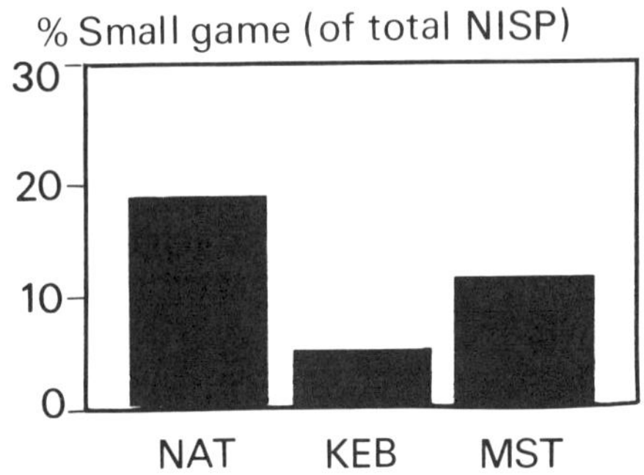

Figure 4. Percentages of small game in Natufian (NAT), Kebaran (KEB), and Mousterian (MST) in piece-plotted faunal sample (total NISP), from stratigraphically intact deposits.

Table 2. Taxonomic representation in the Natufian, Kebaran, and Mousterian of Hayonim Cave

	Natufian	Kebaran	Mousterian
a. P-P NISP percentages for general taxon groups:			
Small game	19%	5%	12%
Ungulate	79	93	87
Carnivores	2	2	1
Total P-P NISP:	*220*	*706*	*3127*
b. P-P NISP percentages for small game categories:			
Reptiles	9%	60%	83%
Birds	56	34	8
Lagomorphs	35	6	1
Ostrich egg	0	0	7
Small game P-P NISP:	*43*	*35*	*375*
c. P-P NISP percentages for ungulate remains identified to species:			
Horse	0%	0%	«1%
Onager	0	1	1
Roe deer	5	2	1
Fallow deer	3	13	38
Red deer	4	6	9
Aurochs	3	2	6
Wild pig	5	4	3
Ibex/wild goat	3	5	«1
Gazelle	77	65	41
Species-specific ungulate P-P NISP:	*74*	*201*	*684*
d. P-P NISP percentages by ungulate body size categories:			
Small ungulate	70%	54%	40%
Medium ungulate	22	36	48
Large ungulate	7	10	12
All ungulate P-P NISP:	*173*	*658*	*2729*

Ungulates always predominate in these assemblages (79–93%), whereas carnivores are always rare (1–2%). The overall proportions of small game vary among the three periods, ranging between 5% and 19% of total P-P NISP and showing no clear trend in this regard.

Within the small game fraction, however, an opposing theme is evidenced for three broadly defined taxonomic categories—large reptiles, birds, and lagomorphs (hares) (Figure 5a and Table 2b). Large reptiles dominate in the Mousterian (83%): the great majority of them are tortoises; most of the remainder legless lizards. Proportionally fewer reptiles are present in the Kebaran (60%), and they are much less common in the Natufian (9%). By contrast, birds (thrush-sized and larger) are prevalent in the Natufian (56%), somewhat fewer in the Kebaran (34%), and rare in the Mousterian (8%). Indeed, the few bird remains present in the Mousterian layers are mainly from owls and probably were not food to these early humans. Hares were an outstanding feature of the diet only by Natufian times (35%).

A Chi-square statistic for the reptile-bird-lagomorph comparison indicates that the noted contrasts are highly significant ($\chi^2 = 145.0$, $p < 0.005$, $df = 4$) despite the small size

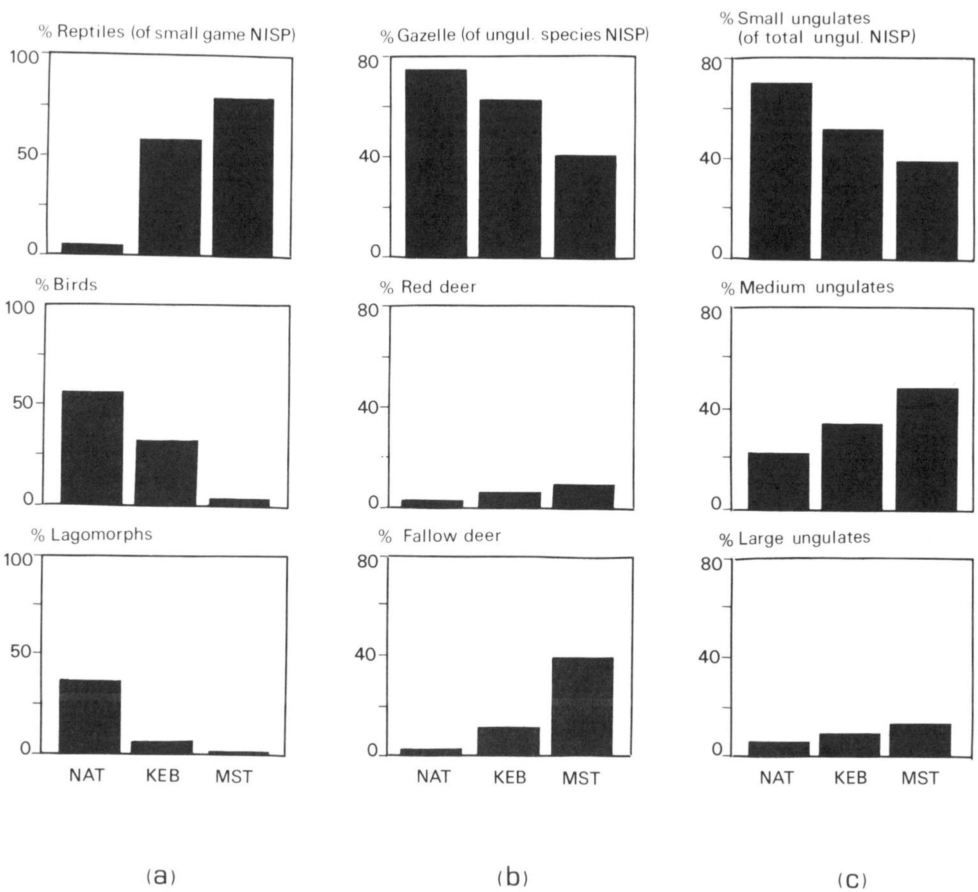

Figure 5. Percentages of (a) reptiles, birds, and lagomorphs in Natufian (NAT), Kebaran (KEB), and Mousterian (MST) in piece-plotted small game NISP fraction from stratigraphically intact deposits; (b) percentages of gazelles, red deer, and fallow deer (% of ungulate NISP that could be identified to the species level); (c) percentages of small, medium, and large ungulates (% of total ungulate NISP).

of the Natufian sample. The trend is interesting from a human behavioral standpoint because all of these taxonomic groups were present in the Upper Galilee throughout the three culture periods (Tchernov 1988, 1989, 1992). Hominids reliance on tortoises, birds, and lagomorphs nonetheless changed a great deal over time. Not illustrated is a largely Mousterian item at Hayonim, unworked ostrich eggshell fragments, which constitute 7% of small game P-P NISP (see Table 2b).

The tortoise remains are from a single species of *Testudo*, probably *T. graeca*, which continues to be widely distributed in Mediterranean ecosystems (Pope 1956). Some large partial specimens were recovered in the lower Mousterian layers, owing to excellent (macroscopic) preservation conditions in some areas of the main trench. In other instances substantial portions of the carapaces or plastrons could be reconstructed from fragments found together (e.g., Figure 1c). The tortoise shells and upper limb bones show every indication of human processing, usually with the aid of fire. The animals appear to have been roasted; carapace and plastron fragments often display burning damage, and many of the specimens occur in close spatial proximity to hearth areas or in cemented ashy concentrations. Percussion fractures (Fig. 1e) on some specimens indicate that the tortoises' shells were split open by placing them on edge on an anvil and striking along the opposing margin with a hard hammer (Figure 6). Upper limb bones (humerus and femur) of the tortoises were frequently damaged by tearing of the proximal epiphysis and/or sectioning them at mid- or upper-shaft (Figure 1d); lower limb bones are less abundant despite thorough recovery procedures, either because they were destroyed by fire during roasting or microfaunal recognition criteria tend to overlook them. Legless lizard remains are much less common, but exceptional burning frequencies and damage to mandibles (now under study) show that they too were also occasionally eaten by humans at Hayonim Cave. Large snakes of the genus Colubra may also have been eaten, if burning frequencies may be taken as fair indication (but see Stiner et al. 1995), but the association between the snake remains and human agencies is less clear.

Another trend in species representation across the three cultural layers is found within the ungulate fraction (Table 2c). The differences, though less marked than those for small game, involve the most common ungulate taxa—gazelles, red deer, and fallow deer. Figure 5b shows that gazelles are proportionally fewer in the Mousterian as a whole (41%), and highest in the Natufian (77%). Fallow deer show essentially the opposite pattern, being highest in the Mousterian (38%) and lowest in the Natufian (3%). The story is less clear for red deer because they are uncommon throughout the Pleistocene sequence of Hayonim, but their abundance appears to track fallow deer in the Hayonim sequence, consistent with the fact that both species are of Eurasian origin. A Chi-square statistic for the comparisons between gazelle, red deer, and fallow deer shows that the contrasts are also highly significant ($\chi^2 = 65.9$, $p < 0.005$, $df = 4$).

Variation in ungulate abundances over time can also be explored in terms of body sizes (Figure 5c), for which sample sizes are greater than those for material securely identified to species (Table 2c-d). The small ungulate category nonetheless is nearly all from gazelle, and wild goat/ibex; large ungulate includes aurochs, horse, and (in this study) red deer;[†] the medium ungulate category is mostly fallow deer; large ungulates are a more even mix of species and few in number overall. It is clear from the size-oriented compari-

† Values in the ungulate body size categories here include specimens identified to the species-specific level as well as those relegated to more general taxonomic levels. The small ungulate category therefore includes gazelle, roe deer, and wild pig; medium ungulate includes onager, fallow deer, and wild goat/ibex; large ungulate includes aurochs, horse, and (in this study) red deer.

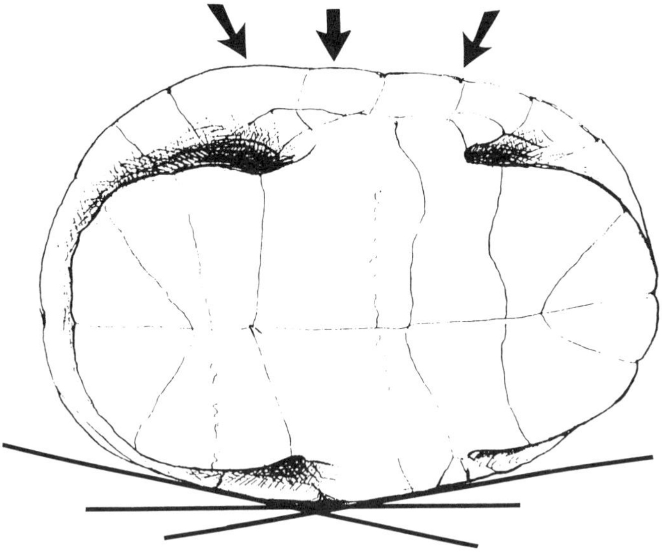

Figure 6. Tortoise shells apparently were split open by placing them on edge on an anvil (base lines) and striking along the opposing margin (arrows) with a hard hammer, as indicated by percussion fractures on some specimens. Not shown but also of interest are damage patterns on upper limb bones (humerus and femur), many of which were sectioned or torn at the mid- or upper-shaft.

son in Figure 5c that the ungulate prey of the Mousterian were considerably larger on average than those emphasized during the later periods. The Kebaran shows intermediate values in this regard, whereas the body sizes of ungulates procured during the Natufian were smallest. Again, the contrasts are highly significant ($\chi^2 = 65.3, p < 0.005, df = 4$).

4.2. Variation in Species Proportions within the Mousterian

Up to this point in the presentation, the Mousterian has been considered as if it were a single monolithic cultural unit. In fact variation also exists within the Mousterian at Hayonim Cave (Table 3), a deep deposit that obviously formed over many thousands of years. The Mousterian sample here is confined to piece-plotted material from the main excavation trench at depths of 345 to 455 cm below datum. Examination of tortoise P-P NISP by 10 cm depth increments (Figure 7) reveals a distinct rise in frequencies just below 404 cm. Ostrich eggshell fragments are found only below this depth as well, and some significant shifts in microfauna species may be indicated within the next 20 cm below 400 cm bd (Tchernov, research in progress). On these grounds one can provisionally divide the Mousterian faunal series excavated from the main trench thus far into chronologically younger and older segments, separated for the sake of this discussion at 404–405 cm below datum (all are included, however, in the Upper Mousterian strata described by Meignen in this volume). Further refinements in the division of Layer E are inevitable as research at Hayonim Cave continues, but this boundary helps establish the existence of significant variation within the Mousterian.

Table 3a shows the proportions of small game (mostly tortoises, see Table 3b), ungulates, and carnivores above and below 404–405 cm below datum in the Mousterian deposits, based on P-P NISP counts. Small game is somewhat more prevalent below 404 cm (15%) than it is above this depth (5%). A Chi-square statistic for the reptile-bird-lagomorph com-

Table 3. Taxonomic representation in the Mousterian of Hayonim Cave

	<405 cm bd	≥405 cm bd
a. P-P NISP percentages for general taxon groups:		
Small game	5%	15%
Ungulate	95	85
Carnivores	«1	«1
Total P-P NISP:	*482*	*1929*
b. P-P NISP percentages for small game categories:		
Reptiles	61%	91%
Birds	26	4
Lagomorphs	13	0
Ostrich egg	0	4
Small game P-P NISP:	*23*	*288*
c. P-P NISP percentages for ungulate remains identifed to species:		
Horse	0%	1%
Onager	0	«1
Roe deer	5	1
Fallow deer	27	48
Red deer	13	5
Aurochs	8	5
Wild pig	2	3
Ibex/wild goat	0	«1
Gazelle	45	36
Species-specific ungulate P-P NISP:	*85*	*412*
d. P-P NISP percentages by ungulate body size categories:		
Small ungulate	40%	38%
Medium ungulate	45	53
Large ungulate	14	9
All ungulate P-P NISP:	*458*	*1632*

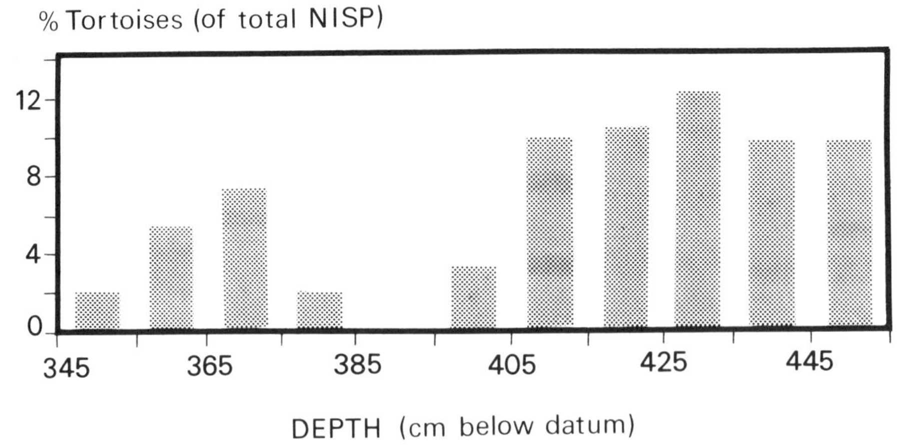

Figure 7. Frequency of tortoise bones (percentage of total P-P NISP) by 10 cm depth increments within the Mousterian deposits of the main excavation trench.

parison shows the contrast between the above defined segments of the Mousterian deposits to be highly significant ($\chi^2 = 54.0$, $p < 0.001$, $df = 2$). The remainder of the samples are taken up by ungulate remains, which reach 95% in the Mousterian deposits above 405 cm, but only 85% below this boundary. These data show that the proportions of gazelle, fallow deer, and red deer were also in flux over the course of the Mousterian (Figure 8, Table 3c). There is considerable temporal variation with respect to predominant species and the overall importance of ungulates in the diet ($\chi^2 = 14.6$, $p < 0.001$, $df = 2$). Specifically, there are more gazelles in the later Mousterian, more deer early on.

5. EXPLANATIONS FOR SPECIES TRENDS IN THE CULTURAL LAYERS

At least two potential explanations may be proposed for trends in the proportions of small to large game in Hayonim Cave, and the species that Natufian, Kebaran, and

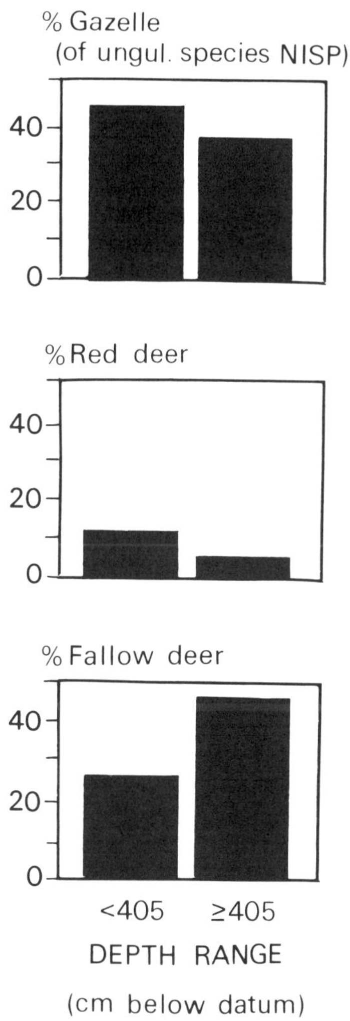

Figure 8. Percentages of gazelle, red deer, and fallow deer in the Mousterian deposits above and below 404–405 cm below datum, based on piece-plotted specimens from the main excavation trench that could be identified to species.

Mousterian foragers emphasized therein. The first explanation emphasizes climate-driven changes in foraging opportunities—or species "availability"—if we accept that hominids should simply respond to the natural abundances of appropriately sized prey species in a given foraging locality. The second type of explanation instead holds that certain species were emphasized in human diets disproportionately to their natural abundances in the environment, due to biases imposed by human behavior or biology. In the latter case the challenges would be to identify those behavioral or mechanical thresholds governing access to prey species, the timing of their appearance in human adaptations, and the causes of change in evolutionary terms.

5.1. Climate-Caused Variation in Large Game Species Harvested

Unless proved otherwise, climate-driven change in the composition of animal communities and foraging opportunities is the more parsimonious of the two explanations above. We know from Tchernov's work (1968, 1988, 1989, 1992), for example, that periodic expansion and contraction of Eurasian versus Afro-Arabian animal communities tracked climatic changes to a large extent, and that the boundary shared by these vast biogeographic zones periodically straddled the Near East. Independently, it has been shown by a variety of wildlife studies that modern carnivores and human hunter-gatherers respond fairly directly to local abundances of prey species, albeit within certain predetermined prey size ranges, activity cycles, access behaviors, and foraging "substrates" (reviewed in Stiner 1992, 1994). In regions where animal diversity is high, several different prey species might fit within the size, behavioral, or geographic criteria that define a predator's foraging window and therefore represent acceptable substitutes from the forager's point of view. Absence or rarity of other species on the menu is also informative.

The proportional differences in ungulate species harvested by humans at Hayonim Cave during the Natufian, Kebaran, and earlier and later Mousterian are comparatively minor (see Table 2) and vary mainly in terms of body size—Eurasian fallow deer and red deer are bigger than West Asian gazelles. Maximum proportional differences between the Natufian and the Mousterian periods in the frequencies of gazelle and fallow deer never exceed 30% in this P-P NISP comparison. Hence there is no basis for arguing that people's access was limited by the body sizes of the ungulate species. Humans of all culture periods were quite able to obtain substantial numbers of both gazelles and deer using the techniques of the times. The species abundance data for ungulates therefore argue for a climatic cause (see also Garrard 1980; Uerpmann 1981; Tchernov 1992), with people simply adjusting to fluctuations in encounter rates over many thousands of years, and continuing to obtain ungulates despite shifts in the composition of nature's larder. It is interesting that Mousterian humans took larger ungulate prey on the average than later Paleolithic peoples, presumably because larger ungulates were more prevalent in the Upper Galilee at the time, and humans' hunting capabilities were up to the task.

We have seen that small game also was an important food supplement to humans at Hayonim Cave during all of the culture periods considered. Mediterranean ecosystems tend to support high levels of species diversity (e.g., Cody 1986), especially among smaller animal taxa (e.g., Bate 1942, 1943; Tchernov 1968; Stuart 1982, 1991). Hence human foragers would frequently have come across a variety of potentially edible small animals. This much certainly is the product of both the climate and the physiographic properties of Mediterranean provinces, as well as the plant communities they support; small animal species should have been important in prehistoric hunter-gatherer economies of West Asia, and indeed they were.

5.2. Human-Caused Variation in Small Game Species Harvested

The second kind of explanation—human-caused biases in species acquisition—implies that some threshold set by early human adaptations in West Asia expanded over time, though not necessarily as a consequence of consciously directed food preferences, since animal food sources have very similar contents if compared to plant food sources. Examples of access thresholds include, among others, prevailing technology for acquiring and processing game, and human demographic conditions in which advantages to intensifying the foraging regimen are minimal or great.

While small game was important to all of the Paleolithic cultures considered, differences in the types of small game species emphasized by humans at Hayonim Cave during the Natufian, Kebaran, and Mousterian periods are not as easy to explain in climatic terms. A first indication that this may be the case arises from the observations that (1) the variation in the relative abundances of three small game categories—tortoises, birds, and lagomorphs—is much more pronounced than it is for large game taxa, and (2) most of the types of small game considered were present in animal communities of the Upper Galilee throughout the three culture periods. Hares, for example, were present in West Asia as early as 250,000 years ago and evidently remained in the region thereafter (Tchernov 1992, 1994). Tortoises also were available throughout this long time span, and, although bird species in the area underwent change, there always have been a variety of local bird species to choose from.

Explanations for human-caused taxonomic biases in small game harvesting across the three Paleolithic periods may be tied to the distinct behavioral and biological properties of the tortoises, birds, and lagomorphs themselves. These categories of small prey can be distinguished in terms of (1) the medium or habitats in which they live and locomote, (2) their escape strategies, and (3) the harvesting intensities that populations of the species can viably sustain due to intrinsic life history and reproductive characteristics.

With respect to habitat, it is clear that flying animals (adult birds) were not important in the Mousterian diet, yet birds such as chukar (*Alectoris chukar*) were quite important to people in later Paleolithic periods, especially the Natufian. Small game items emphasized during the Mousterian of Mediterranean regions instead were obtained by very simple means, primarily on land or along the waters' edge. At Hayonim Cave it was tortoises, along with a few legless lizards, ostrich eggs, and possibly non-poisonous snakes. Elsewhere along the Mediterranean Rim tortoises and littoral shellfish were eaten at coastal caves (Leonardi 1935; Blanc 1958–61; Palma di Cesnola 1965; Vitagliano 1984; Klein and Scott 1986; Stiner 1992, 1993, 1994).

The second property of small prey that is of interest for this study are their escape mechanisms. Although the birds, tortoises, legless lizards, and hares of the Galilee would have yielded grossly comparable food weights, these animals differ greatly in what it takes to catch them. Lagomorphs and birds are fast and maneuverable, whereas the tortoise's strategy is to be cryptic or to bore its opponent if discovered. It is clear that Mousterian small game harvesting focused upon comparatively sessile species, which are gathered as much as hunted; some of these animals are truly immobile whereas others are either slow-moving or possess very small territories. Access to such resources is not normally assisted by technology among modern human foragers—persons of either sex and almost any age can obtain them. By Kebaran and Natufian times the situation was very different in that substantial numbers of lagomorphs and ground and other birds were taken in addition to tortoises, species that are quick, mobile, and perhaps required nets or traps in order to obtain in quantity.

The third and perhaps most intriguing property of the small game categories defined above concerns the harvesting intensities that populations of these prey taxa can viably sustain, given species-specific reproductive and/or development rates. Lagomorphs and many birds reproduce and grow quickly and therefore also exhibit the potential for rapid population growth and resiliency. Tortoises generally do not possess these qualities (Stiner's research in progress), requiring up to a decade or more to reach breeding age and appreciable body size in the case of *Testudo* (Stubbs 1989). Yet tortoises were the small game most commonly eaten by Mousterian hominids at Hayonim Cave, representing at least 12% of total piece-plotted NISP in the Mousterian layers as a whole. In the earlier Mousterian (i.e., ≥405 cm below datum), the percentage of tortoises is even higher—at least 15%. (Note that the percentages of small game will rise for all periods when the screen-recovered samples are integrated with the piece-plotted data.)

Not only are there more tortoises in the Mousterian than in the later cultural layers, but the individual animals collected by Mousterian people were considerably larger on average. Measurements of tortoise humeri demonstrate a size reduction trend in the Hayonim stratigraphic sequence from Mousterian through Natufian times (Figure 9). The dimensions of this weight-bearing limb element in tortoises are directly sensitive to changes in body mass during life. The upper graph in Figure 9 is based on an anterior-posterior measurement (mm) of the proximal humeral epiphysis (the "ball"), and the lower graph is based on a medio-lateral measurement (mm) of the narrowest point in the humeral shaft diaphysis. The Mousterian sequence is subdivided by 100 cm depth increments to compensate for the substantially longer time frame represented in relation to subsequent culture periods. Klein and Cruz-Uribe (1983; see also Klein 1989) have identified a size reduction trend in angulate tortoises from Middle and Late Stone Age deposits in South African caves as well. However, size reduction in the South African tortoises occurs much later in prehistory than in the area of Hayonim Cave in Israel.

All of the tortoises from Hayonim Cave appear to be of a single species (always *Testudo graeca*), and the size reduction trend may reflect an increase in harvesting intensities by humans (see also Klein and Cruz-Uribe 1983). Heavy harvesting (such as in modern Morocco for the pet market) is known to reduce the mean size of individuals in living populations of *Testudo* (e.g., Stubbs 1989). Because tortoises grow slowly even under favorable conditions, the large sizes of the Mousterian tortoises (especially the earliest ones) imply low rates of exploitation by humans at that time. Either exploitation was so irregular during the early Mousterian at Hayonim Cave that tortoises in the area were able to grow to a larger size, or Mousterian foragers simply did not need to glean the vegetation for small ones. The large mean sizes of the Mousterian tortoises are especially striking in light of the fact that a substantial proportion of the game in the Mousterian diet came from tortoises in particular. The size reduction trend for tortoises in Hayonim Cave probably represents a more widespread condition; tortoise measurement samples will increase as more of the Hayonim material comes under study. Samples from other sites in the northern Galilee are currently under study by the first author.

The size reduction trend in the Hayonim tortoises could also be a consequence of climate change, especially if conditions grew increasingly arid and rates of energy flow through the ecosystem declined accordingly; tortoises grow more slowly and may achieve smaller adult sizes on poor feed (Dr. John Behler, New York Zoological Society, personal communication 2/96). It is peculiar, however, that the trend at Hayonim is less responsive to known climatic oscillations of the Upper Paleolithic-Natufian than it was during the long time span represented by the Mousterian (Stiner et al. n.d.). Whether the trend corresponds to shifts in rodent species (Tchernov, research in progress) from Hayonim Cave

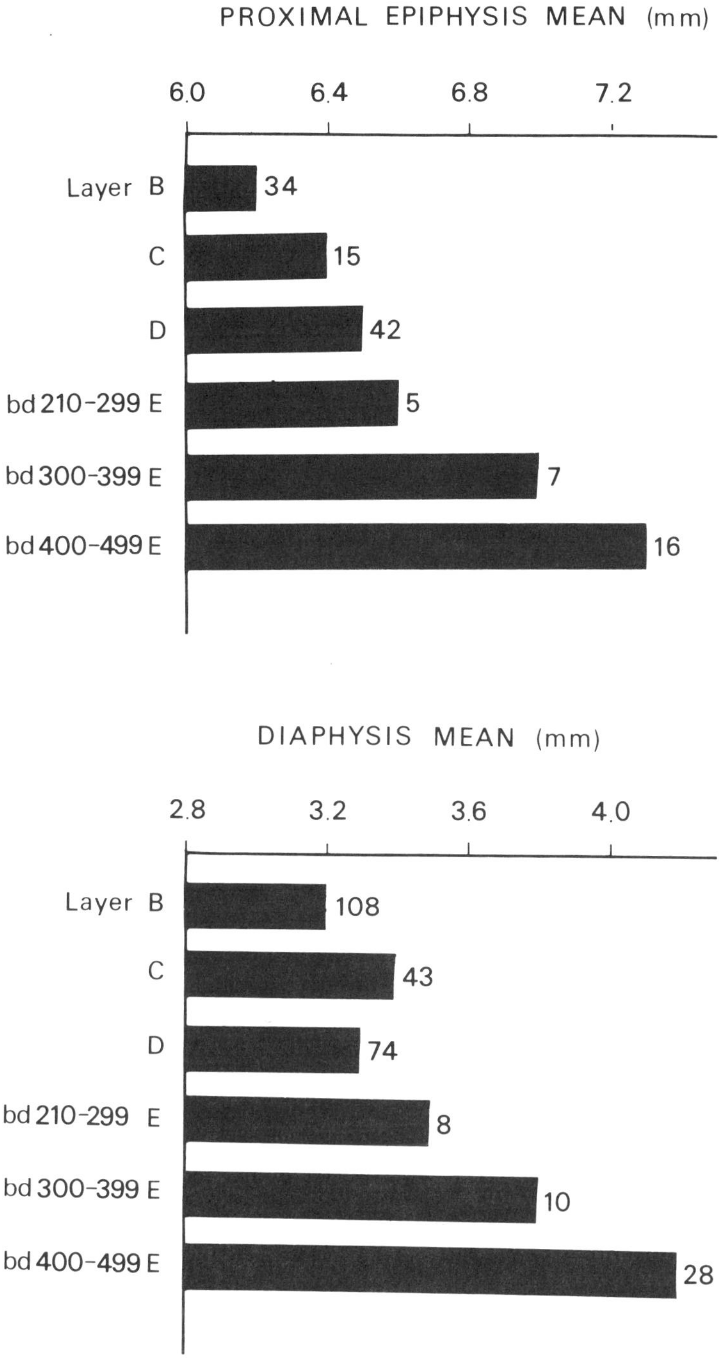

Figure 9. Preliminary results on a size reduction trend for tortoises (apparently always *Testudo graeca*) in the stratigraphic sequence of Hayonim Cave: Natufian (Layer B), Kebaran (C), Aurignacian (D), and Mousterian (E), the latter by 100 cm depth increments below datum (bd). All measurements are for the humerus. The upper graph is based on an anterior-posterior measurement (mm) of the rounded part ("ball") of the proximal humeral epiphysis; the lower graph is based on a medio-lateral measurement (mm) of the narrowest point in the humeral shaft diaphysis. The number of specimens measured, which currently combines right and left sides, is shown at the end of each bar. The combined Mousterian means are 7.1 mm (N=28) and 4.0 mm (N=46) respectively. Note that measurements of tortoises from the Kebaran deposits of Meged Shelter, just uphill from Hayonim Cave, yield a similar if slightly lower mean value (3.1 mm) for the humeral diaphysis (N=24).

and other sites remains to be seen. Regardless, one is at a loss to explain the near absence of birds and hares in Mousterian faunas in strictly climatic terms. Increasing emphasis on these r-selected taxa in human diets over time appears to be linked to the decline in mean tortoise size and the proportion of tortoises in the small game fraction of the Hayonim assemblages.

If the interpretation of a human cause is correct, low harvesting rates of tortoises may undermine explanations citing ineffective foraging by premodern humans (cf. Klein 1989, this volume), pointing instead to the simpler conclusion that human population densities of the Mousterian were especially low by hunter-gatherer standards of subsequent periods in West Asia. Certainly the cultural barriers to humans' access to ungulates were no greater in the Mousterian than they were in later periods. And if technologic innovations are marshalled to account for differences in humans' access to distinct classes of small game, one must also recognize that we are speaking only of innovations specifically geared to obtaining these prey in quantity, rather than a case of general ineptitude. The demographic and ecologic contexts in which these technologic innovations became valuable therefore also merits investigation.

If Mousterian groups were smaller and/or the frequency and duration of visits to Hayonim Cave were significantly lower than in the Kebaran and Natufian, demographically-linked incentives to intensify resource extraction methods, such as by altering the prey types sought to include those that are more challenging to acquire, may not have been as important in the selective milieu of early Mousterian times. The notion that low intensity game exploitation during the Mousterian—especially the earlier Mousterian—at Hayonim Cave reflects very ephemeral human presence is supported by other aspects of the faunal and artifact assemblages. Taking into account problems of in situ decomposition of bone, the rates of faunal and stone artifact accumulation in the Mousterian deposits of Hayonim appear to have been extremely low relative to sediment accumulation. The large samples we are now removing from the site instead owe their appreciable sizes to the long time frame represented and large amounts of sediment excavated; some generally similar patterns are documented in certain early Middle Paleolithic caves in Italy (Grotta dei Moscerini, Kuhn 1995; Stiner and Kuhn 1992; Stiner 1994). Patterns of wood ash accumulation and diagenesis tell a similar story at Hayonim Cave (Schiegl et al. 1994), suggesting that the ash concentrations are the highly compressed remains of what were once much thicker residues. Yet stone tools and bones are sparsely distributed in the Middle Paleolithic layers overall. These and other preliminary findings suggest that the early Mousterian record of Hayonim Cave poses some striking contrasts to the situation in the later Mousterian (<70 ka) of Kebara Cave, both in terms of technology (Meignen, this volume) and hominids use of game (Speth and Tchernov, this volume). The early Mousterian poses an even starker contrast to the foraging systems of later Paleolithic peoples in West Asia.

6. CONCLUSION

In the case of Hayonim, it is the small game fraction that speaks most strongly to questions about constraints upon and trends in Pleistocene human foraging niche. Small game use is a very interesting feature of hominid foraging strategies from an evolutionary perspective, because the small-large dichotomy in prey body size often corresponds to socially significant divisions in modern hunter-gatherer labor networks (e.g., Stiner 1993, 1994; Hawkes et al. 1997). Unfortunately, small game use remains rather poorly documented for the Mousterian as a whole, despite the numerical significance of these taxa in

Paleolithic human diets in some Mediterranean habitats. Small game use informs us about classes of human foraging strategy that are somewhat or wholly unique if compared to those for obtaining large game.

We see at Hayonim and elsewhere along the Mediterranean Rim that most of the small animals utilized by Mousterian humans were of the species most easily gathered. Quite unlike the situation for large mammal procurement, the types of small game exploited at Hayonim Cave are more readily separable in terms of their inherent biological characteristics—especially their escape mechanisms and reproductive and development rates. These are features that may have tested the limits of Pleistocene technology and hunting tactics under conditions of rising human population densities, setting new premiums on the value of certain classes of technological innovation.

It is natural for students of human nature to suppose on the basis of these findings that technological innovations, such as specialized projectile weapons, traps, nets, and snares, made possible the harvesting of quick small prey species in appreciable quantities. Indeed there is a fairly clear if general association in time between profound radiations in the design of tools for capturing small animals and proportional increases in the frequencies of these species in human diets. However, correlation in this case does not isolate cause because it does not demonstrate that the same classes of innovation had not arisen in periods prior to later Paleolithic times, only that no such innovations gained widespread use until a certain time in the past (Kuhn and Stiner 1998).

The view that Middle Paleolithic hominids were generally ineffective or inefficient in their quest for food appears to be widely held, yet the findings of this study argue only for low harvesting rates in conjunction with very low human population densities during the Mousterian, especially the earlier Mousterian. Efficiency and effectiveness are relative concepts by definition, and therefore quite prone to circular reasoning. And important theoretical differences exist between the interpretations of "not able to" and "no need to" in human evolution research, only the latter of which seems testable. The underlying question raised by these findings concerns aspects of the selective milieu that could foster or set in motion a feedback loop between human population stability/growth and resource intensification where it had not done so previously. This question about the Paleolithic sequence is warranted for the simple reasons that the capacity for innovative behavior is widespread among animals that rely heavily on learning, and the transmission of such information between individuals is contingent at least in part on the frequency of social contact and the biological and/or social rewards for doing so (Kuhn and Stiner 1998). What might have altered the value of technical innovations with respect to human foraging agendas? A demographic phenomenon almost certainly forms part of the cycle of cause and effect noted here (see also Keeley 1995), more fundamental than explanations which visualize human evolution as progress toward modern physiological and intellectual ideals.

ACKNOWLEDGMENTS

I am indebted to my colleagues Steven Kuhn, Ofer Bar-Yosef, Anna Belfer-Cohen, Paul Goldberg, Liliane Meignen, Henry Schwarcz, Hélène Valladas, and Stephen Weiner for their help and constructive criticisms throughout this research. Also important to my thinking have been a number of general comments on the implications of small game use shared with me by Liora Horwitz and Patricia Smith some time ago. I am grateful to Professors Takeru Akazawa, Kenichi Aoki, and Ofer Bar-Yosef for organizing the Tokyo University Museum symposium entitled "Neanderthals and Modern Humans in West Asia," a very stimulating forum in which to have first presented these data. Detailed commentary on earlier drafts by Steve Kuhn and Chris Monahan, and new research on the Natufian faunas of Hayonim Cave by Natalie Munro, were a great help for preparing the final version of this manuscript. The research on the Hayonim archaeofaunas is supported by a grant to the first author by the National Science Foundation (SBR-9511894).

REFERENCES

Bar-Yosef, O. 1989. Upper Pleistocene cultural stratigraphy in southwest Asia, 154–79. In *Patterns and Processes in Later Pleistocene Human Emergence*, ed. E. Trinkaus. Cambridge: Cambridge University Press.

Bar-Yosef, O. 1990. The Last Glacial Maximum in the Mediterranean Levant, 58–77. In *The World at 18,000 BP*, Vol. 2, ed. C. Gamble and O. Soffer. London: Unwin Hyman.

Bar-Yosef, O. 1991. The archaeology of the Natufian layer at Hayonim Cave, 81–93. In *The Natufian Culture in the Levant*, ed. O. Bar-Yosef and F. Valla. Ann Arbor: International Monographs in Prehistory.

Bar-Yosef, O., and A. Belfer-Cohen. 1988. The early Upper Palaeolithic in Levantine caves, 23–41. In *The Early Upper Paleolithic in Europe and the Near East*, ed. J. H. Hoffecker and C. A. Wolf. Oxford: British Archaeological Reports International Series 437.

Bate, D. M. A. 1942. Pleistocene Murinae from Palestine. *Annual Magazine of Natural History* 9:465–86.

Bate, D. M. A. 1943. Pleistocene Cricetinae from Palestine. *Annual Magazine of Natural History* 11:813–36.

Belfer-Cohen, A. 1988. The Natufian graveyard in Hayonim Cave. *Paléorient* 14:297–308.

Belfer-Cohen, A., and O. Bar-Yosef. 1981. The Aurignacian at Hayonim Cave. *Paléorient* 7:19–42.

Blanc, A. C. 1958–61. Industria musteriana su calcare e su valve di Meretrix chione associata con fossili di elefante e rinocerante, in nuovi giacimenti costieri del Capo de Leuca. *Quaternaria* 5:308–13.

Cody, M. L. 1986. Diversity, rarity, and conservation in Mediterranean-climate regions, 122–52. In *Conservation Biology: The Science of Scarcity and Diversity*, ed. M. E. Soule. Sunderland, MA: Sinauer Associates.

Davis, S. 1980. Late Pleistocene-Holocene gazelles of northern Israel. *Israel Journal of Zoology* 29:135–40.

Davis, S. 1981. The effects of temperature change and domestication on the body size of Late Pleistocene to Holocene mammals. *Paleobiology* 7:101–14.

Davis, S. 1985. A preliminary report on the fauna from Hatoula: A Natoufian-Khiamian (PPNA) site near Latroun, Israel, 71–98. Appendix B. In *Le Site Natoufien-Khiamien de Hatoula, pres de Latroun, Israel*, ed. M. Lechevallier and A. Ronen. Paris: Association Paléorient.

Gamble, C. 1986. *The Palaeolithic Settlement of Europe. Cambridge: Cambridge University Press.*

Garrard, A. N. 1980. Man-Animal-Plant Relationships During the Upper Pleistocene and Early Holocene of the Levant. Ph. D. diss. Darwin College, Cambridge University. Cambridge.

Garrard, A. N. 1982. The environmental implications of a reanalysis of the large mammal fauna from Wadi el-Mughara caves, Palestine, 165–87. In *Palaeoclimates, Palaeoenvironments, and Human Communities in the Eastern Mediterranean Region in Later Prehistory*, ed. J. L. Bintliff and W. van Zeist. Oxford: British Archaeological Reports International Series 133.

Goldberg, P. 1979. Micromorphology of sediments from Hayonim Cave, Israel. *Catena* 6:167–81.

Goldberg, P., and H. Laville. 1988. Le contexte stratigraphique des occupations paleolithiques de Kebara (Israel). *Paléorient* 14:117–23.

Grayson, D. K. 1984. *Quantitative Zooarchaeology*. Orlando: Academic Press.

Hawkes, K., J. F. O'Connell, and N. Blurton Jones. 1997. Hadza women's time allocation, offspring provisioning, and the evolution of long post-menopausal lifespans. *Current Anthropology* 38. In press.

Henry, D. O., A. Leroi-Gourhan, and S. Davis. 1981. The excavation of Hayonim Terrace: An examination of terminal Pleistocene climatic and adaptive changes. *Journal of Archaeological Science* 8:33–58.

Keeley, L. H. 1995. Protoagricultural practices among hunter-gatherers: A cross-cultural survey, 243–72. In *Last Hunters, First Farms: New Perspectives on the Prehistoric Transition to Agriculture*, ed. T. D. Price and A. B. Gebauer. Santa Fe: School of American Research Press.

Klein, R. G. 1989. Biological and behavioural perspectives on modern human origins in southern Africa, 529–46. In *The Human Revolution*, ed. P. A. Mellars and C. B. Stringer. Edinburgh: Edinburgh University Press.

Klein, R. G., and K. Cruz-Uribe. 1983. Stone Age population numbers and average tortoise size at Bynes Kranskop Cave 1 and Die Kelders Cave 1, Southern Cape Province, South Africa. *South African Archaeological Bulletin* 38:26–30.

Klein, R. G., and K. Scott. 1986. Re-analysis of faunal assemblages from the Haua Fteah and other Late Quaternary archaeological sites in Cyrenaican Libya. *Journal of Archaeological Science* 13:515–42.

Kuhn, S. L. 1995. *Mousterian Lithic Technology: An Ecological Perspective*. Princeton: Princeton University Press.

Kuhn, S. L., and M. C. Stiner. 1998. Middle Paleolithic "creativity": Reflections on an oxymoron? 143–64. In *Creativity and Human Evolution in Prehistory*, ed. S. Mithen. London: Routledge. In press.

Leonardi, P. 1935. *I Balzi Rossi: le faune, 1. I molluschi pleistocenici della Barma Grande*. Firenze: Istituto Italiano di Paleontologia Umana, Anno XIII.

Palma di Cesnola, A. 1965. Notizie preliminari sulla terza campagna di scavi nella Grotta del Cavallo (Lecce). *Rivista di Scienze Preistoriche* 25:3–87.

Pope, C. H. 1956. *The Reptile World: A Natural History of the Snakes, Lizards, Turtles, and Crocodilians*. New York: Alfred A. Knopf.

Rabinovich, R. n.d. Patterns of Animal Exploitation and Subsistence in Israel during the Upper Paleolithic and Epipaleolithic (40,000–12,500 BP), as Studied from Selected Case Studies. Ph. D. diss. (in preparation). Hebrew University. Jerusalem.

Schiegl, S., S. Lev-Yadun, O. Bar-Yosef, A. El Boresy, and S. Weiner 1994. Siliceous aggregates from prehistoric wood ash: A major component of sediments in Kebara and Hayonim Caves (Israel). *Israel Journal of Earth Sciences* 43:267–78.

Stiner, M. C. 1991. A taphonomic perspective on the origins of the faunal remains of Grotta Guattari (Latium, Italy). *Current Anthropology* 32:103–17.

Stiner, M. C. 1992. Overlapping species "choice" by Italian Upper Pleistocene predators. *Current Anthropology* 33:433–51.

Stiner, M. C. 1993. Small animal exploitation and its relation to hunting, scavenging, and gathering in the Italian Mousterian, 101–19. In *Hunting and Animal Exploitation in the Later Palaeolithic and Mesolithic of Eurasia*, ed. G. L. Peterkin, H. Bricker, and P. Mellars. Washington, D.C.: American Anthropological Association.

Stiner, M. C. 1994. *Honor Among Thieves: A Zooarchaeological Study of Neandertal Ecology*. Princeton: Princeton University Press.

Stiner, M. C., and S. L. Kuhn 1992. Subsistence, technology, and adaptive variation in Middle Paleolithic Italy. *American Anthropologist* 94:12–46.

Stiner, M. C., N. Munro, and T. A. Surovell. n.d. The tortoise and the hare: Small game, Paleolithic human demography. In preparation.

Stiner, M. C., S. Weiner, O. Bar-Yosef, and S. L. Kuhn 1995. Differential burning, fragmentation, and preservation of archaeological bone. *Journal of Archaeological Science* 22:223–37.

Straus, L. G. 1982. Carnivores and cave sites in Cantabrian Spain. *Journal of Anthropological Research* 38:75–96.

Stuart, A. J. 1982. *Pleistocene Vertebrates in the British Isles*. London: Longman.

Stuart, A. J. 1991. Mammalian extinctions in the Late Pleistocene of northern Eurasia and North America. *Biological Review* 66:453–562.

Stubbs, D. 1989. *Testudo graeca* spur-thighed tortoise, 31–3. In *The Conservation Biology of Tortoises*, ed. I. R. Wsingland and M. W. Klemens, Occasional Papers of the IUCN Species Survival Commission (SSC), no. 5.

Tchernov, E. 1968. *Succession of Rodent Faunas during the Upper Pleistocene of Israel*. Hamburg and Berlin: Parey.

Tchernov, E. 1981. The biostratigraphy of the Middle East, 67–97. In *Préhistoire du Levant*, ed. J. Cauvin and P. Sanlaville. Paris: CNRS Editions.

Tchernov, E. 1988. Biochronology of the Middle Paleolithic and dispersal events of hominids in the Levant, 153–68. In *L'Homme de Neandertal*, No. 2, ed. M. Otte. Liège: ERAUL.

Tchernov, E. 1989. The Middle Paleolithic mammalian sequence and its bearing on the origin of Homo sapiens in the southern Levant, 25–38. In *Investigations in South Levantine Prehistory (Prehistoire du Sud-Levant)*, ed. O. Bar Yosef and B. Vandermeersch. Oxford: British Archaeological Reports International Series 497.

Tchernov, E. 1992. Biochronology, paleoecology, and dispersal events of hominids in the southern Levant, 149–88. In *The Evolution and Dispersal of Modern Humans in Asia*, ed. T. Akazawa, K. Aoki and T. Kimura. Tokyo: Hokusen-Sha.

Tchernov, E. 1994. New comments on the biostratigraphy of the Middle and Upper Pleistocene of the southern Levant, 333–50. In *Late Quaternary Chronology and Paleoclimates of the Eastern Mediterranean*, ed. O. Bar-Yosef and R. S. Kra. Tucson and Cambridge: Radiocarbon and the Peabody Museum of Archaeology and Ethnography, Harvard University.

Uerpmann, H. P. 1981. The major faunal areas of the Middle East during the late Pleistocene and early Holocene, 99–106. In *Préhistoire du Levant*, ed. J. Cauvin and P. Sanlaville. Paris: CNRS Editions.

Valla, F. R., H. Plisson, and R. Buxo i Capdevila. 1989. Notes preliminaries sur les fouilles en cours sur la terrasse d'Hayonim. *Paléorient* 15:245–58.

Vitagliano, S. 1984. Nota sul Pontiniano della Grotta dei Moscerini, Gaeta (Latina), 155–64. In *Atti della XXIV Riunione Scientifica dell'Istituto Italiano di Preistoria e Protostoria nel Lazio*, October 8–11, 1982.

Weiner, S., and O. Bar-Yosef. 1990. States of preservation of bones from prehistoric sites in the Near East: A survey. *Journal of Archaeological Science* 17:187–96.

Weiner, S., and P. Goldberg. 1990. On-site Fourier Transform Infrared Spectrometry at an archaeological excavation. *Spectroscopy* 5:46–50.

Weiner, S., P. Goldberg, and O. Bar-Yosef. 1993. Bone preservation in Kebara Cave, Israel, using on-site Fourier transform infrared spectrometry. *Journal of Archaeological Science* 20:613–27.

16

NEANDERTAL AND EARLY MODERN HUMAN MOBILITY PATTERNS

Comparing Archaeological and Anatomical Evidence

Daniel E. Lieberman

Department of Anthropology
The George Washington University
Washington, DC 20052

1. INTRODUCTION

The well-known general similarities in the archaeological record associated with Neandertals and early anatomically modern humans in the Levant challenge many long-held theories about human origins. Levantine Mousterian assemblages contain broadly similar lithic industries, hearths, prey species, and burial practices, regardless of whether they are in context with Neandertals or early modern humans. Although there may be some indications for possible behavioral differences between Neandertals and early modern humans (see below), most of their apparent behaviors are unquestionably more similar to each other than either are to those of Upper Paleolithic humans. How do we interpret such similarities? To some paleoanthropologists these similarities suggest that these hominids were behaviorally identical, and perhaps belonged to the same species (e.g., Clark and Lindly 1989; Wolpoff 1989). Others, however, use fossil and archaeological evidence to argue that Neandertal and early modern human behaviors may have differed in some respects (e.g., Shea 1991; Trinkaus 1992; Lieberman and Shea 1993; Klein 1995; Trinkaus et al., this volume). Regardless of which hypothesis is correct, we should expect a broad degree of behavioral similarity between the two taxa. They are both closely-related, large-brained, intelligent hunter-gatherers that evolved during the Middle Paleolithic times and lived at various times in the Levant, sometimes even in the same sites (although probably not concurrently). It is, therefore, unlikely that they would have made very different kinds of tools, used them differently, hunted different species, or even occupied different sites. Most of their similarities, especially those aspects of behavior that relate to subsistence, are probably irrelevant to resolving their taxonomy, but these similarities may have other important implications.

Logically, the only way to test for similarity is to look for difference. The archaeological and fossil records both provide limited data with which to investigate potential

Neandertals and Modern Humans in Western Asia, edited by Akazawa *et al.*
Plenum Press, New York, 1998.

behavioral contrasts between Neandertal and modern human hunter-gatherers in the Levant. Important behaviors that may or may not distinguish these taxa include their cognitive and linguistic capacities, their use of tools, burials and fire, their hunting strategies and ecological niches, and their functional morphological adaptations (summarized in Klein 1995). One dimension that stands out among these possible differences is the energetic workload associated with mobility. Mobility is crucial because, with few exceptions, hunter-gatherers expend a large proportion of their energy and time on their feet, traveling from location to location to acquire and transport resources. Mobility is, thus, directly related to strategies of energy use and resource acquisition (Krebs and Davies 1987).

This paper focuses on evidence for mobility to test the hypothesis that there probably were some important behavioral differences between Neandertals and modern humans. Mobility, however, is difficult to document in the archaeological and fossil records because it leaves few direct traces. We can indirectly test whether these taxa had different mobility patterns using two independent sources of data. First, comparisons of seasonal hunting patterns based on seasonally deposited bands in teeth provide some indication of seasonal mobility patterns. Second, comparisons of systemic cortical bone robusticity between Neandertals and modern humans preserve some indication of overall workload levels.

2. SEASONALITY DATA

Information on the seasonal movements of Levantine Mousterian hominids can potentially reveal much about their mobility strategies. Most hunter-gatherers migrate seasonally, often to different habitats (see Kelly 1992). Seasonal mobility is a sensible, perhaps optimal strategy in most regions because it allows hunter-gatherers to take advantage efficiently of the seasonal appearance of diverse resources in more than one ecological or geographic zone, and it avoids the problems of resource depletion that result from long-term habitation of a single site or region. Ethnographic studies demonstrate that recent hunter-gatherers, with few exceptions, move frequently, rarely occupying a site for more than one season (for examples, see Lieberman 1993a). Although it is almost certain that Levantine Neandertals and early anatomically modern humans were both seasonally mobile (i.e., not sedentary), they could have used the landscape in many possible ways. Therefore, one way to test hypotheses about variability in Mousterian mobility strategies is to examine the seasonal use of sites associated with each taxon in relation to biogeographic zones within discrete time periods.

Estimating the seasonal occupation of Levantine Mousterian sites is a challenge. Although a number of attempts to make inferences about their mobility strategies have been made from differences in the variability and density of lithic assemblages (Henry 1995), from regional variations in site size and distribution (Marks and Friedel 1977; Marks 1988), and from other archaeological traces (summarized in Lieberman 1998), there is little direct evidence for Neandertal and early modern human mobility patterns. Perhaps the most direct archaeological evidence for seasonal mobility comes from data on the seasonal hunting of animals.* If a given layer of a site contains prey that were hunted during just one season, then these data comprise a reasonable minimum estimate of its seasonal use. Of course, absence of prey from a particular season does not prove that the site was then unoccupied, so it is crucial that attempts to estimate seasonality incorporate as many

* Plant remains are an excellent source of seasonality data but these tend to be poorly preserved in most sites, were not recovered from older excavations, and currently remain an underused source of data.

sources of evidence as possible. Nevertheless, seasonal hunting data allow us to build a framework with which to test hypotheses about seasonal site occupation and mobility using other data.

The season of death of hunted animals can be estimated reliably from several kinds of data including juvenile age profiles derived from tooth eruption patterns (e.g., Davis 1983; Klein 1987; Stiner 1995), from the presence of seasonally migratory fauna or commensals, and from the analysis of seasonally-deposited cementum increments in animal teeth (Klevezal and Kleinenberg 1967; Lieberman 1993b, 1994). Unfortunately, most of this potential information is so far unavailable or unanalyzed from Levantine Mousterian sites. Migratory birds are rare, and there have been few attempts to estimate the season of death of mammals using tooth eruption or other techniques (although such estimates should be forthcoming). Cementum increment analysis currently provides one of the best sources of information on seasonal activities in the Levantine Mousterian. These estimates are accurate, reliable, and not directly influenced by human behavior, but it is important to stress they should be considered only provisional, minimum estimates of site occupation. As Rabinovich and Tchernov (1995) note, additional information will be necessary to provide a more complete basis for evaluating seasonal site use and mobility.

Before discussing the data derived from cementum increment analysis, it is worth reviewing its biological basis. Cementum increment analysis has long been available as a potential source of information on seasonality, but it has been used infrequently because of confusion about the data it provides. Cementum is a bone-like tissue that grows around tooth roots. It is deposited by specialized cells, cementoblasts, which derive from the gum (periodontal ligament) and mineralize around bundles of collagen (Sharpey's fibers) that extrude from the gum into the narrow space around the tooth. The tissue, therefore, functions to anchor each tooth to the jaw and skull by connecting the root to the Sharpey's fibers of the gum. In almost all mammals, cementum accrues slowly and continuously after tooth eruption and is rarely remodeled or resorbed. If one examines a thin-section of a tooth root under transmitted polarized light, the cementum appears in alternating bands that have different optical properties and correlate well with patterns of seasonal growth and diet. In general, translucent bands are less mineralized and tend to deposit during seasons of more rapid growth, whereas opaque bands are more densely mineralized and deposit during seasons of slower growth (Lieberman 1993b, 1994). The optical contrasts between cementum increments can also result from variations in collagen organization. The orientation of the Sharpey's fibers (as well as intrinsic collagen) differs between bands to a large extent because the magnitude and frequency of masticatory forces necessary to process food vary from season to season, altering the alignment of seasonally deposited fibers (Lieberman 1993b, 1994).

The most useful animal for cementum increment analysis in the Levantine Mousterian archaeological record is the mountain gazelle, *Gazella gazella*. Gazelle breed rapidly, are easy to hunt, are not migratory (Baharav 1983), and are thus the dominant mammal species found in Upper Pleistocene sites. Gazelle also have distinct seasonal diets that generate clear seasonal cementum bands. During the dry summer, gazelle tend to be primarily browsers, eating mostly Zizyphus lotus (jujube) and other bushes; during the wet and cool winter, gazelle switch to being grazers, with forbs and grasses comprising over 95% of their diet (Baharav 1981). The differences between their summer and winter diets generate histological contrasts in terms of mineral density and collagen organization that result in clear cementum bands. Studies of modern gazelle of known date of death demonstrate that, at a standard region (the lingual surface of M1 near the crown), the rate of growth of the translucent bands between March and September is 4.6 microns/month

(r^2=0.75), whereas the rate of growth of the opaque bands between October and February is 2.7 microns/month (r^2=0.87) (Lieberman 1993c, 1994). The predictable growth rate of these bands allows fairly precise estimates of season of death for recent animals, and presumably for gazelle from archaeological contexts.

Cementum increment analyses of gazelle teeth from archaeological layers associated with anatomically modern humans and Neandertals, therefore, provide preliminary information with which to begin to test whether seasonal hunting and mobility strategies differed between these taxa. Accordingly, gazelle mandibles and maxillae from sites associated with archaic and modern humans (Table 1) were prepared and analyzed to estimate their season of death using techniques described in Lieberman and Meadow (1992) and Lieberman (1994). As many samples as possible were analyzed: for Qafzeh every available sample was used; only a small faunal sample remains for Tabun C and D and Skhul; at Kebara, 20 samples were selected from layer E of Stekelis' excavations (Schick and Stekelis 1977) and 10 samples from a single décapage stratum (Unit X) of the recent excavations. The latter probably represents an occupational layer of limited duration (Bar-Yosef et al. 1992), and thus is especially important for potentially controlling for the effects of time-averaging on the seasonality data. To ensure as much objectivity as possible, all samples were studied in random order so their provenience was unknown at the time of analysis; computer image analysis was also used to ascertain and measure the nature of the outermost cementum increment (Lieberman et al. 1990). Only teeth still encased in alveolar bone were used in order to avoid potentially false readings from exfoliated or otherwise missing outer increments. A few samples from other taxa (*Bos*, *Cervus*, *Dama*, and *Capra*) were also studied (see Lieberman 1993c). Table 1 includes sample sizes, the mean, mode, range, and coefficient of variation of their estimated seasons of death, and an index of seasonality (Is) that summarizes the seasonal diversity of kills. Is (calculated as the ratio of fall–winter kills minus spring–summer kills divided by the total number of kills) is thus −1.0 for sites in which animals were killed solely in the fall–winter; 1.0 for sites in which animals were killed solely in the spring–summer, and 0.0 for sites in which animal deaths are distributed equally throughout the year. Techniques for preparing and analyzing the samples are described elsewhere (Lieberman et al. 1990; Lieberman and Meadow 1992; Lieberman 1994).

The results in Table 1 must be considered preliminary, minimal estimates for seasonal mobility because of the small samples sizes, the lack of stratigraphic control of the samples,[†] and the paucity of sites with associations between fauna and hominids—although an ANOVA indicates that sample size is independent of the estimates of seasonal site occupation (Lieberman, 1998). Note that there appear to be consistent differences in

Table 1. Estimates of seasonal gazelle hunting at Levantine Mousterian sites

Site	Age (ka)	Phase	N	Mean	Range	CV	Mode	Is	**Estimate**
Kebara E	60	Tabun B	17(20)	4.0	1-7	0.45	3/5	-0.06	Multiseasonal
Kebara X Decapage	60	Tabun B	9(10)	4.1	1-7	0.46	5	-0.11	Multiseasonal
Kebara (combined)	60	Tabun B	26(30)	4.0	1-7	0.45	5	-0.08	Multiseasonal
Tabun B	80-100	Tabun B	8(12)	3.9	2-6	0.41	7	0.00	Multiseasonal
Qafzeh XVI-XXVI	100	Tabun C	14(15)	1.4	1-6	0.28	1	-0.86	Winter/Spring
Tabun C	100-120	Tabun C	8(10)	2.3	2-3	0.22	2	1.00	Fall/Winter
Skhul	100-150	Tabun C	7(8)	1.6	1-5	0.93	1	0.71	Fall

† Note that with the exception of the Kebara X samples, most of the samples come from stratigraphic units of considerable and highly variable thickness.

the probable season of death of gazelle hunted at sites associated with Tabun B and Tabun C phases of the Levantine Mousterian. In particular, sites with Tabun C industries have a distinctly single season pattern of gazelle hunting (the same appears to be true for other species) in spite of the lack of stratigraphic control of the samples from Tabun and Skhul and the deep sequence from Qafzeh. These data raise the possibility that the early modern human hunter-gatherers lived in these sites only during certain seasons. Since gazelle were hunted year round in the Levant (see below), hunter-gatherers may have occupied these sites repeatedly during the same season, like modern hunter-gatherers in other Mediterranean, coastal environments (e.g., Kroeber 1925). A likely scenario is a pattern of seasonal transhumance—what Binford (1980) calls logistical foraging—from the inner highlands in the spring and summer to the lowland, coastal region during the fall and winter, a strategy that would have maximized resource diversity (Lieberman 1993a).

Seasonality data for gazelle hunting from the Tabun B phase of the Levantine Mousterian are more problematic to interpret, but appear to present a different picture of seasonal transhumance. The gazelle analyzed from Kebara and Tabun B were hunted during every season of the year in approximately equal percentages, as indicated by the seasonality index (Is) values close to 0. In addition, coefficients of variation (CV) for the estimates of seasonal death are significantly larger ($p < 0.01$) in Tabun B than Tabun C sites. This multi-seasonal pattern does not necessarily mean that these sites were permanently occupied—Middle Paleolithic sedentism may have been unlikely—but it does mean that hunter-gatherers returned to these caves after hunting gazelle and other species during both the wet and dry seasons. This pattern contrasts not only with the Tabun C phase of the Levantine Mousterian, but also with the rest of the Middle Paleolithic, the entire Upper Paleolithic, and the Epipaleolithic until the Natufian period (Lieberman 1993a). Without more data from these and other sites it is impossible to document the specific nature of the resource exploitation strategies during the Tabun B phase, but one possibility is that these sites represent a more regionally-focused system of transhumance in which there was less migration from habitat to habitat on a seasonal basis. Such a mobility strategy would fall into Binford's (1980) category of foraging. In addition, it is likely but not certain that all these Tabun B assemblages were created by Neandertals. There are clear associations between Neandertals and Tabun B levels at Kebara and Amud (whose fauna are not yet available for analysis); the Tabun I Neandertal female was found at the transition between levels B and C (Garrod and Bate 1937) and is possibly intrusive from level B, but her provenience needs to be verified through direct dating.

The above seasonality estimates tentatively suggest that Neandertal and early modern human mobility strategies differed, but are insufficient to test the hypothesis conclusively. They provide, however, an important heuristic message: while there are probably archaeologically observable behavioral contrasts between these taxa, any such differences are likely to be ones of process (how people did things) rather than kind (what people did), making them difficult to document. Unless we look for such differences, we are unlikely to find them. It is, therefore, significant that other sources of data provide some corroboration for the hypothesis that Tabun B (Neandertal) and Tabun C (early modern human) mobility patterns differed (Lieberman and Shea 1994; see, however, Rabinovitch and Tchernov 1995). The density of both the fauna and lithics in recently excavated Tabun B sites appears to be substantially greater than in Tabun C sites, suggesting either more occupational intensity or longer-term occupation (Bar-Yosef, this volume). In addition, Meignen and Bar Yosef (1988) and Shea (1991) have shown that Tabun B sites associated with Neandertals have an approximately five-times greater percentage of points than Tabun C sites associated with modern humans. Impact scars on the points indicate that these points were primarily hunting weapons, suggesting that Neandertals may have hunted more (Shea 1989). Intensive hunting is a predicted consequence of less mobile resource acquisition strategies in which hunter-gatherers exploit a given region for a longer period of time (Lieberman and Shea 1994). In other words, Neandertals may have occupied sites or regions more intensively and for multiple seasons, whereas early modern humans may have been more seasonally mobile. To test this hypothesis properly, however, we need more direct data on seasonality from other faunal and plant evidence.

3. SYSTEMIC BONE ROBUSTICITY

If Neandertal and modern human mobility strategies differed somewhat, as the above data tentatively suggest, were these part of radically contrasting behavioral patterns that differed in terms of energetic efficiency or workload levels? The best data with which to test this are the bones of the hominids themselves. As one might expect from the many other similarities in Neandertal and modern human hunter-gatherer adaptations, these data indicate that their overall workload levels were probably comparable.

Bone records information about life history variables and energetics because it is a dynamic tissue with numerous functions that responds to a variety of genetic and non-genetic stimuli. Although bones protect vital organs, store calcium, and produce red blood cells, their primary function is to provide stiffness and strength in response to mechanical loading (Currey 1984). Activities generate force, which elicits strain (deformation per unit length). Strains are important stimuli for bone growth because, if sufficient in magnitude and frequency, they cause the accumulation of microcracks and fractures that can eventually lead to mechanical failure (Martin and Burr 1989). Bones, thus, respond to loading through a variety of mechanisms. On a local level, bones often respond to force by growing (modeling) to become thicker. Modeling counteracts load-induced strain by distributing more mass in its planes of deformation so that a given force generates less strain. As Trinkaus et al. (this volume) have demonstrated, the thick lower limbs of Levantine Neandertals and early modern humans indicate that local modeling responses to strain were high, suggesting that both taxa engendered higher levels of mechanical loading from walking or running in comparison with recent modern humans.

An additional source of information with which to test hypotheses about workload levels in archaic and modern fossil humans is the degree of overall bone thickness (robusticity) in regions such as the cranial vault that are not directly strained by walking or running. Exercise appears to have significant systemic effects on bone growth, probably through the actions of growth hormone (GH) (Lieberman 1996). GH, which is released in pulsatile bursts by the anterior pituitary every three to four hours, has numerous roles including regulating systemic bone growth. In most cases, GH activates bone modeling, causing individuals to have thicker, and sometimes, longer bones (Brixen et al. 1990). GH influences bone deposition by stimulating the synthesis of local growth factors, especially insulin-like growth factors (IGF-I and II), that mediate many of its effects on bone-producing cells (osteoblasts) and other skeletal cell types (Daughaday 1989). High GH levels thus induce bone modeling throughout the skeleton, including the cranial vault (Vogl et al. 1993). Elevated GH levels in children cause acromegaly, resulting in extreme cortical thickening throughout the skeleton including the cranial vault (Randall 1989). In contrast, GH deficiencies in children lead to abnormally thin skulls and postcrania (Pirinen et al. 1994).

The exact mechanisms for systemic bone responses to exercise are not entirely understood, but most likely occur through the mediating effects of GH. Because GH also stimulates glycolysis, exercise rapidly elevates GH levels significantly in proportion to workload intensity, so that individuals who exercise regularly have higher circulating GH levels than more sedentary individuals (Borer 1980; Naveri 1985; Felsing et al. 1992). GH levels decline rapidly after exercise, but the down-the-line effects of GH on IGF-I appear to have long-term effects on systemic osteogenesis. Yeh et al. (1994), for example, showed IGF-I levels and periosteal bone growth rates to be significantly higher in rats injected daily with GH and in rats who exercised daily compared with controls, but that IGF-I and bone formation rates increased even more dramatically in rats who both exercised and had GH injections. The systemic effects of exercise-induced GH surges, however, are expected to decline with age as GH responses to stimuli such as exercise become less active (Bala et al. 1981; Termine 1990; Benedict et al. 1994).

Evidence for the effects of exercise on systemic bone growth, including the cranial vault, was recently tested in an experiment on swine (for details, see Lieberman 1996). Six same-sex inbred sibling pigs (*Sus scrofa*) were used, of which three ran on a treadmill for 60 minutes a day at 4.8 Kph from the age of one to four months. All pairs grew under otherwise identical conditions and did not differ significantly at any point in the experiment in terms of weight or cranio-dental dimensions. The results, summarized in Table 2, show that significantly more cortical bone growth occurred in the exercised than control animals throughout the skeleton, indicating a systemic response to the exercise. In many regions of the skeleton, such as the tibia, the faster growth of the exercised animals is probably attributable to higher magnitudes and frequencies of strain from the treadmill running. High shear strains recorded in vivo with strain gauges (1,647 ± 391.9 με; n = 45) on the tibia, for example, are sufficient to induce local growth responses (Frost 1986; Martin and Burr 1989). The low shear strains recorded in the cranial vault (155 ± 39.7 με; n = 47), however, are almost certainly too minor to induce any such local bone growth (Frost 1986), but instead may be

Table 2. Systemic robusticity in exercised versus control swine (see Lieberman 1996 for details)

	Controls (n=3)	s.d.	Runners (n=3)	s.d.	P*	% difference
Cranial vault thickness						
Bregma (mm)	6.6	0.8	8.8	0.4	0.05	33.3
Parietal eminence (mm)	6.2	1.2	7.5	0.5	0.19	21.0
Supraoccipital (mm)	8.1	0.9	10.9	0.5	0.05	34.6
Caudal vertebrae						
C1 medio-lateral (mm)	24.5	2.7	31.2	1.5	0.05	27.1
C1 dorso-ventral (mm)	10.5	0.5	11.8	1	0.13	12.7
C4 medio-lateral (mm)	14.4	0.4	16.2	0.9	0.05	12.3
C4 dorso-ventral (mm)	6.4	0.2	7.2	0.2	0.05	11.5
Tibia midshaft						
Polar moment (mm4)	2350.4	180.6	3653.3	408.0	0.05	55.4
Cortical area (mm2)	96.8	3.1	119.5	0.4	0.05	23.4
Dental dimensions						
M1 bucco-lingual (mm)	9.9	0.1	9.8	0.1	0.51	1.0
M1 mesio-distal (mm)	13.5	0.1	13.6	0.1	0.51	0.2
Body weights						
Experiment start (kg)	4.1	0.2	4.1	0.5	0.51	1.5
Experiment middle (kg)	10.8	0.9	10.9	0.9	0.99	0.8
Experiment end (kg)	28	1	26.5	2.3	0.51	5.6

* Mann-Whitney U-test

manifestations of a systemic, hormonal response to exercise. Similarly low strains have also been reported in the cranial vault of dogs and even humans (see Lieberman 1996).

Data on cranial vault thickness in archaic and early modern humans from the Levant and elsewhere may therefore provide some information on exercise-induced systemic growth. The cranial vault in humans, as in other mammals, is known to experience extremely low levels of strain from walking, running, or chewing (Demes 1985; R. Hillam, personal communication). In addition, the upper half of the cranial vault is comprised almost exclusively of depository growth fields on both its internal (endocranial) and external (ectocranial) surfaces. For this reason, the upper half of the cranial vault can only grow thicker. Finally, the above experiments on swine demonstrate that the thickness of the cranial vault has a low degree of genetic heritability, and therefore is not subject to strong natural selection, but instead appears to reflect overall systemic robusticity that occurs in response to exercise.

Figure 1 compares cranial vault thickness for Neandertals and early anatomically modern humans from the Levant and elsewhere. The measurements (from Nawrocki 1991; Lieberman 1996) come from two locations in the cranial vault: bregma, the intersection of the frontal and parietal bones; and the parietal eminences (left and right sides averaged where possible). Mean values for vault thickness at these two locations are also included for a sample of Holocene modern human populations. There is huge range of variability in all taxa, but an ANOVA demonstrates no significant differences in vault thickness between Neandertal and Pleistocene modern humans, both within the Levant and elsewhere. A thick cranial vault is clearly not a derived character of modern humans. Modern Holocene populations, however, have significantly thinner vaults than Pleistocene humans at bregma and the parietal eminences ($p < 0.02$). This decline in robusticity is probably related to changes in subsistence economy since all but one of these populations are either farmers or urban-dwelling populations. Note that the recent decreases in vault thickness are unlikely to result from any deterioration in health or diet since recent, post-industrial era populations have even thinner vaults ($p < 0.05$) than agricultural populations (Lieberman 1996).

If vault thickness is a measure of overall exercise levels, particularly during childhood, than these data suggest that workload levels were probably comparable for early modern and archaic human hunter-gatherers in the Levant. Thus, while Neandertals and modern humans may have used somewhat different mobility strategies, the energetic costs of their mobility patterns appear to have been roughly equivalent. These data agree with the analyses of Trinkaus et al. (this volume) which demonstrate that, when corrected for size and body proportion, the lower limb bones of Levantine Neandertals and early modern humans were similar in strength.

4. DISCUSSION

There is probably no simple distinction that we can draw between Neandertal and early modern human adaptations in the Levantine Mousterian in terms of mobility, resource acquisition strategies, and perhaps overall workload levels. While site seasonality data indicate that there may have been some differences in their mobility strategies, it appears that Neandertals and modern humans had roughly equivalent workload levels. Even if the latter practiced logistical collecting and the former practiced some form of circulating foraging (if, indeed, such terms can be applied to the Levantine Mousterian), such a conclusion makes sense since both strategies require high levels of mobility. In fact, the preponderance of the archaeological evidence suggests that these taxa were more similar to each other behaviorally than either are to more recent modern humans. Such a degree of

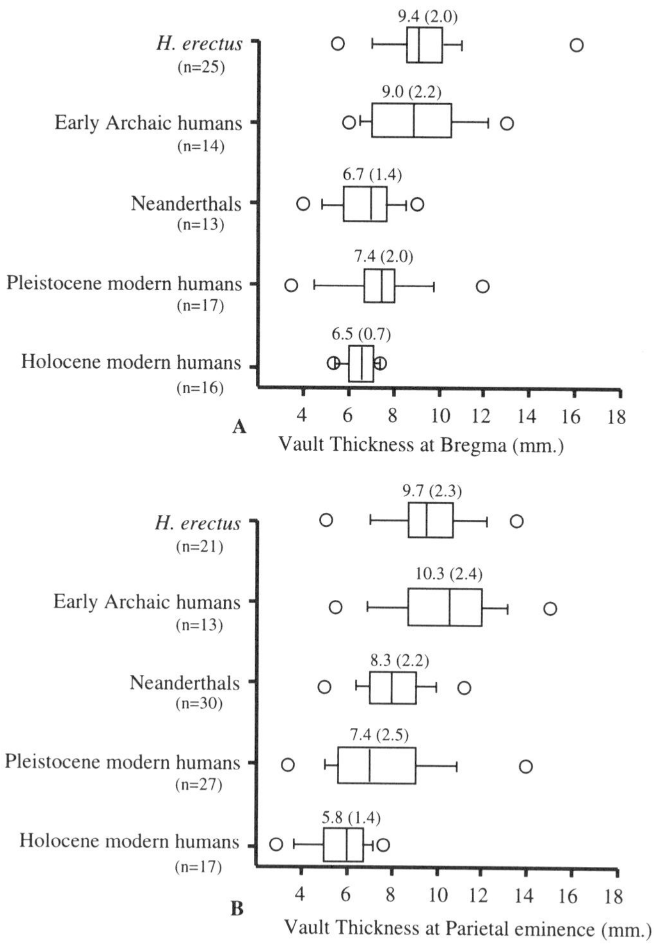

Figure 1. Comparison of vault thickness at bregma and parietal eminences for archaic and modern human taxa, including Levantine fossils. For details on measurements and on the fossils included, see Lieberman (1996).

similarity is exactly what we should expect between two closely-related hominid taxa living in the same geographical region (albeit at different times). Neandertals and early modern humans were both large-brained, intelligent hunter-gatherers who evolved during the Middle Paleolithic/Middle Stone Age. The similarities among the Tabun B, C, and D lithic variants may be a consequence of the fact that they used the Levallois technique to make blanks but used little retouch to modify them (Meignen and Bar-Yosef 1988). Other archaeological similarities are even more predictable: there are limited numbers of Levantine animals and plants to hunt and gather; a limited number of ways to use stone tools; a limited number of caves to occupy; and a limited number of ways to build a simple hearth. So far, archaeological traces of more complex modern behaviors typical of the Upper Paleolithic such as the production of art are rare for both taxa. One must conclude that both Levantine Neandertals and early anatomically modern humans lacked these dimensions of "modern" behavior either because these behaviors had not yet been invented culturally, and/or because they lacked the cognitive capacity (Klein 1995). Such hypotheses are currently impossible to test, and may be based on a spurious notion of "modernity."

The above data, however, are insufficient to argue that Neandertals and early anatomically modern humans lived identically, because any major behavioral differences between these taxa are likely to be ones of process rather than kind. There are numerous morphological differences between these two taxa which support this hypothesis. As Trinkaus (1992), Churchill (1996), Ruff et al. (1994), and others have shown, early modern humans relied less on upper body strength, used their anterior teeth less, and had far fewer incidences of skeletal trauma to their bodies. Archaeological traces of any possible differences in how Neandertals and modern humans acquired resources are likely to be harder to document, but a few are also apparent from the Levantine archaeological record. Of the hominids with undisputed archaeological provenience, Neandertals (Amud, Kebara) have so far been found only in association with Tabun B industries, whereas early anatomically modern humans (Tabun II, Skhul, Qafzeh) are found with Tabun C industries (Tabun I, a female Neandertal, was found at the border between Tabun B and C). Despite their general similarities, Tabun B and Tabun C industries and some of their archaeological associations are not identical. Dorsal scar patterns on flakes indicate that the Tabun B industry was produced using a primarily unipolar convergent core reduction strategy, and that the Tabun C industry was produced using a mostly centripetal core reduction strategy (Bar-Yosef and Meignen 1992). In addition, Neandertals and early modern humans may not have had the same patterns of seasonal mobility, and perhaps relied on hunting to different degrees (Lieberman and Shea 1994). It is important to note that most of these hypotheses are potentially testable with other types of data. Stable isotope analyses have the potential to provide direct evidence of the relative component of meat, grasses, and other foods in their diets (e.g., Ambrose, this volume). In addition, detailed taphonomic analyses from Levantine Mousterian sites will not only provide more information on site seasonality, but will also yield data on hunting strategies, carcass processing techniques, and other aspects of their diet (e.g., Stiner 1995, this volume; Speth and Tchernov, this volume).

One final, related issue is the question of taxonomy. If Neandertals and early anatomically modern humans were behaviorally more similar to each other than either are to recent, modern humans, should we place them in the same taxon as some researchers argue? The answer is clearly no. Although behavior can influence phylogeny by creating reproductive barriers, taxonomic units (including species) are most appropriately defined on the basis of genetically heritable characteristics that reflect ancestry and descent (Simpson 1961). Since genetic data are with one exception unavailable for Neandertals and early anatomically modern humans, morphological characteristics, not archaeological residues, are the next most reliable source of information to assess their systematic relationships. Neandertals and early modern humans most likely belong to separate taxa because each have a set of unique derived characters (autapomorphies) that indicate separate evolutionary histories. Undisputed Neandertal autapomorphies include a suprainiac fossa, a retromolar space, and an expanded midface in which the sagittally-oriented infraorbital margin arises above the M2, forming a transversely and sagittally convex midface (Hublin 1978; Rak 1986; Trinkaus 1987). Anatomically modern human autapomorphies include a short skull base, an orthognathic face tucked underneath the anterior cranial fossa, and a true chin caused by resorption of the superior alveolar margin of the mental symphysis (Lieberman 1995). Logically, these characters comprise sufficient evidence to conclude that Neandertals and early modern humans must belong to separate taxa, although not necessarily to distinct species.

There are additional reasons why we cannot define hominid taxa, especially species, on the basis of their archaeological traces. Complex, intelligent hominids with a high degree of cultural capacity learn how to make fires, retouch stone tools, hunt, and other such tasks.

Although learned traits sometimes correlate well with inherited traits (because parents and kin tend to teach them to us), they are not, strictly speaking, genetically inherited characters and, therefore, cannot provide reliable systematic information. Archaeological modernity (traditionally defined, perhaps mistakenly, as the Upper Paleolithic) is clearly unrelated to anatomical modernity.[‡] Consequently, the archaeological similarities between Neandertals and modern humans do not provide sufficient evidence to combine them in one taxon.

From an evolutionary perspective, behavioral similarities between Neandertals and modern humans may be more important than differences. If they had similar prey species, hunting technology, and other such behaviors, and yet were sufficiently distinct on a taxonomic level to suggest possible behavioral or genetic barriers to interbreeding, than their overlapping niches would have made them direct competitors were they ever sympatric. The archaeological evidence to test this hypothesis is still inadequate, but provides some support. In western Europe, the appearance of modern humans may correlate with the rapid Neandertal extinction that occurred by 30,000 years ago, with late Neandertal populations surviving only in the peripheries (Hublin et al. 1996). The Levantine situation may be even more complicated, however, because Neandertals may have actually replaced or out-competed early anatomically modern humans at about 80,000 years ago, only to be later re-replaced by modern humans (see Bar-Yosef, this volume). Paleoanthropologists will undoubtedly continue to search for behavioral contrasts between these taxa with which to explain such evolutionary scenarios.

ACKNOWLEDGMENTS

I am deeply grateful to Drs. Akazawa, Aoki, and Bar-Yosef for the invitation to contribute to this volume and in the stimulating symposium from which it arose. I thank O. Bar-Yosef, S. Churchill, E. Hovers, and J. Shea for their comments, as well as R. Meadow, D. Pilbeam, J. Shea, E. Tchernov, B. Vandermeersch, the Israel Antiquities Authority, and the Natural History Museum for their support and/or permission to analyze specimens in their care. This research was supported by grants from the Wenner-Gren Foundation, the L. S. B. Leakey Foundation, the National Science Foundation, and the Irene Levi-Sala CARE Foundation.

REFERENCES

Baharav, D. 1981. Food habits of the mountain gazelle in semi-arid habitats of eastern Lower Galilee, Israel. *Journal of Arid Environments* 4:63–9.

Baharav, D. 1983. Observation on the ecology of the mountain gazelle in the Upper Galilee, Israel. *Mammalia* 47:59–69.

Bala, R. M., J. Lopatka, A. Leung, E. McCoy, and R. G. McArthur. 1981. Serum immunoreactive somatomedi levels in normal adults, pregnant women, children at various ages, and children with constitutionally delayed growth. *Journal of Clinical Endocrinology and Metabolism* 52:508–12.

Bar-Yosef, O. 1992. The role of western Asia in modern human origins. *Philosophical Transactions of the Royal Society of London* B 337:193–200.

Bar-Yosef, O., and L. Meignen. 1992. Insights into Levantine Middle Palaeolithic cutural variability, 163–82. In *The Middle Palaeolithic: Adaptation, Behavior and Variability*, ed. H. L. Dibble and P. Mellars. Philadelphia: The University Museum of the University of Pennsylvania.

---

‡ Such culturally-based taxonomic criteria would incorrectly assign Australian aborigines to a separate taxon from other modern humans on the basis of their lithic assemblages.

Bar-Yosef, O., B. Vandermeersch, B. Arensburg, A. Belfer-Cohen, P. Goldberg, H. Laville, L. Meignen, Y. Rak, J. D. Speth, E. Tchernov, A. M. Tillier, and S. Weiner. 1992. The excavations at Kebara Cave, Mt. Carmel. *Current Anthropology* 33:497–550.

Benedict, M. R., S. Adiyaman, D. C. Ayers, F. D. Thomas, J. D. Calore, V. Dhar, and R. A. Richman. 1994. Dissociation of bone mineral density from age-related decreases in insulin-like growth factor-I and its binding proteins in the male rat. *Journal of Gerontology* 49:B224–30.

Binford, L.R. 1980. Willow smoke and dogs' tails: Hunter-gatherer settlement systems and archaeological site formation. *American Antiquity* 45:4–20.

Borer, K. T. 1980. Characteristics of growth-inducing exercise. *Physiology Behavior* 24:713–20.

Brixen, K., M. Kassem, E. F. Eriksen, H. K. Nielsen, A. Flyvbjerg, and L. Mosekilde. 1993. Growth hormone GH and adult bone remodeling: The potential use of GH in treatment of osteoporosis. *Journal of Pediatric Endocrinology* 6:65–71.

Churchill, S. E. 1996. Particulate versus integrated evolution of the upper body in Late Pleistocene humans: A test of two models. *American Journal of Physical Anthropology* 100:559–83.

Clark, G., and J. Lindly. 1989. Modern human origins in the Levant and western Asia: The fossil and archaeological evidence. *American Anthropologist* 91:962–85.

Currey, J. 1984. *The Mechanical Adaptations of Bones.* Princeton: Princeton University Press.

Daughaday, W. H. 1989. Growth hormone: Normal synthesis, secretion, control, and mechanisms of action, 318–29. In *Endocrinology*, Vol. 1, ed. L. J. DeGroot. Phildelphia: W. B. Saunders.

Davis, S. J. M. 1983. The age profiles of gazelles predated by ancient man in Israel: Possible evidence for a shift from seasonality to sedentism in the Natufian. *Paléorient* 9:55–62.

Demes, B. 1985. Biomechanics of the primate skull base. *Advances in Anatomy, Embryology and Cell Biology* 94:1–59

Felsing N. E., J. A. Brasel, and D. Cooper. 1992. Effect of low and high intensity exercise on circulating growth hormone in men. *Journal of Clinical Endocrinology and Metababolism* 75:157–62.

Frost, H. M. 1986. *The Intermediary Organization of the Skeleton*. Boca Raton: CRC Press.

Garrod, D. A. E., and D. M. A. Bate. 1937. *The Stone Age of Mt. Carmel.* Oxford: Clarendon.

Henry, D. O. 1995. *Prehistoric Cultural Ecology and Evolution: Insights from Southern Jordan.* New York: Plenum Press.

Hublin, J. J. 1978. Quelques charactères apomorphes du crane néanderthalien et leur interprétation phylogénetique. *Comptes Rendus de l'Académie des Sciences* 287:923–4.

Hublin, J. J., F. Spoor, M. Braun, F. Zonneveld, and S. Condemi. 1996. A late Neanderthal associated with Upper Palaeolithic artefacts. *Nature* 381:224–6.

Kelly, R. L. 1992. Mobility/sedentism: Concepts, archaeological measures, and effects. *Annual Review of Anthropology* 21:43–66.

Klein, R. G. 1987. Problems and prospects in understanding how early people exploited animals, 11–45. In *The Evolution of Human Hunting*, ed. M. H. Nitecki and D.V. Nitecki. New York: Plenum Press.

Klein, R. G. 1995. Anatomy, behavior and modern human origins. *Journal of World Prehistory* 9:167–98.

Klevezal, G. A., and S. E. Kleinenberg. 1967. *Age Determination of Mammals by Layered Structure in Teeth and Bone.* Moscow: Nauka.

Krebs, J. R., and N. B. Davies. 1987. *An Introduction to Behavioural Ecology*. Sunderland, MA: Sinnauer Associates.

Kroeber, A. L. 1925. *Handbook of the Indians of California.* New York: Dover.

Lieberman, D. E. 1993a. The rise and fall of seasonal mobility among hunter-gatherers: The case of the southern Levant. *Current Anthropology* 34:599–631.

Lieberman, D. E. 1993b. Life history variables preserved in dental cementum microstructure. *Science* 261:1162–4.

Lieberman, D. E. 1993c. Mobility and Strain: The Biology of Cementum and the Evolution of Hunter-Gatherer Mobility in the Southern Levant. Ph. D. diss. Harvard University. Cambridge.

Lieberman, D. E. 1994 .The biological basis for seasonal increments in dental cementum and their application to archaeological research. *Journal of Archaeological Science* 21:525–39.

Lieberman, D. E. 1995. Testing hypotheses about recent human evolution from skulls: Integrating morphology, function, development and phylogeny. *Current Anthropology* 36:159–97.

Lieberman, D. E. 1996. How and why recent humans grow thin skulls: Experimental data on systemic cortical robusticity. *American Journal of Physical Anthropology* 101:217–36.

Lieberman, D. E. 1998. Natufian "sedentism" and the importance of biological data for estimating reduced mobility, 75–92. In *Identifying Seasonality and Sedentism in Archaeological Sites: Old and New World Perspectives*, ed. T. R. Rocek and O. Bar-Yosef. Cambridge: Peabody Museum of Archaeology and Ethnology, Harvard University.

Lieberman, D. E., and R. H. Meadow. 1992. The biology of cementum increments: An archaeological perspective. *Mammal Review* 22:57–77.

Lieberman, D. E., T. W. Deacon, and R. H. Meadow. 1990. Computer image enhancement and analysis of cementum increments as applied to teeth of Gazella gazella. *Journal of Archaeological Science* 17:519–33.

Lieberman, D. E., and J. J. Shea. 1994. Behavioral differences between archaic and modern humans in the Levantine Mousterian. *American Anthropologist* 96:300–32.

Marks, A. E. 1988. The Middle to Upper Paleolithic transition in the southern Levant: Technological change as an adaptation to increasing mobility, 109–23. In *L'Homme de Néandertal*, Vol. 8, ed. M. Otte. Liége: ERAUL.

Marks, A. E., and D. A. Friedel. 1977. Prehistoric settlement patterns in the Avdat/Aqev area, 191–232. *In Prehistory and Paleoenvironments in the Central Negev, Israel*, Vol. 2, ed. A. E. Marks. Dallas: SMU Press.

Martin, R. B., and D. B. Burr. 1989. *Structure, Function, and Adaptation of Compact Bone.* New York: Raven Press.

Meignen, L., and O. Bar-Yosef. 1988. Variabilité technologique au Proche Orient: L'exemple de Kebara, 81–95. In *L'Homme de Neanderthal*, Vol. 4, ed. M. Otte. Liège: ERAUL.

Naveri, H. 1985. Blood hormone and metabolic levels during graded cycle ergometer exercise. *Scandanavian Journal of Laboratory and Clinical Investigations* 45:559–603.

Nawrocki, S. P. 1991. A Biomechanical Model of Cranial Vault Thickness in Archaic Homo. Ph. D. diss. State University of New York. Binghamton.

Pirinen, S., A. Majurin, H. L. Lenko, and K. Koski. 1994. Craniofacial features in patients with deficient and excessive growth hormone. *Journal of Craniofacial Genetics and Developmental Biology* 14:144–52.

Rabinovitch, R., and E. Tchernov. 1995. Chronological, palaeoecological and taphonomical aspects of the Middle Palaeolithic site of Qafzeh, Israel, 5–43. In *Archaeozoology of the Near East*, Vol. 2, ed. H. Buitenuis and H. P. Uerpmann. Leiden: Backhuys.

Rak, Y. 1986. The Neanderthal: A new look at an old face. *Journal of Human Evolution* 15:151–64.

Randall, R. V. 1989. Acromegaly and gigantism, 330–50. In *Endocrinology*, Vol. 1, ed. L. J. DeGroot. Philadelphia: W. B. Saunders.

Ruff, C. B., E. Trinkaus, and A. Walker. 1994. Postcranial robusticity in Homo, III: Ontogeny. *American Journal of Physical Anthropology* 93:35–54.

Schick, T., and M. Stekelis. 1977. Mousterian assemblages from Kebara Cave, Mount Carmel, 97–149. In *Moshé Stekelis Memorial Volume*, ed. O. Bar-Yosef and B. Arensburg. Jerusalem: The Israel Exploration Society.

Shea, J. J. 1989. A functional study of the lithic industries associated with hominid fossils in the Kebara and Qafzeh caves, Israel, 611–25. In *The Human Revolution*, ed. P. A. Mellars and C. B. Stringer. Edinburgh: Edinburgh University Press.

Shea, J. J. 1991. The Behavioral Significance of Mousterian Industrial Variability. Ph. D. diss. Harvard University. Cambridge.

Simpson, G. G. 1961. *Principles of Animal Taxonomy*. New York: Columbia University Press.

Stiner, M. 1995. *Honor Among Thieves: A Zooarchaeological Study of Neandertal Ecology*. Princeton: Princeton University Press.

Termine, J. D. 1990. Cellular activity, matrix proteins, and aging bone. *Experimental Gerontology* 25:217–21.

Trinkaus, E. 1987. The Neanderthal face: Evolutionary and functional perspectives on a recent hominid face. *Journal of Human Evolution* 16:429–43.

Trinkaus, E. 1992. Morphological contrasts between the Near Eastern Qafzeh-Skhul and late archaic human samples: Grounds for a behavioral difference, 277–94. In *The Evolution and Dispersal of Modern Humans in Asia*, ed. T. Akazawa, K. Aoki, and T. Kimura. Tokyo: Hokusen-Sha.

Vogl, C., W. R. Atchley, D. R. Cowley, P. Crenshaw, J. Murray, and D. Pomp. 1993. The epigenetic influence of growth hormone on skeletal development. *Growth Development and Aging* 57:163–82.

Wolpoff, M. H. 1989. The place of Neanderthals in human evolution, 97–141. In *The Emergence of Modern Humans: Biocultural Adaptations in the Later Pleistocene*, ed. E. Trinkaus. Cambridge: Cambridge University Press.

Yeh, J. K., J. F. Aliola, M. Chen, N. Ling, H. C. Koo, and W. J. Millard. 1994. Effect of growth hormone administration and treadmill exercise on serum and skeletal IGF-I in rats. *American Journal of Physiology* 266:E129–35.

17

PROSPECTS FOR STABLE ISOTOPIC ANALYSIS OF LATER PLEISTOCENE HOMINID DIETS IN WEST ASIA AND EUROPE

Stanley H. Ambrose

Department of Anthropology
University of Illinois
Urbana, Illinois 61801

1. INTRODUCTION

The transition from "archaic" to "modern" human behavior, coincident with the transition from the Middle to the Upper Paleolithic (MP/UP) and the Middle to the Later Stone Age (MSA/LSA), may have been accompanied by changes in dietary adaptations. Increased mobility and exchange of exotic lithics and other resources in the UP/LSA may reflect modern human capacities for adjusting social and territorial organization to environments with different food resource structures (Ambrose and Lorenz 1990; Lieberman and Shea 1994). Changes in diet composition may have occurred as well. Middle Paleolithic and MSA humans in Europe and southern Africa have been characterized as less effective predators than those in the Upper Paleolithic and LSA (Binford 1984, 1985; Klein 1989), or to have had predation patterns resembling those of non-human carnivores and scavengers (Stiner and Kuhn 1992; Stiner 1993, 1994). Other studies have concluded that there were no significant differences in faunal exploitation patterns or predator effectiveness (Chase 1989).

Neandertal skeletal morphology and paleopathology indicate they frequently engaged in strenuous activities and suffered traumas like those of rodeo cowboys, suggesting bold confrontational predation patterns (Berger and Trinkaus 1995). Linear enamel hypoplasias, which reflect nutritional stress and/or infection, occur predominantly before or during weaning age in Upper Paleolithic anatomically modern humans, but occur throughout dental development in Neandertals and other archaic humans (Ogilvie et al. 1989; Skinner 1996). The restricted distribution of hypoplasias in modern humans may reflect bouts of infection associated with the introduction of solid foods. Combined with evidence for younger average age at death and rapid dental wear, the prevalence and distribution of enamel defects in Neandertals suggests "periods of poor nutritional quality and/or famine must have been relatively common" (Trinkaus 1995:138). If differences in nutritional

Neandertals and Modern Humans in Western Asia, edited by Akazawa *et al.*
Plenum Press, New York, 1998.

quality are a function of the amount of animal protein, then the difference in incidence of hypoplasias may also reflect differences in trophic level between Neandertals and modern humans. Robust and gracile australopithecines in South Africa have an analogous pattern of differences in incidence and age distribution of linear enamel hypoplasias (White 1978), which may also reflect trophic level differences.

Archaeological and skeletal evidence, particularly greater dental attrition, suggest lower diet quality and thus a lower trophic level in Neandertals, but the evidence is open to alternative interpretations and this stereotyped characterization of diet may be overly simplistic. If MP/MSA humans were less effective predators than their UP/LSA counterparts, did they consume proportionately less meat and more plants? Alternatively, did they consume the same proportions of meat versus plant foods as Upper Paleolithic/LSA humans but simply work harder and take more risks in subsistence? Or could the difference in nutritional stress reflect the presence or absence of food storage (Soffer 1989)?

Answering these questions requires quantitative data on the composition of individual diets. Faunal and floral remains cannot be used for quantitative diet reconstruction because they are never preserved or recovered in the proportions in which they were eaten. Dental microwear and skeletal pathologies also provide only qualitative dietary data at best (Walker 1981; Ogilvie et al. 1989; Trinkaus 1995; Skinner 1996). Chemical and isotopic analyses of bones and teeth can, however, provide quantitative data on some aspects of individual diet (Sillen and Cavanaugh 1982; Ambrose 1993). If there were significant differences in trophic levels between behaviorally and/or anatomically archaic versus modern humans, they could be identified by isotopic analysis of hominid skeletons. Analysis of stable carbon and nitrogen isotopes of bone collagen, combined with carbon isotopes in apatite carbonate, is a powerful tool for dietary reconstruction (Ambrose 1993; Ambrose and Norr 1993). The potentials and problems of diet reconstruction with stable isotopes will be illustrated with analyses of Neandertal and non-hominid mammal bones from Dederiyeh Cave in Syria and Pleistocene and recent bones and teeth from Europe and Africa.

2. DIET RECONSTRUCTION WITH STABLE ISOTOPES OF BONE

Diet reconstruction with stable isotopes is based on the observation that you are what you eat. In other words, the isotopic composition of consumer tissues is a direct and constant function of that of the diet. The tissue-diet function differs for each tissue and must be subtracted from the tissue isotope ratio to estimate that of the diet. Where significant and systematic differences in carbon and/or nitrogen isotope ratios ($\delta^{13}C$ and $\delta^{15}N$)* exist between major classes of dietary resources, proportions consumed can be determined by isotopic analysis of consumer tissues.

Carbon isotope ratios of bone can be used to reconstruct some aspects of diet where the isotopic composition of major classes of dietary resources differ significantly. Non-tropical grasses, most leafy plants, and trees and shrubs use C_3 photosynthesis and have average $\delta^{13}C$ values of –26‰. Tropical grasses and some pioneering weeds adapted to hot sunny habitats and summer rainfall use C_4 photosynthesis. Their $\delta^{13}C$ values average –12‰ (Smith 1972). Marine resources have $\delta^{13}C$ values intermediate between those of C_3 and C_4 plants (Smith and Epstein 1971) and have $\delta^{15}N$ values higher than those of terrestrial ani-

* Stable isotope ratios are expressed using the δ notation in parts per thousand (‰) relative to a standard, calculated as follows: $\delta‰ = (R_{sample}/R_{std} - 1) \times 1000$, where R is $^{13}C/^{12}C$ or $^{15}N/^{14}N$ of the sample and standard.

mals (Schoeninger et al. 1983; Schoeninger and DeNiro 1984). High latitude and Mediterranean (winter rainfall) terrestrial ecosystems, including southern Europe and West Asia, are completely dominated by C_3 plants. In purely C_3 terrestrial foodwebs, carbon isotopic analysis of collagen provides less useful paleodietary information because the range of variation is so small, but when combined with analysis of collagen nitrogen and carbonate carbon isotopes, it can be informative.

Observations of free-ranging animals and controlled diet experiments with rats have been used to estimate the average diet-tissue difference for bone collagen and carbonate (Ambrose 1993, in preparation). For nitrogen isotopes of bone collagen, you are what you eat plus about 3.3‰ (DeNiro and Epstein 1981). There is a ~3‰ step-wise increase in $\delta^{15}N$ values between trophic levels within foodwebs (Minagawa and Wada 1984; Schoeninger and DeNiro 1984). In arid environments there may be a greater difference between trophic levels (Ambrose and DeNiro 1986; Sealy et al. 1987; Ambrose 1991), but this aridity effect is not always observed (Vogel et al. 1990).

Controlled diet experiments with rodents show that carbon isotopes in collagen and other consumer tissue proteins are derived mainly from dietary proteins rather than from the whole diet (Ambrose and Norr 1993; Tieszen and Fagre 1993). When protein and non-protein components of diet have the same $\delta^{13}C$ values, consumer bone collagen is enriched by 5‰ and animal flesh by 1.5‰ (Vogel 1978; Krueger and Sullivan 1984; Ambrose, in preparation). You are what you eat plus 5‰ in collagen or 1.5‰ in muscle, but *only* when dietary protein and non-protein are monoisotopic. When they differ, for example when protein is C_3 and non-protein is C_4, and conversely, the collagen-whole diet difference can range from −2‰ to +10‰ (Ambrose and Norr 1993, Ambrose, in preparation). Carnivore collagen $\delta^{13}C$ is enriched by approximately 1‰ compared to herbivore collagen (Schoeninger 1985; Matheus 1995), probably because of the 1–1.5‰ enrichment between the diet and flesh of their herbivore prey. East African Neolithic pastoralists clearly show this trophic level effect in collagen. They have the highest collagen $\delta^{13}C$ values reported for any human or non-human population, probably reflecting their heavy reliance on milk, meat, and blood of domestic animals that fed on C_4 plants (Ambrose 1986). This enrichment in collagen $\delta^{13}C$ is, however, too small relative to the natural variation within trophic levels to provide accurate estimates of trophic level on its own.

Controlled diet experiments with rats and mice demonstrate that bone apatite carbonate $\delta^{13}C$ values are enriched by around 9.4‰ relative to the diet, even when the protein and non-protein diet components have very different $\delta^{13}C$ values (Ambrose and Norr 1993; Tieszen and Fagre 1993). Carbonate carbon is apparently *always* what you eat plus 9.4‰, even when diets have very low (5%), normal (20%), and very high amounts of protein (70%). Wild carnivore bone carbonate is also enriched by about 9–9.5‰, but that of herbivores is enriched by about 11–13‰ (Krueger and Sullivan 1984; Lee-Thorp et al. 1989). The large difference in herbivores is usually considered a trophic level effect, but another explanation has been proposed by Hedges and van Klinken (in press) based on studies of ruminant digestion. Bone carbonate is derived from dissolved CO_2 and bicarbonate in the blood stream, which ultimately comes from energy metabolism of food. Methanogenic bacteria in ruminant guts produce methane with very low $\delta^{13}C$ values, which is balanced by production of metabolic CO_2 with very high $\delta^{13}C$ values (Metges et al. 1990). This ^{13}C-enriched CO_2 labels the bone carbonate. Controlled diet experiments with rats on very low and very high protein diets failed to produce a trophic level effect (Ambrose and Norr 1993), so the large difference between diet and carbonate in herbivore bone may reflect isotopic enrichment due to methanogenesis in ruminant digestion rather than to eating plants alone.

The difference (Δ) between the $\delta^{13}C$ values of bone carbonate and collagen ($\Delta^{13}C_{carb-coll}$) has been used as an indicator of trophic level. It is larger for herbivores (6–9‰) than for carnivores and omnivores (4–7‰) (Krueger and Sullivan 1984; Lee-Thorp et al. 1989). The large difference in herbivores reflects the high diet-carbonate difference. The small $\Delta^{13}C_{carb-coll}$ in carnivores and omnivores is partly caused by the difference between the $\delta^{13}C$ values of dietary proteins, fats (lipids), and carbohydrates (Krueger and Sullivan 1984). Collagen $\delta^{13}C$ values reflect mainly those of dietary protein, so they are not greatly affected by those of lipids or carbohydrates. Lipid $\delta^{13}C$ values are always 4–6‰ more negative than those of the whole animal (DeNiro and Epstein 1978; Tieszen and Fagre 1993). Lipids lower the $\delta^{13}C$ value of the total diet, which would shift the carbonate $\delta^{13}C$ value closer to that of collagen. If animal flesh $\delta^{13}C$ values are 1–1.5‰ less negative than those of plants, then carnivore and omnivore collagen $\delta^{13}C$ values should be shifted closer to those of their carbonate. Since herbivore diets are typically low in proteins and lipids, the differences in their carbon isotopic composition should have little effect on $\Delta^{13}C_{carb-coll}$ values. When both $\Delta^{13}C_{carb-coll}$ and collagen $\delta^{15}N$ values are used together, clear differences between trophic levels emerge, as will be illustrated below.

3. POST-MORTEM PRESERVATION OF ISOTOPIC SIGNATURES IN BONE

Collagen is rarely preserved for more than 3000 years in hot tropical environments in Africa (Ambrose 1990) and in Pleistocene-aged bones from the Levant (Weiner and Bar-Yosef 1990). In colder environments in Europe, collagen is often preserved through the last Glacial period (Bocherens et al. 1991; Bocherens et al. 1995; Fizet et al. 1995). It should thus be possible to use stable isotopes to determine differences in diet between Middle and Upper Paleolithic humans in Europe. However, it is important to assess the quality of the collagen and carbonate of each sample before using the results for diet reconstruction because preservation can vary substantially within sites due to differences in bone density, rates of burial, exposure to heat, groundwater and soil chemistry (Ambrose and DeNiro 1989; Ambrose 1990; Weiner and Bar-Yosef 1990; Weiner et al. 1993; Stiner et al. 1995).

When bone collagen is not preserved, interpretation of bone and dentine apatite carbonate carbon isotopes is problematic because of the potential for diagenetic alteration. Tooth enamel apatite carbonate is not significantly affected by diagenesis (Lee-Thorp and van der Merwe 1991; Wang and Cerling 1994; Lee-Thorp et al. 1995). Diagenetic alteration of bone and tooth dentine carbonate isotope ratios often occurs in chalky, weathered bones, even if they have not been buried, and in buried bones after collagen has degraded (Koch et al. 1990; Bocherens et al. 1995). Bones of immature individuals also have lower density and crystallinity and higher porosity, rendering them more susceptible to diagenesis than dense cortical bone of adult individuals. Therefore when collagen is not present, and/or where bone and dentine are soft and weathered, bone carbonate $\delta^{13}C$ values should not be used for diet reconstruction. Tooth enamel is not significantly affected by diagenesis and can be used for diet reconstruction for millions of years after bone and dentine have been irrevocably altered (Lee-Thorp and van der Merwe 1991; Wang and Cerling 1994). Simple and reliable methods for characterizing the composition and preservation of organic matter in bones and teeth (Ambrose 1990) can be used to assess the preservation of Later Pleistocene mammal bones from Central Europe and Syria. These criteria will be described in the Results section, below.

4. MATERIALS AND METHODS

4.1. Materials

Modern bones were collected in 1987–1991 in the Kenya Rift Valley and adjacent highlands, using methods employed in a previous study (Ambrose and DeNiro 1986). Olga Soffer provided mammal bones from the Central Gravettian sites of Milovice, Dolni Vestonice II and Bohuslavice, in the Czech and Slovak Republics (Klima 1987; Oliva 1988; Svoboda et al. 1996). They date from 22,000 to 26,000 bp. The soil matrix is calcareous loess. Species analyzed include wolf (*Canis lupus*), arctic fox (*Alopex lagopus*), hare (*Lepus europaeus*, *L. timidus*), reindeer (*Rangifer tarandus*), horse (*Equus caballus*), and woolly mammoth (*Mammuthus primigenius*). Mirko Malez provided one cave bear (*Ursus spelaeus*) bone and two cave bear teeth from Vindija Cave, a very large limestone cave in Croatia (Malez et al. 1980). The age of the Vindija specimens is not precisely known. Level G1, about 2.8 m below the surface, marks the MP/UP transition and dates to 33,000±400 bp (Karavanic 1995). The specimens analyzed come from six meters below the surface in Level J, correlated to climatic stage Würm 1 (early Glacial). They may be close to 70,000 years old. Takeru Akazawa provided skull and other bone fragments of a Neandertal child burial, together with an assortment of unidentifiable fragmentary mammal bones, from Dederiyeh Cave in Syria. The skeleton was associated with a Tabun B Mousterian lithic industry and may date to the early Glacial, between 75,000 and 45,000 BP (Akazawa et al. 1993; 1995).

One modern elephant bone from Kenya was severely weathered. Most prehistoric specimens had dense unweathered bone. Preservation was outstanding at Vindija. Mammal bones from Dederiyeh were dense and unweathered; many were stained dark brown by humic acids and mottled by MnO_2. Neandertal bone fragments from Dederiyeh were softer, as expected for immature bone, and some had been treated with preservatives and glues.

4.2. Methods

Methods for preparation of bone collagen and apatite carbonate for isotopic analysis are described in detail elsewhere (Ambrose 1990; Lee-Thorp and van der Merwe 1991; Ambrose and Norr 1992). All surfaces of prehistoric bones were scraped and discolored portions carved out with a scalpel before sample preparation. Collagen is purified by demineralizing ground bone in 0.2 M HCl. Humic contaminants are removed by treatment with 0.125 M NaOH for 20 hours. Decontaminated collagen is gelatinized by heating (95°C) in acidic water (pH 3) for 10 hours, followed by filtration, condensation, and freeze-drying. Collagen is combusted at 800–900°C in evacuated sealed quartz tubes and CO_2 and N_2 are separated by cryogenic distillation. Carbon and nitrogen contents are measured during distillation. In poorly-preserved samples, the collagen residue may be mainly humic acids, salts, and other inorganic substances, rather than or in addition to collagen. Such samples usually have very low carbon and nitrogen concentrations and C:N ratios outside the acceptable range (2.9–3.6) for collagen, and usually have carbon and nitrogen isotope ratios unlike those of their modern analogs (Ambrose 1990).

Bone carbonate is purified by treatment with NaOCl (50% Clorox) to remove organic matter, followed by 1 M acetic acid to remove diagenetic carbonate contaminants. Samples are then reacted under vacuum with 100% phosphoric acid at 25°C. CO_2 is collected by cryogenic distillation.

The Dederiyeh Cave hominid bones had been treated with PVA (polymerized vinyl acetate) and some fragments had been glued with either Cemedine-C or Kony Bond glue. Visible glue was scraped away. Residual glue and PVA were removed by treatment with acetone for two days before grinding and two days after grinding. Kony Bond did not dissolvc in acctone, but turned bright white when soaked in cold water; contaminated portions were readily identifiable and removed with a scalpel. The non-hominid mammal bone provided for analysis was not glued or treated with PVA. Subsamples of mammal bone were treated with PVA and the acetone-soluble Cemedine C glue in order to determine if they could be completely removed with acetone and if they affected the isotope ratios of collagen.

5. RESULTS AND DISCUSSION

5.1. Criteria for Assessing Preservation of Isotopic Integrity

Collagen presence and purity are determined by three simple criteria: (1) collagen concentrations in bone or tooth, (2) carbon and nitrogen concentrations in collagen, and (3) atomic carbon-to-nitrogen ratios of collagen (Ambrose 1990). Modern bones and whole teeth have more than 20% collagen by weight. Ancient bones with yields above 1–4% by weight usually have collagen (Ambrose 1990; Ambrose and Norr 1992). European bones and teeth (Table 1) have 1.2 to 14.1% collagen, with the exception of one weathered sample from Milovice (ED19), which has 0.8% collagen. The Dederiyeh Cave bones have only 0.6% to 1.4% collagen (Table 2).

Modern bone collagen has 42% carbon and 16% nitrogen by weight. Carbon concentrations below 3–8%, and nitrogen concentrations below 1–3% usually indicate collagen is absent. Figure 1 shows that European bones have relatively high carbon and nitrogen concentrations in collagen, with the exception of ED19 (Table 1). Cave bears from Vindija

Table 1. Collagen concentration, carbon and nitrogen concentration, atomic C:N ratio, carbon and nitrogen isotopic values, apatite concentration, percent carbon, carbon isotope values, and carbonate-collagen $\delta^{13}C$ difference values of Upper Pleistocene mammal bones from Dolni Vestonice II (DV II), Milovice (ML), Bohuslavice (BH), and Vindija Cave (VN)

	Bone Collagen						Bone Apatite Carbonate						
Lab #	%Coll (wt %)	%C (wt %)	%N (wt %)	C:N	$\delta^{13}C$‰	$\delta^{15}N$‰	Lab # (MA)	Apatite (wt %)	%C (wt %)	$\delta^{13}C$‰ carb-coll	$\Delta^{13}C$‰	Species	Site
MC60	6.3	38.8	14.6	3.1	-19.6	6.0	99	42.6	0.89	-15.3	4.3	Canis lupus	DV II
MC61	1.6	20.1	7.5	3.2	-19.0	9.5	100	40.2	0.91	-15.0	4.0	Alopex lagopus	DV II
MC62	14.1	45.8	17.2	3.1	-20.6	4.7	101	32.9	0.85	-13.3	7.3	Equus caballus	DV II
MC63	4.6	35.2	13.3	3.1	-20.1	2.0	102	34.7	0.86	-13.3	6.8	Lepus europaeus	DV II
MC64	9.5	44.9	16.9	3.1	-21.0	8.0	103	32.8	0.86	-12.5	8.5	Mammuthus primigenius	DV II
MC65	4.1	35.9	13.4	3.1	-19.4	9.3	104	39.8	0.86	-13.4	6.0	Rangifer tarandus	DV II
MC67	1.2	22.1	8.1	3.2	-21.0	7.9	106	57.2	1.13	-13.3	7.7	Mammuthus primigenius	BH
MC66	5.2	39.1	14.8	3.1	-20.7	7.0	105	42.0	1.07	-13.0	7.7	Mammuthus primigenius	ML
ED19*	0.8	5.3	1.6	3.9	-21.3		135	52.3	1.23	-12.0	9.3	Mammuthus primigenius	ML
ED20	3.1	32.6	12.3	3.1	-21.0	7.3	136	62.4	1.13	-12.5	8.5	Mammuthus primigenius	ML
ED21	2.1	23.8	9.1	3.1	-20.9	7.8	137	74.1	1.29	-12.1	8.8	Mammuthus primigenius	ML
ED3§	4.0	38.6	14.4	3.1	-19.9	9.7						Canis lupus	ML
ED4	7.8	38.5	14.7	3.1	-20.5	5.9	133	38.3	0.96	-12.8	7.7	Equus caballus	ML
ED5	4.9	40.1	15.2	3.1	-19.8	8.5	134	38.5	1.09	-14.7	5.1	Lepus europaeus	ML
SA1	2.4	9.1	3.1	3.4	-23.0	4.4	128	65.3	0.99	-14.3	8.7	Ursus spelaeus	VN
SA2	1.8	26.1	9.7	3.1	-22.6	4.8	129	66.0	1.07	-13.7	8.9	Ursus spelaeus	VN
SA3	1.2	15.6	5.5	3.3	-21.6	4.3	130	60.5	1.17	-12.8	8.8	Ursus spelaeus	VN

Table 2. Collagen concentration, carbon and nitrogen concentration, atomic C:N ratio, apatite concentration, percent carbon, and carbon isotope ratios of mammal and Neandertal bones from Dederiyeh Cave. There was insufficient organic matter for isotopic analysis in the residue from the collagen extraction procedure

Bone Collagen					Bone Apatite Carbonate				
Lab #	%Coll (wt %)	%C (wt %)	%N (wt %)	C:N	Lab #	Apatite (wt %)	%C (wt %)	$\delta^{13}C$‰	Species
SA328	1.39	0.42	0.05	10.1	SA333	64.0	1.24	-10.8	Mammal (PVA treated)
SA329	0.84	0.38	0.04	10.4	SA334	68.8	1.13	-11.7	Mammal (not treated)
SA330	0.77	0.51	0.05	12.6	SA335	63.4	1.15	-12.4	Human non-skull
SA331	0.57	0.56	0.06	10.7	SA336	63.9	1.11	-12.5	Human skull

have relatively high C and N concentrations despite their low collagen yields. The Dederiyeh Cave samples have virtually no organic carbon or nitrogen in the "collagen" residue.

The atomic C:N ratio of modern collagen is 3.2 (Ambrose 1993). Ancient bones with C:N ratios between 2.9 and 3.6 usually, but not always, have high collagen yields and high C and N concentrations in collagen (Ambrose 1990). Most European specimens have acceptable C:N ratios, with the exception of ED19 from Milovice, which has a C:N of 3.9 (Table 1). The sample from Vindija with the lowest C and N concentrations has a slightly high but acceptable C:N and relatively high C and N concentrations. Bones from Dederiyeh Cave have very low "collagen" concentrations and unacceptably high C:N ratios (Table 2).

5.2. Effects of Modern Contaminants on Dederiyeh Bone Collagen

The elemental and isotopic composition of glues and PVA are listed in Table 3. Differences in carbon and nitrogen concentrations between experimentally contaminated samples treated with acetone and uncontaminated bones from Dederiyeh Cave were insignificant (Table 2). The "collagen" in bones from Dederiyeh Cave did not have enough

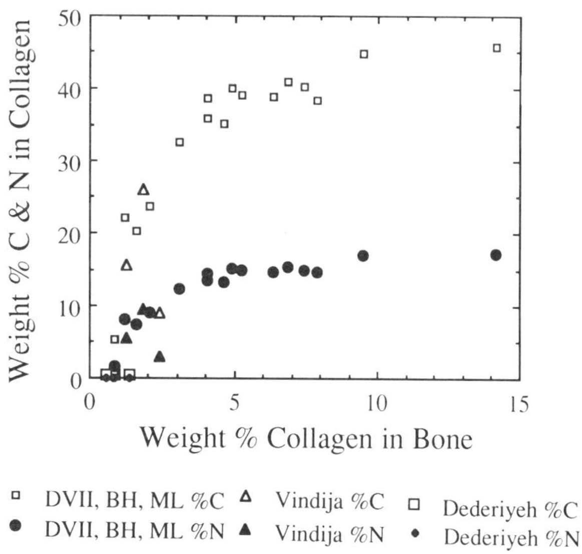

Figure 1. Collagen concentration in whole bone (weight %) versus carbon and nitrogen concentrations (weight %) in collagen of prehistoric bones from Dolni Vestonice II, Milovice, Bohuslavice (Upper Paleolithic), Vindija Cave and Dederiyeh Cave (Middle Paleolithic).

Table 3. Elemental and carbon isotopic composition of PVA preservative and glues used on bones of the Neandertal child from Dederiyeh Cave. PVA and Cemedine C are soluble in acetone; Kony Bond is insoluble in acetone and cold water

Lab #	%C (wt %)	%N (wt %)	C:N	δ^{13}C‰	δ^{15}N‰	Material
SA325	55.5	0.0		-34.0		Polymerized vinyl acetate (PVA)
SA328	37.4	7.2	6.1	-28.9	-1.1	Cemedine C glue
SA327	54.7	0.0		-30.4		Kony Bond glue

carbon or nitrogen for isotopic analysis, suggesting the procedures used to remove glues and PVA were very effective.

5.3. Apatite Carbonate Stable Isotopes

Modern bones have approximately 0.9% carbonate carbon by weight (equivalent to 4.5% CO_3 by weight). The poorly-preserved mammoth from Milovice (Table 1) and all bones from Dederiyeh (Table 2) have carbonate C concentrations ≥1.1%. They may have accumulated small amounts of diagenetic carbonate.

The effect of weathering and collagen loss on the isotopic integrity of bone apatite carbon isotopes is illustrated in Figure 2, which shows the weathered elephant bone has an anomalously high $\Delta^{13}C_{carb\text{-}coll}$ value. This bone had reached an extremely advanced stage of surface weathering, becoming splintered and chalky, resembling a moss-covered log. It has 5.9% collagen and its carbonate C concentration was 1.1%. Its carbonate $\delta^{13}C$ value has shifted toward that of atmospheric carbon dioxide, which has a $\delta^{13}C$ value of around −7‰ (Marino and McElroy 1992). The poorly-preserved mammoth (ED19) from Milovice has the highest $\Delta^{13}C_{carb\text{-}coll}$ value among the European bones (Table 1). Many from Kent's Cavern, England also have relatively high $\Delta^{13}C_{carb\text{-}coll}$ values (Bocherens et al. 1995), but the pattern of lower $\Delta^{13}C_{carb\text{-}coll}$ values for carnivores is maintained.

Bone carbonate carbon isotope ratios have been shown to increase in variance relative to collagen carbon isotopes in weathered bones prior to burial, and also increase in

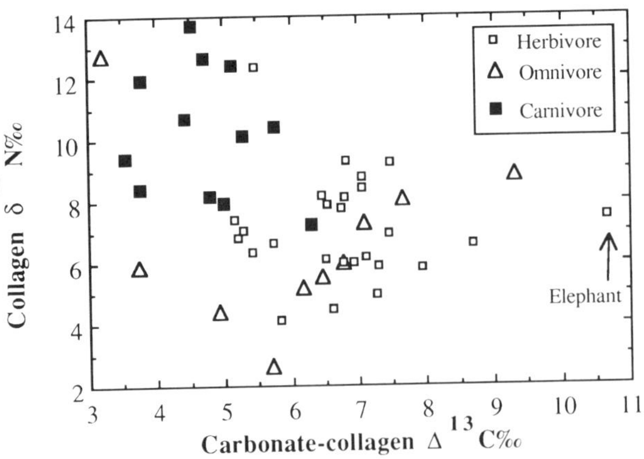

Figure 2. Collagen δ^{15}N and carbonate-collagen δ^{13}C difference values of bones of modern mammals from highland Kenya.

variance relative to bone collagen and tooth enamel carbon isotopes of the same individuals in earlier Holocene humans (Koch et al. 1990). Without collagen or tooth enamel for reference, diagenesis of apatite carbonate $\delta^{13}C$ values of bones from Dederiyeh cannot be evaluated and they cannot be used for diet reconstruction. A survey of the state of preservation of bones from prehistoric sites in the Near East found traces of collagen in only one bone older than 10,000 BP. It is not likely that any bones older than 40,000 BP in this region will be usable for diet reconstruction with stable isotopes, but those from central Europe can be used for paleodietary research.

5.4. Foodweb Stable Isotope Patterns

Bone collagen $\delta^{15}N$ and $\Delta^{13}C_{carb\text{-}coll}$ values of modern Kenyan herbivores, omnivores and carnivores are plotted in Figure 2 and descriptive statistics are presented in Table 4. Carnivores have high $\delta^{15}N$ and small $\Delta^{13}C_{carb\text{-}coll}$, herbivores have low $\delta^{15}N$ and high $\Delta^{13}C_{carb\text{-}coll}$, and omnivores have intermediate values. Similar results have been obtained in modern mammals in southern Africa (Lee-Thorp et al. 1989) and prehistoric mammals in Europe (Bocherens et al. 1995). Trophic levels can be determined when both $\Delta^{13}C_{carb\text{-}coll}$ and collagen $\delta^{15}N$ values are used together.

The isotopic composition of collagen of Pleistocene mammals from central Europe is shown in Figure 3 (Table 1). Carnivores have the highest and herbivores have the lowest $\delta^{15}N$ values. Cave bears from Vindija have very low $\delta^{15}N$ values, like those from sites in western Europe (Bocherens et al. 1994), suggesting a largely herbivorous diet. Carnivore collagen $\delta^{13}C$ values are slightly less negative than those of herbivores, reflecting the slight trophic level effect discussed above and observed in other studies (Schoeninger 1985; Bocherens et al. 1995; Matheus 1995). Trophic level differences are also apparent in low $\Delta^{13}C_{carb\text{-}coll}$ values for carnivores and high values for herbivores. A similar pattern was found at Kent's Cavern, England (Bocherens et al. 1995).

Two specimens from Dolni Vestonice II have anomalous isotopic compositions. The reindeer has a distinctively carnivore-like collagen and carbonate isotopic composition. The wolf has an anomalously low $\delta^{15}N$ value but a carnivore-like $\Delta^{13}C_{carb\text{-}coll}$ (Table 1).

Isotopic analysis of early Upper Pleistocene human diets has only been performed on bone collagen of two Neandertals, from levels 9 and 10 of Marillac Cave, France

Table 4. Means, standard deviations, and sample sizes of bone collagen $\delta^{15}N$ and $\Delta^{13}C_{carb\text{-}coll}$ values of modern herbivores, omnivores, and carnivores collected in natural environments in the central Rift Valley of Kenya in 1987 and 1991[a]

Trophic Level	Collagen $\delta^{15}N$‰	$\Delta^{13}C_{carb\text{-}coll}$‰
Herbivore	7.1 ± 1.7 (26)	6.6 ± 0.9 (25)
Omnivore	6.7 ± 2.8 (10)	6.1 ± 1.8 (10)
Carnivore	10.3 ± 2.1 (12)	4.7 ± 0.6 (12)

[a]Sample sizes are in parentheses. One sample altered by diagenesis has been excluded from the calculation of the mean $\Delta^{13}C_{carb\text{-}coll}$ value. Individual values are plotted in Figure 2.

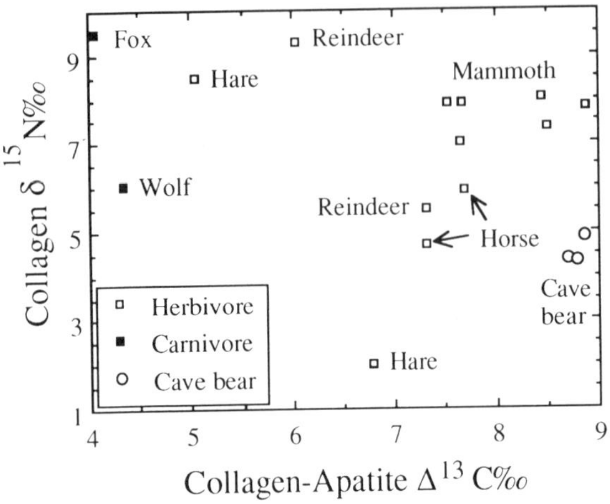

Figure 3. Collagen $\delta^{15}N$ and carbonate-collagen $\delta^{13}C$ difference values of bones from Dolni Vestonice II, Milovice, Bohuslavice (Upper Paleolithic) and Vindija Cave (Middle Paleolithic).

(Bocherens et al. 1991; Fizet et al. 1995). The specimen from Level 9 has a very low collagen yield and a non-collagenous amino acid composition. One from Level 10 has a collagenous amino acid composition. Both individuals have $\delta^{15}N$ values as high as or higher than those of wolves and hyenas, placing them at the top of the Marillac food chain, but only the results on the individual from level 10 can be considered reliable. Apatite carbonate has not yet been analyzed but $\Delta^{13}C_{carb\text{-}coll}$ values would corroborate this interpretation of their high trophic level.

6. CONCLUSIONS

The prospects for diet reconstruction with stable isotopes are good in Europe but poor in Western Asia. A relatively warm climate may have accelerated collagen degradation throughout Western Asia and may account for the rarity of collagen in Upper Pleistocene bones from other sites in this region (Weiner and Bar-Yosef 1990). The warm climate of the last interglacial may have been unfavorable for collagen preservation everywhere. European Upper Pleistocene mammals often have excellent collagen preservation throughout the last Glacial period (70,000 years) and can be used for stable carbon and nitrogen isotope reconstruction of Neandertal and modern human diets.

We may thus be able to address an important aspect of the ongoing debate over the presence or absence of differences in subsistence behavior of Middle versus Upper Paleolithic humans. Faunal evidence suggests that Neandertals were ineffective hunters of large game (Klein 1989), had predation patterns like those of carnivores rather than humans (Stiner 1994), and obtained most meat by scavenging, a view of predation abilities boldly articulated by Lewis Binford (1984, 1985). Paleopathological evidence suggests Neandertals practiced risky, brute force predation strategies and suffered frequent food stress. High rates of dental attrition suggest a tougher, lower quality diet (Ogilvie et al. 1989; Berger and Trinkaus 1995; Trinkaus 1995; Skinner 1996). In contrast, Upper Paleolithic humans are portrayed as more effective predators and paleopathology suggests less frequent and less intense food stress. If Middle Paleolithic and Middle Stone Age humans were ineffective hunters, then they probably ate more plants than Upper Paleolithic and Later Stone Age humans. This should be reflected by lower $\delta^{15}N$ and collagen $\delta^{13}C$ values, and larger differences between carbon isotope ratios of apatite and collagen than in modern humans.

Nitrogen isotopic analysis of one Neandertal seems to contradict this hypothesis, indicating a high trophic level (Fizet et al. 1995), but apatite carbonate isotope analysis of Neandertals has not yet been performed. Moreover, stable isotopic analyses have not yet been reported for any early Upper Paleolithic modern humans, so hypotheses of dietary differences from Neandertals cannot be tested. Isotopic analysis of early European Upper Paleolithic modern humans is eminently feasible and long overdue. Great care should, however, be taken to evaluate the preservation of collagen and apatite before dietary interpretations of stable isotope ratios are made.

ACKNOWLEDGMENTS

I am grateful to Professor Akazawa for giving me the opportunity to perform isotopic analysis on materials from Dederiyeh Cave and to Ofer Bar-Yosef and Ken Aoki for inviting me to participate in the symposium on Neandertals and Modern Humans in West Asia. I thank Olga Soffer and the late Dr. Malez for providing specimens from Central Europe, and an anonymous reviewer for comments and suggestions that have improved the paper. Elizabeth Dinan assisted in the preparation of some samples from Vindija and Milovice. This research was supported, in part, by National Science Foundation grants BNS 9010937 and SBR 9212466 and by grants from the University of Illinois Research Board.

REFERENCES

Akazawa, T., Y. Dodo, S. Muhesen, A. Abdul-Salam, Y. Abe, O. Kondo, and Y. Mizoguchi. 1993. The Neanderthal remains from Dederiyeh Cave, Syria: Interim report. *Anthropological Science* 101:361–87.

Akazawa, T., S. Muhesen, Y. Dodo, O. Kondo, and Y. Mizoguchi. 1995. Neanderthal infant burial. *Nature* 377:585–6.

Ambrose, S. H. 1986. Stable carbon and nitrogen isotope analysis of human diet in Africa. *Journal of Human Evolution* 15:707–73.

Ambrose, S. H. 1990. Preparation and characterization of bone and tooth collagen for isotopic analysis. *Journal of Archaeological Science* 17:431–51.

Ambrose, S. H. 1991. Effects of diet, climate and physiology on nitrogen isotope abundances in terrestrial foodwebs. *Journal of Archaeological Science* 18:293–317.

Ambrose, S. H. 1993. Isotopic analysis: Methodological and interpretive considerations, 59–130. In *Investigations of Ancient Human Tissue: Chemical Analyses in Anthropology*, ed. M. K. Sandford. New York: Gordon and Breach Scientific.

Ambrose, S. H., and M. J. DeNiro. 1986. The isotopic ecology of East African mammals. *Oecologia* 69:395–406.

Ambrose, S. H., and M. J. DeNiro. 1989. Climate and habitat reconstruction using stable carbon and nitrogen isotope ratios of collagen in prehistoric herbivore teeth from Kenya. *Quaternary Research* 31:407–22.

Ambrose, S. H., and K. G. Lorenz. 1990. Social and ecological models for the Middle Stone Age in southern Africa, 3–33. In *The Emergence of Modern Humans. An Archaeological Perspective*, ed. P. Mellars. Edinburgh: Edinburgh University Press.

Ambrose, S. H., and L. Norr. 1992. On stable isotopic data and prehistoric subsistence in the Soconusco region. *Current Anthropology* 33:401–4.

Ambrose, S. H., and L. Norr. 1993. Isotopic composition of dietary protein and energy versus bone collagen and apatite: Purified diet growth experiments, 1–37. *Molecular Archaeology of Prehistoric Human Bone*, ed. J. Lambert and G. Grupe. Berlin: Springer.

Berger, T. D., and E. Trinkaus. 1995. Patterns of trauma among the Neandertals. *Journal of Archaeological Science* 22:841–52.

Binford, L. R. 1984. *The Faunal Remains at Klasies River Mouth.* New York: Academic Press.

Binford, L. R. 1985. Human ancestors: Changing views of their behavior. *Journal of Anthropological Archaeology* 4:292–327.

Bocherens, H., M. Fizet, and A. Mariotti. 1994. Diet, physiology and ecology of fossil mammals as inferred from stable carbon and nitrogen isotope biogeochemistry: implications for Pleistocene bears. *Palaeogeography, Palaeoclimatology, Palaeoecology* 107:213–25.

Bocherens, H., M. Fizet, A. Mariotti, B. Lange-Badre, B. Vandermeersch, J. P. Borel, and G. Bellon. 1991. Isotopic biogeochemistry (^{13}C, ^{15}N) of fossil vertebrate collagen: application to the study of a past food web including Neandertal man. *Journal of Human Evolution* 20:481–92.

Bocherens, H., M. L. Fogel, N. Tuross, and M. Zeder. 1995. Trophic structure and climatic information from isotopic signatures in Pleistocene cave fauna from southern England. *Journal of Archaeological Science* 22:327–40.

Chase, P. 1989. How different was Middle Paleolithic subsistence? A zooarchaeological perspective on the Middle to Upper Paleolithic transition, 321–37. *The Human Revolution*, ed. P. A. Mellars and C. B. Stringer. Edinburgh: Edinburgh University Press.

DeNiro, M. J., and S. Epstein. 1978. Influence of diet on the distribution of carbon isotopes in animals. *Geochimica et Cosmochimica Acta* 42:495–506.

DeNiro, M. J., and S. Epstein. 1981. Influence of diet on the distribution of nitrogen isotopes in animals. *Geochimica et Cosmochimica Acta* 45:341–51.

Fizet, M., A. Mariotti, H. Bocherens, B. Lange-Badre, B. Vandermeersch, J. P. Borel, and G. Bellon. 1995. Effect of diet, physiology and climate on carbon and nitrogen stable isotopes of collagen in a Late Pleistocene anthropic paleoecosystem: Marillac, Charente, France. *Journal of Archaeological Science* 22:67–79.

Hedges, R. E. M., and G. J. van Klinken. In press. "Consider a spherical cow..."—on modelling relationships between isotopic composition of tissues and diet. In *Close to the Bone: Advances in Chemical and Isotopic Approaches to Paleodietary Analysis*, ed. S. H. Ambrose and M. A. Katzenberg. New York: Plenum Press.

Karavanic, I. 1995. Upper Paleolithic occupation levels and late-occurring Neandertal at Vindija Cave (Croatia) in the context of Central Europe and the Balkans. *Journal of Anthropological Research* 51:9–35.

Klein, R. G. 1989. Biological and behavioral perspectives on modern human origins in Southern Africa, 529–46. In *The Human Revolution*, ed. P. A. Mellars and C. B. Stringer. Edinburgh: Edinburgh University Press.

Klima, B. 1987. A triple burial from the Upper Paleolithic of Dolni Vestonice. *Journal of Human Evolution* 16:831–5.

Koch, P., A. K. Behrensmeyer, N. Tuross, and M. L. Fogel. 1990. Isotopic fidelity during bone weathering and burial. *Carnegie Institution, Annual Report of The Director, Geophysical Laboratory*, 1989–1990:105–110.

Krueger, H. W., and C. H. Sullivan. 1984. Models for carbon isotope fractionation between diet and bone, 205–22. In *Stable Isotopes in Nutrition*, ed. J. F. Turnlund and P. E. Johnson. Washington: American Chemical Society.

Lee-Thorp, J. A., J. Sealy, and N. J. van der Merwe. 1989. Stable carbon isotope ratio differences between bone collagen and bone apatite, and their relationship to diet. *Journal of Archaeological Science* 16:585–99.

Lee-Thorp, J. A., and N.J. van der Merwe. 1991. Aspects of the chemistry of modern and fossil biological apatites. *Journal of Archaeological Science* 18:343–54.

Lee-Thorp, J. A., N. J. van der Merwe, and C. K. Brain. 1995. Diet of *Australopithecus robustus* at Swartkrans from stable carbon isotopic analysis. *Journal of Human Evolution* 27:361–72.

Lieberman, D. E., and J. J. Shea. 1994. Behavioral differences between archaic and modern humans in the Levantine Mousterian. *American Anthropologist* 96:300–32.

Malez, M., F. H. Smith, J. Radovcic, and D. Rukavina. 1980. Upper Pleistocene hominids from Vindija, Croatia, Yugoslavia. *Current Anthropology* 21:365–7.

Marino, B. D., and M. B. McElroy. 1992. Isotopic composition of atmospheric CO_2 inferred from carbon in C_4 plant cellulose. *Nature* 349:127–31.

Matheus, P. E. 1995. Diet and co-ecology of Pleistocene short-faced bears and brown bears in eastern Berengia. *Quaternary Research* 44:447–53.

Metges, C., K. Kempe, and H. L. Schmidt. 1990. Dependence of the carbon isotope contents of breath carbon dioxide, milk, serum and rumen fermentation products on the $\delta^{13}C$ value of food in dairy cows. *British Journal of Nutrition* 63:187–96.

Minagawa, M., and E. Wada. 1984. Step-wise enrichment of ^{15}N along food chains: Further evidence and the relationship between $d^{15}N$ and animal age. *Geochimica et Cosmochimica Acta* 48:1135–40.

Ogilvie, M. D., B. K. Curran, and E. Trinkaus. 1989. The incidence and patterning of dental enamel hypoplasias among the Neanderthals. *American Journal of Physical Anthropology* 79:25–41.

Oliva, M. 1988. Discovery of mammoth bone hut at Milovice (Moravia) Czechoslovakia. *Journal of Human Evolution* 17:787–90.

Schoeninger, M. J. 1985. Trophic level effects on $^{15}N/^{14}N$ and $^{13}C/^{12}C$ ratios in bone collagen and strontium levels in bone mineral. *Journal of Human Evolution* 14:515–25.

Schoeninger, M. J.. and M. J. DeNiro. 1984. Nitrogen and carbon isotopic composition of bone collagen from marine and terrestrial animals. *Geochimica et Cosmochimica Acta* 48:625–39.

Schoeninger, M. J., M. J. DeNiro, and H. Tauber. 1983. $^{15}N/^{14}N$ ratios of bone collagen reflect marine and terrestrial components of prehistoric human diet. *Science* 220:1381–3.

Sealy, J. C., N. J. van der Merwe, J. A. Lee-Thorp, and J. Lanham. 1987. Nitrogen isotope ecology in southern Africa: implications for environmental and dietary tracing. *Geochimica et Cosmochimica Acta* 51:2707–17.

Sillen, A. B., and M. Cavanaugh. 1982. Strontium and paleodietary research: A review. *Yearbook of Physical Anthropology* 25:67–90.

Skinner, M. 1996. Developmental stress in immature hominines from Late Pleistocene Eurasia: Evidence from enamel hypoplasia. *Journal of Archaeological Science* 23:833–52.

Smith, B. N. 1972. Natural abundance of the stable isotopes of carbon in biological systems. *BioScience* 22:226–31.

Smith, B. N., and S. Epstein. 1971. Two categories of $^{13}C/^{12}C$ ratios for higher plants. *Plant Physiology* 47:380–4.

Soffer, O. 1989. Storage, sedentism and the Eurasian Paleolithic record. *Antiquity* 63:719–32.

Stiner, M. 1993. Small animal exploitation and its relation to hunting, scavenging and gathering in the Italian Mousterian, 107–25 In *Hunting and Animal Exploitation in the Later Paleolithic and Mesolithic of Eurasia,* ed. G. L. Peterkin, H. M. Bricker, and P. Mellars. Washington, D. C.: American Anthropological Association.

Stiner, M. 1994. *Honor Among Thieves: A Zooarchaeological Study of Neandertal Ecology.* Princeton: Princeton University Press.

Stiner, M. C., and S. L. Kuhn. 1992. Subsistence, technology, and adaptive variation in Middle Paleolithic Italy. *American Anthropologist* 94:306–39.

Stiner, M. C., S. L. Kuhn, S. Weiner, and O. Bar-Yosef. 1995. Differential burning, recrystallization and fragmentation of archaeological bone. *Journal of Archaeological Science* 22:223–37.

Svoboda, J., V. Lezek, and E. Velcek, eds. 1996. *Hunters Between West and East.* New York: Plenum Press.

Tieszen, L. L., and T. Fagre. 1993. Effect of diet quality on the isotopic composition of respiratory CO_2 bone collagen, bioapatite and soft tissues, 121–55. In *Prehistoric Human Bone: Archaeology at the Molecular Level,* ed. J. B. Lambert and G. Grupe. Berlin: Springer-Verlag.

Trinkaus, E. 1995. Neanderthal mortality patterns. *Journal of Archaeological Science* 22:121–42.

Vogel, J. C. 1978. Isotopic assessment of the dietary habits of ungulates. *South African Journal of Science* 74:298–301.

Vogel, J. C., A. S. Talma, A. J. Hall-Martin, and P. J. Viljoen. 1990. Carbon and nitrogen isotopes in elephants. *South African Journal of Science* 86:147–50.

Walker, A. 1981. Dietary hypotheses and human evolution *Philosophical Transactions of the Royal Society, London B* 292:57–64.

Wang, Y., and T. E. Cerling. 1994. A model of fossil tooth and bone diagenesis: Implications for paleodiet reconstruction from stable isotopes. *Palaeogeography, Palaeoclimatology, Paleoecology* 107:281–9.

Weiner, S., and O. Bar-Yosef. 1990. States of preservation of bones from prehistoric sites in the Near East: A survey. *Journal of Archaeological Science* 17:187–96.

Weiner, S., P. Goldberg, and O. Bar-Yosef. 1993. Bone preservation in Kebara Cave, Israel, using on-site Fourier transform infrared spectrometry. *Journal of Archaeological Science* 20:613–27.

White, T. D. 1978. Early hominid enamel hypoplasia. *American Journal of Physical Anthropology* 49:79–84.

PART III

The Human Fossils

It is undoubtedly the human fossils that attract the attention of both the public as well as many students of anthropology to the subject matter of this volume. Since the discovery of the first Neandertals and Cro-Magnons in Europe during the nineteenth century, as well as the finding of *Homo erectus* remains in Java, the number of researchers in the field of human evolution has been growing continuously. Scientists are both building hypotheses concerning the phylogenetics of primates and humans and at the same time searching for more fossils. When prehistoric investigations began, the newly discovered fossils made the international news in almost every country. The same reaction was found when pioneers, including F. Turville-Petre, D. Garrod, R. Neuville, and M. Stekelis, uncovered human remains in Zuttiyeh, Tabun, Skhul, and Qafzeh caves. Similar excitement was brought about when R. Solecki exposed Neandertal skeletons in Shanidar cave. More recently the fossils from Amud, Kebara, and again from Amud raised chronological questions and opened a debate concerning the capacity for language. At the time of this writing, Dederiyeh Cave is the last site to produce important relics.

The taxonomy of the fossils and their phylogenetic relationships constitutes an ongoing discussion. Two general reviews by F. Clark Howell and Chris Stringer opened this volume. The papers in this section, however, widen the geographic scope of the discussion and also introduce particular details of specific hominids. Several authors examine the relationships between prehistoric Europeans and their Levantine contemporaries. The reader will find that there are differing views concerning the interpretation of the fossil record, and we can be certain that the subject is not yet exhausted.

Jean-Jacques Hublin reviews the evolution of human populations in Europe, where for historical reasons there are more fossils relative to the size of that continent. As Neandertals were also found in western Asia, it seems appropriate to examine their definition, which was coined based on the European specimens. The Neandertal special morphological traits are better known from examining the skulls and jaws than from the postcranial elements. The evolutionary process that resulted in what are known today as the classical Neandertals is described by the author within the context of the ecological belts of Europe and the sequence of climatic changes. Hublin begins with the Middle Pleistocene hominids and continues through the various stages to the period of the Last Glaciation. He views the accretion of Neandertal traits as a shift in frequencies, and thus he examines the selective pressures that were placed on the European population through the Glacial-Interglacial cycles. Finally, he examines the eastern expansion of the Neandertals and stresses the fact that fossils found at a distance from the core area do not display the same number of cold-adapted characteristics as those from Neandertals who survived in western Europe. The cold conditions of Isotope Stage 4, which were previously suggested as the trigger for Neandertal expansion into the Near East, are supported by the cultural connections between the Balkanian Mousterian and similar industries in the Taurus-Zagros.

Baruch Arensburg and Anna Belfer-Cohen present a view point which diverges entirely from the one held by most of the participants in the symposium. They see very little evidence, if any at all, for the presence of western Asian Neandertals. The paper opens with the definitions of Neandertals, and the authors point out the historical contradictions in assigning the fossils recovered in Israel to one taxon or another. They cite the original observations of Keith and McCown concerning Tabun I and suggest that the morphological features indicate that it has affinities with an archaic *Homo sapiens* and not with the Neandertals. As for Tabun II, the isolated jaw, they see no difficulty in attributing it to anatomically modern humans (abbreviated as AMHS). They suggest a gradual evolution from Zuttiyeh to the Tabun specimens. The skull of Amud I is examined in detail and the authors conclude that it is more related to the AMHS than to the "classic" Neandertals and the Shanidar skeletal remains. Kebara 2 receives the same treatment, and its morphological traits are compared to a host of other fossils. Contrary to earlier claims, the authors view the pelvis as indicating affinities with AMHS such as Skhul IX and recent humans. In addition, they briefly state that the Shanidar fossils are only partially Neandertals. In their conclusions, the authors accept the African connection for the origins of Near Eastern Upper Pleistocene populations. They ascertain that the earliest distribution of AMHS was in Africa and the Near East and that the morphological variability observed among the Levantine fossils resembles that of the African ones. Thus the latter is seen as the source for the modern Eurasian hominids.

The new discovery of an infant Neandertal skeleton in Dederiyeh Cave (northern Syria) is discussed by Yukio Dodo and his associates. The age (estimated as about two years) and stature (a little over 80 cm) are the first components. The traits of the various cranial elements and post-cranial skeleton are compared to other human fossils. The authors conclude that this infant is closely related to European Neandertals.

Nancy Minugh-Purvis' paper involves a comparison between the juveniles from Krapina and Skhul. The study of immature individuals serves as a means for understanding the development of particular morphology. She begins by outlining the general agreement concerning the modern features of the Qafzeh hominids and proceeds to examine Skhul 1, as the skeletal remains from this site demonstrate a higher degree of variability. She proposes to test the hypothesis that differences and similarities between the Krapina and Skhul relics would indicate a directional gene flow. Skhul 1 was aged as 3.5 to 5 years old and Krapina aged as 1 to between 6 and 8. The author compares the skull components for each of the fossils and concludes that Krapina 1 is a Neandertal and is distinct from the anatomically modern human Skhul 1.

The issue of whether there are two or only one hominid species in a Mousterian cave is tackled by Yoel Rak. Citing the notion the "the skull is the creation of God, while the jaw is the work of the devil," the author proceeds to discuss two mandibles, namely, Tabun I and Tabun II. It is crucial to stress that it is now widely accepted that even the excavator, Dorothy Garrod, could not decide in her final site report (*The Stone Age of Mount Carmel, Vol. 1),* if Tabun I belonged to Layer C, whereas she had no doubt that Tabun II was deeply embedded in this layer. Rak examines the morphological details of the two jaws including the chin, the ramus, the retromolar space and a few other elements. He then tries to figure out how Keith and McCown arrived at the conclusion that the two specimens belonged to the same "race." Rak suggests that it was the chronological assignment that caused them to make this assertion. After mentioning other cases of human relics in which the attribution of the fossils to a certain time span and cultural complex caused scientists to attribute them to the Neandertal group (or grade as was once thought), he concludes that one can achieve the correct cladogenetic position of the fossils only by adopting an anatomical assessment. Hence, the supposed contradiction that emerged from the archaeological European sequence, namely that the human fossils associated with the Mousterian can be only Neandertals, will be abandoned. The Levantine Mousterian demonstrates the presence of two different human populations within apparently similar lithic industries.

For Karen Rosenberg, the postcranial elements serve as a basis for discussing obstetric and locomotion behavior. In spite of the generally poor preservation of pubic bones, the author notes, investigators had previously observed a suite of characteristics among juvenile Neandertals that did not occur among modern humans. Her suggested revision is based on new information such as the Kebara pelvis (dated to about 60 ka by ESR and TL) and the Jinniushan pelvis (dated to about 280 ka by ESR) and the functional approach to these elements. Following a review of the relevant literature, the author examines the finds from the Levant and east Asia. One of her conclusions is that both specimens are different from the pelvic morphology of modern humans. However, her main points are that the elongated pubis of Neandertals is a pleisomorphic condition, and obstetric hypotheses are insufficient to explain the presence of this morphology among both males and females. Finally, it seems that both Neandertals and archaic *Homo sapiens* gave birth in a similar fashion to modern women.

Anne-marie Tillier tackles the ontogenetic variation in the Near East. She accepts the notion of the presence of two populations in the region: AMHS and the others who had Neandertal affinities. In examining the evidence, she relies on the changes in morphology that are related to growth and maturation processes. These can be assessed from the study of juvenile cranial sizes. In addition to the Levantine fossils, her survey includes others from Europe including Upper Paleolithic specimens, and her conclusions discuss the cautions that must be taken when cross-cultural and evolutionary comparisons are made.

A comparison between the loading patterns on upper and lower limb bones is the target of the paper written by Erik Trinkaus and his associates. This study tests the hypothesis that there were biologically reflected behavioral differences between AMHS and Neandertals in the Near East. Earlier analysis already indicated the presence of a mosaic of behavioral similarities and differences. This paper centers on shaft robusticity and shape as exposed in the cross-sectional geometric analysis. Following the description and discussion of humeral, femoral, and tibial diaphyseal strength, the author concludes that there are no differences in lower limb robusticity between the two populations. These conclusions are in line with what is known about earlier Pleistocene hominids. However, there are differences in the upper limb robusticity, especially when the right arm is considered. According to the authors, the latter observations are not supported by the archaeological information. They suggest instead that given the ecological variability over short distances in the Levant, locomotor differences between juveniles and adults reflect social organization.

Historically the two major human remains from Garrod's excavations at Tabun, namely, the woman (Tabun I) and the isolated jaw (Tabun II), were reported as being uncovered in Layer C. Garrod's own hesitations were mentioned earlier and it should be stressed that there were no doubts in her mind and in the minds of careful readers of her publication that Tabun II was clearly embedded in Layer C. Quam and Smith reexamine the Tabun CII jaw in their paper. Their historical review covers the stratigraphy as well as previously published analyses and conclusions concerning this jaw. The possible relationship between the lithic assemblage from Tabun C and similar occurrences in other sites such as Qafzeh is noted. In addition, they also examine the other fragmentary remains from Layer C. The morphological study of the jaw forms the core of the paper. This is followed by a careful test of the possible nature of this specimen as another example of a relic of modern humans. No less important is the authors' assertion that one cannot exclude the possibility that this mandible is the result of hybridization. Accepting the ESR and TL dates would mean that the Tabun CII jaw is the earliest example of modern human morphology, present in the Levant prior to 120 ka.

In sum, there is no doubt in our minds that the debate will continue concerning the physical evidence for the origin of modern humans and the role of the Levantine fossils. The papers in this section reflect the existing divergence of observations and interpretations. They also demonstrate the need to close the apparent interpretational gap between the other archaeological remains (lithics, bones, settlement pattern, etc.) and the skeletal evidence. Finally, this section indicates how the chronological placement of the fossils may impact our conclusions concerning evolutionary trends.

18

CLIMATIC CHANGES, PALEOGEOGRAPHY, AND THE EVOLUTION OF THE NEANDERTALS

Jean-Jacques Hublin

UMR 152 du CNRS
Laboratoire d'Anthropologie
Musée de l'Homme
Paris 75116, France

1. INTRODUCTION

Mainly because of historical reasons, Europe has provided the largest series of Middle and Upper Pleistocene hominids. Many of the sites which yielded these specimens can be placed in a reliable environmental and chronological framework, and they allow us to produce detailed evolutionary scenarios covering the entire last half a million years.

However, although this area was once used as a model for the entire Old World, it is actually marginal when considering the development of hominids using broader chronological and geographical scales. The evolutionary processes of the European population are indeed peculiar. At the center of this peculiarity are the Neandertals, a group of archaic humans that remains unknown outside of Western Eurasia (i.e., Europe and its oriental fringes). The rise, development, and extinction of Neanderthals can be traced in relation to the environmental and cultural changes in Europe.

It is easy to recognize the unique nature of the area that produced this original group. Europe is a long and rather narrow peninsula in the middle latitudes that became isolated by the development of seas and mountain chains. Located at the western extremity of Eurasia, it is far from the intertropical zone where most of the evolutionary processes of the early hominids took place. Another aspect of this European particularism is the dramatic climate changes that it experienced for most of the last million years. These changes caused tremendous contraction of the non-arctic bio-climatic zones and led to the periodic development of periarctic conditions in moderately high latitudes.

It is tempting to imagine a relationship between the peculiar paleobiogeographic history of Europe and the development of an odd lineage of hominids. In this perspective, the following questions arise: What series of phenomena could determine this link? When did the divergence of the European hominids from their Asian and African relatives begin? How fast was it? How far did it go biologically? What is the geographical extension of the process?

Neandertals and Modern Humans in Western Asia, edited by Akazawa *et al.*
Plenum Press, New York, 1998.

The goal of this paper is to review briefly the tempo of the development of European particularism in relation to the geographical and climatic changes, and then to question the boundaries of this process and the relationships between Europe and adjacent areas.

2. DEFINITION OF THE NEANDERTALS

The development in the late 1970s of cladistic methods in human paleontology led to a reassessment of the anatomical definition of the Neandertals. It allowed for a better understanding of the polarity of the anatomical features of the group, sorting out archaic retentions, derived features shared with modern humans, and proper Neandertal apomorphies. One of the main results of this new approach was the refutation of several previous models or hypotheses based on a misunderstanding of the "global morphological pattern." First, the so-called "European pre-*sapiens*," supposedly the ancestors of modern humans in the Middle Pleistocene of Europe, were demonstrated to be genuine pre-Neandertals. They were reminiscent in some respects of the modern morphology only because of symplesiomorphies. The African and Asian "Neandertaloids" appeared to be well-separated from the West Eurasian Neandertals with whom they essentially shared primitive retentions and robusticity.

However, the cladistic approach to the Neandertal problem suffered several criticisms. Phylogenetic analysis was mainly conducted on cranio-mandibular features, while post-cranial morphology did not yield very clear cut evidence. In particular, the limited knowledge of contemporary and immediately preceding out-groups made the phylogenetic interpretation of Neandertal post-cranial morphology difficult. Some, if not most, of the "Neandertal" post-cranial features that are missing in contemporary or later modern humans could be plesiomorphic retentions and thus would be observable in the direct ancestors of modern humans as well. In addition, the degree of *Homo*plasy between closely related groups can be high. This is especially true when dealing with locomotor or climatic adaptations which shaped the morphology, robusticity, and proportion of the limbs and trunk of large bipedal hominids. Finally, the possible use of the cladist methodology within a species (*Homo sapiens*) was questioned on theoretical bases and heavily criticized as typological and non-biological (Trinkaus 1990).

It may be said that dealing with the *Homo*plasy was the fate of cladism from its dawn and that this does not invalidate the great progress that resulted from its use as a theoretical framework. Questioning evolutionary processes in paleobiological terms is certainly one of the major advances of paleoanthropology in the last two decades. Still, the baby should not be thrown out with the bath water. The adaptative perspective does not resolve all the questions raised by the fossil hominids, especially in terms of phylogeny and the pattern of peopling. Furthermore, the features which develop along the evolution of a fossil lineage do not all result from direct adaptative responses. Genetic drift and pleitropic effects are also involved.

Regarding the possible (or impossible) use of the cladistic methodology within the species *Homo sapiens*, it must be said that the subspecific status of the Neandertals is still, at the least, debatable. However, in order to avoid the sin of "typologism," intrapopulational (or intraspecific) variation must always be taken into consideration. When dealing with metric features, clear overlaps are observed between the Neandertals and neighboring groups of hominids. Whenever possible, the occurrences of some morphological features should also be considered in terms of frequency (see for example Franciscus and Trinkaus 1995). Yet there are almost no real problems in distinguishing the Neandertals and their close ancestors from contemporary groups. In a pragmatic way, cladism does work when dealing with the Neandertals. The question of whether cladism works because it is useful at a subspecific level or because Neandertals are a distinct species of the genus *Homo* is important theoretically, but it is also secondary.

The series of cranio-mandibular features presented in Table 1 were operational in distinguishing the Neandertal specimens. Their possible occurrences in the Lower and Middle Pleistocene mark the milestones of Neandertal emergence.

3. STAGES OF THE NEANDERTALIZATION PROCESS

3.1. The Lower Pleistocene Group

Dating the earliest peopling of Europe has been the subject of a long-term debate. The sites of Soleihac and Le Vallonet (France) can be assigned to the Jaramillo event (from 0.99 to 1.07 My) on the basis of paleomagnetic evidence (Thouveny and Bonifay 1984; de Lumley et al. 1988). Still, some doubts have been raised on the human origins of some of the artifacts and structures discovered in these two sites. However, indisputable

Table 1. Neandertal craniomandibular derived features, unique or most frequent in the group

Upper face and mandible:
- Rounded supraorbital torus without distinct elements
- High orbits
- Mid-facial prognathism resulting in low subspinal angle, low nasiofrontal angle, large difference between M1 alveolus and zygomaxillare radii
- Infraorbital area horizontally flat or convex, obliquely receding in alignment with the antero-lateral surface of the zygomatic
- Posterior rooting of the facial crest
- Bucco-lingually expanded anterior dentition
- Extended taurodontism
- Laterally expanded mandibular condyle
- Mental foramina posteriorly set relatively to the dental arcade
- Retromolar space
- Oval horizontal shape of the mandibular foramen
- Large medial pterygoid tubercle

Cranial vault:
- Secondarily increased relative platycephaly
- "En bombe" cranial shape
- Low symmetrically arched temporal squama
- *Meatus acusticus externus* at the level of the posterior zygomatic arch with a strong inclination of the basal groove of this process
- Highly convex upper scale of the occipital

Basicranium:
- Flat articular eminence
- Mediolaterally developed postglenoid process
- Elongated foramen magnum
- Root of the stylomastoidian process medial to the anterior end of the digastric groove and stylomastoidian foramen
- Small and inferiorly situated posterior semicircular canal

Occipito-mastoid area:
- Laterally flattened mastoid process, medially oriented inferiorly
- Tuberculum mastoideum anterius
- Fully developed suprainiac fossa associated with a bilaterally protruding occipital torus

late Lower Pleistocene artifacts and human remains are known in layer TD6 of the Gran Dolina of Atapuerca (Spain) (Carbonell et al. 1995). The area of Orce (Spain) has yielded even older archaeological evidence in two sites: Fuenta Nueva 3 and Barranco Leon 5. A recent reexamination of the biostratigraphic and paleomagnetic framework of the whole sedimentary basin indicates an immediately pre-Jaramillo age for the two sites, which would then be dated between 1.2 and 1 myrs (Turq et al. 1996). 'Ubeidiyeh (Israel) and Dmanisi (Georgia) might only be slightly older (Tchernov 1986; Gabunia and Vekua 1995; Brauer and Schultz 1996). It is likely that the early Lower Pleistocene witnessed the first attempts of colonization of the Mediterranean parts of Europe.

In any case, it is important to distinguish European Mediterranean areas from latitudes above 45 degrees. These were geographically well-separated and represented distinct ecological zones. If the supporters of the so-called "long chronology" recently had more success in the south of Europe, the supporters of the "short chronology" are still not defeated in the higher altitudes. Humans, as tropical animals primarily adapted to warm and open environments, seem to have colonized the temperate to cold environments only relatively recently. There is still no indisputable evidence of more or less permanent human occupation outside the Mediterranean area before 500 ky BP (Roebrooks et al. 1992). However, the paleontological and archaeological records become relatively abundant and continuous after this date.

The morphology of the Dmanisi mandible falls in the wide range of variation of *Homo erectus sensu lato*. Yet some features of this specimen can be interpreted as rather progressive, especially the morphology of the anterior part of the mandibular body and the size and proportion of the cheek teeth. According to Brauer and Schultz (1996), this morphology questions the previously proposed late Pliocene age of the specimen (Gabunia and Vekua 1995), as the occurrence of "unexpectedly derived traits" in a nearly 1.6 or 1.8 my old hominid is deemed unlikely. However, the Dmanisi mandible gives us some indication of the possible nature of the first colonizers of Europe. Given its time range, it does not demonstrate a clear regional pattern.

In the preliminary description of the remains from the TD6 layer of the Gran Dolina of Atapuerca (Spain) (Carbonell and al. 1995), one dental and several mandibular features of this sample were claimed to demonstrate an evolutionary continuity between the TD6 hominids and much (?) later Middle Pleistocene hominids of Atapuerca-Simas de los Huesos (SH). One finds the buccolingual enlargement of the second lower incisors among them. On average, the Neandertals as a group clearly display this enlargement relative to older forms (Semal 1988). Yet such a feature can display strong individual variation and is difficult to assess on isolated specimens. The Dmanisi mandible does not display this pattern and, in contrast, some African individuals assigned to *Homo erectus sensu lato* (e.g., WT 15000, Tighenif 1 and 3) do display buccolingual enlargement of the lower incisors comparable to that of the Neandertals.

3.2. The Early Middle Pleistocene Group

This group includes the specimens from Mauer (Germany), Boxgrove (England), Tautavel (France), Petralona (Greece), and Verteszöllos (Hungary). Regarding the last two specimens, some radiometric ages contradict the paleontological data and suggest a possible younger age (Schwarcz and Latham 1984; Grün 1996). More fragmentary remains from the late Cromerian come from Fontana Ranuccio and Visogliano in Italy. The specimen from Ceprano (Central Italy) is likely to be referred to this group or to be older (Ascenzi et al. 1996).

In this group, and for the first time, we see the development of an incipient but clearly Neandertal morphology. On Arago 21 and the Petralona skull, some facial features are reminiscent of the Neandertals. These include a flat or slightly convex horizontal profile of the maxilla between the nasal opening and the zygomaxillary suture, obliquely oriented posteriorly and followed by an obliquely oriented zygomatic. Some details of the nasal aperture including the shape of the nasal floor in Petralona are also reminiscent of the later Neandertals. The description of the material from the Sima de Los Huesos in Atapuerca, likely dating later (Bischoff et al. 1997) and displaying an even more obvious trend toward a Neandertal facial morphology and its so-called mid-facial prognathism (Arsuaga et al. 1997), have confirmed this observation.

The mandibles also display some Neandertal trends. Mauer, which could be the oldest specimen of the series, weakly displays this tendency. The size of the anterior teeth relative to the cheek teeth has been suggested to be a pre-Neandertal characteristic of the specimen (Wolpoff 1982; Bermudez de Castro 1986). On Arago 2 and 13, which are estimated to be nearly 450 kyrs old (de Lumley et al. 1984), one finds the lateral expansion of the mandibular condyle. In addition, Arago 2, which is much more gracile than Arago 13, displays an enlargement of the anterior part of the dental arcade and a mental foramina located under M1.

It should be emphasized that other anatomical areas in this early Middle Pleistocene group, such as the occipital, the temporal, and the cranial vault as a whole, still display a rather primitive aspect in contrast to the facial area. The skull exhibits a broad, simple, and midprojecting occipital torus quite different from the typical Neandertal aspect. The braincases have a pentagonal outline in posterior view distinctive from the "en bombe" shape characteristic of the Neandertals.

3.3. The "Holstein-Hoxnian" Group

A second step in the development of the Neandertal cranial morphology is marked by the development of a series of unique features on the occipital. The occurrence of a moderate occipital torus displaying a bilateral development instead of a medial maximal point of projection, associated with a horizontally extended well-marked suprainiac depression, is one of the most striking autapomorphic Neandertal features (Hublin 1988a). This suprainiac fossa is first shallow and widely extended and later deep, long, and narrow. A genuine external occipital protuberance is always lacking. In Europe, this set of features appears clearly on specimens such as Steinheim and Swanscombe, which are respectively assigned to the Hoxnian and Holstein "interglacials." They would belong to the isotopic stages 11 or 9 and to a range of time between 430 and 300 ky. The specimens of Reilingen (Germany), Bilzingsleben (Germany), and possibly those from the Atapuerca SH (Spain) series are likely found in the same range of time. Reilingen clearly displays the same morphology as Swanscombe and Steinheim (Dean et al., in press). Atapuerca SH skulls display a rough and/or porous suprainiac surface (Arsuaga et al. 1997). They fill the morphological gap between the middle projected and wide occipital torus of Verteszöllos and the Swanscombe-Steinheim-Reilingen group with their shallow and still ill-defined surpainiac fossa associated with a bilaterally projected occipital torus. In this group, Swanscombe seems to be the more derived, combining a rather Neandertal-like suprainiac area with a highly convex upper occipital scale. This observation calls into question its possible assignment to stage 11 (Bridgland 1994). In contrast, one finds the primitive aspect of the occipital reminiscent of the previous group on one individual in Bilzingsleben, and a suprainiac fossa on another one (Condemi, in press). If a late age for Verteszöllos is retained, this specimen would serve as another example of persistence of the occipital primitive condition at a relatively late period.

At this chronological stage, the trend toward the facial and mandibular Neandertal morphology is more expressed than in the previous group, as documented by SH skull 5. This specimen displays a strong pneumatization of the maxillar and frontal, a Neandertal-like supraorbital torus, midfacial projection, and flattened and receding infraorbital area simultaneously (Arsuaga et al. 1993, 1997). However, a rather primitive facial morphology still exists in Atapuerca 404 . The "primitive" (but sometimes also said to be "modern-like") shape of Steinheim results at least partly from a severe distortion of the specimen.

On these specimens incipient and occasional Neandertal features are observed in the glenoid cavity and in the position of the stylomastoid foramen. A strong juxta mastoid eminence can be observed (Hublin 1988b; Martinez and Arsuaga 1997) but never reaches the Neandertal conditions. Yet many of the typical Neandertal aspects of the temporal are still missing, in particular in the mastoid area. On adult specimens, the mastoid process is usually strong and projects downward when it is preserved. There is no mastoidian anterior tubercle, with perhaps one exception: Atapuerca AT 86 (Elyatqine 1995, but see Martinez and Arsuaga 1997). In the series of Simas de los Huesos, on Steinheim and Reilingen, the outline of the temporal squama does not display the primitive low and angular pattern observed in *Homo erectus*. It is relatively short and high, contrasting also with the Neandertal conditions (relatively low and symmetrical) which appear to be derived. Castel de Guido (Italy) could be the first specimen where the Neandertal outline is observed (Elyatqine 1995). The position of the posterior root of the zygomatic arch is still basically primitive with a weak inclinaison of the basal groove.

Regarding the vault shape, the pattern is still primitive relative to the Neandertals with a pentagonal shape on all the specimens. The lateral walls of the skull are either parallel or converge slightly upward. However, on some specimens (e.g., Swanscombe and Reilingen) the outline of the skull in posterior view is more rounded, forerunning the "en bombe" shape observed in the Neandertals. It should be noted that, as in the previous group, an exceptional angular torus can be observed. This superstructure, frequent in *Homo erectus* but unknown in the Neandertals, is present on Reilingen and some of the SH specimens (4 and 5) (Arsuaga et al. 1993).

The apparent simultaneous occurence of ancestral or derived conditions for the same features, within this group, could sometimes derive from imperfections in our chronological calibration. However, it also clearly results from the actual coexistence of different morphologies in the same populations, the derived one becoming more frequent over time. This model applies to the occipital morphology as well as to the development of facial and mandibular derived features, which started earlier in the pre-Neandertal lineage. It explains the morphological contrasts observed within the series of Arago, Bilzingsleben, and Atapuerca SH.

3.4. The Saalian Group

There are no European specimens assigned clearly to isotopic stage 8. The group of specimens covering stages 7 and 6 includes the material from Biache-Saint-Vaast (France), Ehringsdorf (Germany), La Chaise-Suard (France), Fontechevade 2 (France), Le Lazaret (France), Pontnewydd (Wales), and possibly Montmaurin (France). In this series, it is already possible to recognize individually most if not all of the Neandertal derived features. They are still observed in combination with some primitive retentions on a rather complete individual, but some isolated bones could be practically indistinguishable from a stage 4 "classic" specimen. In particular, the evolution of the occipital morphology seems to be virtually completed. Already in stage 7, we find very derived conditions with the specimen of Biache-Saint-Vaast 1 (Auguste 1995a). The "en bombe" shape of the vault in posterior view is observed for the first time, as well as a Neandertal-like temporal morphology. The mastoid process is small and flattened, obliquely oriented medio-inferiorly, and associated with a juxtamastoid eminence more inferiorly projecting than the mastoid process itself (Stringer et al. 1984). This specimen is not fully adult and the mastoid could have developed further. Yet La Chaise-Suard stage 6 temporal displays indisputable Neandertal conditions in the mastoid area (Condemi and Piveteau 1988). The specimens from La Chaise and Ehringsdorf are the first in Europe to display a well-developed retromolar space on the mandible (Franciscus and Trinkaus 1995).

3.5. The Eemian and Weschelian Groups

During stages 5, 4, and 3 we find Neandertal populations where the expression and frequency of the derived Neandertal features increased. Only with the "classical" (stage 4) specimens are all the details of the temporal morphology completed as well as the Neandertal vault shape (bombe-like in posterior view, strongly elongated antero-posteriorly, and ending with a highly convex upper occipital scale). But the positions of the mental foramen display a higher frequency of the Neandertal conditions in stage 5 than in stage 4 (Trinkaus 1993).

4. ACCRETION OF THE NEANDERTAL FEATURES: A SHIFT IN FREQUENCY

This rapid review demonstrates that the development of the Neandertal morphology results from an *accretion* phenomenon beginning in the middle of the Middle Pleistocene, around 450,000 BP or a bit before. Not all the anatomical areas are effected simultaneously. In a simplistic view, the craniomandibular sequence seems to be: 1) upper facial and mandibular features; 2) occipital area features; 3) temporal area and vault shape features. However, as this accretion proceeds from an increase of the *frequency* of the derived conditions, complex combinations could result from the phenomenon. The mosaic of derived and primitive features may be different in two contemporaneous specimens. But this does not mean that we have different contemporary taxa inside Europe (Vlcek 1993; Otte 1996). As a matter of fact, the mosaic can also be different from one individual to another *within the same population* as documented by the Middle Pleistocene sites of Arago and Atapuerca SH. The change in frequency of the derived conditions is different from one anatomical area to another not only in terms of timing but also in terms of degree. One hundred percent of the known post-Holstein occipitals in Europe display more or less accentuated derived conditions. Yet for other anatomical areas, it is still possible to find the persistence of primitive conditions on some rather late individuals. Although the upper facial and mandibular area seems to be the first to exhibit Neandertal features, the Montmaurin mandible, which could be as late as early stage 5 (Tavoso 1982), or the even younger Bagnoles specimen (Julia and Bischoff 1991), are still very plesiomorphic. Amidst the late Neandertals, one isolated individual at Spy still displays a pentagonal shape of the vault in posterior view (Thoma 1975).

This approach clearly discards the typological views on the Neandertals and their ancestors but its use results in various problems, such as the difficulty of sharply defining grades (Hublin 1988b). Considering the mosaic nature of the accretion phenomenon, tracing clear divisions along the pre-Neandertals/Neandertals lineage is quite artificial. The chronological grades tentatively presented above are affected by consistent morphological overlaps.

Regarding the taxonomical problem, it seems difficult not to include in *Homo neanderthalensis* (or *Homo sapiens neanderthalensis*) all the specimens involved in this Neandertalization process, even if they display only a few derived Neandertal features. Separating the oldest sequence of the series as *Homo heidelbergensis* would lead to significant problems in the anatomical definition of the two taxa, especially if some African specimens are also included in *Homo heidelbergensis* (Rightmire 1990; Stringer 1991). If, in Europe, a taxon anatomically distinguishable from *Homo erectus sensu lato* was present before the development of the Neandertal lineage, its hypodigm should be restricted to the populations anterior to the development of the first Neandertal apomorphies. In this case, the term *Homo heidelbergensis* itself, with the Mauer mandible as a type specimen, would be inappropriate. Considering its fragmentary nature, this specimen is at least to be regarded as an *insertae sedis*, if it is not one of the first representatives of *Homo neanderthalensis*.

5. ENVIRONMENTAL CHANGES AND EVOLUTIONARY PROCESS

Adaptation to a new environment or ecological niche led the Neandertal ancestors to evolve under the pressures of selection. In particular, the body proportions of the European Neandertals are demonstrated to be hyper-arctic, contrasting with the conditions observed in the Pleistocene representatives of the genus *Homo* in lower latitudes and in the early modern humans in the Levant and in Europe (Trinkaus 1981; Holliday 1995). This adaptation clearly resulted from the colonization of higher latitudes but not necessarily from the constant exposure to peri-arctic conditions, if one considers the likely limited ability of Middle Pleistocene hominids to resist technically the climatic stress. In addition, moist conditions are demonstrated to result in effects similar to cold conditions (Holliday and Falsetti 1995). As a matter of fact, climatic conditions during the last half million years were always notably colder than the present except during very brief episodes of stages 11, 9, and 5e. Besides the features which developed directly as adaptative responses to the environment, some Neandertal features may have developed secondarily from pleiotropic effects and therefore remain beyond the reach of simple interpretations. For example, similarities in the relative proportions of the phalanxes in the foot and the hand (Trinkaus 1983) could result from such a phenomenon.

In order for the Neandertals to develop as a subspecies of *Homo sapiens*, and even more as a separate species, some level of genetic isolation from adjacent populations of archaic humans was needed. Throughout the evolution of species, this isolation has usually resulted from peculiar geographical and/or ecological conditions. Howell (1960) emphasized the role of paleogeography in the rise of the Middle and Upper Pleistocene groups of hominids, underlining the possible effects of a glacial event on the isolation of Europe. As is best documented by the conditions of the last pleniglacial, the development of an ice sheet over northwestern Europe during a glacial maximum and the occurrence of a wide zone of permafrost all along this ice sheet dramatically reduced the possibilities of humans settling in the mid-latitudes. Very continental environments developed in Eastern Europe over broad areas. A strong glacial system covered the main mountain chains, in particular the Caucasus. In addition, the Caspian sea, which was fed by a powerful glacial fluvial system, developed on the edge of the glacial sheet and was not regressive but rather transgressive and dramatically expanded toward the northwest. It extended beyond 50 degrees of north latitude and approached the southern end of the Ural mountains, reaching a latitudinal limit that was never surpassed in the North by permanent Paleolithic settlements before stage 5. During the Pleistocene, it also periodically developed a channel along the Manytsch depression from its west bank to the Azov and Black seas, thereby doubling the geographic barrier of the Caucasus. While the regression of the Atlantic ocean increased the continental surface of Europe toward the west, the possibilities of exchange were very limited in the east.

When the isotopic climatic curves were first widely used, some attempts were made following Howell's model to match the development of the "Classic Neandertals" with the cold oscillation post-dating stage 5e (Boaz et al. 1982). However, this model can be criticized. Neandertals were known to occur already during this interglacial (Eemian), and the emergence of their lineage is now demonstrated to be much older, as demonstrated above. However, if this early development occurred between 500 and 400 ky BP, the isotopic curves do show a major change in the climatic history of Europe (Figure 1). Throughout the Pleistocene, there is a shift from a pattern of moderate climatic oscillations to alternation between contrasting cold and temperate periods. The isotopic curves established in the mid-Atlantic at latitude c.41 degrees N demonstrate an increase in the amplitude and wave length around 0.7 my. In the Mediterranean sea, the curve obtained by Vergnaud-Grazzini et al. (1990) in the Tyrhenian basin demonstrates that the swing between cold to temperate conditions passed through 2 stages during which it significantly increased. In the first period, around 900,000 BP, a clear system of alternating cold and warm faunas is observed in Europe and glaciations are documented in Alpine deposits below the Matuyama/Bruhnes boundary (Audra and Rochette 1993). A second stage in the increase of amplitude is observed near 450,000 BP. It is contemporary with the development of cold species of mammals and with the emergence of the European hominids' endemicity.

The warmest interglacial in the last million years (stage 11) followed an exceptionally severe glaciation (stage 12) (Shackleton 1987). This increase in the amplitude of the

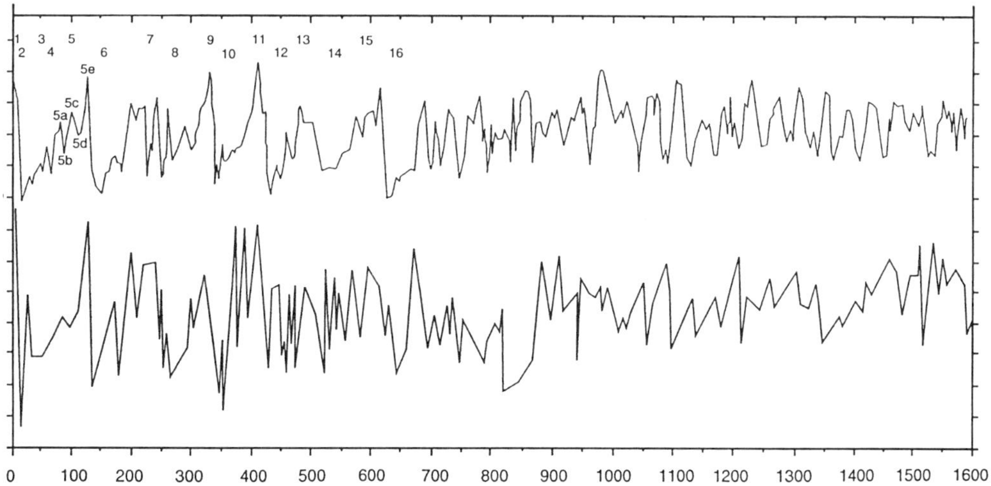

Figure 1. Top: The oxygen isotopic record in borehole (site 607) in the mid-Atlantic at latitude c. 418N (from Ruddiman et al. 1989). Bottom: The ^{18}O record of *G. bulloides* of ODP hole 653A in the Tyrrhenian Basin (Vergnaud-Grazzini et al. 1990).

climatic oscillations is associated with the occurrence of brief warm interglacials (the already mentioned stages as well as 11, 9, and 5e) which may have played a role in the peopling of the mid-latitudes of Europe (but see discussion in Gamble 1986; Roebroeks et al. 1992). Stage 7, while "interglacial," witnessed the persistence of cool conditions while extensive ice caps remained present in the northern hemisphere (Shackleton 1987). The period covering stages 8, 7, and 6 represents virtually a double glaciation of 150,000 years, which witnessed the final development of the Neandertal morphology.

During most of the last half million years, European climatic conditions were regionally cool to cold, limiting both the northern dispersal and the density of human peopling. The main exchanges between Europe and more easterly regions were then likely limited to the corridors between Anatolia and the Balkans. North of the Black Sea, few or no continuous human settlements above 50 degrees of latitude are demonstrated to predate stage 5e (Klein 1973; Soffer 1989). By this time, the differentiation of the European hominids as Neandertals was accomplished virtually entirely (although they are still sometimes called "Proto-Neandertal" or "early Neandertal").

As a whole, the European subcontinent was rather isolated when Neandertals emerged. However, the impact of these cold episodes was not only seen in increased isolation. The extreme conditions, although only found during rather relatively short periods, could also have had a major impact on the distribution, demography, and subsequently the evolutionary processes of European peoples. In addition to a north-to-south change of the human population density, a west-to-east cline may also have developed. The pattern of raw material transport (Feblot-Augustins 1993) suggests quite a different style in land occupation in Western versus Central/Eastern Europe during the Middle Paleolithic. This idea is supported by modern examples showing that the density of hunter-gatherer territorial exploitation in middle latitudes decreased significantly from the oceanic and more humid western areas to the inner and eastern zones that were exposed to continental conditions (Demars 1996).

The conditions of stage 2, when modern humans had already peopled Europe, are better known. Almost a century and a half of archeological surveys and excavations demonstrated that, at the peak of stage 2, most of the European territories north of the Alps were abandoned by humans who were forced to find refuge in small pockets in the Southern part of the continent. In the southwest, glaciers developed in the central sierras of Spain. Most of the Solutrean sites are distributed along the coasts of the Iberic Peninsula within 50 km or less of the present coast. Farther north, other disconnected refuges are represented by the Dordogne and Charente area, the north facade or the Pyrenees Mountains, and the Gard-Ardèche-Provence area more to the east (Straus 1991). Only eight percent of the French Solutrean sites are open air sites, which represents the lowest proportion observed in the local Upper Paleolithic (Demars 1996). Only two sites are known in the Parisian basin (Arcy-sur-Cure and La Montagne de Segrez). The Solutrean is unknown further north. In the eastern areas, although Germany and most of Poland were abandoned, Italy, Greece, and the Balkans offered refuges comparable to those found in southwestern Europe. The clustering effect of this climatic event on the animal populations resulted in faunal differentiation between Eastern and Western Europe (Delpech and Guadelli 1992). It also affected the cultural evolution of Upper Paleolithic with the subsequent separation of Western European cultural lineages (Solutrean and then Magdalenian) from eastern branches (Epigravetian) (Djinjian 1996).

If a pleniglacial episode had such a dramatic effect on Upper Paleolithic populations, who used needles to make clothing and had sophisticated dwellings, one can assume that it may have had a comparable if not stronger impact on Middle Pleistocene archaic humans. During the range of time when Neandertals emerged, the conditions of the last glacial maximum were significantly exceeded by stage 12 and perhaps marginally by stage 6 (Shackleton 1987). Overall the latter witnessed less extreme temperatures than stage 2 but was much longer (Jouzel et al. 1993). Stage 10 was comparable to stage 2 in intensity. Aside from the intensity and length of these major glacial episodes, one can also question the effect of brutal climatic changes on the human settlements. Glaciological data demonstrate that in contrast to the extreme stability of the Holocene, the previous 250 ka have been characterized in the North Atlantic region by marked climatic instability (Dansgaard et al. 1993). Even during the culmination of the Eemian, brief but catastrophic events are recorded in the ice deposits of Groenland (Anklin et al. 1993). They witnessed average temperature fluctuations of 10 to 14°C within the time length of one or a few human generations.

During the second half of the Middle Pleistocene, glacial peaks provoked major contractions of the available European territory and certainly led to a disjunction of discrete populations in separate pockets. As in stage 2, local increases in population density may have occurred, supported by the development of new continental areas (Van Andel 1989). Yet at the scale of the entire continent, it is likely that massive reductions in population size occurred. When they last for several thousand years, these phenomena inevitably lead to the reduction of genetic variability and to the promotion of genetic drift. It should be emphasized that similar effects are observed within the ungulate populations (Delpech, in press). During both stage 2 and stage 6, after having extended their living area towards an extremely southwestern limit, populations of reindeers and saiga antelopes were eventually reduced to isolates in southwestern France. These isolates were apparently totally separated from their more eastern representatives and suffered body size reduction as well as some local morphological change. This evolution certainly affected the available biomass and possibly the density of human peopling. It could also parallel some aspects of human biological evolution.

It has already been argued that the northern extension of the Neandertals and their direct ancestors was limited during the cold stages (Tuffreau 1987; Hublin 1990). Although a series of archeological sites is assigned to stage 6 in middle latitudes (Achenheim, Ariendorf 2, La Cotte St Brelade), none of these rare sites can be securely assigned to the peak of the pleniglacial conditions. In Northern France, very few sites are reported to have been inhabited during the cold conditions of stage 4 (Auguste 1995b). In the late Middle Pleistocene, the most numerous sites and the richest, e.g., Biache (Tuffreau and Sommé 1988), are related to interglacial, interstadial, or early glacial periods. During most of the Lower and Middle Paleolithic, human occupation of European middle latitudes was discontinuous in space and pulsating, developing essentially during temperate to moderately cold climatic periods. In this situation, the difficulty in evaluating possible demographic crashes at the peak of the glacial episodes comes from the limited chronological resolution in the Middle Pleistocene. In fact, a relatively brief but intense cold episode comparable to that of stage 2, which could have resulted in major depopulation, would likely remain undetectable in the archaeological record.

6. EASTERN EXPANSION OF THE NEANDERTALS

As effective as it was, the isolation of the European populations was not complete. Although there is no evidence of significant biological exchange across the straits of Gibraltar during the late Middle Pleistocene (Hublin 1992), we do find Neandertals in the

Near East. However, they are not documented in sites as old as those in Europe. The Zuttiyeh skull, with a possible age estimated between 200,000 and 280,000 years (stage 8 to 7) (Bar-Yosef 1995; Vandermeersch 1995), does not display clear Neandertal affinities (see Sohn and Wolpoff 1993 for discussion and references) at a time when European Neandertal derived features are already observed. In addition, in contrast to the local early modern humans of stage 5, eastern Neandertals display some cold-adapted body proportions, although these are not as extreme as those of the western Neandertals (Trinkaus 1981; Holliday 1995). This is now usually interpreted as resulting from their origin in latitudes higher than those of the Levant area. Still, as documented by the European Upper Paleolithic record, a short period of time (less than 20 ka) is enough to allow significant changes in the climatic adaptation of body proportions (Holliday 1995).

One can consider the occurrence of Neandertals in the Near East as late expansion out of their natural European territory. It has been proposed by Bar-Yosef (1988) that this movement could have been related to environmental pressure resulting from the glacial stage 4. For some authors, this hypothesis would be consistent with overall similarities between Near Eastern Neandertals and European Neandertals from stage 5 (Condemi 1992). One may wonder why such a movement did not happen during an earlier cold stage, such as stage 6. A related question concerns determining the nature of the Mousterian peopling of Anatolia. Where was the southern boundary of the palearctic province peopled by Neandertals during the cold stages of the late Middle Pleistocene: along the paths between the Anatolian plateau and the Balkans, or the Taurus-Zagros line? Supporting the second hypothesis is the evidence for cultural affinities between the interglacial Middle Paleolithic industries of the Balkans and northern Middle East (Zagros Group) (Kozlowski 1992).

Another way to consider the expansion of Neandertal territory toward the southeast and the east is to emphasize the role of temperate rather than cold periods. In this view, the Neandertal peopling of the Near East would not result from environmental pressure during stage 4 but rather from their previous spread into the middle latitudes of Eastern Europe. As a matter of fact, stage 5e is certainly more exceptional as a warm stage than stage 4 is as a cold stage. The new conditions on the eve of stage 4 included the extension of the Neandertals' settlements not only into the Ukrainian and Russian plain but probably even more to the east. In this range of latitudes, Neandertal adaptations allowed their expansion beyond the northern shores of the Caspian sea which, in contrast with the Mediterranean and Black seas, underwent a regression during the interglacial. The occurrence of Neandertals further east at Teshik-Tash (Uzbekistan) could have resulted from the use of the northern and shortest route from Eastern Europe to Central Asia. Future discoveries will tell us how far east the Neandertal extension continued in the middle latitudes. This situation would also explain a later major movement into the Near East during the following cold stage (4).

An early occurrence of the Neandertals, at least in the northern part of the Near East, remains difficult to verify. It would, however, explain the morphological differences observed between them and the Western Neandertals. It would also support the interglacial age of the Shanidar layer D hominid sample proposed by some authors (Trinkaus 1995). However, the chronology of this site is still debated, as is the chronological position of the Tabun C1 hominid in the Levant. An assignment of this specimen to late stage 7 or early stage 6, predating the Skhul/Qafzeh sample (Mercier et al. 1995), would support an early age for the southernmost penetration of the Neandertals in the Near East. Yet Tabun C1 displays one of the most derived morphologies of the occipito-mastoid area amidst the Near East Neandertals. This feature could support a later age and an intrusive burial of the specimen in layer C of Tabun (see Bar Yosef 1995 for further comments). While the mastoid area is one of the last anatomical areas to develop a derived Neandertal morphology, this evolution is already observed in Europe on Saalian, Stage 6, specimens. However, if the oldest estimation is retained for Tabun C1 dating (Vandermeersch 1995), some of its derived temporal conditions would (perhaps surprisingly) predate the earliest indisputable occurrence of these features in Europe.

7. CONCLUSIONS

The first demonstrated attempts at the colonization of Europe are documented around 1 million years or slightly earlier. One or several distinct waves of populations entered the southern part of the subcontinent in the Mediterranean areas. However, it is not before 500 ky BP that a dense record of archeological sites is observed in higher latitudes. Even after this date, humans seem to have been primarily restricted to an area south of 53 degrees north. This limit includes most of the sites dated to cold as well as relatively warm stages prior to the arrival of modern humans in Europe. This limit is effective not only in the west, where the last extension of the ice-sheet could have destroyed hypothetical sites further north, but is also observed in the Russian plain where the oldest significant extension of Middle Paleolithic settlements toward the north occurs during the exceptionally warm stage 5e.

Geographical and climatic constraints resulted in a relative isolation of the European populations, with possible genetic exchange through the Bosphorus corridor and an apparent biological isolation from northwestern Africa. This isolation was intensified periodically by the development of glacial conditions. The development of European endemicity is consistent with the occurrence of major cold stages after 450,000 BP.

Since this period, the European hominids display the development of derived morphological features relating all the known specimens to a unique lineage leading to the Neandertals. This evolution resulted from an *accretion process* characterized by the successive occurrence of new features and by an *increase in their frequency* within the pre-Neandertal populations. Because of this shift in frequency, and although some anatomical areas are affected earlier than others, clear cut stages can hardly be defined. From a taxonomic point of view, integrating all the specimens displaying this trend into a unique taxon (*Homo sapiens neanderthalensis* or *Homo neanderthalensis*) seems the best option.

The major cold stages of the last half million years increased the climatic stress resulting in the isolation of European hominids and the evolution of cold adaptations. These stages also had a major effect on the distribution and size of the Neandertal populations. The model developed by examining the effects of stage 2 on the Upper Paleolithic groups strongly suggests that the most extreme glacial conditions could have lead to short but intense clustering of the European populations along with a dramatic reduction of their size. In particular, the peak of stage 12 (425–480 ky BP), which exceeded the cold conditions of stage 2, may have caused a significant demographic crash in the European populations and triggered their evolutionary divergence. Added to the founder effect following the initial colonization of Europe by small populations, such events would have produced genetic drift episodes resulting in the fixation of derived features. In this view, some of these features could have developed although they do not have a clear adaptative significance. This model would predict a decrease of the variability, as some studies suggest (Maureille 1994; Elyatqine 1995), in the pre-Neandertal/Neandertal populations that experienced major changes of their size.

Regarding the geographical distribution of the Neandertals, the exceptionally warm stage 5e played an important role in the eastern extension of Neandertals, allowing them to occupy territories between the Ural Mountains and the Caspian depression and eventually part of Central Asia. However, it would not be surprising to find earlier occurrences of the Neandertals between the Bosphorus and Levant corridors in cold episodes prior to stage 4.

REFERENCES

Anklin, M., J. M. Barnola, J. Beer, T. Blunier, J. Chappellaz, H. B. Clausen, D. Dahl-Jensen, W. Dansgaard, M. de Angelis, R. J. Delmas, P. Duval, M. Fratta, A. Fuchs, K. Fuhrer, N. Gundestrup, C. Hammer, P. Iversen, S. Johnsen, J. Jouzel, J. Kipfstuhl, M. Legrand, C. Lorius, V. Maggi, H. Miller, J. C. Moore, H. Oeschger, G. Orombelli, D. A. Peel, G. Raisbeck, D. Raynaud, C. Schott-Hvidberg, J. Schwander, H. Shoji, R. Souchez, B. Stauffer, J. P. Steffensen, M. Stievenard, A. Sveinbjrnsdottir, T. Thorsteinsson, and E. W. Wolff. 1993. Climatic instability during the last interglacial period recorded in GRIP ice core. *Nature* 364:203–7.

Arsuaga, J. L., I. Martinez, A. Gracia, J. M. Carretero, and E. Carbonell. 1993. Three new human skulls from the Sima de los Huesos Middle Pleistocene site in Sierra de Atapuerca, Spain. *Nature* 362:534–7.

Arsuaga, J. L., I. Martnez, A. Gracia, and C. Lorenzo. 1997. The Sima de los Huesos crania (Sierra de Atapuerca, Spain). A comparative study. *Journal of Human Evolution* 33:219–81.

Ascenzi A., I. Bidditu, P. F. Cassoli, A. G. Segre, and E. Segre-Naldini. 1996. A calvarium of late *Homo erectus* from Ceprano, Italy. *Journal of Human Evolution* 31(5):409–23.

Audra, P., and P. Rochette, 1993. Premières traces de glaciations du Pleistocène inférieur dans le massif des Alpes. Datation par paléomagnétisme de remplissages à la grotte Vallier (Vercors, Isère, France). *Comptes Rendus de l'Académie des Sciences* 317(II):1403–9.

Auguste, P. 1995a. Cadres biostratigraphiques et paléoécologiques du peuplement humain dans la France septentrionale durant le Pleistocène. Apports de l'étude paléontologique des grands mammifères du gisement de Biache-Saint-Vaast (Pas-de-Calais). Ph. D. diss. Museum Nationale d'Histoire Naturelle. Paris.

Auguste, P. 1995b. Adaptation biologique et culturelle des prénéandertaliens et des néandertaliens aux modifications paléoécologiques et fauniques en France septentrionale, 99–117. In *Nature et Culture*, ed. M. Otte. Liège: ERAUL 68.

Bar-Yosef, O. 1988. The date of Southwest Asian Neandertals, 31–8. In *L'Homme de Neandertal*, No. 3, ed. E. Trinkaus. Liège: ERAUL 30.

Bar-Yosef, O. 1995. The Lower and Middle Palaeolithic in the Mediterranean Levant: Chronology and cultural entities, 247–63. In *Man and Environment in the Palaeolithic*. H. Ullrich ed. Liège: ERAUL 62.

Bermudez de Castro, J. M. 1986. Dental remains from Atapuerca (Spain) I. Metrics. *Journal of Human Evolution* 15:265–87

Bischoff, J. L., J. A. Fitzpatrick, L. Len, J. L. Arsuaga, C. Falguères, J. J. Bahain, and T. Bullen. 1997. Geology and preliminary dating of the hominid-bearing sedimentary fill of the Sima de los Huesos Chamber, Cueva Mayor of the Sierra de Atapuerca, Burgos, Spain. *Journal of Human Evolution* 33:129–54.

Boaz, N. T., D. Ninkovitch, and M. Rossignol-Strick. 1982. Paleoclimatic setting for *Homo sapiens* neanderthalensis. *Naturwissenschaften* 69:29–33.

Brauer, G., and M. Schultz. 1996. The morphological affinities of the Plio-Pleistocene mandible from Dmanisi, Georgia. *Journal of Human Evolution* 30:445–81.

Brigland, D. R. 1994. *Quaternary of the Thames*. London: Chapman and Hall.

Carbonell, E., J. M. Bermudez de Castro, J. L. Arsuaga, J. C. Diez, A. Rosas, G. Cuenca-Bescos, R. Sala, M. Mosquera, and X. P. Rodriguez. 1995. Lower Pleistocene hominids and artifacts from Atapuerca-TD6 (Spain). *Science* 269:826–9.

Condemi, S. 1992. *Les Hommes Fossiles de Saccopastore et Leurs Relations Phylogénétiques*. Paris: CNRS Editions.

Condemi, S. 1996. Une origine européenne pour les néandertaliens du Proche-Orient? L'apport des nouvelles données chronologiques. Paper presented at the *121e Congrès National des Sociétés Historiques et Scientifiques*, Nice, October 26–31, 1996.

Condemi, S., and J. Piveteau. 1988. L'os temporal Riss-Würm (BD 7) provenant de la grotte de La Chaise, abri Bourgeois-Delaunay, 105–10. In *L'Homme de Néandertal*, No. 3, ed. E. Trinkaus. Liège: ERAUL 30.

Dansgaard, W., S. J. Johnsen, H. B. Clausen, D. Dahl-Jensen, N. S. Gundestrup, C. U. Hammer, C. S. Hvidberg, J. P. Steffensen, A. E. Sveinbjörnsdottir, J. Jouzel, and G. Bond. Evidence for general instability of past climate from a 250-kyr ice-core record. *Nature* 364:218–20.

Dean, D., J. J. Hublin, R. Holloway, and R. Ziegler. In press. On the phylogenetic position of the pre-Neandertal specimen from Reilingen, Germany. *Journal of Human Evolution.*

Delpech, F. In press. Biomasse d'Ongulés et peuplement paléolithique au nord de l'arc alpin. In *Archéologie et inférences démographiques*, ed. J. P. Bocquet-Appel. *Bulletins et Mémoires de la Société d'Anthropologie de Paris.*

Delpech, F., and J. L. Guadelli. 1992. Les grands mammifères gravettiens et aurignaciens de la grotte de Temnata, 141–216. In *Temnata Cave. Investigations in Karlukovo Karst Area*, Vol. 1. Krakow: Jagellonia University Press.

Demars, P. Y. 1996. Démographie et occupation de l'espace au Paléolithique Supérieur et au Mésolithique en France. *Préhistoire Européenne* 8:3–26

Elyatqine, M. 1995. Variabilité et evolution de l'os temporal chez Homo sapiens. Comparaison avec Homo erectus. Ph. D. diss. University of Bordeaux I. Bordeaux.

Djindjian, F. 1996. Le cloisonnement du peuplement de Chasseurs-Cueilleurs au maximum glaciaire en Méditerranée occidentale, 161–2. In abstracts volume of the *121e Congrès National des Sociétés Historiques et Scientifiques*, Nice, October 26–31, 1996.

Féblot-Augustins, J. 1993. Mobility strategies in the Late Middle Paleolithic of Central Europe and Western Europe: Elements of stability and variability. *Journal of Anthropological Archaeology* 12:211–65.

Franciscus, R. G., and E. Trinkaus. 1995. Determinants of retromolar space presence in Pleistocene *Homo* mandibles. *Journal of Human Evolution* 28:577–95.

Gabunia, L., and A. Vekua. 1995. A Plio-Pleistocene hominid from Dmanisi, East Georgia, Caucasus. *Nature* 373:509–12.

Gamble, C. 1986. *The Palaeolithic Settlement of Europe.* Cambridge: Cambridge University Press.

Grün, R. 1996. A re-analysis of electron spin resonance dating results associated with the Petralona hominid. *Journal of Human Evolution* 30(3):227–41.

Holliday, T. W. 1995. Body Size and Proportions in the Late Pleistocene Western Old World and the Origins of Modern Humans. Ph. D. diss. The University of New Mexico. Albuquerque.

Holliday, T. W., and A. B. Falsetti. 1995. Lower limb length of European early modern humans in relation to mobility and climate. *Journal of Human Evolution* 29:141–53.

Howell, F. C. 1960. European and Northwestern African Middle Pleistocene Hominids. *Current Anthropology* 1:195–232.

Hublin, J. J. 1988a. Caractères derivés de la region occipito-mastoidienne chez les Neandertaliens, 67–73. In *L'Homme de Néandertal* , No. 3, ed E. Trinkaus. Liège: ERAUL 30.

Hublin, J. J. 1988b. Les plus anciens représentants de la lignée prénéandertalienne, 81–94. In *L'Homme de Néandertal* , No. 3, ed E. Trinkaus. Liège: ERAUL 30.

Hublin, J. J. 1990. Les peuplements paléolithiques de l'Europe, 29–37. In *Paléolithique Moyen Récent et Paléolithique Supérieur Ancien en Europe,* ed. C. Farizy. Nemours: APRAIF.

Hublin, J. J. 1992. Recent human evolution in Northwestern Africa. *Philosophical Transactions of the Royal Society of London* 337B:185–91

Jouzel, J., N. I. Barkov, J. M. Barnola, M. Bender, J. Chapellaz, C. Genthon, V. M. Kotlyakov, V. Lipenkov, C. Lorius, J. R. Petit, D. Raynaud, G. Raibeck, C. Ritz, T. Sowers, M. Stievenard, F. Yiou, and P. Yiou. Extending the Vostock ice-core record of palaeoclimate to the penultimate glacial period. *Nature* 364:407–12.

Julia, R., and J. Bischoff. 1991. Radiometric dating of Quaternary deposits and the hominid mandible of Lake Banyolas, Spain. *Journal of Archaeological Science* 18:707–22

Klein, R. G. 1973. *Ice-Age Hunters of Ukraine.* Chicago: University of Chicago Press.

Kozlowski, J. K. 1992. The Balkans in the Middle and Upper Paleolithic: The gate to Europe or a cul-de-sac? *Proceedings of the Prehistoric Society* 58:1–20.

Lumley, H. de, A. Fournier, J. Krzepkowska, and A. Echassoux. 1988. L'industrie du Pleistocène inférieur de la grotte du Vallonet Roquebrune-Cap-Martin, Alpes-Maritimes. *L'Anthropologie* 92:501–614.

Lumley, H. de, A. Fournier, Y. C. Park, Y. Yokoyama, and A. Demouy. 1984. Stratigraphie du remplissage pleistocène moyen de la Caune de l'Arago à Tautavel. Etude de huit carottages effectués de 1981 à 1983. *L'Anthropologie* 88:5–18.

Martinez, I., and J. L. Arsuaga. 1977. The temporal bones from Sima de los Huesos Middle Pleistocene site (Sierra de Atapuerca, Spain). A phylogenetic approach. *Journal of Human Evolution* 33:283–318.

Maureille, B. 1994. La Face Chez Homo erectus et Homo sapiens: Recherche sur la Variabilité Morphologique et Métrique. Ph. D. diss. University of Bordeaux I. Bordeaux.

Mercier, N., H. Valladas, G. Valladas, and J. L. Reyss. 1995. TL dates of burnt flints from Jelinek's excavations at Tabun and their implications. *Journal of Archaeological Science* 22:495–509.

Otte, M. 1996. *Le Paléolithique Inférieur et Moyen en Europe.* Paris: Armand Colin.

Rightmire, P. 1990. *The Evolution of* Homo erectus. Cambridge: Cambridge University Press.

Roebroeks W., N. J. Conard, and T. van Kolfschoten. 1992. Dense forests, cold steppes, and the Palaeolithic settlement of Northern Europe. *Current Anthropology* 33(5):551–86.

Ruddiman, W. F., M. E. Raymo, D. G. Martinsin, B. M. Clement, and J. Backman. 1989. Pleistocene evolution: Northern hemisphere ice sheets and North Atlantic Ocean. *Paleoceanography* 4:353–412.

Semal, P. 1988. Evolution et variabilité des dimensions dentaires chez *Homo sapiens* neanderthalensis. Viroinval: Editions du Centre d'Etudes et de Documentation Archéologiques.

Schwarcz, H. P., and A. G. Latham. 1984. Uranium series age determinations of travertines from the site of Verteszöllös, Hungary. *Journal of Archaeological Science* 11:327–36.

Shackleton, N. J. 1987. Oxygen isotopes, ice volume and sea level. *Quaternary Science Reviews* 6:183–90.

Sohn, S., and M. H. Wolpoff. 1993. Zuttiyeh face: A view from the East. *American Journal of Physical Anthropology* 91:325–47.

Soffer, O. 1989. The Middle to Upper Paleolithic Transition on the Russian Plain, 714–42. In *The Human Revolution*, ed. P. Mellars and C. B. Stringer. Edinburgh: Edinburgh University Press.

Straus, L. G. 1991. Southwestern Europe at the Last Glacial Maximum. *Current Anthropology* 32(2):189–99.

Stringer, C. B. 1991. *Homo erectus* et "Homo sapiens archaïque." Peut-on définir *Homo erectus*?, 51–74. In *Aux origines d'Homo sapiens*, ed. J. J. Hublin and A. M. Tillier. Paris: Presses Universitaires de France.

Tavoso, A. 1982. Le cadre chronologique de la mandibule de Montmaurin, examen des données disponibles. *First International Congess of Human Paleontology*. Nice: CNRS.

Tchernov, E., ed. 1986. *Les Mammifères du Pleistocène inférieur de la vallée du Jourdain à Oubeidiyeh*. Paris: Association Paléorient.

Thoma, A. 1975. Were the Spy fossils evolutionary intermediates between Classic Neandertal and Modern Man. *Journal of Human Evolution* 4:387–410.

Thouveny, N., and E. Bonifay. 1984. New chronological data on European Plio-Pleistocene faunas and hominid occupation sites. *Nature* 308:355–8.

Trinkaus, E. 1981. Neandertal limb proportions and cold adaptation, 187–224. In *Aspects of Human Evolution*, ed. C. B. Stringer. London: Taylor and Francis.

Trinkaus, E. 1983. Functional aspects of Neandertal pedal remains. *Foot and Ankle* 3(6):377–90.

Trinkaus, E. 1990. Cladistics and the hominid fossil record. *American Journal of Physical Anthropology* 83:1–11.

Trinkaus, E. 1993. Variability in the position of the mental foramen and the identification of Neandertal apomorphies. *Rivista di Antropologia* 71(2):259–74.

Trinkaus, E. 1995. Near Eastern Late Archaic humans. *Paléorient* 21(2):9–24.

Tuffreau, A. 1987. Le Paléolithique Inférieur et Moyen du Nord de la France (Nord, Pas-de-Calais, Picardie) dans Son Cadre Stratigraphique. Ph. D. diss. University of Lille. Villeneuve d'Asq.

Turq, A., B. Martinez-Navarro, P. Palmqvist, A. Arribas, J. Agusti, and J. Rodriguez Vidal. 1996. Le Plio-Pleistocene de la région d'Orce, Province de Grenade, Espagne: Bilan et perspectives de recherche. *Paleo* 8:161–204.

Van Andel, T. H. 1989. Late Quaternary sea-level changes and archaeology. *Antiquity* 63:733–45.

Vandermeersch, B. 1995. Le rôle du Levant dans l'évolution de l'humanité au Pleistocène supérieur. *Paléorient* 21(2):25–34.

Vergnaud-Grazzini, C., J. F. Saliège, M. J. Urrutiaguer, and A. Iannace. 1990. Oxygen and carbon isotope stratigraphy of ODP Hole 653A and site 654: The Pliocene-pleistocene glacial history recorded in the tyrrhenian basin (West Mediterranean), 361–86. In *Proceedings of the Ocean Drilling Program, Scientific Results*, Vol. 107, ed. K. A. Kastens, J. Mascle, J. et al.

Vlcek, E. 1993. Entwicklung der Populationen im Pleistozän Europas, 167–79. In *Man and Environment in the Paleolithic*, ed. H. Ullrich. Liège: ERAUL 62.

Wolpoff, M. H. 1982. The Arago dental sample in the context of hominid dental evolution, 389–410. In "L'*Homo erectus* et la place de l'Homme de Tautavel parmi les Hominidés," *1er Congrès International de Paléontologie Humaine, Prétirage* Nice: CNRS Editions.

19

SAPIENS AND NEANDERTALS

Rethinking the Levantine Middle Paleolithic Hominids

Baruch Arensburg[1] and Anna Belfer-Cohen[2]

[1]Department of Anatomy and Anthropology
Sackler School of Medicine
Tel Aviv University
Ramat Aviv 69978, Israel
[2]Institute of Archaeology
Hebrew University
Mt. Scopus, Jerusalem 91905, Israel

1. INTRODUCTION

The prevailing point of view, until quite recently, was that the Mousterian culture in Israel spanned the period between 75 ka BP (Tabun Cave) and 45 ka BP (Boker Tachtit). A 10 ka interval was posited between the earlier Tabun I skeleton and the later Skhul human remains (Oakley 1969; Masters 1982). This chronological scheme conceived of the alleged Neandertals from Israel as contemporaneous with Mousterian populations in Europe. A common pre-Neandertal origin, or a European-Neandertal migration to the Near East, were proposed as explanations for the morphometric similarities between these two synchronic groups (Vandermeersch 1981; Bar-Yosef 1988, 1989; Trinkaus 1991). However, new dating techniques, additional information furnished by the resumed excavations at certain prehistoric sites, and, to no lesser degree, new approaches to the interpretation of human fossil morphology (Klein 1989; Bar-Yosef et al. 1992) make alternative views of the Levantine Middle Paleolithic more tenable. We present herein some data in support of the view that there is actually little evidence for the existence of a "Southwest Asiatic *Neandertal* population" in Israel during the Middle Paleolithic period.

2. DEFINITIONS OF "NEANDERTAL"

Since the very first stages of the anthropological study of Neandertals in the nineteenth century, opinions have been divided concerning Neandertal phylogenetic status and

Neandertals and Modern Humans in Western Asia, edited by Akazawa *et al.*
Plenum Press, New York, 1998.

their relation to modern humans. Huxley maintained that the Neandertal cranium, rather than constituting an isolated phenomenon, was "the extreme term of a series leading gradually from it to the . . . best developed of human crania" (1863:80). His contemporary, Schwalbe (1899), argued that the Neandertal features suggested a race so morphologically distant from all existing human varieties as to merit the rank of a distinct species.

One hundred years later, the taxonomic position of the Neandertal group is still undecided (Stringer and Gamble 1993). Neandertals' relations to other contemporaneous Mousterian humans as well as to various Upper Paleolithic populations remain to be elucidated (Wolpoff 1992).

Any attempt to evaluate such relationships is necessarily heavily dependent on the definitions of the entities involved. Equally obvious is that such definitions should preferably rest on morphological grounds rather than on cultural or chronological ones. In his listing of various features of the Neandertal hand, Musgrave noted that although "each (of the features) can be found in modern hands . . . they rarely occur combined in one individual hand" (1971:541). He concludes accordingly that it is "the total morphological pattern of the Neandertal hand . . . (that is) unique." Following such an approach, it is not necessary for a specimen to manifest *all* distinctive features to be classified as a Neandertal, but it should display *at least* some of the more diagnostic ones, such as the morphology of the face, the vault or the pelvis, or the limb proportions. These traits must occur "combined in one individual" because it is the totality of the particular morphological pattern which sets the Neandertals apart.

The following discussion examines the distinctive morphological features of the Israel "Neandertals" and evaluates the extent to which they do or do not correspond with the "classic" European Neandertal pattern.

The number of specimens relevant for a Neandertal evaluation in Israel has diminished considerably throughout the years, as a result of a continuous process of "de-Neandertalization." This term refers to the historical process through which human fossils originally labeled as Neandertals were subsequently reevaluated and defined otherwise.

Thus the Zuttiyeh facial fragment, originally considered as proto-Neandertal (Keith 1931), is currently regarded as pre-Sapiens (Vandermeersch 1989), partly because of its pre-Mousterian archaeological context (Gisis and Bar-Yosef 1974) but chiefly because it is devoid of the "typical" Neandertal facial morphology. Others associate it with certain Asian Paleolithic populations (Sohn and Wolpoff 1993), and still others (Smith et al. 1989; Simmons et al. 1991) argue for a pre-Neandertal origin.

Even though the two mandibles recovered from Tabun (the mandible of Tabun I and the isolated mandible, Tabun II) were originally assigned to the same local Neandertal population (McCown and Keith 1939), the morphological differences observed between the two enabled a later reappraisal to consider Tabun II as belonging to an anatomically modern *Homo sapiens* (AMHS) specimen (Vandermeersch 1981).

Similarly, the Skhul remains, suspected from the beginning of possessing modern *Homo sapiens* features and consequently accorded only a "limited" Neandertal status (McCown and Keith 1939), were eventually further dissociated from this category and acknowledged as early anatomically modern *Homo sapiens* (AMHS) (e.g., Santa Luca 1978; Vandermeersch 1982).

The Qafzeh remains, also originally considered as "Neandertaloids" (Vallois and Vandermeersch 1972), were eventually recognized as AMHS (Vandermeersch 1982). This view was later corroborated by the chronological evidence which suggested a much earlier date for these specimens, clearly implying the improbability of any affinity with the later "classic" European Neandertals.

As a result of this "de-Neandertalization" process, Tabun I, the Amud, and the Kebara remains are the only fossils from Israel still regarded as genuine "Near Eastern Neandertals."

2.1. Tabun I

The female skeleton Tabun I (alongside the Tabun II male mandible) was considered by McCown and Keith to represent a Neandertal-related human type. Morphological reexaminations, however, do not seem to corroborate this claim. As noted by McCown and Keith (1939), the specimen's poor state of preservation was a major problem from the start in the evaluation of its morphology. For example, its cranial measurements, such as the basion-bregma height, were estimated from a flattened vertex and a missing basion. Nevertheless, even under these adverse circumstances, McCown and Keith themselves recognized in it "an early stage in the lifting upward of the vault of the skull, a Neanthropic (i.e., modern human) feature which finds a full development in Skhul V" (1939:254). This Neanthropic pattern is similarly apparent in the large mastoid process and in the almost vertical position of the occiput. All these are markedly at odds with the low vault, the protruding shape of the occipital, and the small mastoid bones which are so distinctive of Neandertal morphology.

The claim for Tabun I's Neandertal affinities is not supported by the evidence of mandibular morphology either. Instead, this line of evidence suggests archaic *Homo sapiens* affinities for this specimen. Conversely, the mandibular morphology indicates that Tabun II is rather related to AMHS. Thus, for example, although both specimens retain plesiomorphic traits such as the posteriorly positioned mental foramina and retromolar space, Tabun II has a fairly distinct mental protuberance whereas Tabun I has no trace of it (Vandermeersch 1982).

Whether Tabun I belongs with layer C or B, both layers are considered to date within oxygen isotope stage 5 (Grün et al. 1991). Thus Tabun I should be dated to about 110,000 ka, and possibly even earlier, which seems severely at odds with its identification as Neandertal. In light of the new dates provided for the Upper Pleistocene in Israel, such an identification would constitute evidence for the otherwise highly unlikely contemporaneity of AMHS and Neandertals along some 70 ka at least.

In contrast, the available evidence is altogether compatible with the view that archaic *Homo sapiens* (e.g., Zuttiyeh, Tabun), who inhabited the region as early as 250–130 ka ago, gradually evolved into AMHS. These AMHS, however, featured certain synapomorphies shared with the Neandertals.

Another viewpoint is expressed by Trinkaus (1995). He believes that there is a separate regional lineage in the Near East of Late Pleistocene hominids, the Near Eastern late archaic humans who share affinities with both Near Eastern early modern humans and the European Neandertals while having some unique characteristics of their own.

2.2. Amud I

Suzuki and Takai (1970) maintained that the male skeleton Amud I, although morphologically somewhat more advanced than Tabun I and Shanidar 1, nevertheless shares with them several common characteristics. These features assigned all three specimens into closely-related local Neandertal varieties which may also incorporate the Zuttiyeh fragment. Suzuki and Takai consider Amud I morphologically less advanced than the Skhul population as a whole, yet similar enough in cranial morphology to Skhul IV as to justify their inclusion within the same type.

Reevaluation of the morphological data concerning Amud I reveals that it lacks many of the distinctive Neandertal features. Its cranial height is "greater than any of Middle Palaeolithic man in Europe and Western Asia and is nearly the same as those of the Upper Palaeolithic skulls....(Its) supra orbital torus...is closer to that of Skhul IV than those of Neandertals and Shanidar 1. As far as the frontal angles are concerned, the inclination of (its) frontal bone...is intermediate between Neandertals and Upper Palaeolithic man" (Suzuki and Takai 1970:198).

Amud I's most modern cranial features are found in its temporal bone. The high border of its temporal squama is as curved as in modern man. Its well-developed mastoid process resembles that of AMHS in both shape and size, and differs from that of the "classic" Neandertals and Shanidar 1. Its *fossa mandibularis* is intermediate in depth between those of Neandertals and AMHS.

Amud I's occiput deviates markedly from the Neandertal pattern, and its internal occipital protuberance is on almost the same level as the inion, rather resembling that of AMHS. Its nuchal plane "is swollen much more than is the case in classic Neandertals and even in Shanidar man" (Suzuki and Takai 1970:199).

Some of Amud I's "archaic" cranial features (e.g., the morphology of the tympanic plate, malar bone, and dental arcade) also appear within modern populations. It shares other features (for example, the mental foramen position and retromolar space) not only with Neandertals and with AMHS, but with *Homo erectus* as well (Franciscus and Trinkaus 1995).

Suzuki and Takai (1970) argued that Amud I's facial features are its most "classic-Neandertal" ones. This claim is somewhat at variance with their acknowledgment that in certain facial features, particularly the malar bone, Amud I is intermediate between Neandertals and AMHS. In other features (e.g., maxillary height; nasomalar and zygomaxillary angles), Amud I is definitely different from both European Neandertals and Shanidar 1. The problematic nature of Suzuki and Takai's claim is further illustrated by the fact that Amud I's high, large orbits, which are claimed to be "common to Neandertals and Shanidar 1," are in fact not exclusive to Neandertals but occur in other humans types as well (Table 1).

Apart from this, even a *cursory* examination of Amud I's face would call into question any claims based on its properties, since its incompleteness, especially the lack of contact between the maxilla and the neuro-cranium and zygomatic bones, is most obviously a potential source for misinterpretation. Thus, for instance, its nasal cavity, of which only the inferior part has been preserved, cannot be reconstructed; all measurements per-

Table 1. Orbital and nasal dimensions of Middle Paleolithic specimens[a]

Specimen	Orbit (L/B) (in mm)	Nasal cavity (L/B) (in mm)
Amud I	49 × 38	65 × 34
Broken Hill	51 × 39	51 × 31
Qafzeh VI	48 × 35	
Jebel Irhoud	43 × 41	
Petralona		64 × 32
Saccopastore II		60 × 34
Tabun I		58 × 34
La Chapelle		61 × 34
Mount Circeo		66 × 36

[a]After Suzuki and Takai 1970.

taining to it must be approximated. Nevertheless, even according to these approximate data, Amud I's nasal aperture is just as similar to those of various archaic *Homo sapiens* specimens (Broken Hill, Petralona, Saccopastore II, and Tabun I) as to those of certain classic Neandertal ones (La Chappelle, Mount Circeo). The reported measurements seem to correspond better with a view of these traits as plesiomorphic rather than derived, Neandertal, ones.

Amud I's post cranial skeleton is as distinct as its cranium and it differs from those of both the Neandertals and AMHS. Its upper and lower limb bones are long, in marked contrast to the shorter bones and typical proportions of the allegedly cold-adapted Neandertals (Feldesman et al. 1990). Its estimated stature is quite tall compared with that of "classic" Neandertal males and it is close to that of the Skhul males and the Qafzeh population (see Table 2). The curvature of its radius shaft is slight—definitely not as marked as in the European Neandertals, and its upper limb bones are gracile rather than robust (Hambücken 1995). Its scapula has a dorsal sulcus on its axillary border which does indeed characterize 55.6 percent of the Neandertal specimens but also occurs in many Upper Paleolithic ones as well as in some modern humans (Frayer 1992; Frayer et al. 1993).

In sum, an examination of Amud I's morphology, rather than lending any substantial support to the claim that it corresponds to the Neandertal pattern, clearly affirms its dissociation from this category.

2.3. Kebara 2

Tillier (1991) and Tillier et al. (1989) recognize two major classes of features which characterize the mandible of Kebara 2: plesiomorphies and Neandertal apomorphies (Table 3).

Inspection of the available data reveals that most of the listed apomorphies are not in fact exclusive to Neandertals but occur also within other human groups, including recent populations. Moreover, they are not found in all Neandertals. Thus a posterior position of the mental foramen (under M1), which occurs in 65 percent of the "classic" Neandertals, also characterizes Ehringsdorf, Arago 2 and 13, Sinanthropus AI, BI, III, and Skhul IV. In modern human groups, its rate of incidence is 14.3 percent among African-Americans, 2.7 percent among Australian aborigines, and 4.0 percent among Melanesians (Suzuki and Takai 1970; Trinkaus 1983).

A horizontal mandibular foramen, which characterizes 61.5 percent of the Neandertals, is absent in Amud I, Tabun I, and Shanidar 1. It occurs, however, in 44.0 percent of the European Upper Paleolithic specimens, in 6.0 percent of European Mesolithic remains,

Table 2. Stature measurements of some Middle Paleolithic specimens[a]

Specimen	Gender	Stature (cm)
Amud I	male	177.8
Spy	male	165.3
La Chapelle	male	168.9
Skhul VI	male	176.2
Skhul V	male	185.7
Qafzeh (min.)	female	169.0
Qafzeh (max.)	male	188.0
Kebara	male	173.0

[a]After Suzuki and Takai 1970; Vandermeersch 1991.

Table 3. Plesiomorphic and Neandertal apomorphic traits[a]

Plesiomorphic traits
1. Mandibular body thickness
2. Large digastric fossae
3. Presence of genioglossal fossae
4. Thickness of the retromolar trigone
5. Considerable distance between ramii
6. Absence of menton
Neandertal apomorphic traits
1. Posterior position of mandibular foramen
2. Horizontal shape of the mandibular foramen
3. Retromolar space
4. Large anterior teeth

[a]After Tillier et al. 1989.

and in the majority of an Easter Island skull sample (N=35) (Trinkaus 1983; Frayer 1992; B. Arensburg, personal observation; E. Aspillaga, personal communication on Southern Chile aborigines).

A retromolar space does indeed characterize all "classic" Neandertals with the exception of La Quina 9, but it also occurs in AMHS (Skhul IV) as well as in living human populations, as is frequently observed in dental clinics (Crubezy 1988).

Neandertal tooth morphology has received much attention in the literature. Smith (1989) claims that while early *Homo sapiens* from Israel and the European Upper Paleolithic are markedly similar with regard to the size of their permanent anterior dentition, both types differ significantly in this respect from the Neandertals, both the European and the alleged Near Eastern ones. However, Smith's data do not fully support her contention. Only in four out of 12 measurements (buccolingual dimensions of upper and lower incisors) are the Israel "Neandertals" metrically close to the European Neandertals, and only in a single feature (mesiodistal length of the lower central incisor) are the Israel early *Homo sapiens* and European *Homo sapiens* similar.

In contrast to Smith, Wolpoff (1989) detects a close similarity in tooth size between early Upper Paleolithic European populations and "classic" Neandertals, as well as marked differences between the latter and the Skhul-Qafzeh *Homo sapiens*, especially with regard to their anterior dentition. These contradictory views seem to cast serious doubts on the validity of the exclusive use of metric dental data in establishing associations between Eurasian "Neandertals" and Eurasian *Homo sapiens*.

Tillier (1989) claims that only a few of the Neandertals display extreme anterior tooth wear. She also argues that a large anterior dentition occurs not only among the prognathous Neandertals but also among other Middle Paleolithic hominids in both Southwest Asia and North Africa. Thus it is the combination of extreme anterior tooth wear, large anterior teeth, and midfacial prognathism which forms the unique Neandertal facial trait pattern. The presumed role of the anterior dentition in the reshaping of the Neandertal face (Rak 1986; Trinkaus 1987; Smith and Paquette 1989) is not supported by evidence from the Israel "Neandertals," because the average size of their anterior dentition falls within the range of variation of early anatomically modern *Homo sapiens* (see Smith 1989; Anton 1990) and their dental wear is not pronounced.

Among the Near Eastern Paleolithic hominids, the elongated facial features (e.g., very large nasal aperture, a more anteriorly placed dentition) are most probably plesiomorphic. Moreover, in most if not all cases, the facial features are approximated reconstructions. It seems therefore that the morphometric characteristics of the Kebara 2 mandible are not sufficiently specific to account for its affiliation to any particular human subgroup, let alone the Neandertal one.

It has been suggested that the pelvis of Kebara 2 provides "additional support from yet another anatomical region that two hominid lineages (represented by Qafzeh 9 and Kebara 2) were evolving in parallel during the Middle Paleolithic" (Rak 1990:331). Rak recognizes in this pelvis several Neandertal features, especially the very long and unusually thin superior pubic ramus, the obtuseness of the subpubic angle, and the position of the internal obturator groove in relation to the ischial tuberosity.

Rak's argument for the distinctiveness of the Kebara 2 pelvis is not easily reconciled with his acknowledgment that "it is surely possible that, given the mosaic nature of evolution, the pelvis from Kebara manifests a primitive morphology, whereas the Qafzeh pelvis, like that of modern humans, exhibits the derived form" (1990:331). Furthermore, the very evidence brought in support of this claim seems to us highly questionable. Human pelves of prehistoric age are notable for their inadequate preservation and insufficient representation in the fossil record. In fact, the pelves of Kebara 2 and Qafzeh 9 are the only Middle Paleolithic specimens that are almost completely preserved (see Rak and Arensburg 1987). Like other parts of the Kebara 2 skeleton, the pelvis displays characteristics common to Mousterian AMHS (e.g., Skhul IX) and to recent humans.

According to McCown and Keith (1939), the ischial tuberosity in Neandertals as compared to modern humans is located further up towards the ischial spine, and the internal obturator groove is situated on its medial and posterior border instead of above it, as in modern pelves. In Skhul IV, Tabun I, and the Krapina right fragment (Sp. 208), the position of the groove is intermediate between those in Neandertals and in *Homo sapiens*. An intermediate position of the groove was also detected in the pelves of an Australian aborigine and a Bushman (McCown and Keith 1939). In Skhul V, on the other hand, the ischial tuberosity groove is positioned as in the Neandertal, the La Chapelle, and the Krapina (left) fragments. All this is a clear indication that diversity in the position of the internal obturator groove is in itself a plesiomorphic feature, common in early AMHS as well as in Neandertals, and present in at least some modern humans as well.

Rak argues that the ratio of pubic ramus length to maximum pubic inlet breadth in Kebara 2 sets it "dramatically apart from the distribution of the modern sample" (1990:329). In fact, the value he furnishes for ramus length in Kebara 2 (Table 4), which expresses the distance from the acetabulum to the superior edge of the symphysis, reflects

Table 4. Ramus length in Middle Paleolithic specimens[a]

Specimen	Ramus length (mm)
Skhul IV	72.0
Skhul IX	83.0
Tabun I	80.0–81.0
Shanidar 1	93.0
Shanidar 3	80.0–90.0
La Ferrassie 1	93.0
Kebara 2 (Rak)	89.0
Kebara 2 (Arensburg)	85.0
Modern human[b]	41.0–86.5

[a]After McCown and Keith 1939; Trinkaus 1983; Rosenberg 1986; Rak 1990.
[b]Arensburg, in preparation (N=60 Males and 50 Females).

the actual length of the ramus in this specific specimen less reliably than the measurable distance from the acetabulum to the middle symphysis, a value which we found to be very close to the one reported for Skhul IX (Table 4). The properties of the superior pubic ramus's cross-section have been studied quite extensively in both ancient and modern humans (Stewart 1960) and were reported to show a wide range of variability. In this respect, Kebara 2 falls midway between the thinnest ramii (e.g., Tabun I, Amud I) and the most developed ones (e.g., Skhul IV, La Ferrassie 1, Obercassel, and recent humans).

In our opinion, the plesiomorphic features of the Kebara 2 pelvis, which occur also in some of the anatomically modern Skhul specimens, not only reflect the great diversity found among the various Mousterian remains in Israel but suggest their common origin as well.

In both the vertebral column and the thoracic skeleton, Kebara 2 seems to fall morphologically and metrically within the range of variation observable in modern humans (Arensburg 1989, 1991). The same can be said for Kebara's hyoid bone, which is discussed at length elsewhere (Arensburg et al. 1990).

As indicated in the preceding sections, the existing evidence does not seem to us to warrant assigning Kebara 2 as Neandertal.

2.4. The Shanidar "Neandertals"

Although we shall not discuss here in detail the human remains from Shanidar, Iraq, some comments are relevant. Stringer and Trinkaus maintained at one time that the presence of certain Neandertal features in the Shanidar remains "unquestionably aligns them with other Neandertals," although they were aware that the Shanidar sample "does not display all the characters commonly or exclusively found in European Neandertal crania" (1981:158, 161). Recently, Trinkaus (1991) divided the Shanidar sample into two groups, an ancient "pre-Neandertal" one and a more recent Neandertal one. The alleged Neandertal sample from Shanidar is accordingly only partially Neandertal, whereas another part of it can be safely assumed to have developed from a local hominid population of uncertain origin which manifests more generalized characteristics, or in other words, a more Zuttiyeh-like morphology.

3. DISCUSSION

Various lines of evidence suggest a close evolutionary relationship between African and Levantine Middle Paleolithic hominids. An "African connection" therefore seems to be a plausible hypothesis for approaching the problem of the origin of the Near East Upper Pleistocene hominids and their relation to other hominid populations.

According to Brauer, "the evidence currently available from sub-Saharan Africa has strengthened the case for an early appearance of modern humans at around 100,000 years ago" (1989:148). Thus the chronological evidence relevant for the Zuttiyeh fragments, which expands the time span postulated for the appearance of AMHS in the Near East (see above), corroborates Brauer's view. It also lends substantial support to the hypothesis that AMHS appeared in the Near East and possibly in Africa during oxygen-isotope stage 5 or 6, some 50,000 years earlier than the dates (100,000–80,000 BP) published recently for both Klasies River Mouth Cave in South Africa (Grün et al. 1990) and for Skhul and Qafzeh (Valladas et al. 1988; Stringer et al. 1989). (It should be pointed out that the plesiomorphic traits of the Zuttiyeh remains, for example, the shape of the zygomatic bone and the supra-orbital torus, indeed relate them to AMHS). A similar notion is implied in

Stringers' and Andrews' view that "a South African origin (of *Homo sapiens*) as recent as 100,000 years ago is unlikely," and that "the adjacent area of the Levant cannot be excluded as a possible source area for *Homo sapiens*" (1988:1267).

The evidence for the presence of AMHS in Africa and in the Near East at an early date and the lack of such evidence from other parts of the world suggest that the distribution of AMHS at that time was restricted to these two regions. Indeed, the Middle Paleolithic human remains from sub-Saharan and North Africa and the Mousterian remains (both AMHS and "Neandertals") from Israel show a distinct resemblance: both groups display a similar pattern of marked morphological variability. In both groups, specimens display numerous plesiomorphic traits as well as many that are common to both archaic and modern *Homo sapiens*.

Thus among the African *Homo sapiens* remains, such a variability characterizes both the sub-Saharan fossils (i.e., Omo 1 and 2, Laetoli 18, the Klasies Cave and Border Cave fossils; Rightmire 1983) and the North African specimens. In the latter, this variability occurs among the more ancient remains (ca. 100,000 BP) recovered in Mousterian contexts at Haua Fteah and Jebel Irhoud, as well as in the more recent ones (ca. 40,000 BP) found in the Aterian sites of Temara, Dar es-Sultan, and Harhoura (Stringer et al. 1984; Ferembach 1989; Hublin 1989, 1992). Such marked morphological variability is equally characteristic of the entire body of the Mousterian fossils from Israel. As noted above, Tabun I and II are vastly different from each other in mandibular morphology, and major differences can be also detected among Skhul IV, V, and IX and between Qafzeh 6 and 9 (McCown and Keith 1939; Vallois and Vandermeersch 1972; Vandermeersch 1982). Against this background, it is doubly significant that none of these specimens manifests the total morphological pattern characteristic of the "classic" West European Neandertals.

We suspect that the differences between the Paleolithic populations from Africa and from Israel are due to the decisive role of local evolutionary circumstances in addition to a continuous Afro-Asiatic gene flow, the origin and direction of which we cannot evaluate at present. However, the morphological similarity between these two populations is not really surprising. Even if an "African Eve" hypothesis, positing a common progenitor of all *Homo sapiens*, is rejected, most anthropologists would probably agree that the African Upper Pleistocene population played a major role in the emergence and development of modern Eurasian hominids. The Near East is bound to have been an inevitable land corridor in their long voyage to the West.

3.1. On Morphological Variability and Near East "Neandertals"

The extensive degree of morphological variability characteristic of the Middle Paleolithic sample from Israel is similarly apparent in the Shanidar sample. Thus putative progressive features which "suggest closer affinities with samples of early anatomically modern crania" (Stringer and Trinkaus 1981:161), such as high cranial vault, height of mastoid process, and absence of horizontal mandibular foramen pattern, are present in the Shanidar sample alongside plesiomorphic traits, classic-Neandertal synapomorphic traits, or symplesiomorphic ones. It seems no accident that Tabun I, Amud I, and Kebara 2 likewise manifest diversity of morphological traits as in the preceding specimens, comprising archaic and shared-derived Neandertal features as well as sapiens-like ones.

We believe that regional evolution of Upper Pleistocene *Homo sapiens*, manifesting generalized as well as derived traits, characterizes the Near Eastern Middle Paleolithic hominids. In this region it is apparently possible to recognize a gradient in the expression of certain morphological features, ascending from the more generalized to the more derived.

Within the Levantine sample, the Neandertal morphological traits never occur combined in one individual as has been found in the European Neandertals. Thus the sample is lacking the total morphological pattern that would affiliate it with the "classic" European Neandertals.

Both the regionality hypothesis (e.g., Trinkaus 1983) and the migration hypothesis for the origin of the Near Eastern 'Neandertals' appear to be dominated by the belief that any specimen manifesting any "classic" Neandertal features must indeed be a Neandertal, though possibly not of the "classic" variety. It is our opinion that the emergence of Neandertal features in the course of human evolution may well reflect a genetically inherent human variability. This variability is clearly expressed in the wide range of morphological diversity observed among the early AMHS specimens from the Near East.

Indeed as previously noted, Neandertal autapomorphic traits occur among the Skhul and Qafzeh *Homo sapiens*, while modern human traits can be found among the alleged Near Eastern 'Neandertals.'

The preceding discussion attempts to outline an evolutionary scenario for the Near East that we believe is beneficial to a better understanding of hominid development in this region during the Upper Pleistocene. Instead of the regionality or migration hypotheses for the presumed origin of 'Neandertals' in this region, we advocate the view of a regional evolution of *Homo sapiens* evincing highly variable morphometric traits. The "Neandertal" features observed among the Near Eastern hominids may be interpreted as synapomorphic characteristics that evolved parallel to those of the "classic" Neandertals. For example, it seems that large anterior teeth and extreme anterior tooth-wear are not indicators of Neandertal affiliation *per se*. In fact, biomechanical studies of Neandertal bite-force production suggest that the great significance attributed to heavy anterior dental loading in the evolution of the Neandertal face is apparently unwarranted (Anton 1990). Moreover, the appearance of a gradient of Neandertal traits from Europe through the Near East to Africa would also seem to point to a decreasing north-south or west-east trend in the process of development of Neandertal-derived or autapomorphic-adaptive features.

In sum, analyses of the Israel Middle Paleolithic human fossils reveal numerous incongruences, such as assumed 'Neandertals' lacking specific Neandertal traits and AMHS manifesting Neandertaloid features. Large morphological variability is observable within both early sapiens and "Neandertal" groups and the evidence suggests that "Neandertals" and sapiens were living in the same territory, sometimes in the very same caves. All these, alongside the generally accepted fact of Upper Pleistocene gene flow in an African-Eurasian direction and the logistical problems implicit in an assumption of a Neandertal migration, at whichever date, from Europe to the Near East make the hypothesis of a strong African influence on the south Levantine Middle Paleolithic human groups highly probable. Accordingly, a gene-flow from the northern provinces seems to play a minor role in shaping the particular morphology of that population.

REFERENCES

Anton, S. C. 1990. Neanderthals and the anterior dental loading hypothesis: A biomechanical evaluation of bite force production. *Kroeber Anthropological Society Papers* 71–72:67–76.

Arensburg, B. 1989. New skeletal evidence concerning the anatomy of Middle Palaeolithic populations in the Middle East: The Kebara skeleton, 165–71. In *The Human Revolution*, ed. P. A. Mellars and C. B. Stringer. Edinburgh: Edinburgh University Press.

Arensburg, B. 1991. The vertebral column, thoracic cage and hyoid bone, 113–146. In *Le Squelette Mousterien de Kebara 2*, ed. O. Bar-Yosef and B. Vandermeersch. Paris: CNRS Editions.

Arensburg, B., L. A. Schepartz, A. M. Tillier, B. Vandermeersch, and Y. Rak. 1990. A reappraisal of the anatomical basis for speech in Middle Palaeolithic hominids. *American Journal of Physical Anthropology* 83:137–46.

Bar-Yosef, O. 1988. The date of south-west Asian Neandertals, 39–48. In *L'Homme de Neanderthal*, No. 3, ed. M. Otte. Liège: ERAUL.

Bar-Yosef, O. 1989. Geochronology of the Levantine Middle Palaeolithic, 589–610. In *The Human Revolution*, ed. P. A. Mellars and C. B. Stringer. Edinburgh: Edinburgh University Press.

Bar-Yosef, O., B. Vandermeersch, B. Arensburg, A. Belfer-Cohen, P. Goldberg, H. Laville, L. Meignen, Y. Rak, J. D. Speth, E. Tchernov, A. M. Tillier, and S. Weiner. 1992. The excavations in Kebara Cave, Mt. Carmel. *Current Anthropology* 33(5):497–550.

Brauer, G. 1989. The evolution of Modern Humans: A comparison of the African and non-African evidence, 123–54. In *The Human Revolution*, ed. P. A. Mellars and C. B. Stringer. Edinburgh: Edinburgh University Press.

Condemi, S. 1992. *Les Hommes Fossiles de Saccopastore*. Paris: CNRS Editions.

Crubezy, E. 1988. Le dolmen des places 1, les liquisses hautes à Nant (Aveyron) II: Première approache anthropologique étude paléopathologique. *Gallia Préhistoire* 30:175–86.

Feldesman, M. R., J. G. Kleckuer, and J. K. Lundy. 1990. Femur/stature ratio and estimates of stature in Mid- and Late-Pleistocene fossil hominids. *American Journal of Physical Anthropology* 83:359–72.

Ferembach, D. 1989. L'emergence d'*Homo sapiens*, 265–91. In *Evolution Biologique, Quelques Donnes Actuelles*, ed. J. Bons and M. Delsol. Paris: Edition Boubee.

Franciscus, R. G., and E. Trinkaus. 1995. Determinants of retromolar space presence in Pleistocene genus Homo mandibles. *Journal of Human Evolution* 28:577–95.

Frayer, D. W. 1992. The persistence of Neanderthal features in post-Neanderthal Europeans, 171–80. In *Continuity or Complete Replacement: Controversies in* Homo sapiens *Evolution*, ed. G. Brauer and F. H. Smith. Rotterdam: A. A. Balkema.

Frayer, D. W., M. H. Wolpoff, A. G. Thorne, F. H. Smith, and G. G. Pope. 1993. Theories of modern human origins: The paleontological test. *American Anthropology* 95(1):14–50.

Gisis, I., and O. Bar-Yosef. 1974. New excavation in Zuttiyeh Cave, Wadi Amud, Israel. *Paléorient* 2:175–80.

Grün, R., P. B. Beaumont, and C. B. Stringer. 1990. ESR dating evidence for early modern humans at Border Cave in South Africa. *Nature* 344:537–9.

Grün, R., C. B. Stringer, and H. P. Schwarcz. 1991. ESR dating of teeth from Garrod's Tabun Cave collection. *Journal of Human Evolution* 20:231–48.

Hambucken, A. 1995. Etude du degré de robustess des os longs du membre supérieur des Néandertaliens. *Bulletins et Mémoirs de la Société d'Anthropologie de Paris* 7(1–2):37–47.

Hublin, J. J. 1989. Les origines de l'Homme Moderne: Europe Occidentale et Afrique du Nord, 423–30. In *Homoidae: Proceeding of the 2nd. International Congress of Human Paleontology*, ed. G. Giacobini. Milano: Jaca Book.

Hublin, J. J. 1992. Recent human evolution in northwestern Africa. *Philosophical Transactions of the Royal Society of London* B 337:185–91.

Huxley, T. H. 1863. *Evidence as to Man's Place in Nature*. London: Williams and Norgate.

Keith, A. 1931. *New Discoveries Relating to the Antiquity of Man*. London: Williams and Norgate.

Klein, R. G. 1989. *The Human Career*. Chicago: University of Chicago Press.

Masters, P. M. 1982. An amino acid racemization chronology for Tabun, 43–54. In *The Transition from Lower to Middle Palaeolithic and the Origin of Modern Man*, ed. A. Ronen. Oxford: British Archaeological Reports International Series 151.

McCown, T., and A. Keith. 1939. *The Stone Age of Mount Carmel, Vol. 2, The Fossil Human Remains from the Levalloiso-Mousterian*. Oxford: Clarendon Press.

Musgrave, J. H. 1971. How dexterous was Neanderthal man? *Nature* 233:538–41.

Oakley, K. 1969. *Frameworks For Dating Fossil Man*. London: Weidenfeld and Nicolson.

Rak, Y. 1986. The Neandertal: A new look at an old face. *Journal of Human Evolution* 15:131–64.

Rak, Y. 1990. On the differences between two pelvises of Mousterian context from Qafzeh and Kebara Caves, Israel. *American Journal of Physical Anthropology* 81:323–32.

Rak, Y., and B. Arensburg. 1987. Kebara 2 Neandertal pelvis; First look at a complete inlet. *American Journal of Physical Anthropology* 73:227–31.

Rightmire, G. P. 1983. The Lake Ndutu cranium and early *Homo sapiens* in Africa. *American Journal of Physical Anthropology* 61:245–54.

Rosenberg, R. K. 1986. The Functional Significance of Neandertal Pubic Morphology. Ph. D. diss. University of Michigan. Ann Arbor.

Santa Luca, A. P. 1978. A re-examination of presumed Neanderthal-like fossils. *Journal of Human Evolution* 7:619–36.

Schwalbe, B. 1899. Studien uber *Pithecanthropus erectus. Zeitschrift fur Morphologie und Anthropologie* 1:16–240.

Simmons, T., A. B. Falsetti, and F. H. Smith. 1991. Frontal bone morphometrics of Southwest Asian Pleistocene hominids. *Journal of Human Evolution* 20:249–70.

Smith, P. 1989. Dental evidence for phylogenetic relationships of Middle Palaeolithic hominids, 111–20. In *L' Homme de Neandertal*, No. 7, ed. M. Otte. Liège: ERAUL.

Smith, F. H., and S. P. Paquette. 1989. The adaptive basis of Neandertal facial form, with some thoughts on the nature of modern human origins, 181–210. In *The Emergence of Modern Humans: Biocultural Adaptations in the Later Pleistocene*, ed. E. Trinkaus. Cambridge: Cambridge University Press.

Smith, F. H., A. B. Falsetti, and S. M. Donnelly. 1989. Modern human origins. *Yearbook of Physical Anthropology* 32:35–68.

Sohn, S., and M. H. Wolpoff. 1993. Zuttiyeh face: A view from the east. *American Journal of Physical Anthropology* 91:325–47.

Stewart, T. D. 1960. Form of the pubic bone in the Neandertal man. *Science* 131:1437–8.

Stringer, C. B., and E. Trinkaus. 1981. The Shanidar Neandertal crania, 129–65. In *Aspects of Human Evolution*, ed. C. B. Stringer. London: Taylor and Francis.

Stringer, C. B., J. J. Hublin, and B. Vandermeersch. 1984. The origin of anatomically Modern Humans in Western Europe, 51–135. In *The Origins of Modern Humans; A World Survey of the Fossil Evidence*, ed. F. H. Smith and F. Spencer. New York: Alan R. Liss.

Stringer, C. B., and P. Andrews. 1988. Genetic and fossil evidence for the origins of Modern Humans. *Science* 239:1263–8.

Stringer, C. B., R. Grun, H. P. Schwarcz, and P. Goldberg. 1989. ESR dates for the hominid burial site of Es Skhul in Israel. *Nature* 338:756–8.

Stringer, C. B., and C. Gamble. 1993. *In Search of the Neandertals*. London: Thames and Hudson.

Suzuki, H., and F. Takai. 1970. *The Amud Man and His Cave Site*. Tokyo: Academic Press.

Tillier, A. M. 1989. The evolution of Modern Humans: Evidence from young Mousterian individuals, 286–97. In *The Human Revolution*, ed. P. A. Mellars and C. B. Stringer. Edinburgh: Edinburgh University Press.

Tillier, A. M. 1991. La mandibule et les dents, 97–111. In *Le Squelette Mousterien de Kebara 2*, ed. O. Bar-Yosef and B. Vandermeersch. Paris: CNRS Editions.

Tillier, A. M., B. Arensburg, and H. Duday. 1989. La Mandibule et les Dents du Neanderthalien de Kebara (Homo 2), Mont Carmel, Israel. *Paléorient* 15:39–58.

Trinkaus, E. 1983. *The Shanidar Neanderthals*. New York: Academic Press.

Trinkaus, E. 1987. The Neanderthal face—Evolutionary and functional perspectives on a recent hominid face. *Journal of Human Evolution* 16:429–43.

Trinkaus, E. 1991. The fossil human remains from Shanidar Cave, Iraq: Evolution and continuity among Near East Late Archaic Humans. *L'Anthropologie* 95(2/3):535–72.

Trinkaus, E. 1995. Near Eastern Late Archaic Humans. *Paléorient* 21(2):9–24.

Valladas, A., J. L. Reyes, J. L. Joron, G. Valladas, O. Bar-Yosef, and B. Vandermeersch. 1988. Thermoluminescence dating of Mousterian Proto-Cro-Magnon remains from Israel and the origin of Modern Man. *Nature* 331:614–16.

Vallois, H. V., and B. Vandermeersch. 1972. Le crane Mousterien de Qafzeh (Homo VI). Etude anthropologique. *L'Anthropologie* 76:71–96.

Vandermeersch, B. 1981. Les premiers *Homo sapiens* au Proche Orient, 97–100. In *Les Processus de l'Hominisation*, ed. D. Ferembach. Paris: CNRS Editions.

Vandermeersch, B. 1982. The first *Homo sapiens sapiens* in the Near East, 297–9. In *The Transition from Lower to Middle Palaeolithic and the Origin of Modern Man,* ed. A. Ronen. Oxford: British Archaeological Reports International Series 151.

Vandermeersch, B. 1989. The evolution of modern humans: Recent evidence from southwest Asia, 155–64. In *The Human Revolution*, ed. P. A. Mellars and C. B. Stringer. Princeton: Princeton University Press.

Vandermeersch, B. 1991. La ceinture scapulaire et les membres superieurs, 157–78. In *Le Squelette Mousterien de Kebara 2*, ed. O. Bar-Yosef and B. Vandermeersch. Paris: CNRS Editions.

Wolpoff, M. H. 1989. The place of the Neandertals in human evolution, 97–141. In *The Emergence of Modern Man,* ed. E. Trinkaus. Cambridge: Cambridge University Press.

Wolpoff, M. H. 1992. Theories of modern human origins, 25–63. In *Continuity or Replacement? Controversies in Homo sapiens Evolution*, ed. G. Brauer and F. H. Smith. Rotterdam: A. A. Balkema.

20

ANATOMY OF THE NEANDERTAL INFANT SKELETON FROM DEDERIYEH CAVE, SYRIA

Yukio Dodo,[1] Osamu Kondo,[1] Sultan Muhesen,[2] and Takeru Akazawa[3]

[1]Department of Anatomy
Tohoku University School of Medicine
Sendai 980, Japan
[2]Director-General of Antiquities and Museums
Damascus, Syrian Arab Republic
[3]International Research Center for Japanese Studies
Kyoto 610-11, Japan

1. INTRODUCTION

A human infant skeleton was discovered in the Mousterian formation at Dederiyeh Cave, northern Syria, in 1993 by the team of the Japan-Syrian joint expedition (Akazawa et al. 1995a,b). The skeleton was lying in the expected anatomical position embedded in a stratigraphic entity that contained a "Tabun B-type" Mousterian industry. The preservation of the bones was remarkably good, and all the vertebrae and the left ribs were identifiable (Figure 1).

This paper will focus only on the anatomical aspects of the skeleton. The good preservation will allow us to assign the specimen to a taxon fairly easily despite its immature state.

2. AGE ESTIMATION

Table 1 contains the deciduous and permanent dentitions of the Dederiyeh infant. Of the deciduous teeth, the mandibular lateral incisors, canines, and second molars have not fully erupted. All the permanent teeth were originally within the jaw bones, including the maxillae and mandible. Some of them, however, could be observed with the naked eye (Figure 2). According to Moorrees et al. (1963), the formation stages of these permanent teeth correspond to the average for modern two-year-old white North American children.

Neandertals and Modern Humans in Western Asia, edited by Akazawa *et al.*
Plenum Press, New York, 1998.

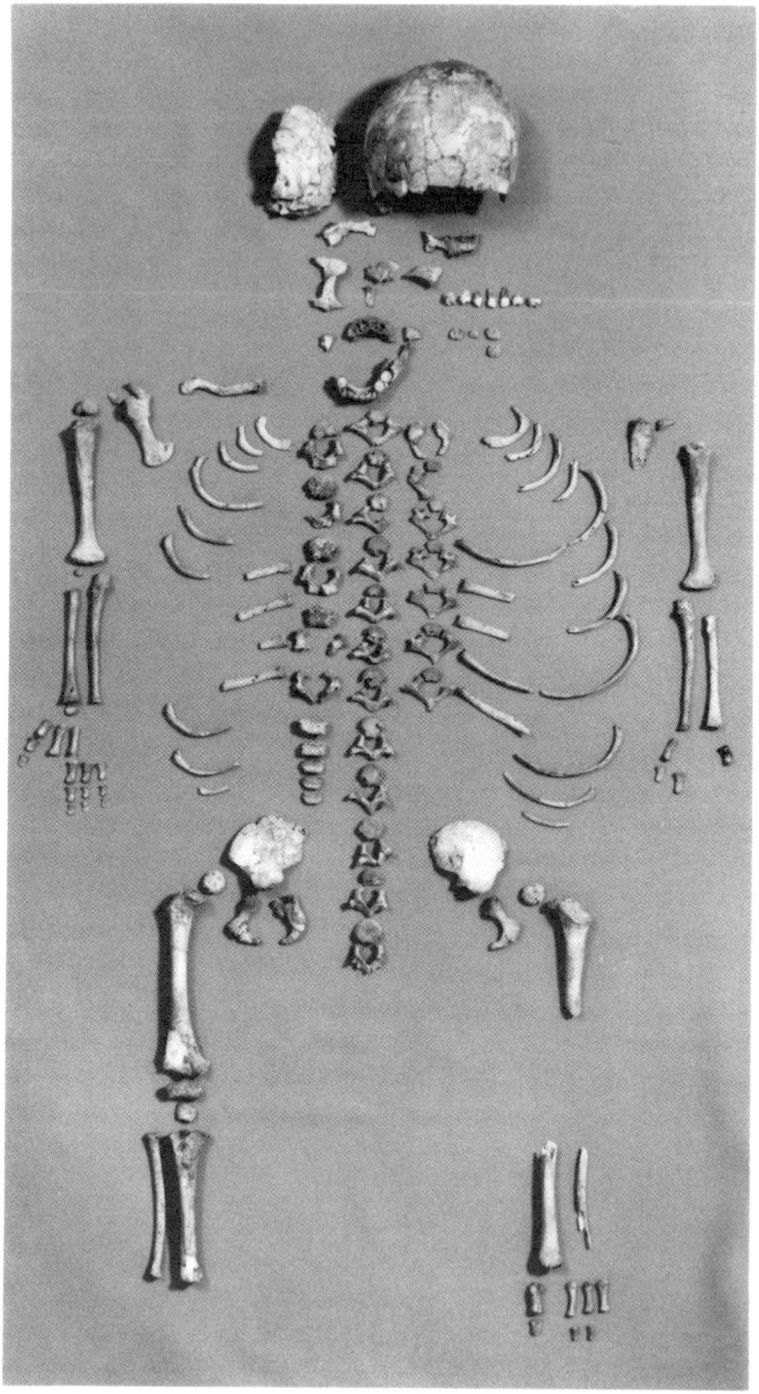

Figure 1. Preservation of the Dederiyeh infant skeleton. Although the greater part of the frontal and facial bones, left clavicle, left hand bones, and right foot bones are missing, many other bones of the skeleton are well-preserved.

Table 1. Deciduous and permanent dentitions of the Dederiyeh infant

M1	—	—	C	—	I1	I1	—	C	—	—	M1
	m2	m1	c	—	i1	i1	—	—	m1	m2	
	—	m1	c	i2	i1	i1	i2	c	m1	m2	
—	—	—	C	I2	I1	I1	I2	C	—	—	M1

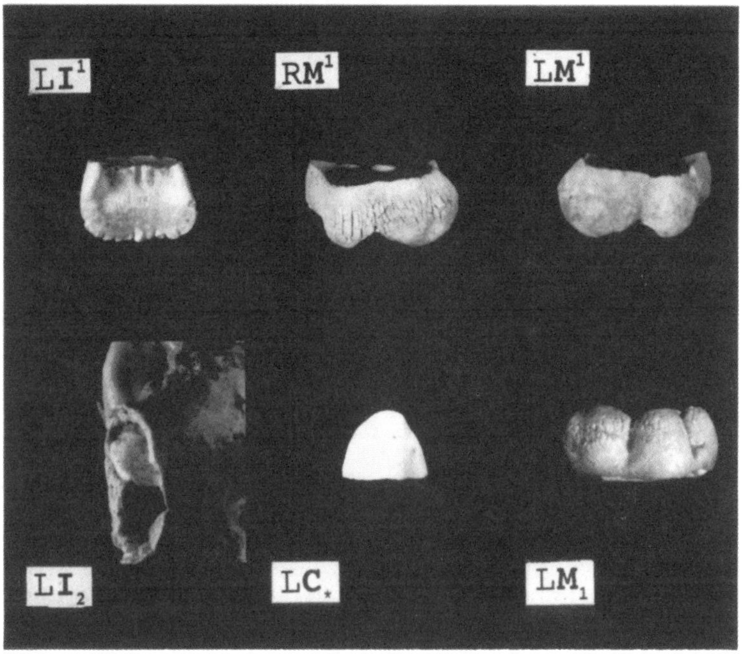

Figure 2. Permanent teeth of the Dederiyeh infant which could be observed by the naked eye. The maxillary left first molar was sectioned for microscopic examination.

A microscopic examination of enamel cross striations on the section of the maxillary first permanent molar indicated that the age of the Dederiyeh infant at death was two years, at the most (Sasaki et al. 1995).

Table 2 shows a comparison of the developmental stages of mandibular teeth between Dederiyeh and Pech de l'Azé, the youngest Neandertal specimen possessing a relatively complete cranium found thus far. Apparently, the developmental stages of both the deciduous and permanent teeth are earlier in Dederiyeh than in Pech de l'Azé, whose age at death is considered a little less than two to three years (Legoux 1970; Minugh-Purvis 1988, Stringer et al. 1990; Faerman et al. 1994). Taking these observations into account, the age at death of the Dederiyeh infant is considered to be about two years.

3. STATURE ESTIMATION

The stature of the Dederiyeh infant was estimated on the basis of the length of the long bones. Telkkä et al. (1962) devised regression formulas for estimating the stature of children using radiographs of their long bones. When the maximum diaphysial lengths of

Table 2. Comparison of developmental stages of mandibular teeth between Dederiyeh and Pech de l'Aze[a,b]

	Dederiyeh		Pech de l'Azé
Permanent teeth			
I1	Cr1/2	<	Cr3/4
I2	Cr1/2	<	Cr3/4
C	Coc	<	Cr1/2
PM1	?		Cco
PM2	?		—
M1	Cr3/4	=	Cr3/4
Deciduous teeth			
i1	A1/2	<	Ac
i2	Rc	<	A1/2
c	R1/2	<	Rc
m1	Rc	<	Ac
m2	R1/4	<	Rc

[a]Cco, coalescence of cusps; Coc, cusp outline complete; Cr1/2, crown 1/2 complete; Cr3/4, crown 3/4 complete; R1/4, root length 1/4; R1/2, root length 1/2; Rc, root length complete; A1/2, apex 1/2 closed; Ac, apical closure complete. Standards for the stages of calcification are given in Moorrees et al. (1963).
[b]The developmental stages in Pech de l'Azé are cited from Faerman et al. (1994).

the humerus, radius, ulna, femur, tibia, and fibula of the Dederiyeh infant were applied to these formulas, the stature was estimated, on average, to be 83.1 cm for a boy and 83.5 cm for a girl.

Direct measurements of the stature were carried out by reconstructing the standing posture of the skeleton (Figure 3; Kondo et al. submitted). Since the obtained height (81.7 cm) is essentially similar to those estimated by Telkkä's formulas, the stature of the Dederiyeh infant is considered to have been 82 to 83 cm.

4. DESCRIPTION AND DISCUSSION

4.1. Calvaria

The neurocranium is round when viewed from behind, and the maximum cranial breadth (132 mm) is very large for its age (Figure 4; Akazawa et al. 1995a,b).

Figure 5 shows a scatter diagram of the maximum cranial breadth by age for the Neandertal and modern Japanese children. The averages of age groups in the modern Japanese are connected by the solid line. The Neandertal children, with the exception of La Quina 18, tend to have wide neurocrania for their ages. In particular, the maximum breadths of Pech de l'Azé and Devil's Tower are notably large for their ages. The Dederiyeh cranium is also wide and its maximum breadth corresponds to the average of a modern Japanese six-year-old, while the width of the Skhul 1, an anatomically modern child, is considerably smaller than the modern Japanese average for about the same age.

The wide neurocranium of the Dederiyeh infant is also indicated by the ratio of the maximum breadth to the sagittal parietal arc. Figure 6 shows the arc-breadth index for Neandertal and modern Japanese children. Apparently, Neandertals have a relatively wide neurocranium compared to modern Japanese. The neurocranium of the Dederiyeh infant is extremely wide in contrast to those of Skhul 1 and Qafzeh 11, whose indices fall within the range of variation of modern Japanese.

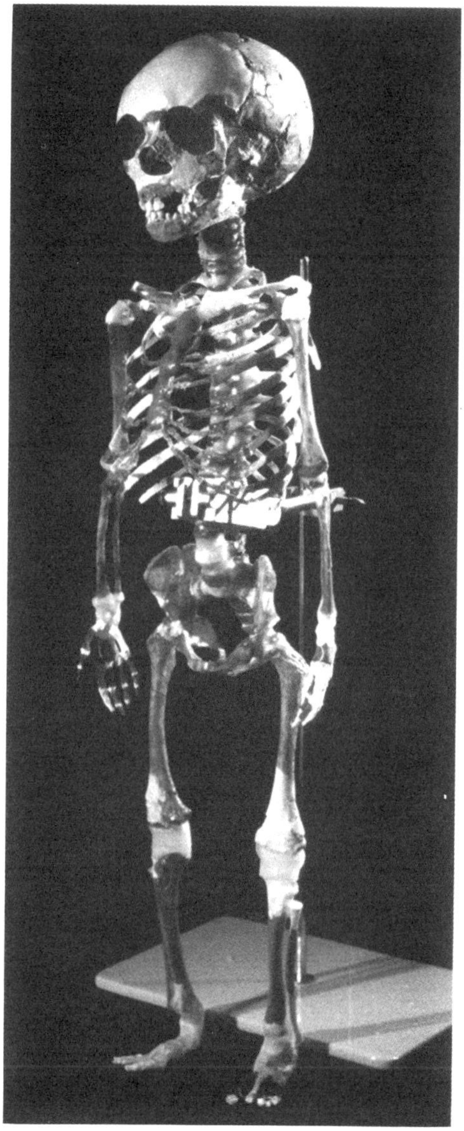

Figure 3. Reconstructed standing posture of the Dederiyeh skeleton. For details of the methods, see Kondo et al. (submitted).

Coexistence of the suprainiac fossa and occipital torus is observable in the reconstructed occipital bone (Figure 7a), although they are not as conspicuous as in adult Neandertal skulls.

The Dederiyeh infant has a considerably large nuchal plane for its age. The distance from the medial end of the suture between the lateral and squamous parts of the occipital bone to the superior nuchal line was measured to estimate the degree of development of the nuchal plane. The scatter diagram of this distance against age for the Neandertal and modern Japanese children shows that Neandertals tend to have a larger nuchal plane than modern Japanese (Figure 8).

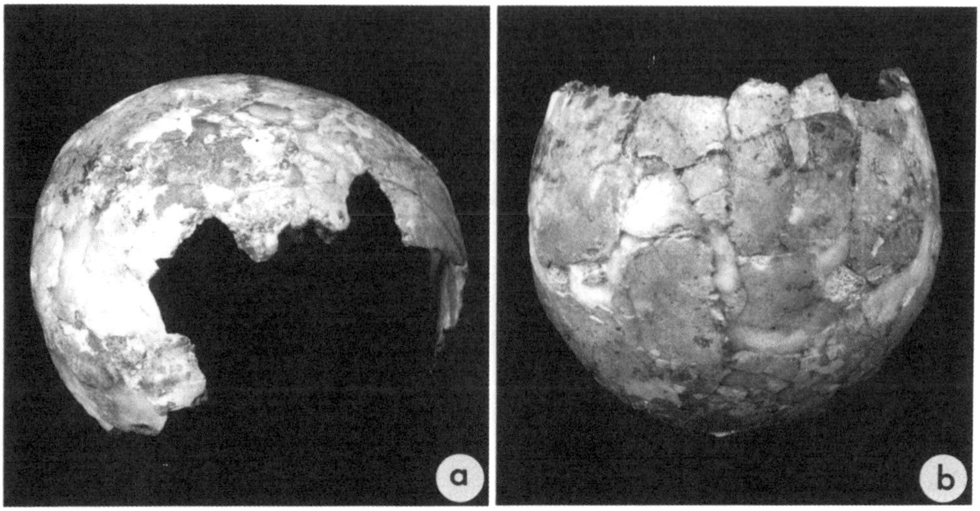

Figure 4. a: Posterior view of the neurocranium of the Dederiyeh infant. The right half of the occipital bone is not glued. b: Superior view of the neurocranium. The squamous part of the frontal bone is missing.

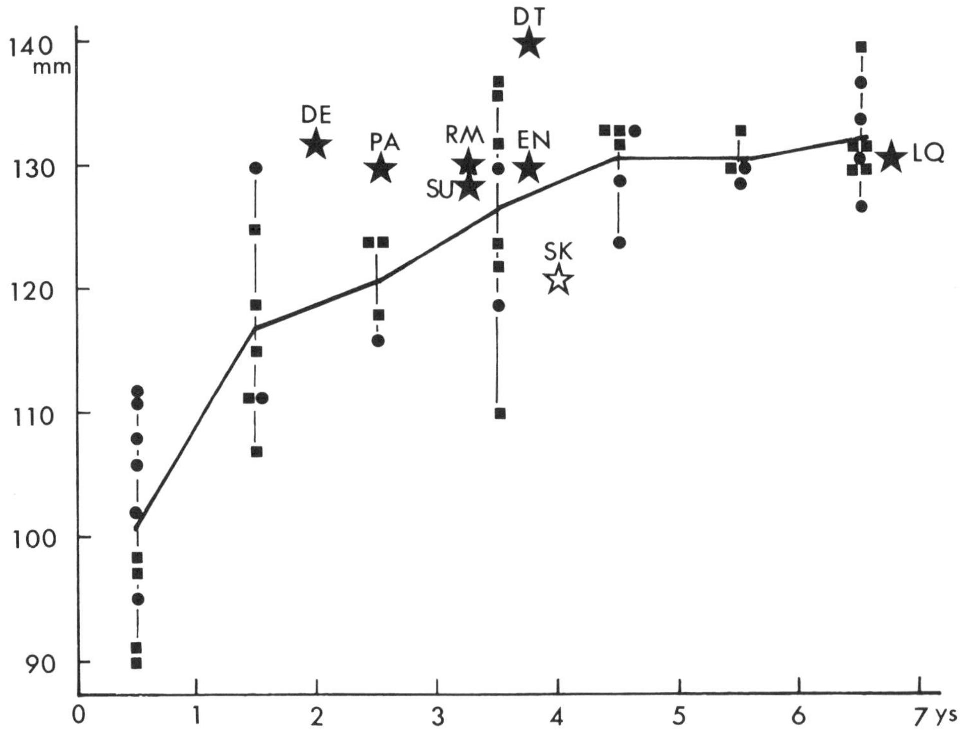

Figure 5. Scatter diagram of the maximum cranial breadth against age for Neandertal (solid star) and modern Japanese (solid square: male; solid circle: female) children. The averages of respective age groups in the modern Japanese children are connected by the solid line. Ages and maximum cranial breadths of the Neandertal children were assessed by consulting Faerman et al. (1994), Legoux (1970), Madre-Dupouy (1992), Minugh-Purvis (1988), Patte (1957), Stringer et al. (1990), Thoma (1963), Tillier (1982, 1983, 1992), Trinkaus and Tompkins (1990). Ages, sexes, and measurements of the modern Japanese children were cited from Wakebe (1990 and personal communication). Age decision and measurement of Skhul 1 were due to McCown and Keith (1939), Minugh-Purvis (1988), and Tillier (1989a). DE, Dederiyeh; PA, Pech de l'Azé; SU, Subalyuk 2; RM, Roc de Marsal; DT, Devil's Tower; EN, Engis 2; LQ, La Quina 18; SK, Skhul 1.

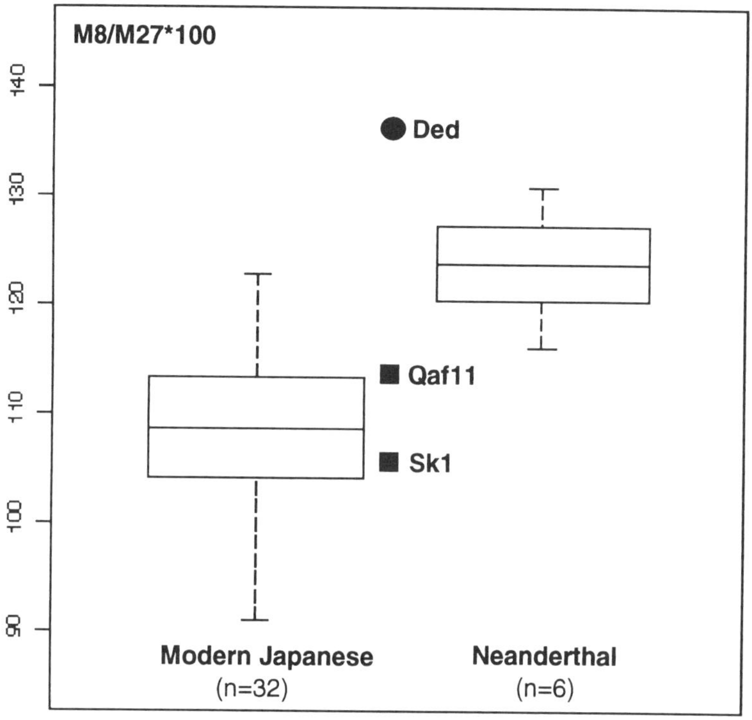

Figure 6. Comparison by box and whisker chart of the parietal arc-breadth index* between Neandertal and modern Japanese children. For ages and measurements of the Neandertal and modern Japanese children, see the legend of Figure 5. The indices of Skhul 1 (Sk1) and Qafzeh 11 (Qaf11) were calculated from the data given by McCown and Keith (1939) and Tillier (1992). *maximum cranial breadth (M8)/median sagittal parietal arc (M27) × 100.

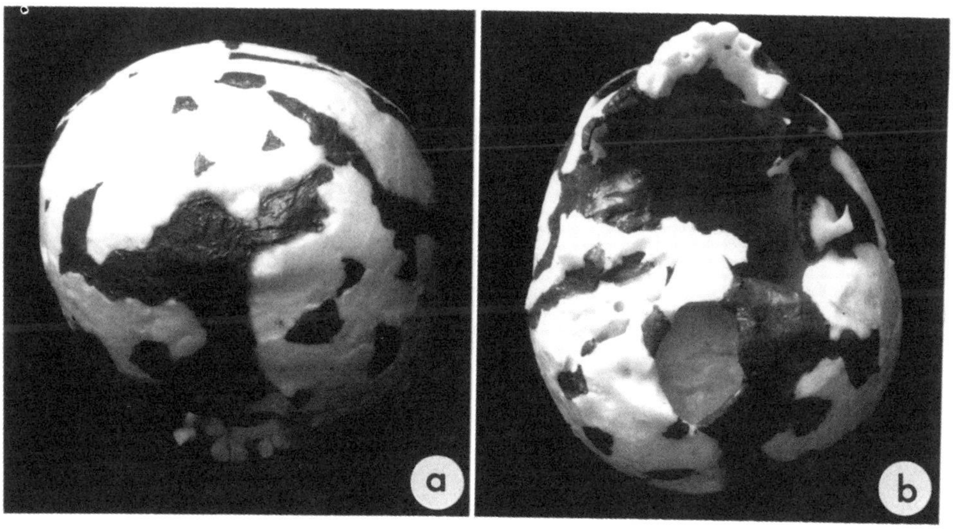

Figure 7. Posterior view (a) and inferior view (b) of the reconstructed cranium of the Dederiyeh infant. The white portions are replicas of the cranial bones and the black ones are the fillings with paraffin wax.

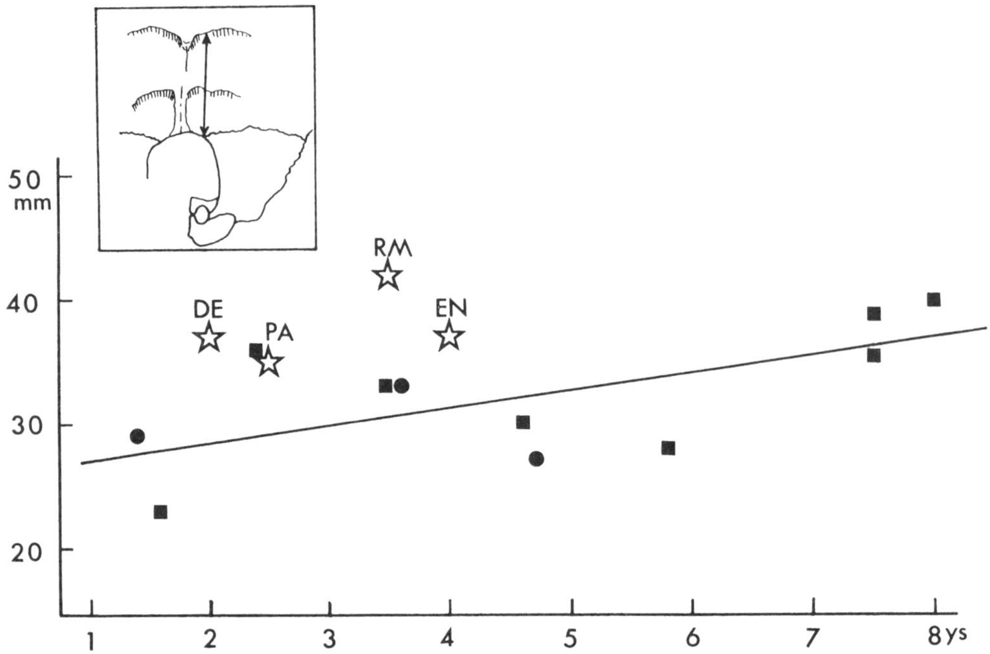

Figure 8. Scatter diagram against age of the distance from the medial end of the posterior intraoccipital synchondrosis* to the superior nuchal line for Neandertal and modern Japanese children. Solid square indicates males and solid circle indicates females of modern Japanese children. White star indicates DE, Dederiyeh; PA, Pech de l'Azé; RM, Roc de Marsal; EN, Engis 2. *The suture between the squamous and lateral parts of the occipital bone.

The stylomastoid foramen and the base of the styloid process are arranged almost medio-laterally. A medio-lateral arrangement was also observed in the European Neandertal children such as Pech de l'Azé, Engis 2, La Quina 18, and La Ferrassie 3, but was never seen in the 29 skulls of Japanese children that were examined.

As for the foramen magnum, the curvature of its right lateral margin is very weak (Figure 7b), suggesting a long oval shape of the foramen as has been pointed out in the Amud 7 Neandertal infant (Rak et al. 1994). In the tympanic plate of the temporal bone, the anterior and posterior limbs have not united with each other and a large defect remains in the tympanic plate (Figure 9). Developmental stages of the tympanic plate were classified into five categories and the occurrences of each category against age were compared between Neandertal and modern Japanese children (Figure 10). The development of the tympanic plate is apparently retarded in the Neandertal children.

4.2. Facial Skeleton

The zygomatic bone of the poorly preserved facial skeleton is of considerable interest. The long frontal process of the zygomatic bone of the Dederiyeh infant suggests that its orbital openings are high. The ratio of the frontal process height to the total zygomatic height was defined as the length index of the frontal process. As illustrated in Figure 11,

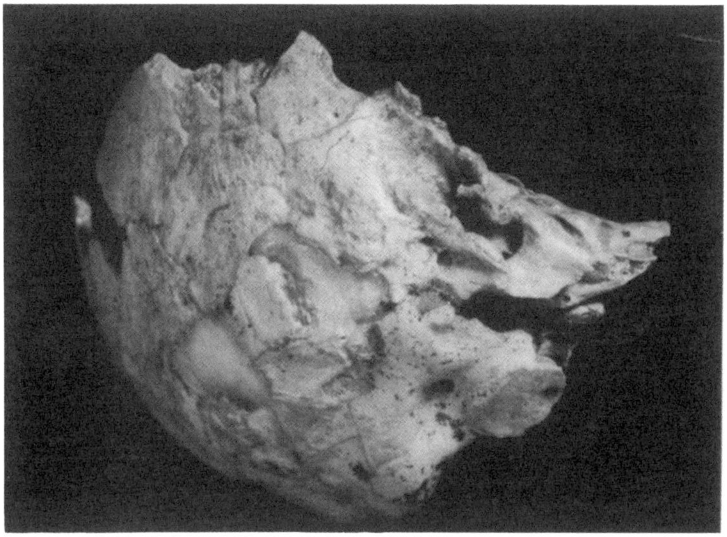

Figure 9. Right tympanic plate. The anterior and posterior limbs have not united with each other and a large defect is seen in the tympanic plate.

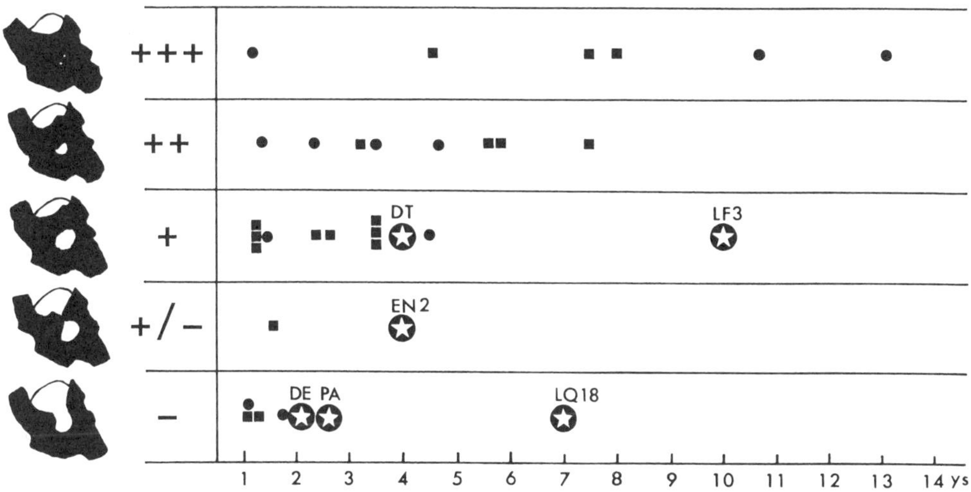

Figure 10. Scatter diagram of developmental stages of the tympanic plate against age for Neandertal and modern Japanese children. The developmental stages were classified into five categories as shown in the left column of the figure. Solid square indicates males and solid circle indicates females of modern Japanese children. White star indicates DE, Dederiyeh; PA, Pech de l'Azé; DT, Devil's Tower; EN2, Engis 2; LQ18, La Quina 18; LF3, La Ferrassie 3. The age of La Ferrassie 3 is cited from Heim (1982).

the length index of Dederiyeh as well as Pech de l'Azé and La Quina 18 exceeds the range of variation of modern Japanese children.

The right nasal bone is extremely wide (7 mm in minimum breadth) for the two-year-old infant and the profile of the anterior surface is strongly curved, suggesting a prominent nasal bridge.

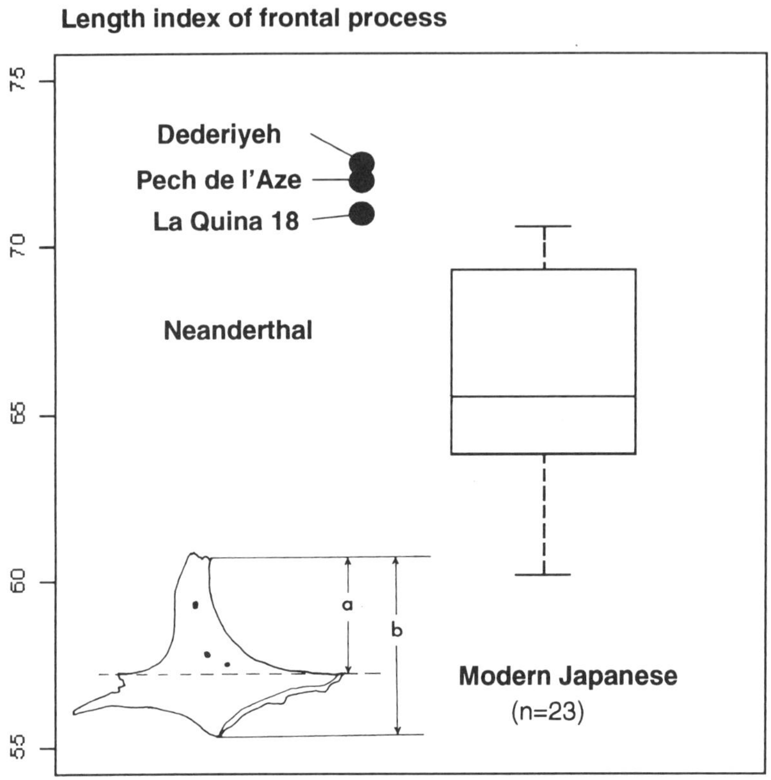

Figure 11. Comparison of length index of the frontal process of the zygomatic bone between three Neandertal and 23 modern Japanese children. The length index of the frontal process was defined as a/b × 100 as shown in the figure.

4.3. Mandible

Although the mental fossae and a faint manifestation of the mental trigone are observable, the mandible of the Dederiyeh infant is characterized by such Neandertal features as receding symphysial profile with the chin angle of 78 degrees, presence of a submandibular notch, anterior marginal tubercle, and double mental foramina (Figure 12a; Akazawa et al. 1995a,b).

The posterior margin of the mandibular notch is attached not to the lateral end of the mandibular condyle but to a point about 5 mm medial to the lateral end. A similar configuration has been described in both infant and adult Neandertals such as Amud 7 (Hovers et al. 1995), Pech de l'Azé (Patte 1957), Roc de Marsal (present study), Amud I (Suzuki 1970), and Kebara 2 (Tillier 1989b).

The first deciduous molar of the mandible shows a strong tendency for taurodontism (Figure 12b; Akazawa et al. 1995b). Taurodont deciduous molars were also observed in the Neandertal infant mandible from Barakai Cave in the Western Caucasus (Faerman et al. 1994).

4.4. Postcranial Skeleton

The postcranial bones of the Dederiyeh infant also exhibit many anatomical features suggesting its Neandertal affiliation. Nineteen postcranial features of the Dederiyeh infant

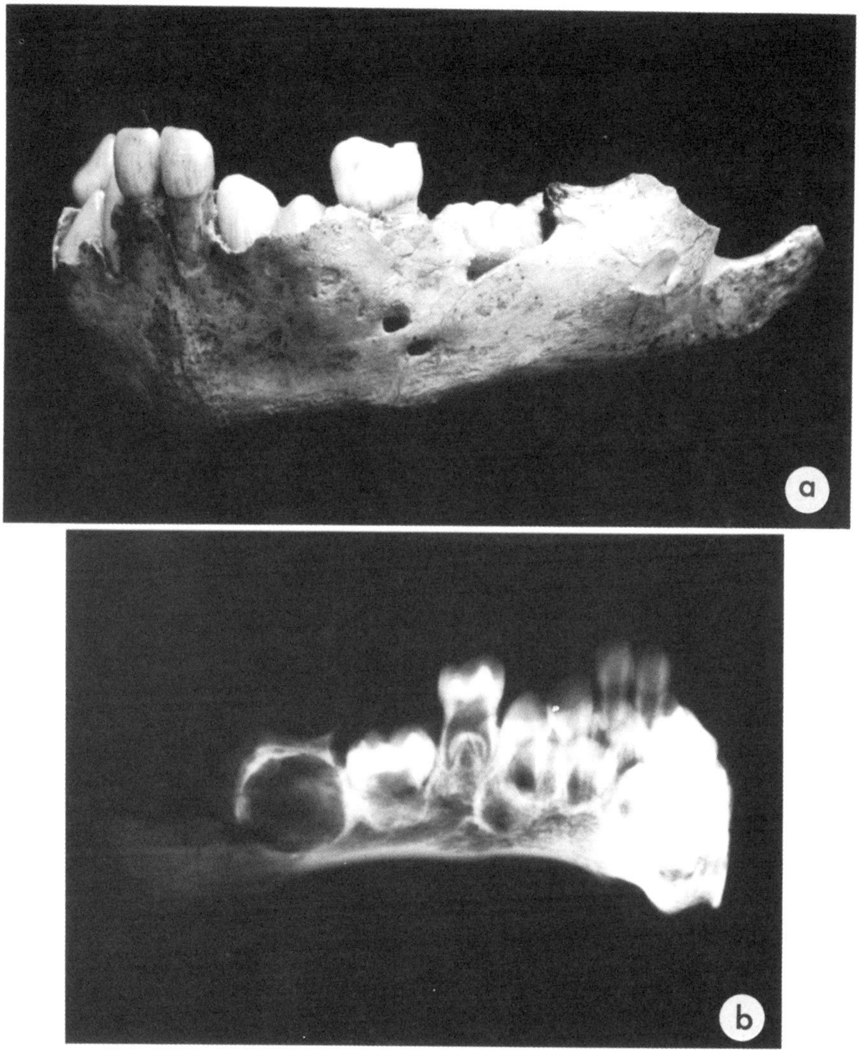

Figure 12. a: Antero-lateral view of the mandible of the Dederiyeh infant. The symphysial region recedes inferiorly with a slight indication of the mental trigone. The mental foramen is duplicated. The lateral deciduous incisors, deciduous canines and second deciduous molar have not fully erupted. b: Radiograph of the left mandibular body. The first deciduous molar shows a strong tendency to taurodontism.

were compared with those of two Neandertal (Roc de Marsal and La Ferrassie 6) and one anatomically modern *Homo sapiens* (Skhul 1) children (Table 3).

The main anatomical features shared with the European Neandertals are the following: (1) relatively long clavicle; (2) medially-oriented tuberosity of the radius; (3) acetabulo-symphysial elongation of the pubis (Figure 13a); (4) relatively large diaphysial ends of the humerus, femur, and tibia; (5) thick cortical layer of the long bones (Figure 13b); (6) strong antero-posterior bowing of the shafts of the femur and tibia; (7) relatively high claviculo-humeral index and considerably low crural index (Akazawa et al. 1995b).

Proportional characteristics of the superior pubic ramus and the limb segments were compared between Neandertal and modern children. The indices compared were

Table 3. Comparison of postcranial anatomical features[a]

	Ded[b]	Roc[b]	Fe6[b]	Sk1[b]
Clavicle:				
Relative length	+	+	/	–
Supero-inferior flattening	+	+	/	+
Scapula:				
Direction of glenoid cavity (dorso inferior)	++	++	/	/
Humerus:				
Medial butress of olecranon fossa	–	–	–	+
Radius:				
Medially oriented radial tuberosity	+	+	++	–
Laterally facing of proximal facet	–	+	+	+
Lateral bowing	–	+	+	–
Ulna:				
Antero-posterior curvature	–	–	+	/
Interosseous crest	–	–	–	/
Ilium:				
Orientation of anterior inferior spine (vertical)	+	++	+	–
Supraacetabular fossa	+	+	+	–
Pubis:				
Elongation of superior ramus	++	+?	+	/
Pecten ossis pubis	+	/	+	/
Femur:				
Antero-posterior curvature	+	+	+	–
Linea aspera	–	–	–	+
3rd trochanter	+	+	+	–
Size of both diaphysial ends	++	+	+	–
Tibia:				
Antero-posterior curvature	++	/	+	/
Retroversion of proximal end	+	/	/	/

[a]+, strong manifestation; –, weak manifestation; /, unobservable.
[b]Ded, Dederiyeh; Roc, Roc de Marsal (Madre-Dupouy 1992 and present study); Fe6, La Ferrassie 6 (Heim 1982 and present study); Sk1, Skhul 1 (McCown and Keith 1939 and present study)

acetabulo-symphysial length (ASL)/iliac breadth (IB); claviculo-humeral index; brachial index (radius/humerus); and crural index (tibia/femur). The results are given in Figure 14.

As has been demonstrated with the La Ferrassie 6 infant (Tompkins and Trinkaus 1987) and various adult Neandertals (McCown and Keith 1939; Stewart 1960; Endo and Kimura 1970; Rak and Arensburg 1987), the ASL of the Dederiyeh infant is significantly greater than that of the modern sample ($T = 3.54$, $P = 0.0008$). Although the reason for the elongation of the superior pubic ramus remains unclear (Rak and Arensburg 1987; Rosenberg 1988), this morphology is probably one of the derived traits of Neandertals.

Suggestive Neandertal features are seen in the claviculo-humeral and crural indices, although they are not significantly different from those of the modern children. As in the Dederiyeh infant, the clavicle is relatively long in the Roc de Marsal and the tibia is relatively short in the Roc de Marsal and La Ferrassie 6. Skhul 1, an anatomically modern child, shows a striking contrast to the Neandertal children with respect to these features. The claviculo-humeral index is near the lower limit of variation of moderns and the crural index is beyond the upper limit.

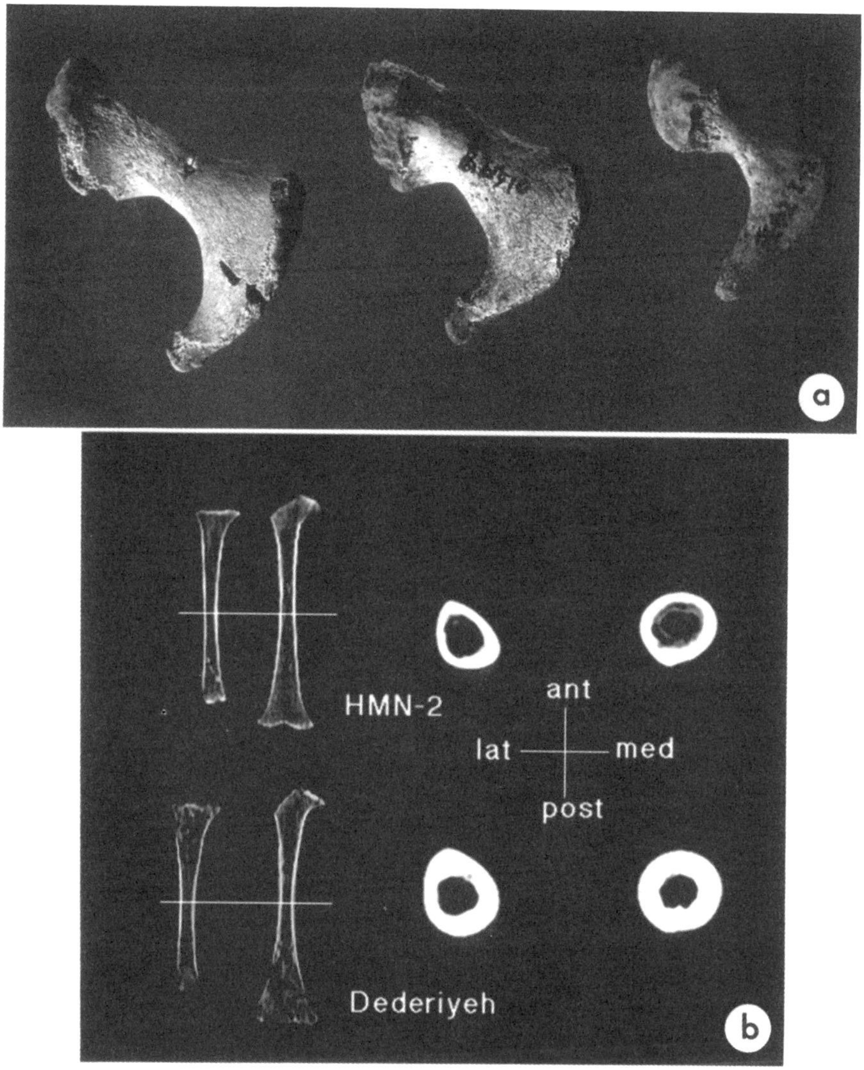

Figure 13. a: Medial view of the left pubic bones of the Dederiyeh infant (left) and two archaeological specimens from Japan. Takasago 10 (middle) is a prehistoric Jomon specimen of about four years, and Hamanaka 2 (right) is a prehistoric Okhotsk specimen of about three years. b: Cross-sectional image by CT-scan of the tibiae and femora of the Dederiyeh infant (below) and prehistoric Okhotsk Hamanaka 2 specimen (above). The tibia is shown on the left side of the figure and the femur on the right side.

5. CONCLUSION

Anatomical features of the infant skeletal remains discovered in the Mousterian formation at Dederiyeh Cave, northern Syria, were described and discussed. The age at death was considered about two years based on dental development. The stature was estimated at about 83 cm from the lengths of long bones and about 82 cm from the reconstructed posture of the skeleton. Despite the immature state of the skeleton, both the cranial and postcranial anatomical features indicated that the Dederiyeh infant was certainly a Nean-

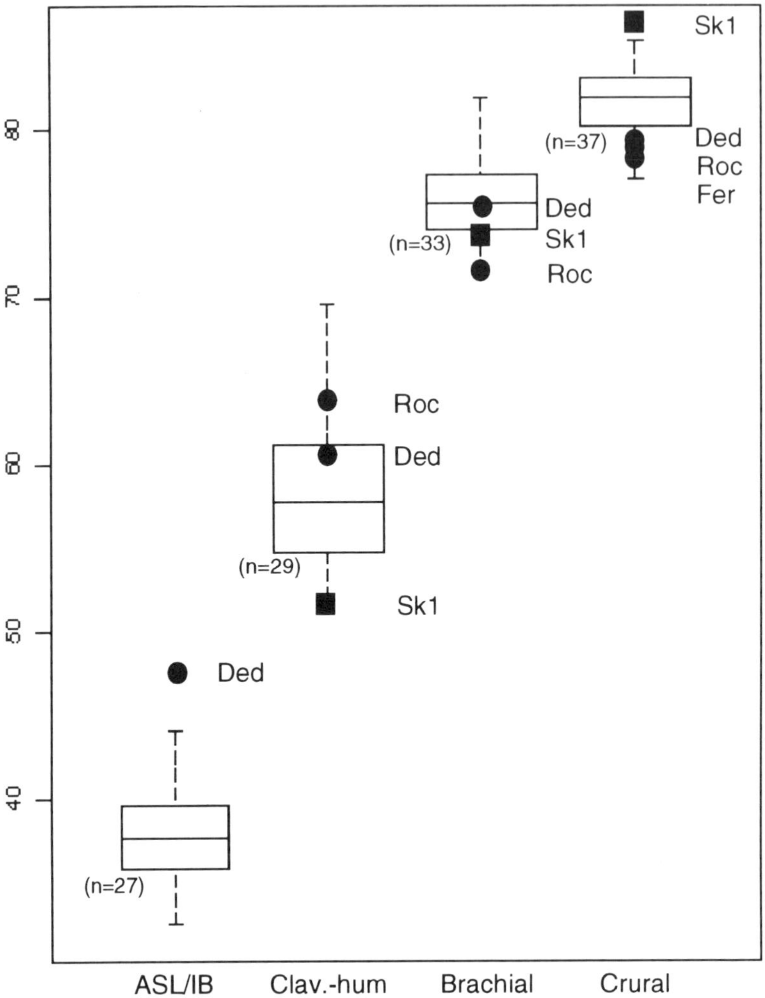

Figure 14. Comparisons of acetabulo-symphysial length index (ASL/IB), claviculo-humeral index, brachial index, and crural index between Neandertal, anatomically modern Middle Paleolithic and modern Japanese/European children. Ded, Dederiyeh; Roc, Roc de Marsal; Fer, La Ferrassie 6; Sk1, Skhul 1. The data of La Ferrassie 6 are cited from Tompkins and Trinkaus (1987) and those of Skhul 1 are from McCown and Keith (1939).

dertal, or more strictly speaking, a Levantine Middle Paleolithic infant very closely related to the European Neandertals.

ACKNOWLEDGMENTS

We would like to thank the following scientists and curators who have permitted us to examine original materials or casts under their care: Drs. B. Vandermeersch and A. M. Tillier, Université Bordeaux I; Drs. J. J. Hublin and M. Chech, Musée de l'Homme; Dr. H. Marino, Musée des Antiquites Nationales; Dr. Madeleine, Musée National de Préhistoire; Dr. T. Kamiya, University Museum, University of Tokyo. The research behind this paper has been supported, in part, by Grant-in-Aid for Overseas Scientific Research, The Ministry of Education, Science, and Culture, Japan.

REFERENCES

Akazawa, T., S. Muhesen, Y. Dodo, O. Kondo, and Y. Mizoguchi. 1995a. Neanderthal infant burial. *Nature* 377:584–5.

Akazawa, T., S. Muhesen, Y. Dodo, O. Kondo, Y. Mizoguchi, Y. Abe, Y. Nishiaki, S. Ohta, T. Oguchi, and J. Haydal. 1995b. Neanderthal infant burial from the Dederiyeh Cave in Syria. *Paléorient* 21:77–86.

Endo, B., and T. Kimura. 1970. Postcranial skeleton of the Amud man, 231–406. In *The Amud Man and His Cave Site*, ed. H. Suzuki and F. Takai. Tokyo: Academic Press of Japan.

Faerman, M., U. Zilberman, and P. Smith. 1994. A Neanderthal infant from the Barakai Cave, Western Caucasus. *Journal of Human Evolution* 27:405–15.

Heim, J. L. 1982. *Les Enfants Néandertaliens de La Ferrassie.* Paris: Masson.

Hovers, E., R. Rak, R. Lavi, and W. H. Kimbel. 1995. Hominid remains from Amud Cave in the context of the Levantine Middle Paleolithic. *Paléorient* 21:47–61.

Kondo, O., Y. Dodo, T. Akazawa, and S. Muhesen. n.d. Estimation of stature from the skeletal reconstruction of an immature Neanderthal from Dederiyeh Cave in Syria. (submitted).

Legoux, P. 1970. Etude odontologique de l'enfant néandertalien du Pech-de-l'Azé , 53–87. In *L'enfant du Pech-de-l'Azé*, ed. D. Ferembach, P. Legoux, R. Fenart, R. Empereur Buisson, and E. Vleck. Archives de l'Institut de Paléontologie Humaine Memoire 33. Paris: Masson.

Madre-Dupouy, M. 1992. *L'enfant du Roc de Marsal. Etude analytique et comparative.* Paris: CNRS Editions.

McCown, T. D., and F. R. S. Keith. 1939. *The Stone Age of Mount Carmel.* Vol II. *The Fossil Human Remains from the Levalloiso-Mousterian.* Oxford: Clarendon.

Minugh-Purvis, N. 1988. Patterns of Craniofacial Growth and Development in Upper Pleistocene Hominids. Ph. D. diss. University of Pennsylvania. Philadelphia.

Moorrees, C. F. A., E. A. Fanning, and E. E. Hunt. 1963. Age variation of formation stages for ten permanent teeth. *Journal of Dental Research* 42:1490–502.

Patte, E. 1957. *L'enfant Néanderthalien du Pech de L'Azé.* Paris: Masson.

Rak, Y., and B. Arensburg. 1987. Kebara 2 Neanderthal pelvis: First look at a complete inlet. *American Journal of Physical Anthropology* 73:227–31.

Rak, Y., W. H. Kimbel, and E. Hovers. 1994. A Neandertal infant from Amud Cave, Israel. *Journal of Human Evolution* 26:313–24.

Rosenberg, K. R. 1988. The functional significance of Neanderthal pubic length. *Current Anthropology* 29:595–617.

Sasaki, C., K. Suzuki, H. Takagi, Y. Kozawa, and Y. Dodo. 1995. Age determination with enamel cross striation of fossil man. *Acta Anatomica Nipponica Supplement* 70:171.

Stewart, T. D. 1960. Form of the pubic bone in Neanderthal man. *Science* 131:1437–8.

Stringer, C. B., M. C. Dean, and R. D. Martin. 1990. A comparative study of cranial and dental development within a recent British sample and among Neandertals, 115–52. In *Primate Life History and Evolution*, ed. C. J. DeRousseau. New York: Wiley-Liss.

Suzuki, H. 1970. The skull of the Amud man, 123–206. In *The Amud Man and His Cave Site*, ed. H. Suzuki and F. Takai. Tokyo: Academic Press of Japan.

Telkkä, A., A. Palkama, and P. Virtama. 1962. Estimation of stature from radiographs of long bones in children, Part I: Children aged from one to nine. *Annales Medicinae Experimentalis et Biologiae Fenniae* 40:91–6.

Thoma, A. 1963. The dentition of the Subalyuk Neandertal child. *Zeitschrift für Morphologie und Anthropologie* 54:127–50.

Tillier, A. M. 1982. Les enfants néanderthaliens de Devil's Tower (Gibraltar). *Zeitschrift für Morphologie und Anthropologie* 73:125–48.

Tillier, A. M. 1983. Le crâne d'enfant d'Engis 2: Un exemple de distribution des caractères juvéniles, primitifs et néanderthaliens. *Bulletin de la Societé Royale Belge d'Anthropologie et de Préhistoire* 94:51–75.

Tillier, A. M. 1989a. The evolution of modern humans: Evidence from young Mousterian individuals, 286–97. In *The Human Revolution*, ed. P. A. Mellars and C. B. Stringer. Edinburgh: Edinburgh University Press.

Tillier, A. M. 1989b. La mandibule et les dents, 97 111. In *Le Squelette Moustérien de Kébara 2*, ed. O. Bar-Yosef and B. Vandermeersch. Paris: CNRS Editions.

Tillier, A. M. 1992. The origins of modern humans in southwest Asia: Ontogenetic aspects, 15–28. In *The Evolution and Dispersal of Modern Humans in Asia*, ed. T. Akazawa, K. Aoki, and T. Kimura. Tokyo: Hokusen-sha.

Tompkins, R. L., and E. Trinkaus. 1987. La Ferrassie 6 and the development of Neandertal pubic morphology. *American Journal of Physical Anthropology* 73:233–9.

Trinkaus, E. 1983. *The Shanidar Neandertals.* New York: Academic Press.

Trinkaus, E., and S. E. Churchill. 1988. Neandertal radial tuberosity orientation. *American Journal of Physical Anthropology* 75:15–21.

Trinkaus, E., and R. L. Tompkins. 1990. The Neandertal life cycle: The possibility, probability, and perceptibility of contrasts with recent humans, 153–80. In *Primate Life History and Evolution*, ed. C. J. DeRousseau. New York: Wiley-Liss.

Wakebe, T. 1990. A morphological study of cranium of infants and children in the Japanese. *Nagasaki Igakkai Zasshi* 65:805–24.

21

THE SEARCH FOR THE EARLIEST MODERN EUROPEANS

A Comparison of the Es-Skhul 1 and Krapina 1 Juveniles

Nancy Minugh-Purvis[1–3]

[1]Department of Neurobiology and Anatomy
Allegheny University of the Health Sciences
Philadelphia, Pennsylvania 19129
[2]University Museum of Archaeology and Anthropology
University of Pennsylvania
Philadelphia, Pennsylvania 19104
[3]Division of Plastic Surgery
The Children's Hospital of Philadelphia
Philadelphia, Pennsylvania 19104

1. INTRODUCTION

Sorting out the population affinities of late Pleistocene circum-Mediterranean peoples is key to understanding the evolutionary significance of the appearance of modern humans in Europe and West Asia. Traditionally, most of this research on fossil remains, usually comprising analyses of morphology, has depended heavily upon the study of adult material. This is primarily because we lacked sufficient experience in placing immature specimens within a developmental context when assessing their morphology. Fortunately, this situation is improving and it is now possible to evaluate the remains of children within appropriate ontogenetic frameworks for specimens of all age groups.

Investigations in recent years have begun to demonstrate the value of immature remains for understanding the developmental basis of morphology. In this regard, the fossil remains of children promise to clarify the significance of much morphological variation seen in the paleoanthropological record, as well as other aspects of the biological and behavioral adaptations of earlier hominids. Moreover, because analyses of developmental biology have the potential to reveal genetically-based unique growth signatures, they may ultimately prove to be of considerable importance in hominid systematics. For these reasons, attempts to understand the late Pleistocene populations of Europe and West Asia will obviously benefit from investigations of their ontogeny.

Neandertals and Modern Humans in Western Asia, edited by Akazawa *et al.*
Plenum Press, New York, 1998.

Crucial to the question of modern human origins are the Levantine sites of Qafzeh and es-Skhul, which have yielded among the earliest evidence of modern human morphology. Qafzeh, dated by TL to approximately 92±5 ka (Valladas et al. 1988 summarized in Bar-Yosef 1994) and by ESR to between 96±13 ka and 115±15 ka (Schwarcz et al. 1988; Stringer et al. 1989; Grün and Stringer 1991 summarized in Bar-Yosef 1994), has an average age of perhaps 110 ka. Es-Skhul has been dated by TL on flints in Layer B at approximately 119±18 ka (Mercier et al. 1993) and by ESR at 81±15 ka to 101±12 ka (Schwarcz et al. 1988; Stringer et al. 1989; Grün and Stringer 1991 summarized in Bar-Yosef 1994) suggesting an average age of perhaps 100 ka. However, Stringer (personal communication) suspects that an even wider span of dates might be present in the hominid-bearing deposits at es-Skhul and has suggested, as have Ronen (1976) and Wolpoff (1995), that the unusual amount of morphological heterogeneity at Skhul may be due to chronological differences within the sample.

Among those working with the Levantine material there has been consistent agreement over the relatively modern appearance of the Qafzeh hominids. In contrast, differences of opinion continue to surround interpretations of the morphology at es-Skhul, which many describe as overlapping with the Neandertal range of variation while also including rather modern appearing individuals (McCown and Keith 1939; Wolpoff 1995). However, one of the most modern appearing Skhul specimens is the Skhul 1 child which, based on its supposed location within the deposits, is most probably dated with the later specimens such as Skhul V, at 73.5–83.5 ka. This date, while considerably later than the average for Qafzeh, nevertheless remains far earlier than the appearance of modern appearing hominids in the European fossil record, and predates the presence of Neandertals at such nearby Levantine sites as Kebara at 60 ka (Valladas et al. 1987; Grün and Stringer 1991). Thus assessing the population affinities of both the earlier and later Skhul specimens is of utmost importance for understanding the pattern of a possible modern human diaspora, and particularly for better resolving issues regarding the directionality of gene flow during late Pleistocene dispersals.

Dates recently obtained for the well-known late Pleistocene sample from Krapina, Croatia (Rink et al. 1995), if correct, place Krapina earlier than Levantine early modern humans and most other known European fossils. The proximity of Krapina to the Adriatic, at a time when sea levels would have been similar to those of today, clearly places it on one of the most likely routes into Europe from the Levant (and vice versa). In this regard, two specimens from Krapina are of particular interest: the juvenile Krapina 1 cranium (formerly Krapina Skull A) and Krapina 11, an adult occipital. Both fossils were recovered from Level 8—the highest hominid-bearing stratum at the site. Furthermore, they are noteworthy in having been described (Skerlj 1958; Smith 1976; Wolpoff 1980; Minugh-Purvis 1988a,b; Caspari 1991; Minugh-Purvis and Radovčić 1991, 1995; Minugh-Purvis et al., in press; and others) as possibly more modern than the balance of the clearly Neandertal remains from deeper in the Krapina deposits. In view of these new, earlier dates, the Krapina specimens warrant examination for possible resemblances to early modern Levantine peoples. With this in mind, this paper is devoted to comparing the most complete of the Level 8 specimens, the Krapina 1 child, and the most complete of the es-Skhul juveniles, Skhul 1, with other late Pleistocene hominids from Europe and the Near East as well as with recent human children.

This study addresses three questions: (1) Do any traits found in Krapina 1 suggest the presence of Levantine genes in this European specimen or, conversely, do any traits in Skhul 1 provide evidence of European genes in this Levantine child? (2) Are populational differences apparent between the European and other samples examined? (3) If in evidence, are any populational differences or similarities more probably attributable to subspecific (i.e., regional) or phylogenetic (i.e., evolutionary) levels of variation?

2. MATERIALS AND METHODS

Utilizing qualitative observations and metric data collected by myself from the originals, comparisons were made between common preserved portions of Krapina 1 and Skhul 1: the frontal, parietal, and temporal bones (Figure 1a,b). These were, in turn, compared with other European and Levantine late Pleistocene juveniles, including Neandertals and early Upper Paleolithic associated modern Europeans (details regarding the source of most of the late Pleistocene specimens are available in Minugh-Purvis 1988a; nd). Although the Qafzeh juveniles are not yet generally available for study, comparisons using pertinent data published by Tillier (1992) have been included in the present study whenever possible.

Skhul 1 has been superbly described by McCown and Keith (1939) although not within a comprehensive developmental context. More recent assessments of Skhul 1 within an ontogenetic framework are to be found in Minugh-Purvis (1988a) and Tillier (1992). A comprehensive description of the newly reconstructed Krapina A specimen is currently underway (Minugh-Purvis et al., in press) although earlier descriptions of this individual, prior to the addition of further bone fragments in the 1980s and 1990s, are available (Gorjanović-Kramberger 1906; Smith 1976; Wolpoff 1980; Minugh-Purvis 1988a; and others).

Fossil metric data were compared with measurements from two samples of recent *Homo sapiens sapiens* children: the 26th to 30th Dynasty Gizeh Egyptian E Series (N=72) (described in Pearson and Davin 1924 and for which I am indebted to Dr. Mark Skinner), spanning the growth period from early childhood to late adolescence, or from approximately 2.5 to 18 years; and a late 19th and early 20th century European sample (N=20), published by Madre-Dupoy (1992), spanning the very youngest age intervals of infancy and early childhood, from birth to approximately 6 years of age. The late Pleistocene remains were divided into two separate samples: (1) Neandertals and (2) Upper Paleolithic associated modern Europeans believed to pre-date 18,000 BP.

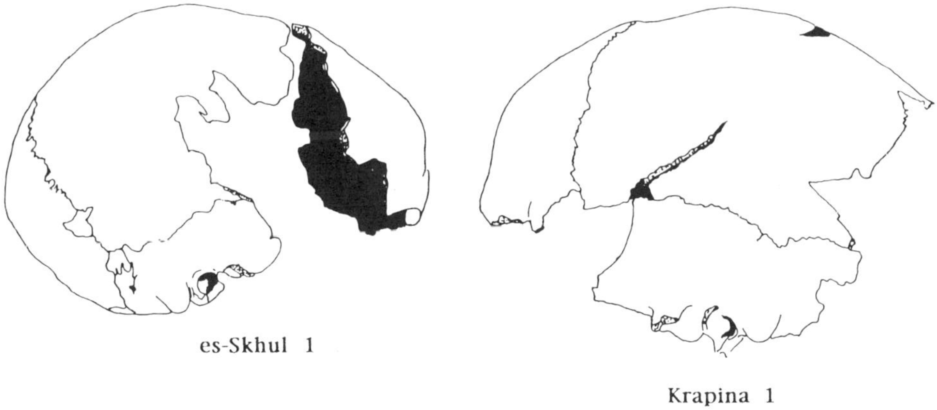

Figure 1. Lateral aspect of es-Skhul 1 and Krapina 1. (Skhul 1 after McCown and Keith 1939).

3. AGES AT DEATH

As outlined in Minugh-Purvis (1988a), Skhul 1 preserves a developing dentition and dental calcification scoring and yields an approximate age of 4–5, or an average of 4.5 years of age at death, which is consistent with the age determination arrived at by McCown and Keith (1939) utilizing dental formation and indicators of skeletal maturity. It should be noted, however, that Skinner (1985) arrived at a younger age range of 3.2–3.6 years at death using dental calcification scoring and Tillier (1992) considers the child approximately 3 to 4 years at the time of death. Thus plotted dimensions for Skhul 1 on graphs are depicted as 3.5–5 years of age, spanning the age range estimated by all of the above investigators on this specimen.

Estimating the approximate age at death for Krapina 1 is more problematic owing to its lack of any associated dentition. Some developing features of the Krapina 1 cranium, including frontal sinus development and calvarial thickness (Minugh-Purvis et al. nd), point to an approximate age of 6 years or greater at the time of death. However, the articular surface of the glenoid fossa is still angulated slightly off the horizontal, suggesting it is unlikely that this child was older than 8 years when it died (see Minugh-Purvis 1988a,b). Nevertheless, because these are extremely variable criteria, it is important to recognize that age at death cannot be determined with much precision for Krapina A. Thus plotted measurements for Krapina 1 are depicted on graphs as spanning the mid-childhood age interval of 6.0 to 7.9 years.

The recent Egyptian comparative material was aged by M. Skinner (1985), utilizing a system of dental calcification scoring similar to that employed by Minugh-Purvis (1988a). The recent European sample was aged by Madre-Dupoy (1992) using an unspecified method of dental aging.

4. COMPARISONS

4.1. Frontal

Among the most interesting qualitative features available for comparison between Skhul 1 and Krapina 1 is the supraorbital region. Both Skhul 1 and Krapina 1 possess supraorbitals defined superiorly by a slight recess in the supraglabellar region, which runs laterally to end just above the temporal lines on either side. In both specimens, clearly distinguishable medial and lateral supraorbital elements are present although the medial portion is only very slightly developed in the ontogenetically young Skhul 1. In this regard, both fossil children differ from modern children, who commonly lack any development of the medial supraorbital portion (Minugh-Purvis, personal observation). The Krapina 1 supraorbitals are larger than those of Skhul 1 in the lateral as well as the medial portion although this is, at least in part, attributable to an older age at death. Smith and his coworkers (Smith and Ranyard 1980; Wolpoff et al. 1981; Smith 1982, 1984, 1985, 1992) have convincingly demonstrated both a size reduction and morphological changes in the supraorbitals through late Pleistocene hominid evolution. These changes have been best documented in south central Europe: the browridge changes from a continuous, typically Neandertal torus by an initial reduction in overall dimension while a differentiation begins between the medial and lateral elements. The mid-superior orbital rim thins or becomes "pinched" in appearance while the more central median browridge increases in vertical height (Wolpoff 1989) as the height of the lateral segment shortens. Previous studies have found that neither Neandertal (Tillier 1986; Minugh-Purvis 1988a) nor early modern European supraorbitals (Minugh-Purvis 1988a) begin to experience any major growth in size until the appearance of the mixed dentition, although their adult shape configuration appears as early as 2.5 years (Vlček 1970; Minugh-Purvis 1988a). If Skhul 1 followed a similar developmental timetable to that known for Neandertals and early modern Upper Paleolithic associated Europeans, its browridge form would accurately predict an adult morphology distinct from that of European Neandertals.

The similarities between the Skhul 1 and Krapina 1 browridges are even more interesting, however, in view of the differences between the supraorbital region of Krapina 1 and many juvenile Neandertals. The majority of Neandertal supraorbitals, including those of children, comprise a continuous incipient bar over the orbits without marked differentiation of the medial and lateral elements (Vlček 1970; Minugh-Purvis 1988a). The presence of a considerable range of morphological variation in the relative development of the medial and lateral supraorbital elements among juvenile Neandertals suggests the possibility that the bipartite supraorbitals in Krapina 1 might simply reflect one extreme of the Neandertal range and not necessarily a modern feature. At the same time, the reduction of the lateral supraorbitals in Krapina 1 is far more dramatic than in any other Neandertal child of comparable developmental age or of any other individual from the Krapina sample (Minugh-Purvis et al., in press). Moreover, because Krapina 1 was clearly old enough at the time of death to have possessed a supraorbital form indicative of its adult browridge morphology, there is no question that it would have differed substantially—in the direction of early modern humans—in the form of this feature had it lived to adulthood.

Other perhaps less significant qualitative features of the frontal include the orbital rim. Here there is a contrast between the specimens: Skhul 1 presents a squared-off superior orbital margin, similar to the squared, angular shape found in some early modern Europeans; Krapina 1 displays a clearly rounded superior orbital margin and superomedial orbital angle, in the form most commonly encountered in European Neandertals. The specimens likewise differ in their frontal eminence morphology, with those of Skhul 1 within the modern range of variation (Minugh-Purvis, personal observation contra McCown and Keith 1939:310), while those of Krapina 1 are not quite so prominent in their development. In addition, the Krapina 1 forehead does not rise nearly as vertically as Skhul 1 to metopion.

Metric comparisons between the frontals of the two specimens reveal marked differences between Krapina 1 and Skhul 1 only in measurements of frontal breadth with Krapina 1 being considerably wider than Skhul 1 in both minimum (see Figure 2) and maximum (see Figure 3) frontal breadths. However, comparisons with the other samples indicate that wide frontal breadths are found in all three European groups: the Neandertals, early Upper Paleolithic associated Europeans, and the recent European sample. In contrast, cranial breadths for the recent Egyptians are considerably narrower than all of these European samples. Skhul 1 falls near the mean for the recent Egyptian children both in minimum and maximum frontal breadth. Although slightly higher in the recent Egyptian range, Qafzeh 11 is also smaller in frontal breadth than most of the European children. Thus these differences in frontal breadth measurements between Krapina 1 and Skhul 1 appear to be regional rather than evolutionary patterns of variation.

4.2. Parietal

One striking qualitative feature of the Krapina 1 parietal region is the presence of marked lambdoid flattening (see Figure 1). While this trait is fairly well-developed on

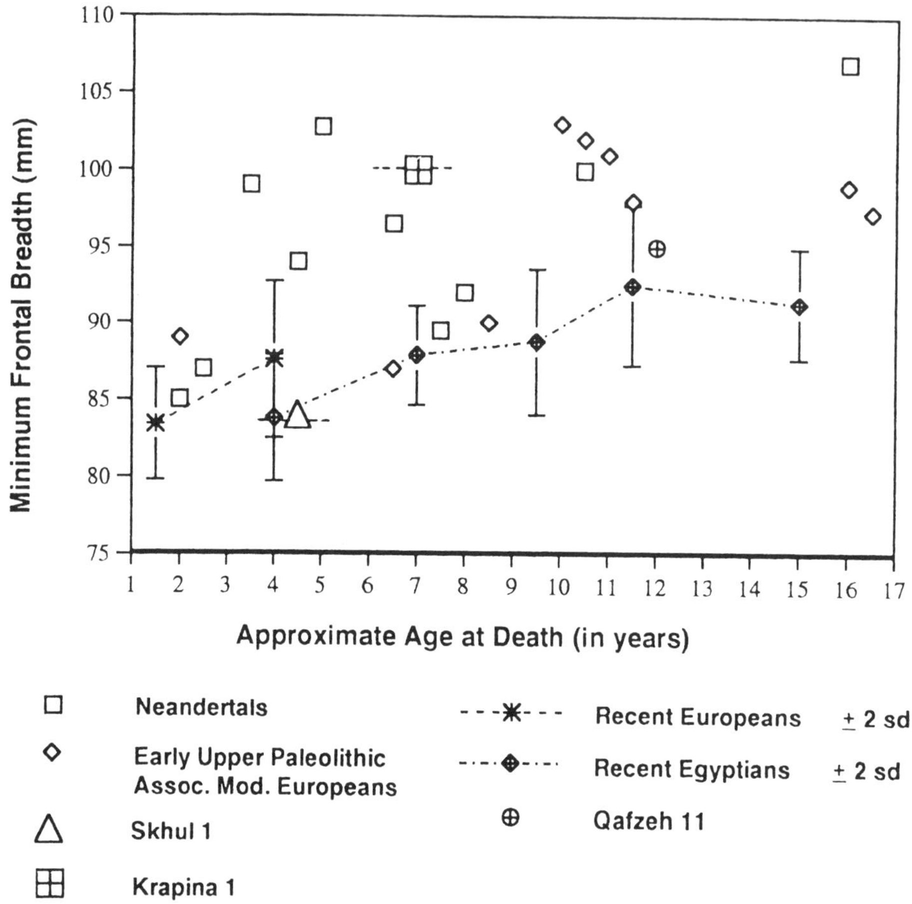

Figure 2. Regional growth comparisons: minimum frontal breadth.

some early Upper Paleolithic associated European juveniles, such as Mladeč 40 (Frayer et al. nd), strong lambdoid flattening is consistently found in Neandertals, including children as young as the 4.5 year old from Engis, Belgium (Tillier 1983; Minugh-Purvis 1988a). No lambdoid flattening is present in Skhul 1.

Another noteworthy qualitative contrast of the posterior parietal is the presence of a well-developed parietal eminence in Skhul 1. This is particularly obvious when viewed from a posterior aspect in which the lateral prominence of the parietal tubers creates a distinctly pentagonal form typical of modern humans. Krapina 1 also possesses a somewhat pentagonal posterior profile, although not of the same modern form as Skhul 1 (see Figure 4). Specifically, in Krapina 1, the parietal bosses are located more superiorly, at the top of the vault, so that the mid-vault segment is extremely flattened above at that level. In contrast, the Skhul 1 vault continues to rise well above the level of the parietal eminences as in modern humans.

Metric comparisons for the parietal bone reveal clear differences in mid-vault length between the two specimens. For bregma-lambda chord length, Krapina 1 falls into the range of the Neandertal children while Skhul 1 falls at the modern mean (see Figure 5). I have shown elsewhere (Minugh-Purvis, nd) that length of the mid-vault segment is

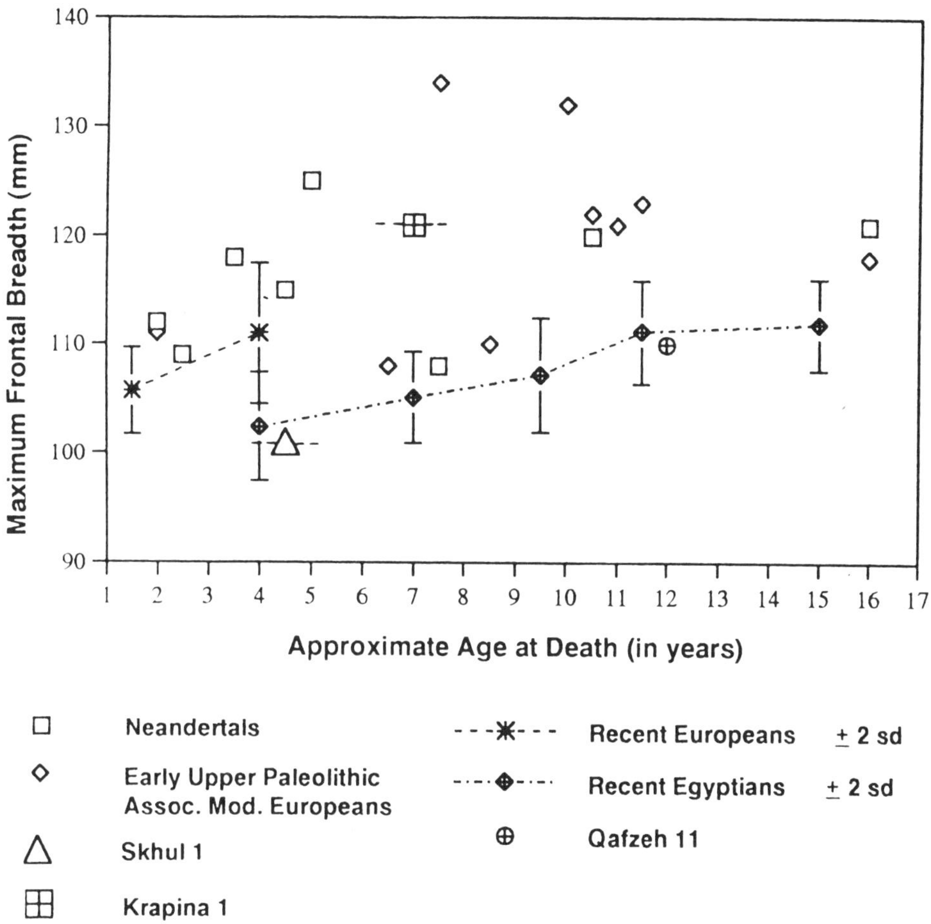

Figure 3. Regional growth comparisons: maximum frontal breadth.

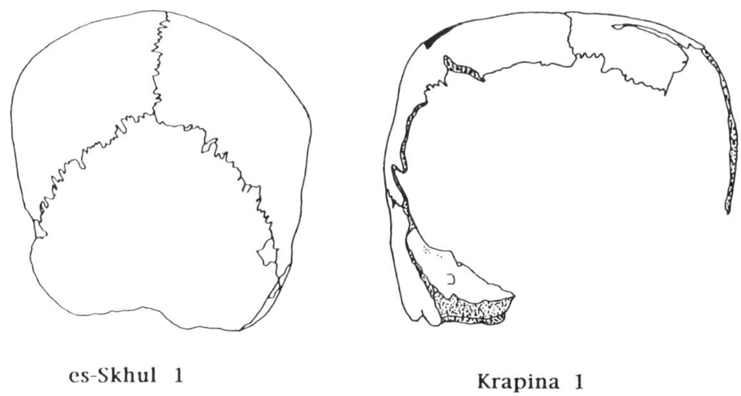

Figure 4. Posterior aspect of es-Skhul 1 and Krapina 1 (Skhul 1 after McCown and Keith 1939).

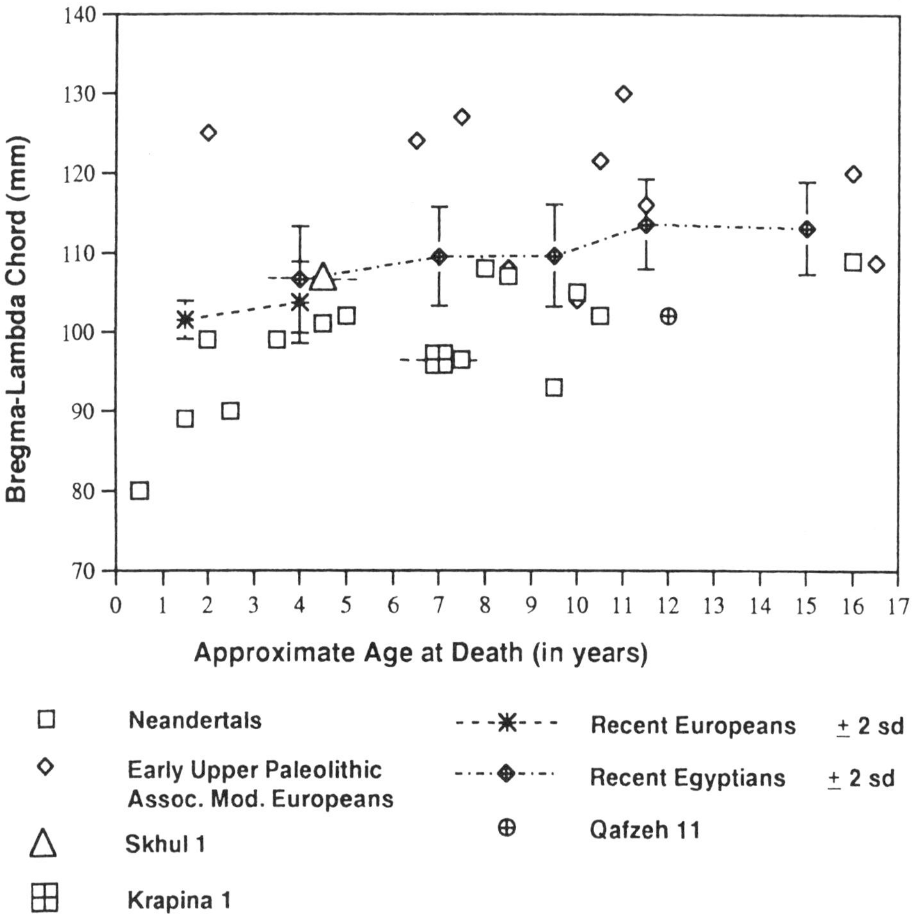

Figure 5. Regional growth comparisons: bregma-lambda chord.

possibly the most significant dimension for differentiating between juvenile archaic and early modern human neurocrania at any ontogenetic age. Here, this is dramatically illustrated by the wide gap between the early Upper Paleolithic associated modern Europeans and Neandertals. The recent European children, while somewhat smaller than most of the early Upper Paleolithic Europeans, are consistently longer than the Neandertals in the parietals. This contrast between the Neandertals and other Europeans as well as the recent Egyptian children (Figure 5) clearly demonstrates that these bregma-lambda chord length differences cannot be attributed to regional differences, suggesting an evolutionary or phylogenetic component to this pattern of variation. Interestingly, the Qafzeh 11 bregma-lambda chord fails to group it with the modern children, indicating that, at least in the mid-vault, this specimen was not anatomically modern.

When compared for bregma-lambda arc length, Skhul 1 falls into the modern range while Krapina 1 falls below it, again well within the range of variation displayed by the immature European Neandertals (see Figure 6). It is noteworthy, however, that although Skhul 1 falls within the modern human range, it does not even closely approach the range of variation for the extremely large vaulted early Upper Paleolithic associated modern European children who contrast with both the European Neandertals as well as the two re-

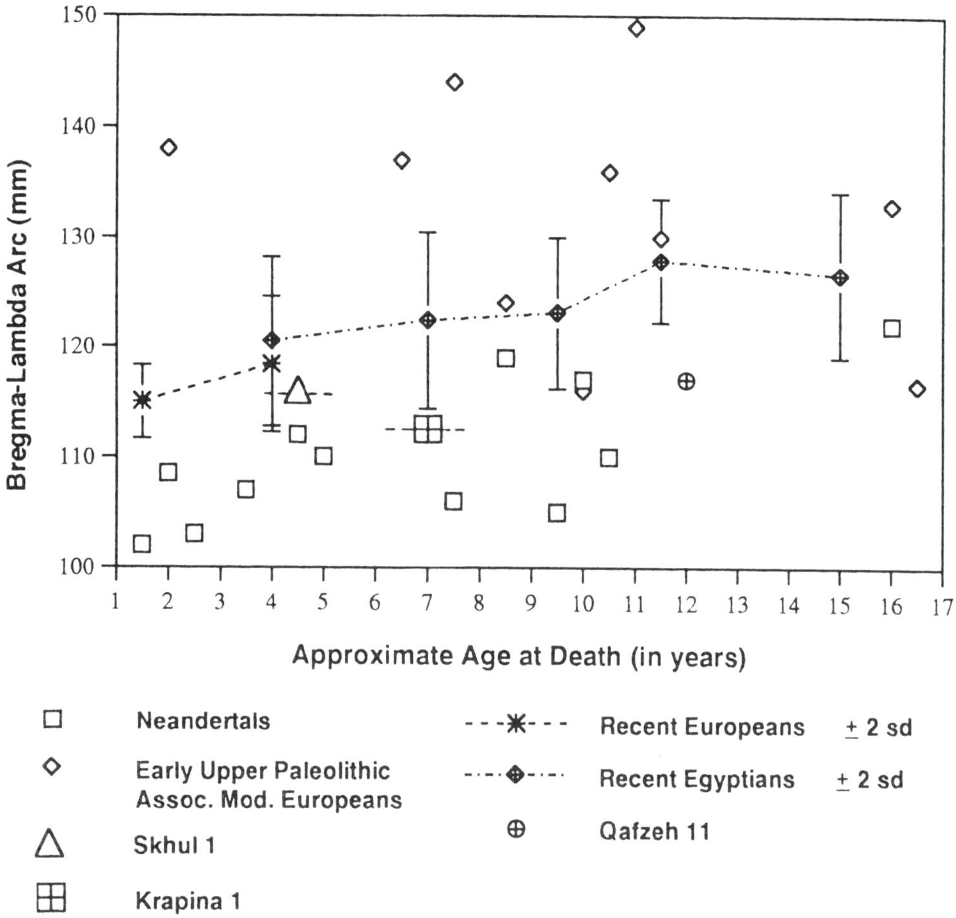

Figure 6. Regional growth comparisons: bregma-lambda arc.

cent comparative human samples for this dimension. As with the bregma-lambda chord, growth differences in bregma-lambda arc do not follow a regional pattern. Rather, a huge disparity is evident between the Neandertal children and the early Upper Paleolithic associated modern Europeans who followed them in time.

Posteriorly on the parietal, both Skhul 1 and Krapina 1 possess relatively long lambdoid chords (Figure 7), although Krapina 1 falls within the range of Neandertal children and outside of the modern range while Skhul 1, on the other hand, falls at the uppermost end of the recent Egyptian range.

4.3. Temporal

Three complexes of the temporal bone present qualitative traits providing interesting comparisons between Skhul 1 and Krapina 1. For example, the Skhul 1 mastoid is well-defined for a young child, protruding laterally from below the squamous temporal as in modern children of comparable developmental age. Krapina 1 also possesses a fairly well-developed mastoid process, which is more strongly developed than is usual in Neandertals of mid-childhood age. However, in other aspects of the mastoid region Krapina 1 differs

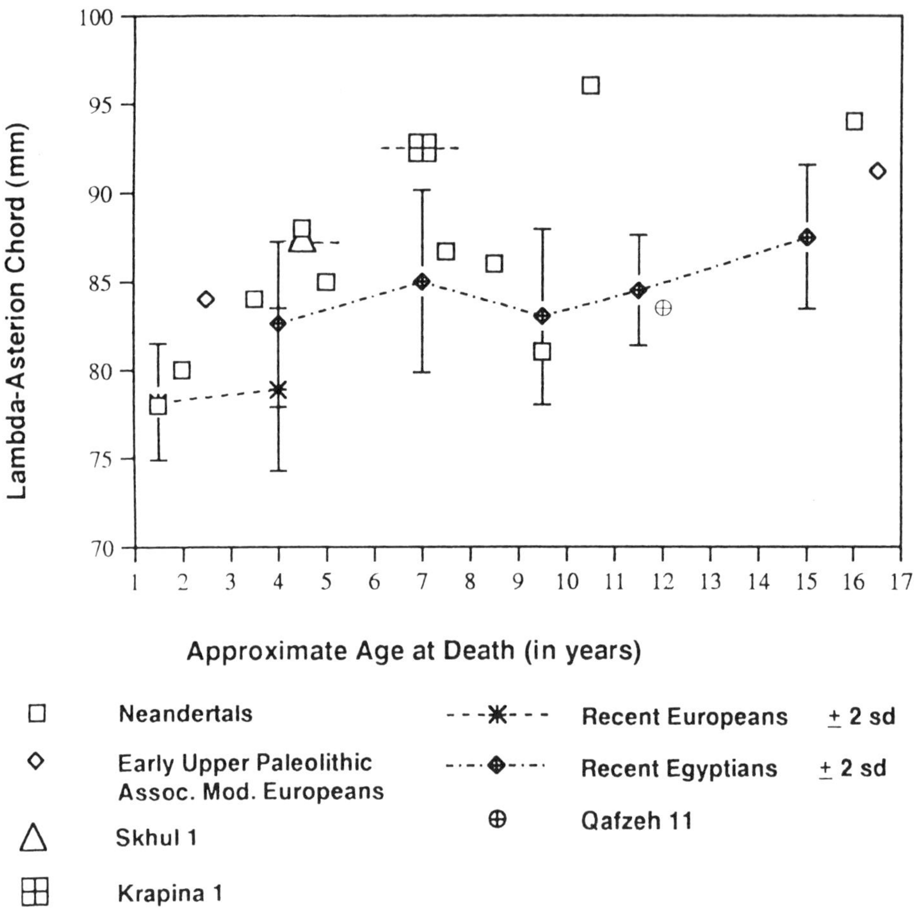

Figure 7. Regional growth comparisons: lambda-asterion chord.

from Skhul 1, particularly in possessing a well-developed juxta-mastoid eminence as is typical of Neandertals. As noted by Tillier (1992), this latter feature is absent in Skhul 1 whose temporal bone levels off medial to the digastric groove as in modern children. A second area of interest is the tympanic plate which in Skhul 1 is also modern in form. Its inferior margin forms the lower or posterior border of the vaginal process in contrast to Krapina 1, which possesses an inferior surface of the tympanic plate showing some division into anterior and posterior areas by a vaginal ridge as is typical of European Neandertals. Thirdly, in the tempro-mandibular region, both Skhul 1 and Krapina 1 possess well-defined glenoid fossae with an articular eminence of the form found in modern children. Although not regarded as typically Neandertal, this morphology is seen in other specimens at Krapina and thus falls within the morphological range of variation from that site (Minugh-Purvis et al., nd).

Comparison of temporal dimensions between Skhul 1, Krapina 1, and other late Pleistocene specimens illustrates the fairly short squamous temporal height of Skhul 1. Krapina 1 and the Neandertal children examined (Figure 8) share this pattern with Skhul 1, as do the recent European children. In this single instance then, Skhul 1 more closely resembles the Europeans. Temporal bone comparisons between the two specimens thus

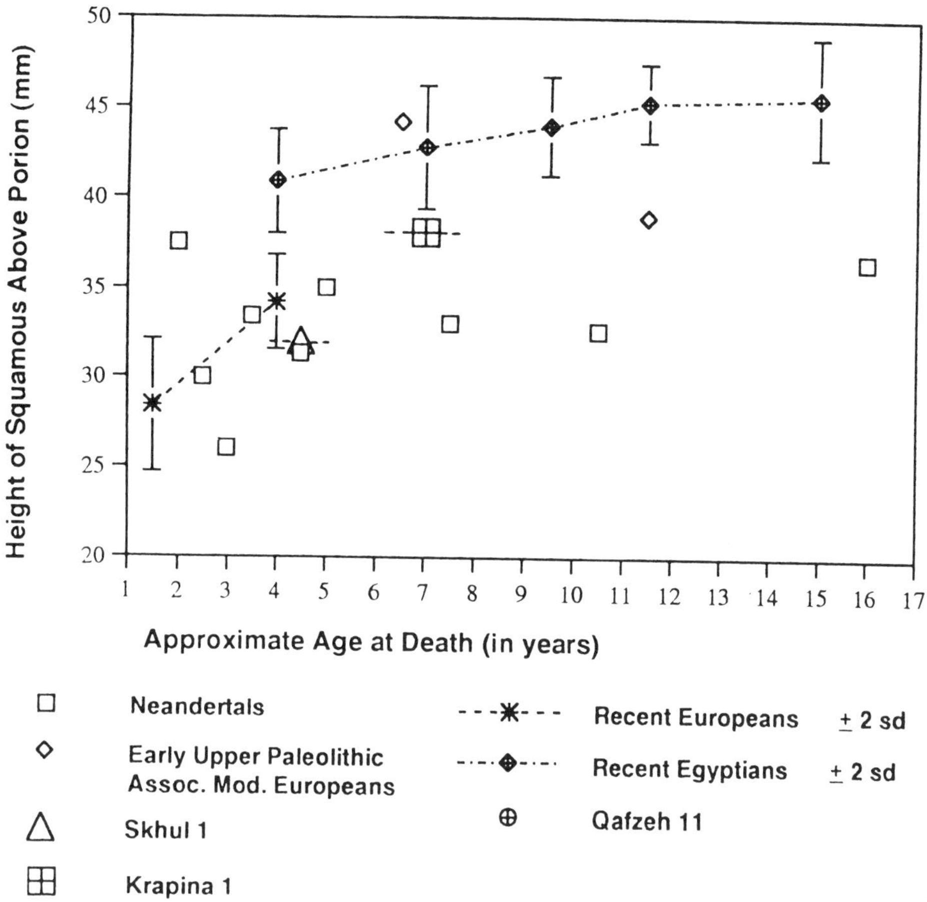

Figure 8. Regional comparisons: squamous temporal height.

reveal a mixture of archaic and modern European features in Krapina 1 but a clearly modern temporal morphological pattern in Skhul 1, although metrically Skhul 1 is closer to the European regional pattern than to the recent Egyptian children for squamous temporal height.

5. DISCUSSION/CONCLUSIONS

Results obtained from comparisons of these neurocranial dimensions and features reveal numerous differences between Krapina 1 and Skhul 1 (see Table 1). Both the qualitative and metric comparisons place Krapina 1 solidly within the European range of variation and, with the exception of its tempro-mandibular joint morphology and the possible exception of its supraorbital form, within the Neandertal range of variation. On the other hand, there are three important points concerning Skhul 1: (1) It is not European in morphology, although it does possess a low squamous temporal height in common with the European samples. However, patterns of growth in squamous temporal height in other populations require further investigation before this single resemblance should be re-

Table 1. Comparison of non-metric traits in Skhul 1 and Krapina 1

Traits in common	Traits that differ
Frontal	
1. Elevation at bregma from the frontal to parietal	1. Skhul 1: modern frontal eminences
2. Bipartite browridge	2. Krapina 1: bossing not modern in extent
	3. Skhul 1: squared off superior orbital margin
	4. Krapina 1: rounded orbital margin
Parietal	
1. Pentagonal posterior profile	1. Skhul 1: modern posterior aspect
	2. Krapina 1: parietal bosses nearly at height of vault
	3. Skhul 1: lambdoid flattening absent
	4. Krapina 1: lambdoid flattening
Temporal	
1. Projecting mastoid process	1. Krapina 1: juxta-mastoid eminence
2. Articular eminence present	2. Skhul 1: absence of juxta-mastoid eminence
	3. Skhul 1: modern petro-tympanic crest
	4. Krapina 1: bipartite petro-tymnpanic crest
	5. Skhul 1: double digastric groove
	6. Krapina 1: single digastric groove

garded as a European trait. (2) Skhul 1 possesses a derived parietal segment in mid-sagittal length and posterior parietal breadth and vaulting which groups it with modern humans (although it does not exhibit the very large sagittal parietal length of early modern Europeans). (3) Skhul 1 exhibits distinctly non-European narrow minimum and maximum frontal breadths and, for both dimensions, falls into the range, approaching the means of the recent Egyptian comparative sample. Thus in those portions of its morphology examined here, Skhul 1 exhibits little, if any, evidence of European genes.

These comparisons between Skhul 1 and Krapina 1 reveal morphological differences strongly supporting the contention that Skhul 1 and Krapina 1 were the products of late Pleistocene populations with fairly distinctive gene pools. Moreover, although it appears that both phylogenetic and regional differences separate these two specimens, the present analysis does not support the notion that Neandertals, such Krapina 1, are distinct beyond the subspecies level from early modern humans such as Skhul 1. Despite the differences identified here, continuity in qualitative aspects of morphology and also in growth patterns—evident in late Pleistocene to present-day European frontal breadths presented here and other features which have been extensively documented (Smith 1976, 1982, 1984, 1985, 1992; Frayer 1978, 1980, 1984, 1992; Smith and Ranyard 1980; Wolpoff 1980, 1989, 1995; Minugh-Purvis 1988a, nd; Frayer et al. 1993; and others)—speak strongly to the preservation of distinctive and unique parts of the European genome from the Neandertal populations present at 130 ka to today. Finally, despite the presence of a supraorbital with modern morphological tendencies, in conjunction with an otherwise strongly Neandertal phenotype in Krapina 1, there is no substantial evidence that modern genes, such as those present in the Levant during the early Upper Pleistocene, had yet arrived on the European continent by 130 ka.

ACKNOWLEDGMENTS

I thank Professor Takeru Akazawa for inviting me to participate in the symposium "Neanderthals and Modern Humans in West Asia" and for his gracious generosity and hospitality during my visit to Tokyo in November, 1995. I also thank the many individuals who kindly permitted me access to the fossil materials in their care, especially Dr. Jakov Radovčić (Croatian Natural History Museum, Zagreb, Croatia) and Mr. Joe Zias (Department of Antiquities, Rockefeller Museum, Jerusalem, Israel). Their assistance is deeply appreciated and gratefully acknowledged. Also, thanks to Doug Purvis for his continuous support, assistance, and encouragement throughout the course of this work. Supported in part by the Department of Neurobiology and Anatomy, Allegheny University of the Health Sciences; the LSB Leakey Foundation; Foundation for Research on the Origins of Man; Richard D. Irwin Foundation; and the Department of Anthropology, University of Pennsylvania.

REFERENCES

Bar-Yosef, O. 1994. The contributions of southwest Asia to the study of the origin of modern humans, 23–66. In *Origins of Anatomically Modern Humans*, ed. M. H. Nitecki and D. V. Nitecki. New York: Plenum Press.

Caspari, R. E. 1991. *The Evolution of the Posterior Cranial Vault in the Central European Upper Pleistocene*. Ann Arbor: University Microfilms.

Frayer, D. W. 1977. Metric dental change in the European Upper Paleolithic and Mesolithic. *American Journal of Physical Anthropology* 46:109–20.

Frayer, D. W. 1978. Evolution of the dentition in Upper Paleolithic and Mesolithic Europe. *University of Kansas Publications in Anthropology* Vol. 10. Lawrence: University of Kansas.

Frayer, D. W. 1984. Biological and cultural change in the European late Pleistocene and early Holocene, 211–50. In *The Origins of Modern Humans: A World Survey of the Fossil Evidence*, ed. F. H. Smith, and F. Spencer. New York: A. R. Liss.

Frayer, D. W. 1992. The persistence of Neanderthal features in post-Neanderthal Europeans, 179–88. In *Continuity or Replacement: Controversies in* Homo sapiens *Evolution*, ed. G. Braüer and F. H. Smith. Rotterdam: A. A. Balkema.

Frayer, D. W., M. H. Wolpoff, A. G. Thorne, F. H. Smith, and G. G. Pope. 1993. Theories of modern human origins: The paleontological test. *American Anthropologist* 95:14–50.

Frayer, D. W., J. Jelínek, N. Minugh-Purvis, F. H. Smith, and M. H. Wolpoff. nd. *Upper Pleistocene Human Remains from Mladec Cave, Moravia.*

Gorjanovič-Kramberger, D. 1906. *Der Diluviale Mensch von Krapina in Kroatien*. Wiesbaden: C. W. Kreidel.

Grün, R., and C. B. Stringer. 1991. Electron spin resonance dating and the evolution of modern humans. *Archaeometry* 33:153–99.

Madre-Dupoy, M. 1992. *L'enfant du Roc de Marsal*. Paris: CNRS Editions.

McCown, T. D., and A. Keith. 1939. *The Stone Age of Mount Carmel*. Vol. 2, *The Fossil Human Remains from the Levalloiso-Mousterian*. Oxford: Clarendon Press.

Mercier, N., H. Valladas, O. Bar-Yosef, B. Vandermeersch, C. B. Stringer, and J. L. Joron 1993. Thermoluminescence date for the Mousterian burial site of es-Skhul, Mt. Carmel. *Journal of Archaeological Science* 20:169–74.

Minugh-Purvis, N. 1988a. *Patterns of Craniofacial Growth and Development in Upper Pleistocene Hominids*. Ann Arbor: University Microfilms.

Minugh-Purvis, N. 1988b. Interpreting immature craniofacial morphology at Krapina: Problems and possibilities. Paper presented at the XII ICAES, Zagreb, Croatia (Yugoslavia), July, 1988.

Minugh-Purvis, N. nd. Metric aspects of ontogenetic patterning in late Pleistocene hominids: I. Neurocranium. Submitted to *American Journal of Physical Anthropology*.

Minugh-Purvis, N., and J. Radovčić. 1991. Krapina A: Neandertal or not? Paper Presented at the 61st Annual Meeting of the American Association of Physical Anthropologists. Milwaukee, April, 1991.

Minugh-Purvis, N., J. Radovčić, F. H. Smith. In press. Krapina 1: A juvenile Neandertal from the late Pleistocene of Croatia. *American Journal of Physical Anthropology*.

Pearson, K., and A. G. Davin. 1924. On the biometric constants of the human skull. *Biometrika* 16:328–53.

Rink, W. J., H. P. Schwarcz, F. H. Smith, and J. Radovčić. 1995. ESR dates for Krapina hominids. *Nature* 378:24.

Ronen, A. 1976. The Skhul burials: An archaeological review. In *Les Sépultures Néandertaliennes*, ed. B. Vandermeersch. Paris: CNRS Editions.

Schwarcz, H. P., R. Grün, B. Vandermeersch, O. Bar-Yosef, H. Valladas, and E. Tchernov. 1988. ESR dates from the hominid burial site of Qafzeh in Israel. *Journal of Human Evolution* 17:733–7.

Škerlj, B. 1958. Were Neanderthalers the only inhabitants of Krapina? *Bulletin Scientifique* (Yugoslvia) 4:44.

Skinner, M. 1985. Dental wear in immature Upper Pleistocene hominids of Europe. Report to Social Sciences and Humanities Research Council of Canada.

Smith, F. 1976. *The Neandertal Remains from Krapina. A Descriptive and Comparative Study.* Report of Investigations Number 15. Department of Anthropology, University of Tennessee.

Smith, F. H. 1982. Upper Pleistocene hominid evolution in South-Central Europe: A review of the evidence and an analysis of trends. *Current Anthropology* 23:667–703.

Smith, F. H. 1984. Fossil hominids from the Upper Pleistocene of Central Europe and the origins of modern humans, 137–209. In *The Origins of Modern Humans: A World Survey of the Fossil Evidence*, ed. F. H. Smith and F. Spencer. New York: A. R. Liss.

Smith, F. H. 1985 Continuity and change in the origin of modern *Homo sapiens. Zeitschrift für Morphologie und Anthropologie* 75:197–222.

Smith, F. H. 1992. The role of continuity in modern human origins, 145–56. In *Continuity or Replacement? Controversies in* Homo sapiens *Evolution*, ed. G. Braüer and F. H. Smith. Rotterdam: A. A. Balkema.

Smith, F. H., and G. Ranyard. 1980. Evolution of the supraorbital region in Upper Pleistocene fossil hominids from South Central Europe. *American Journal of Physical Anthropology* 53:589–610.

Stringer, C. B., R. Grün, H. P. Schwarcz, and P. Goldberg. 1989. ESR dates for the hominid burial site of es Skhul in Israel. *Nature* 338:756–8.

Tillier, A. M. 1983. Le crâne d'enfant d'Engis 2: Un example de distribution des caractères juvéniles, primitifs et néanderthaliens. *Bulletin de la Sociéte Royale Belge d'Anthropologie et de Préhistoire* 94:51–75.

Tillier, A. M. 1986 Quelques aspects de l'ontogenèse du squelette cranien des Néanderthaliens. *Fossil Man, New Facts, New Ideas* 23:207–18.

Tillier, A. M. 1992. The origins of modern humans in Southwest Asia: Ontogenetic aspects, 15–28. In *The Evolution and Dispersal of Modern Humans in Asia.* ed. T. Akazawa, K. Aoki, and T. Kimura. Tokyo: Hokusen-Sha.

Valladas, H., J. L. Joron, G. Valladas, B. Arensburg, O. Bar-Yosef, A. Belfer-Cohen, P. Goldberg, H. Laville, L. Meignen, Y. Rak, E. Tchernov, A. M. Tillier, and B. Vandermeersch. 1987. Thermoluminescence dates for the Neanderthal burial site at Kebara in Israel. *Nature* 330:159–60.

Valladas, H., J. L. Reyss, J. L. Joron, G. Valladas, O. Bar-Yosef, and B. Vandermeersch. 1988. Thermoluminescence dating of Mousterian "Proto-Cro Magnon" remains from Israel and the origin of modern man. *Nature* 331:614–16.

Vlček, E 1970. Étude comparative onto-phylogénetique de l'enfant du Pech de l'Azé par rapport à d'autres enfants néanderthaliens. *Archives d'Institute du Paléontologie Humaine* 33:149–80.

Wolpoff, M. H. 1980. *Paleoanthropology*. New York: Knopf.

Wolpoff, M. H. 1989. Multiregional evolution: The fossil alternative to Eden, 62–108. In *The Human Revolution*, ed. P. A. Mellars and C. B. Stringer. Edinburgh: Edinburgh University Press.

Wolpoff, M. H. 1995. Preliminary publication of *Paleoanthropology*. Second Edition. New York: McGraw Hill, Inc.

Wolpoff, M. H., F. H. Smith, M. Malez, J. Radovcic, and D. Rukavina. 1981. Upper Pleistocene human remains from Vindija Cave, Croatia, Yugoslavia. *American Journal of Physical Anthropology* 54:499–545.

22

DOES ANY MOUSTERIAN CAVE PRESENT EVIDENCE OF TWO HOMINID SPECIES?

Yoel Rak

Department of Anatomy and Anthropology
Sackler Faculty of Medicine
Tel Aviv University
Ramat Aviv 69978, Israel

1. INTRODUCTION

"While the skull is the creation of God, the jaw is the work of the devil."* Be the source of this quote as it may, it is undoubtedly expressive of the frustration that researchers feel when, as so often happens, they examine a mandible that is disarticulated from the skull. Phrased in scientific terms, this quotation simply states that the diabolic mandible bears fewer taxonomic characters—is less diagnostic—than the cranium. Consequently, it lends itself less readily to taxonomic evaluation, with the result that the ensuing taxonomic decision is less secure.† Two mandibles were found in Tabun Cave, both reported as originating in layer C, as described in the field report.

> From the point of view of human remains C was the most interesting of all the layers in Tabun, since it contained a nearly complete skeleton (Tabun I). This lay with head pointing nearly due west, just outside the cave mouth, on the west side of the talus, at 1.35 m below the datum (0.35 m below the surface of C at this point) (Pl. XXXII)....
>
> The skeleton lay so near the surface of C that the question must arise whether it does not represent a burial from Layer B. There was no obvious sign of disturbance, but since the earth of B in this area was also red, and the industry so far in the same tradition, that the presence of a few intrusive flints in C could not possibly be detected, I feel that this must remain an open question....
>
> On the east side of the talus, at 2.20 m below datum (1.20 m below the surface of C at this point), six fragments of a lower jaw (Tabun II) were found scattered over a small area outside the east alcove.... (Garrod and Bate 1937:64).

* Source unknown. I cannot identify the precise origin of this saying which I heard in my student years. Clark Howell (personal communication) believes it can be attributed to William Straus.

† This statement is limited in its validity and, as demonstrated in these pages, I have never espoused it. Making a taxonomic decision based on the mandible of a *generalized* animal is, indeed, problematic. However, when the mandible derives from an animal with a specialized masticatory system, the determination remains relatively easy. But does this not hold true for the cranium as well?

Neandertals and Modern Humans in Western Asia, edited by Akazawa *et al.*
Plenum Press, New York, 1998.

It is clear from this description written by Garrod (Garrod and Bate 1937) that the vertical distance between the two specimens was only about 85 cm. Although the exact location of the second mandible is not marked on any of the maps, my calculations suggest that the two specimens were found about 7 m apart, with the Tabun II mandible lying almost due east of the skeleton, "outside the east alcove" (Garrod and Bate 1937:64).

The depth at which the Tabun II specimen was found in layer C (1.20 m below the top of the layer) and the layer's relative horizontality (according to the published drawings) ensure that the specimen does indeed belong to this layer. Because of the layer's horizontality—untainted by the stratigraphic problems emanating from the inclination of the lower layers—we can also be confident that the stratigraphic relationship between the two specimens remained constant despite the significant horizontal distance between them. In contrast to the attribution of Tabun II to layer C, the Tabun I skeleton, though clearly marked on Garrod's map, was uncovered close to the top of this layer. Therefore, it is not inconceivable that this specimen originated from the layer above, layer B. If any substantial chronological difference did exist between Tabun I and II, Tabun I would, in any case, be the younger of the two.

Tabun I is readily defined taxonomically, as diagnostic anatomical elements from the skull and the skeleton are present along with the mandibular morphology. Together, these characters lead to almost universal agreement that the specimen is a Neandertal. The Tabun II specimen has always been much more problematic, as it consists solely of the mandible, which, as stated above, is merely the work of the devil!

Although the mandible undoubtedly bears fewer taxonomic characters than the skull, one cannot help but be amazed at the way some of these characters, though they may be familiar, have been sorely overlooked. Thus it appears that the main obstacle to the taxonomic identification of the mandible has not been the paucity of diagnostic taxonomic characters at all. These, we shall see, were found in more than sufficient abundance to allow for a precise taxonomic decision. The real obstacle is entirely different and turns the discussion of the anatomy of the Tabun mandibles and their taxonomic status into a fascinating lesson whose moral does not lie so much in the realm of systematics but rather in the history of science.

The purpose of this paper, more than to clarify the taxonomic status of Tabun II, is to show that McCown and Keith's specific paradigm when they studied the mandibles for publication in 1939, rather than the paucity of taxonomic characters, led to difficulties in reaching a taxonomic definition of the specimen. Only by identifying the preconceived ideas to which McCown and Keith were hostage can we explain their difficulties, hesitancies, and erroneous (as I see it) taxonomic determination. To this day several investigators have retained this paradigm, to a greater or lesser degree, and it is noticeable every time the subject of Neandertals is raised for discussion.

2. THE CHIN

First and foremost, the mandible of Tabun II bears the hallmark of *Homo sapiens*: a real chin, which contains all the typical elements that join to form this structure, as described by McCown and Keith:

> The chief differences are to be noted in the region of the chin. In the Heidelberg jaw, as in that of Tabun I, there is no chin; alveolar and basal elements of the symphysis continue in the same oblique plane, the alveolar being more massive than the basal element. The basal element of the symphysis of Tabun II (Figure 145) has become bent forward on the alveolar, an open angle being formed in front between the two. In addition, a bony eminence or prop is be-

> ing built up in front to support the overhanging alveolar shelf. Thus is the chin evolved. In the Tabun man [Tabun II] the teeth and the alveolar element of the mandible seem to have retained their full development; they are not yet reduced. The floor of the mouth is being expanded by a forward movement of the basal element of the symphysis, and new bone is being formed in front.... (1939:218).

The authors are referring to what is, indeed, a mental protuberance, a true chin as classically described in modern humans: two mental tubercles that are clearly separated from the superior part of the mandible by the mental sulcus and merge at the midline to form the prominent mental, or trigonal, eminence. The latter is separated at the midline from the more superior part of the mandible (the alveolar element) by a typical depression—*impressio subincisiva externa* (see, for example, Inke 1967).

Despite this topography and the substantial differences in the symphyseal region of Tabun I and II, of which McCown and Keith are well aware, they choose to focus on the trivial difference between the angle of the Tabun II mandible and that of the sole modern *Homo sapiens* mandible that they use for comparison (80° in Tabun II versus 86° in their Australian aboriginal specimen). Strangely enough, the 6° difference in the symphyseal axis, which should be expected in any population, leads them to conclude that "Tabun II shows an *early* stage in the evolution of the chin" (McCown and Keith 1939:225; my emphasis). This conclusion, based primarily on the inclination of the symphyseal axis, is a persistent theme throughout McCown and Keith's discussion of the Tabun mandibles. Emphasizing that Tabun II is only "an early stage" prepares the way for the authors' inevitable taxonomic unification of the two Tabun specimens, as necessitated by their bias.

I would like to stress, therefore, that my insistence on recognizing a true chin in Tabun II is based on the topography of the anterior section of the mandible, which consists of the unique morphological components defined above as constituting a true chin, and not on the inclination of the symphyseal axis. The 6° difference that presumably exists between the symphyseal axis in the Tabun II mandible and that of modern humans is meaningless in the identification of a true chin. The steepness of the symphyseal axis is a variable parameter in hominid evolution. On the basis of symphyseal steepness alone, one can easily reach absurd conclusions—for example, that specimens bearing a "chin" of this type would include not only the chinless Amud 1 and Spy 1 but also specimens of *Australopithecus* such as the *A. boisei* mandible from Peninj.

Tabun II, in short, is a mandible with a true chin (full-fledged and modern looking), which is actually also observed by those who first described it. In addition, McCown and Keith fully recognize the absence of a true chin in Tabun I. However, despite the fact that these investigators are aware of the substantial difference between the specimens with respect to the chin, they persist in stating that the two mandibles are the same. The substantial difference between Tabun I and Tabun II in the topography of the front of the symphysis led them to view Tabun II as slightly more advanced ("an early stage in the evolution of the chin") than Tabun I in the evolutionary sequence of the chin's formation. The impression obtained from the text is that they believed that it would be natural to find, within a living population, individual specimens that are morphologically more "advanced" than others. The presence of "advanced" specimens would imply that part of the population consists of intermediate types, of which Tabun II is an example. These intermediate specimens are said to "bridge the structural hiatus lying between the Neandertal and the Neanthropic types" (McCown and Keith 1939:372).

Nevertheless, additional significant anatomical differences that are no less important than the chin morphology exist. Suprisingly, these diagnostic elements were either missed or ignored by the original investigators and many others who followed them.

3. THE ASCENDING RAMUS OF THE MANDIBLE

In a lateral view, the similar size of the condylar and coronoid processes in the upper part of the ascending ramus gives this structure a symmetrical appearance. The mandibular notch connecting the processes dips deeply between them, with its deepest point near the midpoint between the two processes (Figure 1, top). This configuration is clearly not what characterizes the Neandertal's mandibular ramus. Its coronoid process is much larger than its condylar process. The outline of the mandibular notch, therefore, is very shallow, and its deepest point (which is not deep at all) is located very close to the

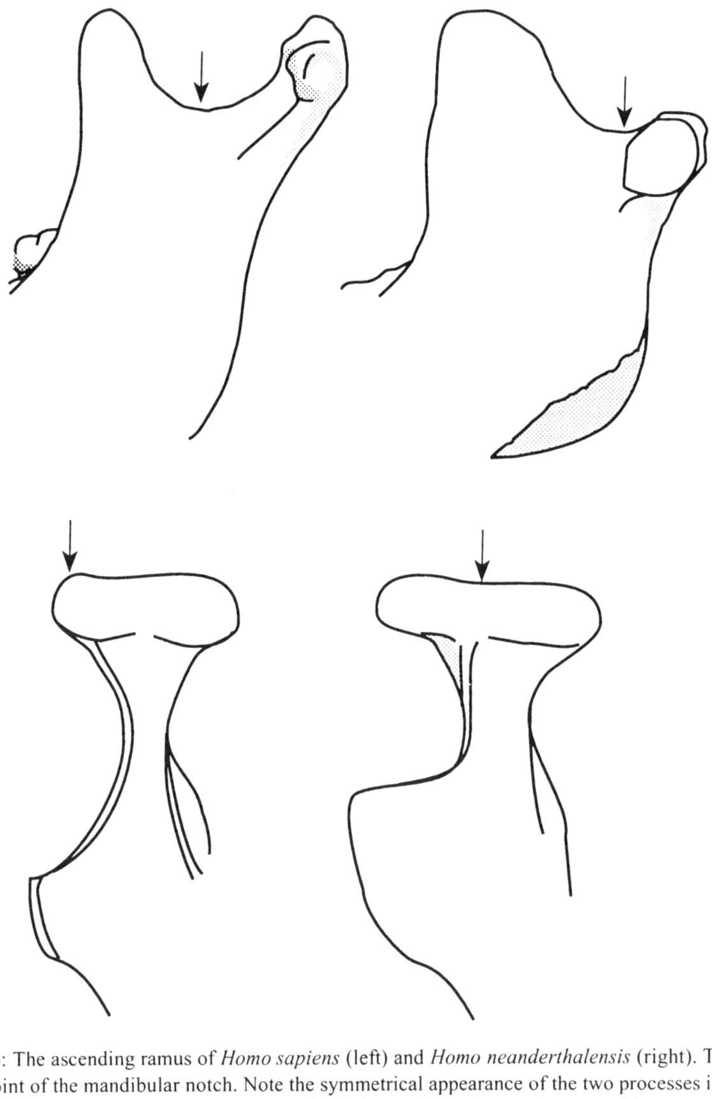

Figure 1. Top: The ascending ramus of *Homo sapiens* (left) and *Homo neanderthalensis* (right). The arrow marks the deepest point of the mandibular notch. Note the symmetrical appearance of the two processes in *Homo sapiens* and their asymmetrical appearance in *Homo neanderthalensis*. In the latter, the coronoid process towers over the condylar process, and the notch's deepest point is close to the condylar process. Bottom: The relationship between the crest of the mandibular notch and the condyle. The arrow indicates the point where the crest meets the condyle. Note that in *Homo sapiens* (left) the crest extends to the lateral end of the condyle, whereas in the Neandertal, the crest bisects the condyle and is perpendicular to the latter's long axis.

condyle (Figure 1, top). This configuration provides the upper part of the ramus with an extremely asymmetric appearance, in which the overwhelmingly large coronoid process towers over the condyle. All Neandertal mandibles appear to share this morphology to some degree (Figure 2). Among them, Régourdou 1 and Amud 1 certainly display it most

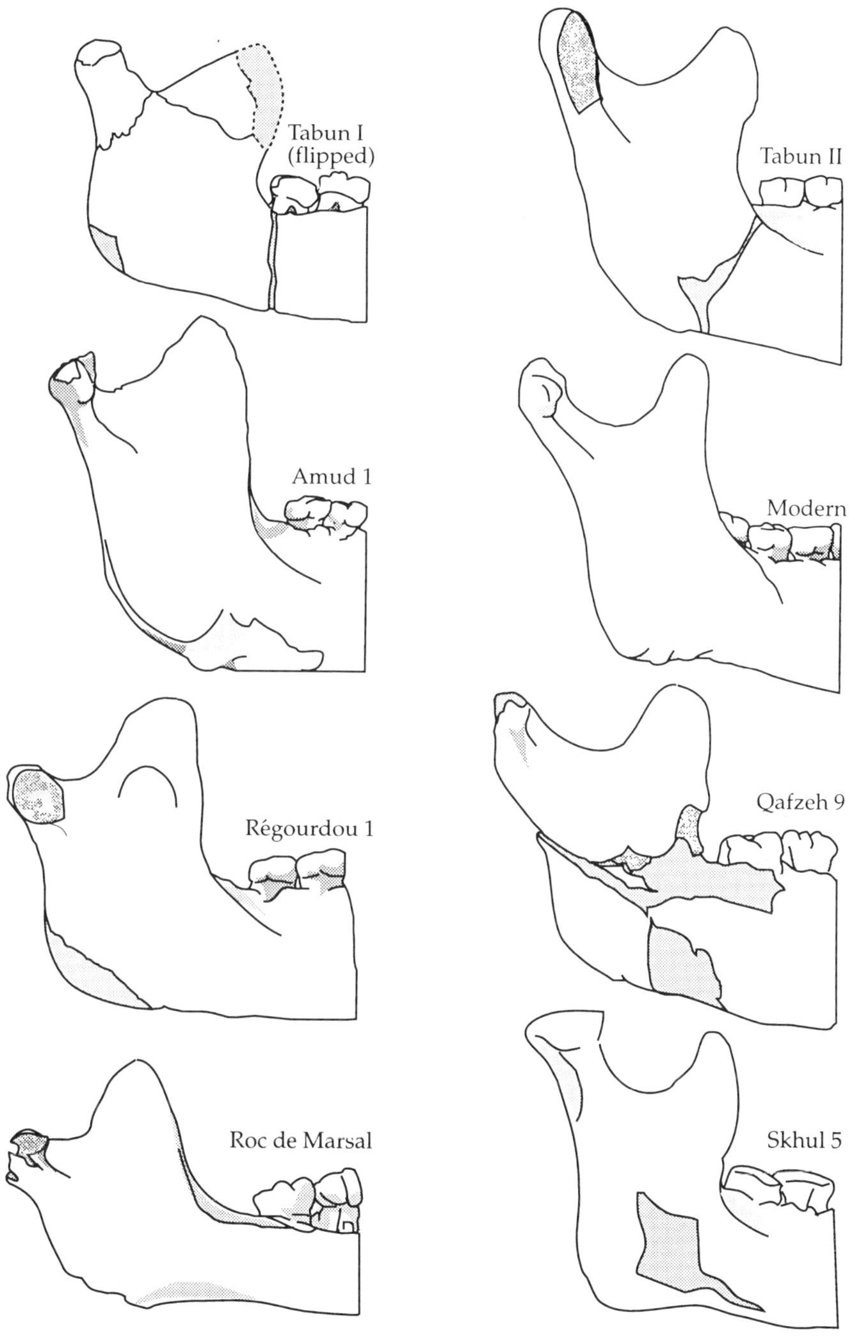

Figure 2. A side view of the ascending ramus of some Neandertal specimens (left) and some *Homo sapiens* specimens (right).

dramatically, and it is prominent even in juveniles (such as Roc de Marsal, Figure 2). It is not surprising to find that Tabun I, as a Neandertal, also exhibits this asymmetric appearance, which is readily discernible despite some destruction of the coronoid process.

Tabun II, on the other hand, displays the arrangement that characterizes *Homo sapiens*, as described above: the condylar process and the coronoid process are of similar size, and a deep, concave notch connects the two. The deepest part of the mandibular notch is found at a point equidistant to the processes, thereby making them equal in size. This morphology is also seen in all the specimens from Skhul and Qafzeh, for which there appears to be a consensus regarding their taxonomic status as early *Homo sapiens* (Figure 2).

In light of the fact that *Homo sapiens* shares this configuration with other hominids and primates, we might conclude that it is apomorphic in the Neandertal, although the rarity of complete rami in the hominid fossil record precludes a definitive determination. However, since the Ternifine III specimen represents a variation of this Neandertal configuration, perhaps the anatomy described as characterizing the Neandertal ramus is better viewed as a synapomorphic feature of a particular clade as opposed to an autapomorphic feature of just the Neandertal. In any event, since our primary concern is the distinction between *Homo sapiens* and *Homo neanderthalensis*, what really matters at this point is not whether the observed morphology represents the primitive or the derived morphology but that the Tabun II specimen displays the *Homo sapiens* pattern whereas Tabun I displays the pattern of *Homo neanderthalensis*.

Besides the shallow outline of the Neandertal's mandibular notch (as viewed from the side) and the posterior location of its deepest point, the manner in which the sharp crest of the notch is connected to the condyle as seen in a superior view (Figure 1, bottom) is unique among hominids. In modern humans as well as in many other hominids and primates, the crest deviates sharply laterally as it approaches the condyle and eventually joins the condyle's lateral part. This configuration undoubtedly represents the plesiomorphic arrangement (see Figure 3). In the Neandertal, however, the crest of the mandibular notch runs directly to the condyle, almost perpendicular to the condyle's bilateral axis. The crest, therefore, reaches the condyle almost at the middle of the latter (nearly half the condylar breadth protrudes laterally from the sagittal plateau of the crest). Although this description represents how I prefer to view the configuration, researchers have long recognized it as characteristic of Neandertal mandibles. However, they have usually emphasized the "lateral tubercle" on a Neandertal mandible. In other words, they use the expression "lateral tubercle" for the entire lateral portion of the condyle, the part that is lateral to the crest of the mandibular notch. (But see, for example, Weidenreich's description [1936], in which he also chooses to portray the ordinary—plesiomorphic—condylar morphology of *Homo erectus* in terms of the relationship between the crest of the mandibular notch and the condyle.)

All of the Neandertals exhibit the unusual relationship between the crest of the mandibular notch and the condyle. This group includes the hominids that were found in Israeli caves and are universally viewed as Neandertals (Amud 1 and Kebara 2, Figure 3). The destruction of the condyles in the Tabun I specimen makes it harder to discern the feature unequivocally, although from the little that remains it is possible to deduce that it portrays a Neandertal configuration.

All the specimens of Skhul and Qafzeh clearly demonstrate the generalized configuration, that which is seen in today's modern humans (Figure 3). In all the early *Homo sapiens* specimens that permit this observation, the crest deviates laterally, as in modern humans, and reaches the lateral end of the condyle; the front of the medial part of the condyle is slightly deepened as the lateral pterygoid muscle's insertion site. In this respect, Tabun II is identical to the Skhul-Qafzeh group and hence to today's modern humans. The

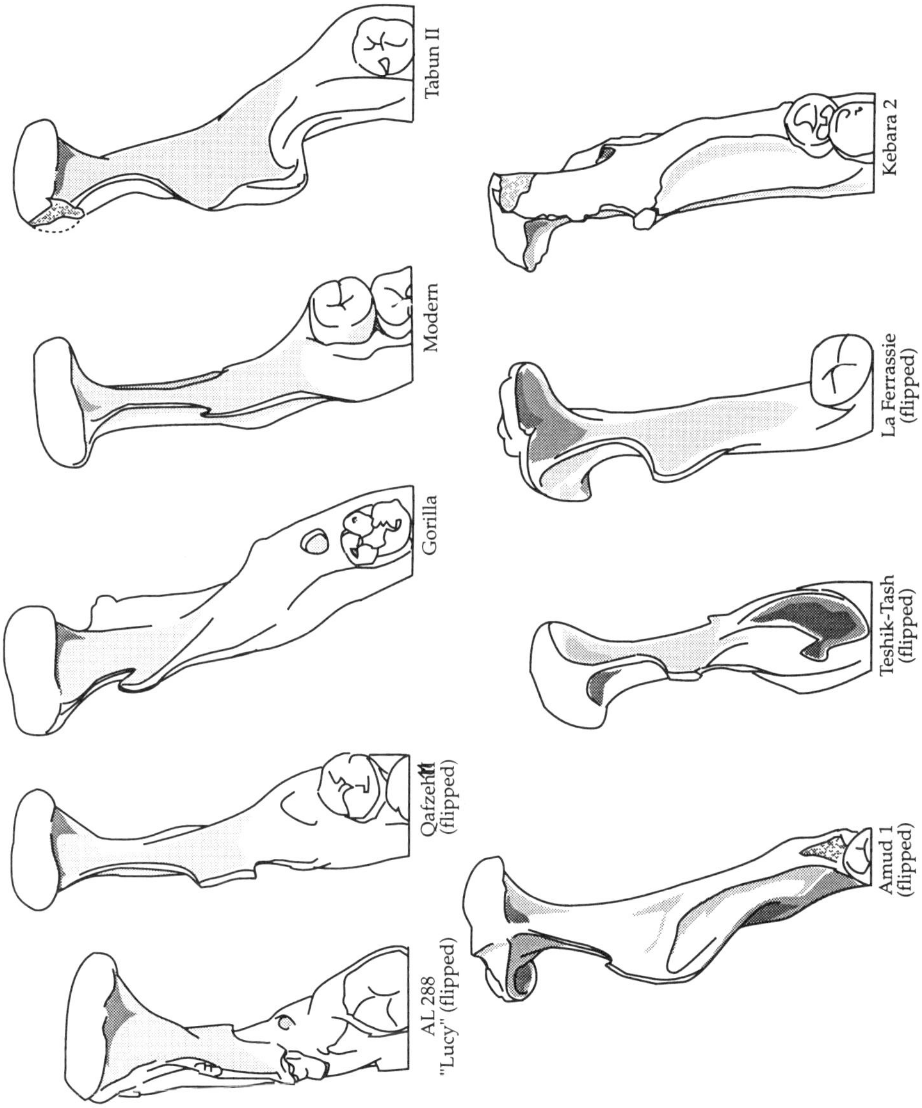

Figure 3. The relationship between the crest of the mandibular notch and the mandibular condyle. The top row shows the generalized configuration, and the bottom, the Neandertal morphology.

relationship between the crest of the mandibular notch and the condyle is clearly an autapomorphic Neandertal feature.

4. THE RETROMOLAR SPACE

A discussion of the features of the Tabun II mandible would not be complete without reference to the retromolar space. In a recent paper, Franciscus and Trinkaus (1995) provide a comprehensive review of the historical literature as well as a thorough discussion of the trait's frequency in archaic and modern humans. These authors generally confirm what had already been observed in the past, namely, that the retromolar space is, indeed, larger and present in a higher frequency among Neandertals and is relatively less frequent in chronologically pre- and post-Neandertal hominids. Nevertheless, the authors emphasize that "the contrast between these late archaic humans [i.e., Neandertals] and early humans (especially from the Near East) is less pronounced, removing this feature from the list of possible autapomorphies" (Franciscus and Trinkaus 1995:577).

However, I tend to see things differently. Although formally the term "retromolar space" is defined as the space between the anterior margin of the ascending rami and the third molar, I believe that the retromolar space in *Homo neanderthalensis* and *Homo sapiens* (such as the Tabun II, Skhul, and Qafzeh specimens) is the product of an entirely different anatomical configuration. In Neandertals, M3 is located farther anteriorly (incidental to the shortening of the anteroposterior length of the dental arcade), thus creating a large gap between the molar and the ramus. In early *Homo sapiens* specimens such as Tabun II, the retromolar space results not from an anterior position of M3 but from a deep, local indentation, called the pre-angular notch, in the contour of the anterior margin of the ascending rami. The deeper the notch at the front of the ramus root, the greater the distance seems to be between the ramus and the third molar. In Tabun II, this notch is especially deep.

This hypothesis is easy to test. Instead of measuring the retromolar space itself (in the traditional manner followed by Franciscus and Trinkaus), one can measure the distance between the rear margin of the ramus (at the level of the alveolar plane) and M3 (see Figure 4, top). In this way, a measurement is obtained that neutralizes the influence of the depth of the pre-angular notch's indentation and acknowledges only the effect of M3's position in creating the gap. Figure 4 demonstrates the differences between two mandibles—Tabun II and the La Ferrassie Neandertal, which, except for those differences, are almost of the same size (Franciscus and Trinkaus 1995). Note in the lower drawing that the position of La Ferrassie's third molar overlaps that of the anterior half of the second molar in the Tabun II mandible; the La Ferrassie molar is situated much farther anteriorly to the rear ramus margin than the same molar in Tabun II (the difference is 16 mm, the length of one and a half molars). In other words, whereas the retromolar space stems from the anterior position of M3 in La Ferrassie, the narrow ramus in the form of the pre-angular notch is what produces the space in Tabun II. The proportionately narrow ramus in the latter is acknowledged by Franciscus and Trinkaus (1995).

The pre-angular notch is normally found to some extent in every modern human population. It seems to be more common in early *Homo sapiens* populations, as it can be identified in all the Skhul and Qafzeh specimens and in the later Natufians, as well as in the Cro-Magnon specimens, Ohalo 2, Choukoutien Upper Cave, and others. The notch's depth in these individuals is closely related to the size of the retromolar space. As mentioned above, in Tabun II, the pre-angular notch is particularly deep, and therefore the gap

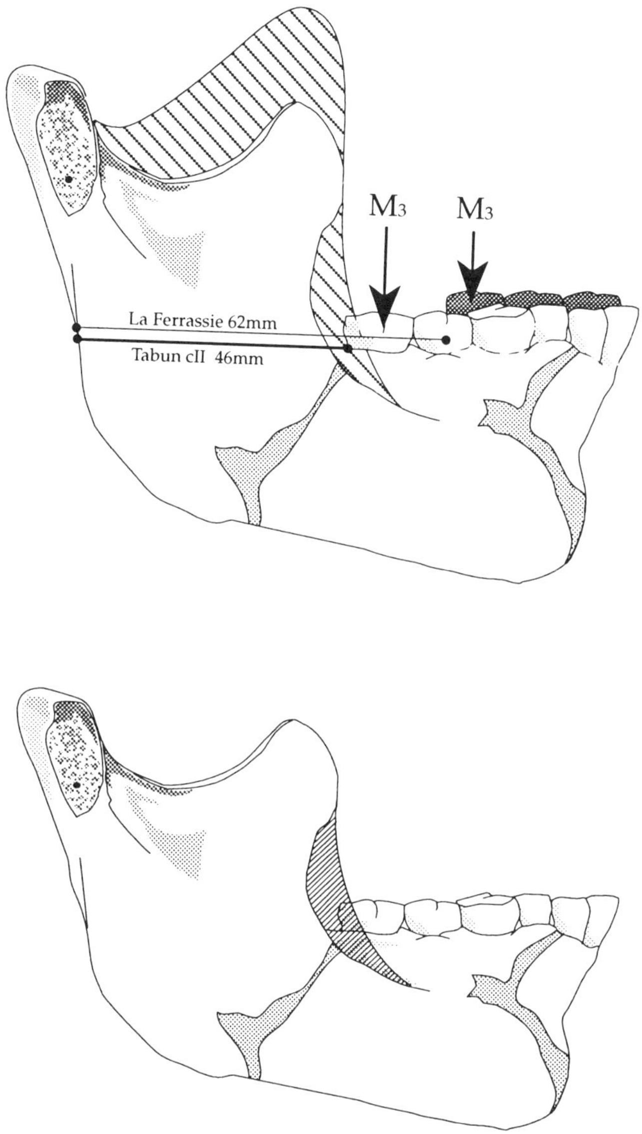

Figure 4. Top: The position of M3 in La Ferrassie and Tabun II relative to the posterior margin of the ramus. Note the considerably anterior position of M3 in La Ferrassie. The hatched area represents the shape of a Neandertal's ascending ramus. Bottom: The pre-angular notch in Tabun II and its effect in creating the retromolar space. Note that when the pre-angular notch (the shaded area) is eliminated, M3 is hidden behind the ramus.

between M3 and the anterior ramus margin is noticeably larger. There is no doubt that the presence of the notch makes the ramus disproportionately narrow. In Tabun II the ramus' breadth constitutes only 32.7 percent of the total length of the mandible, versus the mean of 37.4 percent that I calculated from all the archaic hominids on Franciscus and Trinkaus's list except for the early *Homo sapiens* specimens from Skhul, Qafzeh, and Tabun II.

To demonstrate the effect of the pre-angular notch in creating the retromolar space, we can eliminate the outline of the notch artificially, as shown in Figure 4, by increasing the breadth of the ramus by 5.6 mm. This increase is the equivalent of adding 4.7% to the ramus breadth in Tabun II, which would give the specimen the same breadth measurement as the average of the other hominids. (To conceal part of M3 behind the ramus, the breadth must be increased by approximately 10 mm, which would render it slightly greater than the mean hominid breadth). Canceling out the effect of the notch on the narrowing of the ramus and thereby on the ramus's shape transforms the ramus to its "ordinary" shape and dimensions. Once the notch is eliminated, the retromolar space disappears. Such an exercise cannot be performed on a typical Neandertal mandible since a pre-angular notch is not usually present in these hominids. Even when Neandertal specimens do display a notch, it is modest (e.g., Régourdou and La Ferrassie). In any case, canceling out the notch in such Neandertal specimens does not eliminate the retromolar space, as its size is determined by the anteriorly situated M3.

We submit, therefore, that it is not the space itself that is an autapomorphic feature of *Homo neanderthalensis* (since it appears also in early *Homo sapiens*) but, rather, the unusual anterior position of M3, which also results in the shortening of the dental arcade at the expense of the latter's rear portion. In other words, a distinction must be made between a retromolar space of the type that appears in the Neandertal and that which *sometimes* appears in *Homo sapiens* and other hominids. This distinction is apparently not made by Franciscus and Trinkaus (1995), as can be seen from their sweeping conclusion that "in sum, it is clearly a combination of relatively short dental arcade lengths and *small ramus breadths* in the context of maintained (or only slightly reduced) mandibular lengths which best accounts for the presence of retromolar spaces in our sample" (Franciscus and Trinkaus 1995:591; my emphasis).

It is surprising that McCown and Keith do not take the retromolar space into account in their discussions. In their attempt to demonstrate that the two specimens are of a single "race," they could have well used this feature.[‡] At first, one might assume that they ignore it because they are unaware of its formal diagnostic importance. However, this anatomy had already been described and proposed as a Neandertal trait well before their study (see Martin 1923). It is more reasonable to assume that they disregard this particular trait because the retromolar space exists also in the Skhul specimens and thus is of no use when one wishes to distinguish the Skhul from the Tabun hominids. In any case, the retromolar space is not the point. The real issue lies in the fact that the researchers overlook the major differences, mentioned above, that are eminently diagnostic—differences that they clearly noticed (e.g., see their Figure 149)—and instead they recruit a bizarre reasoning in order to demonstrate the taxonomic proximity between the two Tabun specimens: "Any lingering doubt [the authors are quick to point out, preparing their defense] as to the rightness of our opinion that the Tabun man, represented solely by the mandible, is of the same race as the chinless Tabun woman is dispelled when we note that in both there is the same short-rooted first molar (pl. XVI A, D [sic]). The first molar roots are of normal length in both Skhul IV and V (pl. XXVI C, E)" (McCown and Keith 1939:227).

No less surprising is the discussion of the trabecular pattern revealed in McCown and Keith's X-ray photographs. Even though they maintain that the two Tabun mandibles are similar in their trabecular pattern (1939:217), this is not at all the impression given by the discussion on page 228 or the relevant plates. The authors do not provide any X-rays of modern humans, not even of the single Sikh or Australian specimens they generally used for comparison, to illustrate the pattern in modern *Homo sapiens* (not to mention the range of variation of this feature in a population).

‡ It should be mentioned, however, that only a rather modest retromolar space can be found in Tabun I.

5. DISCUSSION

The real question we are left with, therefore, is how McCown and Keith arrive at the conclusion that the two Tabun specimens are of the same "race." How are what appear to be such clear-cut morphological differences overlooked in their interpretation? Undoubtedly the key to this riddle lies in the following statement: "In spite of the fact that Tabun II is much more robust than that of the woman (Tabun I) and shows other distinguishing features, notably an early stage of chin development, we are of the opinion that both male and female are probably of the same race, as well as being of the same time" (1939: 217). I daresay that in these researchers' minds, the last part of the sentence dictates the first part. In other words, because the specimens are "of the same time" (a fact provided by the archaeologist), they have been bulldozed into the same category—"the same race"—despite the differing anatomical features.

McCown and Keith readily deal with the fact that the hominids from the adjacent Skhul Cave are more "advanced" than those from Tabun. They simply treat the Skhul specimens as later, that is, more modern-like. This approach is expressed in the names they gave to the different groups of hominids: Neanthropic—the "new ones"—to which the Skhul group is closer; and Paleoanthropic—the "old ones"—to which they ascribe both Tabun specimens. The convoluted reasoning that we witness in the discussion of the Tabun mandibles and, in fact, throughout the volume (McCown and Keith 1939) stems directly, I believe, from the fact that the investigators are hostage to the anagenetic—unilinear—concept that the Neandertal is just a link in the evolutionary chain, or a "phase in human evolution," as epitomized by Hrdlička's famous title. Consequently, McCown and Keith cannot imagine that two apparently different hominids, one of which (Tabun I) so obviously belongs to their Paleoanthropic group and the other of which (Tabun II) is plainly closer to hominids of their Neanthropic group, might have come from the same layer. They clearly see the similarity between Tabun II and Skhul, but given the two Tabun specimens' contemporary existence as dictated by the archaeological context, it was the clear-cut anatomy that paid the price.

As stated earlier, McCown and Keith are by no means the only ones biased by this notion. Reactions to the view of the Neandertal as representing an evolutionary phase preceding that of modern humans may be found in many studies on the hominids discovered in the caves of the Near East. Precisely because of this bias and the allegedly young age of the Mousterian layers in the Amud Cave, Suzuki (1970) and Watanabe (1970), as indicated elsewhere (Rak et al. 1994; Hovers et al. 1995), attribute both the lithic assemblage and the Neandertal anatomy of the Amud specimen to a transitional stage that bridged between the classic Neandertals and early *Homo sapiens* as well as between the Middle and Upper Paleolithic industries. Later studies demonstrate that neither the lithic assemblage, which turned out to be much earlier than the original investigators thought (Ohnuma 1992; Hovers et al. 1995), nor the Neandertal anatomy of the Amud specimen constituted an intermediate stage (see Lavi 1994).

Jelinek (1982a,b), in contrast, views the hominids from Amud as full-fledged representatives of the Neandertal taxon, and the Amud lithic assemblage as not at all intermediate but actually similar to Tabun layer D. On the basis of these views and the evolving cultural sequence of Tabun, represented by the ratio of the flakes' thickness to their breadth, he suggested that a continuum of biological evolution coincides with the cultural development (herein the similarity to the approach of Suzuki and Watanabe). In his view, there is a local evolution of the Neandertals residing in the early Mousterian layers of Tabun. He states: "Thus our current evidence from Tabun suggests an orderly and continuous progression of industries in the southern Levant, paralleled by a morphological progression from Neandertal to modern man.... The continuous sequence of cultural and biological development supported by the Tabun sequence appears to be a more economical and firmly based interpretation" (1982b:1374). Jelinek did not even experience the quandaries and hesitations of McCown and Keith. In his opinion, both Tabun I and Tabun II must be Neandertals since, in light of the linear technological continuum, the Neandertal—which was responsible for the thick-flaked, earlier portion of the Tabun sequence—and *Homo sapiens*—which was responsible for the slender-flaked, later portion—cannot possibly be contemporary. Jelinek believes that "the implications of this record of continuity in human biological development in this region are very important. A strong case can be presented for the concomitant development of manipulative and conceptual abilities of the hominids, as shown in the progression toward broader and thinner flakes and in the increasing specialization in assemblages, and for a chronological succession of fossil men that places the *Tabun I, Tabun II, and Amud Neandertals antecedent in time to the more modern Skhul and Qafzeh hominids"* (1982a:99; my emphasis).

Later, but before the TL dates became available, Trinkaus (1984) states a view similar to that of the original Amud excavators in reference to the Qafzeh hominids: "All of the techniques that have been used to date the Qafzeh remains have limitations, and each is insufficient to confirm a date. However, the younger date would alleviate the need to account for how two distinct groups of *Homo sapiens* [Neandertals and early *Homo sapiens*] could have coexisted for several millennia in the small area of northern Israel, using the same cultural adaptive complex and yet remaining biologically distinct. For the purposes of this discussion, I will employ Occam's Razor, use the date for the Qafzeh remains that creates the *fewest problems*, and consider them to date to around 40 ky BP, about the same age as the Skhul layer B remains" (1984:258; my emphasis). Here, it is clearly the anatomy that leads Trinkaus to place the Qafzeh specimens later in time. One wonders why Qafzeh and not the Neandertals.

The assumption that the Neandertals are the forerunners of *Homo sapiens* as promoted by Trinkaus, for example, in his work on facial development (1987) demands that the chronological gap between them be long enough to accommodate the transition from one form to the other, a solution that introduces "the fewest problems." This gap can be short or long, depending on the rate at which the transition occurs, but it must exist, as the unilinear concept does not allow for the coexistence of two forms, a coexistence that is apparently the source of all the problems.

Of great significance to this discussion, in both historical and systematic terms, are the two specimens discovered in the Haua Fteah Cave in Libya, southeast of the Mediterranean Sea. Both specimens consist solely of the mandibular ramus, and both originate in the Mousterian layers of the cave. (These isolated specimens are, therefore, ideal taxonomic test cases.) Despite their fragmentary nature, their anatomy is diagnostic enough to allow us to confidently rule out the possibility that they are Neandertals and strongly suggests that they are *Homo sapiens.* Both specimens have a ramus with a symmetric appearance: The coronoid and condylar processes are equal in size, the notch between them is deep, and the deepest point of the notch occurs halfway between the processes. The crest between the processes also deviates to the side, in a superior view, and merges with the lateral portion of the condyle. Nevertheless, Tobias (1967) draws the following conclusion:

> When all the evidence is taken into account, it seems reasonable to infer that the Haua Fteah remains—pitifully slight as they are—represent individuals who formed part of an advanced Neandertaloid population. This people extended at least from Shanidar in the east, possibly as

> far as Jebel Irhoud (Morocco) in the west. They seem to have supplanted in time the more archaic, *erectus-neanderthalensis*, transitional population of Africa, but carried forward a number of their morphological traits. It is not far-fetched to suggest that this advanced, Afro-Asian Neandertaloid population had at least some genetic roots in Africa and indeed well south of the Sahara. (1967:349)

This paragraph echoes not only Hrdlička's "Neandertal Phase of Man" but also his scandalous definition of the Neandertal: "The only workable definition of the Neandertal man and period seems to be, for the time being, the man and period of the Mousterian culture" (Hrdlička 1927:251). This phase of human evolution, according to Tobias, includes all the specimens from Irhoud to Shanidar, despite the significant anatomical differences between them.

It is almost superfluous to emphasize that only in the anagenetic (a single species lineage) concept of evolution is chronology attributed such great importance. Since evolution in this view constitutes a successive continuum of stages, time necessarily plays a central role. In cladogenesis (a model of two or more lineages), on the other hand, chronology is of little significance, as stated in the principles of the cladistic method. The Neandertals, for all that matters, could have existed until the present day or become extinct long before they actually did without the basic geometry of the phylogenetic tree being changed in any way.

The notion of the Neandertals with their distinctive face—including the specialized mandibular elements—as representing a side branch in the evolution of modern humans is by no means the product of chronological considerations, such as the new Qafzeh dates that place its hominids earlier than the younger Neandertals of the Levant. Contemporaneity alone does not exclude an ancestral-descendant relationship. The inevitable conclusion that the Neandertals are in many ways derived vis-à-vis modern humans and thus represent a side branch that has nothing to do with the evolution of modern *Homo sapiens* results from their comparison to an outgroup along with simple considerations of parsimony. Their position on a side branch, or, in other words, the basic phylogenetic geometry, does not depend on when they lived or became extinct. Neandertals cannot serve as our ancestors because that would constitute a violation of the principle of parsimony. The TL and ESR dates of Qafzeh and Skhul only confirm this geometry of the phylogenetic tree.

From a historical point of view, we should keep in mind that the reasoning of many scholars who do not accept the evolution of modern humans directly from Neandertals and who view the Neandertals as a side branch is not rooted in parsimony. Even though they consider the Neandertals anatomically primitive, not derived, they maintain that the chronological gap—again, chronology!—was too short to allow one form to develop into the other. The most prominent proponent of this view is Boule (1908), who claimed that in light of the anatomy and the dates of the modern specimens from Grimaldi, the Neandertals could not be the forerunners of *Homo sapiens* because of the chronological discrepancy. Boule arrived at this conclusion in spite of his view of the Neandertals as anatomically primitive enough to serve as the ideal intermediate stage between *Pithecanthropus* and modern humans.

Adopting a cladogenetic interpretation and an anatomic assessment based on the two mandibles from Tabun—that is, recognizing through standard paleontological criteria the presence of two hominid species, *Homo neanderthalensis* and *Homo sapiens*, in the Mousterian layers of the cave—presents a simple, logical solution. This interpretation leads to the "fewest problems," or, to borrow a phrase from Jelinek, "appears to be a more economical and firmly based interpretation" (Jelinek 1982b:1374).

REFERENCES

Boule, M. 1908. L'homme fossile de la Chapelle-Aux-Saints. *L'Anthropologie* 19:519.

Franciscus, R., and E. Trinkaus. 1995. Determinants of retromolar space presence in Pleistocene Homo mandibles. *Journal of Human Evolution* 28:577–95.

Garrod, D. A. E., and D. M. A. Bate. 1937. *The Stone Age of Mount Carmel, Vol. 1, Excavations at the Wady El-Mughara*. Oxford: Clarendon Press.

Hovers, E., Y. Rak, R. Lavi, and W. H. Kimbel. 1995. Hominid remains from Amud cave in the context of the Levantine Middle Paleolithic. *Paléorient* 21(2):47–61.

Hrdlička, A. 1927. The Neanderthal phase of man. *Journal of the Royal Anthropological Institute* 57:249–74.

Inke, V. G. 1967. Eigenschaften und Bedeutung der Grenzflache zwischen Basal und Alveolarteil des menschlichen Unterkiefers. *Nova Acta Leopoldina* 33:161–95.

Jelinek, A. J. 1982a. The Middle Palaeolithic in the southern Levant, with comments on the appearance of modern *Homo sapiens*, 57–104. In *The Transition from Lower to Middle Palaeolithic and the Origin of Modern Man*, ed. A. Ronen. Oxford: British Archaeological Reports International Series 151.

Jelinek, A. J. 1982b. The Tabun Cave and Paleolithic man in the Levant. *Science* 216:1369–75.

Lavi, R. 1994. The Amud I Skull: Taxonomic Assessment and Phylogenetic Status (in Hebrew). M. A. thesis. Tel Aviv University. Tel Aviv.

Martin, H. 1923. *L'Homme Fossile de La Quina*. Paris: Librairie Octave Doin.

McCown, T., and A. Keith. 1939. *The Stone Age of Mount Carmel*, Vol. 2, *The Fossil Human Remains from the Levalloiso-Mousterian*. Oxford: Clarendon.

Ohnuma, K. 1992. The significance of layer B (square 8–19) of the Amud Cave (Israel) in the Levantine Levalloiso-Mousterian: A technological study, 83–100. In *The Origins and Dispersal of Modern Humans in Asia*. ed. T. Akazawa, K. Aoki, and T. Kimura. Tokyo: Hokusen-Sha.

Rak, Y., W. H. Kimbel, and E. Hovers. 1994. A Neandertal infant from Amud Cave, Israel. *Journal of Human Evolution* 26:313–24.

Suzuki, H. 1970. The skull of the Amud man, 123–206. In *The Amud Man and His Cave Site*, ed. H. Suzuki and F. Takai. Tokyo: Academic Press of Japan.

Tobias, P. 1967. The hominid skeletal remains of Haua Fteah, 338–52. In *The Haua Fteah (Cyrenaica) and the Stone Age of the South-East Mediterranean*, ed. C. McBurney. Cambridge: Cambridge University Press.

Trinkaus, E. 1984. Western Asia, 251–93. In *The Origin of Modern Humans*, ed. F. H. Smith and F. Spencer. New York: Alan R. Liss.

Trinkaus, E. 1987. The Neandertal face: Evolutionary and functional perspective on a recent hominid face. *Journal of Human Evolution* 16:429–43.

Watanabe, H. 1970. A Paleolithic industry from the Amud cave, 77–114. In *The Amud Man and His Cave Site*, ed. H. Suzuki and F. Takai. Tokyo: Academic Press of Japan.

Weidenreich, F. 1936. The mandibles of *Sinanthropus pekinensis*: A comparative study. *Palaeontologia Sinica Series D* 7:1–162.

23

MORPHOLOGICAL VARIATION IN WEST ASIAN POSTCRANIA

Implications for Obstetric and Locomotor Behavior

Karen R. Rosenberg

Department of Anthropology
University of Delaware
Newark, Delaware 19716

1. INTRODUCTION

For paleoanthropologists interested in behavioral questions, studying the pelvis provides a rare opportunity to look at morphology associated with two vitally important human functions: walking and having babies. Structurally, the bony pelvis is of obvious significance to both of these functions which place important, albeit often conflicting, selective constraints on its form (Lovejoy 1988). Aspects of birth are also studied through an analysis of other parts of the skeleton, such as neonatal morphology (Minugh-Purvis 1988; Tillier 1992a,b; Akazawa et al. 1995; Tillier et al. 1995). This allows us to make inferences about early growth and development and possibly maternal-infant size relationships (with implications for neonatal development and parental care). Lower limb morphology allows us to reconstruct habitual locomotor behavior (Lovejoy and Trinkaus 1980; Ruff et al. 1993; Trinkaus 1983, 1992). Since the first fossils preserving pelvic material were discovered, Neandertals have been described as unusual in several aspects of their pelvic morphology, specifically the form of the superior pubic ramus, which forms the anterior wall of the bony birth canal and the front of the pelvic girdle. Interpretations of the significance of this distinctiveness are changing as the result of both an expanded fossil record and a recent emphasis on a functional understanding of skeletal morphology in fossil hominids.

Unfortunately, our knowledge of the pelvic morphology of Neandertals and other archaic *Homo sapiens* is fairly limited due to poor archaeological preservation of this skeletal element. Many of those specimens that are known were discovered early in this century, however, so debate about this portion of the skeleton has a long history. In considering the variation between the two neighboring sites of Tabun and Skhul on Mt. Carmel, McCown and Keith (1939) concluded that "the Mount Carmel people were in the

Neandertals and Modern Humans in Western Asia, edited by Akazawa *et al.*
Plenum Press, New York, 1998.

throes of evolutionary change," describing "anatomical features of the Tabun woman which demand particular consideration . . . [including] the peculiar form taken by the pubic part of her pelvis" (McCown and Keith 1939:373). With the later discovery of fragmentary pubic bones at Shanidar Cave in Iraq, a number of workers emphasized the distinctiveness of the pubis in these Neandertals compared to anatomically modern humans (Stewart 1960, 1977; Trinkaus 1976). In these initial discussions of Neandertal pubic morphology, attention was focused on the unusual cross-section of the pubic bones at the highest point of the obturator foramen in the fragmentary specimens. In these specimens, the superior pubic ramus is thin in the antero-superior to postero-inferior direction compared to modern humans. When an elongation of the pubis was noted, it was assumed to be related to the unique cross-section (Smith 1976; Trinkaus 1976). These characteristic aspects of morphology appear early in the ontogeny of Neandertals, as they can be seen in the La Ferrassie 6 child, aged 6–8 years (Tompkins and Trinkaus 1987) and the recently discovered Dederiyeh specimen from Syria, aged no more than two years old (Akazawa et al. 1995). These distinctive aspects of Neandertal pubic morphology were seen as one of a suite of derived features which occurred in Neandertals but not in modern humans, and which were interpreted by many scholars as possible evidence of the phylogenetic position of Neandertals off the line of ancestry leading to anatomically modern humans.

2. BACKGROUND

The last fifteen years have seen two important developments which affect our understanding of the pelvic morphology of Neandertals and other archaic humans: one is a change in data and the other is a change in approach. First, our knowledge of the morphology of the pelvic region has been substantially improved by the discovery and analysis of two very well-preserved specimens of archaic *Homo sapiens* from opposite ends of the Asian continent (Kebara from Israel and Jinniushan from China). Second, many scholars now look at the morphology of these archaic humans in an explicitly functional and comparative perspective to better understand their behavioral implications. These developments make this a timely moment to revisit the hypotheses of the 1980s in light of the new discoveries.

In the 1980s, several workers turned their attention to the functional significance of Neandertal pelvic morphology. In modern humans, females have elongated pubic bones compared to males as a concomitant of birth canal expansion (the pubis forms the anterior wall of the bony birth canal), so it was logical for the functional explanations of pubic elongation in Neandertals to center on obstetric constraints (Figure 1). Several workers therefore proposed hypotheses for the typically elongated Neandertal pubic bone that depended on the assumption that the elongation was related to birth canal enlargement. They did not deal directly with the shape of the superior pubic ramus in cross section. These hypotheses all assumed that Neandertal birth canals were large to accommodate large babies, but they differed in the reasons given for the hypothetically large Neandertal babies. Trinkaus (1984) proposed that Neandertals had an elongated gestation period compared to modern humans. As a result, Neandertal babies were further along in the developmental process at birth and were therefore larger than modern babies. Wolpoff (1980) and Rosenberg (1986, 1988) suggested that Neandertal babies were large because of maternal-infant size relationships. Since maternal weight is an important determinant of birthweight and since Neandertal women were heavy relative to their stature, maternal birth canals were

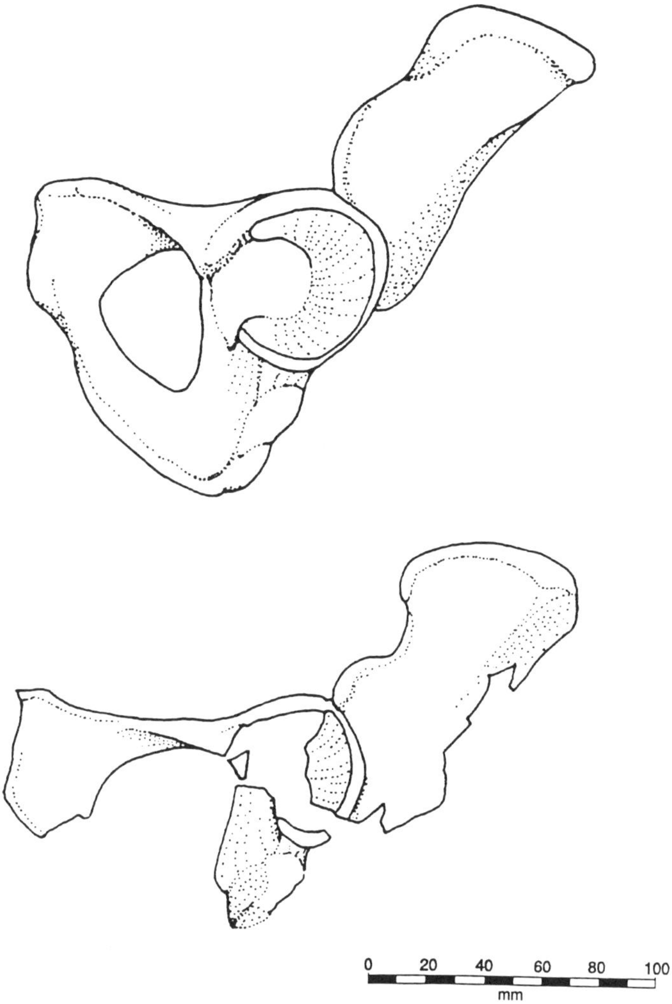

Figure 1. Left innominates of a modern human female (above) compared to a Neandertal female, Tabun C1 (below). The pubic elongation described for Neandertals can be seen in the medial-lateral elongation of the superior pubic ramus (a greater acetabulo-symphyseal length) from the medial edge of the acetabulum to the most superior point on the pubic symphysis as described in Rosenberg (1988). Figure redrawn from Rosenberg (1988) and McCown and Keith (1939).

large relative to maternal stature. Dean, Stringer, and Bromage (1986) suggested that Neandertals had accelerated fetal growth. Although babies were born after the same gestation time as modern humans, this accelerated growth *in utero* meant that they were larger on average than modern newborns. Finally, Friedlander and Jordan (1994) and Minugh-Purvis (1988) suggested that the assumed need for a relatively large birth canal in Neandertals was related to bone robusticity and density. They suggested that denser neonatal cranial bone and a less flexible pelvic aperture in Neandertals may have necessitated a relatively large birth canal. All of these hypotheses had intriguing behavioral implications in the areas of birth, pregnancy, and parental care of neonates. When these hypotheses were proposed, the known sample of archaic *Homo sapiens* (including Neandertals) pelvic bones was both small and fragmentary.

It is worth noting that two significant characteristics of Neandertal pelvic morphology remained unexplained by all of these hypotheses. First, the unusual cross-section of the superior pubic ramus was not addressed in a satisfactory way by any of these models (Figure 2). Although it was assumed to be related to the elongation of the ramus, no plausible mechanical (or other) models were proposed to explain why these two characteristic features should occur together in Neandertals. Second, although these explanations were all basically obstetrical in origin, they did not address the fact that male Neandertals had even longer pubic bones than females, a situation which is unknown for any modern human populations and which would clearly not be the expected pattern if an obstetric explanation accounted for the morphology. Although selection on one sex (in this case, females) often involves a correlated response in the other sex (Lande 1980), "dragging it along," one would certainly not expect that the "non-selected" sex would show the morphology to a more extreme degree than the sex under direct selection. Yet this is apparently the case for pubic elongation in Neandertals and perhaps other archaic humans as well.

3. NEW SPECIMENS FROM THE LEVANT AND EAST ASIA

Since the time these hypotheses were proposed, two important and more complete pelvic specimens of archaic *Homo sapiens* were discovered on opposite sides of the Asian continent. The Kebara specimen was excavated from Mousterian deposits at Kebara Cave in Israel in 1983 (Rak and Arensburg 1987; Rak 1991a; Bar Yosef et al. 1992). Though it lacked the most commonly studied portion of Neandertal anatomy, namely the skull, the Kebara specimen had a well-preserved complete pelvis and was assumed to be a male Neandertal. Shortly after, in 1984, the Jinniushan specimen was discovered in Liaoning Province, in Northeastern China (Lü 1990). This specimen was the first archaic *Homo sapiens* to preserve a pelvis from East Asia.

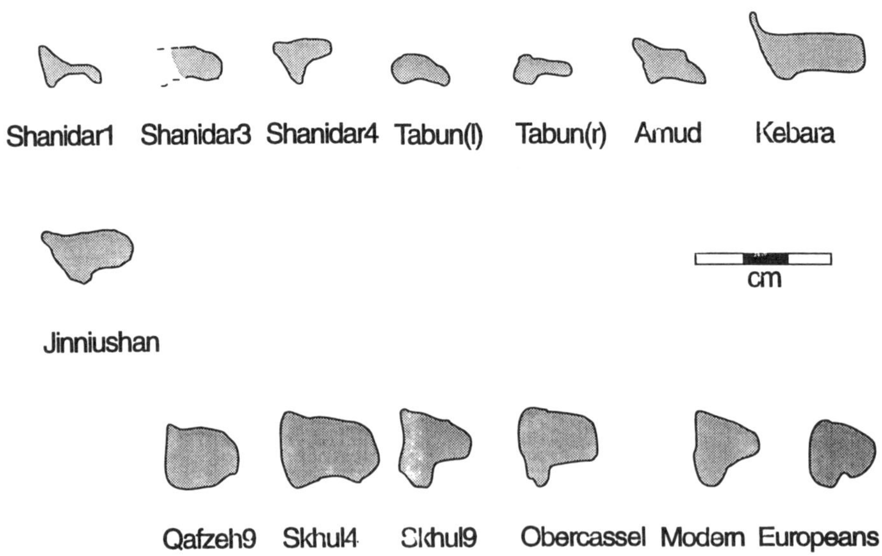

Figure 2. The cross-section of the superior pubic ramus at the highest point of the obturator foramen in Neandertals (top of figure), "anatomically modern humans" (bottom of figure), and the Jinniushan specimen (middle). Anterior is to the right of the figure, posterior to the left; drawings are done looking at the cross-section from the medial aspect towards the lateral portion of the pelvis.

3.1. Kebara

The Kebara 2 skeleton is the remains of a mature male, excavated from Mousterian layers dated to about 60,000 BP (Bar Yosef 1992). Because the discovery of this specimen came at a time when attention was focused on the obstetrical constraints of the Neandertal pelvis, there has been considerable discussion in the literature about its pelvic aperture (Rak and Arensburg 1987; Rak 1990, 1991). Although the pubis is long, as in other specimens designated Neandertals, the pelvic inlet may not be as wide as would be predicted by pubic length, compared to a modern sample. Rak suggested therefore that obstetric demands were not sufficient to explain the distinctive aspects of Neandertal pelvic morphology and argued instead that the pelvis was altered as a result of "posture and locomotor differences." Specifically, he argued that in Kebara, and presumably other Neandertals, the inlet is situated more anteriorly in relation to the acetabulum than in modern humans.

Early studies of the Kebara pelvis focused on the dimensions of the pelvic inlet, the first of the obstetrical planes that a fetus passes through during the birth process. However, at least three obsteterical planes are of importance to the birth process (Walrath and Glantz 1996): the inlet, midplane, and outlet. These form a tunnel of changing cross-sectional shape through which the fetus passes during parturition. Tague (1992) compared the pelvic aperture dimensions of the Kebara 2 to those of males from a number of modern human populations not only at the pelvic inlet but at the other obstetrically important planes of the pelvic aperture. Tague showed that compared to modern males, Kebara 2 is spacious in transverse and anterior-posterior diameters and circumference of the pelvic inlet (greater than 94.1, 95.8, and 91.5% of the males in his sample respectively). However at the midplane and outlet, Kebara 2 is actually constricted compared to modern males. For the index of pelvic funneling [(outlet circumference/inlet circumference) × 100], Kebara 2 (74.9) is considerably more funneled than modern males (87.6 ± 4.8). The index of outlet shape [(anterior-posterior diameter/transverse diameter) × 100] of Kebara 2 is 77.2, whereas for modern males it is 114.6 ± 14.4 (N = 181) and for modern females 102.2 ± 13.8 (N = 154) (Tague 1992) making Kebara much more flattened (short in the anterior-posterior dimension) than modern humans. Because Kebara 2 is a male specimen, it was necessary for Tague to assume that Neandertals had a similar type and degree of sexual dimorphism as modern humans. If Neandertal females were more elliptical at this plane than Neandertal males as is the case for modern humans, it is unlikely that they would have been capable of giving birth to a large brained, fully-developed infant.

Because the Kebara 2 specimen is male, it is not ideally suited for drawing direct conclusions about birth in an extinct human population (Rosenberg 1988). Nevertheless, Kebara 2 is the most complete archaic *Homo sapiens* specimen we have to date and provides an opportunity to formulate hypotheses about female pelvic morphology in archaic *Homo sapiens* from the Near East if certain assumptions are made (see below). Tague (1992) suggested that while Neandertal males and females probably had spacious pelvic inlets, it is likely that at the outlet, both sexes were at the lower end of the modern range of variation. If the discovery of pelvic remains of Neandertal females confirms that they did have relatively constricted pelvic outlets, then all the previously mentioned explanations of Neandertal pubic morphology which predict larger newborns in Neandertals would have the implication of less favorable cephalo-pelvic relations in Neandertals than in modern humans, suggesting more difficult births (Tague 1992). This seems unlikely and is, in fact, the opposite of what Trinkaus (1984) suggested when he hypothesized that the origin of modern humans might have been accompanied by an increase in cephalo-pelvic constriction and a compensatory improvement in obstetrical and neonatal care to take care of even more helpless and vulnerable newborns. Given the high risks of mortality or injury for both mothers and newborns, associated with birth difficulties today, a situation with even greater difficulties in the past appears unlikely.

It is useful to remember, however, that we do not know what pelvic sexual dimorphism was like in Neandertals. Hager (1989, 1991) cautioned against assuming that modern pelvic patterns of sexual dimorphism characterized early stages of human evolution. Though she was generally referring to earlier time periods, there is reason to think that modern patterns of sexual dimorphism in the pelvis (though not necessarily in other parts of the skeleton) might not apply to Neandertals. Assuming that our sample is representative, Neandertals differed from living humans in that (at least for the very small sample of individuals who preserve a pubic bone) females actually had shorter pubic bones than did males (Rosenberg 1986, 1988).

Obviously, hypotheses about birth and about obstetrically related anatomy in Neandertals must await the discovery of more complete female remains. Assuming that Kebara has been sexed correctly, although it has a pelvic inlet, it did not have a birth canal. This is not a trivial distinction. Washburn (1948, 1949; Hanna and Washburn 1953) pointed out in the 1940s that the degree of sexual dimorphism in the pelvis is itself a variable quantity among living human populations and it is possible that sexual dimorphism among Neandertals was different from that in any living human group. For example, I have recently found considerable variation in the sciatic notch among modern human populations. Although it is the case in all populations that the sciatic notch is wider on the average in females than in males, the "cut-point" between the sexes varies considerably from one population to another. Wider sciatic notches seem to occur in east Asian populations (Wu et al. 1982; Rosenberg 1993) but very narrow ones in Australians (Davivongs 1963). The fact that male Neandertals had a pubis longer than females suggests that pelvic sexual dimorphism at least was different from that known for any living human population.

An interesting question from the perspective of modern human origins is the nature of the Skhul and Qafzeh specimens, long regarded as anatomically modern but, like the Neandertals from Kebara and Tabun, associated with a Mousterian tool industry. Although Rak (1990) has argued that Qafzeh 9 differs fundamentally from Kebara in its pubic morphology, it is also relevant to examine the Skhul 9 specimen. If Skhul 9 is male, as it has traditionally been considered by virtually everyone since the time of McCown and Keith (Wolpoff 1999), then it also has a longer pubis than would be predicted for its body size based on modern males (it falls directly on the female regression line). This suggests that it may have some Neandertal features and if not transitional, given the dates, is at least intermediate in its pubic morphology between Neandertals and anatomically modern humans in the Levant. Arensburg and Belfer-Cohen (this volume) make this point as well, arguing that many of the supposedly distinctive features of the Kebara pelvis are plesiomorphic and also occur in other Mousterian remains from Israel such as specimens from Skhul.

3.2. Jinniushan

Although we are limited in our knowledge of female Neandertal pelvic anatomy, a recently discovered specimen from China gives us important information about the pelvic anatomy of female archaic *Homo sapiens* from East Asia. The hominid specimen from Jinniushan (Golden Ox Mountain) was discovered in 1984 in a fissure in Pleistocene cave deposits in an isolated karst prominence in the south-west part of the district of Yinkou in the province of Liaoning (Lü 1990, 1995). The site was excavated by Professor Lü Zune of

Beijing University, who has allowed me to collaborate with him in the description of this pelvis. Jinniushan is the most northern of all pre-modern Chinese hominids. The individual is represented by a cranium, with most of the maxillary dentition, a left innominate, six vertebrae (one cervical and five thoracic), two left ribs, a complete left ulna, a complete left patella, and numerous well-preserved bones of both hands and both feet.

The Jinniushan specimen has been dated by uranium series and ESR to about 280,000 BP and this is consistent with the associated fauna from the site (Chen et al. 1994). Other later pelvic material is available from East Asia, including Liujiang, a late Pleistocene specimen (with a radiocarbon date of >40,000 years BP) (Woo 1959; Wu and Zhang 1985; Rosenberg 1993), and five individuals which preserve some or all of the pelvis from the Okinawan site of Minatogawa, dating to about 18,000 years ago (Baba and Endo 1982). There is another recently excavated but undescribed Chinese specimen from Laishui dated at about 28,000 years ago which also preserves a pelvis (it will be described by Professor Lü of Beijing University). These later specimens certainly appear well within the range of modern humans in all aspects of their pelvic anatomy.

Obviously, the interpretation of human pelvic morphology always depends heavily on the sex of skeletal remains. Because features which distinguish males and females also differ from region to region and through time, it is necessary to determine sex using diagnostic features other than the pelvic morphology under analysis. Although little has been written about the Jinniushan specimen to date, most paleoanthropologists who have worked on this specimen have assumed it was male. However, our recent work (Lü 1995; Rosenberg and Lü 1997 and in preparation) indicates that it is probably female. Although the cranium is robust relative to modern human crania, and the joint surfaces (such as the hip) are large, it has a number of features which are gracile in comparison to its contemporaries. The temporal lines are weakly developed; in fact, they are indistinct posteriorly and the face is relatively small. The supra-orbital torus is quite robust, but the arches thin over the middle portion of the orbit. Although the Jinniushan cranium has the same outer dimensions as the Dali specimen (Wu and Wu 1985), its cranial vault is very thin, much thinner than in Dali, and, as a result, its cranial capacity is slightly larger. Turning to the pelvis, the pubis is long (87.4 mm), but since it is long in archaic *Homo sapiens* from the western portion of the Old World (i.e., Neandertals), it would be circular to make a sex determination of this specimen based on the length of the pubic bone. It is, however, possible to examine the lower area of the anterior and medial portion of the pubis for features described by Phenice (1969) as being good indicators of sex. The area where the ventral arc would be is somewhat abraded so that it is not possible to make a statement about the presence or absence of a ventral arc. However, there is a subpubic concavity and a ridge on the ischio-pubic ramus, both female traits. The sciatic notch, which is usually quite narrow in Neandertals, is intermediate in breadth. Compared to the Kebara specimen from later in time, Jinniushan is gracile in such features as the iliac buttress and ischial tuberosity. The iliac blade, while curved, is not nearly as sinuous in shape as Kebara. The individual was about 165 cm tall based on the length of the ulna, the only long bone preserved. Based on the most appropriate available comparisons of the skull with Dali and the pelvis with Kebara, we feel confident that the specimen is female.

The Jinniushan specimen represents the earliest pelvis in the human fossil record and the only female specimen outside of Africa and Europe. For those aspects of pelvic morphology which are preserved on the Arago XLIV individual (e.g., morphology of the iliac blade, ischial tuberosity), Jinniushan is almost identical to that specimen. The superior pubic ramus of the Jinniushan specimen, like the Neandertals, appears to be elongated mediolaterally and as a result the obturator foramen looks large. The superior pubic ramus itself is fairly thin in cross-section, again, like Neandertals. The presence of an elongated pubis in a 280,000 year old hominid from East Asia supports the notion that this is the plesiomorphic or primitive condition at least for *Homo* and not a derived feature of Neandertals. Additional support for this comes from the australopithecines who, although they have a different pelvic morphology from any archaic *Homo sapiens*, also have a pubic bone which is long relative to body size (Lovejoy 1975; Berge et al. 1984; Tague and Lovejoy 1986; Frayer 1988; Rak 1991b). Indeed, Rak (1993) has argued that while Neandertal facial morphology is derived, its pubic morphology represents the primitive condition for *Homo* and Arensburg and Belfer-Cohen make the same point about the pubis in this volume. It is worth noting, however that also like the female Neandertals, the length of the superior pubic ramus is not very different from what one would expect from a modern female of that body size (which would be a very large female by modern human standards). Figure 3 shows acetabulosymphyseal length (from the edge of the acetabulum to the top of the pubic symphysis) versus a correlate of body size, namely femoral head diameter squared. Of course females have a pubis which is longer than males relative to body size. The so-called early anatomically modern humans from Europe, the Near East, and East Asia all fit with the modern human sample

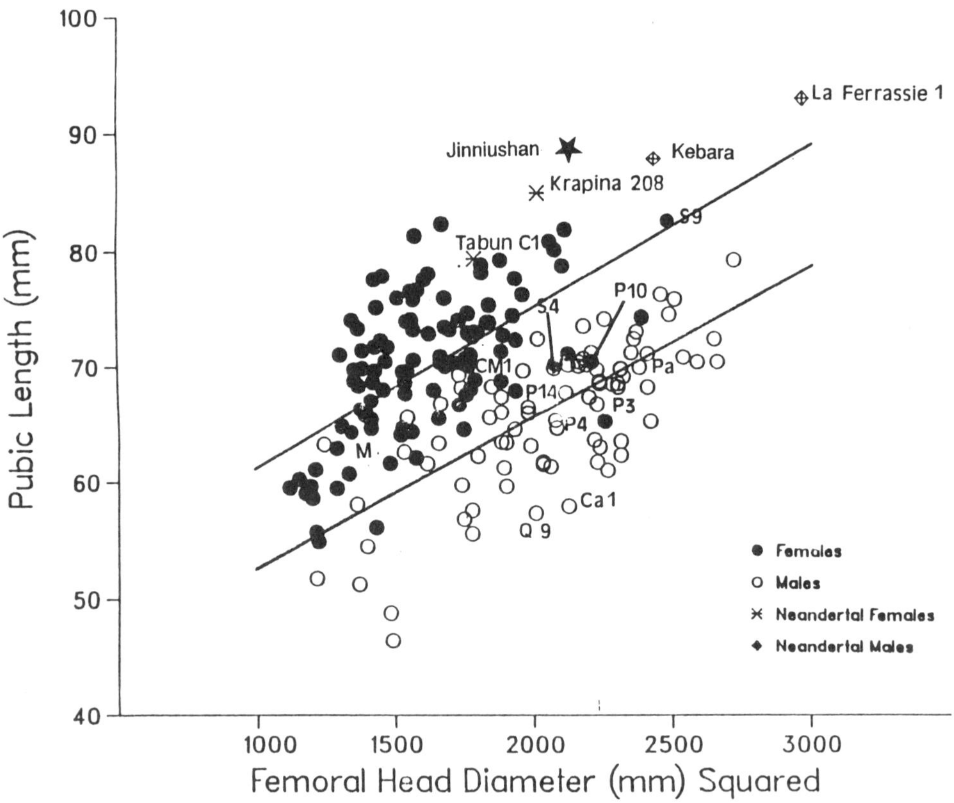

Figure 3. Acetabulosymphyseal length (mm) versus femoral head diameter squared in recent and archaic humans. Black dots represent females, open circles are males. Least squares regression lines are plotted separately for males and females. Abbreviations for early anatomically modern humans are as follows: Ca: Barma del Caviglione; CM: Cro-Magnon; P: Predmosti; Pa: Paviland; Q: Qafzeh and S: Skhul). When femoral head diameter was not available for a fossil specimen it was estimated for acetabular diameter using a regression equation derived from modern humans and discussed in Rosenberg (1988).

(with the exception of Skhul 9, see above). The female archaic humans (Tabun C1, Krapina 208, and Jinniushan) are all at or just outside the upper end of the modern human range of pubic length as well as of body size. Although the Jinniushan specimen has a longer pubic length than any modern human female in this sample, it also has a larger acetabulum and femoral head than any of the modern human females and, like Krapina and Tabun, follows the same general relationship between these two dimensions as the wide ranging modern sample (i.e., it is above the line, but within the expected size range for this relationship).

The Jinniushan specimen does not preserve a sacrum. This means that determinations of the size and shape of the complete pelvic aperture are not possible, though dimensions of the innominate (such as pubic length and the length of the arcuate line) are certainly useful indicators of the relative size of the birth canal.

What stands out on Figure 3, of course, and what still remains unexplained, is the morphology of the pubis in Neandertal males (Kebara and La Ferrassie 1), who have pubic lengths well above what one would predict based on their body size.

The cross-sectional morphology of the superior pubic ramus of Neandertals is another feature that has been claimed to be unique. Figure 2 shows the shape of the superior pubis ramus at the highest point of the obturator foramen in the available Neandertals, anatomically modern humans, and Jinniushan. Clearly there is a striking difference in this morphology between these groups, with Jinniushan not following clearly with either group. Again, we can see that it is the archaic males who are most different from modern humans. In this feature, as in pubic elongation, the female archaic *Homo sapiens* from Jinniushan, like the Neandertal females who come later in time, is not as different from modern humans as the males.

4. DISCUSSION

In general, the pubic morphology of archaic females, either Neandertals or Jinniushan, seems to fit the modern human pattern. This has important implications for our understanding of obstetric demands in these hominids. If, as these data suggest, the relationship between maternal body size and birth canal capacity was approximately the same in female archaic *Homo sapiens* and in modern humans, and if archaic *Homo sapiens* infants were born at approximately the same stage of their development as modern human infants, then it is likely that the mechanism of birth was similar in archaic *Homo sapiens* and in modern humans.

Birth in modern humans is significantly different from birth in non-human primates. In apes, the birth canal is quite spacious relative to the size of the newborn infant and the shape of the birth canal is the same (oval with the long axis in the anterior posterior direction) at all planes (inlet, midplane, and outlet) of the birth canal. This means that the fetus can enter the inlet with its head facing anteriorly and aligned with the anterior-posterior axis of the pelvis. It continues through the bony birth canal without rotation and emerges from the mother's body face up. In contrast, living humans give birth in a more complex manner. Because the dimensions of the infant's head are close to those of the birth canal and because the shape of the birth canal is different at each of the different planes (widest transversely at the inlet, widest in the anterior-posterior dimension at the midplane, and approximately round at the outlet), the infant head and shoulders must rotate to the widest dimension at each level. In addition, the infant emerges from the birth canal in the "occiput anterior" position since the anterior portion of the birth canal is wider than the posterior portion and can better accommodate the infant's occiput. This means that the infant emerges facing away from the mother. Trevathan (1987, 1988) has argued that this unique aspect of modern human birth has an important behavioral consequence for mothers. Whereas ape and monkey mothers (as well as virtually all other mammals) give birth in seclusion, away from other members of their group, humans are distinctive in seeking assistance during this time. Unlike non-human primate mothers who reach down during birth and clear their infants breathing passage or lift them up as they emerge, human mothers are unable to assist their babies during birth. Trevathan (1987, 1988) suggests that because of the way the baby emerges, human mothers benefit from assistance during birth in ways that other primates would not. She argues that "obligate midwifery" has probably characterized humans for some time and that women who seek assistance during birth would be at an advantage during this time.

If archaic *Homo sapiens* females gave birth with a modern human birth mechanism as the pelvic morphology suggests, Trevathan's argument for "obligate midwifery" would apply to them (Rosenberg and Trevathan 1996). Such an interpretation clearly has important behavioral and particularly social implications for such issues as how archaic *Homo sapiens* women formed social ties with one other. Although obstetric hypotheses cannot provide an explanation for the unusual morphology seen in male archaic *Homo sapiens*, the morphology of their female counterparts (which is similar to modern humans) has obstetric implications. As I have suggested, the evidence of female archaic pelvic morphology more closely resembles the human pattern and clearly implies a similarity in pattern of birth and associated behaviors. The evidence from the neonatal and infant morphology of Neandertals which suggests a similar degree of development at birth (Minugh-Purvis, unpublished; Tillier et al. 1995) also supports the idea that they required similar degrees of maternal care at birth.

However, the morphology in the males still remains to be explained. A clear conclusion from the foregoing discussion is that although they are linked, male and female pelvic morphology must be addressed separately. It is the male pelvic pattern which differs most significantly from the modern human condition and which obviously can not be accounted for by parturition. Suggestions have been made that this morphology relates to differences between Neandertals and modern humans in locomotion, posture, or trunk shape, but to date formal hypotheses have not been tested. The extreme thinning of the superior pubic ramus has never been satisfactorily addressed by any of the workers who have examined pelvic morphology in Neandertals. Evidence from other parts of the skeleton, such as the long bones, support the idea that there may be subtle differences in locomotor function, posture, or trunk shape. I would hope that an explanation of the extreme elongation of the pubis in Neandertal males should also attempt to explain the unusual cross-section of the ramus and other aspects of the postcranial skeleton such as lower limb robusticity. Analysis of the pelvis however, must take into account all of the important functions of this region of the skeleton.

5. CONCLUSIONS

The sample available for a study of Neandertal or other archaic *Homo sapiens* pelvic morphology is, as has often been pointed out, extremely small and inadequate. At the moment however, several points seem clear.

1. The elongated pubis which seems to characterize Neandertals is not a derived condition unique to that group but rather the primitive or plesiomorphic condition for *Homo sapiens* and quite probably for *Homo* and *Australopithecus*.

2. The distinctive characteristics of the cross-sectional shape of the superior pubic ramus remain unexplained. Functional and mechanical models must be developed and the implications of pubic elongation for forces acting on the superior pubic ramus must be fully evaluated and tested.
3. Obstetric hypotheses are not sufficient to explain the occurrence of elongated pubic morphology in Neandertals given its presence in males as well as females.
4. Nevertheless, it is clear from the pelvic morphology that Neandertal women (and other archaic Homo sapiens) almost certainly gave birth in a similar fashion to modern human women, with the baby undergoing the same series of rotations as it passes through the birth canal and emerging in the same position (occiput anterior). This would have had the same implications for Neandertals that it does for modern humans, namely that they were characterized by "obligate midwifery," i.e., that they gave birth with the assistance of their close relatives or associates. This suggests that social ties between women would have had great importance for them.

ACKNOWLEDGMENTS

I am very grateful to Professors Takeru Akazawa and Ofer Bar-Yosef for the invitation to participate in the Tokyo University Museum Symposium on "Neanderthals and Modern Humans in West Asia." For access to fossil specimens, I thank: Baruch Arensburg (Tel Aviv), Jean-Louis Heim (Paris), Lü Zune (Beijing), Jakov Radovčić (Zagreb), Yoel Rak (Tel Aviv), Chris Stringer (London), Erik Trinkaus (Albuquerque), Joe Zias (Jerusalem), and Zhang Yinyun (Beijing). For comments on the manuscript, I am grateful to Tracey Crummett, Nancy Minugh-Purvis, and Tom Rocek.

REFERENCES

Akazawa, T., S. Muhesen, Y. Dodo, O. Kondo, and Y. Mizoguchi. 1995. Neanderthal infant burial. *Nature* 377:585–6.

Baba, H., and B. Endo. 1982. Postcranial skeleton of the Minatogawa Man, 61–195. In *The Minatogawa Man*, ed. H. Suzuki and K. Hanihara. Tokyo: University of Tokyo Press.

Bar Yosef, O., B. Vandermeersch, B. Arensburg, A. Belfer-Cohen, P. Goldberg, H. Laville, L. Meignen. Y. Rak, J. D. Speth, E. Tchernov, A. M. Tillier, and S. Wiener. 1992. The excavations in Kebara Cave, Mt. Carmel. *Current Anthropology* 33:497–550.

Berge, C., R. Orban-Segebarth, and P. Schmid. 1984. Obstetrical interpretation of the australopithecine pelvic cavity. *Journal of Human Evolution* 13:573–87.

Chen, T., Q. Yang, and E. Wu. 1994. Antiquity of *Homo sapiens* in China. *Nature* 368:55–6.

Davivongs, V. 1963. The pelvic girdle of the Australian Aborigine: Sex differences and sex determination. *American Journal of Physical Anthropology* 21:443–5.

Dean, M. C., C. B. Stringer, and T. G. Bromage. 1986. Age at death of the Neanderthal child from Devil's Tower, Gibraltar, and the implications for studies of general growth and development in Neanderthals. *American Journal of Physical Anthropology* 70:301–9.

Frayer, D. 1988. Response to "The functional significance of Neandertal pubic morphology." *Current Anthropology* 29:608.

Friedlander, N. J., and D. K. Jordan. 1994. Obstetric implications of Neanderthal robusticity and bone density. *Human Evolution* 9:331–42.

Hager, L. 1989. The Evolution of Sex Differences in the Hominid Bony Pelvis. Ph. D. Diss. University of California. Berkeley.

Hager, L. 1991. The evidence for sex differences in the hominid fossil record, 46–9. In *The Archaeology of Gender: Proceedings of the Twenty-second Annual Conference of the Archaeological Association of the University of Calgary*, ed. D. Walde and N. D. Willows. Calgary: University of Calgary Archaeological Association.

Hanna, R. E., and S. L. Washburn. 1953. The determination of the sex of skeletons, as illustrated by a study of the Eskimo pelvis. *Human Biology* 25:21–7.

Lande, R. 1980. Sexual dimorphism, sexual selection, and adaptation in polygenic characters. *Evolution* 34:292–305.

Lovejoy, C. O. 1975. Biomechanical perspectives on the lower limb of early hominids, 291–326. In *Primate Functional Morphology and Evolution*, ed. R. H. Tuttle. The Hague: Mouton Publishers.

Lovejoy, C. O. 1988 Evolution of human walking. *Scientific American* 259(5):118–25.

Lovejoy, C. O., and E. Trinkaus. 1980. Strength and robusticity of the Neandertal tibia. *American Journal of Physical Anthropology* 53:465–70.

Lü, Z. 1990. La découverte de l'homme fossile de Jing-niu-shan: Premiere étude. *L'Anthropologie* (Paris) 94:899–902.

Lü, Z. 1995. Study of the hip bone of Jinniushan Man. *Quarterly Journal of Historical Relics* (english summary) 2:1–9.

McCown, T. D., and A. Keith. 1939. *The Stone Age of Mount Carmel.* Vol. 2, *The Fossil Human Remains from the Levalloiso-Mousterian*. Oxford: Clarendon Press.

Minugh-Purvis, N. 1988. Patterns of Craniofacial Growth and Development in Upper Pleistocene Hominids. Ph. D. Diss. University of Pennyslvania. Philadelphia.

Minugh-Purvis, N. n.d. The neurocranial evidence for Neandertal birth. Unpublished manuscript.

Phenice, T. D. 1969. A newly developed visual method of sexing the os pubis. *American Journal of Physical Anthropology* 30:297–302.

Rak, Y. 1990. On the differences between two pelvises of Mousterian context from the Qafzeh and Kebara Caves, Israel. *American Journal of Physical Anthropology* 81:323–32.

Rak, Y. 1991a. The pelvis, 147–56. In *Le Squelette Moustérien de Kébara*, ed. O. Bar-Yosef and B. Vandermeersch. Paris: CNRS Editions.

Rak, Y. 1991b. Lucy's pelvic anatomy: Its role in bipedal gait. *Journal of Human Evolution* 20:283–90.

Rak, Y. 1993. Morphological variation in *Homo neanderthalensis* and *Homo sapiens* in the Levant: A biogeographic model, 523–36. In *Species, Species Concepts, and Primate Evolution*, ed. W. H. Kimbel and L. B. Martin. New York: Plenum Press.

Rak, Y., and B. Arensburg. 1987. A new Neanderthal pelvis: First look at a complete inlet. *American Journal of Physical Anthropology* 73:227–31.

Rosenberg, K. R. 1986. The Functional Significance of Neandertal Pubic Morphology. Ph. D. Diss. University of Michigan. Ann Arbor.

Rosenberg, K. R. 1988. The functional significance of Neandertal pubic length. *Current Anthropology* 29:595–617.

Rosenberg, K. R. 1992. The evolution of modern human childbirth. *Yearbook of Physical Anthropology* 35:89–124

Rosenberg, K. R. 1993. Regional features in an upper Pleistocene Chinese pelvis: Implications for sex diagnosis. *American Journal of Physical Anthropology, Supplement* 16:168 (abstract).

Rosenberg, K. R., and Z. Lü. 1997 The pelvis of the Jinniushan specimen: An East Asian perspective on Neandertal morphology. *American Journal of Physical Anthropology, Supplement* 24:199–200 (abstract).

Rosenberg, K. R., and W. Trevathan. 1996. Bipedalism and human birth: The obstetrical dilemma revisited. *Evolutionary Anthropology* 4:161–8.

Ruff, C. B. 1995. Biomechanics of the hip and birth in early *Homo*. *American Journal of Physical Anthropology* 98:527–74.

Ruff, C. B., E. Trinkaus, A. Walker, and C. S. Larsen 1993. Postcranial robusticity in Homo I: Temporal trends and mechanical interpretation. *American Journal of Physical Anthropology* 91:21–53.

Smith, F. H. 1976. *The Neandertal Remains from Krapina: A Descriptive and Comparative Study*. University of Tennessee Department of Anthropology Reports of Investigations 15.

Stewart, T. D. 1960. Form of the pubic bone in Neanderthal man. *Science* 131:1437–8.

Stewart, T. D. 1977. The Neanderthal skeletal remains from Shanidar Cave, Iraq: A summary of findings to date. *Proceedings of the American Philosophical Society* 121:121–65.

Tague, R. G. 1992. Sexual dimorphism in the human bony pelvis, with a consideration of the Neandertal pelvis from Kebara Cave, Israel. *American Journal of Physical Anthropology* 88:1–21.

Tague, R. G., and C. O. Lovejoy. 1986. The obstetric pelvis of A. L. 288–1 (Lucy). *Journal of Human Evolution* 15:237–55.

Tillier, A. 1992a. The origins of modern humans in southwest Asia: Ontogenetic aspects, 15–28. In *The Evolution and Dispersal of Modern Humans in Asia* ed. T. Akazawa, K. Aoki, and T. Kimura. Tokyo: Hokusen-sha.

Tillier, A. 1992b. Reproductive patterns in Neanderthals: A source for critical analysis. *Collegium Anthropologium* 16:53–8.

Tillier, A., T. Majo, and J. Bružek. 1995. Limites d'une approche "bio-culturelle" des populations anciennes: L'exemple du complexe femoro-pelvien des Neanderthaliens, 579–95. In *Nature et Culture, Colloque de Liège*. ed. M. Otte. Liège: Université de Liège.

Tompkins, R. L., and E. Trinkaus. 1987. La Ferrassie 6 and the development of Neandertal pubic morphology. *American Journal of Physical Anthropology* 73:233–9.

Trevathan, W. 1987. *Human Birth: An Evolutionary Perspective.* New York: Aldine de Gruyter.

Trevathan, W. 1988. Fetal emergence patterns in evolutionary perspective. *American Anthropologist* 90:674–81.

Trinkaus, E. 1976. The morphology of European and southwest Asian Neandertal pubic bones. *American Journal of Physical Anthropology* 44:95–104.

Trinkaus, E. 1983. Neandertal postcrania and adaptive shift to modern humans, 165–200. In *The Mousterian Legacy: Human Biocultural Change in the Upper Pleistocene*, ed. E. Trinkaus. Oxford: British Archaeological Reports International Series 164.

Trinkaus, E. 1984. Neandertal pubic morphology and gestation length. *Current Anthropology* 25:509–14.

Trinkaus, E. 1992. Morphological contrasts between the Near Eastern Qafzeh-Skhul and late Archaic human samples: Grounds for a behavioral difference?, 277–94. In *The Evolution and Dispersal of Modern Humans in Asia*, ed. T. Akazawa, K. Aoki, and T. Kimura. Tokyo: Hokusen-sha.

Walrath, D. E., and M. M. Glantz. 1996. Sexual dimorphism in the pelvic midplane and its relationship to Neandertal reproductive patterns. *American Journal of Physical Anthropology* 100:89–100.

Washburn, S. L. 1948. Sex differences in the pubic bone. *American Journal of Physical Anthropology* 6:199–207.

Washburn, S. L. 1949. Sex differences in the pubic bone of Bantu and Bushmen. *American Journal of Physical Anthropology*, 7:425–32.

Wolpoff, M. 1980. *Paleoanthropology*. New York: Knopf.

Wolpoff, M. 1999. *Paleoanthropology*, 2nd edition. New York: McGraw Hill.

Woo, J. 1959. Human fossils found in Liukiang, Kwangsi, China. *Palaeovertebrata et Palaeoanthropologica* 1:97–104.

Wu, X., and M. Wu. 1985. Early *Homo sapiens* in China, 91–106. In *Palaeoanthropology and Palaeolithic Archaeology in the People's Republic of China*, ed. R. Wu and J. W. Olsen. Orlando: Academic Press.

Wu, X., and Z. Zhang. 1985. *Homo sapiens* remains from Late Paleolithic and Neolithic China, 107–33. In *Palaeoanthropology and Palaeolithic Archaeology in the People's Republic of China*, ed. R. Wu and J. W. Olsen. Orlando: Academic Press.

Wu, X., X. Shao, and H. Wang. 1982. Sex differences and sex determination of the innominate bone of modern Han nationality. *Acta Anthropologica Sinica* 1:118–31.

24

ONTOGENETIC VARIATION IN LATE PLEISTOCENE *HOMO SAPIENS* FROM THE NEAR EAST

Implications for Methodological Bias in Reconstructing Evolutionary Biology

Anne-marie Tillier

URA 376 CNRS
Laboratoire d'Anthropologie
Université de Bordeaux I
Talence 33405, France

1. INTRODUCTION

Since the last conference held in Tokyo on the subject of "The Evolution and Dispersal of Modern Humans in Asia" (Akazawa et al. 1992), recent discoveries in Middle Paleolithic caves such as Amud in Israel (Rak et al. 1994) and Dederiyeh in Syria (Akazawa et al. 1993, 1995) have refocused attention on the significance of studies devoted to immature individuals. Ontogenetic studies are thus reintroduced into the debate on the origins of modern humans and their evolutionary relationship to Neandertals.

The location of the Dederiyeh and Amud Caves in the Near East reinforced the significance of the discoveries with regard to the uniqueness of this geographic region. Indeed, the history of human populations in this region during the Middle Paleolithic is complex, as it was occupied by two morphologically different populations. Early anatomically modern humans are known from the Skhul and Qafzeh Caves while other hominids with European Neandertal affinities were the inhabitants of sites such as Amud, Tabun, and Kebara in Israel and Shanidar in Iraq. Because the archaeological records left by the two human groups are not significantly distinct, many scholars propose a model based on biological differences such as reproductive patterns, infant mortality, and life expectancy in order to explain geographical dispersal and prevalence of early modern humans. Most of the ontogenetic studies attempt to establish different parameters of maturation between early modern humans and Neandertals, yet few of them pay attention to methodological problems involved in conducting paleoauxological investigations.

Neandertals and Modern Humans in Western Asia, edited by Akazawa *et al.*
Plenum Press, New York, 1998.

2. METHODOLOGICAL BIASES IN PALEOAUXOLOGY

One major problem with studies that reconstruct ontogeny and growth patterns of past populations is that they use data collected from children whose early age of death indicates that they were the non-survivors of their populations (Saunders and Hopa 1993). Other kinds of problems follow when studies are based on single bones or single specimens, as well as in comparative analysis that is restricted to recent European populations. The interpretation of morphological characteristics of *Homo sapiens* juveniles requires a prior evaluation of the ontogenetic variation within early modern humans and a better understanding of the differences between these fossil hominids and Neandertals (Tillier 1989, 1992a,b). Finally, changes in morphology related to the growth and maturation processes have to be taken into account when a taxonomic identification is attempted. These considerations highlight a few problems that scholars face in attempting paleoauxological studies.

Ontogenetic studies are extremely important for building our knowledge of the site inhabitants in the case of single discoveries of non-adult specimens, as illustrated recently by the Dederiyeh find. I would like to approach the methodological aspects involved in such studies through two examples devoted to skull morphometrics.

3. SIZE VERSUS PATTERN IN SKULL MORPHOLOGY OF IMMATURE LATE PLEISTOCENE HOMINIDS

Cranial size is a primary argument in support of a biological shift in infant maturation between early modern humans and Neandertals. In the last decade, it has been suggested that a decrease in cranial size can be predicted for early modern human babies claimed to be "unusually altricial for anthropoid primates" (Trinkaus 1984:510), while large brain size in Neandertal juveniles is seen as a secondary reflection of an accelerated maturation during gestation and postnatal growth. Such a suggestion has been already discussed elsewhere (Tillier 1989, 1992a,b) with regard to evidence from the present fossil record.

Within the context of this discussion, we document the cranial growth variation found in a recent sample. Given the lack of large-size reference series of known age and sex, a sample of 426 non-adult individuals from a Middle Age cemetery located in Slovakia, the Dukove site (Hanakova et al. 1984), was recently employed to investigate age-related changes of several cranial measurements (Juricic et al. 1994). The selection of this sample was justified by the morphometric criteria displayed by the archaeological population supporting similarities in cranial breadths with archaic populations such as Neandertals. Figure 1 illustrates the range of variation for the maximum cranial breadth during growth. The smooth centile growth curves provide evidence of differences between infancy and childhood. They document more rapid increase in cranial size during infancy than later in ontogeny, in contrast with observations for body length.

The first well-documented representatives of modern humans consist of the immature Qafzeh and Skhul individuals. However, the entire Near Eastern sample might reflect regional differences and chronological trends. The metric and morphological characteristics of the Near Eastern fossil sample are thus evaluated in the context of the Upper Paleolithic immature specimens found in Europe. A similar pattern of cranial size growth emerges from the comparative analysis of data collected from both early modern and Neandertal immature individuals. Among the sample of early modern humans from the Near East, the youngest individual available for analysis is represented by the Skhul 1 child (about 3 years at death). The analysis focuses on the developmental period from 24

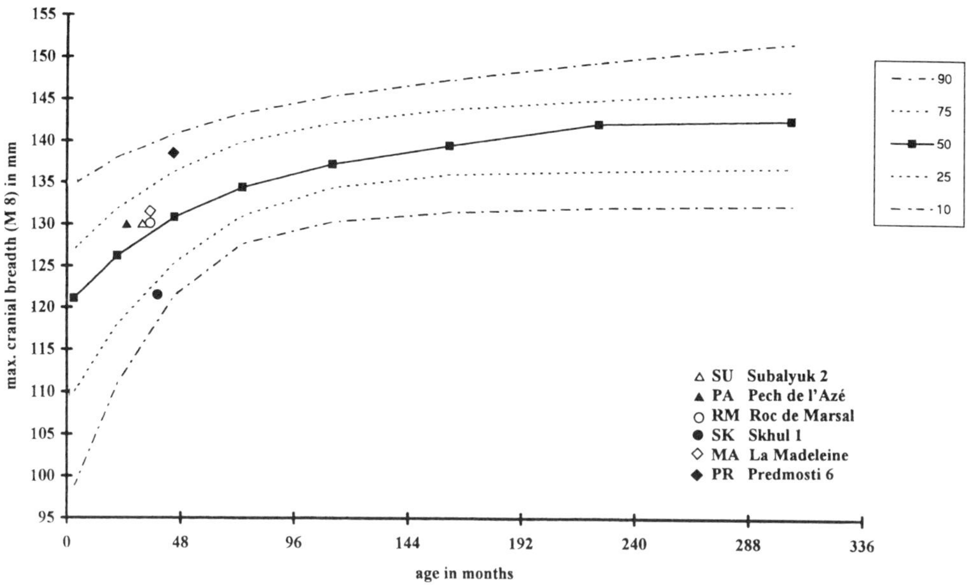

Figure 1. Smooth centile growth curves for the maximum cranial breadth in a modern sample from Slovakia (n = 426). Comparison with early modern humans from the Near East (Skhul 1), Europe (La Madeleine, Predmosti 6), and Neandertals (Pech de l'Azé, Subalyuk 2, and Roc de Marsal). All the fossil children, except Skhul 1, are in the upper limit of the distribution for the recent sample.

to 48 months in order to compare this skull to both recent and prehistoric children. The fossil sample includes three European Upper Paleolithic specimens of similar dental ages (i.e., Le Figuier, La Madeleine, and Predmosti 2, as found in Matiegka 1934; Billy 1979; Heim 1991) and three European Neandertals (i.e., Pech de l'Azé, Subalyuk 2, and Roc de Marsal, as found in Ferembach et al. 1970; Tillier 1983; Madre-Dupouy 1993; Pap et al. 1996). The examination of the recent and fossil data confirms that all children show similarities in cranial growth size. Late Pleistocene children from both Europe and the Near East are within the range of distribution of 80 percent of the recent sample (Figure 1).

Figure 1 demonstrates that the Skhul 1 child can be separated from both Neandertals (Tillier 1992b) and European Upper Paleolithic individuals. Located in the lower limit of the distribution of the recent sample (Figure 1), Skhul 1 exhibits a fully modern braincase morphology which, despite the difference in cranial size, indicates that this child is similar to the Upper Paleolithic children from Europe (Table 1).

4. DEVELOPMENTAL MOSAIC IN CRANIAL ONTOGENY OF EARLY MODERN HUMANS

The Skhul 1 individual permits us to illustrate another aspect of skull ontogenetic variation aside from cranial size. The facial features are also an important morphometric distinction between Skhul 1 and European early modern humans. The face of Skhul 1 is represented exclusively by its mandible. In their original description, McCown and Keith (1939:308–9) mentioned the morphological similarities between Skhul 1 and the only archaic child available at this time for comparative analysis, that is, the Devil's Tower Neandertal child.

Table 1. Comparative cranial morphometrics for Skhul 1 and early modern humans from Europe (Le Figuier, La Madeleine, Predmosti 6). All specimens between 3 and 4 years of age at death

Martin's measur.	Skhul1	Le Figuier	La Madeleine	Predmosti 6
	(1)	(2)	(3)	(4)
Max. cranial breadth M8	121	130 (?)	131	138
Max. cranial length M1	167	174 (?)	169	179
Cranial index M8/M1	72.5	–	77.5	77.1
Biasterionic Breadth M12	106	84	105	–
Max. frontal breadth M10	100	106	100	111

all measurements in mm;
(1) McCown and Keith, 1939
(2) Billy, 1979; (?) estimated
(3) Heim, 1992
(4) Matiegka, 1934

An expanded comparison between Skhul 1 and other immature archaic individuals of similar dental ages at death demonstrates similarities between the Skhul 1 child and archaic *Homo sapiens*. The Skhul 1 mandible lacks the inferior portion of the anterior symphyseal body, and measurements of the symphyseal height and thickness are incomplete. In addition conclusions are possible on the presence and/or the development of a chin eminence. But the Skhul 1 mandible manifests the same enlargement of the dental arch as specimens classified as Neandertals, such as Roc de Marsal from France (Tillier 1983; Madre-Dupouy 1992) and Barakai from Crimea (Faerman et al. 1994), which are similar in dental ages (Table 2). In addition, the three mandibles share archaic features such as the anterior alveolar flattening, the transverse straightness of the mandible across the front teeth, and, as already mentioned by McCown and Keith (1939), a strong development of the attachment for the medial pterygoid muscle on the internal surface of the ramus. A limited comparison with European Upper Paleolithic juveniles of similar dental ages at death (Le Figuier and La Madeleine) provides evidence of their distinction from the Skhul 1 mandible. Unlike Neaderthals and Skhul 1, the two European specimens lack the anterior widening and flattening of the dental arcade.

The Skhul 1 child illustrates a developmental mosaic in the cranial ontogeny of early modern humans dated to about 100 ka BP. Thus, a question can be raised concerning the taxonomic assignment of the Skhul 1 mandible, if it was indeed a case of isolated discovery. Should we consider the mixture of archaic and modern features displayed by the Skhul 1 child as unique to this individual or rather as a fair reflection of the skeletal morphology displayed by anatomically modern humans in the Near East dated to the Middle Paleolithic?

Table 2. Comparative mandibular morphometrics for Skhul 1, early modern humans from Europe (Le Figuier, La Madeleine) and Neandertals (Pech de l'Azé, Roc de Marsal, and Barakai). All specimens between 2 and 4 years of age at death

	Skhul 1	Le Figuier	La Madeleine	Barakai	Pech de l'Aze	Roc de Marsal	Modern sample
measurements in mm	(1) *	(2)	(3)	(4) *	(5)	6	*
Symphyseal height	(20,1)	20.3		24.1	17.8	20.4	19,8 ± 1,6
Symphyseal thickness	(11,0)	10.1		12.5	10.0	12.5	10,9 ± 0,8
Corpus height m1 - m2	16.4	16.5	18	20.1	14.2	17	16,9 ± 1,7
Corpus thickness m1-m2	(11,1)	11.2	12.5	14.2	11.5	13.1	10,3 ± 0,9
Bicanine Breadth	(30,1)	27,9(a)		36.3	30,5(a)	32.5	27,0 ± 1,7
Bimolar-m1 Breadth	43	35.1		49	41,1(a)	43.1	35,3 ± 1,9
Bimolar-m2 Breadth	52.7	37,6(a)		59	49,3(a)	50.8	41,9 ± 2,4

(a) measurements of the dental arcade provided are twice the distance from the preserved side to the midline. Modern sample of known age and sex (n=15) from Spitalfields
(1) author's data; (2) Billy, 1979; (3) Heim, 1992; (4) Faerman et al., 1993;.
(5) Ferembach, 1972; (6) Tillier 1983. * author's data added.

Within the Skhul hominid sample, the other immature individuals are too incomplete for documenting the potential individual variation in cranial ontogeny. More information comes from the Qafzeh immature sample, which provides reasonable evidence for differences between individuals with regard to the midfacial features (Tillier 1989, 1992b). Age-related patterning in the facial morphology cannot be totally excluded. Archaic retentions in the mandibular corpus morphology, including a wide anterior dental arch, presence of genioglossal pit and planum alveolare, and lack of a truly prominent chin bone, can be identified on a child of about 7 years of dental age (Qafzeh 4 as found in Tillier 1979). These traits are lacking on an older individual, such as Qafzeh 11 (approximately 13 years old), which exhibits a full development of the chin accompanied by a vertical symphyseal profile (Tillier 1984). In contrast, the Qafzeh 4 mandible, which manifests the same symphyseal profile, retains the plesiomorphous condition of the posterior corpus morphology.

Perhaps the developmental pattern involved in the ontogeny of early modern humans depends upon the antiquity of the specimens and reflects an evolutionary trend. For this reason it is of interest to examine the morphology displayed by Late Paleolithic children. The immature individual recovered from Upper Paleolithic layers in the Near East is an Aurignacian child from Ksar 'Akil (Lebanon), about 8 years old (Ewing 1963; Bergman and Stringer 1989). The Ksar 'Akil immature mandible seems to exhibit an anterior flattening of the dental arcade and lacks the development of a prominent chin, while the cranial morphology appears fully modern (author's observation on a cast, as the original is not to be found). Given the chronological and geographical isolation of this fossil, it is important to examine other Upper Paleolithic individuals.

In Europe, several Upper Paleolithic children were described from the Predmosti site in the Czech Republic (Matiegka 1934) and Sungir in Russia (Trofimova 1984). Although there is a chronological gap between the Near Eastern and the European sites, the latter being dated to the Gravettian, individual variation with respect to the midfacial development can be observed. Between the two Sungir individuals, for instance, the anterior mandibular corpus morphology is more receding on Sungir 2, a child of about 12–14 years, than on Sungir 3, about 9–10 years old at death (author's observation).

This individual variation in mandibular morphology can be seen as an illustration of either age- or sex-related changes in the expression of the cranial patterning during ontogeny. The fact that some specimens lack archaic features commonly associated with bone robusticity might be a reflection of sexual dimorphism. For instance, Sungir 3 is described as a girl while Sungir 2 is described as a boy (Trofimova 1984). Sex-related patterning in mandibular features among immature recent humans was discussed by Schutkowski (1993), who examined a reference sample of known age and sex from the Spitalfieds collection. The prominence of the chin region as well as the size and shape of the anterior dental arcade were among the sex-typical features selected. However, the author remarked that the discrete traits "fail to allocate girls reliably" in the sample used for testing (Schutkowski 1993:202).

The possible utility of mandibular morphology for sex estimation among immature fossil individuals currently remains unclear. It is therefore premature to propose a solution in terms of sexual dimorphism for single cranial bones, especially as the question of sex estimation for immature skeletons is not accurately solved (Majo et al. 1992).

5. THE FORAMEN MAGNUM SHAPE IN UPPER PLEISTOCENE CHILDREN: THE STATE OF THE ART

A point worthy of attention is therefore that taxonomic assignment of immature Late Pleistocene specimens can be complicated by the "mosaic," relative to the pattern in living humans, of features that they may exhibit. It might be suggested that variation in facial skeleton patterning among early modern humans can be explained by differences in masticatory requirements during life. To illustrate the difficulty in selecting suitable morphological traits on non-adult skeletons, another example regarding the skull base can be discussed.

A taxonomic distinction between modern humans and Neandertals based upon the configuration of the foramen magnum was recently proposed by Rak et al. (1994). A rounded shape of the foramen magnum is described as a modern feature while an elongated and oval-shaped foramen (low value of the breadth/length ratio) is seen as a derived Neandertal feature. One paleontological implication is the reconstruction by these scholars of a so-called modern shaped foramen for Skhul 1, which contrasts sharply with the shape present on the young Neandertal Amud 7 child (Rak et al. 1994:318).

The conclusions of Rak and his associates can be discussed in light of the data collected in a sample of recent children from western Europe. From the available information, it appears that the foramen magnum can be either rounded or elongated in recent children, as shown by the range of variation of the width/length ratio values (Table 3).

In addition, the foramen magnum index values calculated in a sample of Middle and Upper Paleolithic individuals, which includes children and adults, is employed for comparative analysis. The comparison yields results different from those proposed by Rak et al. (Table 3), namely that both Neandertal and other fossil individuals exhibit a certain range of variation in the foramen magnum's proportions.

Table 3. Comparative analysis of the foramen magnum index values (M.16 × 100 / M.7)

NEANDERTHALS		
children		
AMUD 7	57.1	(Rak *et al.*, 1994)
ENGIS 2	57.5	(Tillier's data)
TESHIK TASH	79.5	(after Gremiatski, 1949)
adults		
LA FERRASSIE 1	79.0	(after Heim, 1976)
LA CHAPELLE-AUX-SAINTS	68.8	(after Heim, 1989)

EARLY MODERN HUMANS		
children		
LA MADELEINE	85.7	(after Heim, 1991)
SUNGIR 3	84.3	(after Trofimova, 1984)
PREDMOSTI 22	60.0	(after Matiegka, 1934)
SUNGIR 2	73.6	(after Trofimova, 1984)
QAFZEH 11	79.9	(after Tillier, 1984)
adults		
SKHUL V	79.7	(after McCown & Keith, 1939)
PREDMOSTI 3	77.7	(after Matiegka, 1934)
SUNGIR 1	65.9	(after Bunak &Guerassimova, 1984)

RECENT CHILDREN			
EUROPEAN (Spitalfields) (known age and sex)	n = 33	mean = 66.7 SD = 19.9	(Tillier's data)
LEVANTINE and ASIAN INDIAN	n = 89	mean = 78.6 SD = 6.3	(Rak *et al.*, 1994)

Furthermore, an age-related change of the foramen magnum's proportions cannot be totally excluded in immature Neandertals, as suggested by the examination of the two unique adult Neandertals with a preserved foramen magnum (Table 3). This certainly limits the acceptance of a taxonomic value for the foramen magnum shape as it is presently proposed with regard to the Near Eastern juveniles.

6. CONCLUSION

Many investigations of the ontogeny of Late Pleistocene hominids are qualitative with observations restricted to single individuals. When a study comparing fossil and recent children is carried out, the results of the statistical analysis depend directly on the selection of the recent reference collection.

A basic aspect of comparative analysis of morphometrical data from fossil and recent individuals is the desire to draw conclusions about the potential distinctiveness of the fossil population concerned. Indeed, in our study of the ontogenetic variation in cranial size, a comparative analysis based on a sample of dolichocephalic children provided results different from those previously obtained in demonstrating the overlapping distribution of the cranial sizes with the Dukove sample.

When reconstructing paleoauxology of past populations, we must be aware that the growth of infants and children is influenced by the interaction of various variables and we need to recognize the potential bias inherent in studies devoted to single individuals. The developmental mosaic involved in the ontogeny and the age-related patterning of features must be taken into consideration. The possible interpretation of the morphometric differences in terms of sex-related criteria is still a matter of much debate.

Among early modern humans, the immature Near Eastern individuals from the Skhul and Qafzeh Middle Paleolithic Caves deserve special attention. They can help us to clarify several aspects of the debate regarding possible contrasts in the adaptative biology of different groups of Upper Pleistocene hominids.

ACKNOWLEDGMENTS

I would like to thank Professor Takeru Akazawa for inviting me to participate in this conference on "Neandertals and modern humans in West Asia." I wish to thank all the members of the Organizing Committee. I am grateful to the Department of Antiquities in Jerusalem and to many colleagues for allowing me to study the fossil remains they have in their charge (in Les Eyzies, London, Budapest, Paris, and Moscow). This research is financially supported by the CNRS (URA 376) and the French Foreign Ministry. Thanks are due to N. Ornstein for the editorial work.

REFERENCES

Akazawa, T., K. Aoki, and T. Kimura, eds. 1992. *The Evolution and Dispersal of Modern Humans in Asia*. Tokyo: Hokusen-Sha.

Akazawa, T., Y. Dodo, S. Muhesen, A. Abdul-Salam, Y. Abe, O. Kondo, and Y. Mizoguchi. 1993. The Neanderthal Remains from Dederiyeh Cave, Syria: Interim Report. *Anthropological Science* 101(4):361–87.

Akazawa, T., Y. Dodo, S. Muhesen, A. Abdul-Salam, Y. Abe, O. Kondo, and Y. Mizoguchi. 1995. Neanderthal infant burial. *Nature* 377:585–6.

Billy, G. 1979. L'enfant Magdalénien de la grotte du Fifuier (Ardèche). *L'Anthropologie* 83(2):223–52.

Ferembach, D. 1970. Le crâne de l'enfant du Pech de l'Azé, 13–49. In *L'enfant du Pech de l'Azé,* ed. D. Ferembach, P. Legoux, R. Fenart, R. Empereur Buisson, and E. Vlček. Archives de l'Institute de Paléontologie Humaine Memoire 33. Paris: Masson.

Hanakova, H., A. Sekacova, and M. Stloukal. 1984. *Pohrebiste v Ducovem. I. Soupis materialu a paleodemograficky nozbor, II. Metricka data a jejich strucna analyza* (in Czech). Prague: Narodni Museum.

Heim, J. L. 1991. L'enfant magdalénien de la Madeleine. *L'Anthropologie* 95(2–3):611–38

Juricic, M., J. Bruzek, J. P. Charlet, and A. M. Tillier. 1994. Dynamique de la croissance du crâne. L'exemple de la population médiévale de Ducove (Slovaquie). Application à l'échantillon fossile de Predmosti (Paléolithique supérieur, République Tchèque). *Bulletins et Mémoires de la Societé d'Anthropologie de Paris* 6(1–2):85–99.

Madre-Dupouy, M. 1992. *L'enfant du Roc de Marsal: Etude Analytique et Comparative.* Cahiers de Paléoanthropologie. Paris: CNRS Editions.

Majo, T., A. M. Tillier, and J. Bruzek. 1993. Test des fonctions discriminantes de Schutkowski impliquant l'ilium pour la détermination du sexe dans les séries d'enfants de sexe et d'âge connus. *Bulletin et Mémoires de la Societé d'Anthropologie de Paris* 5:61–8.

Matiegka, J. 1934 *Homo predmostensis. Fosilni Clovek z Predmosti na Moravé. I. Lebky* (in Czech). Prague: Nakladem Ceske Akademie ved a Umeni.

McCown, T. D., and A. Keith. 1939. *The Stone Age of Mount Carmel, Vol. 2, The Fossil Remains from the Levaloiso-Mousterian*. Oxford: Clarendon.

Pap, I., A. M. Tillier, B. Arensburg, and M. Chech. 1996. The Subalyuk Neanderthal remains: A re-examination. *Annales Historico-Naturales Musei Nationalis Hungarici* 88:233–70.

Rak, Y., B. Kimbel, and E. Hovers. 1994. A Neandertal infant from Amud Cave, Israel. *Journal of Human Evolution* 26:313–24.

Saunders, R. S., and R. D. Hoppa. 1993. Growth deficit in survivors and nonsurvivors: Biological mortality bias in subadult skeletal samples. *Yearbook of Physical Anthropology* 36:127–56.

Schutkowski, H. 1993. Sex determination of infant and juvenile skeletons. *American Journal of Physical Anthropology* 90:199–205.

Tillier, A. M. 1979. Restes craniens de l'enfant moustérien Homo 4 de Qafzeh (Israël): La mandibule et les maxillaires. *Paléorient* 5:67–85.

Tillier, A. M. 1983. L'enfant néanderthalien du Roc de Marsal (Campagne du Bugue, Dordogne). Le squelette facial. *Annales de Paléontologie* 69(2):137–49.

Tillier, A. M. 1984. L'enfant Homo 11 de Qafzeh (Israël) et son apport à la compréhension des modalités de croissance des squelettes moustériens. *Paléorient* 10(1):7–49.

Tillier, A. M. 1989. The evolution of modern humans: Evidence from young Mousterian individuals, 286–97. In *The Human Revolution*, ed. P. A. Mellars and C. B. Stringer. Edinburgh: Edinburgh University Press.

Tillier, A. M. 1992a. Reproductive patterns in Neanderthals: A source for critical analysis. *Collegium Anthropologium* 16(1):53–8.

Tillier, A. M. 1992b. The origins of modern humans in southwest Asia: Ontogenetic aspects, 16–28. In *The Evolution and Dispersal of Modern Humans in Asia*, ed. T. Akazawa, K. Aoki, and T. Kimura. Tokyo: Hokusen-sha.

Trinkaus, E. 1984. Neandertal pubic morphology and gestation length. *Current Anthropology* 25(4):509–14.

Trofimova, T. A. 1984. Cherepa deteï epochi bernevo paleolita iz Sungiria (in Russian), 144–55. In *Sungir, anthropologiceskoie issledlovanie*, ed. A. A. Zubov and V. M. Kharitonov. Moscow: Nauka.

25

UPPER LIMB VERSUS LOWER LIMB LOADING PATTERNS AMONG NEAR EASTERN MIDDLE PALEOLITHIC HOMINIDS

Erik Trinkaus,[1] Christopher B. Ruff,[2] and Steven E. Churchill[3]

[1]Department of Anthropology
Washington University
St. Louis, Missouri 63130
[2]Department of Cell Biology and Anatomy
Johns Hopkins University School of Medicine
Baltimore, Maryland 21205
[3]Department of Biological Anthropology and Anatomy
Duke University
Durham, North Carolina 27708

1. INTRODUCTION

In a previous paper concerned with functionally significant aspects of Near Eastern Middle Paleolithic hominids, one of us posed the question whether the biological differences between late archaic and early modern humans from the Near Eastern Middle Paleolithic were sufficient "grounds for a behavioral difference" between these two regional hominid lineages (Trinkaus 1992). At that time, the issue was whether one could argue for significant behavioral differences between these lineages on the basis of functionally relevant aspects of their skeletal and dental morphology. As a result of comparisons of dental attrition, diaphyseal external size and shape, muscle attachment areas, and articular orientations, the answer was *yes*.

However, since that analysis was intended primarily to establish the possibility and probability of biologically reflected behavioral differences between these two hominid groups, it emphasized differences, rather than similarities, between the samples. In the intervening years, we and our colleagues have pursued the issue, focusing on both degenerative aspects and morphological features to elucidate the pattern of functional/behavioral similarities/differences between these two groups. This work has brought to light a suite of similarities and contrasts in functionally relevant aspects of their paleontologically preserved biologies, suggesting a mosaic of behavioral similarities and differences between these lineages (see Trinkaus 1995 for a review). We focus here on one aspect of this

Neandertals and Modern Humans in Western Asia, edited by Akazawa *et al.*
Plenum Press, New York, 1998.

mosaic: upper and lower limb long bone shaft robusticity and shape as revealed through cross-sectional geometric analysis.

This analysis is based on the conclusion that the Middle Paleolithic of the Near East was associated with two hominid lineages, one of late archaic humans (frequently called Near Eastern Neandertals) and one of cranio-facially robust early modern humans. The late archaic humans are best known from the sites of Amud, Dederiyeh, Kebara, Shanidar, and Tabun (McCown and Keith 1939; Suzuki and Takai 1970; Trinkaus 1983; Bar-Yosef and Vandermeersch 1991; Akazawa et al. 1995), whereas the early modern humans are known from the sites of Qafzeh and Skhul (McCown and Keith 1939; Vandermeersch 1981). This division of the Near Eastern Middle Paleolithic hominids into two late Middle Pleistocene and Late Pleistocene hominid lineages is supported by abundant paleontological data, none of which is used here in functional analyses. Whether these groups represent the same or different species is irrelevant given the porous nature of reproductive boundaries between closely-related species. Even if they did exchange individuals and genes, which remains a possibility, the fossil record confirms that they remained largely distinct as biological lineages.

Consequently, the issue of concern here is the nature and the extent of functional, and by extension behavioral, similarities and contrasts between these two lineages of Middle Paleolithic hominids, as indicated by the distributions of bone in their long bone shafts. Shaft (or diaphyseal) morphology is ideal for such an analysis, since it remains plastic throughout the life of an individual, and hence it reflects the habitual mechanical loads on their limbs from activity patterns (see Cowin 1989 and Trinkaus et al. 1994 for reviews and additional references).

2. MATERIALS AND METHODS

In the cross-sectional geometric analysis of long bone diaphyses, each diaphysis is viewed as though it were a hollow beam subjected to axial, bending, and torsional forces. As such it can be treated mechanically by quantifying both the quantity and distribution of bone in a cross-section of the shaft at a specified percentage of bone (diaphyseal) length. Although cortical bone cross-sectional area is relevant as a measure of strength with respect to axial loading, better measures of bone strength are its second moments of area. They combine the quantity of bone with its squared distance from the geometric centroid of the cross-section, integrated across the full cross-section, and hence they measure resistance to bending, as well as axial and shear stresses (Lovejoy et al. 1976; Ruff and Hayes 1983; Kimura and Takahashi 1992).

Five measures are employed. The first four are second moments of area (I_i), which measure bending strength either relative to anatomically defined planes (I_x and I_y for the antero-posterior and medio-lateral planes, respectively) or as maximum and minimum values (I_{max} and I_{min}). The last measure, the polar moment of area, measures torsional rigidity and serves as an overall measure of diaphyseal strength. In this analysis, the polar moment is employed as a measure of overall strength, and specific second moments of area are used to assess diaphyseal shape and strengths in relevant anatomical planes.

For the femur, I_x and I_y are employed to measure antero-posterior and medio-lateral rigidity, since even incomplete femoral diaphyses can be oriented by placing the linea aspera in the mid-posterior diaphysis. Moreover, the near circularity of many late archaic human femoral midshafts means that small variations in cortical bone distribution can have major effects on the orientation of the major axis of the section. For the tibial midshaft, it is more difficult to anatomically orient diaphyses lacking their epiphyses; however, since the major axis is predominantly antero-posterior near midshaft in human tibiae, I_{max} and I_{min} are employed to assess tibial diaphyseal shape.

At the same time, these values must be scaled to body mass and the length of the beam involved, given variation in bone length and body mass. For the upper limb, the relevant scaling factor is bone length, since humans rarely support their body mass with their upper limbs. For the lower limb, a combination of body mass and beam length is appropriate. For the humerus, the polar moment of area scales to approximately $(\text{length})^4$ (Churchill 1995). For the femur, the polar moment of area scales to $(\text{length})^{5.33}$ (Ruff et al. 1993), whereas for the tibia it should scale to $(\text{femur length})^4 \times (\text{tibia length})^{1.33}$ (following Ruff et al. 1993). Since these values are non-linear, appropriate scaling involves either the use of log-log plots or powers of bone length for appropriate linear treatment of the non-linear relationships between variables.

The analysis here is restricted to Near Eastern Middle Paleolithic specimens which preserve sufficient and appropriate portions of the adult diaphyses of at least one of the three major long bones: the humerus, femur, and tibia. For the humerus, the sample includes Amud 1; Kebara 2; Shanidar 1, 3, 4, and 6; Qafzeh 8 and 9; Skhul 2, 4, 5, and 7a; and Tabun 1. For the femur, the sample includes Amud 1; Qafzeh 8 and 9; Shanidar 4, 5, and 6; Skhul 3, 4, 5, 6, and 7; and Tabun 1 and 3. For the tibia, it includes Amud 1; Qafzeh 8 and 9; Shanidar 2 and 6; Skhul 4, 5, and 6; and Tabun 1.

Of these long bones, Skhul 3 and Tabun 3 (5.9%, N=34) retain only isolated shaft sections and hence are used only in the cross-sectional shape analysis. Of the remainder, 78.1% (N=32) preserve sufficient portions of the limb segment to provide an accurate length estimate for body size and beam length scaling. The lengths for the remainder of the specimens were estimated using either the consistent relationship between humeral and femoral lengths across later Pleistocene and Holocene humans (Trinkaus 1981), or using regressions of proximal versus distal limb segment lengths for the sample (Near Eastern early modern versus late archaic) from which the specimen derives. In addition, since tibial diaphyseal scaling requires femoral length in addition to tibial length, the femoral length of Shanidar 2 was estimated using Near Eastern late archaic limb segment proportions. Given the contrasts in limb segment proportions for these two samples, especially for the lower limb (Trinkaus 1981; Holliday 1995), the use of sample-specific proportions to estimate lengths for missing or incomplete elements will tend to minimize the differences between the two groups, and hence their use is a conservative assumption in the analysis.

The majority of the cross-sections were generated using silicone putty molds to transcribe the subperiosteal contours and cortical thicknesses from biplanar radiography to interpolate the endosteal contours. Of the remaining sections, the Shanidar 6 humerus and Shanidar 4 femur had natural breaks at appropriate levels, and the Shanidar 5 and 6 femoral midshafts were reconstructed from casts and cortical thicknesses on the original specimens. Given fossilization damage and matrix, the Qafzeh 9 femoral midshaft represents a combination of data from the right and left bones. The resultant sections were digitized twice and the geometric properties computed using a PC DOS version of SLICE (Eschman 1992), with the resultant values averaged. For the humeri, given natural asymmetry, right and left values as available are plotted separately. For the femora and tibiae, right and left values as available were averaged, given low and randomly distributed levels of lower limb asymmetry in Recent and later Pleistocene humans (Trinkaus et al. 1994).

The Amud 1 and Shanidar 3, 4, and 6 humeri lack 35% sections or have data directly on them unavailable. To maximize sample size, their 35% cross-sections were predicted from their 65%, 50%, or 43% sections using regressions based on later Pleistocene human humeri.

Given these considerations, the resultant cross-sectional second and polar moments of area are compared for Near Eastern Middle Paleolithic humeri, femora, and tibiae, using mid-distal humeral shaft (35% of length) and femoral and tibial midshaft (50% of length) using bivariate data plots. The figures include distributions of values for the Near Eastern late archaic and early modern humans, against a background distribution of moderately robust horticultural Native Americans, all late prehistoric Puebloan Amerindians from New Mexico (Ruff and Hayes 1983; Churchill 1994). The Native American sample is provided as a general reference and is not intended to characterize all of Recent humans, given considerable variation in upper and lower limb robusticity across Recent humans (Ruff et al. 1993; Churchill 1994).

To statistically compare the distributions of the two Near Eastern Middle Paleolithic hominids, the standard (or raw) residuals of the fossil specimens relative to the Recent human least squares regression lines were compared using t-tests assuming heteroscedasticity (see Trinkaus and Hilton 1996 for a detailed explanation and justification). Even though the use of residuals has limitations (Jungers et al. 1995), it allows two-sample shape comparisons in a strictly bivariate framework without the more severe limitations of ratios. All computations and data plots were done using NCSS (Hintze 1991, 1995).

3. HUMERAL DIAPHYSEAL STRENGTH

Near Eastern Middle Paleolithic humeri exhibit considerable length and diaphyseal size variation. However, if their diaphyseal strengths are scaled to bone length, a clear pattern emerges (Figure 1). There are large and small clusters for both samples, but in each

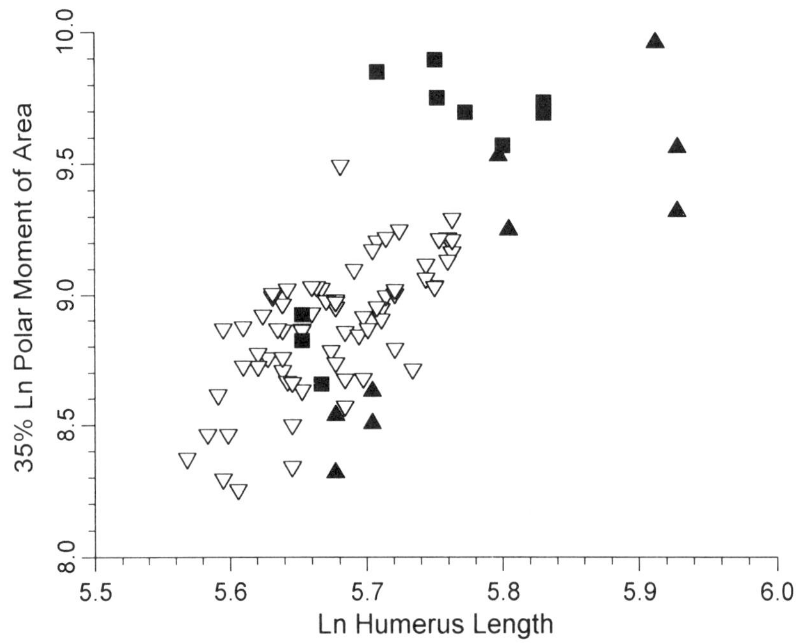

Figure 1. Mid-distal (35%) right and left humeral ln polar moment of area versus ln humeral length for recent Amerindians (▽), Qafzeh-Skhul humans (▲), and Near Eastern late archaic humans (■). For the fossils, all right humeri are more robust than their antimeres and therefore fall slightly above the left ones for the same (or nearly identical) lengths.

cluster there is a separation of the Qafzeh-Skhul sample on the lower, more gracile side and the late archaic humans on the higher, more robust side. There is close proximity only of Qafzeh 8 and 9 to Shanidar 1 and Amud 1, but no overlap between the samples. When the fossil samples are compared statistically, the right humeri are highly significantly different ($p = 0.007$) and the left humeri are moderately significantly different ($p = 0.019$). There is therefore clearly a difference in upper limb robusticity between these samples.

Since this upper limb contrast must be related to levels of manipulative behavior, it is relevant that the more consistent functional contrasts between these samples involve relative rates of anterior dental wear and reflections of upper limb muscularity, habitual grip positions, and mechanical advantages (Vandermeersch 1981, 1991; Trinkaus 1983, 1992, 1995; Churchill 1994). In all of these comparisons, as with humeral diaphyses, the late archaic humans fall on the side of the distribution indicating greater use of the body for manipulative tasks.

4. FEMORAL AND TIBIAL DIAPHYSEAL STRENGTH

In the lower limb, contrasts in patterns of robusticity between the Near Eastern late archaic and early modern human samples have been noted (e.g., Vandermeersch 1981; Trinkaus 1983, 1992). However, given the compounding effects of body mass and bone length in the weight-bearing legs, the contrasts may be more apparent than real.

The plot of femoral midshaft polar moment to length partially separates the Near Eastern late archaics from the early moderns, with $p = 0.106$ (Figure 2). There is some overlap between the samples, with Amud 1 and Tabun 1 being relatively gracile for the late archaics and Qafzeh 3 and 9 being relatively robust for the early moderns. However, such a straight-forward bone shaft to length comparison assumes similar scaling of body mass to bone length in the two samples.

It is well documented, however, that there is considerable ecogeographically-patterned variation in human body form, exemplified across modern humans by the extremes of equatorial Africans and arctic Inuits, Siberians, and Lapps (Trinkaus 1981; Ruff 1994; Holliday 1995). The contrasts are multiple, but they are largely contained in differences in relative trunk breadth and limb length. Limb length is quantified by femoral length, and body breadth is best measured in the living and the dead by comparisons of bi-iliac breadth to stature (or to femoral length). In this, modern arctic populations have high ratios, tropical populations have low ones, and cold temperate ones similar to our Native American reference sample have intermediate ones (Table 1). Moreover, the two Near Eastern Middle Paleolithic specimens preserving these measures, Kebara 2 and

Table 1. Mean bi-iliac breadth to maximum femoral length ratios for Holocene samples and Near Eastern Middle Paleolithic hominids[a]

	Bi-iliac breadth/femoral length
Holocene humans:	
Arctic (Inuits and Aleuts)	0.65
Temperate (Puebloan Amerindian)	0.63
Tropical (east African)	0.53
Near Eastern Middle Paleolithic:	
Kebara 2	0.70
Skhul 4	0.57

[a]Data from Ruff et al. (1993) and Ruff (1994).

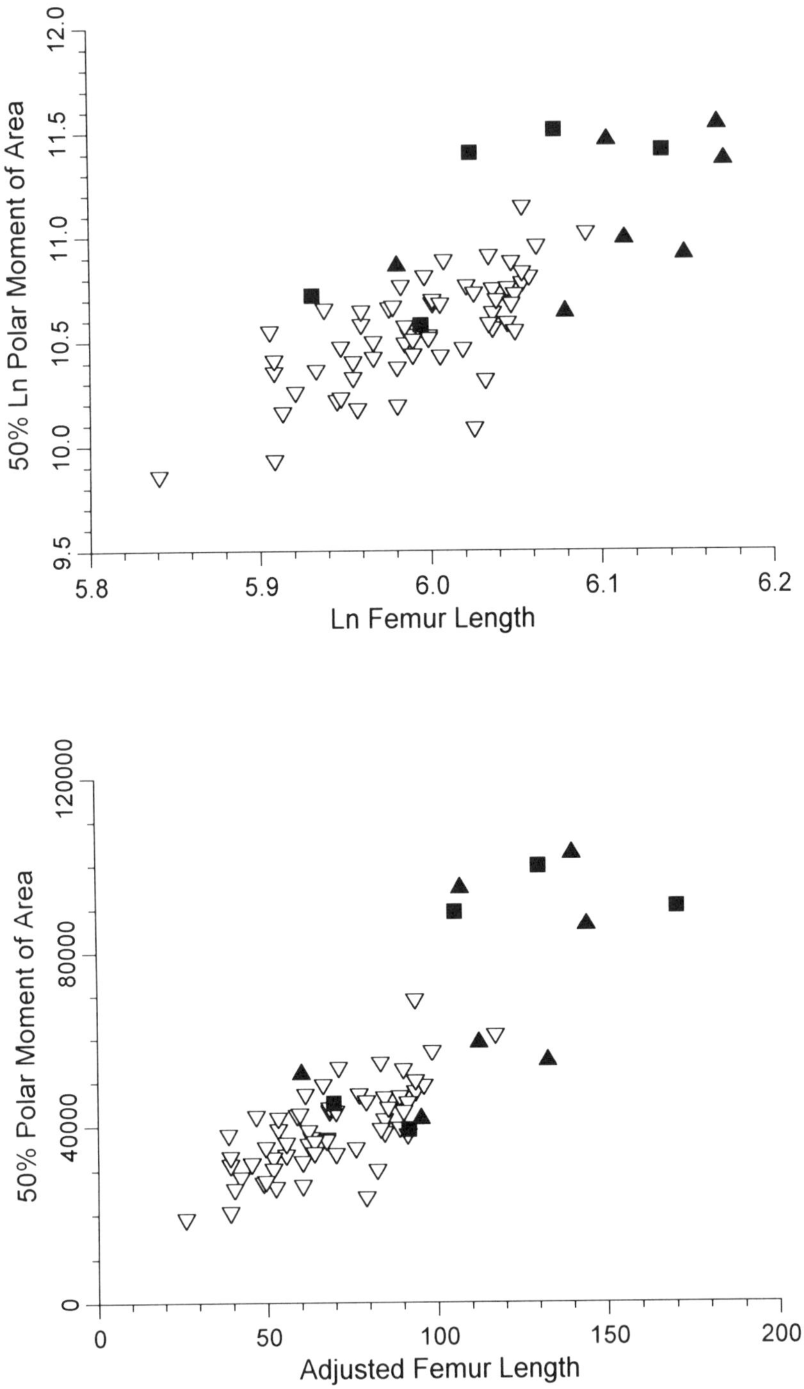

Figure 2. Midshaft (50%) femoral ln polar moment of area versus ln femoral length (above) and polar moment of area versus femoral $(\text{length})^{5.33}$ adjusted for body breadth (below), for recent Amerindians (▽), Qafzeh-Skhul humans (▲), and Near Eastern late archaic humans (■).

Skhul 4, have highly contrasting values. Kebara 2, a late archaic human, has a hyper-polar body breadth, whereas Skhul 4, an early modern human, has a highly tropical shape (Table 1).

Consequently, we have algebraically adjusted our combined measure of body size and beam length, femoral length, for the arctic and equatorial body proportions of these Near Eastern hominids, taking Kebara 2 and Skhul 4 to be representative of their respective samples. Since bi-iliac breadth varies relative to femoral length across Recent human samples to the average slope of 0.237, bi-iliac breadths have been estimated for the less complete fossil specimens based on their femoral lengths. Based on this, a percent correction from the Recent human reference sample has been added to each (femoral length)$^{5.33}$ (see Ruff et al. 1993 for the algebraic details), against which the polar moments of area are plotted.

The addition of body breadth adjustments provides a different pattern (Figure 2). The contrast in femoral robusticity largely disappears, with $p = 0.829$, with the late archaic femora being slightly more robust, on average than those of the early modern humans. In this, Qafzeh 8 and 9 are the most robust, closely approximated by Shanidar 4 and 5, whereas Skhul 4 and 7, Tabun 1, and to a lesser extent Skhul 6 and Amud 1, are relatively gracile.

Similar patterns emerge for tibial robusticity comparisons between the Near Eastern Middle Paleolithic samples. If tibial midshaft polar moments are compared to bone length (Figure 3), the late archaic human sample is consistently more robust than the early modern humans, with $p = 0.024$. Only Amud 1 even approaches the early modern human sample. However, both body breadth and limb segment lengths need to be taken in account.

In addition to contrasts in trunk breadth, these Middle Paleolithic samples also contrast in relative tibial to femoral lengths. If tibial length is plotted against femur length (Figure 3), the early modern humans fall on the high side of the Recent human sample and the late archaic humans all fall well below it. The fossil samples differ at $p = 0.002$. Consequently, tibial midshaft polar moments were then plotted against the product of powers of femoral length, representing body size, and tibial length, representing beam size, both adjusted for body breadth (Figure 3). The resulting scatter plot eliminates any difference between the samples, producing a $p = 0.936$. In this, Shanidar 2 remains highly robust and is approached by Qafzeh 9 and Skhul 5, whereas Amud 1, Tabun 1, Qafzeh 8, and Skhul 4 and 6 are moderately gracile.

Interestingly, these similarities in femoral and tibial robusticity are echoed in recent histological work on the Shanidar, Skhul, and Tabun femora (Abbott et al. 1996). All of them have similarly sized osteons, a reflection of osteoclastic activity levels normally regulated by habitual mechanical load levels on the diaphyses. It is also supported, to a lesser extent, by comparisons of their talar proportions and pedal proximal phalangeal diaphyseal proportions (Trinkaus and Hilton 1996).

5. FEMORAL AND TIBIAL DIAPHYSEAL SHAPE

At the same time, there are clear femoral and tibial diaphyseal cross-sectional shape differences. These are apparent in early modern human pilastric versus late archaic human non-pilastric femoral shafts, and in early modern human tibial cross-sections that are sub-quadrilateral with distinct crests versus late archaic human amygdaloid ones (Figure 4). Yet even though some details of the diaphyseal shapes may be genetically determined and reflect details of muscle attachment patterns, the overall distributions of bone should reflect mechanical loading patterns.

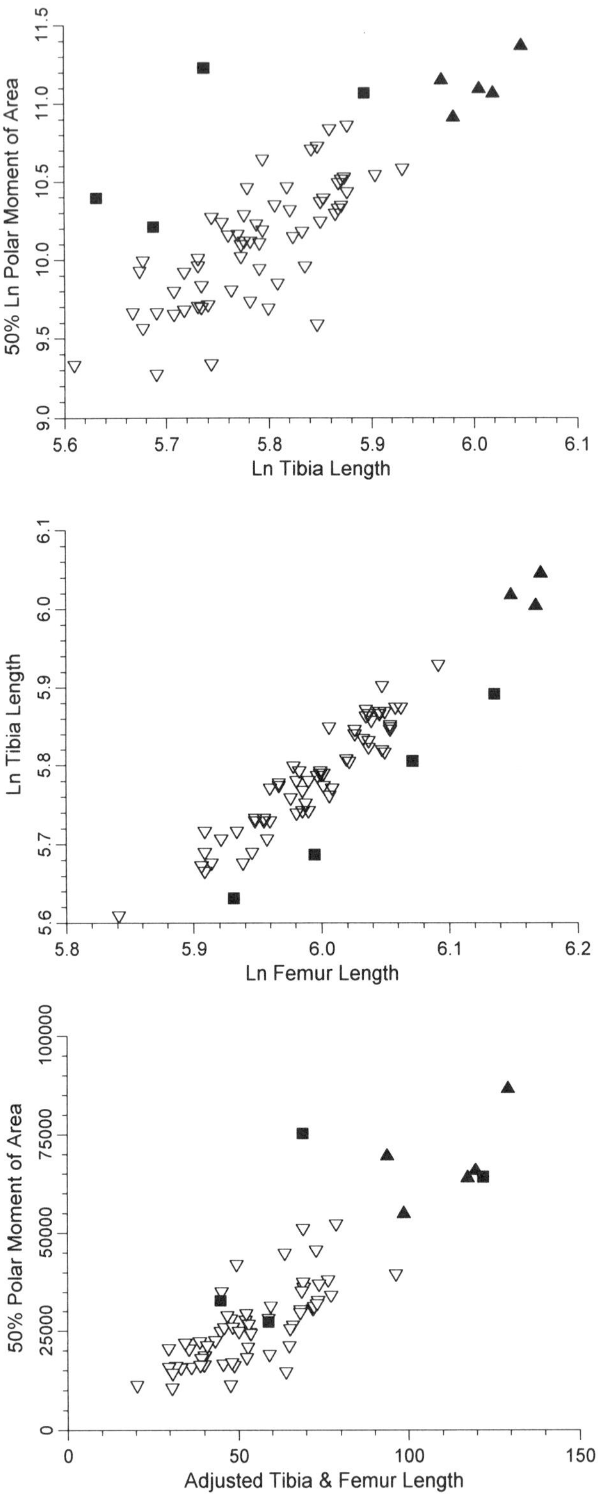

Figure 3. Midshaft (50%) tibial ln polar moment of area versus ln tibial length (top), ln tibial length versus ln femoral length (middle), and midshaft (50%) tibial polar moment of area versus (femoral length)^4 × (tibial length)$^{1.33}$ adjusted for body breadth (bottom), for recent Amerindians (▽), Qafzeh-Skhul humans (▲), and Near Eastern late archaic humans (■).

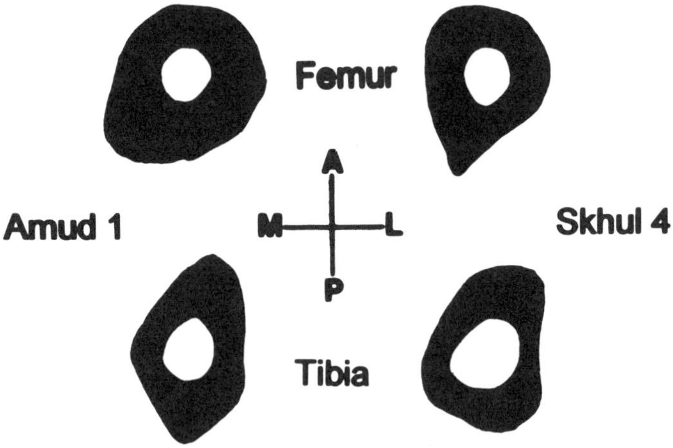

Figure 4. Reconstructed femoral and tibial midshaft (50%) cross-sections for the Amud 1 late archaic human and the Skhul 4 early modern human. The AP and ML lines are 2 cm long.

For the femur, this is reflected in the comparison of antero-posterior versus medio-lateral second moments of area (Figure 5). Not surprisingly given their high pilastric indices (Vandermeersch 1981; Trinkaus 1983, 1992), the Qafzeh-Skhul femora are significantly stronger in the antero-posterior direction relative to the medio-lateral one, at $p = 0.021$. Only Skhul 7 overlaps the late archaic human range of variation in this feature.

If the separate antero-posterior and medio-lateral second moments are compared to femoral length (Figure 5), the nature of the contrast becomes more apparent. The comparison of the antero-posterior second moments to femoral length reveals no difference between the samples, with $p = 0.827$. This should not be surprising, since the femoral midshaft is primarily subjected to antero-posterior bending forces during normal locomotion. However, when the medio-lateral second moments are compared to femoral length, there is a highly significant difference, with $p = 0.002$. In this, the Qafzeh-Skhul sample with its relatively narrow trunks appears more gracile than the relatively wide late archaic humans, with only Qafzeh 3 and 9 being close to the relatively gracile Amud 1 and Tabun 1.

However, as in previous analyses, these comparisons need to be adjusted for differences in body breadth and hence body mass. When this is done (Figure 6), the differences between the samples in the mechanical strengths in the two planes become non-significant. The plot of the antero-posterior second moment to adjusted femur length places the Qafzeh-Skhul sample non-significantly above the late archaic sample ($p = 0.490$), with Qafzeh 8 and 9, and to a lesser extent Qafzeh 3, Shanidar 4 and 5, and Skhul 5, having relatively high values, and Amud 1, Skhul 4 and 7, and Tabun 1 being relatively gracile. The plot of medio-lateral second moments to adjusted femur length places the two samples further apart, with the late archaic humans non-significantly ($p = 0.145$) above the early modern human sample. In this Qafzeh 9 and Shanidar 4 and 5 maintain their high robusticity level, and Skhul 4, 5, and 6 are the most gracile.

In the tibial midshaft maximum to minimum second moment comparison (Figure 7), there is no meaningful difference between the samples, with $p = 0.807$. Relative antero-posterior to medio-lateral load levels in these two hominid samples are therefore similar, all falling at or below the line for the relatively platycnemic Amerindian sample.

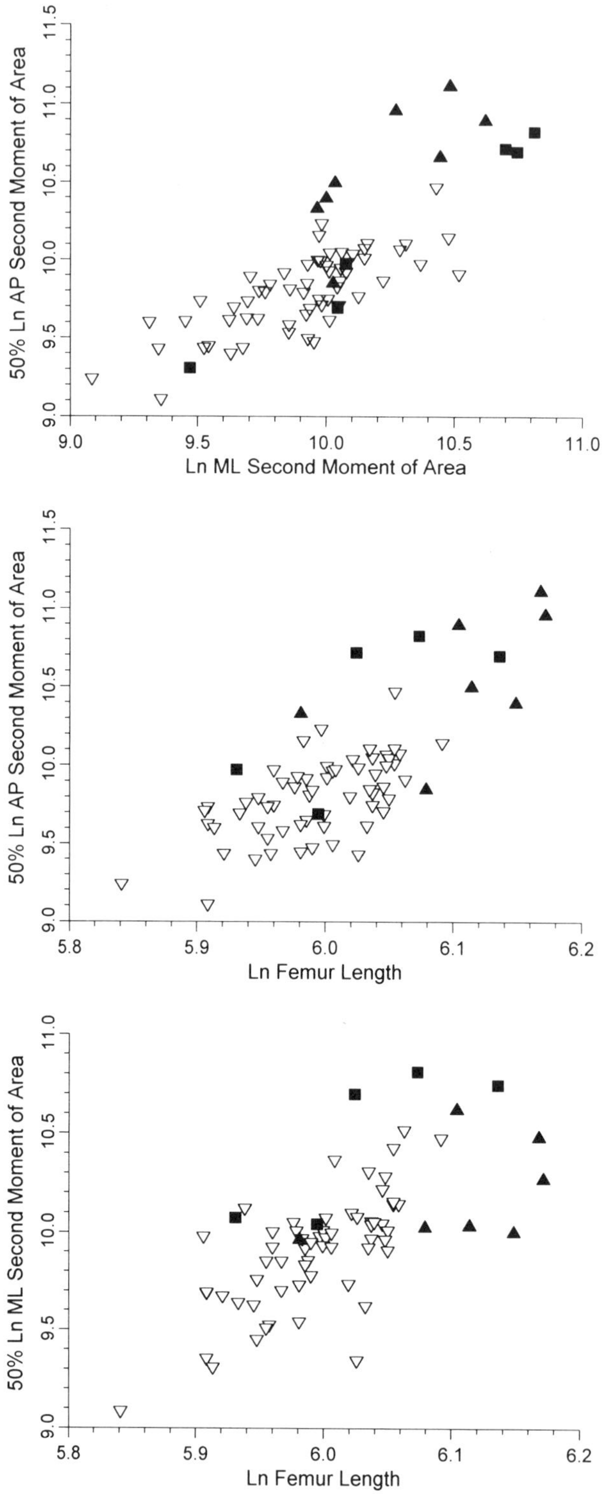

Figure 5. Femoral midshaft (50%) ln antero-posterior second moment of area (I_x) versus ln medio-lateral second moment of area (I_y) (top), ln antero-posterior second moment of area versus ln femoral length (middle), and ln medio-lateral second moment of area versus ln femoral length (bottom), for recent Amerindians (▽), Qafzeh-Skhul humans (▲), and Near Eastern late archaic humans (■).

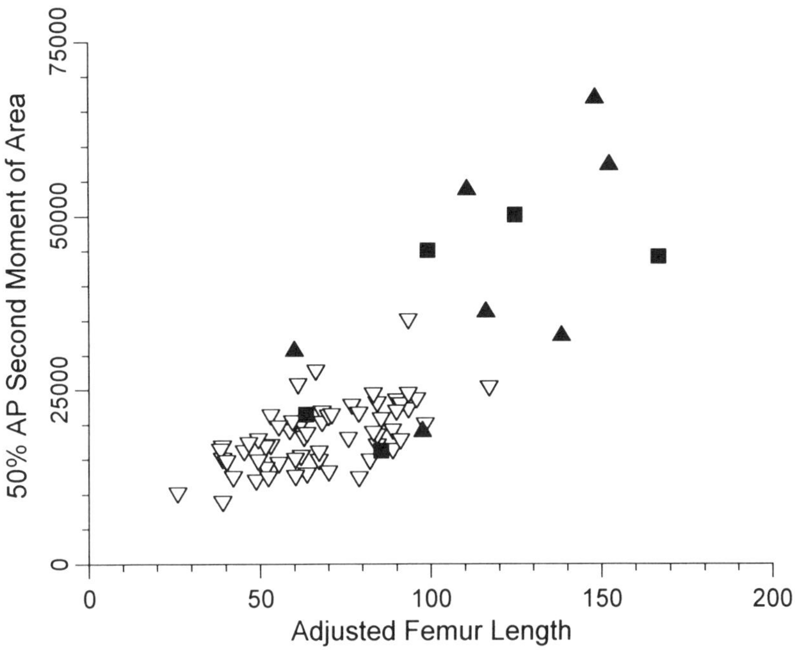

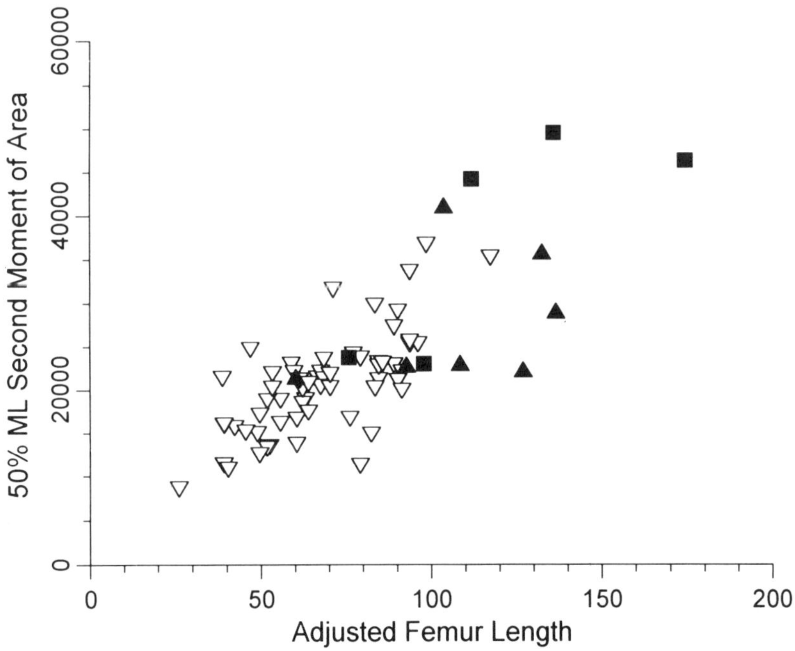

Figure 6. Midshaft (50%) femoral antero-posterior (above) and medio-lateral (below) second moments of area versus femoral $(\text{length})^{5.33}$ adjusted for body breadth for recent Amerindians (▽), Qafzeh-Skhul humans (▲), and Near Eastern late archaic humans (■).

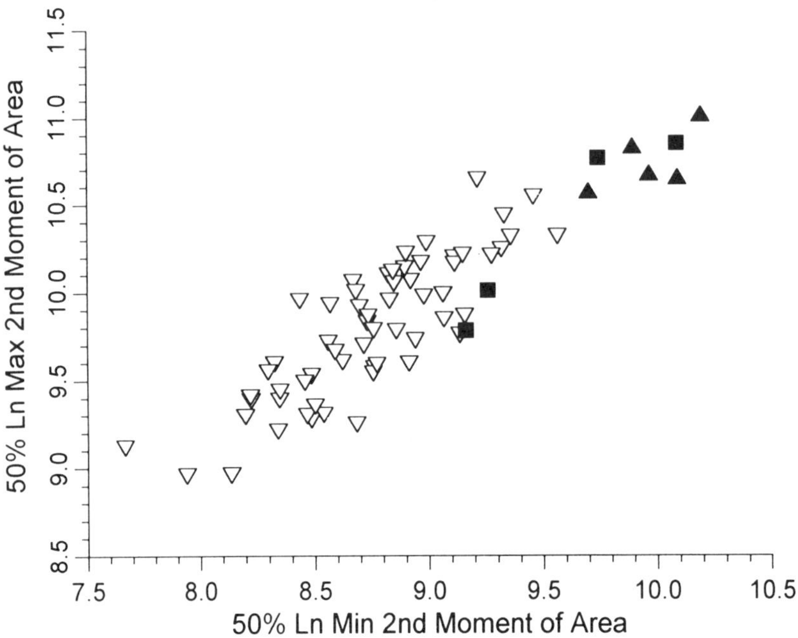

Figure 7. Tibial midshaft (50%) ln maximum (antero-posterior) second moment of area (I_{max}) versus ln minimum (medio-lateral) second moment of area (I_{min}) for recent Amerindians (▽), Qafzeh-Skhul humans (▲), and Near Eastern late archaic humans (■).

6. DISCUSSION AND CONCLUSIONS

Previous analyses of Near Eastern later Pleistocene hominid postcrania have included measures of cross-sectional geometry of their humeri (Smith et al. 1983; Ben-Itzhak et al. 1988; Churchill 1994; Trinkaus et al. 1994), femora (Trinkaus and Ruff 1989; Ruff et al. 1993) and tibiae (Endo and Kimura 1970; Lovejoy and Trinkaus 1980; Trinkaus and Ruff 1989). However, none of these previous analyses of humeri and tibiae have included the full available sample and/or addressed the question of body mass and beam length scaling. The analysis here of all of the available specimens with considerations of scaling factors provides a more complete picture of diaphyseal loading levels for these Middle Paleolithic hominids.

First, once differences in body shape are taken into account, there are no significant differences in lower limb robusticity between these two samples. This is echoed in comparisons of femoral cortical bone histology, and in comparisons of foot articular and phalangeal robusticity. In other words, all of these hominids reflect the pattern of elevated robusticity of the locomotor anatomy evident in Pleistocene *Homo* (Ruff et al. 1993). This conclusion, however, applies only to adolescents and adults, since the consistent contrast in their femoral neck shaft angles implies different levels of locomotor activity between juvenile members of these lineages (Trinkaus 1993a).

Second, there are clear and marked contrasts in humeral diaphyseal robusticity, especially in the right arm. Similar contrasts have been noted by Smith et al. (1983) and Churchill (1994), and they are reinforced by increased sample sizes. These differences are evident in both large and small individuals in the samples and in both arms. They are also echoed in a suite of features from the scapula to the terminal phalanges reflecting contrasts in muscular size, muscle mechanical advantages, and articular resistance to force (Trinkaus 1995). They also correspond to contrasts in relative anterior dental wear levels and patterns.

In light of this, it is interesting that a series of archeological studies has failed to demonstrate major behavioral differences between these samples, ones greater than those expected between similar foraging groups exploiting the Near East across seasons and through a series of regional climatic and ecological shifts (see Lieberman and Shea 1994 for a recent review, despite difficulties in assigning Levantine archeological assemblages to hominid lineages [Trinkaus 1993b]). We do not propose to resolve this apparent contradiction here. However, a couple of suggestions are in order. First, most of the archaeological analyses to date have not been aimed at or sufficient to detect such behavioral contrasts (but see Lieberman and Shea 1994 and other papers in this volume). Second, the Levant is one of the richest regions of the world for high quality lithic raw material. Hence, selective pressures related to raw material acquisition, transport, processing, and use are likely to have been low, and the resultant manufacture and discard pattern may not reveal significant shifts in adaptively important manipulative or mobility behavior. Third, despite ecological diversity within the region, most ecozones are likely to have remained within reasonable distances of the hominids in question, making the exploitation of fauna appear similar for all of these Middle Paleolithic hominids. And fourth, as suggested by the differential contrasts in locomotor robusticity between juvenile and older members of these groups, the real behavioral contrasts may have been social and organizational, and not directly linked to the technology of food acquisition and processing.

ACKNOWLEDGMENTS

We would like to thank the organizers of this symposium, Profs. T. Akazawa, K. Aoki, and O. Bar Yosef, for inviting one of us (ET) to contribute in this forum to the ongoing debate concerning Near Eastern Middle Paleolithic hominids. Access to original Near Eastern fossils and permissions to mold and x-ray them has been kindly granted by B. Arensburg, Isa Salman, H. de Lumley, Muayed Sa-id al-Damirji, D. Pilbeam, Y. Rak, C. B. Stringer, B. Vandermeersch, and J. Zias, and assistance was provided by B. Arensburg, L. Beck, R. A. Bloom, M. Chech, J. Clark, D. George, D. Grimaud-Hervé, R. Kruzynski, D. E. Lieberman, P. Smith, J. Wangermez, and I. Zohar. This research has been supported by NSF and the L. S. B. Leakey Foundation (to ET) for the analysis of the Near Eastern fossil record, and by NSF, the L. S. B. Leakey Foundation, and CNRS (to ET, CBR, and/or SEC) for collection and analysis of the comparative fossil and Recent human data.

REFERENCES

Abbott, S., E. Trinkaus, and D. B. Burr. 1996. Dynamic bone remodeling in later Pleistocene fossil hominids. *American Journal of Physical Anthropology* 99:585–601.

Akazawa T., S. Muhesan, Y. Dodo, O. Kondo, and Y. Mizoguchi. 1995. Neanderthal infant burial. *Nature* 377:585–6.

Bar-Yosef, O., and B. Vandermeersch, eds. 1991. *Le Squelette Moustérien de Kébara 2*. Paris: CNRS Editions.

Ben-Itzhak, S., P. Smith, and R. A. Bloom. 1988. Radiographic study of the humerus in Neandertals and *Homo sapiens sapiens*. *American Journal of Physical Anthropology* 77:231–42.

Churchill, S. E. 1994. Human Upper Body Evolution in the Eurasian Later Pleistocene. Ph. D. diss. University of New Mexico. Albuquerque.

Churchill, S. E. 1995. Humeral strength to bone length scaling relationships in recent humans (abstract). *American Journal of Physical Anthropology Supplement* 20:76–7.

Cowin, S. C. 1989. The mechanical properties of cortical bone tissue, 97–127. In *Bone Mechanics*, ed. S. C. Cowin. Boca Raton: CRC Press.

Endo, B., and T. Kimura. 1970. Postcranial skeleton of the Amud man, 231–406. In *The Amud Man and His Cave Site*, ed. H. Suzuki and F. Takai. Tokyo: Academic Press of Japan.

Eschman, P. N. 1992. *SLCOMM Version 1.6*. Albuquerque: Eschman Archeological Services.

Hintze, J. L. 1991. *Number Cruncher Statistical System Version 5.03 9/91*. Kaysville, UT: NCSS.

Hintze, J. L. 1995. *NCSS 6.0.8 Statistical System for Windows*. Kaysville, UT: NCSS.

Holliday, T. W. 1995. Body Size and Proportions in the Late Pleistocene Western Old World and the Origins of Modern Humans. Ph. D. diss. University of New Mexico. Albuquerque.

Jungers, W. L., A. B. Falsetti, and C. E. Wall. 1995. Shape, relative size, and size-adjustments in morphometrics. *Yearbook of Physical Anthropology*. 38:137–61.

Kimura, T., and H. Takahashi. 1992. Cross sectional geometry of the Minatogawa limb bones, 305–20. In *The Evolution and Dispersal of Modern Humans in Asia*, ed. T. Akazawa, K. Aoki, and T. Kimura. Tokyo: Hokusen-sha.

Lieberman, D. E., and J. J. Shea. 1994. Behavioral differences between archaic and modern humans in the Levantine Mousterian. *American Anthropologist* 96:300–32.

Lovejoy, C. O., A. H. Burstein, and K. G. Heiple. 1976. The biomechanical analysis of bone strength: A method and its application to platycnemia. *American Journal of Physical Anthropology* 44:489–506.

Lovejoy, C. O., and E. Trinkaus. 1980. Strength and robusticity of the Neandertal tibia. *American Journal of Physical Anthropology* 53:465–70.

McCown, T. D., and A. Keith. 1939. *The Stone Age of Mount Carmel, Vol. 2, The Fossil Remains from the Levalloiso-Mousterian*. Oxford: Clarendon.

Ruff, C. B. 1994. Morphological adaptation to climate in modern and fossil hominids. *Yearbook of Physical Anthropology* 37:65–107.

Ruff, C. B., and W. C. Hayes. 1983. Cross-sectional geometry of Pecos Pueblo femora and tibiae—A biomechanical investigation: I. Method and general patterns of variation. *American Journal of Physical Anthropology* 60:359–81.

Ruff, C. B., E. Trinkaus, A. Walker, and C. S. Larsen. 1993. Postcranial robusticity in Homo, I: Temporal trends and mechanical interpretations. *American Journal of Physical Anthropology* 91:21–53.

Smith, P., R. A. Bloom, and J. Berkowitz. 1983. Bone morphology and biomechanical efficiency in fossil hominids. *Current Anthropology* 24:662–3.

Suzuki, H., and F. Takai, eds. 1970. *The Amud Man and His Cave Site*. Tokyo: Academic Press of Japan.

Trinkaus, E. 1981. Neanderthal limb proportions and cold adaptation, 187–224. In *Aspects of Human Evolution*, ed. C. B. Stringer. London: Taylor and Francis.

Trinkaus, E. 1983. *The Shanidar Neandertals*. New York: Academic Press.

Trinkaus, E. 1992. Morphological contrasts between the Near Eastern Qafzeh-Skhul and late archaic human samples: Grounds for a behavioral difference?, 277–94. In *The Evolution and Dispersal of Modern Humans in Asia*, ed. T. Akazawa, K. Aoki, and T. Kimura. Tokyo: Hokusen-Sha.

Trinkaus, E. 1993a. Femoral neck-shaft angles of the Qafzeh-Skhul early modern humans, and activity levels among immature Near Eastern Middle Paleolithic hominids. *Journal of Human Evolution* 25:393–416.

Trinkaus, E. 1993b. Comment on: D. E. Lieberman, "The rise and fall of seasonal mobility among hunter-gatherers: The case of the southern Levant." *Current Anthropology* 34:620–2.

Trinkaus, E. 1995. Near Eastern late archaic humans. *Paléorient* 21:9–23.

Trinkaus, E., S. E. Churchill, and C. B. Ruff. 1994. Postcranial robusticity in Homo, II: Humeral bilateral asymmetry and bone plasticity. *American Journal of Physical Anthropology* 93:1–34.

Trinkaus, E., and C. E. Hilton. 1996. Neandertal pedal proximal phalanges: Diaphyseal loading patterns. *Journal of Human Evolution* 30:399–425.

Trinkaus, E., and C. B. Ruff. 1989. Diaphyseal cross-sectional geometry and biomechanics of the Fond-de-Forêt 1 femur and the Spy 2 femur and tibia. *Bulletin de la Société Royale Belge d'Anthropologie et de Préhistoire* 100:33–42.

Vandermeersch, B. 1981. *Les Hommes Fossiles de Qafzeh (Israël)*. Paris: CNRS Editions.

Vandermeersch, B. 1991. La ceinture scapulaire et les membres supérieurs, 157–78. In *Le Squelette Moustérien de Kébara 2*, ed. O. Bar Yosef and B. Vandermeersch. Paris: CNRS Editions.

26

A REASSESSMENT OF THE TABUN C2 MANDIBLE

Rolf M. Quam[1] and Fred H. Smith[2]

[1]Department of Anthropology
State University of New York
Binghamton, New York 13902
[2]Department of Anthropology
Northern Illinois University
DeKalb, Illinois 60115

1. INTRODUCTION

As recently as just over a decade ago, a reasonable case could be made that modern humans initially appeared throughout the Old World, essentially contemporaneously, between 40 ka and 30 ka (e.g., Wolpoff 1980; Smith 1985). Objections to such an interpretation were numerous and grounded in various lines of evidence (e.g., Howells 1976; Stringer 1978; Butzer et al. 1978; Stringer et al. 1979, 1984; Bar-Yosef and Vandermeersch 1981a; Tchernov 1981; Vandermeersch 1981; Bräuer 1984; Howell 1984; Nei 1985). To many these objections were convincing, but the fact is that no chronometric data existed at that time offering unequivocal evidence anywhere of an emergence of modern humans earlier than about 40 ka. However, by the late 1980s, the chronological picture changed, primarily as the result of the application of thermoluminescence (TL) and election spin resonance (ESR) dating to critical sites in western Asia, Europe, and Africa. These new dates indicate that modern humans appeared appreciably earlier in some areas of the western Old World than in others.

2. DATING AND MODERN HUMAN ORIGINS IN SOUTHWEST ASIA

In 1988, Valladas and colleagues published an age of 92 ka based on TL dating for the Qafzeh hominids, broadly recognized as early modern humans from the Levant (Vandermeersch 1981b). ESR (early uptake) and uranium series dating generally corroborated this, indicating an age of 110 ka to 90 ka for these remains (Schwarcz et al. 1988; McDer-

Neandertals and Modern Humans in Western Asia, edited by Akazawa *et al.*
Plenum Press, New York, 1998.

mott et al. 1993). In addition, uranium series and ESR dating have established an age of well over 40 ka for the Skhul remains (Grün and Stringer 1991; McDermott et al. 1993; Bar-Yosef 1994), which are also generally regarded as early modern humans from the Levant, but the precise chronology of these burials is a matter of some debate (see Wolpoff 1995; Stringer, this volume). Whatever the resolution of the Skhul situation, current dating presents strong evidence that modern humans were established in western Asia significantly earlier than in Europe, where the earliest modern human fossils are dated to about 35 ka (Smith 1991).

The ancestral source of the Skhul/Qafzeh people is a vigorously debated question, with significant implications for the broader issue of modern human origins. In their 1939 analysis of the Skhul and Tabun remains, McCown and Keith identified a series of characters that linked the skeletons from Skhul with the more archaic Tabun remains and concluded that they represented a single, extremely variable population in the "throes of evolutionary change" (1939:14). Specifically, McCown and Keith believed the Skhul/Tabun people were evolving into Neandertals and that the origin of modern peoples was to be found "still further to the East" (1939:17). On the other hand, Coon (1939) and Dobzhansky (1944) considered the Mt. Carmel (Skhul and Tabun) people to be the result of hybridization between Neandertals and modern forms. More recently, Wolpoff, Arensburg, Kramer, Frayer, and colleagues have reasserted the McCown/Keith position that the Skhul (and Qafzeh) people and the Levantine "Neandertals" (Tabun, Amud, and Kebara) are members of a single, variable population which existed in the Near East for several tens of thousands of years (Arensburg 1989; Frayer et al. 1993; Wolpoff 1995; Kramer et al., in press). In the context of multiregional evolution, this population is considered ancestral to recent people in the Near East, as a part of the interconnected web of regional lineages that forms the basis of the multiregional perspective.

Other workers also have viewed the Skhul/Qafzeh people as deriving from indigenous western Asia sources, but not everyone agrees on the source! Citing the chronological evidence available in the 1950s, Howell argued that Tabun was temporally earlier than, and ancestral to, the Skhul people, who were in turn ancestral to modern humans in western Eurasia (Howell 1958, 1959). This view gained considerable support (see Howells 1976; Trinkaus 1984) prior to the new dating of the late 1980s. Alternatively, Vandermeersch (1981a, 1989, 1992) derives the Skhul/Qafzeh people from ancestors represented by the Zuttiyeh specimen, which he connects to the Skhul/Qafzeh sample by a series of facial features (but see Simmons et al. 1991). For Vandermeersch, the Near Eastern Neandertals are distinct from this Zuttiyeh/Skhul/Qafzeh clade and represent relatively late immigrants from the colder climates of Europe. Bar-Yosef (1989, 1993; Bar-Yosef and Vandermeersch 1993) has proposed that Neandertals were pushed into the Levant during the harsh, sometimes arctic conditions of the early part of the last glaciation. This "late migration hypothesis" would see the first Neandertal arriving in the Levant sometime after oxygen isotope stage 5c, somewhat less than 100 ka.

A rather different opinion has been expressed by a number of other researchers, who view the Skhul/Qafzeh people as immigrants into the Near East. Bräuer (1984, 1992), Stringer (1990, 1994; Stringer and Andrews 1988), Rightmire (1984, 1989), and others have focused attention on Africa as the initial region of origin for modern humans. Both Bräuer and Stringer specifically note that African moderns moved into the Levant before spreading to the remainder of Eurasia. Pointing particularly to patterns of body proportions (Trinkaus 1981; Holliday and Falsetti 1995; Holliday, 1997), it is argued that Skhul/Qafzeh, as well as early modern Europeans, are likely tropical in origin and thus were the result of a migration of modern populations from their African homelands.

Proponents of both these views tend to consider Neandertals and modern humans as different species, overlapping each other at various times during the late Pleistocene in the Levant (Rak 1993). Most paleoanthropologists are willing to accept the possibility of some hybridization between these species. Indeed Bräuer (1989, 1992) has referred to his model of the origin of modern Eurasians as "replacement with hybridization." However, it is also clear from Bräuer's writings, and from the work of Stringer and others, that genetic contributions of Neandertals to early modern human populations would have been negligible at most.

One potentially critical specimen to the debate on the origin of the Skhul/Qafzeh people is the large, virtually complete mandible from Tabun, designated Tabun 2 or, more appropriately, Tabun C2 (Figure 1). A careful reading of McCown and Keith reveals that the symphysis and dentition of Tabun C2 played a significant role in connecting the more archaic Tabun C1 specimen with those from Skhul, resulting in their interpretation that the Mt. Carmel skeletons represented "a single people" (1939:12). While most researchers have typically considered Tabun C2 as a Neandertal (e.g., Wolpoff 1975; Trinkaus 1983, 1984; Smith 1985; Smith et al. 1989; Frayer et al. 1993), many of these workers have noted that the specimen's chin is better developed than is typical for Neandertals. In an early paper on the Tabun specimens, McCown (1934) implied that Tabun C1 and C2 represented two different human varieties, a view he subsequently abandoned (McCown and Keith 1937). Grün et al. (1991) mention the Tabun "Neandertals" but refer to Tabun C2 as enigmatic. As Wolpoff puts it: "Tabun 2, a male, has a prominent mental eminence and shows other characteristics that according to some much more closely align it with the Qafzeh sample than with the Levantine 'Neandertals'" (1995:606).

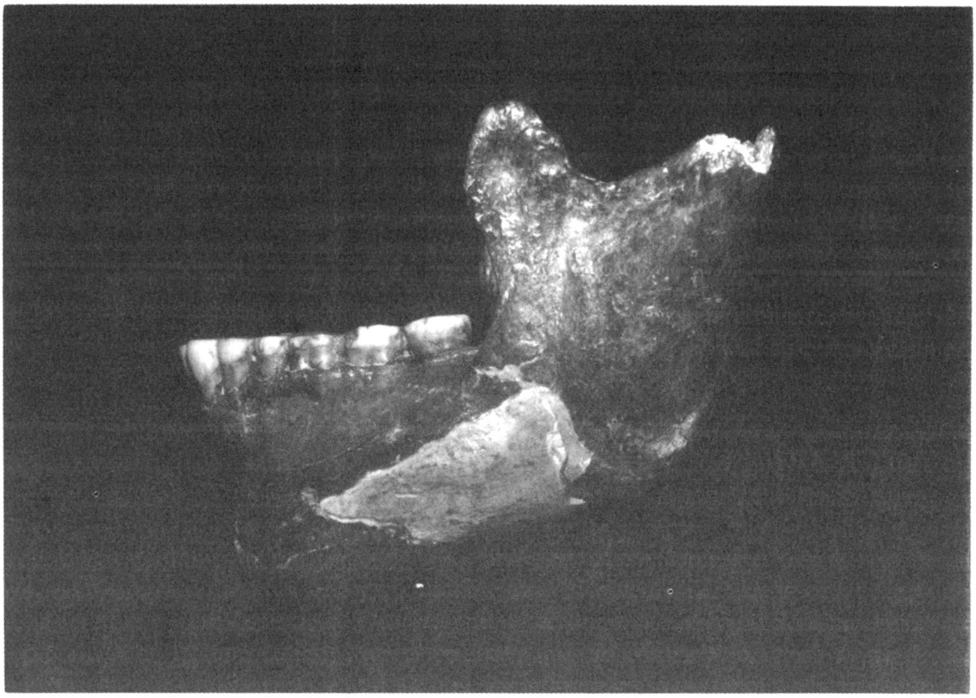

Figure 1. Left lateral view of Tabun C2. Note the symphyseal contours, retromolar space, and missing left condyle.

Vandermeersch (1981b) emphasizes the modern features of Tabun C2 and suggests that, in fact, it may represent an early modern human. Hublin and Tillier (1981) list it as an early modern Southwest Asian. More recently, Bar-Yosef and Pilbeam (1993), Tattersall (1995), and Rak (this volume) have specifically labeled C2 as an early modern human. On the other hand, Trinkaus (1995) notes that some features purported to align C2 with moderns, specifically a prominent mental trigone and mandibular incisure crest position relative to the condyle, are ambiguous and that the specimen exhibits clear archaic affinities in other morphological aspects. In his opinion, Tabun C2 should be viewed "as a late archaic human but one not necessarily aligned with the European Neandertals" (Trinkaus 1995:10).

Despite the critical importance of Tabun C2 and the recent renewed interest in it, the mandible continues to receive only relatively brief mention in most discussions of late Pleistocene human evolution, and no recent detailed studies (except for Stefan and Trinkhaus 1998 and Rak, this volume) have been focused on delineating the affinities of this specimen. While recognizing the limitations of putting too much weight on a single mandible, we present a morphometric assessment of Tabun C2 and an interpretation of its significance for understanding the origin of the Skhul/Qafzeh people.

3. HUMAN REMAINS FROM TABUN AND LATE PLEISTOCENE HUMAN EVOLUTION IN SOUTHWEST ASIA

The site of Tabun was excavated under the direction of D. A. E. Garrod from 1929 until 1934, yielding a complex sequence of archaeological assemblages ranging from the Lower Paleolithic through the Bronze Age (Garrod and Bate 1937). The portion of the sequence spanning the Lower and Middle Paleolithic has emerged as the reference sequence for this time span in Near Eastern prehistory. Changes in the chronology of the Tabun Paleolithic sequence have had significant impacts on perceptions of human evolutionary patterns, both biological and cultural, in this part of the world (Jelinek 1982, 1994; Lieberman 1993; Bar-Yosef 1989, 1992a, 1994). For example, Bar-Yosef (1989, 1992b, 1993, 1994), Lieberman (1993), and Lieberman and Shea (1994) have identified differences in lithic features and resource exploitation between the Mousterian assemblages in Tabun levels C and B. "Tabun C-type" Mousterian is found in Tabun level C and is also associated with the early modern Skhul/Qafzeh remains, while "Tabun B-type" is associated with Tabun level B and the Neandertal-bearing levels at Amud and Kebara (Bar-Yosef 1993, 1994). The suggestion that "C-type" Mousterian is the product of the Skhul/Qafzeh people, while the more recent "B-type" was made by Neandertals is a cornerstone of the "late migration hypothesis" of Near Eastern Neandertal origins.

Human remains were recovered from Tabun levels E (associated with the Acheulo-Yabrudian), C, and B/chimney (associated with the Mousterian). The remains unequivocally derived from level B consist of teeth and a maxillary fragment (McCown and Keith 1939) and are not diagnostic. The level E remains are Middle Pleistocene in age (Trinkaus 1984, 1995; Bar-Yosef 1994) and not pertinent to this discussion. It is the remains from (or purportedly from) Tabun level C that are critical to evaluating the pattern of late Pleistocene human evolution in the Levant.

Garrod and Bate (1937) and McCown and Keith (1939) attributed several human specimens to level C (Table 1). In her discussions of the excavation of this level, Garrod notes without equivocation that specimens Tabun C2 through C7 derive from level C (Garrod and Bate 1937). However for Tabun C1, she states that "the skeleton lay so near

Table 1. Human remains from Tabun level C

Specimen	McCown and Keith Designation (1939)	Howell and Fritz Designation (1975)
Partial skeleton	Tabun I	Tabun C1
Mandible	Tabun II	Tabun C2
Distal right femoral shaft	Tabun 1	Tabun C3
Distal right radius	Tabun 2	Tabun C4
Right hamate	Tabun 3	Tabun C5
Right pisiform	Tabun 4	Tabun C6
Proximal left pollical phalanx	Tabun 5	Tabun C7

the surface of C that the question must arise whether it does not represent a burial from level B" (Garrod and Bate 1937:64). Proponents of the late migration hypothesis have argued that a level B origin for this specimen is likely since "other archaic hominids have been found only in Tabun B contexts [at Amud and Kebara] and modern humans have been found only in Tabun C contexts—including a modern mandible from Tabun C" (Lieberman 1993:603). Acceptance that Neandertals are always associated with "Tabun B-type" and not with the earlier "Tabun C-type" Mousterian certainly would support the late migration hypothesis. Alternatively, Jelinek (1992) has suggested that, since the strata were dipping in the area of the cave where C1 was excavated, the specimen may have come even from the older level D!

Garrod and Bate (1937) note that the Tabun C3 through C7 human remains were excavated from the talus of level C. Trinkaus (1993a) has recently pointed out that the C4 right radius and C5 right hamate are virtually perfect mirror images of the left bones preserved in C1 and almost certainly derive from this specimen. Assuming that the C4 and C5 elements are indeed from within level C, there are two possible explanations for this situation. First, the original attribution of the C1 skeleton to level C is correct, and the C4 and C5 specimens were separated from the skeleton *before* level B was deposited. Second, the skeleton was either from level B proper or intrusive into C, and the C4 and C5 elements were later scattered on the surface of C and "worked in" to the top of the C deposits. Possible support for the first explanation comes from the Tabun C3 right femoral shaft. This specimen is morphologically similar to the Tabun C1 right femur and distinct from the Skhul/Qafzeh femora (Trinkaus 1976). Again assuming that the attribution of this second right femur to level C is correct, it would seem less likely (albeit certainly not impossible) that two individuals were introduced into level C in this way. Thus there is still potential evidence, apart from the C1 skeleton and C2 mandible, that archaic humans were present in the Near East during "Tabun C-type" Mousterian times. This interpretation would be invalidated if the attribution of the Tabun C3 through C7 specimens to level C is shown to be incorrect.

During the early 1980s, radiocarbon dates indicated an age for Tabun level C on the order of 50 ka (Jelinek 1982). More recently, several chronometric techniques have suggested a much greater age, but exactly how much greater is not yet clear. Early and linear uptake ESR dates of 102 ka and 119 ka, respectively, have been obtained from level C (Grün et al. 1991). McDermott et al. (1993) dated level C to 101.6 ka using mass-spectometric uranium series dating, basically corroborating the ESR (especially EU) age estimates. However, recent TL analysis by Mercier et al. (1995) estimate the level's age at 170 ka. This latter date would place the Tabun humans in the Near East long before the Skhul/Qafzeh people.

4. TABUN C2—A MORPHOLOGICAL ASSESSMENT

Tabun C2 was recovered in six pieces, "scattered over a small area outside the east alcove," and at a depth of 1.2 m below the upper limit of level C (Garrod and Bate 1937:64). The six pieces comprise a virtually complete mandible (Figure 1), lacking only the left condyle, a portion of the basilar symphysis (Figures 2 and 3), and a section of the left exterior corpus extending from the mental foramen to the anterior ramus. Other areas of damage are minor and do not affect morphometric assessment of the specimen. The dentition is complete except for the left lateral incisor.

As McCown and Keith (1939), Howell (1958), Trinkaus (1984, 1995), Wolpoff (1995), and others have noted, Tabun C2 is a large mandible with rugose, well-defined areas of muscle attachment for the masseter, temporalis, medial, and lateral pterygoid muscles. It is almost certainly a male. The anterior teeth are large, both absolutely and relative to posterior tooth size, and exhibit an accelerated rate of dental attrition (McCown and Keith 1939; Wolpoff 1970; Bytnar et al. 1994). These features tend to align the C2 mandible with archaic, rather than modern, humans. Our morphological assessment does not include the dentition and focuses on the development of the mental eminence (chin) and six other features (Table 2) that have figured prominently in attempts to distinguish Neandertals from modern humans. The late Pleistocene comparative samples used for our analysis are listed in Table 3.

The morphological feature of Tabun C2 which has been most influential in the fossil's assessment is clearly its degree of mental eminence (chin) development. Weidenreich (1904, 1936) noted that three anatomical characteristics must be evaluated in order to

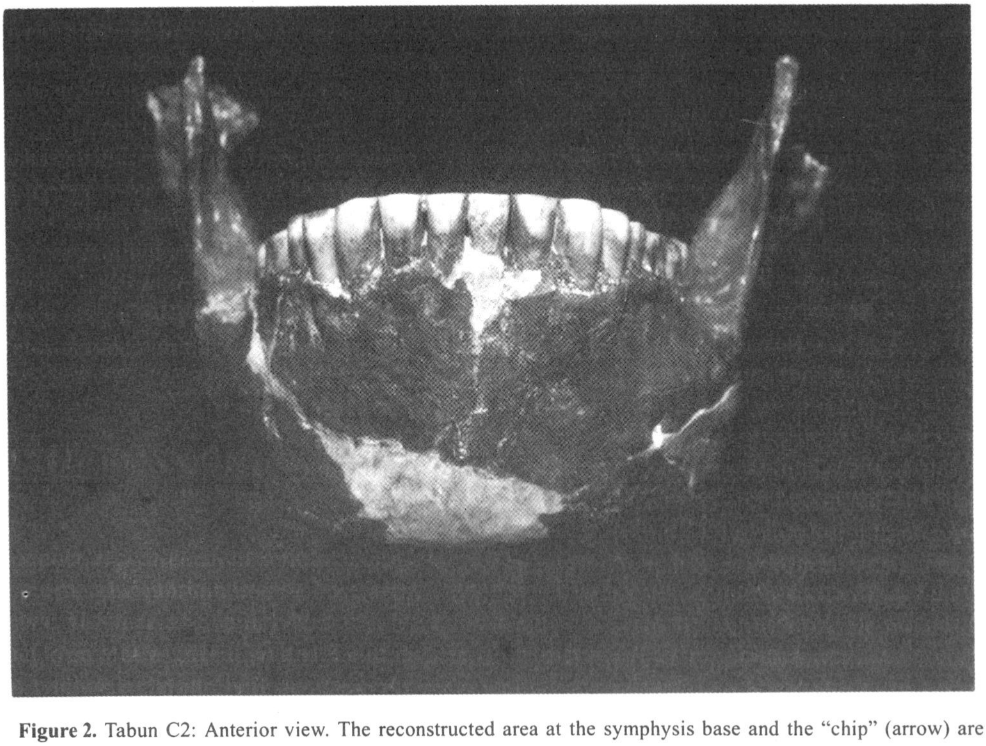

Figure 2. Tabun C2: Anterior view. The reconstructed area at the symphysis base and the "chip" (arrow) are evident.

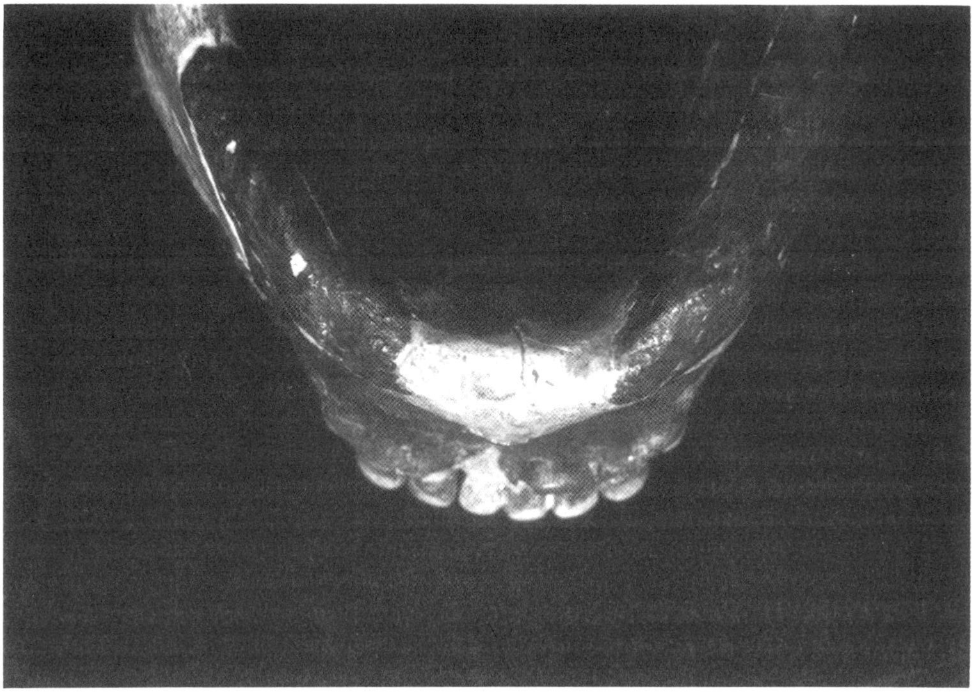

Figure 3. Basilar view of Tabun C2 showing the reconstruction and the bone contours superior to the reconstructed area.

judge the extent of "chin" development: the *incurvatio mandibulae* (im), *mentum osseum* (mo), and *trigonom mentale* (tm) (Figure 4; see also Virchow 1920). The *incurvatio* (im) is the concavity between the alveolar process margin and the basilar portion of the external symphysis (Figure 1). The *mentum osseum* (mo) is the anterior projection of the basilar symphysis that is always associated with formation of the im. According to Weidenreich, a *mentum osseum* exists when the basilar portion forms an equally rounded swelling across the symphysis (Figure 4). By contrast, a *trigonum mentale* occurs when the anterior projection of the basal symphysis exhibits a distinctly triangular shape when viewed from the front, with a vertex that extends superiorly well above the base at the symphysis midline (Figure 4). The central portion of the tm usually projects the most anteriorly, creating the triangular shape of the anterior mandibular corpus when viewed in basilar perspective (see below and Table 2). With the development of a tm, distinct depressions are present on

Table 2. Morphological features

- Anterior Basal Corpus Shape: (square or triangular) when viewed from the basal aspect; this feature is also a reflection of basilar symphysis development.
- Intersection of the Mandibular Incisure Crest at Condyle: (central or lateral).
- Retromolar Space: (present or absent).
- Position of the Mental Foramen Relative to the Tooth Row: (foramen located under P3, P3/P4 septum, P4, P4/M1 septum, or M1).
- Medial Pterygoid Tubercle: (present or absent) attachment area for the medial pterygoid muscle.
- Mandibular Foramen Shape: (horizontal–oval or normal) probably reflects attachment variants of the stylomandibular ligament (see Smith 1978).

Table 3. Early Modern Human and Neandertal samples examined for this study

Early Modern Humans (Europe)	Neandertals (Europe)
Cro-Magnon 1	La Ferrassie 1
La Chancelade	La Chapelle-aux-Saints
Combe-Capelle	La Quina H9
Předmostí 3	St. Césaire
Předmostí 4	La Naulette
Dolní Věstonice 3	Regourdou 1
Pavlov	Guattari 2
Brno 3	Guattari 3
Early Modern Humans (Southwest Asia)	Spy 1
Skhul 4	Subalyuk
Skhul 5	Krapina 57
Qafzeh 9	Krapina 58
Neandertals (Southwest Asia)	Krapina 59
Amud 1	Vindija 206
Kebara 2	Vindija 226
Tabun C1	Vindija 231

either side of the triangular structure above the basilar portion. Weidenreich (1936:37) termed these depressions the *fossa mentalis*. These fossae are delineated inferiorly by development of the lateral tubercles of the tm, as well as the anterior marginal tubercles, and medially by the tm itself. Weidenreich (1904, 1936) posited that an mo can exist without a tm, but that both will be associated always with an im.

Unfortunately Tabun C2 does not preserve the complete anterior base of the symphysis (Figure 2), so that the exact morphology of the specimen in this critical region is difficult to determine. However, the preserved bone just superior to the missing portion gives a reliable indication of some aspects of C2's morphology. The bone along the inferior margin curves sharply anteriorly just before the break, indicating strong development of the anterior marginal tubercle and a well-developed mental fossa on the left side. The bone is not preserved inferiorly enough on the right side to be certain, but such features tend to be bilaterally developed.

At the midline, there is a small "chip" of bone just superior to the break (Figure 2) that is displaced slightly anteriorly relative to the surrounding bone, particularly on the right side. This leads to a potential interpretation, especially based on available casts of

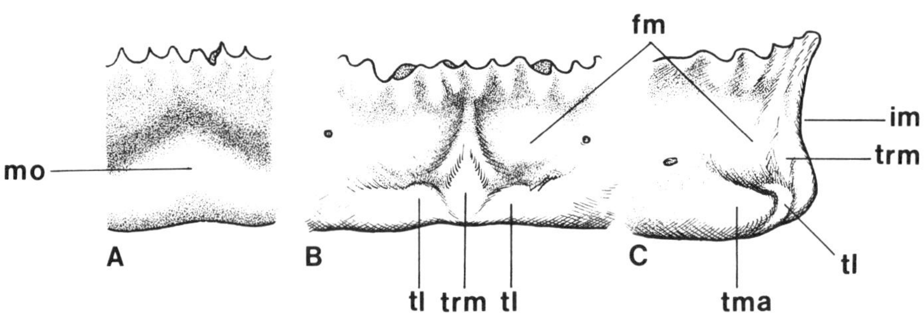

Figure 4. Features of the mandibular symphysis (adapted from Weidenreich 1936). A: *mentum osseum* (anterior view); B: *trigonum mentale* (anterior view); C: Lateral symphyseal view with a mental trigone. Abbreviations: mo - *mentum osseum*; trm - *trigonom mentale*; tl - *tuberculum laterale*; im - *incurvatio mandibulae*; tma - *tuberculum marginale anterius*; fm - *fossa mentalis*.

C2, that a distinct mental trigone is present; but the medial ridge that extends toward the alveolar process in an unequivocal tm (Figure 4) is not detectable on C2. However, when viewed from the basilar perspective, and taking into account the distortion introduced by the "chip," the bony contours still appear to come to an anterior point (Figure 3) that is more consistent with a mental trigone than an mo in Weidenreich's scheme. Also, well-developed marginal fossae and anterior marginal tubercles, like those exhibited by Tabun C2, appear to be consistently associated with mental trigones in other specimens. Be this as it may, the preserved morphology of C2 does not, in our opinion, allow an unequivocal claim for the presence of a mental trigone on this specimen.

The contours preserved on the existing aspects of Tabun C2's symphysis demonstrate the following: (1) a distinct anterior projection of the symphysis base at least consistent with Weidenreich's *mentum osseum*; (2) a well-developed *incurvatio mandibulae*; and (3) indirect indications of a mental trigone. In each of these features, Tabun C2 approaches the condition characteristic of early modern mandibles, including those from Skhul and Qafzeh, to a greater extent than the Neandertal mandibles with which we are familiar, including Vindija 206 and 231, La Quina H9, and Guattari 2 and 3. The external mandibular symphysis of Tabun C2 is consistently more like comparable areas of Skhul 4, Skhul 5, and Qafzeh 9 than are those of Amud 1, Kebara, or Tabun C1.

With respect to other features (Table 3), none is an unequivocal indicator of Neandertal versus modern human affinity. For anterior basal corpus shape (viewed from the basilar perspective), it is true that Neandertals tend to have a blunted or squared contour while modern specimens tend to be more triangular, due to the anterior projection of the mental trigone. But there is some overlap in both directions with the Guattari mandibles exhibiting a triangular profile and Qafzeh 9 a square profile. Furthermore, 6 of 42 recent Iraqi mandibles examined exhibited a square profile (Quam 1996). Similarly, while possession of horizontal-oval (H-O) mandibular foramina, presence of retromolar spaces, and the more posterior position of mental foramina (e.g., under M1 or P4/M1 septum) have clearly higher frequencies among Neandertals (Quam 1996), there is considerable overlap in both directions for each characteristic (Smith 1976, 1978; Frayer 1992; Trinkaus 1993b; Franciscus and Trinkaus 1995). Thus none of these features, particularly when considered independently, can unequivocally indicate a Neandertal or an early modern human affinity for any specimen. Nonetheless, results of Chi-square and Fisher's exact tests (Quam 1996) on the samples in Table 3 support the contention that Neandertals tend to have square anterior basal contours, more posteriorly-placed mental foramina, retromolar spaces, and H-O mandibular foramina.

Tabun C2 appears to exhibit a rather triangular basal corpus profile at the symphysis, similar to the condition more typical of modern humans. On the other hand, C2 also has clear retromolar spaces and H-O mandibular foramina, both of which are more common in Neandertals.

Two additional features have been claimed to represent Neandertal apomorphies by Rak and his colleagues: the medial pterygoid tubercle and a central intersection of the mandibular incisure on the condyle (Rak et al. 1994; Rak and Kimbel 1995; Rak, this volume). In our study, 7 of 10 Neandertals have the central intersection, while all of the early modern and 35 of 38 recent Iraqi mandibles exhibit the lateral intersection. Results of Chi-square ($\chi^2 = 23.4$, $p < .001$) and Fisher's exact test ($p < .0001$) support this distinction. However, Trinkaus (1995) notes that if crest position is considered in terms of the articular head only, a clear distinction between Neandertals and modern humans does not exist. Suzuki (1970), Smith (1976), and Trinkaus (1995) suggest that the central crest position is due to hypertrophy of the lateral tubercle of the fibrous capsule of the temporomandibular joint in Neandertals. This hypertrophy is likely related to the rather unique masticatory apparatus of the Neandertals.

Medial pterygoid tubercles are also variable in their development but tend to be clearly present in Neandertals (7 of 9 in our study) and much less common in modern humans (1 of 10 early modern and 10 of 41 recent Iraqi mandibles). Richards and Plourde (1995) found similar instances in five comparative samples of recent humans. Chi square ($\chi^2 = 12.3$; $p < .005$) and Fisher's exact ($p < .005$) support the distinction of Neandertal and modern humans based on this feature but less robustly than for some other features (see above).

With respect to these two characters, Tabun C2 exhibits a "mixed" pattern. The specimen lacks the left condyle but has a lateral mandibular incisure intersection on the right side. The lateral aspect of the right condyle is also missing, but we do not believe the missing portion affects the assessment of this feature. On the other hand, C2 possesses distinct medial pterygoid tubercles on both sides.

From a morphological standpoint, comparison of the C1 and C2 mandibles from Tabun is particularly interesting. While C1 conforms to the "Neandertal" pattern, C2 obviously does not exhibit so consistent a pattern. Taken together C1 and C2 represent a sample with a range of morphological variation that is not found at other middle or late Pleistocene sites where multiple mandibles are preserved. This is the case, for example, in the much earlier sample from Atapuerca (Rosas 1995), the penecontemporaneous Krapina mandibles (Smith 1976), and the late Neandertals from Vindija (Wolpoff et al. 1981). Even the Klasies River Mouth sample of mandibles, though morphologically quite variable (Rightmire and Deacon 1991; Smith 1993; Lam et al. 1996), do not match Tabun in morphological range. Thus whatever the explanation for the pattern at Tabun, it would not appear to result from sexual dimorphism, as suggested by McCown and Keith (1937).

5. TABUN C2—A METRICAL ASSESSMENT

A series of 42 measurements were obtained on the specimens listed in Table 3, and six indices were calculated for these specimens using the appropriate measurements. Measurements were either derived from the literature or taken directly on original fossils and casts. Cast measurements were subsequently checked on original fossils. Because not all of these specimens are complete, some measurements and indices could not be taken on some specimens. No estimates of missing data were made. T-tests were applied to variables conforming to a normal distribution, and a non-parametric version of the t-test, the Wilcoxon rank transformation, was applied to variables not conforming to a normal distribution. Only the 15 variables (Table 4) that demonstrated a significant difference ($p < .05$) between the Neandertal and early modern samples were used in subsequent analyses.

These 15 variables were subjected to a series of multiple descriminant function analyses. Discriminant function analysis uses continuous variables to classify observations into two or more groups and can be used with both normally and non-normally distributed data. The groups are defined beforehand and each observation (specimen) is placed into one of the groups on the basis of these descriminant functions. Some observations may be misclassified into the wrong group, and the number of misclassifications is one measure of the quality of the discriminant function.

Ideally, a separate data set is used to determine the discriminant functions which are then applied to the data set under study. Since only one data set (the fossils listed in Table 3) was available in the present study, the discriminant functions were determined directly

Table 4. Variables exhibiting statistically significant difference at the $p<.05$ level between Neandertals and early moderns in western Eurasia[a]

1.	Height of mental foramen above basal corpus (M>N)
2.	Bi-mental foramen breadth (N>M)
3.	Bi-condylar breadth (N>M)
4.	Angle of the symphysis to the alveolar plane (M>N)
5.	Angle of the chin to the alveolar plane (M>N)
6.	Infradental projection from C/P3 septum plane (N>M)
7.	Pogonion projection (M>N)
8.	Maximum depth of the *incurvatio mandibulae* (M>N)
9.	Corpus breadth at M1/M2 septum (N>M)
10.	Corpus breadth at M2/M3 septum (N>M)
11.	Corpus thickness at I2/C septum (N>M)
12.	Corpus thickness at P3/P4 septum (N>M)
13.	Mandibular index [total mandibular length:bi-condylar breadth] (M>N)
14.	Pogonion index [pogonion projection:infradental projection] (M>N)
15.	Length of retromolar space at occlusal plane (N>M)

[a]N=Neandertals, M=early moderns.

from it. This has the unfortunate result of producing optimistic error rates. There are certain techniques that can correct for this, such as cross-validation, which compares each observation in the data set to all the other observations in the data set and produces a nearly unbiased error estimate. The error estimates derived in cross-validation are generally higher, and more specimens are misclassified. Since it is more conservative, this technique will be used here.

Discriminant function analysis also takes into consideration the prior probability of a certain observation belonging to a particular group. This is determined by the sample sizes of each group. For example, if the Neandertals were represented by 15 observations and the early moderns by only five, then, by default, there is a greater chance that any particular observation is a Neandertal. Considering the prior probabilities of each group produces a more accurate discriminant function.

When generating several discriminant functions for comparative purposes, one factor which could bias the results is the actual specimens which comprise the data set. This is especially important in paleontology since more often than not, paleontological specimens are fragmentary. Some specimens may preserve the symphyseal area while others may only be represented by the ramus. If one is creating a discriminant function based on the symphyseal area, the actual specimens used may be different from those used to calculate a discriminant function based on the ramus.

In our study, six multiple-variable discriminant functions were calculated, using only those variables which showed a significant difference between the Neandertals and early moderns. The formulation of these discriminant functions proceeded in a series of steps, beginning with each variable being considered separately. In general, single variables are not powerful discriminant functions, and this was done to get a preliminary idea of the relative strength of each variable. These single variables were then combined into logical groupings to form multiple-variable discriminant functions. Finally, all the variables which showed significant differences between the Neandertals and early moderns were combined into one discriminant function which reflects the gross differences between these two groups. For purposes of brevity, only the two most inclusive discriminant functions (i.e., those based on the most variables) will be presented here. The other, less robust functions are discussed in Quam (1996).

Table 5. Discriminant functions

Function	Variables (Table 2)	No. of specimens	Error rate (%, N) (crossvalidation)	Tabun C2?	Misclassified specimens
Symphysis	4, 5, 6, 7, 8, 14	17 Neandertals 10 E. Moderns Tabun C2	Neandertals (11.8%, 2) E. Moderns (20%, 2) Total (14.8%, 4)	E. Modern	La Ferrassie 1 Guattari 3 Předmostí 3 Combe Capelle
Combined	All but 3, 13	8 Neandertals 7 E. Moderns Tabun C2	None	E. Modern	None

Results of the two most inclusive discriminant functions, the symphysis and the combined variables discriminant functions, are given in Table 5. Of the two, the symphyseal discriminant function is the less robust, misclassifying four specimens (14.8%) using the conservative cross-validation method. This is not surprising, given that Neandertals exhibit variability in terms of symphyseal morphology, with specimens from the sites of Vindija, Guattari, and others having aspects of more modern human symphyseal morphology (Smith 1984, 1991; Wolpoff 1995). This discriminant function classifies Tabun C2 as an early modern human (see Table 5), but this could be questioned due to the incompleteness of C2's anterior basal symphysis. Specifically it might be argued that an inaccurate reconstruction makes the region more anteriorly projecting than it actually was. As we note above, however, the basal symphysis seems very accurately reconstructed, and the reconstruction is fully commensurate with the preserved bone contours. If anything, the degree of anterior projection of the basal symphysis is underestimated in the reconstruction, which could foster classification of the specimen as a Neandertal.

The combined variable discriminant function uses fewer specimens because inclusion requires greater completness of each individual specimen. This discriminant function does not misclassify a single specimen, indicating that it is a robust, effective discriminator. This function also includes all of the symphyseal variables that form the basis of the symphysis discriminant function. The specimens from Vindija are not used for the combined discriminant function because the specimens are not complete enough. As with the symphyseal discriminant function, the combined variable function also classifies Tabun C2 as an early modern human (Table 5).

6. TABUN C2 AND THE ORIGIN OF MODERN EURASIANS

Both the metric and morphological aspects of our study show that Tabun C2 has distinct affinities with early modern western Eurasians. These affinities are most evident in the symphysis, as Tabun C2 has a more pronounced anterior projection of the basal symphysis and development of an *incurvatio mandibulae* than any other western Eurasian archaic human mandible. Whether the former feature comprises a *mentum osseum* or a mental trigone is debatable, because much of the critical region is reconstructed or obscured by the previously-discussed bone "chip." While there are significant indirect indications that a trigone is present, it is impossible to be certain.

Regardless, the symphyseal region of Tabun C2 is surprisingly like early modern humans considering its age of 120 ka or older. The only other western Old World specimen with a possibly comparable antiquity and similar degree of a "chin" development is the

Omo Kibish 1 skeleton from East Africa, whose fragmentary mandible exhibits a distinct mental trigone (Bräuer 1984). However, the date of 130 ky on Omo 1 is problematic (Smith 1993). The Klasies River Mouth Cave mandibles also have been claimed to be fully anatomically modern (Rightmire and Deacon 1991) but are maximally about 100 ka and perhaps younger (Smith et al. 1989). Thus they post-date Tabun C by at least 20 ka.

In addition to the symphysis, the lateral intersection of the mandibular incisure crest on the condyle and the combined variable discriminant function also indicate early modern affinities for Tabun C2. Thus we feel that Tabun C2 is the most convincingly modern-like mandible from anywhere in the western Old World with an unquestioned ancestry of 120 ka or more.

Despite the results of our study, designating this specimen as a modern human ignores the archaic morphology it exhibits. Of particular note in this context are the morphology and metric patterning of the dentition (Bytnar et al. 1994), features reflecting a more anterior placement of the alveolar process relative to the basal corpus (i.e., presence of retromolar spaces and placement of the mental foramen), and the presence of medial pterygoid tubercles. In fact, Trinkaus (1995) and Stefan and Trinkhaus (1998) argue that Tabun C2 should continue to be considered an archaic human specimen. While we would not necessarily disagree with Trinkaus' position, we believe the implications of Tabun C2's similarity to definite early modern mandibles in western Eurasia merits additional consideration.

One possibility is that Tabun C2's similarity to early modern humans represents a parallelism and is not relevant to the issue of the origin of modern people. However, since Tabun C2 is the earliest unequivocally dated specimen exhibiting this morphology, the parallelism argument is suspect. Such an argument is normally not used for the earliest appearance of a morphological pattern unless there is strong corroborating evidence for parallelism. No such evidence exists in this case.

A second possibility is that Tabun C2 represents a "stage" in the transition to modern humans in the western Old World. In this case, this "stage" might be descended from an essentially regional ancestor or the result of influence (even migration) from elsewhere, probably Africa. In our opinion, it is difficult to decide between these two alternatives given the available evidence. For example, our knowledge of pre-Mousterian human morphology in west Asia is limited, and there are no diagnostic mandibles. Could the Tabun C people have evolved from something like Zuttiyeh and represent a population "in the throes" of evolving into modern humans, rather than Neandertals as McCown and Keith suggested? Alternatively, does Tabun C2 represent the extension of an "African transitional group" (cf. Bräuer 1984; Smith et al. 1989) into the Near East as a part of a recent African origin model? Because there are no adult mandibles (but see below) unequivocally associated with the African transitional group, this possibility is difficult to evaluate objectively.

Finally, the possibility that Tabun C2's morphology is the result of hybridization cannot be completely dismissed. It is difficult, perhaps impossible, to distinguish between a transitional specimen and a hybrid specimen in many situations. However, given that there are no earlier unequivocally more modern source populations known, this possibility becomes less viable as an explanation for the more modern features seen in Tabun C2. Should it be established in the future that earlier modern populations do exist in Africa or Eurasia, hybridization would need to be reconsidered as a distinct possibility.

Whichever of these explanations for the origin of Tabun C2's morphology is preferred, two things are apparent to us. First, the distinct similarity of Tabun C2 to the Skhul and Qafzeh mandibles provides a strong argument for a phylogenetic connection between them. Second, it is possible to present a case that some of the archaic aspects of Tabun C2 reflect a Neandertal contribution to its ancestry. For example, the only humans known to exhibit retromolar spaces are either Neandertals or humans possibly descendant from them (Franciscus and Trinkaus 1995). Earlier human mandibles (e.g., Ternifine, Mauer, Arago) lack them, as does the only adult mandible *possibly* associated with the African transitional group, the Mousterian-associated Haua Fteah 1 speciman from Libya (Tobias 1967). Thus whatever the factors responsible for the occurrence of the retromolar space (Trinkaus 1987; Franciscus and Trinkaus 1995), its presence in non-Neandertals can be interpreted as a *possible* indicator of Neandertal contributions to their ancestry. A similar argument also can be made for the presence of the medial pterygoid tubercle in Tabun C2, if Rak et al. (1994) are correct that it is a Neandertal marker.

Because Tabun C2 is represented only by a single mandible and its morphological pattern is essentially unique, controversy is certain to continue regarding its phylogenetic significance. However, the morphological pattern of this specimen is yet another example of the apparent complexity of population relationships during the late Pleistocene, especially in the circum-Mediterranean region (Smith et al. 1995). In our opinion, Tabun C2 is currently the most modern-like human fossil known with a geological age unequivocally older than 120 ka and is a much more critical piece in the puzzle of the origin of modern western Eurasians than has been generally recognized.

ACKNOWLEDGMENTS

We thank the several museums and other institutions that have allowed analysis of the specimens used for this study. We are especially grateful to Y. Rak, B. Arensburg, and J. Zias for access to particularly critical specimens and Northern Illinois University for financial support.

REFERENCES

Aiello, L. C. 1993. The fossil evidence for modern human origins in Africa: A revised view. *American Anthropologist* 95:73–96.

Arensburg, B. 1989. New skeletal evidence concerning the anatomy of Middle Paleolithic populations in the Middle East: The Kebara skeleton, 165–71. In *The Human Revolution*, ed. P. A. Mellars and C. B. Stringer. Edinburgh: Edinburgh University Press.

Bar-Yosef, B. 1989. Geochronology of the Levantine Middle Paleolithic, 589–610. In *The Human Revolution*, ed. P. A. Mellars and C. B. Stringer. Edinburgh: Edinburgh University Press.

Bar-Yosef, O. 1992a. Middle Paleolithic chronology and the transition to the Upper Paleolithic in Southwest Asia, 261–72. In *Continuity or Replacement. Controversies in* Homo sapiens *Evolution*, ed. G. Bräuer and F. H. Smith. Rotterdam: A. A. Balkema.

Bar-Yosef, O. 1992b. Middle Paleolithic adaptations in the Mediterranean Levant, 189–215. In *The Evolution and Dispersal of Modern Humans in Asia*, ed. T. Akazawa, K. Aoki, and T. Kimura. Tokyo: Hokusen-Sha.

Bar-Yosef, O. 1993. The role of western Asia in modern human origins, 132–47. In *The Origin of Modern Humans and the Impact of Chronometric Dating*, ed. M. Aitken, C. Stringer, and P. Mellars. Princeton: Princeton University Press.

Bar-Yosef, O. 1994. The contributions of southwest Asia to the study of the origin of modern humans, 23–66. In *Origins of Anatomically Modern Humans*, ed. M. H. and D. V. Nitecki. New York: Plenum Press.

Bar-Yosef, O., and D. Pilbeam. 1993. Dating hominid remains. *Nature* 366:415.

Bar-Yosef, O., and B. Vandermeersch. 1981. Notes concerning the possible age of the Mousterian layers in Qafzeh cave, 281–6. In *Préhistoire du Levant*, ed. J. Cauvin and P. Sanlaville. Paris: CNRS Editions.

Bar-Yosef, O., and B. Vandermeersch. 1993. Modern humans in the Levant. *Scientific American* 268:94–100.

Bräuer, G. 1984. A craniological approach to the origin of anatomically modern humans in Africa and implications for the appearances of modern Europeans, 327–410. In *The Origins of Modern Humans: A World Survey of the Fossil Evidence*, ed. F. H. Smith and F. Spencer. New York: Alan R. Liss.

Bräuer, G. 1989. The evolution of modern humans: A comparison of the African and non-African evidence, 123–54. In *The Human Revolution*, ed. P. A. Mellars and C. B. Stringer. Edinburgh: Edinburgh University Press.

Bräuer, G. 1992. Africa's place in the evolution of Homo sapiens, 83–98. In *Continuity or Replacement: Controversies in* Homo sapiens *Evolution*, ed. G. Bräuer and F. H. Smith. Rotterdam: A. A. Balkema.

Butzer, K., P. Beaumont, and J. Vogel. 1978. Lithostratigraphy of Border Cave, Kwa Zulu, South Africa: A Middle Stone Age sequence beginning ca. 195,000 BP. *Journal of Archaeological Science* 5:317–41.

Bytnar, J., E. Trinkaus, and A. B. Falsetti. 1994. A dental comparison of Middle Paleolithic Near Eastern hominids. *American Journal of Physical Anthropology Supplement* 18:63 (abstract).

Coon, C. S. 1939. *The Races of Europe*. New York: Macmillan.

Dobzhansky, T. 1944. On species and races of living and fossil man. *American Journal of Physical Anthropology* 2:251–65.

Franciscus, R., and E. Trinkaus. 1995. Determinants of retromolar space presence in Pleistocene Homo mandibles. *Journal of Human Evolution* 28:577–95.

Frayer, D. W. 1992. Evolution at the European edge: Neanderthal and Upper Paleolithic relationships. *Préhistoire Européenne* 2:9–69.

Frayer, D. W., M. H. Wolpoff, A. G. Thorne, F. H. Smith, and G. G. Pope. 1993. Theories of modern human origins: The paleontological test. *American Anthropologist* 95:14–50.

Garrod, D. A. E., and D. M. A. Bate. 1937. *The Stone Age of Mount Carmel*, Vol. 1, *Excavations at the Wady el-Mughara*. Oxford: Clarendon Press.

Grün, R., C. B. Stringer, and H. Schwarcz. 1991. ESR dating of teeth from Garrod's Tabun cave collection. *Journal of Human Evolution* 20:231–48.

Grün, R., and C. B. Stringer. 1991. Electron spin resonance dating and the evolution of modern humans. *Archaeometry* 33:153–99.

Holliday, T. W. 1997. Body proportions in late Pleistocene Europe and Modern human origins. *Journal of Human Evolution* 32:423–48.

Holliday, T. W., and A. B. Falsetti. 1995. Lower limb length of European early modern humans in relation to mobility and climate. *Journal of Human Evolution* 29:141–53.

Howell, F. C. 1958. Upper Pleistocene man of the southwest Asian Mousterian, 185–98. In *Hundert Jahre Neanderthaler*, ed. G. H. R. von Koenigswald. Utrecht: Kemink en Zoon.

Howell, F. C. 1959. Upper Pleistocene stratigraphy and early man in the Levant. *Proceedings of the American Philosophical Society* 103:1–65.

Howell, F. C. 1984. Introduction, xiii–xxii. In *The Origins of Modern Humans: A World Survey of the Fossil Evidence*, ed. F. H. Smith and F. Spencer. New York: Alan R. Liss.

Howell, F. C., and M. C. Fritz. 1975. Israel, 125–54. In *Catalogue of Fossil Hominids. Part III: Americas, Asia, Australia*, ed. K. P. Oakley, B. G. Campbell, and T. I. Molleson. London: Trustees of the British Museum (Natural History).

Howells, W. W. 1976. Explaining modern man: Evolutionists versus migrationists. *Journal of Human Evolution* 5:477–95.

Hublin, J. J., and A. M. Tillier. 1981. The Mousterian juvenile from Irhoud (Morocco): A phylogenetic interpretation, 167–85. In *Aspects of Human Evolution*, ed. C. B. Stringer. London: Taylor and Francis.

Jelinek, A. 1982. The Tabun cave and Paleolithic man in the Levant. *Science* 216:1369–75.

Jelinek, A. 1992. Problems in the chronology of the Middle Paleolithic and the first appearances of early modern Homo sapiens in southwest Asia, 253–75. In *The Evolution and Dispersal of Modern Humans in Asia*. Tokyo: Hokusen-sha.

Jelinek, A. 1994. Hominids, energy, environment, and behavior in the late Pleistocene, 68–89. In *Origins of Anatomically Modern Humans*, ed. M. H. Nitecki and D. V. Nitecki. New York: Plenum Press.

Kramer, A., T. Crummett, and M. Wolpoff. In press. Morphological diversity in the Upper Pleistocene hominids of the Levant: Two species? *L'Anthropologie*.

Lam, Y. M., O. M. Pearson, and C. M. Smith. 1996. Chin morphology and sexual dimorphism in the fossil hominid mandibular sample from Klasies River Mouth. *American Journal of Physical Anthropology* 100:545–57.

Lieberman, D. E. 1993. The rise and fall of seasonal mobility among hunter-gatherers. *Current Anthropology* 34:599–631.

Lieberman, D. E., and J. J. Shea. 1994. Behavioral differences between archaic and modern humans in the Levantine Mousterian. *American Anthropologist* 96:300–32.

McCown, T. 1934. The oldest complete skeletons of man. *Bulletin of the American School of Prehistoric Research* 10:12–19.

McCown, T., and A. Keith. 1939. *The Stone Age of Mount Carmel*, Vol. 2, *The Fossil Human Remains from the Levalloiso-Mousterian*. Oxford: Clarendon.

McDermott, F., R. Grün, C. B. Stringer, and C. J. Hawkesworth. 1993. Mass-spectrometric u-series dates for Israeli Neanderthal/early modern hominid sites. *Nature* 363:252–4.

Mercier, N., H. Valladas, G. Valladas, J. L. Reyss, A. Jelinek, L. Meignen and J. L. Joron. 1995. TL dates of burnt flint from Jelinek's excavations at Tabun and their implications. *Journal of Archaeological Science* 22:495–509.

Nei, M. 1985. Human evolution at the molecular level, 41–64. In *Population Genetics and Molecular Evolution*, ed. T. Ohta and K. Aoki. Tokyo: Japan Science Society.

Quam, R. 1996. *Tabun Too? A Metric Comparison of Upper Pleistocene Mandibles*. MA Thesis. Northern Illinois University. DeKalb, IL.

Rak, Y. 1993. Morphological variation in *Homo neanderthalensis* and *Homo sapiens* in the Levant: A biographic model, 523–35. In *Species, Species Concepts and Primate Evolution*, ed. W. H. Kimbel and L. B. Martin. New York: Plenum Press.

Rak, Y., and W. Kimbel. 1995. Diagnostic Neandertal characteristics in the Amud 7 infant. *American Journal of Physical Anthropology Supplement* 20:177–8 (abstract).

Rak, Y., W. H. Kimbel, and E. Hovers. 1994. A Neandertal infant from Amud cave, Israel. *Journal of Human Evolution* 26:313–24.

Richards, G., and A. Plourde. 1995. Reconsideration of the "Neandertal" infant, Amud 7. *American Journal of Physical Anthropology Supplement* 20:180–1 (abstract).

Rightmire, G. P. 1984. *Homo sapiens* in Sub-Saharan Africa, 295–325. In *The Origins of Modern Humans. A World Survey of the Fossil Evidence*, ed. F. H. Smith and F. Spencer. New York: Alan R. Liss.

Rightmire, G. P. 1989. Middle Stone Age human remains from eastern and southern Africa, 109–22. In *The Human Revolution*, ed. P. A. Mellars and C. B. Stringer. Edinburgh: Edinburgh University Press.

Rightmire, G. P., and H. Deacon. 1991. Comparative studies of late Pleistocene human remains from Klasies River Mouth, South Africa. *Journal of Human Evolution* 20:131–56.

Rosas, A. 1995. Seventeen new mandibular specimens from the Atapuerca/Ibeas Middle Pleistocene hominid sample. *Journal of Human Evolution* 28:533–59.

Schwarcz, H., R. Grün, B. Vandermeersch, O. Bar-Yosef, H. Valladas, and E. Tchernov. 1988. ESR dates for the hominid burial site of Qafzeh in Israel. *Journal of Human Evolution* 17:733–7.

Simmons, T., A. B. Falsetti, and F. H. Smith. 1991. Frontal bone morphometrics of southwest Asian Pleistocene hominids. *Journal of Human Evolution* 20:249–69.

Smith, F. H. 1976. *The Neandertal Remains from Krapina: A Descriptive and Comparative Study*. Knoxville: Department of Anthropology, University of Tennessee.

Smith, F. H. 1978. Evolutionary significance of the mandibular foramen area in Neandertals. *American Journal of Physical Anthropology* 48:523–31.

Smith, F. H. 1984. Fossil hominids from the Upper Pleistocene of Central Europe and the origin of modern Europeans, 137–209. In *The Origins of Modern Humans: A World Survey of the Fossil Evidence*, ed. F. H. Smith and F. Spencer. New York: Alan R. Liss.

Smith, F. 1985. Continuity and change in the origin of modern *Homo sapiens*. *Zeitschrift für Morphologie und Anthropologie* 75:197–222.

Smith, F. H. 1991. The Neandertals: Evolutionary dead ends or ancestors of modern people? *Journal of Anthropological Research* 47:219–38.

Smith, F. H. 1993. Models and realities in modern human origins: The African fossil evidence, 234–48. In *The Origin of Modern Humans and the Impact of Chronometric Dating*, ed. M. Aitken, C. Stringer, and P. Mellars. Princeton: Princeton University Press.

Smith, F. H., A. B. Falsetti, and S. M. Donnelly. 1989. Modern human origins. *Yearbook of Physical Anthropology* 32:35–68.

Smith, F. H., A. B. Falsetti, and T. Simmons. 1995. Circum-Mediterranean biological connections and the pattern of late Pleistocene human evolution, 197–207. In *Man and Environment in the Paleolithic*, ed. H. Ullrich. Liège: ERAUL 62.

Stefan, V., and E. Trinkhaus. 1998. Discrete trait and dental morphometric affinities of the Tabun 2 mandible. *Journal of Human Evolution* 34:443–68.

Stringer, C. B. 1978. Some problems in Middle and Upper Pleistocene hominid relationships, 395–418. In *Recent Advances in Primatology*, Vol. 3, ed. D. J. Chivers and K. Joysey. London: Academic Press.

Stringer, C. B. 1990. The emergence of modern humans. *Scientific American* 263:98–104.

Stringer, C. B. 1994. Out of Africa—A personal history, 150–72. In *Origins of Anatomically Modern Humans*, ed. M. H. and D. V. Nitecki. New York: Plenum Press.

Stringer, C. B., and P. Andrews. 1988. Genetic and fossil evidence for the origin of modern humans. *Science* 239:1263–8.

Stringer, C. B., F. C. Howell, and J. Melentis. 1979. The significance of the fossil hominid skull from Petralona. *Journal of Archaeological Science* 6:235–53.

Stringer, C. B., J. J. Hublin, and B. Vandermeersch. 1984. The origin of anatomically modern humans in western Europe, 51–136. In *The Origins of Modern Humans: A World Survey of the Fossil Evidence*, ed. F. H. Smith and F. Spencer. New York: Alan R. Liss.

Suzuki, H. 1970. The skull of the Amud man, 123–206. In *The Amud Man and His Cave Site*, ed. H. Suzuki and F. Takai. Tokyo: Academic Press of Japan.

Tattersall, I. 1995. *The Last Neanderthal*. New York: MacMillan.

Tchernov, E. 1981. The biostratigraphy of the Middle East, 67–98. In *Préhistoire du Levant*, ed. J. Cauvin and P. Sanlaville. Paris: CNRS Editions.

Tobias, P. 1967. The hominid skeletal remains of Haua Fteah, 338–52. In *The Haua Fteah (Cyrenaica) and the Stone Age of the South-East Mediterranean*, ed. C. B. M. McBurney. Cambridge: Cambridge University Press.

Trinkaus, E. 1976. The evolution of the hominid femoral diaphysis during the Upper Pleistocene in Europe and the Near East. *Zeitschrift für Morphologie und Anthropologie* 67:291–319.

Trinkaus, E. 1981. Neanderthal limb proportions and cold adaptation, 187–224. In *Aspects of Human Evolution*, ed. C. B. Stringer. London: Taylor and Francis.

Trinkaus, E. 1983. *The Shanidar Neandertals*. New York: Academic Press.

Trinkaus, E. 1984. Western Asia, 251–93. In *The Origin of Modern Humans: A World Survey of the Fossil Evidence*, ed. F. H. Smith and F. Spencer. New York: Alan R. Liss.

Trinkaus, E. 1987. The Neandertal face: Evolutionary and functional perspectives on a recent hominid face. *Journal of Human Evolution* 16:429–43.

Trinkaus, E. 1993a. Comment. *Current Anthropology* 34:620–2.

Trinkaus, E. 1993b. Variability in the position of the mandibular mental foramen and the identification of Neandertal apomorphics. *Revisita di Anthropologia* 71:259–74.

Trinkaus, E. 1995. Near Eastern late archaic humans. *Paléorient* 21:9–23.

Valladas, H., J. L. Reyss, J. L. Jeron, G. Valladas, O. Bar-Yosef, and B. Vandermeersch. 1988. Thermoluminescence dating of Mousterian "Proto-Cro-Magnon" remains from Israel and the origin of modern man. *Nature* 331:614–16.

Vandermeersch, B. 1981a. Les premiers *Homo sapiens* au Proche-Orient, 97–100. In *Les Processus de l'Hominisation*, ed. D. Ferembach. Paris: CNRS Editions.

Vandermeersch, B. 1981b. *Les Hommes Fossiles de Qafzeh (Israel)*. Paris: CNRS Editions.

Vandermeersch, B. 1989. The evolution of modern humans: Recent evidence from southwest Asia, 155–64. In *The Human Revolution*, ed. P. A. Mellars and C. B. Stringer. Edinburgh: Edinburgh University Press.

Vandermeersch, B. 1992. The Near Eastern hominids and the origins of modern humans in Eurasia, 28–39. In *The Evolution and Dispersal of Modern Humans in Asia*, ed. T. Akazawa, K. Aoki, and T. Kimura. Tokyo: Hokusen-sha.

Virchow, H. 1920. *Die Menschlichen Skeletreste aus dem Kämpfe'schen Bruch im Travertin von Ehringsdorf bei Weimar*. Jena: Fischer.

Weidenreich, F. 1904. Die Bildung des Kinns und seine angebliche Beziehung zur Sprache. *Anatomischer Anziger* 24:545–5.

Weidenreich, F. 1936. The mandibles of *Sinanthropus pekinensis*: a comparative study. *Paleontologia Sinica (Series D)* 7:1–132.

Wolpoff, M. H. 1971. *Metric trends in hominid dental evolution*. Cleveland: Press of Case Western Reserve University.

Wolpoff, M. H. 1975. Some aspects of human mandibular evolution, 1–64. In *Determinants of Mandibular Form and Growth*, ed. J. A. McNamara. Ann Arbor: University of Michigan Press.

Wolpoff, M. H. 1980. *Paleoanthropology*. New York: A. A. Knopf.

Wolpoff, M. H. 1995. *Human Evolution*. New York: McGraw Hill.

Wolpoff, M. H., F. H. Smith, M. Melez, J. Radovčić, and D. Rukavina. 1981. Upper Pleistocene human remains from Vindija cave, Croatia, Yugoslavia. *American Journal of Physical Anthropology* 54:499–545.

PART IV

Cultural and Human Evolution: Views from Neighboring Regions

This part opens with a general paper by Kevin Laland, who examines cultural evolution in a comparative perspective. He notes that the concept of culture is often seen as characteristic of humans, with social learning and transmission at the heart of this definition. Culture as recorded in the animal kingdom is then referred to as "protoculture." According to Laland, the archaeological record indicates that most members of the genus *Homo* had cognitive capabilities and cultural manifestations that were intermediate between animals and modern humans. Social learning through horizontally and vertically transmitted traits could have evolved under different ecological circumstances. Hence, a comparative investigation of social learning among vertebrates may produce a special perspective on the evolution of hominid learning capacities. From an evolutionary standpoint, horizontal transmission evolved first and was later followed by vertical transmission. Importantly, primate societies do not lack evidence for technical innovations, but the dynamics of vertical transmissions within these groups are either poorly understood or limited by population size. Laland concludes by indicating that the simple definition of culture as the social transmission of collected knowledge is not solely human. Social learning evolved over time among both animals and humans, and it will be impossible to explain human cultural evolution without understanding animal behavior.

Avraham Ronen employs the phases and entities from the Oldowan, to the Acheulian, to other modes of tool manufacturing to caution that these are not the best avenues for finding out about degrees of human cognition and language. Examining the evidence for burials, he concludes that these activities reflect a group's perception of differences between life and death. He emphasizes the implications of the known Middle Paleolithic graves, such as those in Qafzeh Cave. Ronen then examines food provisioning, but he pays more detailed attention to fire, the starting and maintenance of which requires knowledge and cooperation. The domestication of fire demands communication, which was achieved with the evolution of language and symbolic behavior and could have emerged during the Acheulian times.

The wider geographic perspective is enhanced by Kathy Schick's survey of Paleolithic cultural patterns. Her review begins with the early Paleolithic technologies known from Africa. Although those core and flake industries lack stylistic patterns, experiments have indicated that human manufacturing processes differ from those of apes in other important ways. For instance, the role of the raw material in these early lithic assemblages is considered crucial in dictating the final forms. Acheulian technology is thought to reflect planned basic tool forms, which raises the possibility of the presence of cultural norms. The dispersal of early humans into Eurasia is surveyed briefly and the technological variants in the various major regions are described. The author comments on the lack of good correlation between environmental variability and the distribution of the Acheulian, and she then discusses the ensuing Middle Paleolithic industries. She stresses the lack of Levallois-based entities in East Asia and suggests examining the technological stasis in Middle Paleolithic technologies across the Old World. She concludes by expressing the need to resolve the meaning of vague verbal designations such as "technological complexity."

Janusz Kozlowski set out to examine the sequences of some lesser known territories, namely those around the Black Sea. Any discussion of possible dispersals of Cro-Magnons from the Near East into Europe should take into account the evidence from the Middle and Upper Paleolithic sequences of this region. His review, accompanied by a series of maps, reports the distribution of the various cultural entities from Isotope Stage 6 through Isotope Stage 3. It clearly indicates that one can trace the cultural continuities in most areas. Shifts within the Mousterian industries are seen as possibly reflecting the expansion of Neandertals eastward, driven by the increasingly harsh climatic conditions. The early appearance of the Aurignacian in the Balkans marks a change and heralds the penetrations of the Cro-Magnons into western Europe.

Marcel Otte offers a different view. He views the formulation of several hypotheses concerning the emergence of modern humans as merely the creation of "myths." Otte suggests that this is due in part to the interpretation prevailing in Europe that the transition from Middle to Upper Paleolithic was abrupt and was related to the biological shift from Neandertals to Cro-Magnons. He does not see such a watershed in other parts of the Old World. Otte presents his views concerning the current models of "out of Africa," examining the archaeological information concerning the transition from the Middle to the Upper Paleolithic in various Asiatic regions, North Africa, and parts of Europe. His discussion centers, however, on Anatolia. He concludes that the evidence from this region is crucial for testing contradictory models.

The European world is reviewed by Paul Mellars, who examines the impact of climatic changes on the demography of late Neandertals and early anatomically modern humans. He summarizes the evidence for the dispersal of the latter. Mellars views almost all aspects of archaeologically recognizable remains that reflect human behavior as the result of the progressive move of AMHS from east to west. He finds support for the proposal that views this change in somewhat similar fashion to the advent of the agricultural revolution in Europe. Undoubtedly, understanding the impact and significance of the climatic fluctuations during Isotope Stage 3 would assist us in explaining the eventual failure and success of both Neandertal and Cro-Magnon populations. Mellars describes the process of replacement, which took a very long time, especially in the Iberian peninsula. In comparing and analyzing the process, he discusses low population densities, subsistence systems, and successive occupation of the same territory. He then proceeds to look at Neandertal extinctions. It is obvious that Neandertals, who were well adapted to the climatic conditions of glaciated Europe, ultimately did not survive the competition with the Cro-Magnons.

Richard Klein, returning to the African continent, begins by presenting the two competing hypotheses concerning the origins of modern humans. Acceptance of the African origins theory is complicated by the fact that early or near modern humans did not leave behind evidence for modern behavior. This behavior is generally expressed by the extensive use of a variety of raw materials aside from lithics, such as antler, bone, and ivory, by discrete tool types, and by evidence for ritual or ceremony when burying the dead. Klein suggests that the stimulus for modern behavior was a neurological change that perhaps enabled them to articulate phonemic speech. This mutation could have been a random event, which makes it difficult to trace in the fossil skulls. He then proceeds to examine the MSA (equivalent to Middle Paleolithic) and LSA (synonym of Upper Paleolithic) in South African sites. Examining the animal remains from both MSA and LSA assemblages, he demonstrates how the limited MSA technology affected the range of hunted game, whereas the evidence indicates that the LSA people were more competent hunters. A similar explanation is suggested for the lack of seasonality in culling seals among the MSA inhabitants of Klasies River Mouth Cave, while the LSA foragers clearly exploited the peak season of seal hunting. On the whole, the densities of MSA people seem to have been lower than those of LSA populations. In sum, Klein sees the major behavioral shift as taking place in sub-Saharan Africa around 50–40 ka, and it was this watershed that provided modern humans with the capacity for culture and led to the rapid replacement of nonmodern populations in Eurasia.

We end this part and the volume itself with a solicited paper written by David Pilbeam, who joins the others with an appreciation of the contribution of Southwest Asia to the study of the origins of modern humans. That this is the most debated topic in paleoanthropology is no surprise. The relative richness of the archaeological and fossil record only enhances the argument. Pilbeam essentially deals with the newly accumulated knowledge in the genetic domain and stresses the importance of learning from this field about the quantitative differences between species and subspecies. While comparing the situation among chimpanzee species, he concludes that the high degree of genetic homogeneity among currently living humans is best explained by their recent origins. He then examines the fossil record and concludes that both the morphological and archaeological evidence is sufficiently ambiguous to enable various scholars to present persuasive yet diverging, and even contradictory, hypotheses. He ends not only by stating his view but also by positing a series of problems for future research, including the patterns of behavior of nonmodern and modern humans in the less studied parts of the world, and the dynamics of Eurasian population interactions from 45 ka to 25 ka.

27

THE EVOLUTION OF CULTURE

A Comparative Perspective

Kevin N. Laland

Sub-Department of Animal Behavior
University of Cambridge
Cambridge CB3 8AA, United Kingdom

1. INTRODUCTION

'Culture' is commonly regarded as a feature of human existence which sets us apart from other species and which, in consequence, defines humanity. This perspective can be found in the writings of many evolutionary biologists (e.g., Simpson 1949; Dobzhansky 1955) and human scientists (e.g., Montagu 1968; Geertz 1973; Sahlins 1976; Washburn 1978). Archaeologists, anthropologists, and evolutionary biologists also concur that a, if not the, definitive characteristic of culture is the social transmission of information (Durham 1990). To qualify as cultural, a given unit of information must be learned from other individuals, not transmitted genetically or acquired from isolated individual experience. Culture is a system for the inheritance of acquired information.

Yet this widespread conception of culture as characteristically human, with social learning and transmission lying at the heart of culture, is somewhat paradoxical when we consider that many other species have their own, albeit more primitive, forms of culture. That is to say that individuals in many animal species are capable of learning patterns of behavior or acquiring adaptive information from other animals. This paradox is usually resolved by emphasizing the differences between human culture and animal protoculture. This includes distinctions based on the central role of symbolic encoding of information in human culture or the social history of accumulated tradition, as well as those expressed in terms of the nature and mechanisms of information transmission.

If we are interested in the roots of human culture, and if a system of social learning and transmission is regarded as central to that culture, then there are several reasons why it may pay to consider the nature of social learning in extant animal populations. First, although there is nothing in our understanding of evolutionary biology that precludes the possibility that some animal species might have evolved more sophisticated cognitive capabilities than recent human ancestors, the wealth of archaeological evidence, together

Neandertals and Modern Humans in Western Asia, edited by Akazawa *et al.*
Plenum Press, New York, 1998.

with comparative analyses of relative brain size, renders this possibility somewhat implausible. It therefore seems reasonable to assume that hominid populations, certainly those of the genus *Homo*, would have had cognitive capabilities and a manifest culture intermediate in sophistication to humans and animals. By dwelling on what animals can and cannot do we can set the lower bounds to what recent hominids were probably capable of doing.

Second, studies of social learning and transmission in animal populations have shed light on the processes of social transmission which were almost certainly the same processes that operated in hominid populations. For example, empirical studies have established that social transmission can have frequency-dependent properties; that learned information is sometimes transmitted horizontally (among peers) and sometimes vertically (from parents to offspring) by different mechanisms; and that there are many reasons why adaptive innovations do not spread.

Third, field and laboratory findings have informed the development of mathematical models which explore the evolution of social learning (Cavalli-Sforza and Feldman 1983a,b; Boyd and Richerson 1985; Aoki and Feldman 1987, 1989; Vickery et al. 1991; Giraldeau et al. 1994; Laland et al. 1996). These analyses suggest that horizontally and vertically transmitted traits may have evolved under very different ecological circumstances as solutions to quite different adaptive problems (Laland et al. 1996).

A comparative approach, which considers the nature of social learning across many vertebrate species, suggests a particular perspective on the evolution of the capacity for social learning and sheds light on human origins.

2. SOCIAL LEARNING IN ANIMALS

Social learning occurs when an animal learns a behavior pattern or a dietary preference as a consequence of observing or interacting with a second animal. The term 'social learning' was introduced as a general term to represent learning that is influenced socially, in contrast to instances of individual learning in which behavior acquisition is not influenced by interaction with others (Box 1984). This general term should not be confused with terms such as 'imitation,' or 'local enhancement,' which loosely describe the mechanism by which social learning takes place (or at least categorize social learning phenomena into 'types'). 'Imitation' refers to the process in which observation of another individual performing an act allows an animal to learn the consequences of the act or the context in which it can be enacted (in other words, to form a response-reinforcement association through observation) (Galef 1988a; Heyes and Dawson 1990). In contrast, 'local enhancement' is a process in which one animal directs another animal's attention to some aspect of the environment or stimulus (Thorpe 1956). If, as a consequence of this tip off, the observer subsequently learns something about the environmental stimulus, then a simple form of social learning has occurred (Galef 1988a). In fact, terms describing these 'types' of social learning have very little real explanatory power. They do not, for example, explain how social and environmental stimuli are analyzed, representations are constructed, associations are formed between events, and like motor acts produced.

Perhaps the most celebrated case of animal social learning is the washing of sweet potatoes by Japanese macaques at the Japanese Monkey Centre at Koshima. Kawai (1965) described how a two-year-old female, called Imo, discovered that she could wash the sand grains off her sweet potatoes in water. This habit spread throughout the troop as 90% of its members had acquired the behavior within ten years. The younger individuals were thought to learn from the older animals and this transmission was referred to as a preculture or protoculture, the implication being that it is a primitive homologue of human culture.

Jane Goodall (1964) reported that chimps will poke twigs and stalks into the openings of termite nests, pull them out, and lick off the termites. Goodall describes how she observed infants watching adults manipulating tools and then imitating them. She refers to the spread of this behavior as a primitive cultural tradition passed on from one generation to the next. While nearly all members of the chimpanzee population in the Gombe National Park in Tanzania fish for termites, chimpanzees in other populations do not exhibit this behavior, and this has led to the suggestion that primitive cultural traditions account for the between-population variation in this behavior (Bonner 1980).

These high-profile, textbook cases present a slightly misleading picture of animal social learning. Most animal social learning is not from parents to offspring, does not involve cognitively demanding transmission mechanisms, and will not maintain alternative traditions between populations. Indeed, there is some doubt whether even the potato washing and termiting traditions have all these qualities (Green 1975; Galef 1992). A more prototypical example is the acquisition of dietary preferences by rats that attend to cues on the breath of conspecifics (Galef 1988b). In general, rats prefer foods that other rats have eaten rather than alternative novel diets, and this simple mechanism probably maintains short term dietary traditions in rat populations (Galef 1988b). Another more representative example is the spread of milk bottle top opening in British tits (Fisher and Hinde 1949). These birds learned to peck open the foil cap on milk bottles and to drink the cream. This habit started in a single location and spread rapidly throughout Britain and into Europe. Hinde and Fisher (1951) found that this behavior probably spreads by local enhancement, where the tits' attention is drawn to the milk bottles by a feeding conspecific and, after this initial tip off, they subsequently learn on their own how to open the tops.

Most animal social learning involves the short-term transmission of feeding and foraging information among populations of social foragers by very simple mechanisms such as local enhancement. A considerable amount of empirical evidence has been accumulating for this subtle kind of animal social transmission. Lefebvre and Palameta (1988) document one hundred studies reporting socially transmitted foraging behaviors among natural populations of animals. It occurs among populations of rats, pigeons, starlings, bats, gulls, and literally hundreds of other species that acquire up-to-date foraging and feeding information from unrelated conspecifics. This is what animal social learning is all about. Animals learn what to eat and where to find it by very simple mechanisms. These social foragers typically live in patchy and highly-variable environments and hence the information they acquire through social learning is only of transient value.

At this point in time, it would appear that most animal social learning disseminates foraging information among unrelated individuals in a social group (Lefebvre and Palameta 1988). It relies on simple mechanisms such as local enhancement, where all that is transmitted is a tip off, mediated by very simple forms of communication (Galef 1988a). And it generates ephemeral, non-accumulatory traditions which track changing environmental resources (Lefebvre and Palameta 1988). In contrast, human culture is principally vertically (or obliquely) transmitted, that is information flows from the parental to the offspring generation (although some information is also transmitted horizontally). This can be seen by focusing on the correlations between parent and offspring for personality traits and behavior patterns (Cavalli-Sforza et al. 1982; Boyd and Richerson 1985). Human culture relies on complex social learning mechanisms such as imitation and teaching, facilitated by complex forms of communication, such as language (Aronson 1995). Human culture also generates stable traditions in which information accumulates from generation to generation.

Of course, the distinction made here between human culture and animal protoculture is not absolute. There are several cases of animal behavior patterns which are transmitted vertically, from parents to offspring, or obliquely, from the parental to the offspring generation. For instance, the traditional use of migratory routes, and the return to specific locations for breeding in birds and fish (Galef 1976; Bonner 1980); the transmission of information concerning territories, territory boundaries, and pathways (Galef 1976); the identification of predators (e.g., the development of a fear of snakes by rhesus monkeys that perceive conspecifics acting fearfully in the presence of a snake, Cook et al. 1985); and the social transmission of bird song (Marler and Tamura 1964; Nottebohm 1972). Moreover, animals sometimes exhibit sophisticated forms of social learning. There is clear evidence for imitation in rats (Heyes and Dawson 1990) and budgerigars (Galef et al. 1986), and although the experimental evidence for imitation in apes (Tomasello et al. 1993; Russon and Galdikas 1993; Byrne and Byrne 1993) is not quite as satisfactory, few doubt that imitation in apes will eventually be clearly demonstrated. Although the current view of animal social learning implies that such cases are exceptions (Laland et al. 1996), with the vast majority of animal cases possessing the archetypal qualities described above, the composite picture is changing rapidly and this conclusion should at best be regarded as tentative.

Most animal social learning can thus be thought of as an adjunct to individual learning, which allows animals to enhance their foraging efficiency in rapidly changing unpredictable environments by transmitting dietary preferences and simple patterns of behavior. In contrast, the 'traditional' transmission which characterizes human culture must address a different kind of adaptive problem germane to features of the environment which are stable for long periods of time relative to the lifetime of an individual organism. There is no adaptive advantage to organisms adopting the behavior patterns of their parents if the environment is likely to change and render the behavior inappropriate.

3. THE EVOLUTION OF CULTURE

Highly-horizontal forms of transmission are phylogenetically much more widespread than vertical forms. Most of the cases of vertical transmission involve traits such as bird song, sexual imprinting, or habitat selection, which are highly constrained and have a very specific function. Such traits cannot be regarded as homologues of the flexible and information-rich vertical transmission system that characterizes human culture. Thus a comparative perspective suggests that it was horizontal social learning that evolved first, and that in hominids, and in a relatively restricted number of other species, a general capacity for vertical transmission subsequently evolved. Thus the problem of how human culture evolved has been transformed into two quite separate questions: (a) how did social learning evolve? (where the appropriate focus is on the adaptive advantages of a highly-horizontal transmission system); and (b) how did a highly-horizontal system of social learning evolve into a traditional system with the characteristics of human culture? Both of these questions have been addressed using mathematical models.

Figure 1, taken from Laland et al. (1996), is an illustration of the results of a mathematical analysis which used population genetics models to explore when natural selection should favor horizontal social transmission in a population already capable of learning. Two key variables were the probability that a naive forager will successfully locate a highly nutritional, but patchily distributed food source (ε), and the rate at which the environment varies (e), for example, reflecting a change in the location of that food source. Note that this

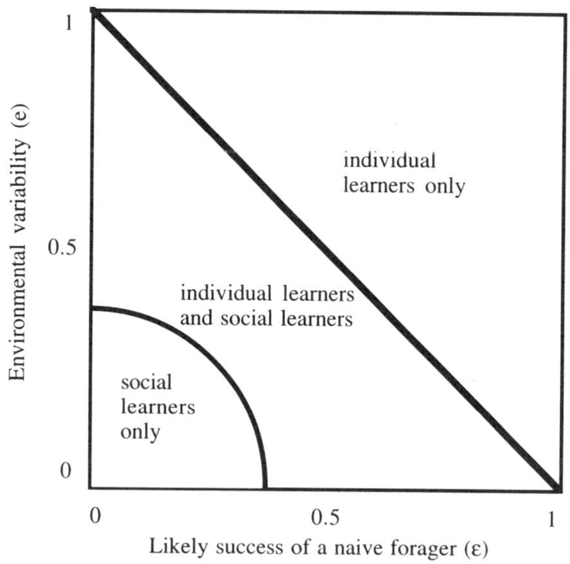

Figure 1. The figure shows the circumstances under which social learning is expected to evolve. Social learning will be favored by selection in a changing environment when the probability of individual foraging success is low, for example, where food sources are difficult to locate or sparsely dispersed.

measure of environmental variability is relative to daily, or more frequent, foraging bouts, and hence even when (e) is small it can represent rapid environmental change relative to the lifetime of an organism. When the probability is high that (individual) foraging will allow a naive animal to locate the nutritional diet, or when the environment changes very rapidly, social learning is of no adaptive advantage and cannot evolve. Where the likely success of a naive forager is low, and when the rate of change in the environment is fast but not too fast, all individuals in the population should be social learners.

There are also circumstances where populations should be polymorphic for individual and social learners. This means that social learning can evolve in a rapidly changing environment when the probability of individual foraging success is low, either because individuals can only search a small section of the population range or because food sources are sparsely dispersed. These are precisely the kinds of environments in which most animal social learning is found. A similar conclusion was reached by Giraldeau et al. (1994) in a related analysis which explored the advantages of social learning in a population competing for limited food resources. Giraldeau et al. hypothesized that a further adaptive function of social learning may be that it allows individuals to circumvent some of the inhibitory effects that scrounging (exploitation of food discovered by others) has on the individual learning of a foraging skill.

The second question raised by the comparative perspective is how did a highly-horizontal system of social learning evolve into a traditional system with the characteristics of human culture. To address this question it is imperative to conduct a similar mathematical analysis of the adaptive advantages of a system of vertical cultural transmission (Cavalli-Sforza and Feldman 1983a,b; Boyd and Richerson 1985; Aoki and Feldman 1987, 1989). Boyd and Richerson (1985) explored the circumstances under which natural selection should favor a system of vertical cultural transmission in a population of individuals that can learn about their environment on their own but not from each other. They concluded that when environments change very slowly, all information should be transmitted genetically, since the modest demands for updating are easily met by the genetic system responding to selection. When environmental change is very rapid, tracking by individual learning is favored (although the analysis presented above suggests that under certain conditions horizontal social transmission will be beneficial). But it is at intermediate rates of change that vertical social learning is an advantage, when changes are not so fast that parents and offspring experience different environments, but not so slow that appropriate genetically transmitted behavior could evolve.

4. TRANSMISSION DYNAMICS

The conclusion that seeming broad environmental conditions favor vertically transmitted cultural traditions was regarded by Boyd and Richerson (1988) as "troubling" when they were faced with the apparent rarity of such traditions in animal populations. It is entirely possible that future research will find that vertical transmission is not quite as infrequent among animals as is currently conceived. However, until such evidence is forthcoming it remains a mystery why vertical cultural transmission is apparently rare among animals. If animals can learn from unrelated individuals, sometimes by quite cognitively demanding mechanisms such as imitation (Heyes and Dawson 1990), one would think they ought to be able to learn from their parents. Animals may not have language, but their communication systems have been proven to be sophisticated enough to transmit simple behavior patterns (Aitchison 1989). Kummer and Goodall (1985) found no shortage of innovation in primate populations, but these innovative behaviors rarely spread. Recent research suggests the dearth of examples of vertically transmitted traditions in animals is probably more a consequence of the complexities of transmission dynamics in a social group than a reflection on the cognitive capabilities of animals. It seems that on many occasions when ecological and technical innovations occur, the innovator is alone or is at least freed from social distractions (Kummer and Goodall 1985). The innovating animal is rarely the dominant individual in the social group (Kummer and Goodall 1985), which may in part explain why innovations are rarely adopted by other animals. The diffusion of information through a population may be biased towards particular pathways or directed from some kinds of individual to others (Coussi-Korbel and Fragaszy 1995). Moreover, frequency-dependent effects, where individuals do what the majority are doing, create a conservative force in transmission (Sugita 1981; Beck and Galef 1989; Lefebvre and Giraldeau 1994) because the innovation has to reach a threshold frequency in the population before it can spread. A focus on the population level dynamics of social transmission will do most to explain the enigma of animal traditions (Laland et al. 1993; Coussi-Korbel and Fragaszy 1995).

Traditional field and laboratory approaches to the study of social learning in animals have recently been supplemented by diffusion studies (which monitor the spread of a novel behavior pattern through a population) and transmission chain studies (which investigate the spread of a behavior along a chain of animals). Diffusion studies have shown that scrounging can be an important factor affecting the diffusion of food-related behavior, and they illustrate that social interaction can interfere with, as well as mediate, social transmission (Palameta and Lefebvre 1985; Lefebvre 1986; Giraldeau and Lefebvre 1987). Such studies have also found that migration between groups can increase the rate of diffusion of socially transmitted traits within groups. The enhanced diffusion may result because migration destabilizes the frequency-dependent equilibrium between the number of animals that have learned socially to produce food and the number of scroungers (Lefebvre and Palameta

1988). Transmission chain studies have demonstrated that diet preferences, foraging, and predator information can be transmitted along chains of individuals and these shed light on factors that affect the stability of transmission (Curio et al. 1978; Laland and Plotkin 1990, 1992, 1993; Galef and Allen 1995). For instance, transmission chain studies have taught us that foraging information can build up slowly throughout transmission (Laland and Plotkin 1990), suggesting that traditions may develop in the absence of a 'clever' or 'innovative' individual who invents the transmitted behavior.

5. MENTAL ABILITIES

The literature on human evolution is rife with claims that there is some key mental ability or behavioral attribute which distinguishes humans from other animals, with the emergence of this ability regarded as a momentous event in human evolution. As understanding of animal behavior grows, these distinctions appear increasingly untenable, since rudimentary forms of that which has traditionally been regarded as exclusively human are consistently found in animal populations. This is not to claim that animals and humans are alike: clearly animal behavior, 'culture,' and communication are simple by human standards. However, there are often continuities between ourselves and other animals with respect to these 'human' abilities, rather than essential and defining differences.

Language is commonly paraded as the attribute which most clearly distinguishes humans from other animals. Whether or not animal species can be regarded as possessing language hangs critically on the definition of language adopted (Aitchison 1989). Since most definitions are manifestly anthropocentric (i.e., that language is the means by which human beings communicate), the linguistic abilities of animals are often automatically excluded. The most promising approach to comparing human and animal communication has been to describe a number of design features which characterize human language, such as arbitrariness, semanticity, cultural transmission, spontaneous usage, displacement, etc., and to assess whether animal communication possesses any or all of these features (Aitchison 1989). Rudimentary forms of most of these features can be found in animal populations. For instance, vervet monkeys use arbitrary vocalizations to represent the impending threat of different predators; apes have been taught sign language and have invented new word combinations; bird dialects are often socially transmitted; and honey bees communicate information about a food source distant in time and space. However, the communication of no single non-human species has all of the characteristics of human language, even in a rudimentary form. Whether or not it is deemed appropriate to describe the communication of non-human species as language, it is important to acknowledge that some animals possess sophisticated communicatory repertoires. Chimpanzee communication is complex, integrating vocalizations, gestures, and facial signs in a manner dependent on the context and the individuals present (Goodall 1965; Mitani 1994). At the very least, studies of teaching apes sign language have established that apes can comprehend large numbers of symbols, respond appropriately to novel sentences, generate signs or symbols for novel objects, and use these skills to communicate information about past and future events or objects distant in space and time (Rumbaugh et al. 1984). A study of ape communication implies, minimally, that hominids could communicate through a complex series of expressive gestures and sounds.

The assumption that humans are conscious while animals are unthinking machines has a long history dating back to Descartes. This position is not in accord with modern conceptions of consciousness (Dennett 1991). A single mental attribute labeled "consciousness" cannot be clearly delineated or assigned to particular species with any precision (Dennett 1991). Experimental studies demonstrate that animals readily combine information from different senses into some form of representation, which is integrated with memories to guide behavior (Dickinson 1988). The "consciousness" of a snake, bat, or honey bee is likely to be very different from that of a human, since the sensory input, and the ability to integrate this information, is very different in each case. What should not be in doubt is that many animals have a concept of 'self' and can distinguish their own activities from the events of the external world. For example, rats can learn to press one of a series of levers according to the activity in which they are currently engaged, showing that they are aware of their own actions (Passingham 1982). This should not be surprising. As Heyes aptly puts it, "If an animal could not distinguish extraneous visual input from visual and kinesthetic feedback from its body, then it would constantly bump into things" (1994:915). That rats, or even many invertebrates, might be described as "conscious" is only surprising when this term implies more than self-conceptualization or when consciousness is regarded as a unitary concept. Self-conceptualization does not imply the ability to recognize oneself in a mirror or photograph (more aptly described as the use of displaced visual feedback to guide action), or the ability to attribute mental states to others, and the concept of consciousness becomes onerous if it prevents such distinctions from being made (Heyes 1994).

Tool use is commonly found in animals and birds (Beck 1980; McGrew 1994). That is, animals frequently manipulate inanimate objects with the effect of improving the animal's efficiency in altering the position or form of some separate object. Well-known examples include the use of twigs and cactus spines by the woodpecker finch to probe for insects in the bark of trees, the use of stones as an anvil to crack open mussels by the sea otter, and termiting in chimpanzees. A recent study reported that crows manufacture and use two different types of hook tools to aid prey capture (Hunt 1996). Wright (1972) and Toth et al. (1993) were able to teach a captive orangutan and bonobo chimpanzee, respectively, to make stone tools and to use them to open a box containing fruit. Given their natural proclivity for tool use, we might expect common chimpanzees to surpass the performance of other apes (humans aside) in such tasks. Most free-ranging chimpanzee populations that have been intensively studied have been found to use tools (McGrew [1994] counted 36 such populations). In addition to termiting, chimpanzees 'fish' for ants and honey. They also use leaves as sponges, napkins for cleaning the body, sticks as clubs and levers, stones as anvils and hammers to crack nuts and as missiles to throw, and several other tools (McGrew 1994). However tool-use repertoires vary widely across geographical regions. Some evidence for forward planning in tool manufacture is evident. For example, chimpanzees will carefully select and shape twigs for termite fishing some distance (up to 100 yards) from the nest, and will sometimes select several tools to carry to the termite heap which they will use one at a time (Goodall 1965). Chimps can also employ a combination of tools, learning to manipulate each in an appropriate manner and temporal order. In using stones to crack nuts, chimps will select appropriate stones to act as anvil and hammer, and they have been regularly observed to use a third stone as a wedge to stabilize the anvil. In most cases of animal tool use, it is not clear to what extent the behavior develops as a result of social learning, although this has often been implicated or assumed. Nonetheless, it is clear that tool exploitation should not be regarded as exclusively human.

According to Dobzhansky, "Man alone adapts himself by actively or even deliberately changing the environment, and by inventing and creating new environments" (1995:339). This claim has been made repeatedly over recent years, despite the fact that it is demonstrably untrue. Many if not most organisms construct their environments in part

(Lewontin 1983; Odling-Smee 1988; Odling-Smee et al. 1996). To varying degrees, organisms choose their own habitats and resources, and construct important components of their local environments such as nests, holes, burrows, paths, dams, and chemical environments, as well as provisioning "nursery" environments for their offspring (von Frisch 1975; Hansell 1984). There is also considerable evidence of evolutionary responses to self-induced selection pressures, found in plants and invertebrates as frequently as in 'higher' vertebrates (Odling-Smee et al. 1996). Niche construction is a general phenomenon, and although humans may be regarded as primate champions, no essential distinction between humans and animals can be made on this ground.

6. IMPLICATIONS FOR THE STUDY OF HUMAN ORIGINS

The preceding sections establish a number of important points that are relevant to the study of prehistory. If culture is defined, pragmatically, as the social transmission of acquired information, it cannot be restricted to humans or hominids. Socially learned behavior patterns are found across a broad range of vertebrate species and even among some invertebrates. Social learning probably originated as an adjunct to individual learning, which allowed animals to enhance their foraging efficiency in rapidly changing unpredictable environments by transmitting dietary preferences and simple patterns of behavior. Vertical transmission probably evolved subsequently, in a more restricted range of species, and facilitated the diffusion of traits that are of adaptive value for long periods of time relative to the lifetime of an individual organism. Culture, like other behavioral and anatomical traits, has a substantial evolutionary history before its appearance as a human specialty, and it is counterproductive to conceive of it as a unitary phenomenon that is either present or absent. While there is probably no homology between human and animal traditions, some of the cognitive abilities which mediate the transmission of learned information (such as a capacity for local enhancement, emulation learning, creative innovation, and perhaps imitation) may well be homologous.

Rudimentary forms of behavioral characteristics and capabilities, often employed to delineate humans from other animals (consciousness, communication, tool use, imitation, niche construction), do not separate humans from other species. No special mental qualities need to be postulated as causes of a cultural tradition. The evidence of tool use in free-living chimpanzees raises the possibility that a growing dependence on tool use may have become an established means of adaptation in ape populations several million years ago. In terms of the cognitive demands of the task, there is no great intellectual leap in going from selecting and modifying a twig for termite fishing, or selecting an anvil and modifying it with a wedge for optimal nut cracking, to making a primitive stone tool. As the comparative evidence is considered, it becomes clear that our current behavioral abilities were pioneered by different species of ancestors over millions of years. I reiterate, this is not to say that there are no distinctive or important differences between the mental abilities of humans and other animals; clearly there are. An understanding of animal behavior will never explain away the problems of human evolution, but it may provide a yardstick to the likely cognitive sophistication and social dynamics of hominid populations.

The pre-eminent mode of thinking in many related disciplines concerned with human evolution is to argue that there was some key characteristic that emerged which was central to the evolution of human culture. The most popular hunches are that human culture 'took off' when hominids evolved language or more sophisticated cognitive capabilities, or when they took up hunting (Mellars 1989). There are two difficulties with such hypotheses which detract from their utility. First, it is not obvious how hypotheses such as "a more sophisticated cognitive capability evolved" can be tested, and if they cannot be tested then they should be regarded as plausible stories rather than scientific postulates. Second, such hypotheses focus almost exclusively on the assumed capabilities of hominids, while at best ignoring, and at worst drawing attention away from, the processes of evolution and cultural transmission.

As illustrated in the earlier section on transmission dynamics, careful consideration of the dynamics of the transmission process might prove a more fruitful approach to archaeological problems than speculation over abilities. If it is the case that lithic technology changes comparatively slowly throughout the Middle Paleolithic, the possibility that this reflects a lack of cognitive sophistication in archaic humans cannot be excluded. However, there are other explanations and it may serve us to consider whether a frequency-dependent conformist tradition might have been in operation, to ask what processes might prevent innovations from spreading, or to reflect on what problem the innovation was required to solve.

Similarly, if we want to understand why it is that hominid cultural elaboration exhibits an accelerating function over the last two million years, instead of arguing that this reflects a series of independent advances in the cognitive and linguistic architecture of hominid populations, rendering them better and better able to acquire and transmit culture, we could consider the possibility that the acceleration in hominid cultural elaboration reflects a single, acceleratory process. Many of the changes that organisms bring about in their world effectively damp out environmental variability. For example, as far as the organism is concerned, migration cancels out seasonal changes in temperature. Culture is a particularly effective means of adaptation, and because, according to the theory presented above, vertical cultural transmission is an adaptation to slowly changing environments, it will create uniformity in other aspects of the population's life history from one generation to the next. In other words, the theoretical analyses imply that vertical transmission may create the conditions where further vertical transmission is favored. The transmission process may itself be intrinsically acceleratory, with the population adapting to the social environment at each stage. If this is indeed a general property of vertical cultural transmission, then it should still be true and could be investigated in contemporary populations.

ACKNOWLEDGMENTS

This work was supported in part by a Royal Society University Research Fellowship. I am grateful to Jeff Galef and John Odling-Smee for helpful comments. Figure 1 is from Laland, Richerson, and Boyd (1996) with permission from Academic Press.

REFERENCES

Aitchison, J. 1989. *The Articulate Mammal: An Introduction to Psycholinguistics*. London: Unwin Hyman.

Aoki, K., and M. W. Feldman. 1987. Toward a theory for the evolution of cultural communication: Coevolution of signal transmission and reception. *Proceedings of the National Academy of Science USA* 84:7164–8.

Aoki, K., and M. W. Feldman. 1989. Pleiotropy and preadaptation in the evolution of human language capacity. *Theoretical Population Biology* 35:181–94.

Aronson, E. 1995. *The Social Animal*. New York: Freeman.

Beck, B. B. 1980. *Animal Tool Behavior: The Use and Manufacture of Tools by Animals*. New York: Garland.

Beck, M., and B. G. Galef, Jr. 1989. Social influences on the selection of a protein-sufficient diet by Norway rats (*Rattus norvegicus*). *Journal of Comparative Psychology* 103:132–9.

Bonner, J. T. 1980. *The Evolution of Culture in Animals*. Princeton: Princeton University Press.

Box, H. 1984. *Primate Behavior and Social Ecology*. London: Chapman and Hall.

Boyd, R., and P. J. Richerson. 1985. *Culture and the Evolutionary Process*. Chicago: University of Chicago Press.

Boyd, R., and P. J. Richerson. 1988. An evolutionary model of social learning: The effects of spatial and temporal variation, 29–48. In *Social Learning: Psychological and Biological Perspectives*, ed.T. Zentall and B. G. Galef, Jr. Hillsdale: Erlbaum.

Byrne, R. W., and J. M. E. Byrne. 1993. Complex leaf-gathering skills of mountain gorillas (*Gorilla g. beringei*): Variability and standardization. *American Journal of Primatology* 31:241–61.

Cavalli-Sforza, L. L., M. W. Feldman, K. H. Chen, and S. M. Dornbusch. 1982. Theory and observation in cultural transmission. *Science* 218:19–27.

Cavalli-Sforza, L. L., and M. W. Feldman. 1983a. Paradox of the evolution of communication and of social interactivity *Proceedings of the National Academy of Science USA* 80:2017–21.

Cavalli-Sforza, L. L., and M. W. Feldman. 1983b. Cultural versus genetic adaptation. *Proceedings of the National Academy of Science USA* 80:4493–6.

Cook, M., S. Mineka, B. Wolkenstein, and K. Laitsch. 1985. Observational conditioning of snake fear in unrelated rhesus monkeys. *Journal of Abnormal Psychology* 94(4):591–610.

Coussi-Korbel, S., and D. M. Fragaszy. 1995. On the relationship between social dynamics and social learning. *Animal Behaviour* 50:1441–53.

Curio, E., U. Ernst, and W. Vieth. 1978. The adaptive significance of avian mobbing II. Cultural transmission of enemy recognition in blackbirds: Effectiveness and some constraints. *Zeitschrift Tierpsychologie* 48:184–202.

Dennett, D. C. 1991. *Consciousness Explained*. New York: Little Brown.

Dickinson, A. 1988. Intentionality in animal conditioning, 305–25. In *Thought Without Language*, ed. L. Weiskrantz. Oxford: Clarendon.

Dobzhansky, T. 1955. *Evolution, Genetics, and Man*. New York: Wiley.

Durham, W. H. 1990. *Coevolution: Genes, Culture and Human Diversity*. Stanford: Stanford University Press.

Fisher, J., and R. A. Hinde. 1949. The opening of milk bottles by birds. *British Birds* 42:347–57.

Galef, Jr., B. G. 1976. Social transmission of acquired behavior: A discussion of tradition and social learning in vertebrates, 77–100. In *Advances in the Study of Behavior*., ed. J. S. Rosenblatt, R. A. Hinde, E. Shaw, and C. Beer. New York: Academic Press.

Galef, Jr., B. G. 1986. Social identification of toxic diets by Norway rats (*Rattus norvegicus*). *Journal of Comparative Psychology* 100(4):331–4.

Galef, Jr., B. G. 1988a. Imitation in animals: History, definitions and interpretation of the data from the psychological laboratory, 3–28. In *Social Learning: Psychological and Biological Perspectives*. ed. T. Zentall and B. G. Galef, Jr. Hillsdale: Erlbaum.

Galef, Jr., B. G. 1988b. Communication of information concerning distant diets in a social, central-place foraging species: *Rattus norvegicus*, 119–39. In *Social Learning: Psychological and Biological Perspectives*. ed. T. Zentall and B. G. Galef, Jr. Hillsdale: Erlbaum.

Galef, Jr., B. G. 1992. The question of animal culture. *Human Nature* 3(2):157–78.

Galef, Jr., B. G., and C. Allen. 1995. A new model system for studying behavioural traditions in animals. *Animal Behaviour* 50(3):705–17.

Geertz, C. 1973. *The Interpretation of Culture*. New York: Basic Books.

Giraldeau, L. A., and L. Lefebvre. 1987. Scrounging prevents cultural transmission of food-finding behavior in pigeons. *Animal Behaviour* 35:387–94.

Giraldeau, L. A., T. Caraco, and T. J. Valone. 1994. Social foraging: Individual learning and cultural transmission of innovations. *Behavioral Ecology* 5(1):35–43.

Goodall, J. 1964. Tool-using and aimed throwing in a community of free-living chimpanzees. *Nature* 201:1264–6.

Goodall, J. 1965. Behaviour of Free-Living Chimpanzees. Ph. D diss. University of Cambridge. Cambridge.

Hansell, M. H. 1984. *Animal Architecture and Building Behaviour*. New York: Longman.

Heyes, C. M. 1994. Reflections on self-recognition in primates. *Animal Behaviour* 47(4):909–19.

Heyes, C. M., and G. Dawson. 1990. A demonstration of observational learning in rats using a bidirectional control. *Quarterly Journal of Experimental Psychology* 42(B):59–71.

Hinde, R. A., and J. Fisher. 1951. Further observations on the opening of milk bottles by birds. *British Birds* 44:393–96.

Hunt, G. R. 1996. Manufacture and use of hook-tools by New Caledonian crows. *Nature* 379:249–51.

Kawai, M. 1965. Newly-acquired pre-cultural behavior of the natural troop of Japanese monkeys on Koshima Islet. *Primates* 6(1):1–30.

Kummer, H., and J. Goodall. 1985. Conditions of innovative behaviour in primates. *Philosophical Transactions of the Royal Society of London B* 308:203–14.

Laland, K. N., and H. C. Plotkin. 1990. Social learning and social transmission of digging for buried food in Norway rats. *Animal Learning and Behavior* 18(3):246–51.

Laland, K. N., and H. C. Plotkin. 1992. Further experimental analysis of the social learning and transmission of foraging information amongst Norway rats. *Behavioural Processes* 27:53–64.

Laland, K. N., and H. C. Plotkin. 1993. Social transmission in Norway rats via excretory marking of food sites. *Animal Learning and Behavior* 21:35–41.

Laland, K. N., P. J. Richerson, and R. Boyd. 1993. Animal social learning: Toward a new theoretical approach. *Perspectives in Ethology* 10:249–77.

Laland, K. N., P. J. Richerson, and R. Boyd. 1996. Developing a theory of animal social learning, 129–54. In *Social Learning in Animals: The Roots of Culture*, ed. C. M. Heyes and B. G. Galef, Jr. New York: Academic Press.

Lefebvre, L. 1986. Cultural diffusion of a novel food finding behavior in urban pigeons: An experimental field test. *Ethology* 71:295–304.

Lefebvre, L., and L. A. Giraldeau. 1994. Cultural transmission in pigeons is affected by the number of tutors and bystanders present. *Animal Behaviour* 47(2):331–7.

Lefebvre, L., and B. Palameta. 1988. Mechanisms, ecology and population diffusion of socially learned food finding behavior in feral pigeons, 141–64. In *Social Learning: Psychological and Biological Perspectives*. ed. T. Zentall and B. G. Galef, Jr. Hillsdale: Erlbaum.

Lewontin, R. C. 1983. Gene, organism, and environment, 273–85. In *Evolution from Molecules to Men*, ed. D. S. Bendall. Cambridge: Cambridge University Press.

Marler, P., and M. Tamura. 1964. Culturally transmitted patterns of vocal behavior in sparrows. *Science*. 146:1483–6.

McGrew, W. C. 1994. Tools compared: The material of culture, 25–40. In *Chimpanzee Cultures*, ed. R. W. Wrangham, W. C. McGrew, F. B. M. de Waal, and P. G. Heltne. Cambridge: Harvard University Press.

Mellars, P. 1989. Major issues in the emergence of modern humans. *Current Anthropology* 30(3):349–85.

Mitani, J. C. 1994. Ethological studies of chimpanzee vocal behaviour, 195–210. In *Chimpanzee Cultures*, ed. R. W. Wrangham, W. C. McGrew, F. B. M. de Waal, and P. G. Heltne. Cambridge: Harvard University Press.

Montagu, A. 1968. *Sociobiology Examined*. New York: Oxford University Press.

Nottebohm, F. 1972. The origins of vocal learning. *The American Naturalist* 106:116–70.

Odling-Smee, F. J. 1988. Niche constructing phenotypes. In *The Role of Behavior in Evolution*, ed. H. C. Plotkin. Cambridge: MIT Press.

Odling-Smee, F. J., K. N. Laland, and M. W. Feldman. 1996. Niche construction. *The American Naturalist* 147(4): 641–48.

Palameta, B., and L. Lefebvre. 1985. The social transmission of a food finding technique in pigeons: What is learned? *Animal Behaviour* 33:892–96.

Passingham, R. 1982. *The Human Primate*. Oxford: Freeman.

Rumbaugh, D. M., E. S. Savage-Rumbaugh, and R. A. Sevcik. 1994. Biobehavioral roots of language: A comparative perspective of chimpanzee, child, and culture, 319–34. In *Chimpanzee Cultures*, ed. R. W. Wrangham, W. C. McGrew, F. B. M. de Waal, and P. G. Heltne. Cambridge: Harvard University Press.

Russon, A., and B. M. F. Galdikas. 1993. Imitation in free-ranging rehabilitant orangutans (*Pongo pygmaeus*). *Journal of Comparative Psychology* 107:147–61.

Sahlins, M. D. 1976. *The Use and Abuse of Biology: An Anthropological Critique of Sociobiology*. Ann Arbor: University of Michigan Press.

Simpson, G. G. 1949. *The Meaning of Evolution*. New Haven: Yale University Press.

Sugita, Y. 1980. Imitative choice behaviour in guppies. *Japanese Psychological Research* 22:7–12.

Thorpe, W. H. 1956. *Learning and Instinct in Animals*. London: Methuen.

Tomasello, M., E. S. Savage-Rumbaugh, and A. C. Kruger. 1993. Imitative learning of actions on objects by children, chimpanzees, and enculturated chimpanzees. *Child Development* 64:1688–1705.

Toth, N., K. D. Schick, E. S. Savage-Rumbaugh, R. A. Sevcik, and D. M. Rumbaugh. 1993. Pan the tool-maker: Investigations into the stone tool-making and tool-using capabilities of a bonobo (*Pan paniscus*). *Journal of Archaeological Science* 20:81–91.

Vickery, W. L., L. A. Giraldeau, J. J. Templeton, D. L. Kramer, and C. A. Chapman. 1991. Producers, scroungers and group foraging. *The American Naturalist* 137:847–63.

von Frisch, K. 1975. *Animal Architecture*. London: Hutchinson.

Washburn, S. L. 1978. The evolution of man. *Scientific American* 239:146–54.

Wright, R. V. S. 1972. Imitative learning of a flaked stone technology—The case of an orangutan. *Mankind* 8:296–306.

28

DOMESTIC FIRE AS EVIDENCE FOR LANGUAGE

Avraham Ronen

Zinman Institute of Archaeology
University of Haifa
Haifa 31905, Israel

1. INTRODUCTION

Human characteristics seem to have emerged gradually from the ape grade. The great apes have the preadaptations that parallel, in embryonic form, those of the human lineage—manipulating objects with hands, binocular color vision, cause-and-effect-oriented cognitive abilities—but they did not evolve along the hominid line (Tooby and DeVore 1987:211). 'Ape-grade' problem solving is not sharply distinguished from that of early hominids (Parker and Milbrath 1993:323). Lateralization, a precondition for human speech (Lancaster 1975:68–9) was attained very early, as seen by handedness in Oldowan tool making (Toth 1985). The human ability to manufacture and use tools, social complexity, the capacity to process information, and language are all linked and increased simultaneously since the initial divergence of hominids from the apes. Cooperative efforts of small human groups would have been an important factor in this trend (Gibson 1993). The key word in the evolution of human cognition and language is *representation* (Donald 1991:3), attained at the stage of "Homo depictor" (Hacking 1983). Thus human language is a symbolic representation of reality. Accordingly, the use of language in prehistory, untraceable on its own, was sought through symboling behavior which could perhaps be traced in the archaeological record. As we shall see, however, identifying clues for symbolic behavior prior to the Upper Paleolithic is problematic. A different approach will therefore be taken here: the attempt to identify the oldest activity reflected in the archaeological record which could not have been performed without language.

Symboling behavior was sought in acts which involve the dual presence of signifier/signified, or a model. Art is the clearest symbol because an art object by definition represents something other than itself, be it the immediate subject model or a more abstract notion. All scholars seem to concur, then, that a fully evolved human language must have existed in the Upper Paleolithic (Marshack 1976; Goodenough 1981; Chase 1991). Art objects claimed to be older than the Upper Paleolithic (e.g., the Acheulean figurine in

Neandertals and Modern Humans in Western Asia, edited by Akazawa *et al.*
Plenum Press, New York, 1998.

Goren-Inbar 1994) are too far apart both spatially and temporally to be considered reliable cognitive markers (e.g., Chase 1991).

2. TOOL MAKING

Beside art, two other types of evidence could have involved symboling: tool manufacturing and burials. It was assumed that clearly standardized tool forms may indicate the existence of a mental template, or an "ideal" type-model, that is, a symbol. There is wide agreement that Oldowan tools do not involve models. Oldowan technology was apparently governed by functional constraints, and if there are apparent tool "classes," they are due to the fact that the job dictated to some extent the artifact's shape, size, and weight (Toth and Shick 1993). Oldowan tool manufacture could have been achieved with an ape's intelligence, communication, and social organization (Wynn and McGrew 1989). Yet even within the "stagnant" tradition which the Oldowan techno-complex seems to represent, some "progress" is seen between assemblages of 2 my ago and those dated to 1.5 my (Texier 1995), possibly indicating the development of somewhat improved technical, conceptual, and/or organizational capacities.

The manufacturer of Acheulean bifaces had a far greater cognitive complexity (Wynn 1991) and could have used more complex organizational schemes in his daily behavior than his Oldowan predecessor (Wynn 1993). The cognitive abilities of the Acheuleans were judged to be rather close to those of modern humans, without specific reference to language (Belfer-Cohen and Goren-Inbar 1994). Toth and Schick (1993), on the other hand, argue that Acheuleans (*Homo erectus*) had a limited capacity to communicate ideas, hence their technology was easily lost when populating new lands and adapting to new environmental conditions, e.g., across the "Movius line." Another claim for a seemingly poor cognitive capacity in the Acheulean was recently advanced by Davidson and Noble (1993), suggesting that bifaces were actually cores and not tools. The same claim is made for the Middle Paleolithic leaf-shaped points, thereby reducing the claimed cognitive capacity of Middle Paleolithic humans as well. However, for some unspecified reason, Solutrean leaf points are recognized as standardized tools, possibly in respect of the Upper Paleolithic artist. Whether bifaces were tools or cores matters little: the important fact is that there is a very clear increase of symmetry, standardization of shapes, quality of finish (Wynn 1993), and technological variability (Belfer-Cohen and Goren-Inbar 1994) between the Early and Late Acheulean, which may be taken to indicate an increase in cognitive abilities.

Cognition and language in the Middle Paleolithic is not less debated than in the Lower Paleolithic. Most students accept that Middle Paleolithic tools are well-standardized and, together with the widespread pre-planned Levallois technnique, they would indicate an advanced cognitive capacity. A few, however, argue that this standardization is more apparent than real, and that it would stem from functional constraints, especially from a sequence of reduction which transforms one type of tool to another (Dibble 1984; Chase and Dibble 1987). The proponents of the functional approach deny that Middle Paleolithic humans possessed mental templates for tools, symboling behavior, and, by implication, language.

On the whole, lithic technology does not seem to be a reliable lead on cognitive and linguistic abilities given that "standardization can appear in the absence of symboling" (Chase 1991:209). Furthermore, even if some relation may have existed between standardization and symboling, this relation is not recognizable by the archaeologist (Chase

1991:209). Wynn has reached a similarly negative conclusion: "the tools themselves cannot answer this question...the evidence for language must come from elsewhere" (1993:404).

3. BURIALS

Does burial of the dead constitute symboling behavior? Although this behavior is unique to humans, the meaning of burial is debated. The oldest clear example of treatment of the dead comes from the Middle Paleolithic. Gargette has argued (1989) that these ancient interments are natural phenomena and not the intentional features they seem to be. Although this extreme view is not generally accepted, the precise meaning of prehistoric burials and their potential symbolic connotations remain unknown (Ucko 1969; Pader 1982). I believe, in accordance with Ullrich (1995), that at the very least burials are a reflection on the difference between life and death. Chimpanzees appear to lack an implicit conceptualization of death, and of killing as the immediate cause of death (Boehm 1992:165). Thus prey animals are killed incidentally in the process of being consumed, while conspecific strangers savagely attacked are abandoned wounded (Boehm 1992:165). Middle Paleolithic burials represent a considerable investment of time and energy, probably by the group or at least a part of it (kin?). This investment carries no sense if a knowledge of death, and awesome feelings toward it, are denied. If this is accepted, then the abstract notions life and death must have been added to the human cognitive experience, as well as a symbolic enactment of separation in general. In my opinion, the symbolism imbued in some Middle Paleolithic burials is even more complex than the minimum suggested here, expressed, for example, in the antlers held in the hand of the juvenile from Qafzeh (Vandermeersch 1970). The presence of language some 100 ka ago has to be accepted.

4. ACTIVITY ANALYSIS

Which activity would have required a human type of communication? The earliest regularly performed activities discernible in the archaeological record are a food provision system followed by a tool provision system (Figure 1). Human food provision is not different from that of monkeys and apes. Its repetitive elements are search, harvest, and consumption. It is an integrated system in the sense that all the elements are clearly linked and clearly perceived by everyone. This system is maintained on the level of the individual. No collaboration is necessary most of the time, and of course no language is required to carry out this activity.

The tool provision system, approximately 2.6 my old, consists of procuring a suitable raw material, followed by manufacture, use, and discard. Like the former, this too is an integrated system with the various steps clearly linked and easily perceived by each individual. Every individual is capable of performing the step-wise, repetitive system when the need is felt. Although human tool production is more varied and more complex than that of non-humans, the provision system consists of the same steps. In both animal and human tool provision systems the job is done by the individual, although we may note that humans tend to carry out tool manufacture activities in small groups of two or three individuals. Tool manufacture can be learned by observation, imitation, and experimentation (Wynn 1991) with no instruction nor language required.

There is, though, a fundamental difference between the above-mentioned provision systems in humans and non-humans: the former carry out these activities in specific locations, a behavior never observed in non-human primates (Dominguez-Rodrigo 1994).

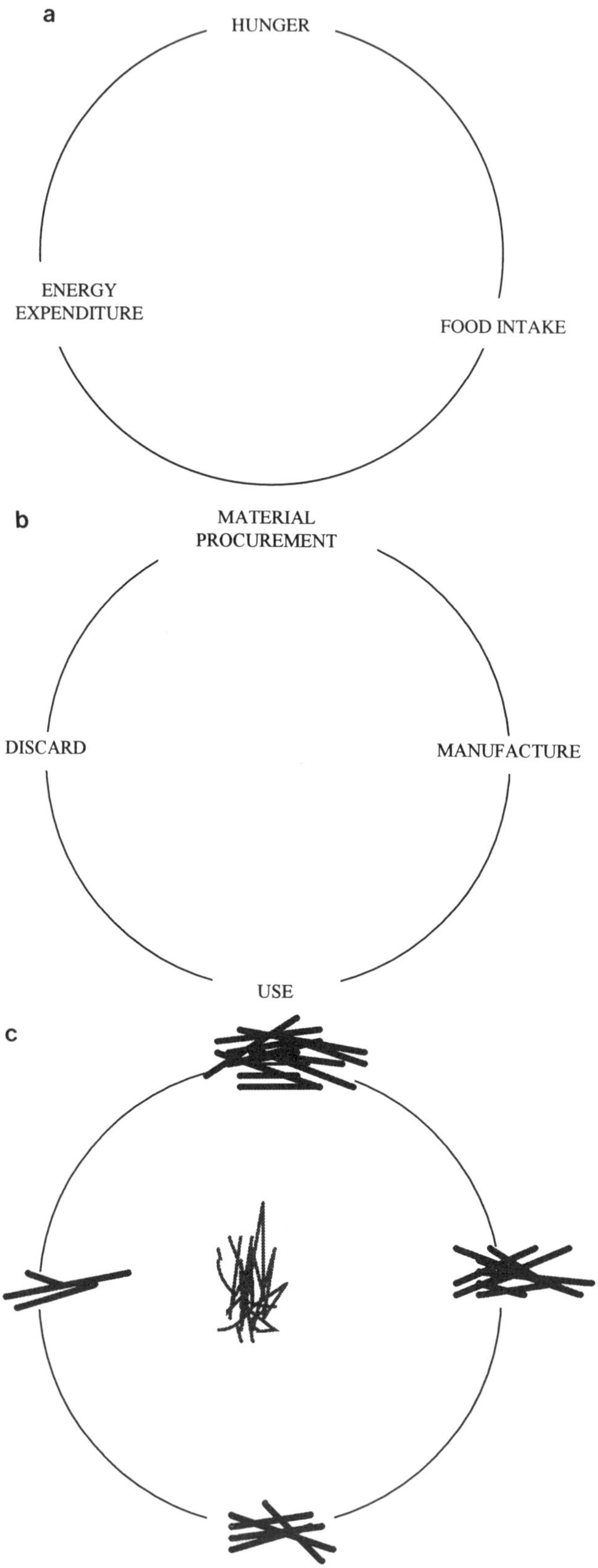

Figure 1. a: Food provision system; b: Tool provision system; c: Fire provision system.

Whether we call such sites "home base" (Isaac 1983) or "preferential location" (Dominguez-Rodrigo 1994), the important fact is that such concentrations in space reflect a qualitative behavioral leap between humans and non-humans. The leap is linked to the capacity for environmental reference, which is similar to the well-known human capacity of name giving (Reiss 1989). Thus the early appearance of this specifically human behavior may perhaps hint at an early appearance of a human system of communication, probably rudimetary.

5. DOMESTIC FIRE

After the food and tool provision systems, a third provision system is introduced to the archaeological record: domestic fire. The precise time of the domestication of fire is still debated, but there is good evidence that it was during the Lower Paleolithic (Ronen and Tsatskin 1994) and preceded the oldest symboling behavior presently recognized, which is manifested in burials. The fire provision system differs in some important aspects from the two former ones. First, it is strictly and uniquely human, not shared in any form by any other species. Hence domestic fire is not a difference in grade, as for example in the case of tool production, but in quality. The implications of such species-specific activities as the maintenance of fire or burials are difficult to evaluate. Nevertheless, the very fact of their uniqueness may perhaps hint at the presence of unique social/cognitive capacities which are not directly evidenced in the archaeological remains.

Secondly, unlike the provision systems discussed formerly, providing for fire is not an integrated system in that the element "fire" and the accumulation of fuel are only indirectly linked. The supply of fuel does not immediately affect the fire. In this provision system the repetitive steps are independent of the subject of concern, the fire. Need and trigger are not as closely perceivable by each group member as in the case of food or tools. The fire provision system is thus a detached system, involving planning to a greater extent than in the integrated systems formerly discussed.

In my view, the initial phase of domestic fire calls for the following assumptions:

1. Fire had to be kept with no knowledge of starting it.
2. Introducing fire to the dwelling place required the consent of all group members.
3. Keeping and supplying the fire were group activities.

The first point above seems justified in that we may expect the novel acquisition of fire to go through similar stages as other inventions, namely from "primitive" to "advanced." Hence it may be expected that it took some time before the technology to start a fire was invented. Before that happened, fire had to be kept alive and every time it was extinguished, it had to be recaptured from some natural source. Keeping the fire means that when the time came, new fuel had to be brought in as a response to some alert, which had to be given by sound to the members of the group. The second and third points relate to a group cooperation and social complexity which, I will argue, exceed those of non-humans. I will argue also that the level of communication required for the communal keeping of the domestic fire also exceeded non-human communication systems. A brief examination of non-human and human coalition and communication systems follows.

6. GROUPS AND COALITIONS

Life in groups is common in the animal world. The main reason for life in groups seems to be an increased foraging capacity (Dunbar 1988:108). Life in groups has its price: "The interests of different individuals often will be in conflict. Therefore, larger patterns of social behavior . . . may be the emergent result of the conflicting interests of interacting individuals. Frequently, therefore, the behavior of an individual . . . will be the mutual result of its interests and the counter-strategies of those with whom the individual is associated" (Dunbar 1988:96). Animals in groups have to coordinate their activities so as to remain together (Dunbar 1988:113), which entails dominance, submission, and high levels of aggression.

Subgroups of varying size and composition are formed for specific activities among humans, chimpanzees, spotted hyenas, lions, and wolves (Manson and Wranglar 1991:372). Primate and non-primate coalitions are used mainly to (1) protect others, especially kin; (2) allow others to gain access to resources; (3) raise rank of kin; (4) improve own access to resources (Harcourt 1992:465). Functionally-oriented coalitions among male chimpanzees change over time depending on gain prospects. Female coalitions, mainly built upon sympathy, are stable over longer periods (de Waal 1992). Thus ape cooperation forms most frequently for immediate purposes, involves only a few individuals, and is based on the evaluation of given and received behavior like grooming, sex, and food sharing (Noe 1992; de Waal 1992). On the other hand, the human coalition involved in keeping fire must have included the entire group and lasted for a long time, while the burden of supplying fuel brought only delayed benefits. This type of group activity appears to be peculiar to humans alone, and seems to indicate a type of social organization unseen among non-humans.

7. COGNITION AND LANGUAGE

Human communication is based upon sound signals, whereas among monkeys and apes, sound is but an addition to other, more important means of communication—olfactory signals and visual gestures (Lancaster 1975:59). Most non-human calling is reserved for information about resources and predators. Thus, chimpanzees converge on particularly good food resources as a result of special calls given by individuals who locate the resource (Reynolds and Reynolds 1965; Tanner 1987). Individuals who spot a predator alert the group by cries (de Waal 1992). While primate calls communicate the emotional situation of the caller, human language also carries reference to the environment. This is done by naming and by numerical abstractions (Lancaster 1975:74; Reiss 1994), that is, by symbols.

Some primate calls, however, may also seem to convey a reference to the environment, for example in the case of cries differentiating between an eagle and a snake (Lancaster 1975:64), or the cry of vervets indicating "leaving the safety of trees" (Cheney and Sefarth 1985). Hamadryas decide on the day's route according to food availability (Dunbar 1988:111), hence they seem to be able to process information pertaining to the environment. But even if these instances indeed testify to some rudimentary reference to the environment, it is clear that primate communication does not allow instruction, discussion of previous experience, or future plans, as human language does. Social learning among apes is done by gestures and observations, mainly between mothers and their offspring (Loy and Peters 1991:110). Among non-humans, information is transmitted mainly to peers (horizontally) rather than across generations (vertically), while humans use both horizontal and vertical information transmission (Laland, this volume).

Tamed fire versus the roaring and terrifying natural mass is not analogous to a stone tool versus its parent rock. In the latter, the potential enclosed in a harmless matter was realized; in the former, a reversed potential was realized, that is, the benefits hidden in a dangerous matter. Such a realization implies the ability to perceive a thing differently than its apparent nature, in other words, to recognize oppositions. The case of fire is rather analogous to transforming a wolf into a domestic dog, an achievement only 12,000 years old.

Keeping and feeding fire must have involved the notion of object permanence and the notion of disappearance, as well as that of success and failure (Gopnik and Meltzoff 1986). Successfully keeping fire required testing cause and effect relationships (as between air, humidity, and fuel), learning from past experience and planning (Parker and Milbrath 1993:329), instructing prohibition, especially to children, and systematically using units of measurement (fuel quantity), hence, mathematical abstraction. Providing for the fire requires yet another feature which is absent in all communication systems but the human: displacement (Hockett and Ascher 1964). Displacement means a message with no immediate sensory contact with the event to which the message refers, as is the case of fuel which is only indirectly related to the fire (the "detached provision system" discussed above). All these mean a human communication system.

8. SYMBOLISM

There is yet another reason for connecting fire with the ability for language. Fire is not merely a "Paleolithic tool and weapon" (Oakley 1956:36). In a recent volume entitled *Tools, Language and Cognition*, fire is hardly mentioned, and then not once as a tool (Gibson and Ingold 1993). This is no accident, and it stands in accordance with folklore and myth where fire plays a role like no other tool: it was hidden from humans and had to be stolen by force or guile from some powerful owner (Frazer 1930). Beyond being a tool, fire is a symbol. At present it symbolizes safety, passion, and excitation (Grinstein 1952; Pyne 1993), life itself.

When did fire become a symbol? A symbol stands for itself and some additional notion, while an ordinary object stands for itself and nothing else. Fire cannot be in this category, for it has no simple self. It is permanently moving and ever changing: it emits warmth and sounds; it consumes food; it is a red living substance, or Life; and, when extinguished, it is Death. Fire is the only substance which humans can kill and revive at will. Its symbolic potential must have emerged, in all probability, fairly early after domestication. Conversely, it cannot be excluded that fire was domesticated precisely because it had acquired some special meaning. This possibility could perhaps explain the otherwise surprising and hardly thinkable introduction of such dangerous and frightening matter into the home (Bachelard 1964:23), within the reach of young children. But, whether before or after domestication, fire must have come to symbolize its most obvious representation: human singularity. Humans stand alone among the myriad of creatures around in possesing fire, a fact which could not have passed unnoticed. It is the ultimate weapon which no Other can use or defend itself against. If there had been a trigger to arouse self consciousness and the ultimate sense of "otherness," it was fire. Given the high cognitive capacity of Acheulean humans (Wynn 1993; Belfer-Cohen and Goren-Inbar 1994), it is unlikely that these humans were unaware of their difference from other animals and their unique position in the world. Symbolic processes express meaning based upon cognitive processes (Langer 1993:308), and it is suggested here that in the course of the Lower Paleolithic, domestic fire could have come to symbolize the unbridgeable gap separating humans from animals, and the human condition as a whole. Through language, then, fire "became paradigm for all of humanity's interaction with nature" (Pyne 1993:248).

ACKNOWLEDGMENTS

Dr. Dan Kaufman read an earlier draft of this paper and his critical remarks were helpful. Dr. Nira Reiss commented on some linguistic aspects and brought to my attention the issue of displacement, for which I am very grateful.

REFERENCES

Bachelard, G. 1964. *The Psychoanalysis of Fire*. London: Routledge and Kegan Paul.

Belfer-Cohen, A., and N. Goren-Inbar. 1994. Cognition and communication in the Levantine Lower Palaeolithic. *World Archaeology* 26:144–57.

Boehm, C. 1992. Segmentary 'warfare' and the management of conflict: comparison of East African chimpanzees and patrilineal-patrilocal humans, 137–73. In *Coalitions and Alliances in Humans and Other Animals*, ed. A. H. Harcourt and F. B. M. de Waal. Oxford: Oxford University Press.

Chase, P. G., and H. L. Dibble. 1987. Middle Paleolithic symbolism: A review of current evidence and interpretations. *Journal of Anthropological Archaeology* 6:263–96.

Chase, P. G. 1991. Symbols and Paleolithic artifacts: Style, standardization, and the imposition of arbitrary form. *Journal of Anthropological Archaeology* 10:193–241.

Cheney, D. L., and R. M. Sefarth. 1985. Vervet monkey alarm calls: Manipulation through shared information? *Behaviour* 94:150–66.

Davidson, I., and W. Noble. 1993. Tools and language in human evolution, 363–88. In *Tools, Language and Cognition in Human Evolution*, ed. K. R. Gibson and T. Ingold. Cambridge: Cambridge University Press.

Dibble, H. L. 1984. Interpreting typological variation of Middle Paleolithic scrapers: Function, style, or sequence of reduction? *Journal of Field Archaeology* 11:431–6.

Dominguez-Rodrigo, M. 1994. The origins of human behaviour: A question of gradual shift or qualitative leap? *Anthropologie* (Brno) 32:33–44.

Donald, M. 1991. *Origins of the Modern Mind*. Cambridge: Harvard University Press.

Dunbar, R. I. M. 1988. *Primate Social Systems*. London: Croom Helm.

Frazer, J. G. 1930. *Myths of the Origin of Fire*. London: McMillan.

Gargett, R. H. 1989. Grave shortcomings: The evidence for Neandertal burial. *Current Anthropology* 30:157–90

Gibson, K. R. 1993. Tool use, language and social behavior in relationship to information processing capacities, 251–69. In *Tools, Language and Cognition in Human Evolution*, ed. K. R. Gibson and T. Ingold. Cambridge: Cambridge University Press.

Gibson, K. R., and T. Ingold, eds. 1993. *Tools, Language and Cognition in Human Evolution*. Cambridge: Cambridge University Press.

Goodenough, W. 1981. *Language, Culture and Society*. Menlow Park: Benjamin Cummings.

Gopnik, A., and A. N. Meltzoff. 1986. Relations between semantic and cognitive development in the one-word stage: The specificity hypothesis. *Child Development* 57:1040–53.

Goren-Inbar, N. 1994. The Lower Paleolithic of Israel, 93–109. In *The Archaeology of Society in the Holy Land*, ed. T. E. Levy. London: Leicester University Press.

Grinstein, A. 1952. Stages in the development of control over fire. *International Journal of Psychoanalysis* 33:416–20.

Hacking, I. 1983. *Representing and Intervening*. Cambridge: Cambridge University Press.

Harcourt, A. H. 1992. Coalitions and alliances: Are primates more complex than non-primates?, 445–71. In *Coalitions and Alliances in Humans and other Animals*, ed. A. H. Harcourt and F. B. M. De Waal. Oxford: Oxford University Press.

Hockett, C., and R. Ascher. 1964. The Human Revolution. *Current Anthropology* 5:135–47.

Isaac, G. L. 1983. Aspects of human evolution, 509–43. In *Evolution from Molecules to Men*, ed. D. S. Bendall. Cambridge: Cambridge University Press.

Lancaster, J. B. 1975. *Primate Behavior and the Emergence of Human Culture*. New York: Holt, Rinehart and Winston.

Langer, J. 1993. Comparative cognitive development, 300–13. In *Tools, Language and Cognition in Human Evolution*, ed. K. R. Gibson and T. Ingold. Cambridge: Cambridge University Press.

Loy, J. D., and C. B. Peters. 1991. *Understanding Behavior*. Oxford: Oxford University Press.

Manson, J. H., and R. W. Wranglar. 1991. Intergroup aggression in chimpanzees and humans. *Current Anthropology* 32:369–90.

Marshack, A. 1976. Some implications of the Paleolithic symbolic evidence for the origins of language. *Current Anthropology* 17:274–82.

Noe, R. 1992. Alliance formation among male baboons: Shopping for profitable partners, 137–73. In *Coalitions and Alliances in Humans and Other Animals*, ed. A. H. Harcourt and F. B. M. de Waal. Oxford: Oxford University Press.

Oakley, K. 1956. Fire as Palaeolithic tool and weapon. *Proceedings of the Prehistoric Society* 21:36–48.

Pader, E. J. 1982. *Symbolism, Social Relations and the Interpretation of Mortuary Remains*. Oxford: British Archaeological Reports International Series 130.

Parker, S. T., and C. Milbrath. 1993. Higher intelligence, propositional language, and culture as adaptations for planning, 314–33. In *Tools, Language and Cognition in Human Evolution*, ed. K. R. Gibson and T. Ingold. Cambridge: Cambridge University Press.

Pyne, S. J. 1993. Keeper of the flame: A survey of anthropogenic fire, 245–66. *Fire in the Environment*, ed. P. J. Crutzon and J. G. Goldammer. New York: Wiley.

Reiss, N. 1989. Speech act taxonomy, chimpanzee communication, and the evolutionary basis of language, 282–304. In *Studies in Language Origins*, Vol. 1, ed. J. Wind, E. G. Pulleyblanc, E. de Grolier, and B. H. Bichakjian. Amsterdam: John Benjamins.

Reynolds, V., and F. Reynolds. 1965. Chimpanzees of the Budongo Forest, 368–424. In *Primate Behaviour*, ed. I. DeVore. New York: Holt, Rheinhart and Winston.

Ronen, A., and A. Tsatskin. 1994. New interpretation of the oldest part of the Tabun cave sequence, Mount Carmel, Israel, 265–81. In *Man and Environment in the Palaeolithic*, ed. H. Ullrich. Liège: ERAUL 62.

Tanner, N. A. 1987. The chimpanzee model revisited and the gathering hypothesis, 3–27. In *The Evolution of Human Behavior: Primate Models*, ed. W. G. Kinzey. Albany: SUNY Press.

Texier, P. J. 1995. The Oldowan assemblage from NY 18 site at Nyabusosi (Toro-Uganda). *Comptes Rendus de l'Académie de Sciences* 320:647–53.

Tooby, J., and I. DeVore. 1987. The reconstruction of hominid behavioral evolution through strategic modeling, 183–237. In *The Evolution of Human Behavior: Primate Models*, ed. W. G. Kinzey. Albany: SUNY Press.

Toth, N. 1985. Archaeological evidence for preferential right-handedness in the Lower and Middle Pleistocene, and its possible implications. *Journal of Human Evolution* 14:607–14.

Toth, N., and K. Schick. 1993. Early stone industries and inferences regarding language and cognition, 346–62. In *Tools, Language and Cognition in Human Evolution*, ed. K. R. Gibson and T. Inghold. Cambridge: Cambridge University Press.

Ucko, P. J. 1969. Ethnography and archaeological interpretation of funerary remains. *World Archaeology* 1:262–80.

Ullrich, H. 1995. Mortuary practices in the Palaeolithic—Reflections on human-environment relations, 363–78. In *Man and Environment in the Palaeolithic*, ed. H. Ullrich. Liège: ERAUL 62.

Vandermeersch, B. 1970. Une sepulture mousterienne avec offrandes decouverte dans la grotte de Qafzeh. *Comptes Rendus de l'Académie des Sciences* 270D:298–301.

de Waal, F. B. M. 1992 Coalitions as part of reciprocal relations in the Arnhem chimpanzee colony, 233–57. In *Coalitions and Alliances in Humans and Other Animals*, ed. A. H. Harcourt and F. B. M. de Waal. Oxford: Oxford University Press.

Wynn, T. 1991. Tools, grammar and the archaeology of cognition. *Cambridge Archaeological Journal* 1:191–206.

Wynn, T. 1993. Layers of thinking in tool behavior, 389–406. In *Tools, Language and Cognition in Human Evolution*, ed. K. R. Gibson and T. Inghold. Cambridge: Cambridge University Press.

Wynn, T., and W. C. McGrew. 1989. An ape's view of the Oldowan. *Man* 24:383–98.

A COMPARATIVE PERSPECTIVE ON PALEOLITHIC CULTURAL PATTERNS

Kathy D. Schick

Anthropology Department and CRAFT Research Center
Indiana University
Bloomington, Indiana 47405

1. INTRODUCTION

Many important questions surround the place of Neandertals and early modern humans in the course of human evolution. Prominent among these are questions concerning the extent of biological and cultural connections among synchronous populations in various parts of Asia, Europe, and Africa throughout the Pleistocene, and the biological and cultural links between any regional population and those that precede or succeed it in time. In addition to exploring such questions through morphological and taxonomic studies of hominid remains, it is also useful to examine the archaeological evidence for broader technological patterns over time and space during the Paleolithic and, particularly, to examine these for possible cultural affinities among hominid populations.

Technological continuities or discontinuities may reflect the nature of cultural links among populations across time or space that, in turn, can bear on questions of biological connections in terms of gene flow. Furthermore, the larger patterns observable in technological activities and products can provide information regarding aspects of cognitive abilities and cultural processes. In order to look at these issues in this perspective, it is useful to examine the broader details of Paleolithic technological developments throughout hominid dispersals out of Africa into Europe and Asia.

This study will take a broad, comparative look at possible archaeological indications of connections among Paleolithic populations in different parts of the Old World throughout the course of early hominid dispersals, migrations, or expansions of population. Questions concerning the earlier Paleolithic will be of central interest, but issues concerning general technological patterns during later phases of the Paleolithic will also be addressed. Of concern here will be evident continuities and/or discontinuities in technological patterns over relatively broad reaches of time and of space. Of primary concern here will be the possible significance of archaeological patterns with regard to cultural connections among hominid populations during the Early and Middle Paleolithic.

Neandertals and Modern Humans in Western Asia, edited by Akazawa *et al.*
Plenum Press, New York, 1998.

2. CULTURAL ASPECTS OF EARLY PALEOLITHIC TECHNOLOGIES

2.1. Oldowan Technology: Minimal Evidence of Culture

The earliest definite evidence of stone tool-making and an archaeological record is in Africa. Simple stone tool industries appear in East Africa well before 2 million years ago (with the oldest evidence yet known at the site of Gona, currently dated to 2.5 million years ago) (Semau et al. 1997). Such industries, often referred to as "Oldowan" technologies, consist basically of flaked cobbles and chunks of stone and the angular debris (flakes and fragments) that have been removed from them by percussive blows with battered hammerstones (Leakey 1971; Isaac 1976, 1982; Harris 1983; Schick and Toth 1993). Such stone tool-making behaviors, archaeologically visible and practiced by at least some early hominid groups by at least two and a half million years ago, become relatively widespread and sufficiently consistent to build up regional concentrations of sites, particularly in various sedimentary basins of East Africa, starting between 2.0 and 1.5 million years ago (Leakey 1971; Isaac 1976).

As such tool-making would appear to represent behaviors shared and learned in a social group and ultimately passed on from one generation to the next, they have often been recognized as a simple yet persistent cultural tradition (Isaac 1976). Although experiments have shown that the flakes prevalent in such assemblages make extremely useful and efficient cutting tools and likely were a main goal of Oldowan stone working (Toth 1985), typological emphasis has traditionally been placed on the cores from which they have been flaked. Thus comparisons and contrasts among sites have often been made on the basis of morphological types, usually with the implicit assumption that similarities may denote or reflect a greater degree of shared culture.

Oldowan technology does indicate a sharing of knowledge of tool manufacturing and tool-using processes, and would appear to indicate a cultural linking in at least a very minimal sense. Although it is not impossible that stone tool-making was invented or re-invented more than once, the continuity of stone tools over long reaches of time and tremendous geographic extent as hominids expanded their range would argue for the regular cultural transmission of stone tool-making as a learned behavior acquired within the social group, just as chimpanzees in the wild show a variety of tool-making and tool-using behaviors that are culturally transmitted within groups or even large-scale communities (McGrew 1992).

A minimal cultural aspect of Oldowan artifacts is, in fact, highlighted by systematic flaking showing intuitive understanding of the physical problems of stone fracture, including assessment of the three-dimensional geometry of the core and controlled placement of the percussive blow. Experimental studies of stone tool manufacture by an ape (*Pan paniscus*) highlight the hidden complexities of the tool-making process for simple Oldowan tools (Toth et al. 1993). Thus despite the lack of stylistic patterning evident in Oldowan tools, knowledge of basic techniques of manufacture as well as involvement of tools in hominid adaptation were most likely culturally transmitted throughout the course of human evolution.

Despite the apparent dependence on social learning for such tool-making behaviors (hence their "cultural" aspect) and their persistence over time (hence their attribution as a "tradition"), however, Oldowan-type or Mode 1 industries (denoting simply flaked cores and flake products) (Clark 1961) do not manifest intricate or elaborate rules governing their manufacture nor distinctive stylistic constraints. Furthermore, experiments demonstrate that the nature of the final products of stone tool-manufacturing in Mode 1 technology depends largely on characteristics of the raw materials used, such as size, shape, and flaking qualities (Toth 1985). Artifact morphologies or "types" can show strong similarities from site to site not due to stylistic rules being shared by the responsible tool-making groups, but due to similar constraints exerted by the raw materials used.

Conversely, Mode 1 industries can show strong differences from site to site not due to different cultural or stylistic rules but largely due to the use of raw materials of different size, shape, or flaking characteristics. Special types of artifacts observed in some assemblages, such as unifacial cores ("choppers") and the heavy duty "Karari" scrapers seen in many Koobi Fora assemblages, could possibly represent local technological styles or trends, but may well be governed by raw material effects, such as the use of small wedge-shaped cobbles influencing the production of unifacial cores. Experiments have also shown that two other Oldowan artifact types often considered deliberately-shaped tool forms, battered quartz spheroids and facetted limestone spheroids, may well be by-products of use of a quartz cobble as a hammerstone (battered spheroids) (Schick and Toth 1994) or the reduction of limestone cobbles to exhaustion (facetted spheroids) (Sahnouni et al. 1997).

As early Mode 1 sites do not tend to show evidence for strong preferential selection of raw material but primarily the ad hoc use of what was available in the region, the influence of raw material on variability in Oldowan or Mode 1 assemblages should not be underestimated. Mode 1 assemblages give no clear indication of close "cultural" relatedness among different tool-making groups throughout space or time. Technological similarities or differences from site to site and time to time do not necessarily reflect the degree of "cultural" similarity or differentiation. Oldowan technology may reveal culturally transmitted tool-making behaviors, but it is unfortunately not a powerful indicator of specific sets of tool-making rules or traditions.

Thus there are no reliable, convincing ways to formulate hypotheses of cultural identity or relatedness based on characteristics of simple Oldowan industries, at least beyond the rudimentary inference that stone tool manufacture and use was habitually learned and passed on through generations and throughout the spread of hominid populations. Mode 1 technology is almost certainly based on basic shared cultural knowledge of toolmaking, but beyond that it yields little information regarding specific cultural rules, norms, or strictures.

Although early tool-makers certainly had culturally patterned behaviors not explicitly reflected in the stone tool record, we unfortunately have no means to ascertain these. For the first million years or so of stone tool-making, simple Oldowan technology prevails in the Early Stone Age with no definite evidence of any substantive technological innovations or culturally-based stylistic variants apparent. Thus Oldowan technology provides minimal, limited evidence of whatever cultural patterns the hominid groups may have had. Therefore, there is little basis to formulate hypotheses of strong cultural affinity among Mode 1 sites in any one region of the world, whether in Africa, Europe, or Asia, or from one of these regions to the next, and of course, even Holocene assemblages can mimic "Oldowan" technology. Oldowan or Mode 1 artifacts may demonstrate a basal level of continuity of a culture of stone-tool making, but beyond this they seem to reveal little regarding cultural rules.

2.2. Acheulean Technology: New Evidence for Cultural Rules

Starting at least 1.5 million years ago with the emergence of Acheulean technology in East Africa, we can identify the first definite evidence of planned, deliberate production of definite target forms in stone tool-making, as in the shaping of large bifacial forms such as handaxes and cleavers. These do not totally replace Oldowan tools, which are generally found alongside handaxes and cleavers at Acheulean sites, and Oldowan-like artifacts also occur at contemporary, handaxe-less sites in Africa, Europe, and Asia (the latter situation is discussed below). Acheulean tools do represent, however, the first good evidence of cultural rules or norms guiding the shaping of standardized forms. As such, they constitute a much more rich and robust indicator of culture and a potential means of tracing cultural connections. Moreover, there can be detected a general temporal trend from cruder to more refined products, as well as refinement in flaking techniques over the span of the Acheulean (e.g., use of the Levallois method to produce flake blanks and "soft hammer" in biface production in some later Acheulean assemblages).

Evident rules at work in Acheulean technology can usefully be broken down into (a) rules of the tool-making procedure, or various operations and technological acts involved in tool manufacture, and (b) rules governing characteristics of the resultant tool, e.g., in standardized characteristics of attributes such as size, shape, symmetry, edge properties (Wynn 1989). (The latter set of rules, those governing the final tool form, depend upon execution of the tool-making procedures, but impose an added set of requirements). The manufacturing process itself shows the operation of a number of standardized procedures starting with the choice or production of the initial form or blank for the tools (either choosing a large, flat nodule or striking off a very large flake), followed by implementation of a number of operations to shape it. Especially in later Acheulean assemblages, this shaping can show simultaneous attention to different concerns, for instance, developing the shape in plan-view, often developing and maintaining bilateral symmetry in plan-view and cross-section, commonly thinning the cross-section of the tools, and developing a sharp and straight or slightly sinuous edge toward the tip, which is usually the most carefully worked and thinned area of a handaxe (see Schick 1994:583–4).

In addition, there are sometimes idiosyncratic technological procedures evident in particular regions, as the minimally modified Tabelbala or Tachenghit cleavers on flakes struck from prepared cores in North Africa (Clark 1970; Isaac 1982); the Kombewa method of producing large, biconvex flake (double-bulb) flake blanks for bifaces (Clark 1970), as seen in northern Africa and now in Israel (Goren-Inbar et al. 1991); the Fauresmith variant of the late Acheulean in South Africa; and the "Victoria West technique" observed in different parts of Africa (Isaac 1982). Thus the basic technological operations involved in producing Acheulean industries show evidence of a prescribed set of rules in tool-making, ones that had to have been culturally transmitted and maintained within certain regions.

In addition to the rules evident in the actual process of tool manufacture, many of the products—the Acheulean tools—often show other signs of strong rules governing their final shape, such that Acheulean bifaces found in one occurrence can show remarkable uniformity in technological procedures or morphological attributes, yet considerable variation from one occurrence to another within a single site-complex (e.g., Kleindienst 1961; Howell et al. 1962; Clark 1969; Isaac 1977). In many Acheulean assemblages, numbers of handaxes and cleavers often show remarkable consistency in any of a number of general characteristics, such as their size or planform (ovate, lanceolate, etc.). A later Acheulean site (approximately 0.5 million years old) studied by J. Desmond Clark and the author in the Middle Awash Valley of Ethiopia, for instance, contained quantities of small, well-made handaxes and cleavers, all between 12 and 13 cm long, and many with a distinctive chisel bit, as just one example of evidence for strong adherence to rules in Acheulean technology.

It is due to the identification of culturally determined or modified patterns in the Acheulean, involving both complicated technological procedures and the imposition of arbitrary patterns of style or design, that we can identify and suggest more precise cultural links from one hominid population to another throughout time and space. It could still be argued that some stylistic and technological trends can emerge by independent invention, and it will be the task of archaeologists to meet this objection squarely and identify patterns that most reasonably appear to represent cultural connections. Nevertheless, the production in the Acheulean of tools finely and proficiently shaped into consistently repeated forms raises some important questions as to the cultural norms governing or influencing this tool-making as well as the cognitive operations of the hominid tool-makers themselves. These questions are addressed further below.

2.3. Technological Thresholds in the Early Paleolithic

As discussed above, in Oldowan technology there is little positive evidence of predetermined forms in the stone tool assemblages. Although stone tool-making shows good understanding of the basic principles of hard hammer and bipolar percussion and was maintained and transmitted for hundreds of thousands of years, undoubtedly through cultural learning, we cannot identify strong rules governing the tool manufacturing process nor recurrent patterns in the artifact assemblages throughout this long time period. It is, of course, impossible to preclude that hominids *could* have made other types of artifacts, ones showing the imposition of elements of style or design, but such patterns are not seen throughout the earliest million years or so of Oldowan tool-making.

Technological procedures observed in Acheulean technology, however, are not ad hoc operations, but show a prescribed set of procedures done in a sequence so as to produce forms with great stylistic consistency at many sites and even within some regions. The rules governing these procedures are certainly cultural in nature, shared among groups and transmitted over vast regions as hominids spread out of Africa and into Europe and Western Asia. Moreover, as discussed below, these tool-making procedures show great tenacity or longevity, as Acheulean traditions are maintained for one to one-and-a-half million years of hominid evolution.

2.4. The Spread to Europe and Western Asia: Technological Variability

As hominid populations spread from Africa into Europe and Asia, tool-making behavior also spread. The precise details of the timing of these excursions from Africa are still somewhat hazy and still need to be worked out with more information from new sites and better chronological information overall. Nevertheless, we can now identify some major areas as important landmarks during these population expansions and establish general time frameworks for this spread.

Sites documented in western Asia represent good evidence of early hominid migrations to this area prior to 1 million years ago. The site of 'Ubeidiya in Israel indicates hominid presence by between 1 to 1.4 million years ago (Bar-Yosef and Goren-Inbar 1993), and the Dmanisi site in Georgia also appears to establish extension of the hominid range further northward during this same general period (perhaps between 1–1.5 million years ago) (Bar-Yosef 1994, 1995). While initial levels at 'Ubeidiya contain Mode 1 artifacts, later ones also have bifacial Acheulean tools. The Dmanisi site contains only simple Mode 1 artifacts (Dzaparidze et al. 1989), establishing variable site patterns early in the course of hominid migrations or expansions in the Early Pleistocene.

The nature and timing of the earliest occupation sites in Europe has been somewhat problematic. Although various sites have been suggested to represent occupation of parts of Europe by the Early Pleistocene, this has been seriously challenged in a study that found that sites pre-dating 500,000 years ago had either dubious artifacts, unreliable chronology, or questionable context (Roebroeks and van Kolfschoten 1994). The lower artifact-bearing level at the Trinchera Dolina site at Atapuerca could represent a good candidate for Early Pleistocene habitation with archaeological evidence predating the Brunhes-Matuyama reversal at 780,000 BP, although questions of chronology have also been raised here (Dennell and Roebroeks 1996). The archaeological materials in these older levels at Atapuerca constitute simple, Oldowan-like cores and flaking debris, although Atapuerca localities dated to the Middle Pleistocene contain tools that have been considered Acheulean. In any case, once habitation of Europe becomes solidly established after 500,000 BP, sites with Acheulean tools and sites with non-Acheulean or simple Mode 1 artifacts are found.

As in Africa, the presence of both types of industries has been a topic of concern and controversy. In Africa, Mode 1 artifacts (i.e., simply flaked cores) are found within Acheulean assemblages and also at contemporaneous non-Acheulean sites in the Early and Middle Pleistocene, as in Beds I and II of Olduvai Gorge (Leakey 1971). Hypotheses have ascribed this variability to possible differences in activities or tool functions or to cultural differences, among different groups, sexes, populations, or even different species. While its overall significance is still unresolved, it should be kept in mind that these two different technological patterns do persist for a very long period in Africa.

Among the various hypotheses that have been proposed to explain this technological differentiation in Europe, Mithen (1994) has suggested that the two different industries might be explained by different circumstances of social learning in different environments. In this construct, Mode 1 industries are found in more closed forests (particularly during interglacials) where smaller groups would have subsisted and relied on more individual learning, while the Acheulean is associated with more open environments of glacial periods, with larger groups involved in group learning and transmission of ideas, including Acheulean technology. Although this model may not apply well to all of Eurasia, the role of cultural learning in the maintenance of Acheulean technologies may have been a critical factor in the course of population movements or expansions out of Africa into unchartered regions of Europe and Asia (Toth and Schick 1993; Schick 1994), and would seem to deserve further thought and attention.

2.5. The Spread into Eastern Asia: Technological Differentiation

The timing of the earliest occupation of eastern Asia has also generated a great deal of research interest as well as some controversy (Ikawa-Smith 1978; Aigner 1981; Wu and Olsen 1985; Schick and Dong 1993). Recent reports have pointed to the possibility of very early occupation of Java (Swisher et al. 1994) and mainland China at Longgupo (Huang et al. 1995). In the case of Java, where dates at Modjokerto and Sangiran have indicated ages of 1.8 and 1.6 million years BP respectively for some hominid finds, further fieldwork and dating would help substantiate the association of the dated sedimentary detritus with the time of fossil deposition. Further information is also needed regarding the artifactual nature of the lithic materials at Longgupo, as well as resolution regarding their context and substantiation of the derived age (1.9 million years BP), which relies on extrapolation from a chronometric date based upon paleomagnetic stratigraphy. If these sites are secured in this early time range, additional fieldwork would be important to investigate if longer-term populations were established by this time.

Relatively firm evidence for hominid occupation and associated stone artifact industries in eastern Asia appears approximately one million years ago. At sites in the Nihewan Basin and at Lantian in north central China, there is evidence for hominid occupation of northern China by at least 800,000 to 1 million years ago (Wei 1988; Schick et al. 1991). Sites from the Early or Middle Pleistocene in China and eastern Asia consistently contain Mode 1 technologies. This conforms to a pattern noted decades ago by Movius (1949), in which he discerned a boundary (the "Movius Line") separating sites containing handaxes in much of Europe and southwestern Asia from sites without handaxes in the area from northeastern Europe to eastern Asia.

Various hypotheses have been offered to explain this technological differentiation between eastern Asia and the Acheulean technologies present in other parts of the occupied Old World during Early and Middle Pleistocene times. These have included models of cultural isolation, environmental barriers to diffusion, functional differences linked to environmental variability, decay and loss of the tradition during migrations, and use of other materials such as wood and bamboo for tools (see overview in Schick 1994). None of these, however, as yet has been universally accepted or embraced.

More recently, it has been suggested that current evidence does not support this differentiation, and that Acheulean biface traditions have been located in eastern Asia. There are some occurrences of large tools in China, but most of these have been either in undated contexts or appear to be after the end of the Acheulean. Furthermore, bifaces found and reported thus far do not show the standardized technological procedures and products observed in vast numbers of Acheulean occurrences in western Eurasia. Some seem to come from later, Upper Pleistocene contexts, although recent reports attribute bifaces in southern China datable to 780,000 years BP (R. Potts, personal communication), and we await further information on these. Most of the bifaces reported thus far are relatively crudely made, and overall lack the technological and stylistic coherence characteristic of many Acheulean assemblages in Africa, Europe, and western Asia. Most of these East Asian biface occurrences appear to be a separate development, perhaps for specific activities in certain circumscribed environmental zones, and mostly much later in time than the Acheulean tradition in the rest of Eurasia (Schick and Dong 1993).

It is curious, then, that Acheulean technology, which successfully transplanted to multiple environments throughout Africa and then ultimately spread to many occupied portions of Europe and western Asia as far eastward as the Indian subcontinent, does not appear to have become well-established in eastern Asia in the course of hundreds of thousands of years of habitation. Decades ago Movius (1949) suggested that early hominids may have entered eastern Asia before the emergence of Acheulean technology, an idea which has reemerged in light of the possible early sites in eastern Asia dating to before 1.5 million years ago. Even if such an early population expansion had occurred, however, it still would not explain the failure for Acheulean technology to diffuse into eastern Asia throughout the entire span of the Early and Middle Pleistocene. The pattern remains, and remains a problem. Why and how was so much technological differentiation maintained over such a long period of time, minimally three-quarters of a million years? This question will be addressed again below.

3. MIDDLE PALEOLITHIC PATTERNS

Distinctive technological changes are seen in various parts of the world in the later Early Pleistocene/early Late Pleistocene. In Africa and western Eurasia, the Acheulean industries and simple core-and-flake assemblages typical of the Early and Middle Pleistocene industries are replaced by industries emphasizing retouched flake tools. Within this general trend, there is a good deal of regional variability from one region to another in tool forms, for example, in the Mousterian assemblage types recognized in western Europe, the Near East, and Mediterranean North Africa; assemblages in North Africa with tanged points typical of the Aterian Industry; and the various Middle Stone Age industries of sub-Saharan Africa.

Levallois technology, which appears in many later Acheulean industries, also is found in many Middle Paleolithic and Middle Stone Age industries in Africa, Europe, and Asia. If Levallois technology is loosely defined as a special prepared core technology, it has been argued that this does not correspond to a "tradition" and that the far-flung occurrences of Levallois-like technology may well be independent invention and devoid of cultural information (Otte 1996). While it is true that preparation or shaping of cores becomes fairly widespread in many flake industries of the Middle Paleolithic, the prevalent, specially prepared Levallois cores with techniques of core shaping and preparation typical of Levallois industries sensu strictu, would still seem to denote some culturally-based rules or norms in their production and the likely diffusion of such ideas from one place to another (although constraints of raw materials necessarily also play a role; Dibble 1991). Intensive study of the Levallois techniques evident at Middle Paleolithic sites in Europe have shown a variety of different Levallois strategies evident in different sites and in different occupations levels of the same site, revealing great detail on trends in the application of the Levallois method (Boëda et al. 1990) that do not seem dictated simply by raw material (Mellars 1996:90) and could possibly be culturally dictated in some way.

Many of the Levallois-like technologies sometimes noted in parts of eastern Asia tend not to show recurrent use of elaborate, stylized core shaping and platform preparation seen in many Middle Paleolithic industries of western Asia, Europe, and Africa. Some assemblages reported in parts of northeastern Asia, as in Mongolia, however, do show specially prepared cores that have been associated with the term "Levallois" in the technical sense, without claims of direct technological or cultural links with the Levallois method evidenced in southwestern Asia and Europe, for example (Derevianko et al. 1996:78). In China technologies in the Early or Middle Paleolithic do not show emphasis on special core preparation or elaborate preparation of platforms for flake removal. Sites attributed to the Middle Paleolithic in China do show an increased emphasis on flake tools with more consistent and regular trimming and retouch of flakes. The artifact assemblages do not, however, show the highly standardized tool types and regional assemblage types observed, for instance, in the Mousterian of Europe and the Levant and in the Middle Stone Age of sub-Saharan Africa. This pattern appears, in fact, to mirror that seen in the Acheulean, that is, there seems to be an absence of Acheulean technology, Levallois technology, and distinct regional styles in Middle Paleolithic flake tool assemblages in much of eastern Asia.

As procedures of Levallois technology can in many ways be seen as an outgrowth of Acheulean procedures, with emphasis on strategic flaking to shape the mass or form of the flakes removed, on platform preparation to help achieve this end, and overall shaping of the mass of the core, it may be significant that both of these technological patterns are lacking in much of eastern Asia. It appears, then, that a number of strong technological differences were maintained over a very long period of time, which may well denote barriers or restrictions in cultural transmission that were maintained over hundreds of thousands of years. If some profound barriers to cultural transmission were maintained over such long periods of time in the Paleolithic, it seems legitimate to ask whether barriers also limited genetic transmission among some of the populations of the world during much of the Pleistocene (Andrews 1982).

4. DISCUSSION AND CONCLUSIONS

Discussion of hominid migrations and the time and place of early occupations of different parts of the world often tend to make implicit assumptions that early occupation in a region denotes continuous population presence from that time onward, that migrations were one-way and did not reverse direction, and that placing two or more dots on a map justifies assuming successful, thriving populations in much of the intervening area. It seems quite possible, however, that hominid populations did not become established in broader Eurasia after the initial "entrance" or "entrances." Initial populations may well have been intermittent, discontinuous, and very possibly quite sparse on the landscape. Climatic factors, for instance, may have made central Asia rather harsh and inhospitable for occupation throughout much of the colder phases of the Pleistocene. There may have been a series of entrances and retreats or population extinctions. Howell (this volume) has suggested a strong likelihood of some degree of isolation of populations during much of this time, and technological evidence may help us identify such patterns.

It has been argued here that the Acheulean tradition is a relatively robust cultural marker, perhaps not an ethnic marker in the modern sense but a sign of a tight package of culturally transmitted rules and norms regarding stone technology. It is important to consider here the processes of cultural transmission of such tool-making traditions. A complicated technology of this sort is potentially very fragile, and depends on consistency and continuity in its transmission. The stone adze-making tradition of New Guinea, for instance, shows similar intricacy and somewhat similar procedures in the initial stages of reduction, but the failure of the younger generation to acquire this information makes the extinction of this tradition appear imminent (Toth et al. 1992). The archaeological evidence would suggest a relatively profound cultural division between eastern Asia and the rest of the Old World, one that appears to be maintained not only over the long time depth of the Acheulean but perhaps well into Middle Paleolithic times. This may well signal the maintenance of important barriers to the delicate balance of this cultural transmission.

The technological evidence can also be a valuable indicator of other aspects of the culture, behavior, and even cognitive characteristics of early hominids, particularly in the tenacity and even rigidity of early cultural traditions of tool-making. One remarkable feature of the Acheulean, for instance, is its amazing longevity, the transmission and maintenance of techniques and procedures of tool-making over a vast period of time and great distances. This sort of continuity of culturally transmitted information and rules is unthinkable now, and as such, the Acheulean technological tradition is not equivalent with a "culture" or ethnic boundary in a more modern sense, though it does show us the repeated use and transmission of information and rules. These hominids are showing us a tremendous capacity for adherence to rules and procedures of manufacture, sometimes bringing about, as Belfer-Cohen and Goren-Inbar have indicated (1994), remarkably similar products by similar or even different procedures. Although we cannot see a universal or unidirectional stylistic change in Acheulean handaxes, by the final stages we often see remarkable technical control in the production of remarkably symmetrical, finely flaked bifaces. This may be an important indicator of cognitive changes by this time, in that the fashioning of these tools was achieving an apparent importance in addition to their functional use.

There is admittedly a good level of technological complexity in the Acheulean tool-making tradition and in Middle Paleolithic industries, as well as evidence of reliable cultural transmission of information and rules. It nevertheless is also evident that some very critical, fundamental changes have occurred throughout human evolution in overall biology, in the brain and its cognitive processes, in behavior, and in capacity for culture. While various cultural traditions have now been well-documented among apes, these seem to have drastic limitations, and tradition extinctions reportedly may be quite common in chimpanzees (Wrangham 1995). The evolution of hominids from a common ancestor with the apes has necessarily involved changes in the cognitive aspects of brain organization which have undoubtedly increased our capacity for cultural learning and perhaps the nature of culture and culture change.

The relative technological stasis or paucity of directional trends of stylistic change within the Middle Paleolithic likewise presents important questions regarding the nature of culture during this time. There is good evidence that more formal, systematic retouch, and recurrent types of flake tools are observed regionally. It remains a nagging question, however, whether the technological patterns observed in Middle Paleolithic times represent just a variant of culture as we know and see it in recent humans, or whether it represents some major departure from or precursor to a more "modern" form of culture, and perhaps even a different capacity for culture in the hominids themselves. Unfortunately, there is no universal acceptance of what constitutes "modern behavior" nor how to detect it in the archaeological record.

While most researchers seem to agree that an apparent explosion of technological diversity, standardization, and inventiveness observed in the Upper Paleolithic—the diverse stylistic strains that seem to denote "ethnicity," coupled with the appearance of representational art, symbolism, and decorative elements—almost certainly represent modern behavior and culture working more as we know it, Middle Paleolithic patterns remain quite controversial in this regard. Ultimately, archaeologists will hopefully develop new ways to deal with and understand the different kinds of variability seen in the Paleolithic record. Terms such as "standardization," "style," "technological complexity," and so on are becoming increasingly clumsy verbal tools to describe changing Paleolithic patterns, and they provide a poor resolution picture of the actual patterns of cultural developments, diversification, and change over time.

REFERENCES

Aigner, J. S. 1981. *Archaeological Remains in Pleistocene China.* Munich: Verlag.

Andrews, P. 1982. An alternative interpretation of the characters used to define *Homo erectus. Courier forschungsinstitut Senckenberg* 69:167–75.

Bar-Yosef, O. 1994. The Lower Paleolithic of the Near East. *Journal of World Prehistory* 8(3):211–65.

Bar-Yosef, O. 1995. The Lower and Middle Palaeolithic in the Mediterranean Levant: Chronology and cultural entities, 247–63. In *Man and Environment in the Palaeolithic*, ed. H. Ullrich. Liege: Université de Liège.

Bar-Yosef, O., and N. Goren-Inbar. 1993. *The lithic assemblages of 'Ubeidiya: A Lower Palaeolithic site in the Jordan Valley.* QEDEM. Jerusalem: The Institute of Archaeology, Hebrew University of Jerusalem.

Belfer-Cohen, A., and N. Goren-Inbar. 1994. Cognition and communication in the Levantine Lower Palaeolithic. *World Archaeology* 26:144–57.

Boëda, E., J. M. Geneste, and L. Meignen. 1990. Identification de chaînes opératoires lithiques du paleolithique ancien et moyen. *Paleo* 2:43–80.

Clark, G. 1961. *World Prehistory: In New Perspective.* Cambridge: Cambridge University Press.

Clark, J. D. 1969. *Kalambo Falls Prehistoric Site*, Vol. 1. Cambridge: Cambridge University Press.

Clark, J. D. 1970. *The Prehistory of Africa.* London: Thames and Hudson.

Dennell, R., and W. Roebroeks. 1996. The earliest colonization of Europe: The short chronology revisited. *Antiquity* 70:535–42.

Derevianko, A. P., J. W. Olsen, D. Tsevendorj, V. T. Petrin, A. N. Zenin, A. I. Krivoshapkin, R. W. Reeves, E. V. Deviatkin, and V. P. Mylnikov. 1996. *Archaeological Studies Carried Out by the Joint Russian-Mongolian-American Expedition in Mongolia in 1995*, ed. A. P. Derevianko, J. W. Olsen, and D. Tsevendorj. Novosibirsk: Izd-vo Instituta arkheologii i etnografii SO RAN.

Dibble, H. L. 1991. Local raw material exploitation and its effects on Lower and Middle Paleolithic assemblage variability, 33–46. In *Raw Material Economies Among Prehistoric Hunter-Gatherers,* ed. A. Montet-White and S. Holen. Lawrence: University of Kansas Publications in Anthropology No. 19.

Dzaparidze, V., G. Bosinski, T. Bugianisvili, L. Gabunia, A. Justus, N. Klopotovskaja, E. Kvavadze, D. Lordkipanidze, G. Majszradze, N. Mgeladze, M. Nioradze, E. Pavlenisvili, H.-U. Schmincke, and D. Sologasvili. 1989. Der Altpalaolithische Fundplatz Dmanisi in Georgien (Kaukasus). *Jahrbuch des Romisch-Germanischen Zentralmuseums Mainz* 36:67–116.

Gabunia, L., and A. Vekua. 1990. L'evolution du paleoenvironement au cours de l'anthropogene en Georgie (Transcaucasie). *L'Anthropologie* 94:643–50.

Goren-Inbar, N., I. Zohar, and D. Ben-Ami. 1991. A new look at old cleavers—Gesher Benot Ya'aqov. *Journal of the Israel Prehistoric Society* 24:7–33.

Harris, J. W. K. 1983. Cultural beginnings: Plio-Pleistocene archaeological occurrences from the Afar, Ethiopia. *African Archaeological Review* 1:3–31.

Howell, F. C., G. H. Cole, and M. R. Kleindienst. 1962. Isimila, an Acheulian occupation site in the Iringa Highlands, 43–80. In *Actes du IVe Congres Panafricain de Prehistoire et de l'Etude du Quaternaire.* Tervuren: Musée Royal de l'Afrique Centrale.

Huang, W., R. Ciochon, Y. Gu, R. Larick, Q. Fang, H. Schwarca, C. Yonge, J. de Vos, and W. Rink. 1995. Early Homo and associated artefacts from Asia. *Nature* 378:275–8.

Ikawa-Smith, F., ed. 1978. *Early Paleolithic in South and East Asia.* The Hague: Mouton.

Isaac, G. L. 1976. Plio-Pleistocene artifact assemblages from East Rudolf, Kenya, 552–64. In *Earliest Man and Environments in the Lake Rudolf Basin: Stratigraphy, Paleoecology, and Evolution*, ed. Y. Coppens. Chicago: University of Chicago Press.

Isaac, G. L. 1977. *Olorgesailie: Archaeological Studies of a Middle Pleistocene Lake Basin in Kenya.* Chicago: University of Chicago Press.

Isaac, G. L. 1982. The earliest archaeological traces, 157–247. In *The Cambridge History of Africa*, Vol. 1, *From the Earliest Times to c. 500 B.C.*, ed. J. D. Clark. Cambridge: Cambridge University Press.

Kleindienst, M. R. 1961. Variability within the Late Acheulean assemblage in eastern Africa. *South African Archaeological Bulletin* 16:35–52.

Leakey, M. D. 1971. *Olduvai Gorge*, Vol. 3, *Excavations in Beds I and II, 1960–1963.* Cambridge: Cambridge University Press.

McGrew, W. 1992. *Chimpanzee Material Culture.* Cambridge: Cambridge University Press.

Mellars, P. 1996. *The Neanderthal Legacy: An Archaeological Perspective from Western Europe.* Princeton: Princeton University Press.

Mithen, S. 1994. Technology and society during the Middle Pleistocene: Hominid group size, social learning and industrial variability. *Cambridge Archaeological Journal* 4(1):3–32.

Movius, H. 1949. The Lower Palaeolithic cultures of southern and eastern Asia. *Transactions of the American Philosophical Society* 38(4):329–420.

Otte, M. 1996. The nature of Levallois, 117–24. In *The Definition and Interpretation of Levallois Technology*, ed. H. L. Dibble and O. Bar-Yosef. Madison: Prehistory Press.

Roebroeks, W., and T. van Kolfschoten. 1994. The earliest occupation of Europe: A short chronology. *Antiquity* 68:489–503.

Sahnouni, M., K. Schick, and N. Toth. 1997. The nature of limestone spheroids in the Lower Palaeolithic: An experimental investigation. *Journal of Archaeological Science* 24:701–13.

Schick, K. D. 1994. The Movius line reconsidered: Perspectives on the earlier Paleolithic of Eastern Asia, 569–95. In *Integrative Paths to the Past: Paleoanthropological Advances in Honor of F. Clark Howell*, ed. R. S. Corruccina and R. L. Ciochon. Englewood Cliffs: Prentice Hall.

Schick, K. D., N. Toth, Q. Wei, J. D. Clark, and D. Etler. 1991. Archaeological perspectives in the Nihewan Basin, China. *Journal of Human Evolution* 21:13–26.

Schick, K. D., and N. Toth. 1993. *Making Silent Stones Speak: Human Evolution and the Dawn of Technology.* New York: Simon and Schuster.

Schick, K. D., and N. Toth. 1994. Early Stone Age technology in Africa: A review and case study into the nature and function of spheroids and subspheroids, 429–49. In *Integrative Paths to the Past: Paleoanthropological Advances in Honor of F. Clark Howell*, ed. R. S. Corruccini and R. L. Ciochon. Englewood Cliffs: Prentice Hall.

Schick, K. D., and Z. Dong. 1993. Early Paleolithic of China and Eastern Asia. *Evolutionary Anthropology* 2:22–35.

Semau, S., P. Renne, J. W. K. Harris, C. S. Feibel, R. L. Bernor, N. Fesseha, and K. Mowbray. 1997. 2.5-million-year-old stone tools from Gona, Ethiopia. *Nature* 385:333–6.

Swisher, C. C. III, G. H. Curtis, T. Jacob, A. G. Getty, A. Suprijo, Widiasmoro. 1994. Age of the earliest known hominids in Java, Indonesia. *Science* 263:1118–21.

Toth, N. 1985. The Oldowan reassessed: A close look at early stone artifacts. *Journal of Archaeological Science* 12:101–20.

Toth, N., J. D. Clark, and G. Ligabue. 1992. The last stone axe makers. *Scientific American* 267:88–93.

Toth, N., K. D. Schick, E. S. Savage-Rumbaugh, R. Sevcik, and D. M. Rumbaugh. 1993. Pan the tool-maker: Investigations into the stone tool-making and tool-using capabilities of a Bonobo (*Pan paniscus*). *Journal of Archaeological Science* 20:81–91.

Toth, N., and K. D. Schick. 1986. The first million years: The archaeology of protohuman culture. *Advances in Archaeological Method and Theory* 9:1–96.

Toth, N., and K. D. Schick. 1993. Early stone industries and inferences regarding language and cognition, 346–62. In *Tools, Language and Cognition in Human Evolution*, ed. K. Gibson and T. Ingold. Cambridge: Cambridge University Press.

Wei, Q. 1988. Le cadre stratigraphique, geochronologique et biostratigraphique des sites les plus anciens connus en China. *L'Anthropologie* 92:931–8.

Wrangham, R. W. 1995. *Ecology and Cultural Diversity in Chimpanzees.* Paper in Invited Session, "Do Apes Have Culture?" American Anthropological Association Annual Meeting, Washington, D.C.

Wynn, T. 1989. *The Evolution of Spatial Competence*. Chicago: University of Illinois Press.

Wu, R., and J. Olsen, eds. 1985. *Palaeoanthropology and Palaeolithic Archaeology in the People's Republic of China.* New York: Academic Press.

30

THE MIDDLE AND THE EARLY UPPER PALEOLITHIC AROUND THE BLACK SEA

Janusz K. Kozlowski

Institute of Archaeology
Jagellonian University
Kraków 31007, Poland

1. INTRODUCTION

The territories around the Black Sea occupy a key position in investigating connections between Asia and Europe. Particular attention should be devoted to land bridges that connect the Near East and Europe given the suggestion that *Homo erectus* spread from Africa via the Near East to Asia and then to Europe and that archaic modern humans migrated along a similar route. Anatolia together with the Balkans, and Transcaucasia with the Kuban basin, were important terrestrial bridges.

Unfortunately, research into the Paleolithic in these territories is poorly developed. Many blank areas exist including Anatolia, a large part of the Balkans, and the eastern and southern part of Transcaucasia. Western Transcaucasia, the Kuban basin, Crimea, and the regions between the lower Danube and the Dneper are better explored.

Studies of the prehistory of the territories around the Black Sea were conducted separately within the various regions attached either to Central-Eastern Europe or to the Near East. As a result, no synthetic work has been written so far that approaches the question of the Paleolithic in the Pontian perspective. The present paper is the first attempt to expound the cultural interrelations which existed around the Black Sea during the Middle and Early Upper Paleolithic.

The territories which construct the littoral around the Black Sea are distinct both ecologically and geomorphologically. In addition, the paucity of research and the few radiometric determinations make it difficult to correlate the paleoclimatic sequences. It should be remembered that employing paleontological criteria (biozonation) for archaeological dating of assemblages, in particular of Lower and Middle Paleolithic age, is impeded by the conservative character of fauna and flora in the Transcaucasian region. Several species which are typical of earlier periods in Europe persisted much longer. This was caused by the isolation of the Pleistocene fauna and the presence of faunal refugia in the Caucasus.

Neandertals and Modern Humans in Western Asia, edited by Akazawa *et al.*
Plenum Press, New York, 1998.

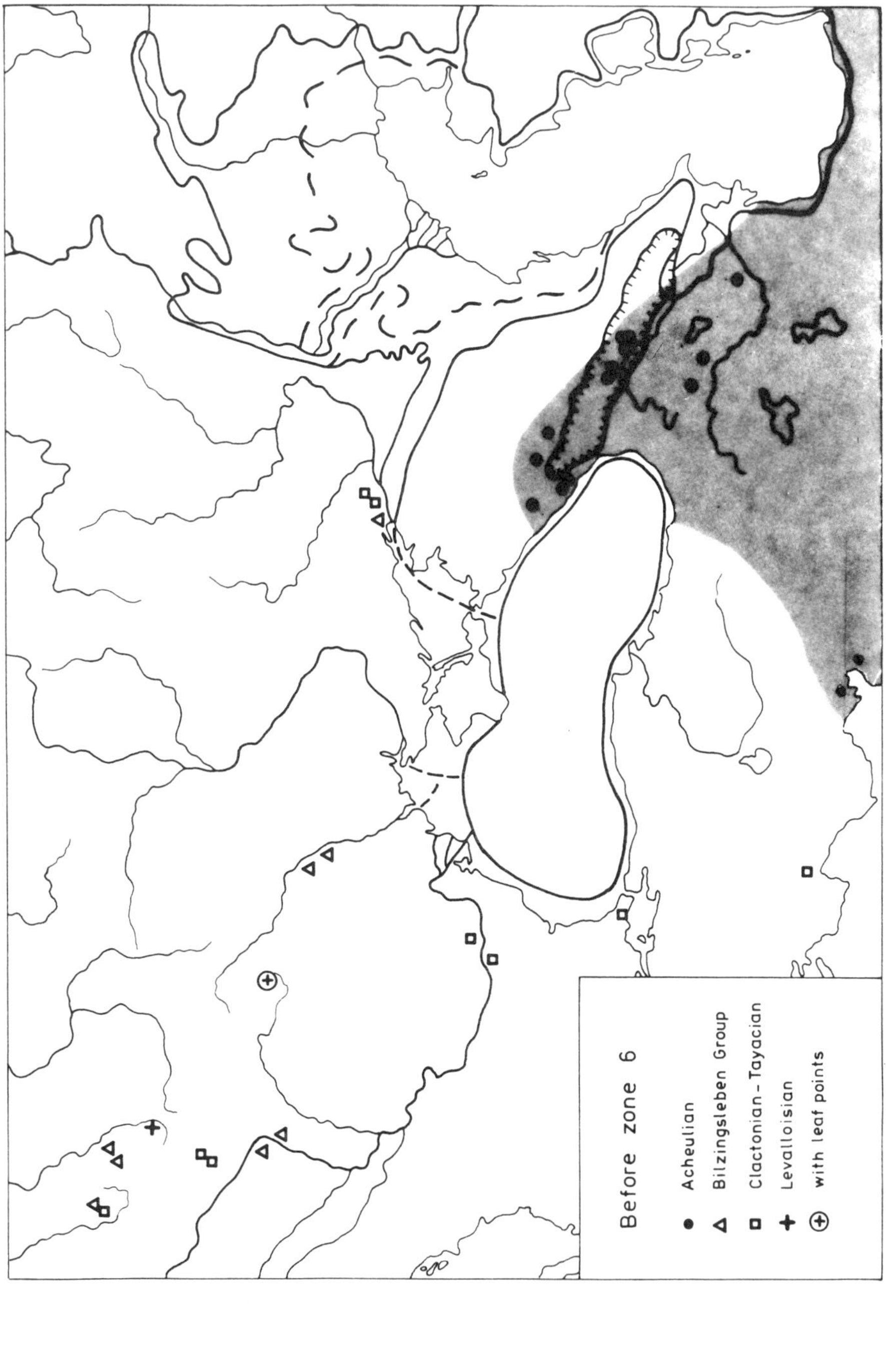

Figure 1. Territories around the Black Sea before isotopic stage 6 (shaded area indicates Near East Acheulean).

2. THE EARLY MIDDLE PALEOLITHIC

Prior to isotope stage 6 (i.e., before 180/200 ka BP), the main cultural boundaries were demarcated by the Caucasus and the Taurus. These mountains constituted the boundary of the western and northern range of the Near East Acheulean province; industries with hand-axes do not occur north of the Caucasian foreland (except for rare sites on the left-side tributaries of the Kuban between Taupase and Maikop). Similarly, hand-axes are known in central and western Anatolia as individual finds without a context. Rich assemblages with hand-axes are recorded in Transcaucasia and in southeastern Anatolia. In these territories, all the phases of the evolution of Acheulean can be traced: from the transitional phase of the "pebble-culture" (e.g., level VI of the Azykh Cave in Azerbaijian dated before the Oka glaciation [=Mindel], i.e., from 0.4 to 0.8 Ma [Guseinov 1985]); through the archaic phase (e.g., the site of Ignatienkov Kutok near Tuapse, on the northwest slopes of the Caucasus dated between the Oka and the Dneper glaciations = isotope stages 13–11 [Velichko et al. 1965]); and continuing until the developed phase represented in the caves of western Georgia. The dating of the Late Acheulean is uncertain as there are no radiometric dates and chronological criteria based on biostages are insufficient. This leaves open the question of whether this phase dates to isotope stage 11, 9, or 7, or possibly even 5. At any rate, as many as four climatic optima were palynologically recorded in the Acheulean layers in the sequence of the Kudaro I Cave (Liubin et al. 1985:178).

In contrast to this stage, where assemblages with hand-axes were presumably connected with the Acheulean of the Near East, the areas west and north of the Black Sea show a distribution of flake industries, possibly with core-chopper tools. The regions between the Carpathians and the Don River belong to the stage of small, almost microlithic industries with denticulated pieces of Bilzingsleben and Vertesszöllös type. Sites belonging to this group were recorded between the Prut and Dnester rivers (Dubossary I and Pogrebi I in the Republic of Moldova; Anisioutkin 1987) and on the lower Don (Mikhaylovskoye; Praslov 1968). These sites are older than the maximum transgression of the Dneper ice-sheet corresponding to isotope stage 8 (300–220 ka). Sites with thicker flake tools, closer to the Clactonian style, although to a large extent based on denticulated-notched types, are also known from the same territory. Among them the site of Khrischtchi in the lower Don basin, whose stratigraphical position is similar to that of the site of Mikhaylovskoye (Praslov 1968). However, it is uncertain whether the individual hand-axes found in Ukraine and Crimea indeed belong to the same period, that is, prior to isotope stage 6.

Industries that developed on the western side of the Carpathians were completely different. Their distribution could have extended as far as the lower Danube. These were industries with the Levallois technique and leaf points. Their stratigraphic position was defined in layers V–Va of the site of Korolevo I in Transcarpathian Ukraine, and dated to isotope stage 7 and is located by TL determinations corresponding to the period from 150 to 220 ka (Gladilin 1989; Gladilin and Sitlivyi 1990). These industries seem to gravitate in the direction of the northwestern European Levalloisian province. At the same time, leaf points are a specific central European and Balkan trait.

In Anatolia, this period is known only from single sites. The most complete sequence before isotope stage 6 is known from Karain Cave E (Yalçinkaya et al. 1995). In this sequence, industries representing the Clactonian style with denticulated-notched tools and steep retouch (complex A) are the earliest. These are followed by "Proto-Charentian" type industries with thick flakes and steep scaled retouch shaping side-scrapers, possibly with rare hand-axes (complexes B–E). The sequence ends with the typical Mousterian with the Levalloisian technique (complexes F–I). The upper part of the sequence has stalagmitic floors and humic soils dated to isotope stage 5 (60/70–110/120 ka) using ESR and TL methods (Rink et al. 1994), whereas the lower layers of Karain Cave can only be dated tentatively using correlations between warmer events recorded in the cave sequence and the isotopic curve (Figure 2). On this basis the "Clactonian" phase could coincide with stage 11 (350–380 ka) and the "Proto-Charentian" complex would correlate with isotope stage 9 (330–300 ka). Consequently, the "Clactonian" would be chronologically contemporary with industries on the northern side of the Black Sea, whereas the "Proto-Charentian" would be even earlier than Acheulo-Yabrudian (the Mugharan tradition) in the Near East. However, these are dated by ESR to about 150 ka, (i.e., within the range of isotope stage 6), but by TL in Tabun Cave to about 300 ka (Bar Yosef, this volume).

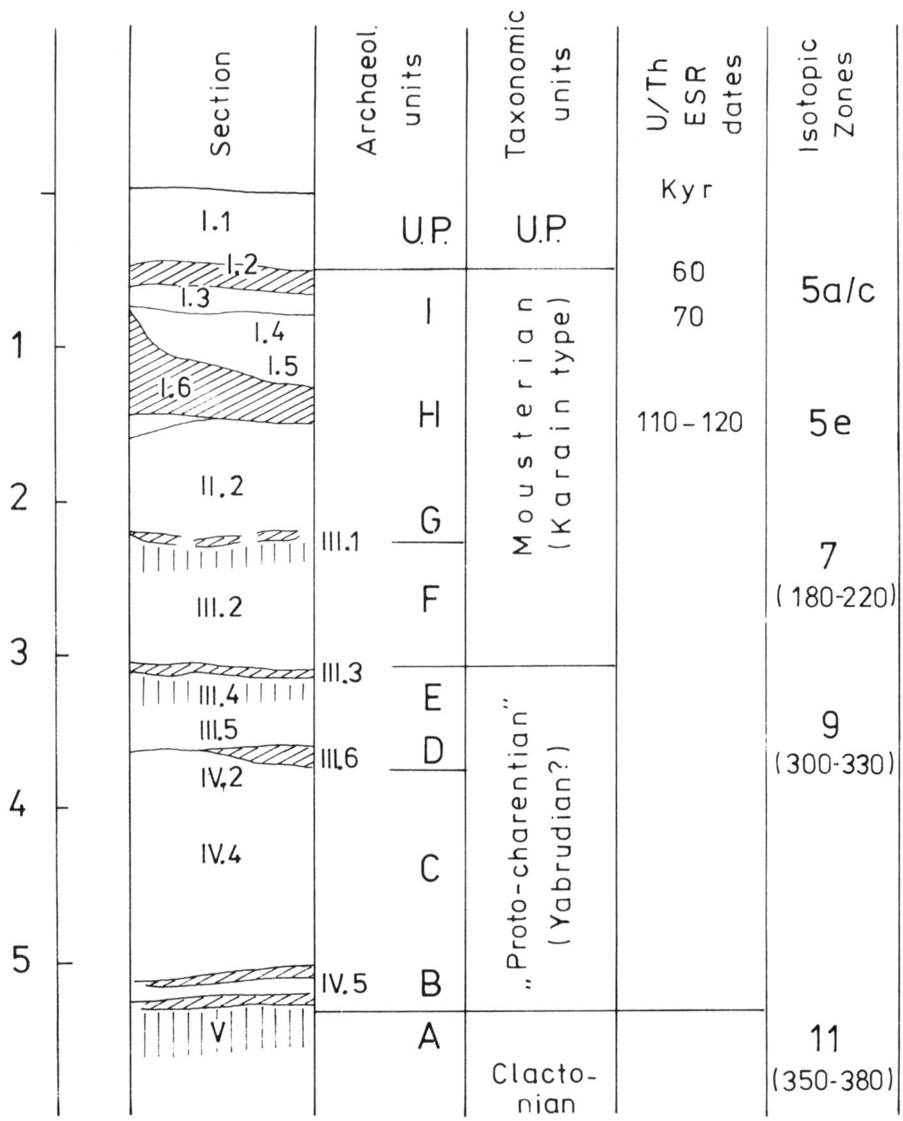

Figure 2. Schematic section of the Karain E Cave (according to Yalçinkaya et al. 1995).

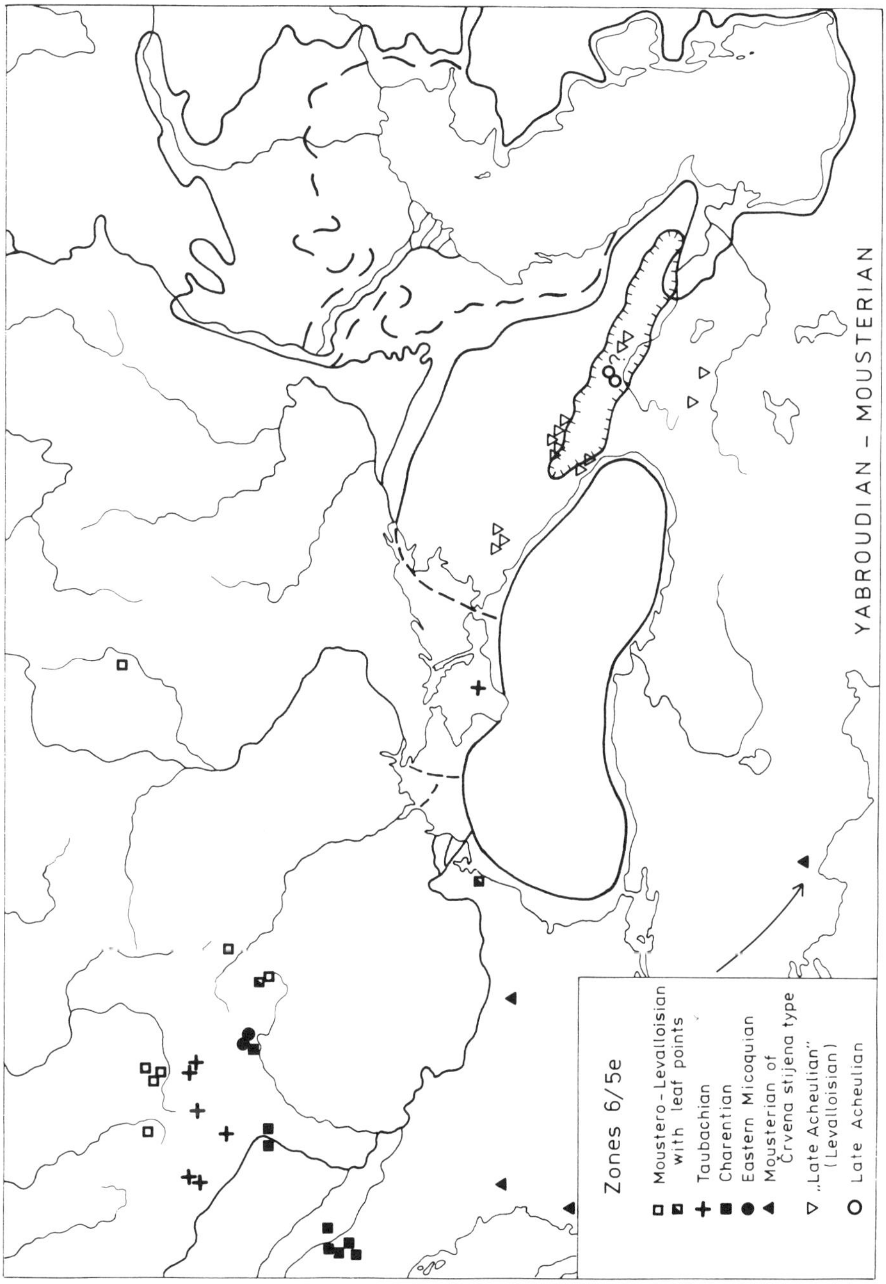

Figure 3. Territories around the Black Sea during isotopic stages 6/5e.

Human remains from the Early Middle Paleolithic were discovered in the Acheulean context in Transcaucasia (a fragment of the lower mandible from layer V in Azykh Cave and a tooth from layer 5b in Kudaro I Cave). These remains were attributed to *Homo erectus*.

3. THE MIDDLE PALEOLITHIC (ISOTOPE STAGES 6 AND 5E)

The emergence of a new cultural stage can be seen in the middle phase of the Paleolithic. This stage covers the area from the Balkans to Anatolia and later to the northern part of the Near East. It is characterized by the presence of the typical Mousterian with the Levallois technique of recurrent type and a high frequency of side-scrapers (Figure 4). Such assemblages embedded in layers dated to isotope stage 6 were recorded in the Črvena Stijena Cave (Figure 5) in Montenegro (Basler 1975), whereas those corresponding to isotope stage 5e were unearthed in fossil beaches in the eastern Peloponese (Reisch 1982). Similar industries occur in complexes F–I in the Karain E Cave where they are dated to 60/70 ka (complex I), 110/120 ka (complex H), and perhaps to the period corre-

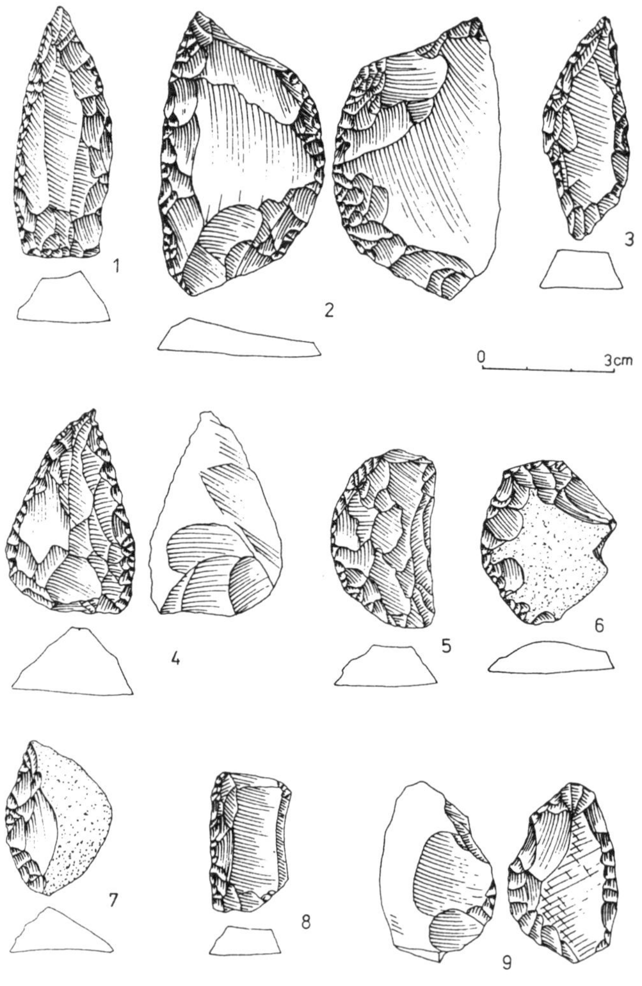

Figure 4a. Crvena stijena (Montenegro) layer XXIV. 1–8: retouched tools (according to Basler 1975).

sponding to isotope stage 6 (F, G; cf. Yalçinkaya et al. 1995). Taking into account the discovery of human teeth attributed to Neandertals (Yalçinkaya 1988) in complex H, in Karain E, I have put forward a hypothesis that this particular facies of the Mousterian was related to the spreading of the Neandertals in the direction of the Near East. This hypothesis is supported by evidence from the early Last Glaciation (Early Würm). The Mousterian of the Zagros type from the northern part of the Fertile Crescent, which is the product of the classical Neandertals, demonstrates a number of parallels with the Mousterian of the Črvena Stijena-Karain E type (Kozlowski 1992).

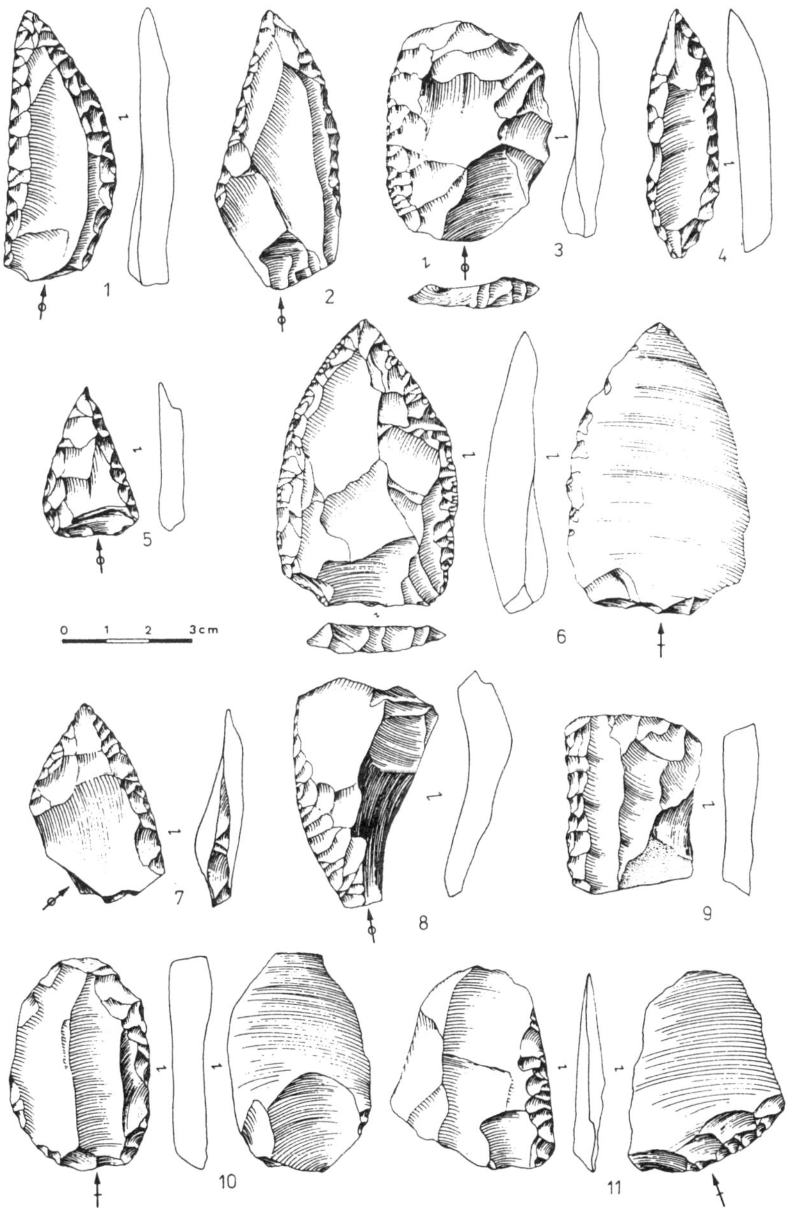

Figure 4b. Karain E (Turkey), Upper Part (Last Interglacial - geological units 1.6–1.7). Lithic artifacts (1–11).

More culturally varied territories on the north side of the Black Sea stand in opposition to the fairly uniform southern sphere. A stage of industries with the Levalloisian technology extended to the northwest of the Black Sea. The industries of this stage dominate the northern belt of Europe along the southern boundary of the European Lowland during the Penultimate Glaciation and the Last Interglacial, and contain products of both Levallois and blade technology. In the Black Sea stage, these industries contain leaf points (e.g., Mamia in Dobrogea [Valoch 1993]).

In Crimea no well-dated sites are known from the Penultimate Glaciation. On the other hand, during the Last Interglacial there were industries with small bifacial and den-

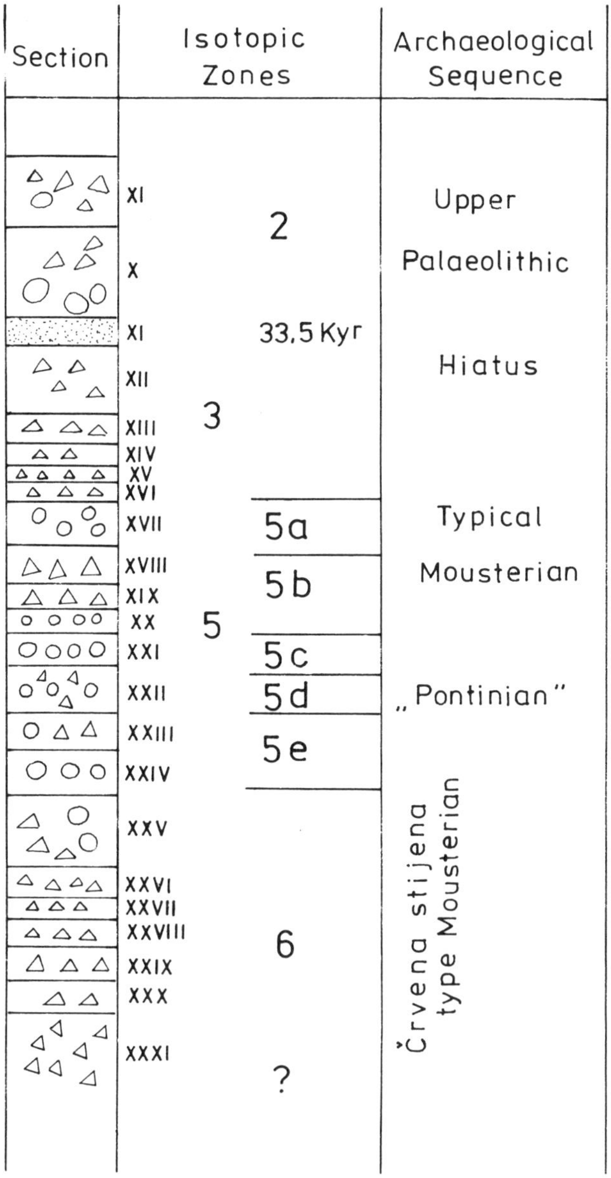

Figure 5. Crvena Stjiena (Montenegro). Schematic sequence of the Cave (according to Brunnacker 1975).

ticulated tools such as those represented in the lower level of the Kiik-Koba Cave (Bonch-Osmolovski 1940; Kolosov et al. 1993). This industry displays some analogies to the Taubachian in the northern part of central Europe, whose chronology is also interglacial. Assuming that the burial from Kiik-Koba can be assigned to the lower layer of this site, then the industry of this layer can be regarded as produced by the Neandertals. This is further confirmed by finds from interglacial travertines in the context of the Taubachian (e.g., Ganovce and Horka in Slovakia).

The situation in the northwestern Caucasus and the Kuban basin is still not very clear. As we have already mentioned, Acheulean industries occupy these territories but their late phase has been dated only by typology (Late Acheulean from the Abadzekh group; Zamiatnin 1961). Recently, Golovanova (1994) proposed the presence of a transitional phase from the Late Acheulean to the Mousterian. This phase is dated to the Last Interglacial and is represented by surface collections or finds from river terraces (Lochkovskoye, Semiyblonovskoye, Urupskoye). These collections show the use of the Levalloisian technique and contain occasional hand-axes, but the association between the two components is uncertain.

Only very few radiometric determinations are available from Transcaucasia. The latest layers of the classical Acheulean sites (e.g., Kudaro I, layer 5a) (Figure 6) may coin-

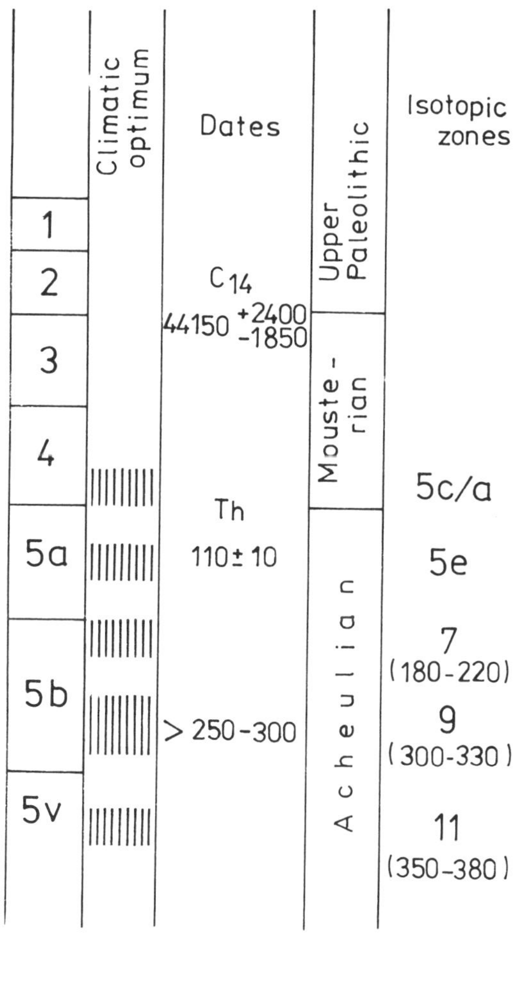

Figure 6. Kudaro I Cave (Osetia). Schematic sequence of the Cave (according to Liubin et al. 1985).

cide with the Last Interglacial as the U/Th date of 110±10 ka BP suggests (Liubin 1989) and not with the earlier interglacials or interstadials. This interpretation is further supported by a distinct lag in the evolution of the Transcaucasian fauna. Baryshnikov has pointed out (1989:826) that *Ursus deningeri* is still present in that region at the beginning of the Holocene and *Rhinoceros merkii* occurs as late as the Last Glaciation. The supposition that the flora of Acheulean layers in Kudaro I is generally concordant with the Karangat transgression of the Black Sea (Liubin et al. 1985) does not resolve the chronological ambiguity. The sediments of this transgression range over a very broad chronological period: from isotope stage 7a to 5a. It incorporates both the warm oscillations of the Penultimate Glaciation as well as the Early Würm. Persistence of the Acheulean in Transcaucasia resulting from its isolation may therefore be real, thereby making this territory different from the Near East where the Acheulo-Yabrudian and early Mousterian already occur in isotope stages 9/8 and 7/6 (Bar Yosef 1996, this volume).

The cultural discontinuity recorded between the regions on the northern side of the Black Sea and Transcaucasia, lasting until isotope stage 5e, may have been caused by the specific paleogeography of the territories between the Azov and the Caspian Sea. These regions were separated by a wide sea inlet called the Manych Inlet, which hampered easy land communication between the lower Don basin and the northern slopes of the Caucasus, including the Kuban basin (Figure 7). The Manych Inlet was alternately flooded by

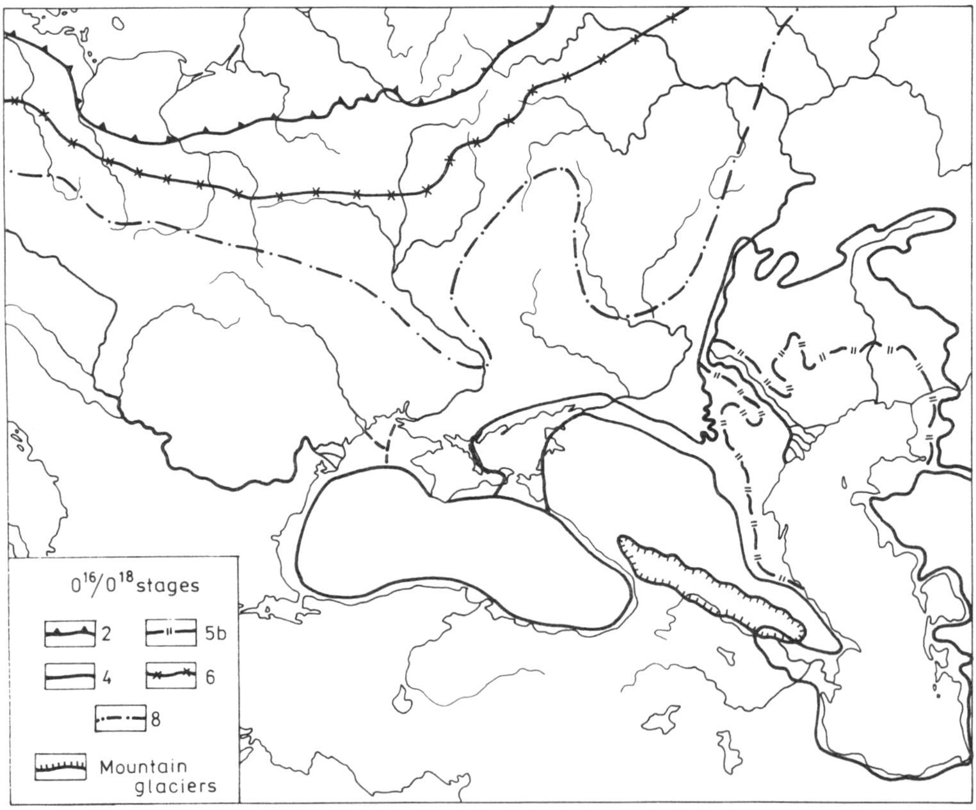

Figure 7. Paleogeographical map of Eastern Europe and Western Asia with ice-sheet transgressions (stages 8, 6, 2) and Caspian/Black Sea regressions/transgressions (isotopic stages 4 and 5b).

waters from the Black Sea or from the Caspian Sea, depending on a transgression phase of one of the two seas. The lack of synchronicity between these transgressions was caused by the fact that the Black Sea transgressions correlated with warm episodes coinciding with the level of the world oceans, whereas the Caspian Sea transgressions occurred when humidity was maximal at high geographic latitudes, i.e., at the beginning of a glacial phase (Figure 8). The transgressions of the Caspian Sea correspond to isotope stages 14, 12, 10, 6, and 5b, while the Black Sea transgressions coincided with isotope stages 9, 7a, and 5e (Zubakov 1992). As a result, the Manych Inlet was repeatedly flooded by waters of one of the two seas, which in effect cut off communication between eastern Europe and the Caucasus until practically isotope stage 5b (i.e., the beginning of Würm). Prior to that time, the main communication routes between Europe and the Near East must have run across the Balkans and Anatolia.

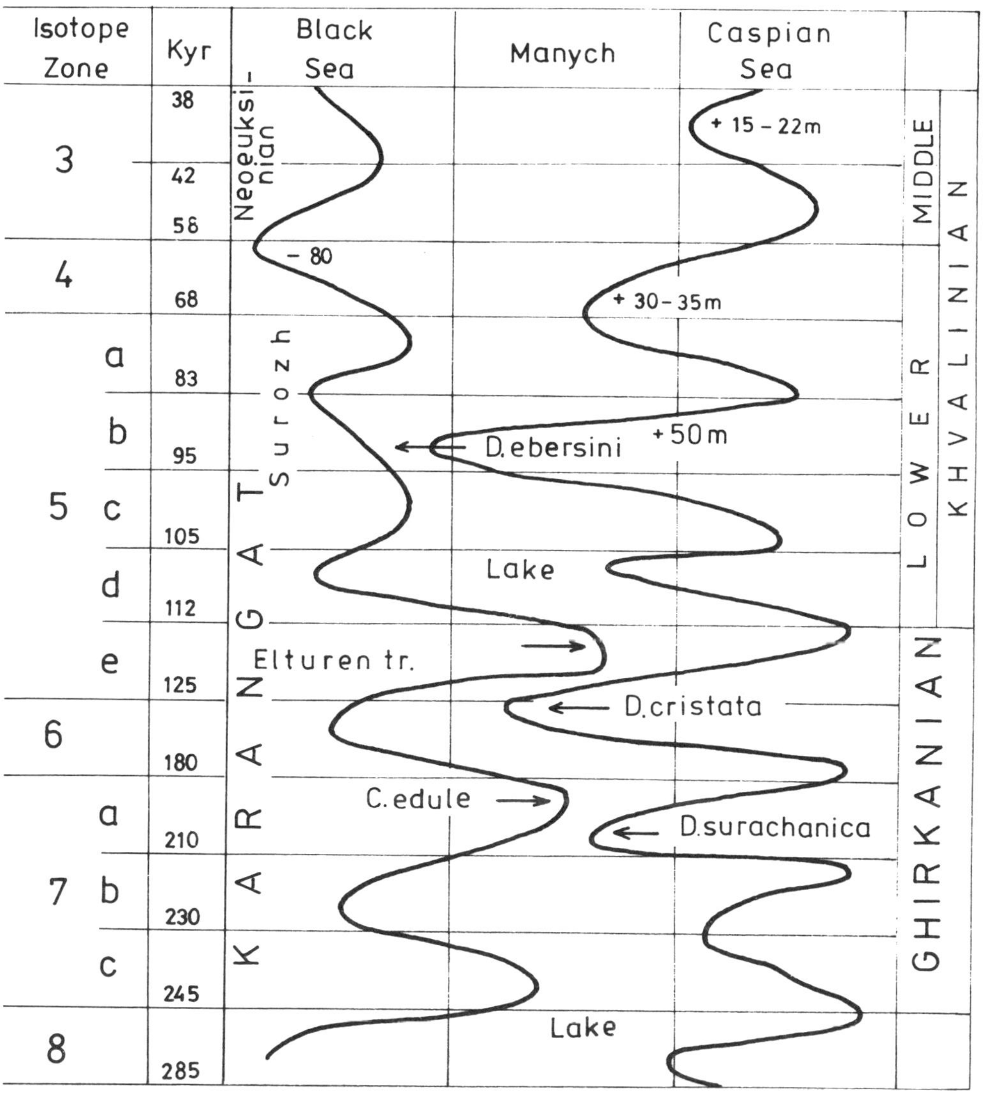

Figure 8. Black Sea/Caspian transgressional and regressional cycles (according to Zubakov 1992).

4. THE LATE MIDDLE PALEOLITHIC (ISOTOPE STAGES 5D–5A AND 4)

During the Early Würm, the eastern Balkans, the lower Danube basin, and the region between the Dnester and the Dneper were the domain of industries with Levalloisian technology. The typological character of this technology was functionally determined in the direction of either the typical Mousterian (enriched sometimes with side-scrapers as a response to conditions of more intensive tool exploitation), or the denticulated Mousterian with leaf-points. The Balkan Moustero-Levalloisian is characterized by a small number of retouched tools. Despite apparent similarities, it is a different entity than the Moustero-Levalloisian in the Levant, from which the former is separated by the whole of central and western Anatolia with the dominant Zagros-type Mousterian.

The stage with industries where the Levalloisian tradition lasted extends over the northern Black Sea coast along with Crimea, where the Kabazi-type Mousterian (Kolosov et al. 1993) is found. Recently called "western Crimean Mousterian" (Chabai et al. 1995), this industry corresponds to the industries with the Levalloisian technology of the Balkan-Lower Danube stage.

On the northern coast of the Black Sea, the region with Levalloisian technology borders the non-Levalloisian Eastern Micoquian, an entity typical of the Eastern European Lowland and the northern part of central Europe (Bosinski 1967; Chmielewski 1969; Kulakovskaya et al. 1993). At the Crimean sites, the Eastern Micoquian as represented by assemblages of the Akkaya (Zaskalnaya), and the Upper Kiik-Koba type, is older than the Kabazi-type Mousterian (Kolosov et al. 1993:43). West of the Black Sea, however, interstratification of both entities provides evidence for contemporaneous development of the Mousterian and the Eastern Micoquian during the Early Würm.

East of the Black Sea, the influence of the Eastern Micoquian industries of the Russian Lowland reaches the Kuban basin. In the Early Würm, these industries spread as far as the Volga river valley. This is demonstrated by the site of Sukhaya Mechetka or Stalingradskaya, dated to 76 ka BP by TL, i.e., to isotope stage 5a (Zamiatnin 1961; Kuznetsova 1989). In the Kuban basin this facies is possibly represented by the lower level of the site of Ilskaya (Zamiatnin 1934; Anisioutkin 1968). The links with Europe were in all likelihood related to the episode of the Caspian Sea regression, which took place at the end of the Surozh transgression of the Black Sea (approximately at the time of isotope stage 5a).

In Transcaucasia, a number of industries affiliated with the Zagros type Mousterian are recorded. The industry from Eryvan I Cave in Armenia (Eritsian 1970) is undoubtedly closest to the classical sites of the northern part of the Near East. This proximity is weaker in the case of the industries from Lusakert I and II Caves in Armenia (Eritsian 1975) and the Taglar Cave in Azerbaidjian (Guseinov 1973). Assemblages in western Georgia (Ortvale, Chakhaty and Sakhazia Caves), described as the Tskhaltsitalskaya group (Nioradze 1990, 1992), and the caves of the Tsutskhavat karstic system, situated farther north (Tushabramishvili 1978; Liubine 1989), are also ascribed to the Zagros-type Mousterian. The northernmost finds resembling this group were registered in the Barakayevska Cave in the Kuban basin (Liubin 1989) and also possibly in the rockshelters of Gubskiy and Monasheskiy (Liubin 1977).

This group of Mousterian assemblages displays characteristic features such as the manipulation of the Levalloisian technique of the recurrent type and the discoidal core technique; a high component of retouched Levalloisian and Mousterian points; a high ratio of side-scrapers; the presence of numerous points and side-scrapers with bases thinned by ventral retouching; a constant component of pieces with Kostenki truncations; and the presence of slender blade points with bilateral retouch (Figure 10).

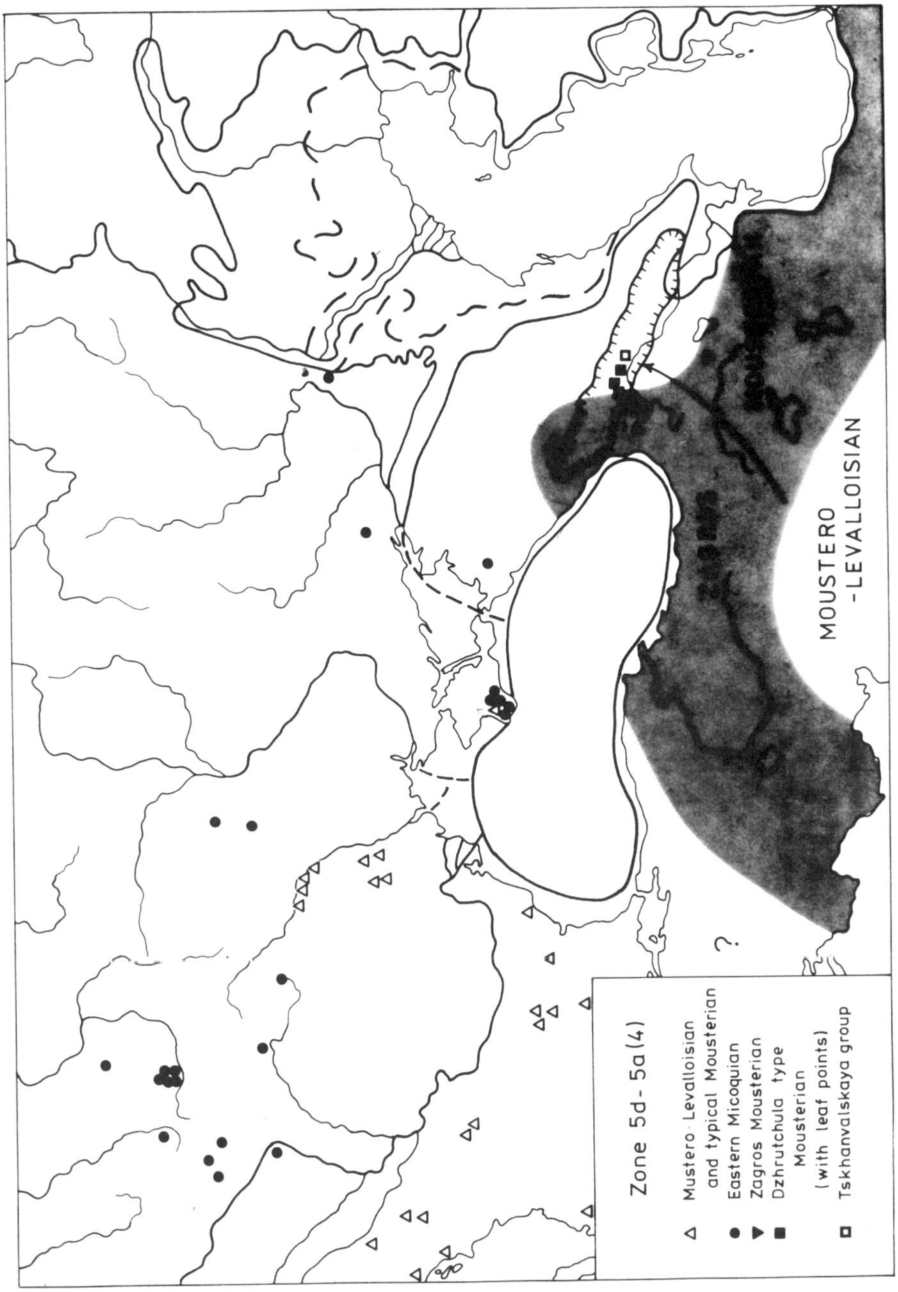

Figure 9. Territories around the Black Sea during isotopic stages 5d–5a–4. Shaded area indicates Zagros Mousterian stage.

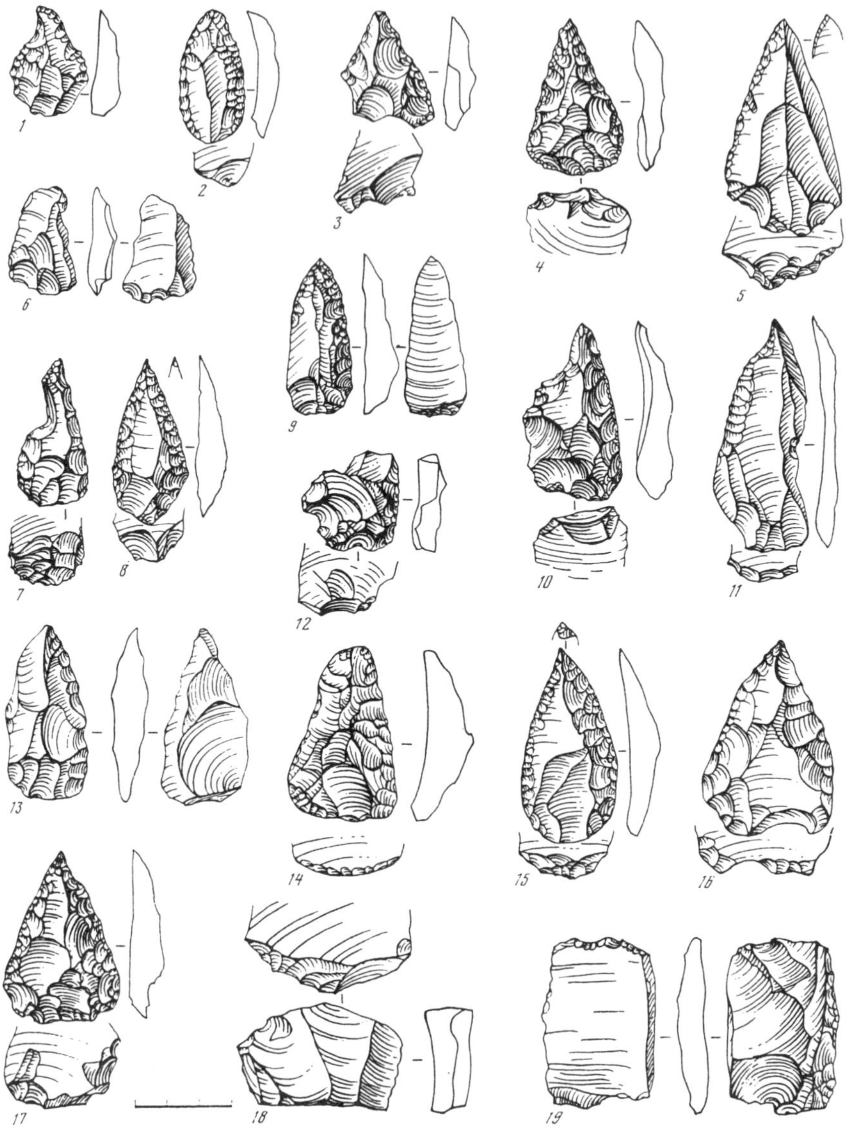

Figure 10. Eryvan Cave (Armenia). 1–19: Artifacts of the Zagros type Mousterian (according to Eritsian 1975).

Aside from those assemblages that are affiliated with the typical Mousterian from the northern part of the Near East, other cultural facies of a much more local character are also recorded in Transcaucasia. The most important among them are assemblages of the Dzhrutchula-Kudaro type, first known from western Georgia (Tsona, Dzhrutchula, Hviraty; Tushabramishvili 1963, 1969; Liubin 1989) and from south Osetia (Kudaro I; Liubin 1989). A typical feature of these assemblages is a highly advanced Levalloisian technique focused on the production of slender points, often blade-like in shape. These points have a flat, marginal retouch, sometimes combined with distal (less often proximal) retouch of the ventral face which shaped *points à face plan* or even leaf points (Figure 11). Such industries have not been recorded in other territories. The finds ascribed to the Tskhanval-

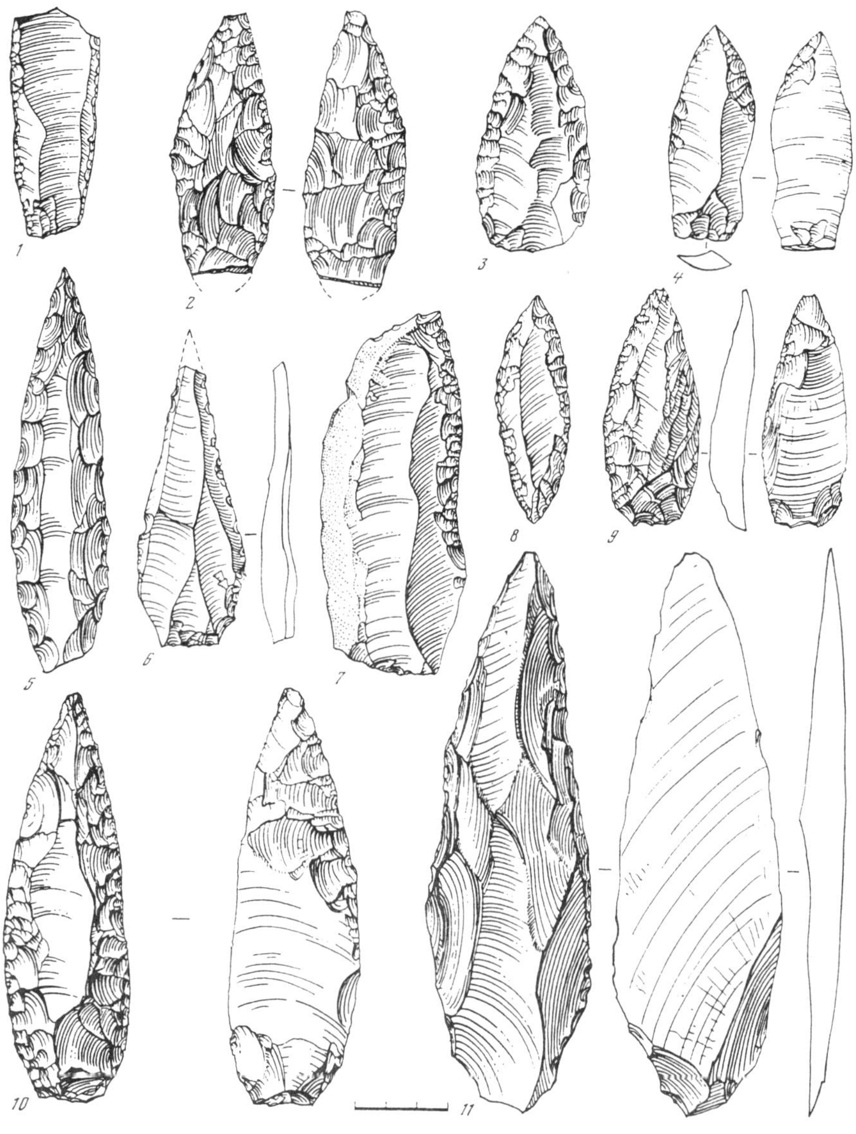

Figure 11. Artifacts of the Dzhrutchula type Mousterian. 1, 3, 4, 7, 8: Dzhrutchula Cave (Georgia); 2, 5: Tsona Cave (Georgia); 6, 11: Kudaro I; 9, 10: Kudaro III Caves (Osetia) (after Liubin 1989).

skaya group in southern Osetia (Liubin 1977) may also be associated with this facies. In southern Osetia, too, a large scale production of Levalloisian points took place but the points are thicker and less often retouched. As a result, they are closer to the Mousterian of the Near East and especially of the Levant.

Obviously, certain Caucasus Mousterian facies reflect functional variability, aside from those facies which we are inclined to interpret in terms of cultural variability. Examples include the Charentian type inventories which resulted from intensity of tool transformation (e.g., at Tsopi) or inventories with denticulated tools (e.g., the Akhschtyr Cave or Malaya Vorontskovskaya in Abkhazia).

In addition to techno-typological variability, we note different faunal compositions related to the diversity of landscapes in Transcaucasia. The correlation of both lithic and faunal variabilities and their expression in socio-economic terms is rather difficult, as demonstrated recently by Doronitchev (1993).

Finds of human remains are usually limited to teeth considered to belong to the Neandertals. A larger fragment of the maxilla was unearthed in Sakhazia Cave and represents a typical Neandertal (Nioradze 1992). It was found within the context of a typical Mousterian assemblage of the Zagros-type Mousterian.

5. TRANSITION TO THE UPPER PALEOLITHIC (ISOTOPE STAGE 3)

During the Interpleniglacial (50 to 30 ka BP), several Mousterian groups continued to survive in most of the territory discussed above. However, the cultural transition to the Upper Paleolithic and the appearance of allochthonous Upper Paleolithic entities created a complex picture. It is interesting that the processes of the technological and typological evolution towards the Upper Paleolithic began earliest in the eastern Balkans. This can be seen in inventories where the Levalloisian single- or double-platform core technique shifts in the direction of the Upper Paleolithic blade technique with centrally positioned crests, as well as in the inventories where end-scrapers, burins, and truncations appear. Such industries have been recorded in Bulgaria (layer VI, the Temnata Cave; Ginter et al. 1996) and in the eastern Carpathians (layer II, Korolevo II; Gladilin and Demidenko 1989). They are dated to between 50 and 45 ka BP. It is noteworthy that these industries coexisted with the Balkan Moustero-Levalloisian and the typical Mousterian, but did not evolve further in the direction of typical Upper Paleolithic cultures. The archaic Aurignacian appears very early in the Balkans (45–40 ka BP; Figure 13) in its form with abundant end-scrapers and retouched blades (e.g., Bachokirian; Kozlowski 1982). It does not show continuation in relation to the "transitional cultures" but occurs as an allochthonous element.

The local development based on the Moustero-Levalloisian base of "transitional cultures" suggests that they were produced by local Neandertal groups, whereas remains of *Homo sapiens sapiens* occur from the very beginning in the Aurignacian context (layer 11, the Bacho Kiro Cave; Glen and Kaczanowski 1982). The association of the Aurignacian and modern humans is best documented in Mladeč Cave (Jelinek 1988).

In this light, the hypothesis linking the Balkan Aurignacian with the migration of modern humans from the Near East to Europe via Anatolia and the Balkans receives further support. Unfortunately, the unsatisfactory state of research in Anatolia, along with erosional processes which removed interpleniglacial sediments from many investigated caves are responsible for the paucity of information from the northern part of the Near East to the Balkans. It should be stressed, however, that both facies of the European early Aurignacian, namely the above-mentioned Bachokirian with numerous end-scrapers and retouched blades and the Mediterranean facies rich in micro-retouched bladelets and Krems-type points (Broglio 1994), have their Near East parallels during the Early Upper Paleolithic. The Baradostian in the northern part of the Near East (Dibble and Olszewski 1994) is the counterpart of the former, whereas some assemblages assigned to the early Ahmarian in the Levant (e.g., levels XI–XIV from Ksar Akil in Lebanon) are similar to the latter.

The hypothesis about the migration of modern humans from the Near East to Europe at about 45 ka cannot be finally verified unless sites are discovered in the territory of Anatolia that would cover the period from 50 to 40 ka.

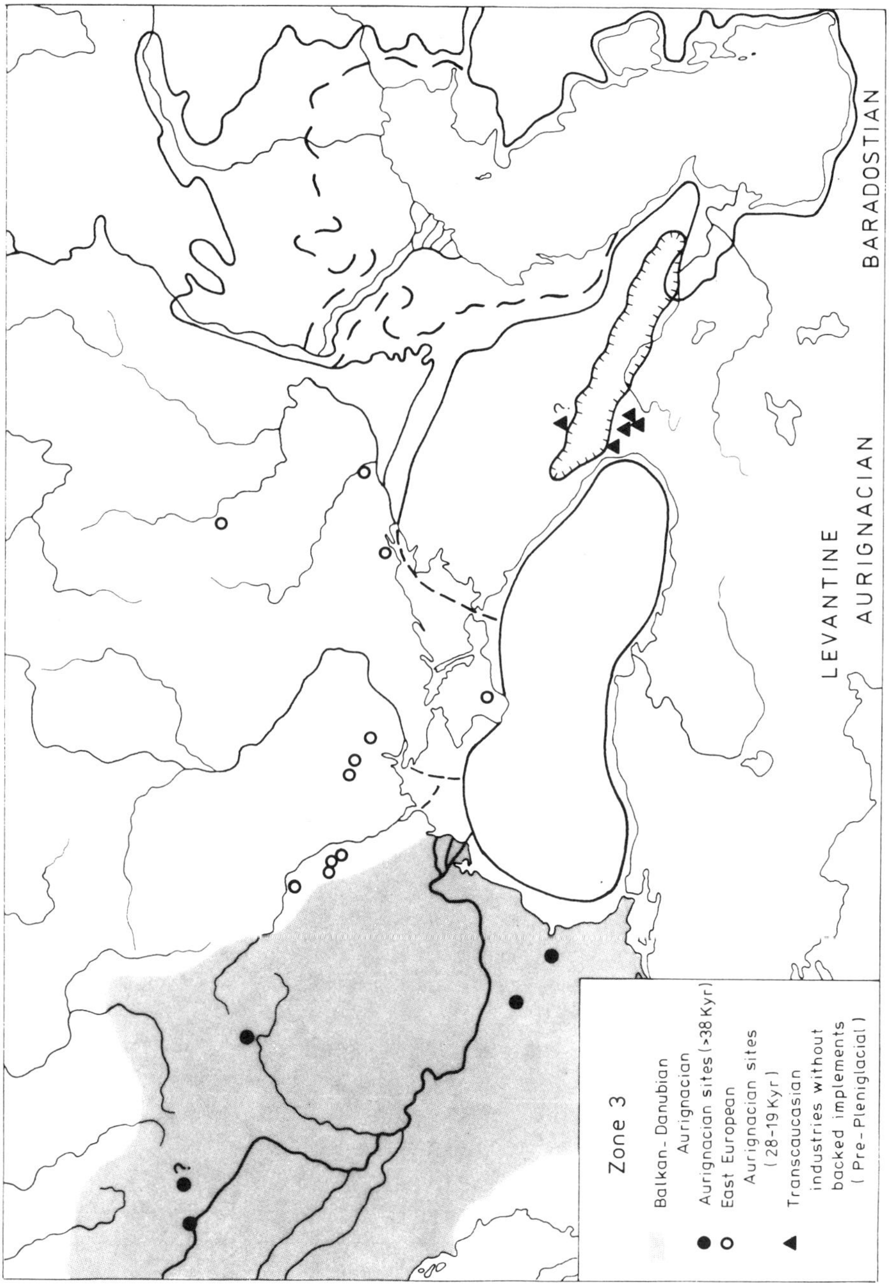

Figure 12. Territories around the Black Sea during isotopic stage 3.

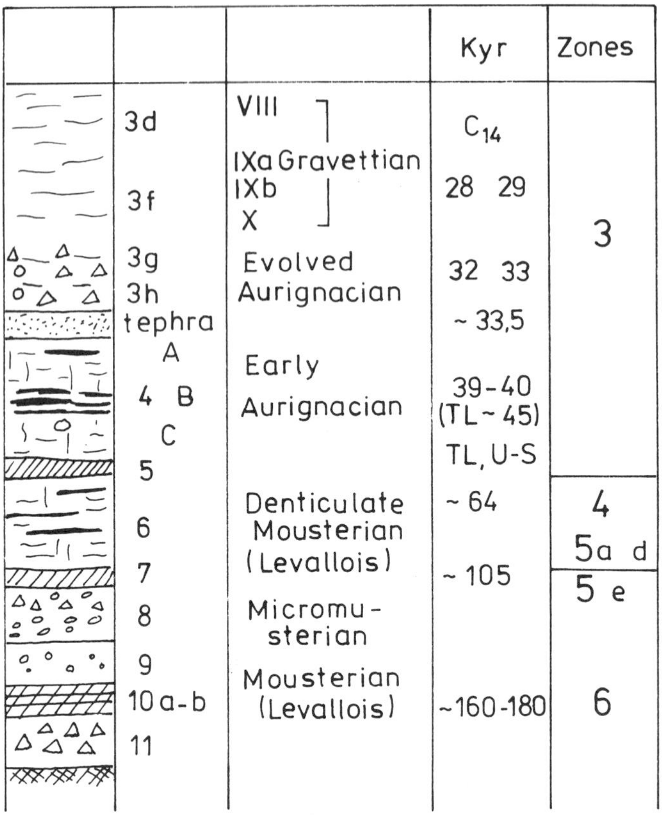

Figure 13. Position of the Early Balkan Aurignacian in the Temnata Cave, Bulgaria.

The situation on the east side of the Black Sea is even less clear than in the Balkans or the lower Danube basin. There is a lack of radiometric dates, and only few interpleniglacial sediments at Transcaucasian sites have been preserved. The dates from Transcaucasia support the interpretation that the local Mousterian of the Kudaro-Dzhrutchula type, e.g., Kudaro, persisted. The Mousterian in layer 3a was dated to $44{,}150^{+2{,}400}_{-1{,}850}$ years BP and the denticulated Mousterian in Mala Vorontsovska Cave to 35,680±480 years BP.

Transcaucasian sites once believed to represent the early phase of the Upper Paleolithic, namely, Taro-klede and Khergulis-klde (Zamiatnin 1957), are a mixture of Middle and Upper Paleolithic inventories (Kozlowski 1969, 1972). The only relatively well-documented inventory of the early phase of the Upper Paleolithic from Sagvardzile Cave (Kozlowski 1969a, Berdzenishvili 1972) contains Upper Paleolithic substratum forms (end-scrapers, retouched blades). At the beginning of the 1970s, I proposed the further evolution of the Upper Paleolithic in that territory as bi-linear, based on a parallel evolution of assemblages with micro-retouched bladelets (e.g., Samertskhle-Klde Cave) and assemblages with backed pieces (most of the so-called Imeretian culture; Kozlowski 1969b). However, that hypothesis was based on correlation between cave sediments and climatic cycles, without taking into account paleontological and palynological data, which were non-existent at that time. Recently, Amirkhanov (1994), relying on the results of sedimentological and palynological analyses carried out in recent years (e.g., Dzudzuana, Apiancha), suggested a division of the Paleolithic of Transcaucasia into two main phases. The first phase would comprise interpleniglacial inventories with micro-retouched bladelets and steep scrapers (Samertskhle-klde, Dzudzuana; Figure 14). The

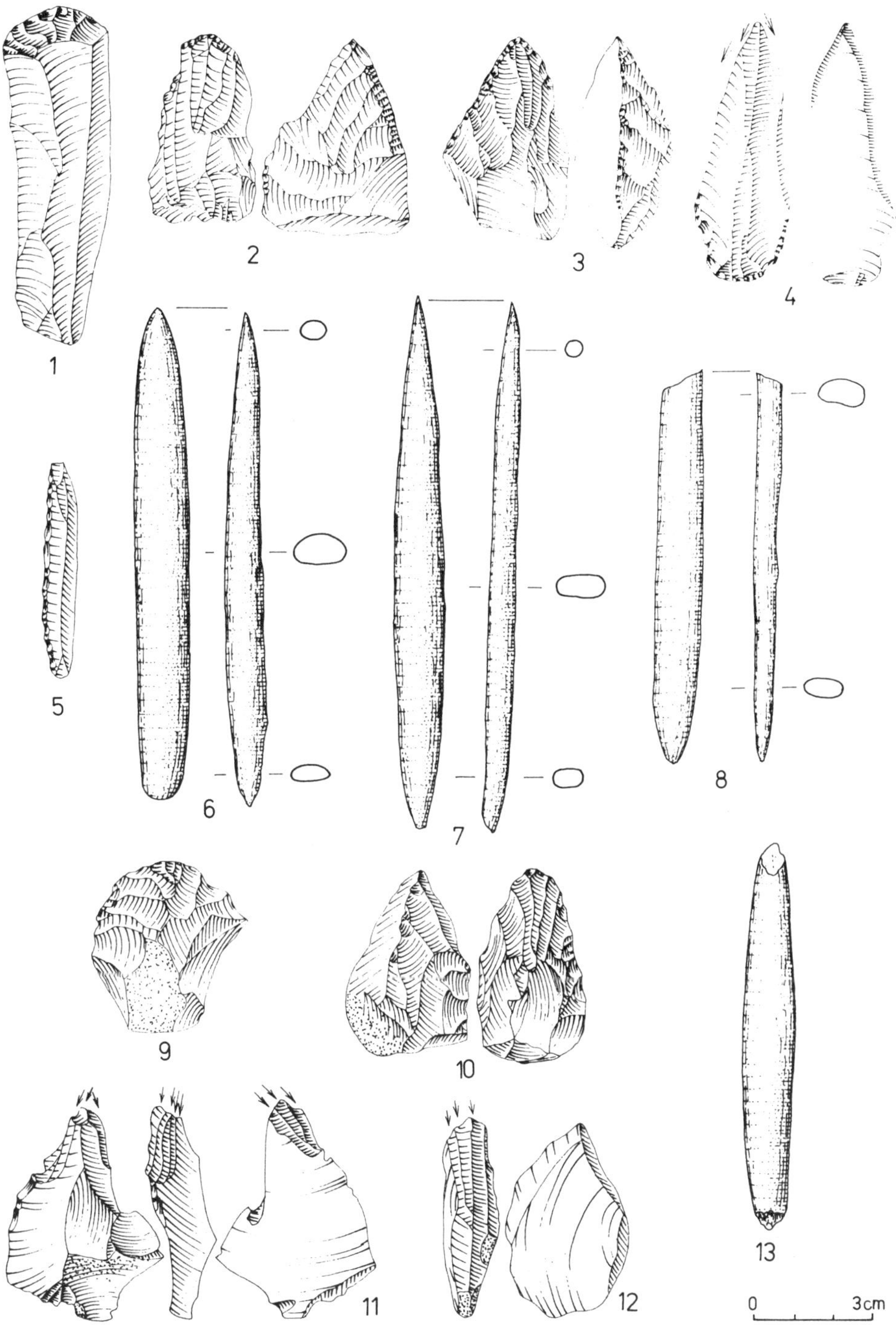

Figure 14. Early Upper Paleolithic lithic artifacts from Transcaucasia. 1–8: Samertskhle Cave (Georgia); 9–13: Dzudzuana Cave (Georgia).

second phase, taking place after the LGM, would comprise all inventories with the various forms of backed pieces. This scheme would be closer to the situation known from the Near East, where the first group would correspond to the Ahmarian and possibly the Levantine Aurignacian, and the second to the Epipaleolithic. Unfortunately, a lack of radiometric dates and an incomplete record of paleoclimatic cycles in cave sediments do not allow us to estimate where to place the inventories of the first group during the Interpleniglacial.

Little is known to date about the early phase of the Upper Paleolithic in the territories on the north side of the Black Sea. The presence of inventories with micro-retouched bladelets and high scrapers in the Siuren Cave in Crimea (Vekilova 1957) may point to an Aurignacian intrusion similar to that in the Russian Lowland (Kostenki I, level 2, 3; Sinitsyn 1993), whose age is placed at about 29 ka BP. There is no evidence to suggest a continuity of local Middle Paleolithic cultures. This continuity has recently been documented in Siuren and Skalistyi for industries with triangular leaf points, but not for the Aurignacian.

6. CONCLUSIONS

Prior to isotope stage 6, the Caucasus and eastern Anatolia form one territory linked with the Acheulean province of the Near East. The regions north and west of the Black Sea are linked with the European province of flake industries whose influence may have also reached western and southern Anatolia.

During isotope stage 6, a new entity of Mousterian industries from the Balkans to the northern part of the Near East can be seen. Presumably, the formation of this cultural stage is connected with the expansion of the Neandertals from Europe to the Near East, which was caused by climatic conditions in eastern and central Europe during the Penultimate Glaciation. During the last Interglacial in the Caucasus, with its Late Acheulean, the paleogeographical conditions in the region of the Manych Inlet constitute a barrier separating Transcaucasia from Europe.

In the Early Würm more intensive circum-Pontinian contacts can be seen, manifested in the influences of the eastern Micoquian province in the direction of the Kuban basin and possibly in contacts between the Balkan Mousterian and Anatolia (e.g., leaf points in the uppermost levels in the Karain E Cave).

The early appearance of the Aurignacian in the Balkans can be related to the migration of modern humans from the Near East to Europe. Transcaucasia at that time was a peripheral stage of cultural influences from the Near East, rather than the bridging territory through which cultural information was disseminated into Europe.

REFERENCES

Amirkhanov, K. A. 1994. K problemie evolutsii i periodizatsii verkhnego paleolita zapadnogo Kavkaza. *Rosiyskaya Archeologia* 1:12–18.

Anisioutkin, N. K. 1968. Dva kompleksa Ilskoy stoyanki. *Sovietskaya Arkheologia* 2:118–26.

Anisioutkin, N. K. 1987. De nouvelles données sur le Paléolithique ancien en Moldavie. *L'Anthropologie* 91(1):69–75.

Baryshnikov, G. 1989. Les mammifères du Paléolithique inférieur du Caucase. *L'Anthropologie* 93(4):813–30.

Basler, D. 1975. *Črvena Stijena*. Niksić: Regional Museum.

Berdzenishvili, N. Z. 1972. K voprosu o natchalnoy stadii verkhnego paleolita Gruzii, 10–11. In *Kamennyi Vek Sredniey Azii i Kazakhstana*. Tashkent: Uzbek Academy of Sciences.

Bonch-Osmolovski, G. A. 1940. *Grot Kiik koba*. Moskva-Leningrad: Izdatel'stvo Akademii Nauk.

Bosinski, G. 1967. *Die mittelpaläolitischen Funde im westlichen Mitteleuropa*. Köln: Bohlau.

Broglio, A. 1994. Il paleolitico superiore del Friuli-Venezia Giulia. Considerazioni sul popolamento umano del territorio tra Penisola Italiana e Penisola Balcanica, 37–56. In *Atti delle XXIX Riunione Scientifica*. Firenze: Instituto Italiano di Preistoria e Protostoria.

Brunnackeř, K. 1975. Die Sedimente der Crvena Stijena, 171–203. In *Crvena Stijena*, ed. D. Basler. Niksic: Regional Musuem.

Chabai, V., A. E. Marks, and A. Yevtuchenko. 1995. Views of the Crimean Middle Palaeolithic past and present. *Préhistoire Européene* 7:59–80.

Chmielewski, W. 1969. Ensembles Micoquo-Prondnikiens en Europe Centrale. *Geographia Polonica* 17:371–86.

Dibble, H., and D. Olszewski. 1994. The Zagros Aurignacian. *Current Anthropology* 35:1:68–75.

Doronitchev, V. B. 1993. Musterskiye industrii bolshogo Kavkaza. *Petersburskiy Arkheologitcheskiy Vestnik* 7:14–23.

Eritsian, V. G. 1970. Erevanskaya pechtchernaya stoyanka i ee mesto sredi drevnieychich pamiatnikov Kavkaza. Avtoreferat kandidatskoy dissertatsii. Moskva.

Eritsian, V. B. 1975. Novaya nizhnepaleolititcheskaya pechtchernaya stoyanka Lusakert I (Armenia). *Kratkiye Soobchtchenia Instituta Archeologii* 141:47–56.

Ginter, B., J. K. Kozlowski, H. Laville, N. Sirakov, and R. E. M. Hedges. 1996. Transition in the Balkans: News from the Temnata Cave, Bulgaria, 164–200. In *The Last Neandertals: The First Anatomically Modern Humans*, ed. E. Carbonell and M. Vaquero. Tarragona: Universitat Rovira i Virgili.

Gladilin, V. N. 1989. The Korolevo Palaeolithic site: Research methods, stratigraphy. *Anthropologie* 27(3–4):5–18.

Gladilin, V. N., and Y. E. Demidenko. 1989. Upper Palaeolithic stone tool complexes from Korolevo. *Anthropologie* 27(3–4):179–222.

Gladilin, V. N., and V. I. Sitlivy. 1990. *Aszel Centralnoy Evropy*. Kiev: Naukova Dumka.

Glen, E., and K. Kaczanowski. 1982. Human remains, 75–80. In *Excavation in the Bacho Kiro Cave (Bulgaria): Final report*, ed. J. K. Kozlowski. Warszawa: Panstwowe Wydawnictwo Naukowe.

Golovanova, L. V. 1994. *Problema perekhoda ot rannego k sredniemu paleolitu na Severno-Zapadnom Kavkaze*. Sankt-Petersbourg: Int material'noi kultury.

Guseinov, M. M. 1973. O paleoliticheskoy stoyanke v pechtchere Taglar. *Materialnaya kultura Azerbeidjana* 7:13–18.

Guseinov, M. M. 1985. Drevniy paleolit Azerbeidjana (po materialam pechtchernoy stoyanki Azykh i dr.) Avtoreferat kandidatskoy dissertatsii. Baku.

Jelinek, J. 1988. Historie, identyfikacie a vyzkum mladecskich anthropologickych nalezu z pocatku mladého paleolitu, 51–70. In *25 lat pavilonu Anthropos 1961–1986.*, ed. L. Seith. Brno: Moravske Muzeum.

Kolosov, Y. G., V. N. Stepanchuk, and V. P. Chabai. 1993. *Ranniy paleolit Kryma*. Kiev: Naukama Dumka.

Kozlowski, J. K. 1969a. Górny paleolit w krajach zakaukaskich i na Bliskim Wschodzie, 121–42. In *Prace Komisji Archeologicznej PAN*, ed. M. L. Bernhard and L. Stefaniak. Kraków: Polska Akademia Nauk.

Kozlowski, J. K. 1969b. Zur Stratigraphie der Grotte Sakazia und ihre Bedentung für die Chronologie des Jungpaläolithikums Georgiens. *Jahresschrift für Mitteldeutsche Vorgeschichte* 59:157–66.

Kozlowski, J. K. 1972. Górny Paleolit w krajach zakaukaskich i na Bliskim Wschodzie: cz.2 - Periodyzacja górnego paleolitu zachodnich krajów zakaukaskich. *Swiatowit* 33:7–47.

Kozlowski, J. K., ed. 1982. *Excavations in the Bacho Kiro Cave (Bulgaria): Final report*. Warszawa: Panstwowe Wydawnictwo Naukowe.

Kozlowski, J. K. 1992. The Balkans in the Middle and Upper Palaeolithic: The Gate to Europe or a cul-de-sac? *Proceedings of the Prehistoric Society* 58:1–20.

Kulakovskaya, L., J. K. Kozlowski, and K. Sobczyk. 1993. Les couteaux micoquiens du Würm ancien. *Préhistoire Européenne* 4:9–32.

Kuznetsova, L. V. 1989. Paleolit srednego i nizhnego Povolzhia. Kuybyshev.

Liubin, V. P. 1977. *Mustierskiye kultury Kavkaza*. Leningrad: Nauka.

Liubin, V. P. 1989. Paleolit Kavkaza, 9–141. In *Paleolit Kavkaza i Centralnoy Azii*, ed. P. Boriskovski. Leningrad: Nauka.

Liubin, V. P., A. G. Tcherniachovski, G. F. Baryshnikov, G. M Levkovskaya, and N. B. Selianova. 1985. La grotte Koudaro I (résultats des recherches pluridisciplinaires). *L'Anthropologie* 89(2)159–80.

Nioradze, M. G. 1990. Pechtchernyie paleoliticheskiye stoyanki uchtchelia reki Tskhaltsitelia, 8–10. In *Paleolit Kavkaza*. Tbilisi: Metsnereba.

Nioradze, M. G. 1992. *Altsteinzeitliche Höhlenfundplatze im Zchalzitela-Tal*. Tbilisi: Metsnereba.

Praslov, N. D. 1968. *Ranni paleolit severo-vostochnogo Priazovia i Nizhnego Dona*. Leningrad: Nauka.

Reisch, L. 1982. The transition from Lower to Middle Palaeolithic in Greece and the Southern Balkans, 223–232. In *The Transition from the Lower to Middle Palaeolithic and the Origin of Modern Man, ed. A. Ronen. Oxford: British Archaeological Reports International Series 151.*

Sinitsin, A. 1993. Les niveaux aurignaciens de Kostenki I, 242–59. In *Aurignacien en Europe et au Proche Orient*, ed. L. Banesz and J. K. Kozlowski. Bratislava: Académie Slovaque des Sciences.

Tushabramishvili, D. M. 1963. Pechtchery Dzhrutchulskogo uchtchelia, 54–58. In *Pechtchery Gruzii*. Tbilisi: Matsnereba.

Tushabramishvili, D. M. 1969. Itogi rabot provedennykh arkheologitcheskoy ekspeditsei Gosudarstvennogo Muzea Gruzii v Kvirilskom uchtchelie. *Arkheologitcheskiye ekspeditsii Gosudarstvennogo Muzea Gruzii* 1:8–9.

Tushabramishvili, D. M. 1978. Arkheologitcheskiye pamiatniki Tsutskhvatskogo mnogoetazhnogo pechtchernogo kompleksa, 127–82. In *Izutcheniye Pechtcher Kolkhidy*. Tbilisi: Matsnereba.

Valoch, K. 1993. Les industries du Paléolithique moyen de Mamaia-Sat, Roumanie. *L'Anthropologie* 97(2–3): 239–64.

Velichko, A. A., I. K. Ivanova, and V. M. Muratov. 1969. Geologitcheskaya istoria Russkoi Ravniniy, Kryma i Kavkaza v pleistocenie i vozrast arkheologitcheskikh kultur, 8–40. In *Priroda i razwitie pervobytnogo obchtchestva Evropeyskoy Chasti SSSR*, ed. I. P. Gerasimov. Moskva: Nauka.

Vekilova, E. A. 1957. Stoyanka Siuren I i ee mesto sredi paleoliticheskikh mestonakhozhdenii Kryma i blizhayshikh territorii. *Materiali i Issledovaniya po Arkheologii SSSR* 59:235–323.

Yalçinkaya, I. 1988. 1987 Yili Karain Kazisi. *Kazi Sonuclari* 10:15–36.

Yalçinkaya, I., M. Otte, J. K. Kozlowski, O. Bar Yosef, H. Taskiran, and P. Noiret. 1995. Evolution technique au Paléolithique ancien de Karain (Turquie). *L'Anthropologie* 99:529–61.

Zamiatnin, S. N. 1934. Itogi poslednikh issledovanii Ilskogo paleoliticheskogo mestonakhozhdenia. In *Trudy II Mezhdunarodnoi Konferencii Associatsii po Izucheniu Chetvertichnogo Perioda*. Leningrad: Nauka

Zamiatnin, S. N. 1957. Paleolit Zapadnogo Zakavkazia. *Sbornik Muzeava Antropologii i Etnografii* 17:463–79.

Zamiatnin S. N. 1961. Stalingradskaya paleoliticheskaya stoyanka. *Kratkiye Soobchtchenia Instituta Arkheologi* 82:14–20.

Zubakov, V. A. 1992. *Lednikovo-mezhlednikovye tsikli plejstocena Russkoi i Sibirskoy Ravniny v pyltsevykh diagramach*. St. Petersbourg: Ministerstro Ekologii i Prirodnych Resursov.

31

TURKEY AS A KEY

Marcel Otte

Service de Préhistoire
Université de Liège
Liège B-4000, Belgium

1. INTRODUCTION

As a geographical intermediary between Asia, Africa, and Europe, Turkey requires special consideration in any theory on the origins of Modern Man and of the Upper Paleolithic (Figure 1). The data available today on this subject are summarized in this paper. However, the real lesson provided by the Anatolian example is theoretical in nature. The meaning of the same expressions used on all sides of Anatolia has to be placed in the historical context that created them. In particular, European Modern Man appears as a biological and behavioral break in an apparently stable substratum. On the other hand, the other great territories (Asia, Africa) display a continuous evolution, with no apparent coincidence between anatomy and behavior.

The author interprets these distinctions as the expression of general evolutionary tendencies that are perpetual and particular to humankind from its origins, more than a million years ago. Consequently, the very relative notion of Modern Man corresponds merely to a state of a universal process, and the search for the moment of origin is just an illusory, fundamentally metaphysical quest.

2. THE CONSTITUTION OF A MYTH

During the eighteenth century, the "Natural Sciences," Geology, Zoology, Botany, developed the notion of transformism, involving the history of the Earth and the beings that occupy it. "Primitive" Man, discovered elsewhere by ethnologists, was himself considered "natural."

At the start of the nineteenth century, the shift from "primitive" beings (of the past) was then transferred from ethnography to paleontology. Soon, Darwin's theory took the place of the Bible in establishing the distinction between present-day Man and his savage ancestor, who was viewed as close to nature.

Neandertals and Modern Humans in Western Asia, edited by Akazawa *et al.*
Plenum Press, New York, 1998.

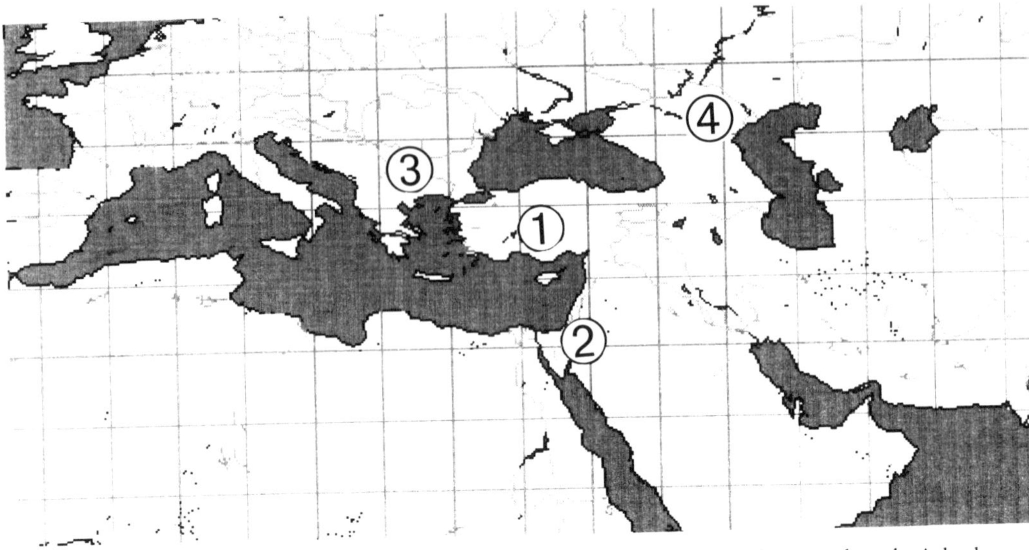

Figure 1. Location of Turkey and the Karain area (1) as a key position between the Levant, where the Acheulean occurs, (2) Balkan countries where "Euroasiatic" traditions are present, (3) and Caucasian ranges where transition from African to Asiatic "people" occurred (4).

The "scientific" explanation through the evolution of species required both the existence of a pre-human stage and the status of accomplished humans, like ourselves. These notions introduced the idea of a vague and variable intermediate stage, or some "missing link." Curiously, this cardinal notion of marking the appearance of our species with a "before" and "after" has been maintained through all the stages of modern paleontology and, today, of microbiology.

The constitution of this new mythology, more powerful than any logic, forms the principal feature of any "human paleontology." This new incarnation of creationist religions has merely changed its name to what are called "Humanities," and displays our obsession about being part of nature (Frank 1990; Cohen 1991; Wright 1994).

The phobia about our own evolutionary flux, which is random, indefinite, and continuous, has inhibited the application of reasoning to the study of the human species and has imposed the notion of "Modern" Men, as opposed to others from before and elsewhere. We claim that this metaphysical phobia has in no way left us, and that it still drives us in the quest for this sacred demarcation.

3. MODERN EUROPE

Borne along by this "atheist mythology," the European discoveries of the nineteenth century crystallized this opposition through the definition of an archetypical primitive fossil man, the "Neandertal." These discoveries, scattered through the century (Engis 1828, Gibraltar 1848, Neandertal 1855, Spy 1886), reinforced the model of the primitive being who was distinct from ourselves. The break was clearly linked to Cro-Magnon, the creator of images, who possessed an advanced technology and was morphologically "identical to us," that is to say, at the pinnacle of evolution.

This coincidence between the quest for a transformist argument and the anatomical discoveries is merely due to the European continent's cul-de-sac location. This geographical disposition of the extreme West accentuates the effect of these differences, which are attenuated everywhere else. Moreover, one cannot rule out the possibility that these ideas themselves became radicalized here through the effect of these extreme concentrations. The myth of Modern Man, together with that of the break that it implies, was nevertheless transposed onto a world scale. This coincided with colonial expansion and was conveyed through the same movement of "faith" in Science as salvation and in civilization as benefactor. The fact that these notions accompanied the conquerors made them all the more readily "accepted" (in the words of La Fontaine, "The strongest is always the most right").*

Since then we have maintained an obsession with finding everywhere the *moment* of coincidence between the appearance of both new behavior and the "modern" form of humankind. Everywhere else, the European models fail to explain the observations, even when the discoveries are made by the most orthodox Europeans. In North Africa, Australia, China, and sub-Saharan Africa, for example, one does not find an abrupt appearance of a Modern Man or of new behavior, but rather *processes* of a slow evolution that are distinct and complex. Hence one is led to wonder why, for a period as short as the Quaternary, human evolution should be considered on a global scale, taking place on all continents, whereas that of other animal species is considered only regionally.

It may seem natural for our own origins to fascinate us more than those of the other species. However, while this may justify obtaining research grants, it should not result in passionate and irrational behavior.

4. EVOLUTIONARY FACTORS PECULIAR TO MAN

The evolutionary biological forces that are particular to every living being also affect our own species. However, with us, culturalized behavior has taken on such importance that it constitutes in itself a powerful evolutionary factor. This is true so much so that whatever the environment, the demographic density, or the period, the cultural component "fashions" the morphological destiny of our species, albeit without breaking its unity.

Hence, in all parts of the world, and quite independently, human populations display analogous evolutionary "tendencies" due, for example, to the manipulation of objects, the development of concepts, group solidarity, or the capacity for running or throwing. These behavioral examples, which are organically linked in our species, *delimit* these evolutionary trends without strictly determining them (Leroi-Gourhan 1964; Chance 1974). This is why all of today's "races" display an analogous morphological outcome although they have emerged from a single *species* that is probably extremely ancient (more than a million years, in our opinion). Furthermore, they have evolved independently, but in parallel, in a variety of environments while respecting the fundamental law of our species, that is, giving priority to conceptual development.

5. THE AFRICAN JOKE

Whether primordial Eve was black or yellow in no way alters the fact that she had numerous distant sisters, cousins, neighbors, and friends who, all on the Earth of the time, produced children who were at an advantage if they were capable of grasping concepts,

* "La raison du plus fort est toujours la meilleure." La Fontaine, Fable I.X, 1668.

manipulating them, transmitting them, and applying them, wherever they were born. It has long been known that Neandertal *capacities* were not inferior to those of Cro-Magnon, but that it was the *accomplishments* which distinguish them. This idea can be likened to comparing an Indian with an eighteenth century European on the basis of the equipment they each possessed and not their capacity for using it. The phantasms from before the eighteenth century still haunt us, and although Indians are no longer commonly accused of being pre-men, this role is filled today by fossil men who were in the process of great change, as indeed we still are.

One can understand that antiracist politics, like those in the United States during the 1960s, aroused tactical ideas among researchers who were constantly seeking funding. But this profoundly respectable approach should not be allowed to impose its truths simply because they appear more "moral," or transpose them to a completely different context. If, for example, the skull of Zuttiyeh, in Israel, dating to about 250,000 years ago (Bar-Yosef 1996), is accompanied by a fauna of African origin (Tchernov 1992), this does not mean that the entire Old World waited for the African reflux before starting to think about its own destiny. And it seems highly improbable and slightly ridiculous that 150,000 years later, through this same Levantine corridor, the same African form should appear, suddenly populating the whole of Eurasia, from the tip of Ireland to the Pacific islands.

6. THE MITOCHONDRIAL GAME

Certainly, microbiology could bring us useful material for reflection. But one can wonder why the usual literature so often imposes on us results obtained from mitochondrial DNA, manifestly on this side of the subspecific definition that interests us. It is obvious that present-day humankind belongs to the same species; all the cross-breeds prove it far better than any confused diagram (Stoneking and Cann 1989 *contra* Darlu 1991). Hence, one sees successive differences converging towards an African nucleus, and nobody doubts that this is the unique and primordial reservoir. The question that needs to be resolved is *when* the diversifications occurred. When one knows that currrent estimates are based on present-day populations, whose migrations are calculated from the data themselves, and when, moreover, the speeds of divergence are unknown and remain inaccessible in the absence of precise palaeodemographic data, one is flabbergasted by the self-assurance of some researchers and the naiveté of many others. Once again, the phenomenon of a mythical fascination has come into play, this time in the hands of the new secular priests who are bereft of any humanist vocation.

Naturally, biochemical analyses can bring finer, more precise results than those evoked above (Tokunga and Yuyi 1992). But one remains staggered at the absence of critical argument between the biological method being used and its relevance to the general anthropological question one has undertaken to resolve. Everything proceeds as if the fascination with a demonstration, simple but wrong, overrides the complex but authentic one that is taken from the archaeological arguments themselves. The behavior of Man, whether modern or not, is infinitely richer than any mitochondrial analysis that has ever been undertaken. The Humanities, and especially History, have for centuries been developing methodological arsenals that are adapted to facing up to such complexities, on the basis of the cultural elements themselves, considered in their evolutionary dynamism.

Once again, the mythical quest for a superior truth, inaccessible and intangible, overrides the values of the scientists themselves (Stengers 1993).

7. A DIFFERENT THOUGHT

It is interesting to note the freedom of thought in those who have largely escaped western influence, such as those of the Chinese, which stands in contrast to these classic dogmas or the excessive fascination with molecular biology (Granet 1968; Needham 1973). The notion of perpetual cycles that is particular to the traditional Chinese thought seems to arouse a far more flexible attitude in relation to evolutionary phenomena, especially those affecting our species (Ximzhi 1991; Pope 1992).

For the Westerners, it would seem that a paradigmatic revolution should come into operation in two stages. First, it is necessary to disengage oneself from the influence of an auxiliary science, such as microbiology, whose sphere of activity does not coincide with the historical processes being sought. Moreover, one should respect the values that are specific to the humanities. They alone are suited to the study of behavior and can achieve, in return, an explanation of biological nature by bringing in some meaning. One hence arrives at the fundamentals of the opposition between the innate and the acquired, which once set Jean-Paul Sartre against Claude Lévi-Straus when they examined the incest taboo (Pouillon 1985). A biological law linked to consanguinity was thus expressed by a universal cultural rule. In our opinion, proceeding from observed behavior towards an eventual biological meaning will afford a greater hope of coming close to these mechanisms than will proceeding from a general, ultimate, and universal law (Lumdsen and Wilson 1984; Samuel 1990).

8. SOME EXAMPLES

The shift between Neandertals and Modern Man spread like wildfire through Europe, where everything (i.e., behavior and anatomy) changes at the same time, giving rise to the dogmatic model mentioned above. Although the respective situations were infinitely more complex everywhere else, the fundamental concepts were developed in Europe and then exported by the Europeans. One should not, therefore, lose sight of the fact that the very questions that were applied to the rest of the world were conceived within western philosophical and scientific premises derived from the Judeo-Christian tradition, as can be observed in the notion of time, for example.

1. The Altaï region, in Central Asia, displays a continuous evolution towards a blade industry from industries with prepared flakes, and this occurs in a completely autonomous way. Neither a cultural break nor an external influence disturb the continuous evolution that can be observed between the early and late Paleolithic (Goebel et al. 1993; Otte and Derevianko 1996).
2. An analogous situation is found in the Negev, where the appearance of blade industries was very early (around 40,000 years ago) but seems neither to mark a break, nor to cause any displacements, nor to correspond to any ethnic—let alone anthropological—modifications (Marks 1988; Bar-Yosef and Vandermeersch 1991).
3. In the Levant, apart from the Ahmarian evoked above, the Aurignacian appears at Hayonim and at Ksar Akil in the form of a late invasion (32,000 years ago) and as a break with the local substratum. If, therefore, a modern type of man appeared there around 115,000–90,000 years ago (Bar-Yosef 1989), he has no link with the far more general movement that affected the whole of Eurasia 40,000 years later

with the Aurignacian. This situation once again simply shows the convergence of anatomical evolutions in different centers such as Asia and Africa.

4. North Africa sees a late evolution in the form of an evolved Middle Paleolithic (Aterian) accompanied by a "primitive" but not Neandertal form of human (Hublin 1989), until around 20,000 years ago. The Upper Paleolithic culture (Iberomaurusian) appears here much later than elsewhere, with modern man. In this extreme west of Africa, a break that is analogous to that of Europe appears late, again under the influence of currents from the east.
5. In the Caucasus, Crimea, and the Balkans, in all the sites where we have worked ourselves on the edge of the Black Sea, a classic model has been found repeatedly in the form of an Aurignacian interruption through a very evolving and innovative Mousterian substratum. The Balkan dates are apparently the oldest (around 44,000 years ago), although most of the eastern sites remain badly dated. This situation allows one to imagine a diffusion at a more or less constant latitude along the Mediterranean route, which was then above sea level (Kozlowski 1990; Otte et al. 1996).
6. The Zagros and the steppes of Central Asia have, until the present, yielded few traces of this change. These are limited to a site at Samarkand and a few sites in the northern foothills of the Zagros (Olszewski and Dibble 1994). However, in our opinion, this is the most promising zone because it is located in an ecological context that is favorable to the kind of change observed in Europe (Otte 1994). It is in these open landscapes that relationships with animals are modified through the use of thrown weapons, the exploitation of natural defenses as attack weapons, and the use of animal images. We have developed this point elsewhere, and it seems crucial to the conceptions that man has of his place in Nature (Otte 1995) and that, in Europe, characterize the mythical break evoked above. One simply needs to recall here that this shift is above all psychic in nature, then economic, and finally demographic and migratory. It thus corresponds to a purely historical phenomenon, like the Barbarian Invasions, the Conquest of the West, or the Battle of the Ardennes. The jolts in the adventure of human paleontology were forgotten long ago. They are no longer anything more than slow transformations, dependent on social behavior, which we pursue constantly even today.

9. ANATOLIA

This immense territory, spread out over two thousand kilometers of latitude, occupies the key intermediary position between the three worlds: Africa, Asia, and Europe. Every theory on the origin of Modern Man must confront the Anatolian argument. After ten years of intense research in Anatolian sites, and despite their richness and their density, no traces have yet been observed of this "mythical shift" at the origins of Modern Man. The Middle Paleolithic, extremely rich and complex (Otte et al. 1995), continues here until quite late. The Epipaleolithic, equally well-documented, only begins around 16,000 years ago. The "Aurignacian" or equivalent phase remains unknown up to the present. The best proof of an African wave for the origin of Modern Man should be provided by Anatolia, the natural bridge to Europe. We believe that rich material will soon replace the existing gaps, especially if one takes into account the Aurignacian sites of the Caucasus, Crimea, and Georgia.

Dates B.P. U/Th-ESR (averages)	O16/O18	Key Humic Horizons (soils and concretions)	Geological Layers	Depths in the Central Sector (cm below datum)	Spits (excavation units)	Archaeological Units	Technical Characteristics	Lithics
			I,1	-500 to -550 cm	1 to 4	**P.S.**	LATE UPPER PALEOLITHIC bladelets; microliths	
60000 to 70000			I.2- .6	-550 to -650 cm	5 to 14	**I**	MOUSTERIAN OF ZAGROS OR KARAIN TYPE Levallois and discoidal technique; many scapers; fine marginally retouched points	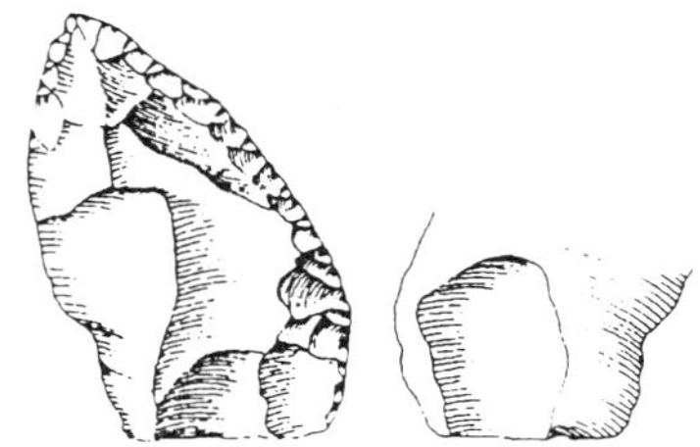
110000 to 120000	5e	1	I.7	-650 to -700 cm	15 to 18	**H**		
			II.1 2,3	-700 to -750 cm	19 to 25	**G**		
195000 to 251000	7	2	III.1	-750 to -770 cm	26	**G**		
			III.2	-770 to -850 cm	27 to 32	**F**		
297000 to 347000	9	3	III.3,4,5	-850 to -880 cm	33 to 37	**E**	"PROTO-CHARENTIAN" thick flakes; hard percussion; centripetal core preparation; heavily retouched notches and denticulates; steep scrapers	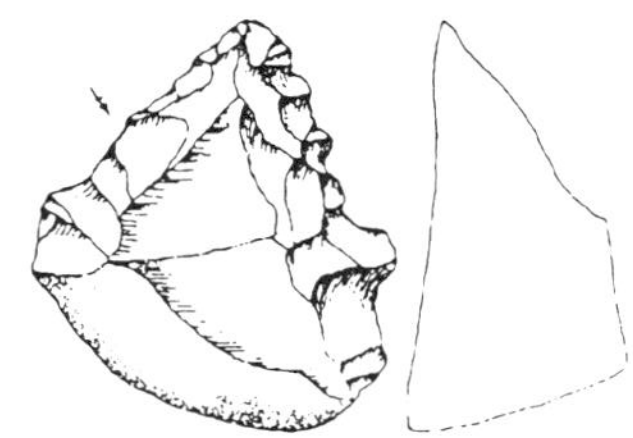
			IV.1	-880 to -900 cm	38 to 39	**D**		
			IV.2,3,4	-900 to -1000 cm	40 to 51	**C**		
			IV.5	-1000 to -1050 cm	52 to 56	**B**		
367000 to 440000	11	4	V	-1050 to 1100 cm	57 to 61	**A**	"CLACTONIAN" notches; no core preparation	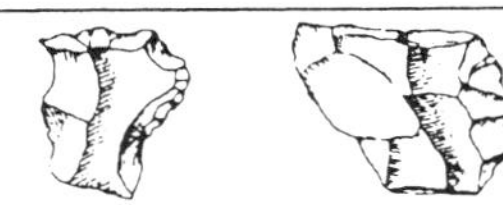

Figure 2. Karain E: Correlation between geological layers and archaeological layers.

One can see a skirting effect here: from the Balkans to the Caucasus, via the north of the Black Sea; and from the Caucasus to the Levant, through upper Mesopotamia (El Kown, for example). A trail of sites appears to correspond to the Aurignacian migrations and looks like a people carrying a specific culture and not a stage of anatomical evolution. Only Europe displays the coincidence of the two simultaneous transformations because they took place here at the extremity of a world in which the Neandertals had been specialized. In Europe, a social behavior accompanies the appearance of a new people, which has remained unchanged since then and which, in our opinion, is the origin of the Indo-Europeans (Otte 1995). Elsewhere, the evolutionary stages merge in the form of cross-breedings or waves overlapping with archaic forms. There is no need at all for a geographical point of origin, because the trend, exclusive and powerful, had already begun. Anatolia proves that there was no passage by land from the Levant to the Balkans at the origins of Modern Man, just as it proves that there was a single date for the transition from the Mousterian to the Upper Paleolithic.

Conversely, the relations between the Anatolian "cultural traditions" and the Zagros or the Balkans display the transversal geographic entity, maintained from the Mousterian onward (Otte et al. 1995). This situation does not correspond to the cultural evolution of the Levant (Bar-Yosef 1996) nor to the anatomical alternation of evolved or archaic forms of fossil human remains (Bar-Yosef and Vandermeersch 1991).

Farther east, the cultural mosaic becomes so complicated that the anatomical criteria cannot be linked to the behavioral components with any regularity (Gabori 1986; Ranov 1991)—so much so that both the notion of modern man and that of the Upper Paleolithic become completely diluted. Between these two worlds, structured and clear in Europe, diffuse and hazy in Central and Eastern Asia, there lies the immense Anatolian plateau whose transitional content forces one to reconsider the very concepts in the premises of Paleolithic research.

From the Acheulian onward, regional transition takes place between the African traditions, made from carved blocks, and those of Asia, made on flakes. In the Mousterian of Karain, one sees the shift from industries made on thick flakes to those made of prepared flakes of Levallois type (Figure 2). In the early Upper Paleolithic, migrations appear peripheral and in no way justify the supposed African origin.

Of course, important new research needs to be undertaken in this immense territory in order to confirm these suppositions, which are based on our present but limited knowledge. However, Anatolia already constitutes the field of research in which we need to test the general theories that are too often restricted to Europe and the Levant. Be that as it may, the Anatolian situation forces one to reconsider the value of the concepts that are used here and there in its territory. It embodies the distance to be established in their meaning: historical in Europe, processual in Asia. Between the two, the notion of Modern Man carries an entirely different meaning.

REFERENCES

Bar-Yosef, O. 1989. Geochronology of the Levantine Middle Palaeolithic, 589–610. In *The Human Revolution*, ed. P. A. Mellars and C. B. Stringer. Edinburgh: Edinburgh University Press.

Bar-Yosef, O. 1996. Modern Humans, Neanderthals and the Middle-Upper Palaeolithic transition in Western Asia, 175–90. In *The Lower and Middle Palaeolithic: Colloquia*, ed. O. Bar-Yosef, L. Cavalli-Sforza, R. March, and M. Piperno. Forli: ABACO Editions.

Bar-Yosef, O., and B. Vandermeersch. 1991. Premiers hommes modernes et néandertaliens au Proche-Orient: Chronologie et culture, 217–50. In *Aux Origines d'Homo sapiens*, ed. J. J. Hublin and A. M. Tillier. Paris: Press Universitaires de France.

Chance, M. 1974. Une dimension absente en biologie: le comportement, 218–27. In *L'Unité de l'Homme, Invariants Biologiques et Universaux Culturels*, ed. E. Morin and M. Piattelli-Palmarini. Paris: Le Seuil.

Cohen, C. 1991. Les races humaines en histoire des Sciences, 9–47. In *Aux Origines d'Homo sapiens*, ed. J. J. Hublin and A. M. Tillier. Paris: Press Universitaires de France.

Darlu, P. 1991. Diversité génétique et origine de l'homme moderne, 367–404. In *Aux origines d'Homo sapiens*, ed. J. J. Hublin and A. M. Tillier. Paris: Press Universitaires de France.

Dobzhansky, T. 1966. *L'homme en Evolution*. Paris: Flammarion.

Frank, M. 1990. *Le Dieu à Venir*. Paris: Actes Sud.

Gabori, M. 1976. *Les Civilisations du Paléolithique Moyen entre les Alpes et l'Oural*. Budapest: Akaddémiai Kiado.

Gibbons, A. 1994. Rewriting and redating prehistory. *Science* 263:1087–8.

Goebel, T., A. Derevianko, and V. Petrin. 1993. Dating the Middle-to-Upper Paleolithic transition at Kara-Bom. *Current Anthropology* 34:452–8.

Granet, M. 1968. *La Pensée Chinoise*. Paris: Albin Michel.

Hublin, J. J. 1989. Les plus anciens représentants de la lignée pré-néandertalienne, 81–94. In *L'homme de Néandertal*, Vol. 3, ed. E. Trinkaus. Liège: ERAUL.

Jacquard, A. 1984. *Inventer l'Homme*. Paris: Edition Complexe.

Jia, L., and C. K. Ho. 1990. Lumière nouvelle sur l'archéologie paléolithique chinoise. *L'Anthropologie* 94(4):851–60.

Kozlowsi, J. 1990. A multi-aspectual approach to the origins of the Upper Palaeolithic in Europe, 419–37. In *The Emergence of Modern Humans: An Archaeological Perspective*, ed. P. A. Mellars. Edinburgh: Edinburgh University Press.

Larichev, V., U. Khol'ushkin, and I. Laricheva. 1988. The Upper Paleolithic of Northern Asia: Achievements, problems and perspectives. I. Western Siberia. *Journal of World Prehistory* 2:(4):359–96.

Leroi-Gourhan, A. 1964. *Le Geste et la Parole. Technique et Language*. Paris: Albin Michel.

Lumsdem, C., and E. Wilson. 1984. *Le Feu de Prométhée, Réflexions sur l'Origine de l'Esprit*. Paris: Mazarine.

Marks, A. 1988. The Middle to Upper Paleolithic transition in the southern Levant, 109–24. In *La Mutation*, ed. J. Kozlowski. Liège: ERAUL.

Mehler, J., and E. Dupoux. 1980. *Naître Humain*. Paris: Odile Jacob.

Needham, J. 1973. *La Science Chinoise et l'Occident*. Paris: Le Seuil.

Olszewski, D., and H. Dibble. 1994. The Zagros Aurignacian. *Current Anthropology* 35(1):68–74.

Otte, M. 1994. Origine de l'homme moderne: Approche comportementale. *Comptes Rendus de l'Académie de Sciences* 318(II):267–73.

Otte, M. 1995. Diffusion des langues modernes en Eurasie préhistorique. *Comptes Rendus de l'Académie de Sciences*.321(IIa)1219–26.

Otte, M., and A. Derevianko. 1996. Transformations techniques au paléolithique de l'Altaï (Sibérie), 131–44. In *Anthropologie et Préhistoire*, ed. I. Jadin. Brussels: Bulletin de la Societé Belge d'Anthropologie et de Préhistoire.

Otte, M., P. Noiret, S. Tatartsev, and I. Lopez-Bayon. 1996. L'Aurignacien de Siuren I (Crimée, Ukraine), 81–92. In *Anthropologie et Préhistoire*, ed. I. Jadin. Brussels: Bulletin de la Societé Belge d'Anthropologie et de Préhistoire.

Otte, M., I. Yalcinkaya, H. Taskiran, J. Kozlowski, O. Bar-Yosef, and P. Noiret. 1995. The Anatolian Middle Paleolithic: New research of Karain Cave. *Journal of Anthropological Research* 51(4):287–99.

Pope, G. 1992. Replacement versus regionally continous models: The paleobehavioral and fossil evidence from East Asia, 3–14. In *The Evolution and Dispersal of Modern Humans in Asia*, ed. T. Akazawa, K. Aoli, and T. Kimura. Tokyo: Hokusen-Sha.

Pouillon, J. 1985. Nature et culture. *Claude Levi-Straus, Le Magazine littéraire* 223:52–3.

Ranov, V. 1991. Les sites très anciens de l'âge de la pierre en URSS, 210–16. In *Les Premiers Européens*, ed. E. Bonifay and B. Vandermeersch. Paris: Congrès des Sociétés Savantes.

Samuel, G. 1990. *Mind, Body and Culture: Anthropology and the Biological Interface*. Cambridge: Cambridge University Press.

Stengers, I. 1993. *L'invention des Sciences Modernes*. Paris: La Découverte.

Stoneking, M., and R. Cann. 1989. African origin and human mitochondrial DNA, 17–30. In *The Human Revolution*, ed. P. A. Mellars and C. B. Stringer. Edingurgh: Edinburgh University Press.

Tchernov, E. 1992. Biochronology, paleoecology, and dispersal events of hominids in the Southern Levant, 149–88. In *The Evolution and Dispersal of Modern Humans in Asia*, ed. T. Akazawa, K. Aoli, and T. Kimura. Tokyo: Hokusen-Sha.

Tokunaga, K., and T. Yuyi. 1992. The migration and dispersal of East Asian populations as viewed from HLA genes and haplotypes, 599–611. In *The Evolution and Dispersal of Modern Humans in Asia*, ed. T. Akazawa, K. Aoli, and T. Kimura. Tokyo: Hokusen-Sha.

Wolpoff, M. H. 1991. "*Homo erectus*" et les origines de la diversité humaine, 99–155. In *Aux Origines d'Homo sapiens*, ed. J. J. Hublin and A. M. Tillier. Paris: Presses Universitaires de France.

Wright, R. 1994. *The Moral Animal: Evolutionary Psychology and Everyday Life*. New York: Pantheon Books.

Ximzhi, W. 1991. Continuité évolutive des hommes fossiles chinois, 159–79. In *Aux Origines d'Homo sapiens*, ed. J. J. Hublin and A. M. Tillier. Paris: Press Universitaires de France.

32

THE IMPACT OF CLIMATIC CHANGES ON THE DEMOGRAPHY OF LATE NEANDERTAL AND EARLY ANATOMICALLY MODERN POPULATIONS IN EUROPE

Paul Mellars

Department of Archaeology
Cambridge CB2 3DZ, United Kingdom

1. INTRODUCTION

One of the most impressive features of research over the past ten years has been the rapid increase in our understanding of the pattern and character of climatic and associated ecological fluctuations throughout the course of the last glaciation. The evidence has come from a variety of sources: high resolution oxygen-isotope studies of deep-sea cores (e.g., Bond et al. 1993); similar high resolution studies of rapidly-accumulating ice core-sequences in Greenland and elsewhere (e.g., Dansgaard et al. 1993; Grootes et al. 1993); and (potentially most important for archaeology) the investigation of a number of exceptionally long and apparently continuous lake pollen sequences in several different regions of Europe and elsewhere (Woillard 1978; de Beaulieu and Reille 1984; Follieri et al. 1990; Reille and de Beaulieu 1990; Guiot et al. 1993; Tzedakis 1994; Watts et al. 1996). As a result of these studies we can now provide a reconstruction of climatic and associated paleoenvironmental fluctuations during the course of the last glaciation, which would have been impossible ten or even five years ago.

The most critical period for studies of the geographical dispersal of anatomically modern populations across Europe and their potential contact and interaction with the local Neandertal populations is that of oxygen-isotope stage 3, extending from about 60,000 to 25,000 BP (Martinson et al. 1987). Sharply focused research has recently been devoted to the precise character of the ecological shifts during this period by the interdisciplinary OIS 3 project recently initiated by the Godwin Institute for Quaternary Research at Cambridge, with the explicit aim of providing the essential climatic and paleoenvironmental

Neandertals and Modern Humans in Western Asia, edited by Akazawa *et al.*
Plenum Press, New York, 1998.

background for studies of contemporaneous human populations (see Van Andel and Tzedakis 1996a,b, 1997). The present paper will take the paleoecological reconstructions provided in these recent studies as a starting point and go on to discuss some of the implications of these ecological patterns for the demographic shifts of anatomically modern and Neandertal populations during the later stages of isotope stage 3.

Although the exact chronology and correlation of climatic oscillations during the course of OIS 3 are still to some extent tentative, the ecological consequences of these changes cannot be disputed. Temperature changes of up to 8°C between major cold and warm episodes in stage 3—apparently in some cases within periods of less than 20 years—have now been documented from several lines of evidence, and the ecological effects of these changes can be seen clearly in the relevant pollen and faunal records (Guiot et al. 1989, 1993; Dansgaard et al. 1993; Van Andel and Tzedakis 1996a, 1997). As an example, it can now be clearly demonstrated from faunal assemblages recovered from cave and rock shelter sequences in southwestern France that faunal communities shifted between those heavily dominated by reindeer to those dominated by bison, horse, or even red deer at several different points during OIS 3 (e.g., Figure 1; Laville et al. 1980; Delpech 1983; Boyle 1988; Mellars 1996a: Fig. 2.22). It is inconceivable that ecological changes of this scale could have occurred without some very significant behavioral adaptations and demographic shifts among the contemporaneous human populations.

In the present paper I look briefly at the impact of these climatic changes from three main perspectives: (a) in terms of the patterns of dispersal of anatomically modern populations across Europe; (b) in terms of possible models for the apparent coexistence and potential interaction of these early anatomically modern and the final Neandertal populations in certain regions of Europe; and (c) in terms of the factors responsible for the eventual decline and extinction of Neandertal populations, in at least the more northerly parts of their range.

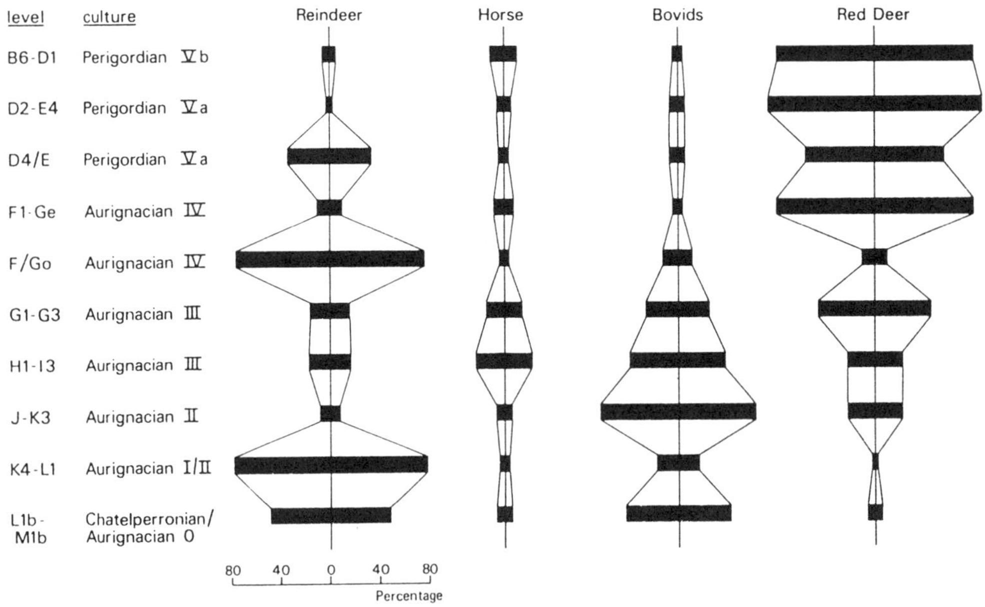

Figure 1. Fluctuations in faunal frequencies in the early Upper Paleolithic levels (about 35–25,000 BP) at La Ferrassie, southwest France. From Mellars 1985.

2. THE DISPERSAL OF ANATOMICALLY MODERN POPULATIONS ACROSS EUROPE

The arguments for and against a large-scale dispersal of anatomically modern (i.e., *Homo sapiens sapiens*, or 'Cro-Magnon') populations across Europe and the associated decline and extinction of the preceding Neandertal populations have been rehearsed at length in the recent literature (see, e.g., papers in Mellars and Stringer 1989; Mellars 1989, 1992a, etc.; Stringer and Gamble 1993; Nitecki and Nitecki 1994; Stringer 1996). Briefly, the arguments in favor of a major population dispersal derive partly from DNA studies of modern human populations (e.g., Stoneking and Cann 1989; Rogers and Jorde 1993; Sykes 1997), partly from studies of the skeletal morphology of the final Neandertal and early anatomically modern populations (Stringer et al. 1984; Gambier 1989, 1993; Hublin 1990; Stringer and Gamble 1993; Hublin et al. 1996a), and partly from independent studies of the archaeological patterns associated with the earliest skeletons of anatomically modern form (Kozlowski 1988, 1992, 1993, 1996; Demars and Hublin 1989; Mellars 1989, 1992a, 1996b; Demars 1990; Banesz and Kozlowski 1993; Djindjian 1993; Broglio 1996). As discussed more fully elsewhere (Mellars 1973, 1992a,b, 1996a,b,c), the archaeological records appear to indicate not only a dramatic transformation in almost all aspects of archaeologically visible human behavioral patterns over this time range (in stone and bone technology, art and ornamentation, and most probably also in several aspects of social organization, demography, communication, and subsistence patterns), but also an apparent cline in the dates for the earliest appearance of these distinctively Upper Paleolithic technologies progressively from east to west across Europe (see Figures 2 and 3; Mellars 1992a). The chronological pattern illustrated in Figure 3 is of course subject to numerous qualifications: the inherent susceptibility of radiocarbon dates to contamination in this age range (Mellars et al. 1987; Aitken 1990); known fluctuations in the radiocarbon activity of the atmosphere over the 30,000–50,000 time range, and the associated need for calibration of the dates to absolute terms (Bard et al. 1990; Laj et al. 1996); and the shortage of reliably dated sites in several critical areas. Nevertheless, the pattern illustrated in Figure 3 does on the face of it provide strong support for the notion of a progressive dispersal of early Upper Paleolithic technology—and by implication the associated anatomically modern populations—essentially from southeast to northwest across Europe over the period (in uncalibrated radiocarbon terms) from about 43,000 to 35,000 BP.

The mechanisms responsible for this dispersal of anatomically modern populations were undoubtedly complex and are likely to have entailed a strong component of internal demographic pressures, quite independent of any effects of contemporaneous climatic changes (Mellars 1996b). As I (1989, 1992a etc.), Bar-Yosef (1992, 1994, 1996) and others have discussed, it is likely that the nature of the various behavioral changes inherent in the so-called 'Upper Paleolithic revolution'—however these may initially have originated beyond the bounds of Europe—would have led to a significant increase in local population numbers, and accordingly to an inherent internal pressure for these expanding populations to colonize new territories progressively to the north and west. As Bar-Yosef (1994, 1996) has recently argued, the process of population expansion could well have had many similarities to that of the subsequent Neolithic revolution, which may well have originated in much the same areas of western Asia or southeastern Europe some 30–35,000 years later. The patterns of population expansion could also have followed broadly similar geographic routes, i.e., along both the Mediterranean coastal zone and north-westwards up the Danube and Rhine valleys into western Europe (Kozlowski 1992, 1993; Djindjian 1993).

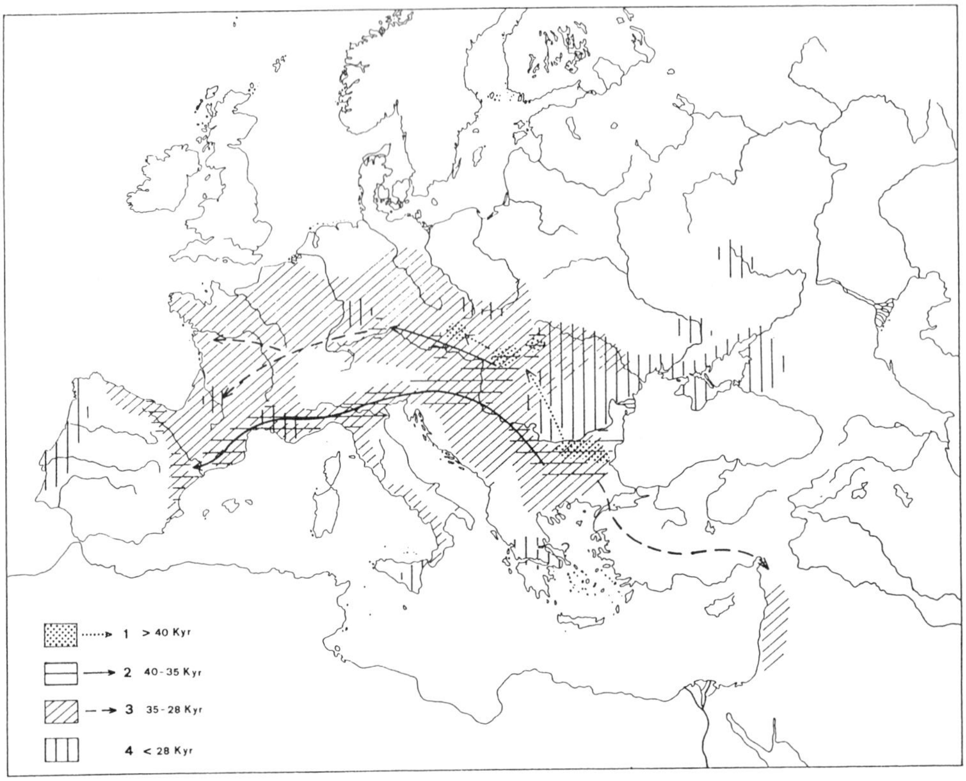

Figure 2. Distribution and spread of Aurignacian technologies across Europe, according to Kozlowski (1992).

The point to be emphasized here is that for populations who originated in the relatively temperate environments of the eastern Mediterranean zone or southwest Asia, any major episode of climatic amelioration would inevitably have made the process of population dispersal and colonization into the periglacial landscapes of eastern, central, and western Europe much easier to achieve. The details of this process would of course have varied between the different regions of Europe in response to local ecological conditions (see Van Andel and Tzedakis 1996a,b, 1997). What can be seen very clearly from Figure 4, however, is that during the major warm episodes of OIS 3 a large area along the southern Mediterranean zone of Europe would have supported various forms of mixed deciduous/coniferous woodland, probably not very different from that to be found in the littoral and immediately inland areas of the southwest Asian region throughout a large part of the last glaciation (Van Andel and Tzedakis 1996a, 1997). Colonization of this region by human populations in direct response to a major climatic amelioration during isotope stage 3 would arguably have required only limited 'adaptation' on their part, either in terms of local climatic and vegetational conditions, or even in terms of the kinds of animal populations available for exploitation. In the latter context it is particularly interesting to see that in many of the earliest settlements of early Aurignacian populations, extending from the Balkans, through Italy to southeast France and northern Spain, a major component in the faunas is frequently red deer (e.g., Straus 1990). Other fairly uniform food resources across this area would probably have come from various plant species found in the mixed deciduous/coniferous forest (though archaeological evidence for this is virtually lacking)

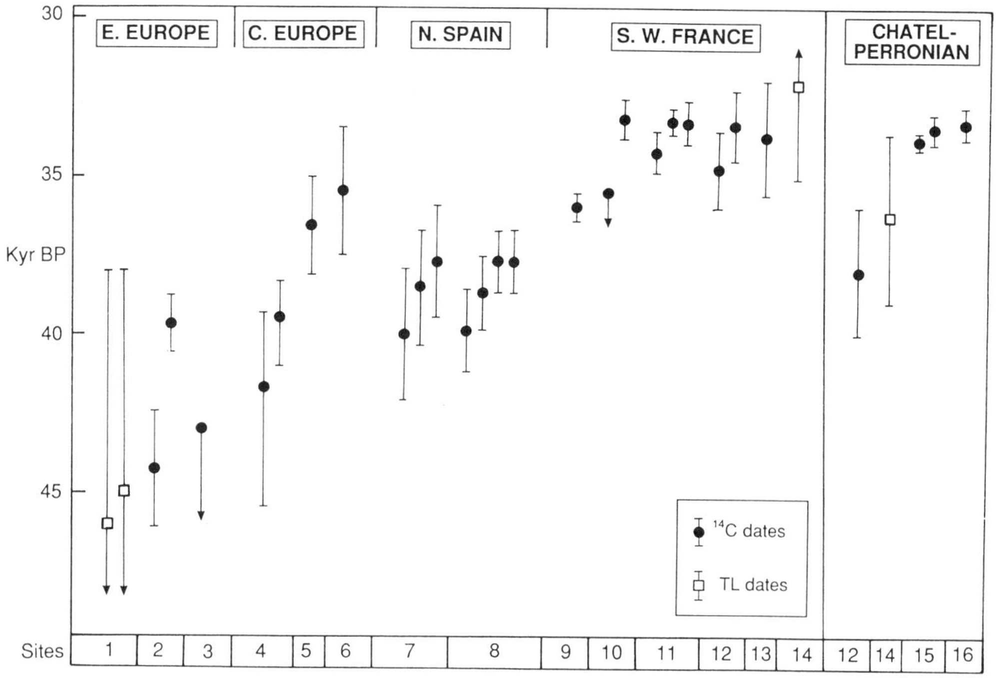

Figure 3. Absolute dates for early Aurignacian industries in Eastern, Central, and Western Europe, and for Chatelperronian industries in France. From Mellars 1992a. The sites shown are: 1. Temnata (Bulgaria); 2. Istállóskö (Hungary); 3. Bacho Kiro (Bulgaria); 4. Willendorf (Austria); 5. Geissenklösterle (Germany); 6. Krems (Austria); 7. Castillo (N.W. Spain); 8. L'Arbreda (N.E. Spain); 9. La Rochette (France); 10. La Ferrassie (France); 11. Abri Pataud (France); 12. Roc de Combe (France); 13. Le Flageolet (France); 14. Saint-Césaire (France); 15. Arcy-sur-Cure (France); 16. Les Cottés (France). Note that radiocarbon dates in this age range are likely to be systematically younger than those produced by other dating techniques by perhaps 3000 years (Bard et al. 1990; Mercier and Valladas 1993; Laj et al. 1996).

as well as from coastal resources distributed along the Mediterranean littoral zone. In short, for early anatomically modern populations to have colonized these more southerly areas of Europe during the course of a major warm episode during OIS 3 may have amounted to little more than 'surfing the ecological tide' in demographic or adaptive terms. This would be especially true if, as discussed below, the same ecological changes would have tended to cause territorial shifts among the late Neandertal populations, which could have left certain geographical habitats along the coastal Mediterranean fringes of Europe at least partially, if not entirely, free of direct demographic competition to colonization by the expanding modern populations (cf Gamble 1986; Soffer 1994).

Some of the current uncertainties over the exact dating of the various climatic ameliorations of OIS 3 call for more caution in the precise correlation of this episode of population dispersal with specific climatic events. Despite the current debates over these chronologies (cf Van Andel and Tzedakis 1996a, 1997; Watts et al. 1996), there is increasing evidence for one very major warm period which immediately follows the sharp glacial episode known as 'Heinrich event 3' in the deep sea core isotope records (Bond et al. 1993; Kerr 1993) and which in terms of the current (uncalibrated) radiocarbon time scale probably centers on about 38,000–41,000 BP. This is the event which Dansgaard et al. (1993) have correlated with the major 'Hengelo interstadial' as represented in the paleobotanical records of northern and western Europe (Behre 1990; Zagwijn 1990). While the

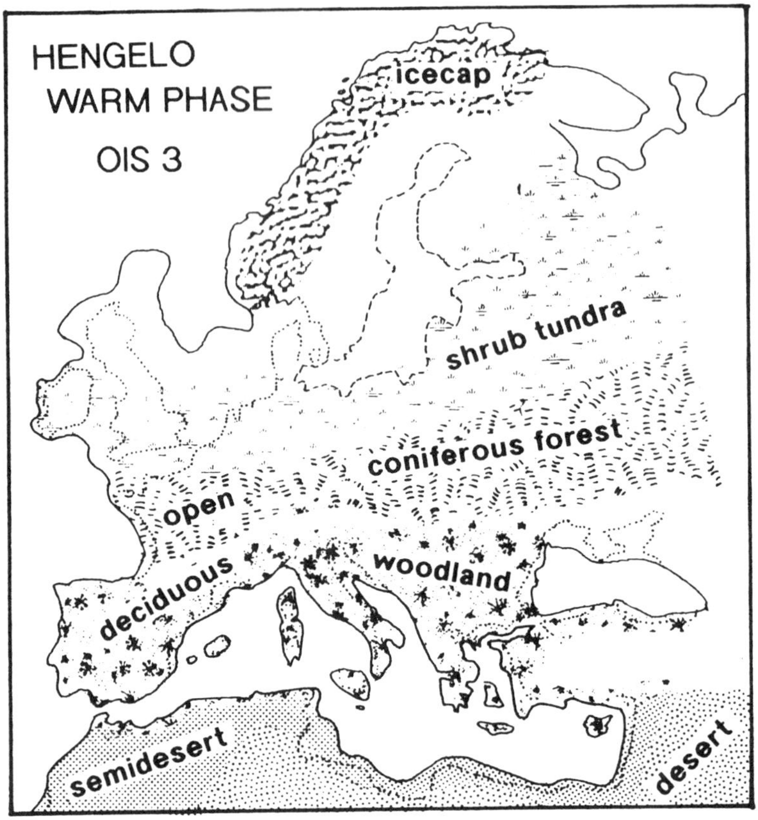

Figure 4. Pattern of vegetational zones in Europe at the time of the Hengelo interstadial, about 40,000 BP (uncalibrated ^{14}C), according to Van Andel and Tzedakis 1997.

problems of precise dating and correlation of the different climatic episodes remain complex, I would suggest that it is this climatic episode which witnessed and facilitated the major episode of demographic expansion and dispersal of anatomically modern populations across Europe, together with the associated patterns of early Upper Paleolithic/ Aurignacian technology.

3. MODELS OF POPULATION COEXISTENCE

Evidence accumulated over the past 15 years shows that the process of eventual demographic replacement of the Neandertal by anatomically modern populations in Europe was by no means an overnight occurrence. While the presence of early Aurignacian industries—and presumed anatomically modern populations—can now be documented in many areas of Europe by around 40,000 BP (Mellars 1992a,b; Djindjian 1993; Kozlowski 1993, 1996; Otte and Straus 1995), there is equally clear evidence that in several areas of the continent local populations of Neandertals with technologies apparently heavily influenced by contact with anatomically modern populations survived down to at least 33,000–34,000 BP. These are perhaps best represented in the characteristically Chatelperronian technologies found in direct association with typically Neandertal fossils at both Saint-Césaire in southwest France (Lévêque and Vandermeersch 1980; Mercier et al. 1991; Lévêque et al. 1993) and Arcy-sur-Cure in northern Burgundy (Farizy 1990; Hublin et al. 1996a). Similar chronological overlap of the final Neandertal and early anatomically modern populations almost certainly occurred in many other regions of Europe, as for example in the Szeletian, Jerzmanovizian, and other leaf-point industries of central and northern Europe, the Italian Uluzzian, and the Streletskayan of southern Russia (Allsworth-Jones 1986, 1990; Soffer 1989; Goia et al. 1990; Mussi 1990; Kozlowski 1992, 1993, 1996; Broglio 1996).

Within southwestern France itself the available stratigraphic and chronological evidence suggests that the time ranges occupied by the final Neandertal (i.e., Chatelperronian) and earliest anatomically modern (i.e., Aurignacian) populations were largely separate, with the Aurignacian populations apparently replacing the Chatelperronian groups at around 33,000–34,000 BP (Mellars et al. 1987; Harrold 1989; Rigaud 1993). The existence of apparently well documented interstratifications of the two industries in at least two French sites (Le Piage and Roc de Combe), however, and the evidence of some degree of chronological overlap in the radiocarbon and paleoclimatic evidence (Figure 3), nevertheless implies that there must have been some degree of closer juxtaposition of the two groups during at least certain parts of this time range (Bordes and Labrot 1967; Champagne and Espitalié 1981; Leroyer and Leroi-Gourhan 1983; Leroyer 1988; Harrold 1989; Demars 1990). What follows is an attempt to identify some of the potential models by which this kind of 'coexistence' has occurred. Essentially, three scenarios can be visualized:

3.1. Low Population Densities

One possibility is that during the time range in question both the final Neandertal populations and the intrusive anatomically modern populations were present in such low population densities in most areas of Europe that the two groups could exploit the same region with only infrequent contacts or direct competition for the use of particular territories or economic resources. The hypothesis is difficult to test directly, but it is often suggested that Neandertal populations in general were living in relatively low population densities throughout Europe (Whallon 1989; Gamble 1993; Soffer 1994), and it could be argued that during the initial, pioneer episodes of population expansion and colonization of new habitats the earliest anatomically modern populations would also have been relatively thin on the ground and widely dispersed. Both groups could well have practiced very mobile annual and seasonal foraging strategies, which could arguably have allowed them to avoid close competition for the use of particular foraging ranges or resources for most of the time.

3.2. Use of Different Economic Resources

The most commonly recorded situation in which two or more species occupy and exploit the same territory arises when the two populations concentrate on different components of the available resources, in such a way that their foraging patterns are largely complementary rather than directly competitive in ecological terms (e.g., Foley 1987). In other words, the separate populations have discrete ecological or foraging 'niches' with only limited overlap.

At present we unfortunately know too little about the precise economic adaptations of the Chatelperronian and earliest Aurignacian groups in western Europe to assess to what extent this can provide a viable explanation for potential population coexistence in the present case. From the faunal assemblages recovered from the Chatelperronian levels at La Ferrassie, Roc de Combe, Le Piage, Les Cottés, Fonténioux, La Chèvre, Pair-non-Pair, Saint-Césaire, Quincay, Arcy sur Cure, and Châtelperron itself, it would appear that these final Neandertal groups practiced a relatively broad-spectrum foraging pattern, usually involving substantial exploitation of at least three or four different species—reindeer, horse, large bovids (either *Bos* or *Bison*), and red deer. By contrast, most of the faunas recovered from early Aurignacian levels in the same region show a striking specialization on reindeer, with reindeer often comprising more than 90% of the documented remains (e.g., at Abri Pataud, Roc de Combe, Le Piage, and La Gravette) (Delpech 1983; Boyle 1988; Mellars 1989; Patou-Mathis 1993). If the Chatelperronian and Aurignacian groups were exploiting different resources then it is likely that they were utilizing to some extent different ranges within the same general territories. Thus the Aurignacians, for example, could have been exploiting predominantly seasonally migrating reindeer herds in the main river valleys of the Perigord (the Vézère, Dordogne, Isle, etc.) while the Chatelperronians were concentrating mainly on more widely dispersed and solitary species (e.g., horse or aurochs) on the intervening plateau and upland areas (Mellars 1989, 1992b). The same patterns could even be tied into more general seasonal contrasts in foraging patterns for these resources, which would have helped further to avoid any direct conflicts for the exploitation of specific resources at the same times and locations.

While it is tempting to suggest that this might provide a very convenient way of proposing distinct foraging specializations for the Chatelperronian and Aurignacian groups (Mellars 1992b), this would overlook the fact that most of the faunal assemblages being compared belong to separate climatic episodes: those of the Chatelperronian belong mainly to the relatively mild conditions of the so-called ‘Les Cottés’ interstadial (about 37,000–34,000 BP), while the Aurignacian faunas date mainly from the succeeding, much colder phase which followed 34,000 BP (Delpech 1983; Leroyer and Leroi-Gourhan 1983; Leroyer 1988; Rigaud 1993). In the adjacent areas of northern Spain, the earlier Aurignacian levels seem to be associated primarily with more temperate woodland forms such as red deer, aurochs, and horse (Straus 1990; Bernaldo de Quiros and Cabrera Valdés 1993). While the possibility of distinct foraging specializations for the Aurignacian and Chatelperronian populations certainly deserves close attention, it might be premature to adopt this too rapidly as an explanation for possible coexistence of the two groups from the evidence presently at hand.

3.3. Successive Rather than Simultaneous Use of the Same Territories

Overall, the most likely explanation for the apparent coexistence of Chatelperronian and Aurignacian populations in western Europe is that at any one point in time the two populations were probably confined very largely to separate economic and demographic territories, which largely avoided any direct competition for the use of specific resources at the same times and places. As discussed above, it is likely that throughout most of the period from about 38,000 to 34/35,000 BP, the greater part of southwestern France (or at least the central Perigord and immediately adjacent areas) was occupied exclusively by Chatelperronian groups, with the contemporaneous Aurignacian groups restricted largely to the areas to the south and east. Essentially this model has in fact been proposed by Leroyer and Leroi-Gourhan (1983), based on paleoclimatic data from various sites, who suggest that there could have been a gradual wave of population displacement extending progressively from southeastern France towards the north and west during the period from about 38,000 to 33,000 BP. Against this, of course, must be set the apparently well-documented instances of direct interstratifications of Chatelperronian and Aurignacian levels recorded at both Roc de Combe and Le Piage (Bordes and Labrot 1967; Champagne and Espitalié 1981; Demars 1990) While these two cases would appear to provide strong *a priori* evidence for the effective coexistence of the two groups within the same general territories, it would be possible to suggest that they reflect simply temporary displacements of the territories occupied by the two populations under the impact of rapidly fluctuating climatic and ecological conditions. Recent studies of the Greenland ice cores have provided dramatic evidence for the highly oscillatory nature of climatic conditions during the later part of isotope stage 3, marked in several cases by shifts in annual temperatures of up to 5–8°C over a period of a few decades (Dansgaard et al. 1993; Van Andel and Tzedakis 1996a, 1997). Climatic oscillations of this scale might well have had a significant effect on the territories occupied by different human groups, especially if the climatic oscillations were accompanied by shifts in the spatial distribution or seasonal migration patterns of certain animal species such as reindeer or red deer (Boyle 1988; Pike-Tay 1991, 1993). Alternatively, or in addition, either major ecological fluctuations or other demographic or health factors might conceivably have led on occasion to the temporary demise of Neandertal populations within specific territories, leaving in effect empty 'ecological space' to be colonized by neighboring groups. At present, it is doubtful whether the degree of temporal resolution of either the archaeological or climatic/ecological data is sufficient to resolve any of these specific issues. We should at least be aware, however, that territorial displacements and demographic oscillations of this kind could well have occurred at many different times during the 5,000 years or so of apparent coexistence between the late Neandertal and earliest anatomically modern populations in western Europe.

4. NEANDERTAL EXTINCTION

As noted above, most of the available evidence points to the conclusion that the period of co-existence of late Neandertal and early anatomically modern populations came to an end in at least most areas of Europe by around 33,000–34,000 BP, as reflected, for example, by the radiocarbon dates for late Chatelperronian levels at Les Cottés and Combe Saunière in southwest France and Arcy-sur-Cure further to the northeast (Figure 3). If the claimed dating evidence for the classic Neandertal mandible from Zafarraya in Andalusia is reliable, this could indicate a rather later survival of Neandertal populations in the more southerly regions of Europe down to about 30,000 BP, as has also been claimed from the southern Balkans (Hublin et al. 1996b; Stringer 1996:105–6). In both cases, however, the exact chronology of the finds remains to be clarified.

In my view, there were probably two critical factors which, acting in combination, led to the eventual extinction and replacement of the final Neandertal populations over at least the greater part of northern and western Europe. The first was the gradual but continuing increase in the total population numbers of the anatomically modern populations, and the consequently increasing competition between the two groups for the occupation and exploitation of specific economic territories. If there is any truth in the assumption that the anatomically modern populations possessed not only more complex and economically efficient technology than the Neandertal groups but probably also more highly structured patterns of social organization and probably more complex and efficient systems of inter-group communication (Mellars 1989, 1991, 1996c; Stringer and Gamble 1993; Bar-Yosef 1994, 1996; Klein 1994), then it seems almost inevitable that in any situation of direct competition of this kind the anatomically modern populations would be more likely to outcompete the Neandertals. The result would be a gradual contraction in the overall territories and economic ranges occupied by the Neandertals, and almost inevitably some fragmentation of the Neandertal groups into smaller and more demographically isolated social units. This process of social and demographic fragmentation alone—with its critical implications for the long-term viability of relatively small, isolated breeding units (cf Zubrow 1989)—could well have been sufficient to ensure the eventual extinction and replacement of Neandertals in most areas of the continent.

The second critical factor in the ultimate extinction of the Neandertals may well have been the climatic events during the later stages of isotope stage 3. Despite all of the caveats over the exact chronology of the events discussed above, there is reason to think that the final demise of the Neandertals in many areas of Europe coincided with an episode of sharply colder climate which most probably coincided with Heinrich event 4 as recorded in the deep-sea cores (Bond et al. 1993; Kerr 1993), and which probably centers on about 33,000–34,000 BP in radiocarbon years (Dansgaard et al. 1993; Van Andel and Tzedakis 1996a, 1997). In southwestern France, for example, this appears to be reflected archaeologically by the final replacement of the late Neandertal Chatelperronian technologies by those of classic early Aurignacian form (Mellars and Bricker 1987; Mellars 1992a, 1996a; see Figure 3), and is marked in the contemporaneous climatic records by a sudden return to extremely cold conditions, associated with exceptionally high frequencies of reindeer (up to 99%) in the associated faunas (Farrand 1975, 1995; Laville 1975; Laville et al. 1980; Delpech 1983; Boyle 1988; Raynal and Guadelli 1990).

While it is true that Neandertal populations had survived climatic and ecological conditions of equal severity at some earlier points in the last glaciation (most notably at the maximum of isotope stage 4), the climatic impact of Heinrich event 4 was not only very severe, but also coincided with increasing pressure of competition between the final Neandertal groups and the expanding anatomically modern populations. For both groups, adaptation to the rapidly deteriorating climatic conditions would have required not only major changes in subsistence strategies (to cope with changing species of animals, and the virtual disappearance of plant food resources in many areas of Europe) but also the capacity to cope with exceptionally severe winters. Winter survival under these conditions would have required not merely a range of specialized subsistence strategies (including strategies to cope with the effects of prolonged snow cover and frozen rivers on the mobility and communication patterns of the human groups) but also the capacity to cope with exceptionally low winter temperatures—probably to well below -20°C in many contexts. The use of skin clothing, artificial living structures, and (above all) abundant and reliable fuel supplies for camp fires would have been essential for winter survival. Indeed, the demands for fuel supplies may well have been the most critical single factor, since the effects of reduced temperatures may have virtually eliminated significant tree growth over large areas of the continent (see Figure 5; Van Andel and Tzedakis 1996a, 1997 and references therein).

The implications of these climatic changes for the survival of the Neandertals seems fairly clear. While the need for adaptation to these climatic changes would be equally essential for both the late Neandertal and early anatomically modern populations, it is reasonable to assume that the more complex and advanced technology and (most probably) patterns of social organization of the anatomically modern populations would have given them powerful adaptive and survival advantages over the Neandertal groups. In short, the combination of these climatic changes with the effects of increasing social fragmentation of the Neandertal groups and increasing competition for social and ecological space from the expanding anatomically modern groups could well have delivered the ultimate *coup de*

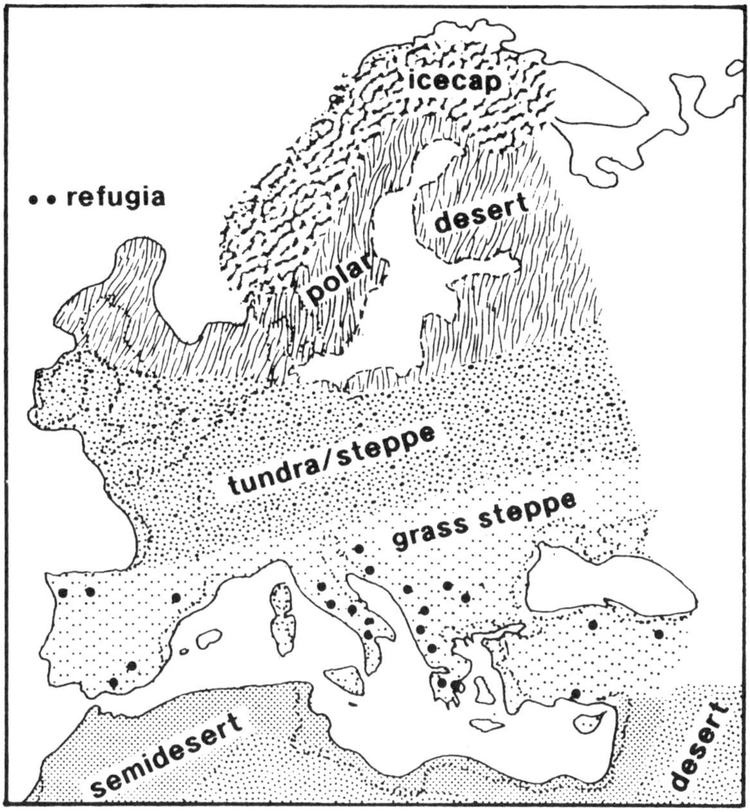

Figure 5. Pattern of vegetation zones in Europe during the major cold episodes of oxygen-isotope stage 3, according to Van Andel and Tzedakis 1997. Note the contrast with the pattern recorded during the Hengelo warm phase, shown in Figure 4.

grace to the final Neandertal populations over large areas of Europe. Heinrich event 4 may, perhaps, have been a final and critical factor in the demise of the European Neandertals, or at least within the more northerly parts of their territorial range.

ACKNOWLEDGMENTS

I am indebted to Tjeerd Van Andel and Chronis Tzedakis for permission to reproduce Figures 4 and 5, and for stimulating discussions. I am also indebted to Janusz Kozlowski for permission to reproduce Figure 2.

REFERENCES

Aitken, M. J. 1990. *Science-based Dating in Archaeology*. London: Longmans.

Allsworth-Jones, P. 1986. *The Szeletian and the Transition from Middle to Upper Palaeolithic in Central Europe*. Oxford: Oxford University Press

Allsworth-Jones, P. 1990. The Szeletian and the stratigraphic succession in Central Europe and adjacent areas: Main trends, recent results, and problems for solution, 160–243. In *The Emergence of Modern Humans: An Archaeological Perspective*, ed. P. Mellars. Edinburgh: Edinburgh University Press.

Banesz, L., and J. K. Kozlowski, eds. 1993. *Aurignacien en Europe et en Proche Orient.* Bratislava: Acts of XII International Congress of Prehistoric and Protohistoric Sciences.

Bar-Yosef, O. 1992. The role of Western Asia in modern human origins. *Philosophical Transactions of the Royal Society of London B* 337:193–200.

Bar-Yosef, O. 1994. The contribution of southwest Asia to the study of the origins of modern humans, 23–66. In *Origins of Anatomically Modern Humans*, ed. M. T. Nitecki and D. V. Nitecki. New York: Plenum Press.

Bar-Yosef, O. 1996. Modern humans, Neanderthals and the Middle/Upper Palaeolithic transition in western Asia, 175–90. In *The Lower and Middle Palaeolithic: Colloquia,* ed. O. Bar-Yosef, L. L. Cavalli-Sforza, R. J. March, and M. Piperno. Forli: ABACO Editions.

Bard, E., B. Hamelin, R. G. Fairbanks, and A. Zindler. 1990. Calibration of the ^{14}C timescale over the past 30,000 years using mass spectrometric U-Th ages from Barbados corals. *Nature* 354:405–10.

Beaulieu, J. L. de, and M. Reille. 1984. A long upper Pleistocene pollen record from Les Echets, near Lyon, France. *Boreas* 13:111–31.

Behre, K. E. 1990. Biostratigraphy of the last glacial period in Europe. *Quaternary Science Reviews* 8:25–44.

Bernaldo de Quiros, F., and V. Cabrera Valdés. 1993. Early Upper Palaeolithic industries of Cantabrian Spain, 57–70. In *Before Lascaux: The Complex Record of the Early Upper Paleolithic*, ed. H. Knecht, A. Pike-Tay, and R. White. Boca Raton: CRC Press.

Bond, G., W. Broeker, S. Johnsen, J. McManus, L. Labeyrie, J. Jouzel, and G. Bonani. 1993. Correlations between climatic records from North Atlantic sediments and Greenland ice. *Nature* 365:143–7.

Bordes, F., and J. Labrot. 1967. La stratigraphie du gisement de Roc-de-Combe (Lot) et ses implications. *Bulletin de la Société Préhistorique Française* 64:15–28.

Boyle, K.V. 1988. *Upper Palaeolithic Faunas from South-west France: A Zoogeographic Perspective*. Oxford: British Archaeological Reports International Series 557.

Broglio, A. 1993. L'Aurignacien au sud des Alpes, 193–203. In *Aurignacien en Europe et au Proche Orient*, ed. L. Banesz and J. K. Kozlowski. Bratislava: Acts of 12th International Congress of Prehistoric and Protohistoric Sciences.

Broglio, A. 1996. The appearance of modern humans in Europe: The archaeological evidence from the Mediterranean regions, 237–49. In *The Lower and Middle Palaeolithic: Colloquia*, ed. O. Bar-Yosef, L. L. Cavalli-Sforza, R. J. March, and M. Piperno. Forli: ABACO Editions.

Cabrera Valdés, V., and J. L. Bischoff. 1989. Accelerator ^{14}C dates for early Upper Palaeolithic (basal Aurignacian) at El Castillo Cave (Spain). *Journal of Archaeological Science* 16:577–84.

Dansgaard, W., S. J. Johnsen, H. B. Clausen, D. Dahl-Jensen, N. S. Gundestrup, C. U. Hammer, C. S. Hvidberg, J. P. Steffensen, A. E. Sveinbjornsdottir, J. Jouzel, and G. Bond. 1993. Evidence for general instability of past climate from a 250-kyr ice-core record. *Nature* 364:218–20.

Delpech, F. 1983. *Les Faunes du Paléolithique Supérieur dans le Sud-ouest de la France.* Paris: CNRS Editions.

Demars, P. Y. 1990. Les interstratifications entre Aurignacien et Châtelperronien a Roc de Combe et au Piage (Lot): Approvisionnement en matières premières et position chronologique, 235–40. In *Paléolithique Moyen Récent et Paléolithique Supérieur Ancien en Europe,* ed. C. Farizy. Nemours: APRAIF.

Demars, P. Y., and J. J. Hublin. 1989. La transition Néandertaliens/Hommes de type moderne en Europe occidentale: Aspects paléontologiques et culturels, 23–7. In *L'Homme de Néandertal*, Vol. 7, ed. M. Otte. Liège: ERAUL.

Djindjian, F. 1993. Les origines du peuplement Aurignacien en Europe, 136–54. In *Aurignacien en Europe et au Proche Orient* , ed. L. Banesz and J. K. Kozlowski. Bratislava: Acts of 12th International Congress of Prehistoric and Protohistoric Sciences.

Farizy, C. 1990. The transition from Middle to Upper Palaeolithic at Arcy-sur-Cure (Yonne, France): Technological, economic and social aspects, 303–26. In *The Emergence of Modern Humans: An Archaeological Perspective*, ed. P. Mellars. Edinburgh: Edinburgh University Press.

Farrand, W. R. 1975. Analysis of the Abri Pataud sediments, 27–68. In *Excavation of the Abri Pataud, Les Eyzies (Dordogne)*, ed. H. L. Movius. Cambridge: Peabody Museum of Archaeology and Ethnology, Harvard University.

Farrand, W. R. 1995. Etude sédimentologique du remplissage de l'Abri Pataud, 31–66. In *Le Paléolithique Supérieur de l'Abri Pataud (Dordogne): Les Fouilles de H. L. Movius Jr.*, ed. H. M. Bricker. Paris: Editions de la Maison des Sciences de l'Homme.

Follieri, M., D. Magri, and L. Sadori, L. 1990. Pollen stratigraphical synthesis of Valle di Castiglione (Roma). *Quaternary International* 3/4:81–4.

Gambier, D. 1989. Fossil hominids from the early Upper Palaeolithic (Aurignacian) of France, 194–211. In *The Human Revolution*, ed. P. A. Mellars and C. B. Stringer. Edinburgh: Edinburgh University Press.

Gambier, D. 1993. Les hommes modernes du debut du Paléolithique supérieur en France: Bilan des donnees anthropologiques et perspectives, 409–30. In *El Origen del Hombre Moderno en el Suroeste de Europa*, ed. V. Cabrera Valdés. Madrid: Universidad Nacional de Educacion a Distancia.

Gamble, C. 1986. *The Palaeolithic Settlement of Europe.* Cambridge: Cambridge University Press.

Goia, P. 1990. La transition Paléolithique Moyen/Paléolithique supérieur en Italie et la question de l'Uluzzien, 241–50. In *Paléolithique Moyen Récent et Paléolithique Supérieur Ancien en Europe*, ed. C. Farizy. Nemours: APRAIF.

Graves, P. 1991. New models and metaphors for the Neandertal debate. *Current Anthropology* 32:513–41.

Grootes, P. M., M. Stuiver, J. W. C. White, S. Johnsen, and J. Jouzel. 1993. Comparison of oxygen isotope records from the GISP2 and GRIP Greenland ice cores. *Nature* 366:552–4.

Guiot, J., A. Pons, J. L. de Beaulieu, and M. Reille. 1989. A 140,000-yr continental climate reconstruction from two European pollen records. *Nature* 338:309–13.

Guiot, J., J. L. de Beaulieu, R. Cheddadi, F. David, P. Ponel, and M. Reille. 1993. The climate in western Europe during the last glacial/interglacial cycle derived from pollen and insect remains. *Palaeogeography, Palaeoclimatology, Palaeoecology* 103:73–94.

Harrold, F. B. 1989. Mousterian, Châtelperronian, and Early Aurignacian in Western Europe: Continuity or discontinuity?, 677–713. In *The Human Revolution*, ed. P. A. Mellars and C. B. Stringer. Edinburgh: Edinburgh University Press.

Hublin, J. J. 1990. Les peuplements paléolithiques de l'Europe: Un point de vue géographique, 29–37. In *Paléolithique Moyen Récent et Paléolithique Supérieur Ancien en Europe*, ed. C. Farizy. Nemours: APRAIF.

Hublin, J. J., F. Spoor, M. Braun, F. Zonneveld, and S. Condemi. 1996a. A late Neanderthal associated with Upper Palaeolithic artefacts. *Nature* 381:224–6.

Hublin, J. J., C. Barroso Ruiz, P. Medina Lara, M. Fontugne, and J. L. Reyss. 1996b. The Mousterian site of Zafarray (Andalucia, Spain): Dating and implications on the Paleolithic peopling of Western Europe. *Comptes Rendus de l'Académie des Sciences de Paris* (in press).

Kerr, R. A. 1993. The whole world had a case of the Ice Age shivers. *Science* 262:1972–3.

Klein, R. G. 1994. The problem of modern human origins, 3–22. In *Origins of Anatomically Modern Humans*, ed. M. T. Nitecki and D. V. Nitecki. New York: Plenum Press.

Kozlowski, J. K. 1988. Transition from the Middle to the early Upper Paleolithic in Central Europe and the Balkans, 193–236. In *The Early Upper Paleolithic: Evidence from Europe and the Near East*, ed. J. F. Hoffecker and C. A. Wolf. Oxford: British Archaeological Reports International Series 437.

Kozlowski, J. K. 1990. A multi-aspectual approach to the origins of the Upper Palaeolithic in Europe, 419–37. In *The Emergence of Modern Humans: An Archaeological Perspective*, ed. P. Mellars. Edinburgh: Edinburgh University Press.

Kozlowski, J. K. 1992. The Balkans in the Middle and Upper Palaeolithic: The gateway to Europe or a cul de sac? *Proceedings of the Prehistoric Society* 58:1–20.

Kozlowski, J. K. 1993. L'Aurignacien en Europe et au Proche Orient, 283–91. In *Aurignacien en Europe et au Proche Orient,* ed. L. Banesz and J. K. Kozlowski. Bratislava: Acts of 12th International Congress of Prehistoric and Protohistoric Sciences.

Kozlowski, J. K. 1996. Cultural context of the last Neanderthals and early modern humans in central-eastern Europe, 205–18. In *The Lower and Middle Palaeolithic: Colloquia*, ed. O. Bar-Yosef, L. L. Cavalli-Sforza, R. J. March, and M. Piperno. Forli: ABACO Editions.

Laj, C., A. Mazaud, and J. C. Duplessy. 1996. Geomagnetic intensity and ^{14}C abundance in the atmosphere and ocean during the past 50 kyr. *Geophysical Research Letters* 23:2045–8.

Laville, H. 1975. *Climatologie et Chronologie du Paléolithique en Périgord: Etude Sedimentologique de Depots en Grottes et sous Abris.* Marseille: Université de Provence.

Laville, H., J. P. Rigaud, and J. Sackett. 1980. *Rock Shelters of the Périgord: Geological Stratigraphy and Archaeological Succession.* New York: Academic Press.

Leroyer, C. 1988. Des occupations castelperroniennes et aurignaciennes dans leur cadre chrono-climatique, 103–08. In *L'Homme de Néandertal*, Vol. 8, ed. M. Otte. Liège: ERAUL.

Leroyer, C., and A. Leroi-Gourhan. 1983. Problèmes de chronologie: Le castelperronien et l'aurignacien. *Bulletin de la Société Préhistorique Française* 80:41–4.

Lévêque, F., A. M. Backer, and M. Guilbaud, ed. 1993. *Context of a Late Neandertal; Implications of Multidisciplinary Research for the Transition to Upper Paleolithic Adaptations at Saint-Césaire, Charente-Maritime, France.* Madison: Prehistory Press.

Lévêque, F., and B. Vandermeersch. 1980. Découverte de restes humains dans un niveau castelperronien à Saint-Césaire (Charente-Maritime). *Comptes Rendus de l'Académie des Sciences de Paris* 291:187–9.

Martinson, D., N. G. Pisias, J. D. Hays, J. Imbrie, T. C. Moore, and N. J. Shackleton. 1987. Age dating and the orbital theory of the ice ages: Development of a high-resolution 0–300,000 year chronostratigraphy. *Quaternary Research* 27:1–29.

Mellars, P. A. 1973. The character of the Middle-Upper Palaeolithic transition in south-west France, 255–76. In *The Explanation of Culture Change: Models in Prehistory*, ed. C. Renfrew. London: Duckworth.

Mellars, P. A. 1985. The ecological basis of social complexity in the Upper Paleolithic of southwestern France, 271–97. In *Prehistoric Hunter-Gatherers: The Emergence of Cultural Complexity*, ed. T. D. Price and J. A. Brown. Orlando: Academic Press.

Mellars, P. A. 1989. Major issues in the emergence of modern humans. *Current Anthropology* 30:349–85.

Mellars, P. A. 1991. Cognitive changes and the emergence of modern humans in Europe. *Cambridge Archaeology Journal* 1:63–76.

Mellars, P. A. 1992a. Archaeology and the population-dispersal hypothesis of modern human origins in Europe. *Philosophical Transactions of the Royal Society of London B* 337:225–34.

Mellars, P. A. 1992b. Archaeology and modern human origins in Europe. *Proceedings of the British Academy* 82:1–35.

Mellars, P. A. 1994. Radiocarbon accelerator dates for El Castillo (Spain), Riparo Mochi (Italy) and Bacho Kiro (Bulgaria). *Archaeometry* 36:345–8.

Mellars, P. A. 1996a. *The Neanderthal Legacy: An Archaeological Perspective from Western Europe.* Princeton: Princeton University Press.

Mellars, P. A. 1996b. Models for the dispersal of anatomically modern populations across Europe: Theoretical and archaeological perspectives, 225–237. In *The Lower and Middle Palaeolithic: Colloquia,* ed. O. Bar-Yosef, L. L. Cavalli-Sforza, R. J. March, and M. Piperno. Forli: ABACO Editions.

Mellars, P. A. 1996c. The emergence of biologically modern populations in Europe: A social and cognitive 'revolution'?, 179–202. In *Evolution of Social Behaviour Patterns in Primates and Man*, ed. W. G. Runciman, J. Maynard Smith, and R. I. M. Dunbar. London: The British Academy.

Mellars, P. A., H. M. Bricker, J. A. J. Gowlett, and R. E. M. Hedges. 1987. Radiocarbon accelerator dating of French Upper Palaeolithic sites. *Current Anthropology* 28:128–33.

Mercier, N., H. Valladas, J. L. Joron, J. L. Reyss, F. Lévêque, and B. Vandermeersch. 1991. Thermoluminescence dating of the late Neanderthal remains from Saint-Césaire. *Nature* 351:737–9.

Mercier, N., and H. Valladas. 1993. Contribution des méthodes de datation par le carbone 14 et de la Thermoluminescence à la chronologie de la transition du Paléolithique moyen au Paléolithique supérieur, 47–60. In *El Origen del Hombre Moderno en el Suroeste de Europa*, ed. V. Cabrera Valdès. Madrid: Universidad Nacional de Educacion a Distancia.

Mussi, M. 1990. Le peuplement de l'Italie à la fin du Paléolithique moyen et au début du Paléolithique supérieur, 251–62. In *Paléolithique Moyen Récent et Paléolithique Supérieur Ancien en Europe,* ed. C. Farizy. Nemours: APRAIF.

Nitecki, M. H., and D. V. Nitecki. 1994. *Origins of Anatomically Modern Humans*. New York: Plenum Press.

Otte, M., and L. G. Straus. 1995. *Le Trou Magrite: Fouilles 1991–1992*. Liège: ERAUL 69.

Patou-Mathis, M. 1993. A taphonomic and paleoethnographic study of the fauna associated with the Neandertal skeleton of Saint-Césaire, 79–102. In *Context of a Late Neandertal; Implications of Multidisciplinary Research for the Transition to Upper Paleolithic Adaptations at Saint-Césaire, Charente-Maritime, France,* ed. F. Lévêque, A. M. Backer, and M. Guilbaud. Madison: Prehistory Press.

Pike-Tay, A. 1991. *Red Deer Hunting in the Upper Paleolithic of Southwest France: A Study in Seasonality*. Oxford: British Archaeological Reports International Series 569.

Pike-Tay, A. 1993. Hunting in the Upper Perigordian: A matter of strategy or expedience?, 85–100. In *Before Lascaux: The Complex Record of the Early Upper Paleolithic*, ed. H. Knecht, A. Pike-Tay, and R. White. Boca Raton: CRC Press.

Raynal, J. P., and J. L. Guadelli. 1990. Milieux physiques et biologiques: Quel changements entre 60 et 30,000 ans a l'ouest de l'Europe?, 53–61. In *Paléolithique Moyen Récent et Paléolithique Supérieur Ancien en Europe,* ed. C. Farizy. Nemours: APRAIF.

Reille, M., and J. L. de Beaulieu. 1990. Pollen analysis of a long upper Pleistocene continental sequence in a Velay maar (Massif Central, France). *Palaeogeography, Palaeoclimatology, Palaeoecology* 80:35–48.

Rigaud, J. P. 1993. La transition Paléolithique moyen/Paléolithique supérieur dans le sud-ouest de la France, 117–26. In *El Origen del Hombre Moderno en el Suroeste de Europa*, ed. V. Cabrera Valdès. Madrid: Universidad Nacional de Educacion a Distancia.

Rogers, A. R., and L. B. Jorde. 1995. Genetic evidence on modern human origins. *Human Biology* 67:1–36.

Soffer, O. 1989. The Middle to Upper Palaeolithic transition on the Russian Plain, 714–42. In *The Human Revolution,* ed. P. A. Mellars and C. B. Stringer. Edinburgh: Edinburgh University Press.

Soffer, O. 1994. Ancestral lifeways in Eurasia—The Middle and Upper Paleolithic records, 101–20. In *Origins of Anatomically Modern Humans*, ed. M. T. Nitecki and D. V. Nitecki. New York: Plenum Press.

Straus, L. G. 1990. The early Upper Palaeolithic of southwest Europe: Cro-Magnon adaptations in the Iberian peripheries, 40,000–20,000 BP, 276–302. In *The Emergence of Modern Humans: An Archaeological Perspective*, ed. P. Mellars. Edinburgh: Edinburgh University Press.

Stringer, C. B. 1996. *African Exodus: The Origins of Modern Humanity*. London: Jonathan Cape.

Stringer, C., and G. Gamble. 1993. *In Search of the Neanderthals: Solving the Puzzle of Human Origins*. London: Thames and Hudson

Stringer, C. B., J. J. Hublin, and B. Vandermeersch. 1984. The origin of anatomically modern humans in Western Europe, 51–135. In *The Origins of Modern Humans: A World Survey of the Fossil Evidence*, ed. F. H. Smith and F. Spencer. New York: Alan R. Liss.

Sykes, B. 1997. The genetic evidence for population movements in the Palaeolithic. In *Science in Archaeology: An Agenda for the Future*, ed. J. Bailey. London: English Heritage (in press).

Tzedakis, P. C. 1994. Vegetation change through glacial-interglacial cycles: A long pollen sequence perspective. *Philosophical Transactions of the Royal Society of London B* 245:403–32.

Van Andel, T. H., and P. C. Tzedakis. 1996a. Palaeolithic landscapes of Europe and environs, 150,000–25,000 years ago. *Quaternary Science Reviews* 15:481–500.

Van Andel, T. H., and P. C. Tzedakis. 1996b. European Palaeolithic landscapes 140–30,000 years ago—A summary, 191–204. In *The Lower and Middle Palaeolithic: Colloquia*, ed. O. Bar-Yosef, L. L. Cavalli-Sforza, R. J. March, and M. Piperno. Forli: ABACO Editions.

Van Andel, T. H. and P. C. Tzedakis. 1997. Priority and opportunity: Reconstructing the European Middle Palaeolithic climate and landscape. In *Science in Archaeology: An Agenda for the Future*, ed. J. Bailey. London: English Heritage (in press).

Watts, W. A., J. R. M. Allen, and B. Huntley. 1996. Vegetation history and palaeoclimate of the last glacial period at Lago Grande di Monticchio, southern Italy. *Quaternary Science Reviews* 15:133–53.

Whallon, R. 1989. Elements of cultural change in the later Paleolithic, 433–54. In *The Human Revolution*, ed. P. A. Mellars and C. B. Stringer. Edinburgh: Edinburgh University Press.

Woillard, G. M. 1978. Grande Pile peat bog: A continuous pollen record for the last 140,000 years. *Quaternary Research* 9:1–21.

Zagwijn, W. H. 1990. Vegetation and climate during warmer intervals in the late Pleistocene of western and central Eurpe. *Quaternary International* 3/4:57–67.

Zubrow, E. 1989. The demographic modeling of Neanderthal extinction, 212–31. In *The Human Revolution*, ed. P. A. Mellars and C. B. Stringer. Edinburgh: Edinburgh University Press.

33

WHY ANATOMICALLY MODERN PEOPLE DID NOT DISPERSE FROM AFRICA 100,000 YEARS AGO

Richard G. Klein

Department of Anthropology
Stanford University
Stanford, California 94305

1. INTRODUCTION

Two competing hypotheses continue to dominate specialist thinking on modern human origins. The first posits that modern people originated in Africa and spread from there to replace non-modern humans in Eurasia (Bräuer and Rimbach 1990; Stringer and Gamble 1993). The second suggests that modern humans originated everywhere—in Africa, but also in Eurasia—where non-modern humans had lived before (Wolpoff et al. 1994; Wolpoff 1996). The controversy will be resolved only with additional fossils and many new, reliable dates, especially in Africa and east Asia. However, the case for replacement is very strong in Europe and west Asia, where relevant fossils and dates are most abundant. In both places, the last non-modern people were the well-known Neandertals, who evolved in Europe before 130,000 years ago (Stringer and Gamble 1993; Hublin 1996) and later spread to west Asia, perhaps at the beginning of the Last Glaciation, about 80,000 years ago (Tchernov 1992, 1994). In both places, fully modern humans were widespread by the 40,000 year limit of conventional radiocarbon dating. Almost everywhere, the Neandertals seem to have disappeared very quickly, perhaps in less than a millennium. Possible exceptions to this rule include western France, where modern humans and Neandertals may have coexisted from 40,000 to 34,000 years ago (Hublin et al. 1996) and southern Spain, where a Neandertal enclave may have survived until 30,000 years ago (Hublin et al. 1995).

For those who favor the replacement hypothesis, a potential difficulty is to explain why anatomically modern or near-modern humans expanded to Eurasia only between 50,000 and 40,000 years ago, more than 50,000 years after they occupied Africa and its immediate southwest Asian margin (Israel) (Rightmire and Deacon 1991; Schwarcz 1994). The reason, however, is probably that early modern or near-modern Africans were not behaviorally modern. In every detectable archeological respect, they were in fact in-

Neandertals and Modern Humans in Western Asia, edited by Akazawa *et al.*
Plenum Press, New York, 1998.

distinguishable from their Eurasian Neandertal contemporaries. Thus, they manufactured relatively informal (unstandardized) stone artifacts that are difficult to subdivide among multiple discrete types; they made few, if any artifacts in plastic raw materials like bone, ivory, antler, and shell; they produced little, if any indisputable art; they built relatively simple fireplaces, around which their living floors were remarkably homogeneous (unorganized); they buried their dead with little or no evidence for ritual or ceremony; and they exploited animal resources relatively ineffectively. It was only when anatomically modern Africans developed modern behavior between 50,000 and 40,000 years ago (Ambrose, 1998) that they gained an undeniable competitive advantage over their non-modern Eurasian contemporaries.

Arguably, the most likely stimulus for modern behavior was a neurological advance, perhaps promoting the fully modern capacity for rapidly articulated phonemic speech (Milo and Quiatt 1994). The neurological hypothesis requires only a random, selectively advantageous mutation like ones that must have occurred many times earlier in human evolution, but the hypothesis is presently impossible to test. This is because humans everywhere had modern-sized brains by 100,000 years ago, and fossil skulls provide no indisputable evidence for differences in brain organization. This is true even when the skulls differ in shape as much as those of Neandertals and modern humans.

Mellars (1996), Bar-Yosef (1994, 1995), and Thackeray (1992) summarize artifactual and other archeological evidence for broadly similar, non-modern behavior before 50,000–40,000 years ago in Europe, west Asia, and Africa. My purpose here is to outline the clearest case for non-modern hunting-gathering. This evidence comes from the west and south coasts of South Africa, at sites that are unique for the diversity and richness of their mixed marine and terrestrial faunas. Much lower diversity in Europe and west Asia may always complicate the development of similar evidence there, though research by Stiner (this volume) suggests that west Asian materials may exhibit some parallels.

2. THE SOUTH AFRICAN SITES

The relevant South African sites represent two major culture-stratigraphic units—the Middle Stone Age (MSA), which spanned the interval between 250,000–200,000 and 50,000–40,000 years ago, and the succeeding Later Stone Age (LSA), after 50,000–40,000 years ago (Volman 1984; Clark 1988; Thackeray 1992; Wadley 1993). Artifactually, the MSA closely resembled the broadly contemporaneous Middle Paleolithic (or Mousterian) of Europe and west Asia. In both the MSA and the Middle Paleolithic, the principal stone blanks were carefully prepared flakes and flake-blades, and the main retouched types were scrapers, points, and denticulates, all of which are difficult to divide into discrete subtypes. In regard to stone artifacts and to other observable archeological characteristics listed above, the MSA and Middle Paleolithic contrast sharply with the LSA and with the Upper Paleolithic, which succeeded the Middle Paleolithic in west Asia and Europe. Artifactually, unlike the MSA and the Middle Paleolithic, the LSA and the Upper Paleolithic diverged sharply from each other, but they shared a greater degree of artifact formalization (standardization), a proliferation of readily discernible discrete artifact types, a much greater tendency for both types and assemblages to vary through time and space, the widespread use of bone and similar organic materials for artifact manufacture, and the production of indisputable art. The sum implies that LSA/Upper Paleolithic people were much more innovative than their predecessors, and it is this above all that ties them more closely to living people.

Figure 1 shows the approximate locations of the South African MSA and LSA sites that have provided most of the hunting-gathering evidence summarized below. The MSA sites include the well-known Klasies River Mouth Cave complex, occupied by MSA people during the Last Interglaciation (global oxygen-isotope stage 5) and the early part of the Last Glaciation (isotope stage 4), between roughly 128,000 and 59,000 years ago (Singer and Wymer 1982; Deacon and Schuurman 1992; Deacon 1995); Die Kelders Cave 1, provisionally dated to the early part of the Last Glaciation, between 74,000 and 59,000 years ago (Tankard and Schweitzer 1976; Grine et al. 1991; Avery et al. 1997); the Sea Harvest and Hoedjies Punt shell middens, tentatively assigned to the earlier part of the Last Interglaciation, between 128,000 and 115,000 years ago (Volman 1978; Parkington and Klein, unpublished); and the Boegoeberg 2 shell midden, also provisionally dated to the Last Interglaciation (Halkett, Hart, Parkington, Klein, and Cruz-Uribe, unpublished).

The most relevant LSA sites for comparison include Nelson Bay Cave (Deacon 1984; Inskeep 1987), Byneskranskop Cave 1 (Schweitzer and Wilson 1982), Die Kelders Cave 1 (Schweitzer 1979), Paternoster Midden (Robertshaw 1977), Kasteelberg A and B (Smith 1987, 1992; Klein and Cruz-Uribe 1989), Elands Bay Cave (Parkington 1987, 1990), and the Dune Field Midden (Parkington et al. 1992). As a group, the LSA sites date partly from the end of the Last Glaciation (= isotope stage 2), between 24,000 and 12,000 years ago, but mainly from the Present Interglaciation (or Holocene) (= isotope stage 1), after 12,000 years ago. MSA and LSA sites from the middle part of the Last Glaciation (= isotope stage 3), between 59,000 and 24,000 years ago, are rare and usually poor in southern Africa, perhaps because conditions were mainly too dry to support substantial human populations

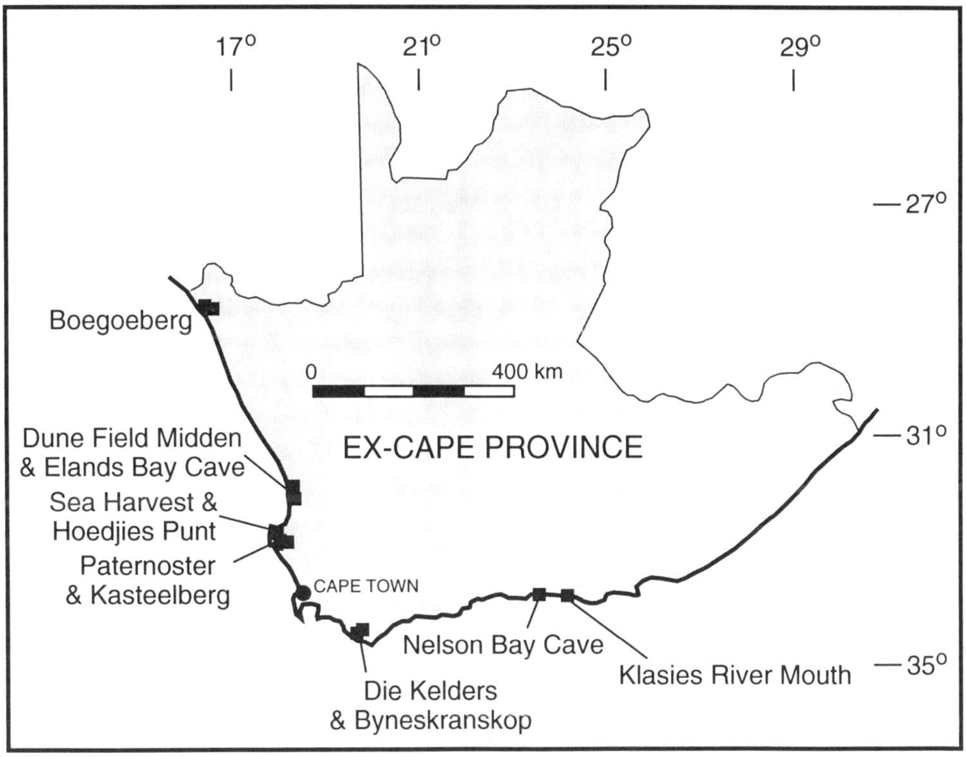

Figure 1. Approximate locations of the sites mentioned in the text.

(Deacon and Thackeray 1984). The long occupation hiatus is vexing because it spans the shift from the MSA to the LSA, which so far is well-documented only in east Africa (Ambrose, 1998). However, the richness of Last and Present Interglacial occupations in South Africa offers a unique opportunity to compare the economies of MSA and LSA people living under very similar conditions. It is on this comparison that I focus here.

3. SPECIES FREQUENCIES

The MSA and LSA sites contain remains of the same species, but often in very different frequencies. The most striking contrast involves fish bones, which are abundant in all coastal LSA sites, but rare or absent in all of the known MSA sites. This is true even though sediments and bones of seals and marine birds indicate that the coastline remained nearby throughout each MSA occupation. Offshore bathymetry further affirms that the coastline was generally within 5 km, even during early last glacial periods of sea level depression (van Andel 1989). The rarity or absence of fish remains in MSA layers is paralleled by the absence of likely fishing implements, which occur only in LSA sites. The sum suggests that only LSA people knew how to fish.

The MSA layers at Klasies River Mouth and Die Kelders Cave 1 also contrast with coastal LSA layers in the relative abundance of various marine birds. Thus, jackass penguins (*Spheniscus demersus*) dominate the MSA layers, while cormorants (*Phalacrocorax* spp.) and other airborne species heavily dominate the LSA sites (Avery 1990). This suggests that only LSA people could routinely catch airborne birds, probably because only they had the bow and arrow and other relevant technology. Recent discoveries at Boegoeberg 1 and 2 may be seen to complicate this interpretation, because Boegoeberg 1, a fossil hyena den, is rich in penguin, with little or no identifiable cormorant, and Boegoeberg 2, an MSA site, is rich in cormorant, with little or no penguin. In the abundance of penguin, the Boegoeberg 1 hyena site is more extreme than the MSA sites at Klasies River Mouth and Die Kelders, while in the rarity of penguin, the Boegoeberg 2 MSA site is actually more extreme than any comparable LSA site. The explanation for the contrast between the two Boegoeberg sites may be environmental, since other faunal observations imply that Boegoeberg 1 formed under remarkably wet (and cool) conditions in what is now a hyper-arid environment. However, even if the Boegoeberg sites suggest that environment may influence the penguin-cormorant ratio, it is difficult to see how environment could explain penguin dominance in the MSA deposits at Klasies River Mouth and Die Kelders 1, where the pattern persisted under both interglacial and glacial conditions.

Finally, the relative abundance of large ungulates differs strongly between MSA and LSA sites where ungulate bones are abundant (Klein and Cruz-Uribe 1996). The clearest example concerns the contrast between the Last Interglacial MSA layers at Klasies River Mouth and the Present Interglacial LSA layers at Nelson Bay Cave. The environs of both sites are very similar today, and sediments and geochemistry indicate that the environs were similar to the present and to each other during the Last and Present Interglaciations. In the historic environment, buffaloes (*Syncerus caffer*) and bush pigs (*Potamochoerus porcus*) significantly outnumbered eland (*Taurotragus oryx*), and this situation is mirrored in the LSA layers at Nelson Bay Cave. However, in the MSA layers at Klasies River Mouth, eland strongly outnumber both buffaloes and pigs, and if the MSA environment truly resembled the historic one, the most likely explanation for eland abundance becomes MSA behavior. Buffaloes and pigs are far more dangerous to hunt than eland, and differences between MSA and LSA artifact assemblages suggest that only LSA people had the bow and arrow, with which they could have obtained dangerous species much more often. Assuming that limited MSA technology explains the relatively small numbers of buffaloes and pigs at Klasies River Mouth, the MSA disadvantage would have been very great, since buffaloes and pigs probably far outnumbered eland nearby.

Eland similarly dominate buffaloes in the MSA layers of Die Kelders Cave 1, while buffaloes strongly dominate eland in the LSA layers of nearby Byneskranskop 1. In this instance, an environmental cause cannot be fully excluded, since the Die Kelders MSA layers formed under a cooler, wetter climate. However, temperature variation had no obvious impact on historic buffalo and eland numbers in Africa; wherever both species occurred together, buffaloes tended to be much more numerous (Skinner and Smithers 1990; Estes 1992). In addition, the eland-to-buffalo ratio remained constant when "glacial" conditions replaced "interglacial" ones at Klasies River Mouth. At least circumstantially then, the much increased abundance of buffaloes at Byneskranskop provides additional support for an advance in LSA hunting competence.

4. SEASONALITY OF OCCUPATION

South African coastal sites invariably contain bones of the Cape fur seal (*Arctocephalus pusillus*), and these afford a special opportunity to determine the season(s) of bone accumulation. The reason is that until 1941, when fur seals were first legally protected from human predation, they bred almost exclusively on offshore rocks (David 1989; Skinner and Smithers 1990). The vast majority of births occurred within a few weeks in late November and early December, and adults forced the young from the rocks about nine months later. In recent times, large numbers of 9–11-month-old seals then washed up ashore, exhausted or dead. It is the short fur seal birth season and the consequent seasonal peak in on-shore availability that allow estimates of when fur seal bones accumulated at fossil sites. In cases where fossil seal ages cluster tightly around 9–11 months, bone accumulation probably centered tightly on the August-October period of 9–11-month-old seal superabundance. Where ages cluster more loosely around 9–11 months, bone accumulation probably included not only the August-October interval, but also other times. And where ages fail to cluster near the 9–11 month average, then accumulation probably fell largely outside the August-October interval. My purpose here is to present evidence that MSA people responded to the peak in fur seal availability differently, and perhaps less rationally, than LSA people did.

Fossil seal bones may be "aged" by comparison to bones from known-age animals in the South African Museum and the South African Department of Sea Fisheries. The known-age collections comprise mainly newborns and individuals about 9–11 months old. Various skeletal elements are useful for age determination, but the distal humerus is especially apposite, because it is relatively durable and thus tends to dominate fossil samples. The most consistently available dimension is the mediolateral diameter or "breadth" of the distal end. The dimensions of other bones, including the mediolateral diameter of the distal femur and the "short length" of the mandible (defined by Woodborne et al. 1995) provide the same basic results, but the patterning is less compelling, because it is based on fewer specimens.

Figure 2 uses a boxplot format to summarize fur seal distal humerus breadths in samples from the MSA and LSA sites listed above, in two samples accumulated by brown hyenas (*Hyaena brunnea*), and in samples of modern newborns and 9-month olds. In each plot, as described by Velleman (1995), the vertical line near the center is the median, the open rectangle encloses the middle half of the data (between the 25th and 75th percentiles), the shaded rectangle is the 95% confidence interval for the median, and the vertical lines at the ends mark the range of more or less continuous data. Circles or star bursts indicate extreme values (points that are far removed from the main body of data). The number of specimens in each sample is given in parentheses. Samples for which the 95% confidence limits do not overlap differ significantly in the conventional statistical sense. To aid visual assessment, a vertical gray bar extends the range for 9-month olds through the figure.

Figure 2 shows that in most samples, median fur-seal-distal-humerus breadth lies within or just outside the range for known 9-month-olds. In addition, median breadths in the LSA samples tend to be tightly clustered around the median, and the sum suggests that LSA people focused their coastal visits on the August-October interval when they could harvest 9–11-month old seals on nearby beaches. The most glaring departure from the LSA pattern involves the MSA sample from Klasies River Mouth Cave 1, where median distal humerus breadths are much more broadly dispersed and the median is substantially and significantly greater than that for 9-month-olds. This implies that MSA people occupied Klasies River Mouth mainly outside the peak period of 9–11-month-old availability, and more generally, that they followed a very different seasonal round than their LSA successors. From the available data, it could even mean that they did not follow a seasonal round at all. Arguably, the MSA sample from Boegoeberg 2 supports the same inference, while the one from Die Kelders Cave 1 suggests an LSA-like pattern. Unfortunately, however, both samples are too small for any firm conclusion.

As an aid to further interpretation, it is instructive to compare the LSA and MSA boxplots to those for the two brown hyena samples at the bottom of Figure 2. The sample labeled "Central Namib" was accumulated between 1990 and 1996 in dens adjacent to a fur-seal breeding colony about 20 km south of Luderitz, Namibia (Skinner and van Aarde 1991; Skinner et al. 1995). The other sample was accumulated at the Boegoeberg 1 fossil hyena den alluded to above. In keeping with hyena access to a breeding colony, the Central Namib sample includes fifty-four humeri from newborn individuals. The abundance of newborn specimens explains why Central Namib median distal-humerus breadth is so small, and the rarity or absence of newborn specimens in the various fossil sites shows that none were located near breeding colonies. When the newborn specimens are removed from the Central Namib sample for heuristic purposes, the median humerus breadth more closely approximates that in the various LSA samples, but the degree of breadth dispersion becomes much greater. This is because the Central Namib brown hyenas remain at the coast throughout the year, including times when the only available seals are older than 9–11 months. The comparable or greater degree of dispersion in the Boegoeberg 1 fossil hyena sample also illustrates non-seasonal bone accumulation.

In sum, the boxplots in Figure 2 imply that LSA people generally timed their coastal visits to include the seasonal peak in young-fur-seal availability, while the MSA occupants of Klasies River Mouth obtained many, if not most of their seals at other seasons. If larger fur seal samples from other MSA sites replicate the Klasies River Mouth pattern, the implication would be that MSA people occupied the coast non-seasonally, perhaps like brown hyenas. MSA people may have differed from LSA people because they failed to perceive seasonal variability in fur seals and other resources, or because they were technologically more limited. In this regard, it may be pertinent that no MSA site has provided evidence for water containers, while LSA sites have repeatedly provided fragmentary or even whole ostrich eggshell canteens (Deacon 1984). The contrast is particularly striking because ostrich eggshell abounds at Boegoeberg 2 and other west coast MSA sites.

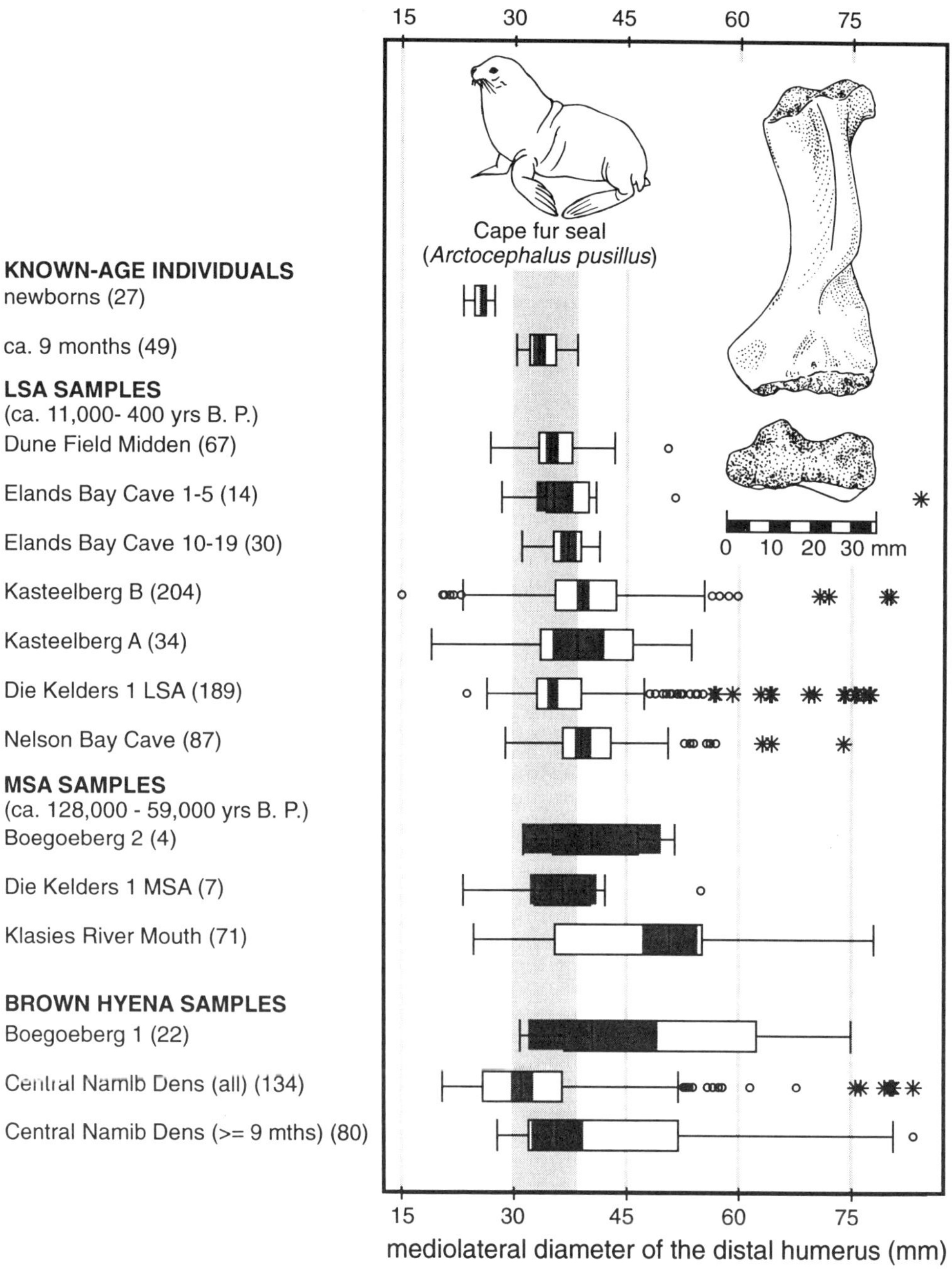

Figure 2. Boxplots summarizing fur seal distal humerus mediolateral diameters in modern and fossil South African samples.

5. HUMAN POPULATION DENSITY

Both MSA and LSA people collected large numbers of tortoises and intertidal mollusks in suitable environments, and this furnishes an opportunity to investigate changes in human population density through time. The underlying assumption is that smaller average tortoise or mollusk size implies more intensive human predation (collection), and that predation pressure in turn was mostly a function of human population density. Archeologically, this hypothesis is supported by the shared pattern of size change in large samples of angulate tortoises (*Chersina angulata*) and granite limpets (*Patella granatina*) from the 13,000-year-long LSA sequence at Elands Bay Cave. These species patently participate in very distinct ecosystems, yet on average, individuals of each were smallest about 10,000 years ago, when other evidence indicates that conditions were especially favorable for hunter-gatherers, and again after 2,000 years ago, when hunting-gathering had been significantly augmented by stock-herding (Klein and Cruz-Uribe, in press; Parkington and Jerardino, in press). Almost surely, the introduction of domestic stock succeeded mainly because it promoted human population growth.

Historically, angulate tortoises were abundant on the western and southwestern coasts of South Africa, but not on the south central coast in the vicinity of Klasies River Mouth and Nelson Bay Cave. This geographic difference appears to have great time depth, since angulate tortoise bones are very rare in the late Quaternary deposits at Klasies River Mouth, Nelson Bay Cave, and all other south-central coast sites, but they abound at Die Kelders Cave 1, Byneskranskop Cave 1, and all sites to the west and northwest (Figure 1). The most plentiful, readily measurable tortoise elements in most sites are distal humeri, and Figure 3 uses the same boxplot format described above to summarize distal humerus mediolateral diameters or "breadths" through the MSA/LSA sequence at Die Kelders Cave 1 (DK1) and Byneskranskop Cave 1 (BNK1).

The figure shows that median breadth differs significantly among successive LSA units beginning nearly 13,000 years ago, and the pattern of size change broadly recalls that through the like-aged deposits at Elands Bay Cave. However, the figure also shows that on average, the MSA tortoises tend to be significantly larger than even the largest LSA specimens. It might be argued that this reflects an especially favorable MSA environment for tortoises, but sedimentological and other faunal evidence indicate cool, moist climatic conditions that might have favored humans and other large mammals, but not tortoises. The most plausible alternative explanation is that the MSA tortoises are large because MSA people preyed on them less intensively, probably because MSA human populations were relatively sparse. The same argument can be made from the very large average size of the tortoises in the MSA layers at Diepkloof Cave near Elands Bay Cave (Yates, personal communication).

The assessment of mollusk size is most straightforward for the west coast of South Africa, where large numbers of intact MSA and LSA specimens of the granite limpet and the granular limpet (*Patella granularis*) have been systematically measured (Buchanan et al. 1978; Parkington and Jerardino, in press; Parkington, Cruz-Uribe, and Klein, unpublished). In every known instance, MSA examples of both limpets tend to be significantly larger than their LSA counterparts. Figure 4 illustrates the pattern, based on the maximum diameter of intact granite limpet shells in the MSA samples from Sea Harvest, Hoedjies Punt, and Boegoeberg 2, in LSA samples from Elands Bay Cave and Paternoster, and finally in seven modern "10-minute" samples. The latter result from exercises in which students were given ten minutes to collect limpets from intertidal rocks that are otherwise unexploited today (Buchanan et al. 1978). For each sample, the vertical line represents the arithmetic mean and the horizontal line is the mean plus or minus one standard deviation. Sample sizes are provided in parentheses. I used this format, as opposed to the boxplots of Figures 2 and 3, because only summary statistics are available for most of the shell samples. The figure shows that on average, granite limpet size in the MSA samples much more closely approaches that in unexploited modern populations, and a reasonable inference is that MSA exploitation was correspondingly light.

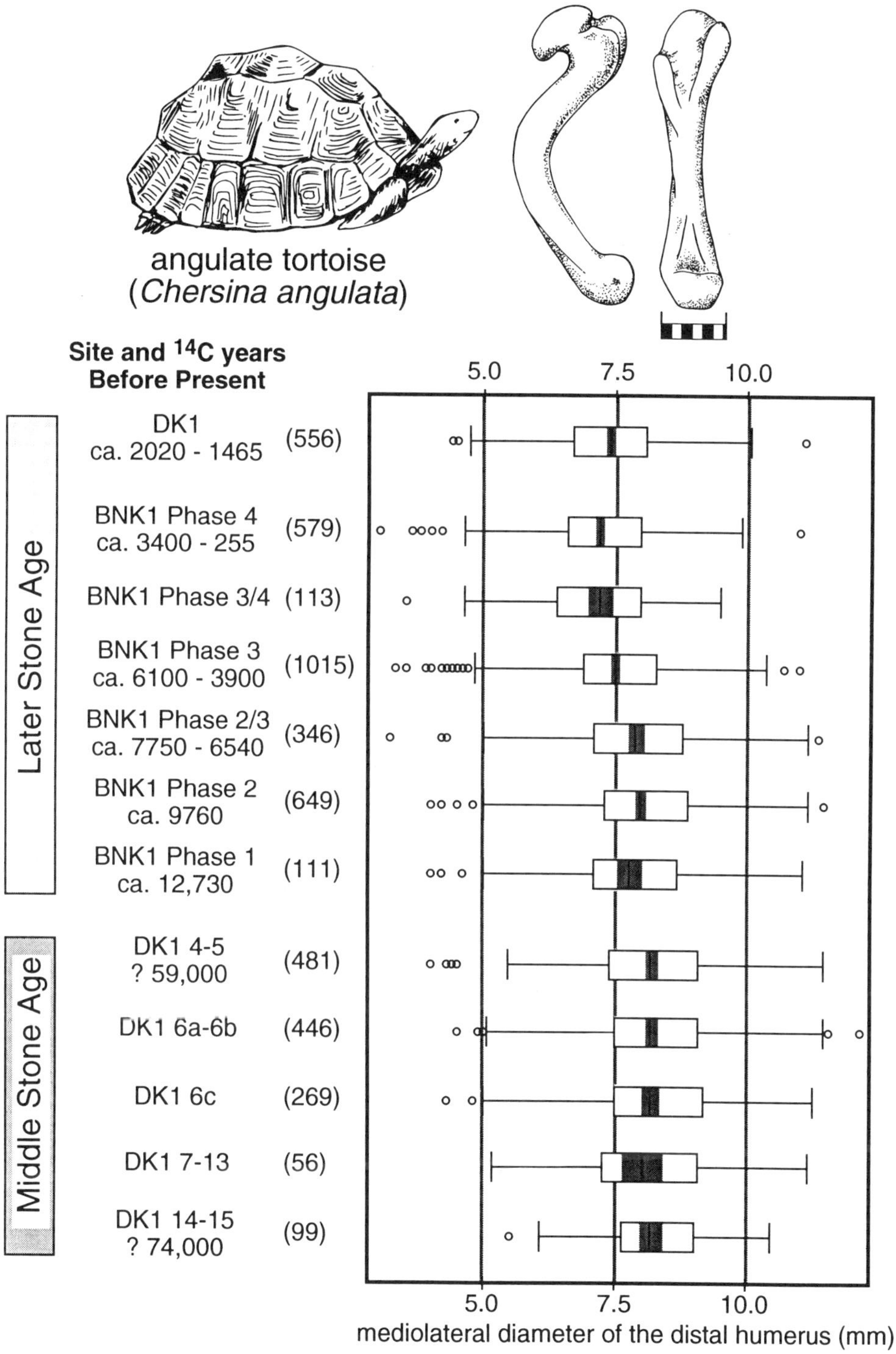

Figure 3. Boxplots summarizing the mediolateral diameter of tortoise distal humeri in fossil samples from Die Kelders Cave 1 (DK1) and Byneskranskop 1 (BNK1).

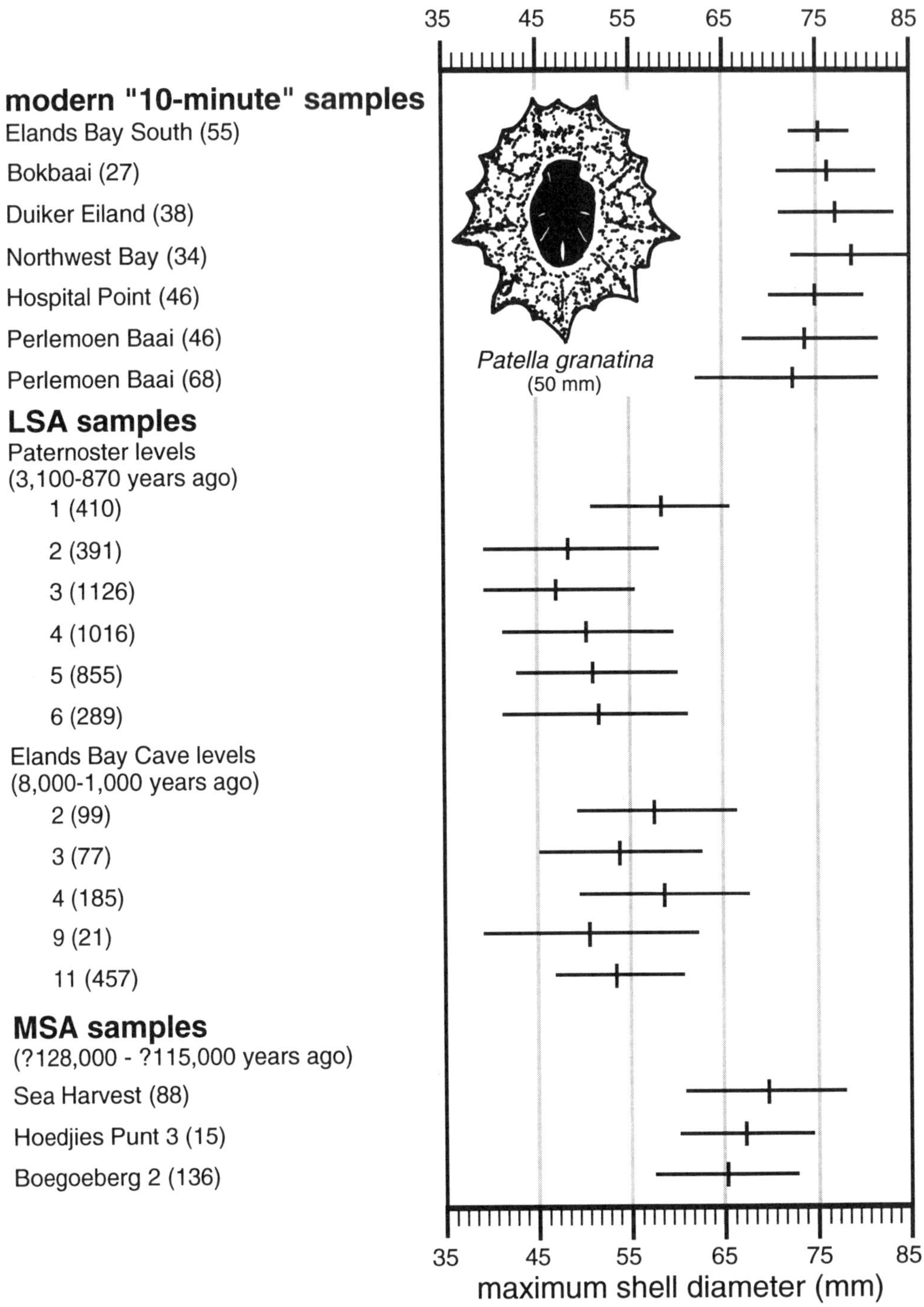

Figure 4. The mean and standard deviation of granite limpet shell length in modern and fossil samples from the west coast of South Africa.

The evaluation of mollusk size is more complicated for the south coast, because shells are too poorly preserved for measurement in the MSA deposits of Die Kelders, and they have not been measured in the large LSA samples from Die Kelders and Nelson Bay Cave. Nonetheless, it is notable that opercula of the giant periwinkle (*Turbo sarmaticus*) tend to be significantly larger in the MSA layers of Klasies River Mouth than in local LSA samples (Voigt 1982; Thackeray 1988). The pattern is the same even when the opercula are separated between a sample excavated in 1967–68 when recovery was comparatively unsystematic and one excavated since 1984 when recovery was much more meticulous. In addition, on average, MSA granite limpets from Klasies River Mouth (Voigt 1982) tend to be as large as those in west coast MSA sites, while granite limpets in the LSA sample from Byneskranskop 1 tend to be much smaller (Schweitzer and Wilson 1982), like those in west coast LSA sites.

In sum, wherever size comparisons are possible between MSA and LSA shells, MSA specimens tend to be significantly larger, and there is no reason to suppose that larger average size reflects more favorable environmental conditions. Independent evidence suggests broadly similar circumstances between, for example, the early Holocene layers at Elands Bay Cave and the probable Last Interglacial levels at Sea Harvest or Boegoeberg 2. The implication is thus that by comparison to LSA people living under similar circumstances, MSA people preyed much less intensively on shellfish. The most parsimonious explanation is that MSA populations were less dense, in keeping with their more limited ability to exploit fish, birds, and mammals.

6. CONCLUSION

The fossil record suggests that modern human morphology evolved in Africa by 100,000 years ago, when the sole inhabitants of Eurasia were the Neanderthals and other equally non-modern people. However, the earliest modern or near-modern Africans were behaviorally (archeologically) indistinguishable from their non-modern, Eurasian contemporaries, and it was only between 50,000 and 40,000 years ago that a major behavioral difference developed. Archaeological indications of this difference include the oldest indisputable ornaments (or art broadly understood); the oldest evidence for routine use of bone, ivory, and shell to produce formal (standardized) artifacts; greatly accelerated variation in stone artifact assemblages through time and space; and, as detailed here, advances in hunting-gathering ability that promoted significantly larger populations. As a complex, the novel traits imply fully modern cognitive and communicative abilities, or more succinctly, the fully modern capacity for Culture. It was the competitive advantage this capacity conferred that allowed physically modern Africans to replace non-modern Eurasians after 50,000 years ago.

ACKNOWLEDGMENTS

I thank the National Science Foundation for financial support, the South African Museum for research facilities, and Kathryn Cruz-Uribe for comments on a draft.

REFERENCES

Ambrose, S. H. 1998. Chronology of the Later Stone Age and food production in East Africa. *Journal of Archaeological Science* 25:377–92..

Avery, G. 1990. Avian Fauna, Palaeoenvironments and Palaeoecology in the Late Quaternary of the Western and Southern Cape, South Africa. Ph. D. diss. University of Cape Town. Cape Town.

Avery, G., K. Cruz-Uribe, P. Goldberg, F. E. Grine, R. G. Klein, M. J. Lenardi, C. W. Marean, W. J. Rink, H. P. Schwarcz, A. I. Thackeray, and M. L. Wilson. 1997. The 1992–93 excavations at the Die Kelders Middle and Later Stone Age cave site, South Africa. *Journal of Field Archaeology* 24:263–91.

Bar-Yosef, O. 1994. The contributions of Southwest Asia to the study of the origin of modern humans, 23–66. In *Origins of Anatomically Modern Humans*, ed. M. H. Nitecki and V. Nitecki. New York: Plenum Press.

Bar-Yosef, O. 1995. The Lower and Middle Palaeolithic in the Mediterranean Levant: Chronology and cultural entities, 247–63. In *Man and Environment in the Palaeolithic*, ed. H. Ullrich. Liège: ERAUL.

Bräuer, G., and K. W. Rimbach. 1990. Late archaic and modern *Homo sapiens* from Europe, Africa, and southwest Asia: Craniometric comparisons and phylogenetic implications. *Journal of Human Evolution* 19:789–807.

Buchanan, W. F., S. L. Hall, J. Henderson, A. Olivier, J. M. Pettigrew, J. E. Parkington, and P. T. Robertshaw. 1978. Coastal shell middens in the Paternoster area, southwestern Cape. *South African Archaeological Bulletin* 33:89–93.

Clark, J. D. 1988. The Middle Stone Age of East Africa and the beginnings of regional identity. *Journal of World Prehistory* 2:235–305.

David, J. H. M. 1989. Seals, 288–302. In *Oceans of Life off Southern Africa*, ed. A. I. L. Payne and R. H. M. Crawford. Cape Town: Vlaeberg Publishers.

Deacon, H. J. 1995. Two late Pleistocene-Holocene archaeological depositories from the southern Cape, South Africa. *South African Archaeological Bulletin* 50:121–31.

Deacon, H. J., and R. Schuurman. 1992. The origins of modern people: The evidence from Klasies River, 121–30. In *Continuity or Replacement: Controversies in* Homo sapiens *Evolution*, ed. G. Braüer and F. H. Smith. Rotterdam: A. A. Balkema.

Deacon, H. J., and J. F. Thackeray. 1984. Late Pleistocene environmental changes and implications for the archaeological record in southern Africa, 375–90. In *Late Cainozoic Palaeoclimates of the Southern Hemisphere*, ed. J. C. Vogel. Rotterdam: A. A. Balkema.

Deacon, J. 1984. Later Stone Age people and their descendants in southern Africa, 221–328. In *Southern African Prehistory and Paleoenvironments*, ed. R. G. Klein. Rotterdam: A. A. Balkema.

Estes, R. D. 1992. *The Behavior Guide to African Mammals*. Berkeley and Los Angeles: The University of California Press.

Grine, F. E., R. G. Klein, and T. P. Volman. 1991. Dating, archaeology, and human fossils from the Middle Stone Age levels of Die Kelders, South Africa. *Journal of Human Evolution* 21:363–95.

Hublin, J. J. 1996. The first Europeans. *Archaeology* 49(1):36–42.

Hublin, J. J., C. Barroso Ruiz, P. Medina Lara, M. Fontugne, and J. L. Reyss. 1995. The Mousterian site of Zafarraya (Andalucia, Spain): Dating and implications on the Palaeolithic peopling processes of Western Europe. *Comptes Rendus de l'Académie des Sciences* 321(IIa):931–7.

Hublin, J. J., F. Spoor, M. Braun, F. Zonneveld, and S. Condemi. 1996. A late Neanderthal associated with Upper Paleolithic artefacts. *Nature* 381:224–6.

Inskeep, R. R. 1987. *Nelson Bay Cave, Cape Province, South Africa. The Holocene Levels*, Vols. 1 and 2. Oxford: British Archaeological Reports International Series 357.

Klein, R. G., and K. Cruz-Uribe. 1989. Faunal evidence for prehistoric herder-forager activities at Kasteelberg, western Cape Province, South Africa. *South African Archaeological Bulletin* 44:82–97.

Klein, R. G., and K. Cruz-Uribe. 1996. Exploitation of large bovids and seals at Middle and Later Stone Age sites in South Africa. *Journal of Human Evolution* 30:315–34.

Klein, R. G., and K. Cruz-Uribe. In press. Large mammals and tortoises. In *Elands Bay Cave: A View of the Past*, ed. J. E. Parkington. Cape Town: University of Cape Town Press.

Mellars, P. A. 1995. *The Neanderthal Legacy: An Archaeological Perspective from Western Europe*. Princeton: Princeton University Press.

Milo, R. G., and D. Quiatt. 1994. Language in the Middle and Late Stone Ages: Glottogenesis in anatomically modern Homo sapiens, 321–39. In *Hominid Culture in Primate Perspective*, ed. D. Quiatt and J. Itani. Niwot (Colorado): University Press of Colorado.

Parkington, J. E. 1987. Changing views of prehistoric settlement in the western Cape, 4–23. In *Papers in the Prehistory of the Western Cape, South Africa*, ed. J. E. Parkington and M. Hall. Oxford: British Archaeological Reports International Series 332.

Parkington, J. E. 1990. A view from the south: Southern Africa before, during, and after the Last Glacial Maximum, 214–28. In *The World at 18,000 BP*, ed. C Gamble and O. Soffer. London: Unwin Hyman.

Parkington, J. E., and A. Jerardino. In press. Shellfish gathering. In *Elands Bay Cave: A View of the Past*, ed. J. E. Parkington. Cape Town: University of Cape Town Press.

Parkington, J. E., P. Nilssen, C. Vermuelen, and C. Henshilwood. 1992. Making sense of space at Dunefield Midden campsite, western Cape, South Africa. *South African Journal of Field Archaeology* 1:63–70.

Rightmire, G. P., and H. J. Deacon. 1991. Comparative studies of late Pleistocene human remains from Klasies River Mouth, South Africa. *Journal of Human Evolution* 20:131–56.

Robertshaw, P. T. 1977. Excavations at Paternoster, south-western Cape. *South African Archaeological Bulletin* 32:63–73.

Schwarcz, H. P. 1994. Chronology of modern humans in the Levant, 21–31. In *Late Quaternary Chronology and Paleoclimates*] *of the Eastern Mediterranean*, ed. O. Bar-Yosef and R. S. Kra. Tucson and Cambridge: Radiocarbon and the Peabody Museum of Archaeology and Ethnography, Harvard University.

Schweitzer, F. R. 1979. Excavations at Die Kelders, Cape Province, South Africa: The Holocene deposits. *Annals of the South African Museum* 78:101–233.

Schweitzer, F. R., and M. L. Wilson. 1982. Byneskranskop 1, a late Quaternary living site in the southern Cape Province, South Africa. *Annals of the South African Museum* 88:1–203.

Singer, R., and J. J. Wymer. 1982. *The Middle Stone Age at Klasies River Mouth in South Africa.* Chicago: University of Chicago Press.

Skinner, J. D., and R. H. N. Smithers. 1990. *The Mammals of the Southern African Subregion*. Pretoria: University of Pretoria.

Skinner, J. D., and R. J. van Aarde 1991. Bone collecting by brown hyenas *Hyaena brunnea* in the Central Namib Desert, Namibia. *Journal of Archaeological Science* 18:513–23.

Skinner, J. D., R. J. van Aarde, and R. A. Goss. 1995. Space and resource use by brown hyenas *Hyaena brunnea* in the Namib Desert. *Journal of Zoology* 237:123–31.

Smith, A. B. 1987. Seasonal exploitation of resources on the Vredenburg Peninsula after 2000 B.P., 393–402. In *Papers in the Prehistory of the Western Cape, South Africa,* ed. J. E. Parkington and M. Hall. Oxford: British Archaeological Reports International Series 332.

Smith, A. B. 1992. Origins and spread of pastoralism in Africa. *Annual Review of Anthropology* 21:125–41.

Stringer, C. B., and C. Gamble. 1993. *In Search of the Neanderthals*. New York: Thames and Hudson.

Tankard, A. J., and F. R. Schweitzer. 1976. Textural analysis of cave sediments: Die Kelders, Cape Province, South Africa, 289–316. In *Geoarchaeology*, ed. D. A. Davidson and M. L. Shackley. London: Duckworth.

Tchernov, E. 1992. Biochronology, paleoecology, and dispersal events of hominids in the southern Levant, 149–88. In *The Evolution and Dispersal of Modern Humans in Asia*, ed. T. Akazawa, K. Aoki, and T. Kimura. Tokyo: Hokusen-Sha.

Tchernov, E. 1994. New comments on the biostratigraphy of the Middle and Upper Pleistocene of the southern Levant, 333–50. In *Late Quaternary Chronology and Paleoclimates of the Eastern Mediterranean*, ed. O. Bar-Yosef and R. S. Kra. Tucson and Cambridge: Radiocarbon and the Peabody Museum of Archaeology and Ethnography, Harvard University.

Thackeray, A. I. 1992. The Middle Stone Age south of the Limpopo River. *Journal of World Prehistory* 6:385–440.

Thackeray, J. F. 1988. Molluscan fauna from Klasies River, South Africa. *South African Archaeological Bulletin* 43:27–32.

van Andel, T. H. 1989. Late Pleistocene sea levels and the human exploitation of the shore and shelf of southern Africa. *Journal of Field Archaeology* 16:133–55.

Velleman, P. 1995. *Data Desk Handbook: Version 5.0*. Ithaca: Data Description, Inc.

Voigt, E. A. 1982. The molluscan fauna, 155–86. In *The Middle Stone Age at Klasies River Mouth in South Africa,* ed. R. Singer and J. J. Wymer. Chicago: University of Chicago Press.

Volman, T. P. 1978. Early evidence for shellfish collecting. *Science* 201:911–3.

Volman, T. P. 1984. Early prehistory of southern Africa, 169–220. In *Southern African Prehistory and Paleoenvironments*, ed. R. G. Klein. Rotterdam: A. A. Balkema.

Wadley, L. 1993. The Pleistocene Later Stone Age south of the Limpopo River. *Journal of World Prehistory* 7:243–96.

Wolpoff, M. H. 1996. *Human Evolution: 1996–97 Edition*. New York: McGraw-Hill.

Wolpoff, M. H., A. G. Thorne, F. H. Smith, D. W. Frayer, and G. G. Pope. 1994. Multiregional evolution: A worldwide source for modern human populations, 175–99. In *Origins of Anatomically Modern Humans*, ed. M. H. Nitecki and V. Nitecki. New York: Plenum Press,

Woodborne, S., K. Hart, and J. E. Parkington. 1995. Seal bones as indicators of the timing and duration of hunter-gatherer coastal visits. *Journal of Archaeological Science* 22:727–40.

34

AFTERWORD

David Pilbeam

Department of Anthropology
Peabody Museum
Harvard University
Cambridge, Massachusetts 02138

I am very pleased to be invited to contribute this brief essay as a closing commentary, having had the pleasure of reading many of these chapters in manuscript. I have been interested for a long time, although as an outsider, in the relationship of the various Middle and Late Pleistocene nonmodern hominids to "modern" humans and to the tempo and pattern of the replacement of the former by the latter. (A note on terminology here: by "modern" humans I mean living humans and their Holocene ancestors, those associated with the European Upper Paleolithic and their approximate contemporaries elsewhere [North, East, and South Africa, Southwest Asia, Australia], as well as the more robust early moderns from Southwest Asia and Africa [Qafzeh, for example] which are older than about 60 ka.)

As we can see from this conference volume, Southwest Asia has become an area of considerable paleoanthropological importance. Indeed, the hominid and archaeological residues from this region have contributed especially significantly over the past decade to progress on modern human origins and the "Neandertal issue." Although it is not abundant, there is good quality fossil material from this region that is well analyzed and increasingly well dated, and the various reviews and discussions here make it clear that a simple "Neandertal-to-modern" transition is not tenable in this region (or in that peripheral northwestern Old World cul-de-sac so beloved of so many paleoanthropologists).

Modern human origins is currently the liveliest topic in paleoanthropology and also the one with the best fossil and archaeological record. But debates still swirl around the same basic topics that they have for years: defining populations (taxonomy), determining their relationships (phylogenetics), discussing adaptive and behavioral differences between earlier and later populations (paleobiology), and coming up with hypotheses, models, scenarios, and explanations for the patterns we observe (the fun stuff). At one level not much has changed in a century: there are still two radically different points of view about relationships. One argues that more-or-less all earlier hominid populations contributed more-or-less all the time to more-or-less all subsequent populations, while the other argues that only some earlier populations did so. There are nuanced intermediate models between the two ends of what is a continuum of favored possibilities. Although there is

Neandertals and Modern Humans in Western Asia, edited by Akazawa *et al.*
Plenum Press, New York, 1998.

still by no means a consensus, there is perhaps more agreement than was the case a decade ago: that at least some populations (in particular the Neandertals of western Eurasia) did not contribute in any significant way to the gene pool of living Homo sapiens.

At other levels there have clearly been some changes over the past century. Perspectives, although still somewhat Eurocentric, are much less so than they were, and there is a general recognition that at any particular time during the Middle and Late Pleistocene most people did not live in northwestern Eurasia, while a majority probably did live in Africa. We see considerably more sophisticated approaches to unravelling the phylogenetic relationships among populations (or at least what passes for populations inferred from an inevitably sparse fossil record). There has been a considerable, and as yet not fully digested, impact of genetic data on issues of population relationships and population history. (Almost as a reprise of the molecules versus morphology arguments over chimpanzee-gorilla-human relationships, there is still reluctance on the part of some paleoanthropologists to come to terms with the new genetic data, and strong temptations to listen to siren voices saying that we really don't need to pay much attention to all this new-fangled stuff.) We have seen very recently the first direct genetic information from a fossil (the original Neandertal specimen) (Krings et al. 1997). And a better archaeological record, more fossils and better interpretations, and better chronological control have allowed us increasingly to ask questions as though we are dealing with extant populations. For example, what was the distribution of hominids across the landscape; where did they live, and not live; what geographical and demographic changes might have been associated with climatic and environmental change?

Genetic data have contributed a great deal to current thinking about modern human origins (e.g., Cavalli-Sforza et al. 1994). First, genetic studies enable us to put humans into a useful comparative framework. It has long been recognized that humans as a species are genetically interesting in showing a pattern of high within- to between-groups variance (Lewontin 1972; Barbujani et al. 1997) (sometimes explicated with an example: if all humans on earth disappeared except for, say, Ethiopians, roughly 80% or more of the current total human genetic variability would still be preserved). Until recently, it has not been clear how this pattern compared with that of our nearest primate relatives, the great apes. But it is now possible to see that the pattern of global, species-wide variation in humans is equivalent, not to variation within other entire ape species but rather to variation at the subspecies level (Ruvolo 1997). For example, the chimpanzee *Pan troglodytes* is divided into three (or perhaps even four) infraspecies groupings, normally recognized formally as subspecies (Groves et al. 1992). The other apes appear similar to chimps in having clearly recognizable subspecies (Ruvolo 1997). Chimpanzee subspecies can be discriminated both morphologically though with some overlap (Groves et al. 1992), and genetically (Morin et al. 1994): based on mitochondrial DNA each subspecies forms a clade. Part—but just part—of one chimpanzee subspecies (*Pan troglodytes schweinfurthii*, the eastern chimpanzee) has a genetic patterning of within- to between-groups variation like living humans (Goldberg and Ruvolo 1997). Putting this another way, humans are genetically equivalent to just a portion of just one of several subspecies of the chimpanzee. The interesting question is, why? Because there has been more gene or population flow in humans? Or because the human species is young and recently expanded geographically?

Which brings us to the second point: the genetic data allow us to say something about population relationships, and a considerable body of evidence and informed interpretation now shows that African populations are more ancient than non-African (e.g., Penny et al. 1995; Zischler et al. 1995; Ruvolo 1996; Krings et al. 1997). That is, extra-African populations have an African derivation. (Considering the size of inhabitable Africa relative to the rest of the Old World, this is perhaps not saying too much!) And third, the same genetic data set in a comparative primate perspective indicates that the last common ancestor of all living humans lived relatively recently (e.g., Goldstein et al. 1995; Ruvolo 1996) (within the later Middle Pleistocene or even Late Pleistocene and not the Early Pleistocene or Late Pliocene). So, regardless of the amount of gene flow, the odd human patterning of genetic variance relative to other hominoids is probably largely a function of the relative youth of our species.

As can be seen from this volume, the fossil record is improving significantly. Neandertals clearly form a separate western Eurasian clade that is identifiable over the last 400 to 500 ka from the expanding fossil record (e.g., Howell, this volume; Hublin, this volume). Within Africa and Southwest Asia regional fossil records are poorer, but in Northwest Africa (Morocco mainly), eastern Africa (Ethiopia, Kenya, Tanzania, Zambia), and South Africa there is now evidence plausibly documenting an evolutionary transition from "archaic" to "modern" human populations. Neandertals are absent. Likewise there is no evidence for Neandertals in South, East, or Southeast Asia. Within Southwest Asia there is a clear pattern of "archaic" Middle Pleistocene hominids followed by robust early modern humans followed by Neandertals, followed again by modern humans (as documented in many papers in this volume). Some aspects of dating, calibration, and correlation still need strengthening in this region, and sampling density is lower than in Europe, but the pattern is clearly not one in which Neandertals precede early modern humans—rather the reverse.

Outside Africa, Europe, and Southwest Asia, the fossil and archaeological record is hardly worth arguing about. Within Europe, the record of population interactions and histories is best represented in the western portion of the continent (the cul-de-sac of the cul-de-sac), making up what would have been at any particular time only a tiny fraction of the inhabited Old World land surface and therefore a similar fraction of the hominid population. In a sense, except for its historical importance (for the shadow it has cast), the western European hominids and their traces are trivial for any consideration of modern hominid origins. I think that present evidence clearly supports population replacement in at least the western part of Europe; even were there interbreeding it had no discernable morphological or genetic impact on later populations (Brauer and Rimbach 1990; Ruvolo 1996; Holliday 1997; Krings et al. 1997; Turbon et al. 1997). Of course, this begs the question of what was going on in the other >95% of the globe/population, although what happened in western Europe probably reflects patterns in perhaps many other currently more poorly sampled parts of the Old World

Returning to genetics, let us imagine taking a time machine and making a sampling trip around the world around 45 ka ago. We would find "modern" populations throughout Africa (based on the admittedly relatively small portions of northwestern, eastern, and southern Africa that have been sampled), Southwest Asia, Australia, and probably some regions between eastern Africa/Southwest Asia and Australia. But in Europe and in many parts of South, East, and Southeast Asia we would find only nonmodern populations. The recent analysis of partial mtDNA of the Neandertal type specimen (Krings et al. 1997), although only a sample of one (and likely to remain so, Cooper et al. 1997), suggests that at least the Neandertal nonmodern population was genetically quite different from modern humans, at least as different as the central and eastern subspecies of chimpanzee are from each other (*Pan troglodytes troglodytes* and *Pan troglodytes schweinfurthii*) (Morin et al. 1994; Goldberg and Ruvolo 1997; Krings et al. 1997). That is, the living human populations (and their "modern" antecedents) would be equivalent genetically to one chimpanzee subspecies (actually, part of one), while the various archaic human populations might have been equivalent genetically to other chimpanzee subspecies.

Using morphological data, there are problems in comparing the degree and patterning of within-species variation between species, but enough has been published now to suggest that the cranial differences between human modern and nonmodern populations (whether or not they are described as species or subspecies) were greater than those between subspecies of the chimpanzee or even the gorilla (Groves 1970; Van Vark 1984; Groves et al. 1992; Turbon et al. 1997). So, apparently paradoxically, we see genetic differences between moderns and nonmoderns (specifically Neandertals) equivalent to those between chimpanzee subspecies (implying equivalent time depths for the distinct clades) (Morin et al. 1994; Krings et al. 1997), whereas the morphological differences are (perhaps considerably) greater implying greater adaptive differences between the hominid populations than between the ape subspecies. Arguing over whether these are or are not species distinctions in either the hominids or the apes is pointless in my opinion, because these are matters of taste and taxonomic philosophy. (But this hominid case presents us with another example of the occasionally marked disjunction between "morphological" and "genetic" differences.)

If we now consider the structure of arguments about modern human origins, although there is still a continuum from "regional continuity" to "replacement" models, with just about all intermediate positions, most would now be clustered toward the "replacement" end of the spectrum. Another interesting and increasingly discussed set of models are those that cover the behavior of these hominids. Were the Neandertals and other nonmoderns "just like us" in their behavioral potential; that is, did their biology allow the full living human range of "cultural" responses? Or was their behavioral capacity different, in that their "biology" was more constraining of behavioral potential, of "culture"? As behavior becomes biologically less constrained (that is, more like living humans) the behavioral repertoire can become more varied. But not all populations of equivalent biologically determined behavioral potential have the same set of adaptive-technological-ecological responses. To illustrate this, a good example would be the adaptive range of human behavior around 7 ka ago when some populations were sedentary agriculturalists, some were hunter-gatherers, and some pastoralists. Another example would be around 35 ka ago when most modern human populations across the globe would not have produced Upper Paleolithic cultures as defined in Europe and western Asia. Yet all these populations were equally human.

The critical issue here is the nature of behavioral potential ("cultural capabilities") in Neandertals and other nonmodern populations versus that (or those) in the various temporal stages of modern humans. The same or different? Were the differences equivalent to those between, say, Near Eastern agriculturalists and European late Paleolithic hunters or between European Upper Paleolithics and North African Aterians? That is, was the degree of biological-constraint on cultural behavior in Neandertals equivalent to that in modern humans? Or were these hominids different in ways analogous to, for example, "common" chimpanzees (*Pan troglodytes*) and "pygmy chimpanzees" or bonobos (*Pan paniscus*)? That is, were the differences more "biologically" based? If Neandertals and moderns were indeed behaviorally different because of some "biological" difference, were they nonetheless equivalent in behavioral complexity (or grade) as are the two chimpanzee species, or chimps and gorillas? Or did they differ in grade, one being more "advanced" than the other, as we see in chimps versus baboons, for example? And what, if any, relevance do these issues have to do with "replacement"? (After all, chimps have "replaced" neither gorillas nor baboons.) Clearly the anatomical and archeological evidence is sufficiently ambiguous that essentially all possible combinations of these model features can be proposed at the moment by competent scholars.

With that said though, I think there have been changes over the past decade that have moved the debate a little toward resolution. We have evidence for the youth and unity of living humans. Human genetic patterning, plus the one genetic fossil datum from the Neandertal type specimen, together with numerous morphological and morphometric studies, suggest that genetically and morphologically living humans are a descendant remnant of what was relatively recently (40 ka ago or even younger) a much more diverse cluster of populations, genetically and morphologically. Without more and different (nuclear as well as mitochondrial DNA) fossil genetic data (which will be difficult indeed to obtain), we shall have to rely on continuing analysis of living populations, but there is little if any sign of the existence of "ancient" lineages of the kind found in nonhuman hominoids (Ruvolo 1997). This is evidence for a "replacement" model. (Or, expressing this conservatively, in the one well-sampled, peripheral, and rather small area—western Europe—for which there is adequate fossil and archeological data, replacement is the more plausible option.).

This still leaves us with some quite major questions to answer. Picking just three as examples: the pattern of evolution from nonmodern to modern morphologies and behaviors away from the periphery; the nature and timing of the behavioral transition to the current strongly culturally inflected biology from one in which the biological constraints on behavioral variation and change were less humanlike; and the dynamics of Eurasian population interactions between 45 ka and 25 ka ago. But a great deal of progress is being made.

REFERENCES

Barbujani, G., A. Magagni, E. Minch, and L. L. Cavalli-Sforza. 1997. An apportionment of human DNA diversity. *Proceedings of the National Academy of Science USA* 94:4516–19.

Brauer, G., and K. Rimbach. 1990. Late archaic and modern Homo sapiens from Europe, Africa, and Southwest Asia: Craniometric comparisons and phylogenetic implications. *Journal of Human Evolution* 19:789–807.

Cavalli-Sforza, L., P. Menozzi, and A. Piazza. 1994. *The History and Geography of Human Genes*. Princeton: Princeton University Press.

Cooper, A., H. N. Poinar, S. Pääbo, J. Radovcic, A. Debénath, M. Caparros, C. Barroso-Ruiz, J. Bertranpetit, C. Nielsen-Marsh, R. E. M. Hedges, and B. Sykes. 1997. Neandertal genetics. *Science* 277:1021–4.

Goldberg, T., and M. Ruvolo. 1997. Molecular phylogenetics and historical biogeography of east African chimpanzees. *Biological Journal of the Linnean Society* 61:301–24.

Goldstein, D. B., A. Ruiz Linares, L. L. Cavalli-Sforza, and M. W. Feldman. 1995. Genetic absolute dating based on microsatellites and the origin of modern humans. *Proceedings of the National Academy of Science USA* 92:6723–7.

Groves, C. 1970. Population systematics of the gorilla. *Journal of the Zoological Society of London* 161:287–300.

Groves, C., C. Westwood, and B. T. Shea. 1992. Unfinished business: Mahalanobis and a clockwork orang. *Journal of Human Evolution* 22:327–40.

Holliday, T. 1997. Body proportions in Late Pleistocene Europe and modern human origins. *Journal of Human Evolution* 32:423–47.

Krings, M., A. Stone, R. W. Schmitz, H. Krainitzki, M. Stoneking, and S. Pääbo. 1997. Neandertal DNA sequences and the origin of modern humans. *Cell* 90:19–30.

Lewontin, R. 1972. The apportionment of human diversity. *Evolutionary Biology* 6:381–98.

Morin, P. A., J. J. Moore, R. Chakraborty, L. Jin, J. Goodall, and D. S. Woodruff. 1994. Kin selection, social structure, gene flow, and the evolution of chimpanzees. *Science* 265:1193–1201.

Penny, D., M. Steel, P. J. Waddell, and M. D. Hendy. 1995. Improved analyses of human mtDNA sequences support a recent African origin for Homo sapiens. *Molecular Biology and Evolution* 12:863–82.

Ruvolo, M. 1996. A new approach to studying modern human origins: Hypothesis testing with coalescence time distributions. *Molecular Phylogenetics and Evolution* 5:202–19.

Ruvolo, M. 1997. Genetic diversity in hominoid primates. *Annual Revue of Anthropology* 26:515–40.

Turbon, D., A. Pérez-Pérez, and C. Stringer. 1997. A multivariate analysis of Pleistocene hominids: Testing hypotheses of European origins. *Journal of Human Evolution* 32:449–68.

Van Vark, G. 1984. On the determination of hominid affinities, 323–50. In *Multivariate Statistical Methods in Physical Anthropology*, ed. G. van Vark and W. W. Howells. Boston: Reidel.

Zischler, H., H. Geisert, A. von Haeseler, and S. Pääbo. 1995. A nuclear "fossil" of the mitochondrial D-loop and the origin of modern humans. *Nature* 378:489–92.

CONTRIBUTORS

Takeru Akazawa
International Research Center for Japanese Studies
Kyoto 610-11, Japan

Stanley H. Ambrose
Department of Anthropology
University of Illinois
Urbana, Illinois 61801

Kenichi Aoki
Department of Biological Sciences
University of Tokyo
Tokyo 113-0033, Japan

Baruch Arensburg
Department of Anatomy and Anthropology
Sackler School of Medicine
Tel Aviv University
Ramat Aviv 69978, Israel

Ofer Bar-Yosef
Department of Anthropology
Peabody Museum
Harvard University
Cambridge, Massachusetts 02138

Anna Belfer-Cohen
Institute of Archaeology
Hebrew University
Mt. Scopus, Jerusalem 91905, Israel

Eric Boëda
Department of Ethnology and Prehistory
Université de Paris X
Nanterre 92001, France

Steven E. Churchill
Department of Biological Anthropology and Anatomy
Duke University
Durham, North Carolina 27708

Jacques Connan
Elf Exploration Production
Direction Exploration, CSTJF
Avenue Larribau, PAU 64018, France

Yukio Dodo
Department of Anatomy
Tohoku University School of Medicine
Sendai 980, Japan

Paul Goldberg
Department of Archaeology
Boston University
Boston, Massachusetts 02215

Naama Goren-Inbar
Institute of Archaeology
Hebrew University
Mt. Scopus, Jerusalem 91905, Israel

Donald O. Henry
Department of Anthropology
University of Tulsa
Tulsa, Oklahoma 74104

Erella Hovers
Institute of Archaeology
Hebrew University
Mt. Scopus, Jerusalem 91905, Israel

F. Clark Howell
Laboratory for Human Evolutionary Studies/Museum of Vertebrate Zoology
University of California
Berkeley, California 94720

Jean-Jacques Hublin
UMR 152 du CNRS
Laboratoire d'Anthropologie
Musée de l'Homme
Paris 75116, France

Jean-Louis Joron
Groupe des Sciences de la Terre
Laboratoire Pierre Süe
CEN Saclay
Gif sur Yvette Cedex 91191, France

Richard G. Klein
Department of Anthropology
Stanford University
Stanford, California 94305

Osamu Kondo
Department of Anatomy
Tohoku University School of Medicine
Sendai 980, Japan

Janusz K. Kozlowski
Institute of Archaeology
Jagellonian University
Kraków 31007, Poland

Kevin N. Laland
Sub-Department of Animal Behaviour
University of Cambridge
Cambridge CB3 8AA, United Kingdom

Daniel E. Lieberman
Department of Anthropology
The George Washington University
Washington, DC 20052

Liliane Meignen
ERA 28 du Centre de Recherches Archeologiques
CNRS, Sophia Antipolis
Valbonne 06560, France

Paul Mellars
Department of Archaeology
Cambridge CB2 3DZ, United Kingdom

Norbert Mercier
Centre de Faibles Radioactivités
Laboratoire Mixte CNRS-CEA
Gif sur Yvette Cedex 91198, France

Nancy Minugh-Purvis
Department of Neurobiology and Anatomy
Allegheny University of the Health Sciences
Philadelphia, Pennsylvania 19129
University Museum of Archaeology and Anthropology
University of Pennsylvania
Philadelphia, Pennsylvania 19104
Division of Plastic Surgery
The Children's Hospital of Philadelphia
Philadelphia, Pennsylvania 19104

Sultan Muheson
Director-General of Antiquities and Museums
Damascus, Syrian Arab Republic

Marcel Otte
Service de Préhistoire
University of Liège
Liège B-4000, Belgium

David Pilbeam
Department of Anthropology
Peabody Museum
Harvard University
Cambridge, Massachusetts 02138

Rolf M. Quam
Department of Anthropology
State University of New York
Binghamton, New York 13902

Yoel Rak
Department of Anatomy and Anthropology
Sackler School of Medicine
Tel Aviv University
Ramat Aviv 69978, Israel

W. J. Rink
Department of Geology
McMaster University
Hamilton, Ontario L8S 4M1, Canada

Avraham Ronen
Zinman Institute of Archaeology
University of Haifa
Haifa 31905, Israel

Karen R. Rosenberg
Department of Anthropology
University of Delaware
Newark, Delaware 19716

Christopher B. Ruff
Department of Cell Biology and Anatomy
Johns Hopkins University School of Medicine
Baltimore, Maryland 21205

Kathy D. Schick
Anthropology Department and CRAFT Research Center
Indiana University
Bloomington, Indiana 47405

Henry P. Schwarcz
Department of Geology
McMaster University
Hamilton, Ontario L8S 4M1, Canada

Fred H. Smith
Department of Anthropology
Northern Illinois University
DeKalb, Illinois 60115

John D. Speth
Museum of Anthropology
University of Michigan
Ann Arbor, Michigan 48109

Mary C. Stiner
Department of Anthropology
University of Arizona
Tucson, Arizona 85721

Chris Stringer
Department of Palaeontology
The Natural History Museum
London SW7 5BD, United Kingdom

Eitan Tchernov
Department of Evolution, Systematics, and Ecology
Hebrew University
Jerusalem 91904, Israel

Anne-marie Tillier
URA 376 CNRS
Laboratoire d'Anthropologie
Université de Bordeaux I
Talence 33405, France

Erik Trinkaus
Department of Anthropology
Washington University
St. Louis, Missouri 63130

Hélène Valladas
Centre de Faibles Radioactivités
Laboratoire Mixte CNRS-CEA
Gif sur Yvette Cedex 91198, France

INDEX